Cellulose and Cellulose Derivatives

Molecular Characterization and its Applications

Cellulose and Cellulose Derivatives

Molecular Characterization and its Applications

Kenji Kamide

*Asukano-kita 1-7-26, Ikoma,
Nara, 630-0134, Japan*

2005

ELSEVIER

Amsterdam – Boston – Heidelberg – London – New York – Oxford – Paris
San Diego – San Francisco – Singapore – Sydney – Tokyo

ELSEVIER B.V.
Radarweg 29
P.O. 211, 1000 AE
Amsterdam, The Netherlands

ELSEVIER Inc.
525 B Street
Suite 1900, San Diego
CA 92101-4495, USA

ELSEVIER Ltd
The Boulevard
Langford Lane, Kidlington,
Oxford OX5 1GB, UK

ELSEVIER Ltd
84 Theobalds Road
London WC1X 8RR
UK

First edition 2005

Library of Congress Cataloging in Publication Data
A catalog record is available from the Library of Congress.

British Library Cataloguing in Publication Data
A catalogue record is available from the British Library.

ISBN: 0-444-82254-2

♾ The paper used in this publication meets the requirements of ANSI/NISO Z39.48-1992 (Permanence of Paper). Printed in The Netherlands

Preface

Over the years, I have had the opportunity to write review papers,[1] book chapters,[2] and to contribute to symposia[3] on the molecular characterization of cellulose and cellulose derivatives. However, I now recognize that it is extremely difficult for students or researchers not immersed in this fascinating and important field to acquire systematic knowledge by referring to original articles written by many different individuals, each presenting detailed, but rather narrow, topics. There are now many highly acclaimed and authoritative books on polymer science where the molecular characterization of polymers is described. Most attention is given to synthetic polymers and, unfortunately, the molecular properties of cellulose and cellulose derivatives are only mentioned in a fragmented way, if at all (see, for example, Section 1.4). However, there is an exception, 'Cellulose and Cellulose Derivatives' (edited by Emil Ott and H.M. Spurlin (for Parts I–III) and N.N. Bikales and L. Segal (for Parts IV and V)) is a magnificent omnibus, published some 20–40 years ago. Even as late as the 1970s there were many unsolved fundamental problems (see Section 1.5) and new challenges have emerged since, most of which are being effectively tackled. However, there is no authoritative text devoted to the achievements during the decades 1970–1990 to which researchers and students can refer to the field for a rapid introduction. It is timely, therefore, for a single authored book, which aims to deal with the field in depth and in a systematic way. It is this belief that has motivated me to try to produce such a text drawing on my many years of considerable involvement with all aspects of cellulose from fundamental science to the industrial application of its many different derivatives. It is hoped that both the experienced cellulose scientist and newcomers to the field will derive as much enjoyment using the text over the coming years as the author has had preparing it.

The book itself aims to examine the recent developments in the areas of structural and molecular characterization of cellulose and cellulose derivatives, with particular reference to cellulose (in aqueous alkali) and cellulose acetate, for which the most reliable experimental data are available. It is not intended to be a handbook or a technical manual, but an in-depth treatment of the fundamental principles of cellulose and its derivatives in solution. Much of the data used stems from work in my own laboratories. The latest theories of polymers in solution are drawn on to achieve a fundamental and quantitative understanding at the molecular level and examples of how such an approach can be applied to problems in the industrial sphere. It is hoped that the approach adopted establishes a platform on which new initiatives can be built.

---

[1] *Senn-I Gakkaishi*, 1977, **33**, 101; *Adv. Polym. Sci.*, 1987, **83**, 5–56 (with Dr M Saito).

[2] *Wood and Cellulose, Characterization of Chemically Modified Cellulose*, **Vol. 17** (eds DN-S Hon and N Shiraishi), Marcel Dekker, New York, 1990, pp. 801–860.

[3] Makromolecular Symposium, *Adv. Polym. Mater.*, 1994, **83**, 233–271 (with Dr M Saito).

The approach taken puts particular emphasis on the following:

(a) The citing of essential references.
(b) Outlining briefly theoretical fundamentals (the detailed explanation of general theories can be found in the books[4] by the author and his colleague together with relevant procedures (conventional analysis of commercial products does not always yield scientifically significant data—many studies have fallen into this trap!)).
(c) Presenting comprehensive and reliable experimental results as figures and tables so that a database as large as possible is established for further analysis by others.
(d) The establishment of accurate and rigorous procedures for data analysis (unless correctly analyzed, the experimental data are of little value).
(e) Presenting as many experimental examples as possible in order to deduce the common nature of cellulose and cellulose derivatives (a single demonstration through one example which favors the theory in question is insufficient and often leads to false conclusions).
(f) Not drawing any conclusions without showing the experimental evidence on which these are based.
(g) The minimization of the number of preconditions or assumptions on which both theory and experiments are based to establish evidence-based cellulose chemistry.
(h) Accepting the complexity of cellulose and cellulose derivatives and the fact that their properties cannot be explained reasonably by simple (or rather empty) theories. For this reason, the models employed deviate from the molecular reality and only the time-average quantities at equilibrium or at some stationary state are considered. The phenomenological diversity of molecular properties of cellulose and cellulose derivatives will undoubtedly ensure their continued fruitful use in the future.
(i) Illustrating industrial (or commercial) applications of fundamental research.
(j) Collecting reliable molecular parameters from reliable experimental data and sound theories.

The content of the book is mainly based on work that I have carried out with the following co-workers to whom I would like to express my sincere gratitude: Prof. Kunihiko Okajima, Dr. Keisuke Kowsaka, Mr. Toshihiko Matsui, Dr. Hideki Iijima, Prof. Sei-ichi Manabe, Miss Etsuko Osafune, Mr. Tetsuro Okada, Mr. Toshikazu Terakawa, the late Prof. Katsumasa Kaneko, Dr. Masanao Ohnishi, Dr. Hidehiko Kobayashi, Mr. Sigeru Nomura, Prof. Kunio Hikichi, Mr. Katsunari Yasuda, Dr. Masatoshi Saito, Prof. Yukio Miyazaki, Mr. Tatsuyuki Abe, Dr. Shigenobu Matsuda, the late Prof. W.R. Moore, Prof. Toru Kawai, Prof. Hidematsu Suzuki, Dr. Yoichiro Muraoka, Dr. Kichizo Ono, Mr. Tei-ichi Shiomi, Mr. Hiroshi Ohkawa, Mr. K. Kishino, the late Prof. Shin-ichiro Ishida, Mr. Hideki Komatsu, Mr. Hiroyuki Katoh, Mr. Takaharu Akedo, Mr. Isamu Itoh, the late Dr. Takashi Yamashiki, Mr. Masatoshi Saitoh, Prof. Taturo Sawada, Miss Miki Inamoto, and Dr. Ikuya Miyamoto.

---

[4] K Kamide and T Dobashi, *Physical Chemistry of Polymer Solutions, Theoretical Background*, Elsevier, Amsterdam, 2000; K Kamide, *Thermodynamics of Polymer Solutions, Phase Equilibria and Critical Phenomena*, Elsevier, Amsterdam, 1990.

I am grateful for the valuable suggestions and critical comments of Prof. H. Suzuki (Nagaoka University of Technology), Prof. T. Dobashi (Gumma University), and Prof. A.F. Johnson (Leeds University), given to the early draft. Of course, I take full responsibility for errors and omissions.

The author is also grateful to the following who gave permission for the reproduction of figures and tables, all of which, except Figures 1.4.1, 3.23.14–3.23.16, and 4.7.15–4.7.25, and Tables 3.23.2, 3.23.3, and 4.7.9–4.7.16 are the work of the authors: (a) Society of Polymer Science, Japan (*Polymer Journal* and *Kobunshi Kagaku*), (b) Society of Fiber Science and Technology, Japan (*Senn-i Gakkaishi*), (c) Society of Textile Machinery, Japan (*Journal of Society of Textile Machinery, Japan*), (d) Society of Economics and Management, Nara Sangyo University (*Journal of Industry and Economics, Nara Sangyo University*), (e) John Wiley & Sons Ltd (*British Polymer Journal, Polymer International, and Journal of Polymer Science*), (f) Pergamon Press plc (European Polymer Journal), (g) Springer Verlag GmbH (*Advances of Polymer Science*), (h) Hüthig & Wept Verlag (*Die Makromolekulare Chemie* and *Makromolekular Symposia*), (i) Academia ROMANA-Filiala (*Cellulose Chemistry & Technology*), (j) Mercel Dekker Inc.

Last, but not least, I would like to express my warmest thanks to my family, in particular, to my wife, Chizuru, who deserves a medal for her great and patient contribution to the drafting of this book. Quite simply, without her this book would not have been possible.

Kenji Kamide
Nara, Japan, September 2004

Contents

CONTENTS

Glossary

A	absorbency; eq. (2.8.5)
A	unperturbed chain dimensions; short-range parameter; eqs. (3.13.8) and (3.13.8$'$) carboxylethyl content; eqs. (2.2.1) and (2.2.2)
ACS	average crystal size; eq. (4.7.2)
$A_{c,w}$	weight-average acetic acid content; eq. (2.5.1)
A_f	A of a hypothetical chain with free internal rotation
$A_{(m)}$	most probable A; Table 3.16.10
$A_r(n)$	relative absorbance
A_{WL}	anticoagulant activity by White–Lee method; Table 2.7.1
A_1	surface area of solution drop; eq. (3.3.4)
A_2	area of contact between drop and thermistor; eq. (3.3.4)
A_2	second virial coefficient; eqs. (3.3.1), (3.4.1), and (3.5.1)
A_2^*	apparent A_2; eq. (3.13.2)
$A_{2,L}$	A_2 by light scattering method; eq. (3.5.1)
$A_{2,o}$	A_2 by membrane osmometry; eq. (3.3.1)
$A_{2,v}$	A_2 by vapor pressure osmometry; eq. (3.3.3)
a	exponent of Mark–Houwink–Sakurada equation; eq. (1.4.1)
a'	length of a link (segment); eq. (3.15.7)
a_D	exponent of molecular weight dependence of D_0; eq. (3.15.3)
B	long-range parameter; eq. (3.13.9): carbamoylethyl content; eq. (2.2.1)
B'	half value width in radian of X-ray diffraction angle; eq. (4.7.3)
B_{WL}	body weight loss ratio; Section 2.7
C	weight concentration
C^*	critical micelle concentration; Figure 2.11.2
C_∞	characteristic ratio; eq. (3.16.2)
C_a	alkali concentration
C_L	lower critical concentration (liquid crystal); Section 3.23
C_p	polymer concentration
C_{sa}	acid concentration of coagulant
c	concentration (weight/volume)
D	diffusion coefficient
D_0	weight-average diffusion coefficient at infinite dilution; eq. (3.15.5$'$)
D_i	D of ith component
D_r	draft ratio
d	hydrodynamic diameter; eq. (3.15.7)
d	diameter of a worm-like touching bead
$d\gamma/dt$	shear rate
E	Young modulus
E_{nb}	energy required for breaking
F	total degree of substitution

F_i^j	degree of substitution of ith glucopyranose unit in jth molecule
$\langle\!\langle F \rangle\!\rangle$	average degree of substitution; eq. (2.1.3)
$\langle\!\langle F \rangle\!\rangle_{\text{chem}}$	$\langle\!\langle F \rangle\!\rangle$ determined by chemical analysis
$\langle F^j \rangle$	average degree of substitution of jth molecule; eq. (2.1.2)
$f_{k,i}^j$	probability of substitution of hyroxyl group at C_k atom in ith glucopyranose unit of jth molecule; eq. (2.1.4)
$\langle\!\langle f_k \rangle\!\rangle$	average degree of substitution of hyroxyl group at $C_{k\ \text{position}}$; eq. (2.1.4)
f_{lmn}	mole fraction of eight (seven substituted and single unsubstituted) glucopyranose units, where l, m, and n mean the existence of substituent group at C_2, C_3, and C_6 positions, respectively, and they can take the value of 0 (unsubstituted) or 1 (substituted); eq. (2.4.6)
$f(X)$	$\equiv f_c$
$f^{\parallel}$	molecular orientation factor in parallel to sheared direction
f_b	molecular orientation by optical birefringence
f_c	crystal orientation factor; eq. (4.7.4)
g	shear velocity gradient
$g(A_c)$	distribution of acetic acid content; eq. (2.5.12)
$g(A_c, M)$	weight fraction of the polymer, whose acetic acid content lies between A_c and $A_c + dA_c$ and molecular weight is between M and $M + dM$; eq. (2.5.1)
$g\langle F^j \rangle$	distribution of $\langle F^j \rangle$; eq. (2.2.6)
G	centrifugal acceleration
G'	storage modulus; dynamic rigidity
h	inhomogeneity parameter for molecular weight distribution; polymolecularity index; eq. (3.5.5)
I	scattering intensity (LS and X-ray); Sections 3.5 and 3.10
I_{am}	integrated peak intensity corresponding to amorphous region in X-ray diffraction curve; eq. (4.7.1)
I_h	integrated peak intensity of higher magnetic peaks in C_4 or C_6 or $C_{2,3,5}$ carbon peak; eqs. (4.7.5), (5.1.1), and (5.1.3)
I^i	integrated peak intensity of ith region of ^{13}C NMR spectrum
I_1	integrated peak intensity of higher magnetic peaks in C_4 or C_6 or $C_{2,3,5}$ carbon peak; eqs. (4.7.5), (5.1.1), and (5.1.3)
I_1	integrated peak intensity of lower magnetic peaks in C_4, C_6 or $C_{2,3,5}$ carbon peak; eqs. (4.7.5), (5.1.1), and (5.1.3)
$I(002)$	peak intensity corresponding to (002) plane in X-ray diffraction curve; eq. (4.7.1)
I_s	scattered light intensity; eq. (3.5.3)
I_0	incident light intensity (LS); eq. (3.5.3)
$I_1(q, z)$	first-order time correlation function (dynamic LS); eq. (3.5.12)
$I_2(q, z)$	second-order time correlation function (dynamic LS); eq. (3.5.5)
K	optical constant (LS); eq. (3.5.2)
K	equilibrium constant; eq. (2.11.6)
K	Flory constant; eq. (3.16.9)
K_m	parameter in Staudinger equation eq. (1.3.4) or MHS eq. (1.4.1)
K_s'	calibration parameter (VPO); eq. (3.3.4)
K_s	parameter in eq. (3.14.1), representing molecular weight dependence of s_0

K_Φ	parameter in eq. (3.15.3)
k	scattering vector; Section (3.10)
k	Boltzmann constant; eq. (3.15.7): constant in Zimm plot; eq. (3.5.1)
k_s	concentration-dependent coefficient; eq. (3.7.2)
k'	Huggins coefficient; eq. (3.6.7)
L	contour length; eqs. (3.13.15) and (3.19.3)
ℓ	projection of pyranose ring onto the chain axis; eq. (3.19.3)
LD_{50}	acute toxicity; Section 2.7
M	molecular weight; eq. (1.3.1)
M_L	shift factor; eq. (3.19.2)
M_n	number-average molecular weight
M_{SD}	sedimentation–diffusion-average molecular weight
M_{SV}	sedimentation–viscosity-average molecular weight
M_w	weight-average molecular weight
m	molecular weight of monomer unit
m_0	molecular weight of segment; eq. (3.13.10)
m_s	molecular weight of solvent; eq. (3.17.1)
N	total number of molecules constituting a chain; number of links connecting segments in one molecule
$N(q, \tau)$	normalized first correlation function; eq. (3.5.13)
N_A	Avogadro's number
N_c	nitrogen content (wt%); eq. (2.5.20)
n	number of solvated solvent molecules per gram of polymer; eq. (3.8.2)
n	refractive index of solution; eq. (3.5.2)
n	number of molecules in micelle; Table 2.12.2
n_0	refractive index of solvent
P	Flory parameter; eq. (3.15.4)
P'	cholesteric pitch; eq. (3.23.1)
P_n	number-average degree of polymerization; eq. (2.1.6)
P_v	viscosity-average degree of polymerization
P_w	weight-average degree of polymerization; eq. (2.1.7)
P_0	saturated vapor pressure; eq. (3.3.5)
p	degree of polymerization
p_j	p of jth molecule
Q_v	volume expansion ratio; Figure 4.6.9
q	persistence length; eq. (3.19.7)
$q_{w,z}$	correction factor for molecular weight distribution of Φ; eq. (3.15.2)
$q_{w,z'}$	polymolecularity correction factor of P; eq. (3.15.4)
q_{BD}^0	q at perturbed state by Benoit–Doty theory; eq. (3.13.5)
q_{CL}	coil limit persistence length; eq. (3.13.23)
q_{CL}^0	coil limit persistence length at unperturbed state; eq. (3.13.23)
R	gas constant
$\langle R^2 \rangle_0$	mean-square end-to-end distance at unperturbed state; eqs. (3.13.8') and (3.16.3)
R_f	rate of flow (TLC); eq. (2.5.4)
$R_{f,w}$	weight-average R_f; eq. (2.5.10)

R_θ	Rayleigh ratio; reduced scattering light intensity; eq. (3.5.3)
S	number of water molecules solvated to an NaOH molecule; eq. (4.3.1)
S	solubility; eq. (2.8.8)
S_a	solubility of cellulose into aq. alkali solution; Section 4.1
$\langle S^2 \rangle$	mean-square radius of gyration; eq. (3.5.1)
$\langle S^2 \rangle_0$	$\langle S^2 \rangle$ of unperturbed chain; eq. (3.13.4)
$\langle S^2 \rangle_w$	weight-average mean-square radius of gyration; eq. (3.5.4)
$\langle S^2 \rangle_{0,\infty,w}$	$\langle S^2 \rangle_w$ of polymer with infinite molecular weight in unperturbed state
$\langle S^2 \rangle_z$	z-average mean-square radius of gyration
s	number of solvated solvent molecules per repeating unit; eq. (3.17.1)
s	sedimentation coefficient; eq. (3.7.1)
s_0	s at infinite dilution; eq. (3.17.2)
s_0	s at infinite dilution
$s_{0,w}$	weight-average s_0; eq. (3.7.2)
T	temperature (°C)
T	temperature (K); eq. (1.3.1)
T_c	critical solution temperature: coagulation temperature; Section 4.7
T_c	crystallization temperature; Section 2.10
T_d	onset temperature of decomposition
TE	tensile elongation
T_g	glass transition temperature
T_m	melting temperature
TS	tensile strength
T_1	spin-lattice relaxation time
T_{1H}	^{1}H spin-lattice relaxation time
t	time
t_c	coagulation time
V	sound velocity of solution; eq. (3.8.1)
V_0	molar volume of solvent; eqs. (1.3.3) and (3.3.2)
V/V_0	volume contraction; Section 4.7
V_m	molar volume of monomer unit; eq. (3.3.2)
V_p	specific volume of polymer; eq. (3.15.5)
v	partial specific volume
v_p	polymer volume fraction; eq. (3.3.2)
v_p	specific volume of polymer; eq. (3.15.5)
w_2	polymer weight fraction; Figure 3.21.2
X	draining parameter; eq. (3.15.7): torsional angle; Section 5.2.1
X_n^*	number-average ratio of polymer to solvent; eq. (3.3.2)
Y	existence ratio of individual coagulation factors; eq. (2.7.2)
Z	point located Z apart from solvent front level (TLC); Section 2.5
Z_f	distance between solvent front and starting point (TLC); Section 2.5
z	excluded volume parameter; eq. (3.13.7)
α	expansion factor of end-to-end distance of chain; eq. (3.15.27): optical rotary angle
α_α	parameter representing selective adsorption; eq. (3.5.9)
α_p	exponent of molecular weight dependence of P; eq. (3.15.6)

α_s	exponent of molecular weight dependence of s_0; eq. (3.14.1)
α_Φ	exponent of molecular weight dependence of Φ; eq. (3.15.3)
α_1	contribution of excluded volume effect to α; eq. (3.15.30)
α_2	contribution of non-Gaussian nature of chain to exponent α; eq. (3.15.5)
α_s	linear expansion factor of radius of gyration; eq. (3.13.4)
$\alpha_s, 3^p$	α_s derived by a third-power law type equation for pearl necklace model; eq. (3.13.21)
$\alpha_s, 3^w$	α_s derived by a third-power law type equation for wormlike chain model (Section 3.13), eq. (3.13.25)
$\alpha_s, 5^w$	α_s derived by a fifth-power law type equation for wormlike chain model (Section 3.13)
α_η	viscosity expansion factor; (Section 3.15)
β	adiabatic compressibility; eq. (3.8.1)
β_s	β of solvent; eq. (3.8.2)
β	binary cluster integral; eq. (3.13.10)
Γ	surface excess concentration; eq. (2.11.1)
Γ'	selective adsorption parameter; eq. (3.5.10)
$\Gamma(x)$	gamma function of x; eq. (3.13.14)
γ	correlation coefficient; Sections 3.16 and 4.1
γ	surface tension; Figure 2.11.1
Δ	contribution of draining effect to exponent α; eq. (3.16.14)
$\Delta_{1/2}$	half value depth
ΔH	heat of dilution; eq. (3.3.5)
ΔI_1	peak intensity at peak 1 ($\theta = 22.9-23.1°$ for Cell I)
$\Delta \mu_1$	chemical potential of solute
Δv	electronegativity; Figure 3.9.4
δ	NMR chemical shift: parameter defined by eq. (2.11.4)
δ^-	electronegativity; Section 3.9
δ_{CH_3}	O-acetyl proton chemical shift; eq. (3.9.1)
δ_H	proton chemical shift
δ_{Na}	^{23}Na chemical shift
δ_{OH}	hyroxyl proton chemical shift; eq. (3.9.6)
$\partial n/\partial c$	refractive index increment; eq. (3.5.2)
$(\partial n/\partial c)_{\Phi_L}$	$\partial n/\partial c$ at constant Φ_L; eq. (3.5.8)
ϵ	dielectric constant
ε	molecular weight dependence of α_s; eq. (3.15.9)
η	viscosity coefficient; viscosity; eq. (1.3.2)
η_0	viscosity of solvent; eq. (1.3.2)
η_r	relative viscosity; eq. (3.6.4)
η_{sp}	specific viscosity; eq. (1.3.3)
$[\eta]$	limiting viscosity number; intrinsic viscosity; eqs. (3.6.2) and (3.6.3)
$[\eta]_H$	$[\eta]$ by Huggins plot
$[\eta]_K$	$[\eta]$ by Kraemer plot
$[\eta]_\theta$	$[\eta]$ in θ solvent
θ	Flory theta temperature; eq. (3.21.6)
θ	scattering angle; eq. (3.5.1)

θ	bond angle
$[\theta]$	specific rotary angle; Section 4.3
κ_0	Flory enthalpy parameter; eq. (3.21.7)
λ_0	wave length of incident light; eq. (3.5.2)
v	molecular weight dependence of A_2; Table 3.13.1: kinematic viscosity; Figure 4.6.2
ξ	friction coefficient; eqs. (3.15.5) and (3.15.5')
π	osmotic pressure; eqs. (3.3.1) and (3.3.2)
ρ	density
ρ_0	density of solvent; eq. (3.15.5)
ζ	friction coefficient between fluid and small particle
σ	steric hindrance factor; eq. (3.16.1): electrical conductivity; Section 4.3, Figure 4.3.2
σ	standard deviation; Table 2.3.1
τ	correlation time; eq. (3.5.11): pulse interval (NMR)
Φ	Flory viscosity parameter; eq. (3.15.1): torsional angle; eq. (5.2.7)
Φ	solute volume fraction eq. (1.3.2)
ϕ	Flory entropy parameter; eqs. (3.20.2) and (3.21.2)
Ψ	penetration function; eq. (3.13.4): torsional angle; eq. (5.2.7)
χ	anticoagurant activity; eq. (2.7.1)
χ	thermodynamic interaction parameter; eq. (2.11.2)
χ	extinction angle (flow birefringence)
$\chi_{am}(C_3)$	degree of breakdown of intramolecular hydrogen bond at C_3 carbon determined by CP/MAS ^{13}C NMR; eqs. (4.7.5) and (5.1.1)
$\chi_{am}(C_k)$	degree of breakdown in intramolecular hydrogen bond at C_k carbon
$\chi_{am}(X)$	amorphous content determined by X-ray method; eq. (4.1.3)
$\chi_c(X)$	X-ray crystallinity; eq. (4.1.2)
$\chi_h(NMR)$	relative amount of the higher field peaks of MNR C_4 peak; eq. (4.1.6)
χ_c	χ at critical solution point
χ_0	anticoagurant activity of heparin standard; eq. (2.7.1)
χ_0	χ at infinite dilution
χ'	aspect ratio; eq. (3.23.2)
ω	angular velocity; eq. (3.7.1)

Abbreviated symbols

ATA, amylose triacetate
ATC, amylose tricarbanilate
CA, cellulose acetate
CCD, chemical composition distribution
CD, circular dichroism
CN, cellulose nitrate
CS, cellulose sulfate
CTA, cellulose triacetate
CTC, cellulose tricarbanilate
CTCp, cellulose tricaproate
CTN, cellulose trinitrate
DCM, dichloromethane;
DMAc, dimethylacetamide
DMSO, dimethyl sulfoxide
DS, degree of substitution
DSC, differential scanning calorimetry
DTG, differential thermogravimetry
EHEC, ethylhyroxyethyl cellulose
FA, fluoroacetic acid
FTNa, iron-sodiumtartrate

GPC, gel permeation chromatography
HEC, hydroxyethyl cellulose
IR, infra-red spectroscopy
LS, light scattering
MC, methylcellulose
MHS, Mark–Houwink–Sakurada equation
NaCMC, sodium carboxymethyl cellulose
NaCX, sodium cellulose xanthate
NMR, nuclear magnetic resonance
SAXS, small angle x-ray scattering
SPF, successive precipitation fractionation
SSF, successive solution fractionation
TCE, tetrachloroethane
TCM, trichloromethane
TFA, trifluoroacetic acid
THF, tetrahyrofuran
TLC, thin-layer chromatography
VPO, vapor pressure osmometry

– 1 –

Introduction

1.1 CELLULOSE: THE MOST ABUNDANT NATURAL ORGANIC COMPOUND

1.1.1 A brief history of the utilization of cellulose

Even before the beginning of hominization, human beings had used cellulose and wood as an extension of their hands for millions of years. After the hominization period, man employed cellulose and wood for tools (primitive hoes, hoe sticks, spear handles (Paleolithic era); primitive ploughs, light ploughs without wheels, spinning sticks, hand looms (Neolithic era); and for buildings, bridges, ships (from the canoe to the wooden steam boat), carriages, furniture (boxes, beds, tables, chairs), utensils and daily necessities (spoons, cups), textiles (cotton, flax), paper, fuel (for warming, cooking, and industrial production such as extracting salt from saline; pottery and metallurgy), charcoal (for the production of iron), and for cattle breeding. Flax and cotton, both almost pure cellulose, have been used as clothing for 10,000 years (since after the transition from gathering-and-hunting economies to agriculture-and-cattle-raising economics).[1]

In the 18th century, the mechanization of spinning and weaving devices (inventions to improve productivity in order to meet market demand) for cotton fabrics formed a strong incentive to the industrial revolution, which brought about mass consumption of chemicals (in particular, inorganic compounds) for scouring, finishing, and dying cotton cloths.[2]

1.1.2 Chemistry of cellulose as an organic compound: survey of chemical structure (1830s–1930s)

Cellulose is a component of at least a third of advanced plants: 40–60% (in weight) of dry wood, and more than 90% of raw cotton (99.9% of purified cotton) and flax. Therefore, cellulose is unquestionably the most abundant naturally occurring reproducible organic compound.

In the second quarter of the 19th century cellulose became recognized as a chemical compound. In the 1830s, the French agriculturist Anselme Payon named the main constituent of cell wall membranes of the plant cellulose.[3] It took some time (until the 1930s) before the molecular structure of cellulose was established. The history of the research on the chemical structure of cellulose is summarized in several books.[4–6]

1

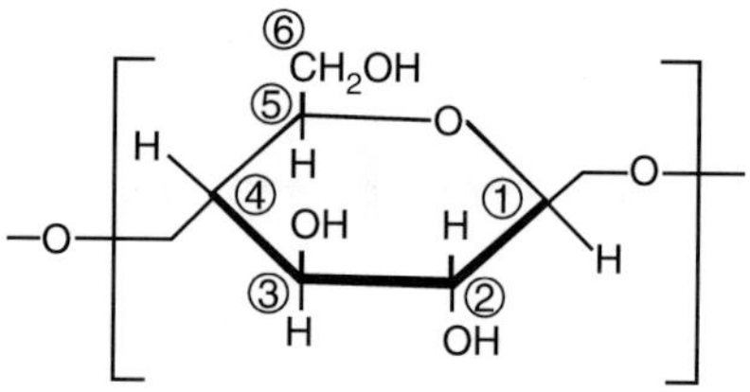

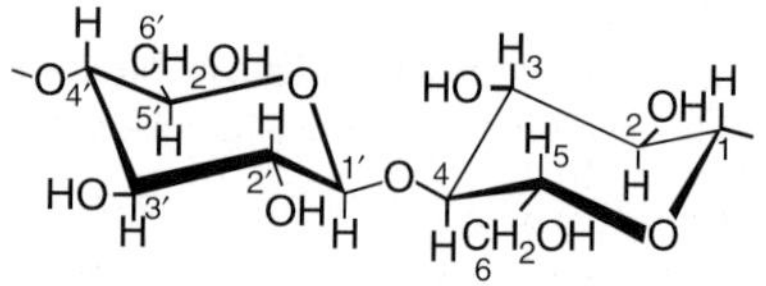

Figure 1.1.1   Glucopyranose unit of cellulose.

Following are some achievements that contributed significantly to the advancement of cellulose chemistry.

(1)   Under severe conditions, cellulose is hydrolyzed by treatment with sulfuric acid quantitatively to saccharides,[7] which was proved glucose.[8] An empirical formula of the chemical composition of cellulose was demonstrated as $(C_6H_{10}O_5)_n$.[9]

(2)   Results of element analysis (C,H,O) of cotton, ramie, and regenerated cellulose fibers coincide excellently with the values, calculated assuming glucose anhydride (anhydroglucose) (C: 44.4%, C: 6.2%, O: 49.4%).[10]

     (1) and (2) strongly suggest that cellulose is composed of a glucose unit only.

(3)   Tri-substitute ester or ether per gluco(pyrano)se units can be prepared from cellulose. Therefore, cellulose has three free hydroxyl groups per glucose unit.[11]

(4)   Cellotriose, cellotetraose, and cellohexaose are isolated during the process of hydrolysis of cellulose, in addition to cellobiose.[12,13] Existence of higher oligosaccharides in a cellulose molecule supports the long-chain theory of cellulose.

(5)   Trimethyl cellulose, obtained by complete methylation of cellulose, is decomposed to give a single kind of trimethylglucose.[14] The yield of trimethylglucose agrees fairly with the calculated value and tetramethyl, dimethyl and monomethyl glucose are not detected in the degradated products. This means that the chemical bonding pattern existing in cellulose is of a single kind.

(6)   Glucose, decomposed from cellulose, is β type (Figure 1.1.1),[15,16] which was proved by β-glucocidase against cellobiose.

(7)   Glucose, decomposed from cellulose, has a six-membered ring structure (pyranoid).[17]

     β-glucopyranose has a steric structure as illustrated in Figure 1.1.2.

(8)   Complete methylation of cellulose yields octametylcellobiose, which is converted by hydrolysis to 2,3,6 trimethylglucose and 2,3,4,6 tetramethylglucose.[18] This suggests the possibility of 1:5 glucoside or 1:4 glucoside bonding. The former was denied because of formation of intra-molecular dehydration of glucose (between 1:5).

Figure 1.1.2   Steric configuration of cellulose.

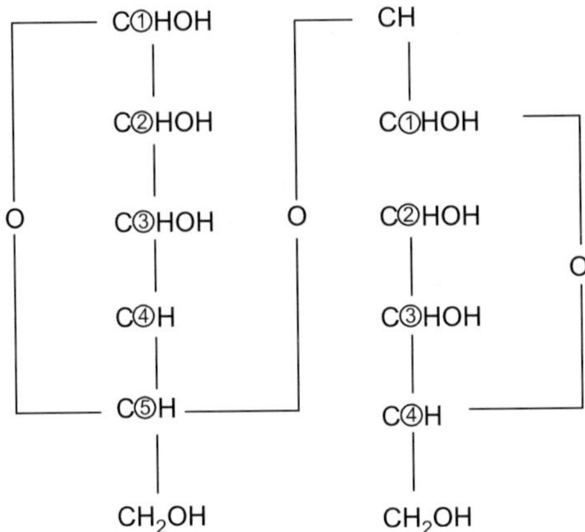

Figure 1.1.3 Cellulose: β-D-glucopyranose.

Cellobiose, obtained from cellulose, is β-glucose(β-D-glucopyranose), which is combined in 1,4-glucoside[19] bonding (Figure 1.1.3). β-glucose has three free hydroxyl groups at the 2, 3, and 6 positions.[20,21] Note that one end unit in methylated cellulose must contain one more methyl group than the units comprising the chain, and then a small amount of tetramethylglucose should be formed on hydrolysis, in addition to trimethylglucose. In the 1920s, the number of glucopyranose units constituting a single cellulose molecule (i.e. degree of polymerization (DP)) was not known.

The molecular weight as estimated by end-group assay is 20,000–40,000.[22]

In 1932, Hawarth[22] isolated an amount of tetra-methylglucose corresponding to one end group in a chain of 100–200 glucose units.

REFERENCES

1. K Kamide, *J. Ind. & Econ., Nara Sangyo Univ.*, 2003, **18**.
2. K Kamide, *History of Textile Industry*, Chapters 4 and 5, Soc. Text. Machin., Japan, 1993.
3. A Payon, *Compt. Rend.*, 1838, **7**, 1052–1125.
4. T Hata, *Polymer Chemistry*, In Fundamentals II, Polymer Structure, Part 4, I. Natural Organic Polymers, Cellulose, (eds K Kanamaru and H Sobue), Shukyo Sha, Tokyo, Chapter 8 Chemical Research of Cellulose, 1944, pp. 1035–1050.
5. CB Purves, A Chemical Structure, Chapter III. In *Cellulose and Cellulose Derivatives, Part I* (eds E Ott and HM Spurlin), Wiley, New York, 1954, p. 54.
6. K Okamura, Wood and Cellulose Chemistry. In *Structure of Cellulose* (eds DN-S Hon and N Shiraishi), Marcel Dekker, New York, 1990, p. 89.
7. H Braconnot, *Ann. Chim. Phys.*, 1819, **12**(2), 172.
8. E Flechsig, *Z. Phys. Chem.*, 1883, **7**, 523.
9. H Ost, *Ann. Chem.*, 1913, **398**, 313.
10. T Hata, *Polymer Chemistry*, In Fundamentals II, Polymer Structure, Part 4, I. Natural Organic Polymers, Cellulose (eds K Kanamaru and H Sobue), Shukyo Sha, Tokyo, Chapter 8 Chemical Research of Cellulose, 1944, p. 1029, Table 4.40.
11. H Ost, *Z. Angew. Chem.*, 1906, **19**, 993.
12. R Willstätter and L Zechneister, *Ber. Dtsch. Chem. Ges.*, 1929, **62**, 722.
13. L Zechneister and G Tóth, *Ber. Dtsch. Chem. Ges.*, 1931, **64**, 854.

14. JC Irvine and EL Hirst, *J. Chem. Soc.*, 1923, **123**, 518.
15. B Tollens, *Ber. Dtsch. Chem. Ges.*, 1883, **16**, 921.
16. E Fischer, *Ber. Dtsch. Chem. Ges.*, 1893, **26**, 2400.
17. W Chalton, WN Haworth and S Peat, *J. Chem. Soc.*, 1926, 89.
18. WN Haworth, CW Long and JH Plant, *J. Chem. Soc.*, 1921, **119**, 193.
19. WS Denham and W Woodhouse, *J. Chem. Soc.*, 1917, **111**, 244.
20. JC Irvine and EL Hirst, *J. Chem. Soc.*, 1922, **121**, 1213.
21. WN Haworth, CW Long and JH Plant, *J. Chem. Soc.*, 1927, 2809.
22. WN Haworth and H Machener, *J. Chem. Soc.*, 1932, **2270**, 2372.

1.2   NEW ERA OF CELLULOSE DERIVATIVES AS ARTIFICIAL MATERIALS IN MIDDLE 19TH CENTURY: CELLULOSE CHEMICAL INDUSTRIES[1]

From the 1830s to the 1850s, the existence of cellulose, with the chemical composition of $(C_6H_{12}O_5)_n$, on the earth glove was recognized as a fundamental component constituting plant tissues On the other hand, in the 1840s and 1850s, medicines and artificial (synthetic) dyestuffs were being developed vigorously for the effective application of dry-distillation residues of coal (coal tar), and as a result, organic chemistry advanced utterly concurrently and remarkably.[2,3] Around this period, the industrial revolution attained a level of maturity, starting first with the mechanization of cotton fiber production and moving on to the utilization of steam engines; and metal machinery in transportation (turnpike roads, before the revolution; water canals and railways).[4] Needless to say, the industrial revolution accelerated the advancement of capitalism. Utilization of wood is a good example. Wood had been extensively employed for the construction of buildings, bridges, and ships, and for fuel. However, there resulted a serious shortage of wood in the whole of Britain, due to the massive usage of charcoal for iron-refining. As a result, the iron-refining industry began importing iron from northern Europe, until the coke method was invented for iron-refining. Tar was the worthless biproduct in the coke production process.

Attempts to convert wood into high-value-added goods were undertaken: The first commercial success was in the paper-manufacturing industry, which used wood pulp as starting material. Until his death, JP Weibel[5]—who was highly successful in wood-pulp manufacture and paper-making—and his associates financed the commercialization of HB de Chardonnet's invention of artificial silk. Prince Guido Henckel von Donnersmarck, an eminent German entrepreneur who succeeded in paper manufacture, financially supported British manufacture of viscose (CF Cross, EJ Bevan and C Beadle) and of regenerated cellulose fiber (CH Stearn and F Topham).[6,7] In addition, Cross and Bevan were classmates and graduates of the paper-making department of Owen's College at Manchester. In the mid-19th century the paper-making industry prospered, particularly in Europe, as a precursor of the polymer chemical industry. The capitalists of this industry were very keen to create additional value-added-goods from wood pulp, other than paper.

In accordance with progress in the industrial revolution, especially the explosive development of the cotton fiber industry (1770–1840), there was massive demand for inorganic chemical compounds used in refining, finishing, and dyeing processes. The prices of these chemicals dropped because of the commercialization of new methods for their production.[8] As a result of the above, study of the chemical reactions of cellulose

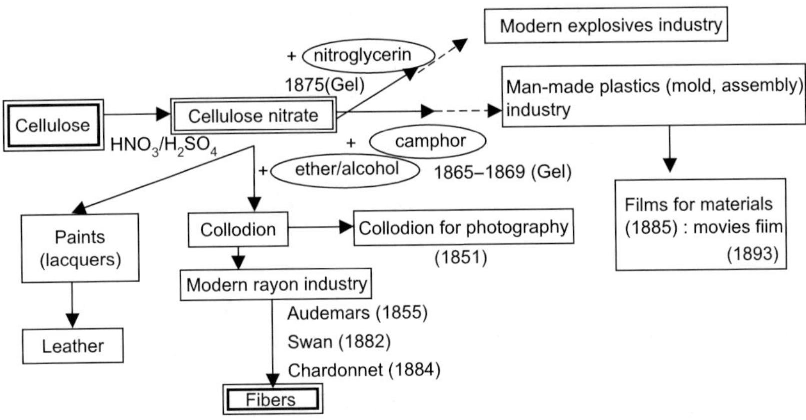

Figure 1.2.1 Genealogy of commercial applications of cellulose nitrate during the 19th century.

became popular between the 1830s and the 1850s. The chemical industry, whose development was induced by the cotton fiber industry (the mechanization of the spinning process (1770–1800) and the weaving process (1830–1840)), motivated the next stage of the development of artificial fiber, which dramatically changed the constitution of the fiber industry thereafter.

Cellulose nitrate, first synthesized in 1833, was the first artificial material, which led to the development of the man-made plastics industry and the modern rayon industry, as illustrated in Figure 1.2.1. Naturally occurring materials, utilized thus far in the traditional industries, were replaced by cellulose nitrate (Table 1.2.1). The products exhibited high performance and the industry expanded quickly.

In addition, cellulose nitrate is readily soluble in various solvents to form gel or solutions, which were used in the commercial processes. For example, until 1895 the following substances were known as solvents for cellulose nitrate:[9] (1) Essig säure, allein order gemischt mit Alkohol order Äther, (2) Schwefel säure, (3) Ätherschwefelsäure, (4) Aldehyde, Anilin, (5) Lösung von Kampfer in Alkohol, Äther, Benzol, Toluol order

Table 1.2.1

Cellulose nitrate as new material in 19th century

Sector	Products used until emergence of cellulose nitrate	Cellulose nitrate products	Remarks
Explosives	Gun powder	Smokeless powder cunamite	Modern explosives
Plastics	Natural resin	Celluloid (first artificial thermoplastics)	Synthetic
Lacquer coating	Oil, natural resin	Lacquer	Modern paints
Fiber	Silk	Artificial silk	Chemical fibers

Tetrachlorkohlenstoff, (6) Essigäther, (7) Aceton, (8) Äther-alkohol, (9) Holzgeist (Methylalkohol), (10) Nitroglyzerin, (11) Nitrobenzol, (12) Amylacetate, (13) Verduennte alkoholische Lösungen verschiedener Salze. These characteristics (i.e. formability of gel and solutions) of cellulose nitrate played a decisive role in the creation of a new industry. It was rather a matter of course that cellulose nitrate was chosen in the late 19th century as a starting material for the production of man-made fibers. Regrettably, no comprehensive historical survey on cellulose nitrate has been undertaken, even in Europe; although the importance of cellulose nitrate from the late 19th century to the early 20th century can never be underestimated. The rise of the cellulose chemical industry during this time saw the true emergence of the polymer chemical industries. From there, the science of cellulose and cellulose derivatives as macromolecules was born.

REFERENCES

1. See, for example, K Kamide, *J. Ind. & Econ., Nara Sangyo Univ.*, 2000, **18**, 157.
2. K Kamide, *History of Textile Industry*, Society of Textile Machinery of Japan, 1993, p. 216.
3. K Kamide, *Polymers*, 2001, **50**, 405–407.
4. K Kamide, *History of Textile Industry*, Chapter 5, pp. 151–183.
5. K Kamide, *History of Textile Industry*, p. 223.
6. K Kamide, *History of Textile Industry*, p. 258.
7. K Kamide, *History of Textile Industry*, p. 248–261.
8. K Kamide, *History of Textile Industry*, p. 212.
9. Th Schlumberger, Deutsche Patent Nr. 93, 009 1895.

1.3 EMERGENCE OF POLYMER SCIENCE IN THE 1920s AND 1930s

1.3.1 Micellar theory

Naturally occurring materials such as cellulose and its derivatives were considered for long low-molecular-weight compounds, whose molecules are associated to form colloid in solutions, and anhydroglucopyranose associates to form a cellulose micelle solid by Van der Waal's force.[1]

Experimental evidence presented from 1925 to 1927 supporting the above theory (micellar or low-molecular-weight compound theory) is as follows:

(1) The molecular weight, estimated by the freezing-point depression method, for cellulose acetate in phenol[2] and cellulose triacetate in glacial acetic acid,[3] coincides approximately with those of an elemental unit constituting the material.

(2) The size of 'micell', as determined by diffusion velocity, of cellulose nitrate solution, agrees with the size of crystallite cellulose nitrate evaluated by X-ray diffraction.[4]

(3) Bonding formation between cellulose and copper in cellulose cuprammonium solution clearly obeys the mass-action law if glucose is taken as unit.[5]

Staudinger and his co-workers strongly insisted that cellulose and cellulose derivatives are macromolecules (macromolecule theory), employing osmometry and viscometry to support their hypothesis,[6] and Staudinger finally gained his point.

1.3.2 Osmotic pressure

In ideal solution, osmotic pressure is related to the molecular weight M of solute through Van't Hoff's equation[7]

$$\pi = RT_c/M \tag{1.3.1}$$

(π, osmotic pressure; R, gas constant; T, temperature; c, concentration; M, molecular weight).

Osmotic pressure of solutions of polymers including dye stuff, gelatin, starch (1910–1916), rubber (1914), cellulose ester (1927–1934), and proteins (1919–1930) were measured as early as 1900, before the establishment of the concept of macromolecules.[8] Wo Ostwald analyzed the literature data on rubber–benzene, gutta percha–benzene, cellulose nitrate–acetone, gelatine–water, and hemoglobin–water systems, showing that Van't Hoff's equation did not hold for all the above systems.[9] Note that Van't Hoff himself described that eq. (1.3.1) is valid only in extremely dilute solution.[10]

It was also shown that Van't Hoff's equation is not valid even for low concentrations for linear chain polymers. Theoretical explanations for the deviation from Van't Hoff's equation were proposed in the 1940s by numerous researchers such as Flory,[11] Huggins,[12,13] Miller,[14] and Guggenheim.[15] Osmometry was first applied for cellulose derivatives in the 1930s: Staudinger, Schulz and their coworkers measured the osmotic pressure for cellulose nitrate in acetone[16,17] and cellulose acetate in acetone.[18] Schulz's results were cited in some references.[19,20] The apparatus and measurement procedure of osmotic-pressure were described in some detail by Schulz,[17] and Fuoss and Mead.[21,22] Dobry demonstrated that the plots of π/c (π, osmotic pressure; c, concentration) versus c for a cellulose nitrate sample in various solvents (ethylbenzoate, ethanol, methylsolicylate $+20\%$ methanol, acetone, acetic acid, methanol, nitrobenzene) yield the same value of intercept at $c = 0$ (i.e. $\lim_{c \to 0} \pi/c$) (the same number-average molecular weight M_n value).[23,24] Doby's results strongly suggest that cellulose nitrate molecularly dissolves in dilute solutions, and these data were cited by Sakurada (1946),[25] Schulz (1953),[26] Doty and Spurlin (1954),[27] and Staudinger (1961).[28]

1.3.3 Solution viscosity

Before the establishment of polymer science, the solution viscosity, measured at the same concentration of natural polymer including cellulose, cellulose derivatives, and starch, had been employed as a measure of quality, representing a degree of degradation.[29]

It was shown that the viscosity of solutions of linear polymer, such as starch and cellulose derivatives, varies with the molecular weight of solute, and cannot be explained by Einstein's rule,[30] derived for solutions in which rigid sphere are suspended.

$$\eta = \eta_0(1 + 2.5\Phi) \tag{1.3.2}$$

where η is the viscosity of solution and η_0, that of solvent, Φ, the solute volume fraction. Equation (1.3.2) is rewritten as

$$\eta_{sp}/c(\equiv \{/\eta_0\} - 1)/c) = 2.5v \tag{1.3.3}$$

η_{sp} is the specific viscosity and c, the concentration (g/cm^3), v, the specific volume of solvent. Equation (1.3.2) predicts the constancy of η_{sp}/c, irrespective of c and the molecular weight.

The experimental values of $\{(\eta/\eta_0) - 1\}/\Phi$ for high-molecular-weight polymer solutions are sometimes 1000 times the theoretical value (2.5).

In 1930, Staudinger and Nozu[31] showed that η_{sp}/c is proportional to the solute molecular weight (by the freezing-point depression method) of linear paraffin fraction solutions in carbon tetrachloride.

Then, Staudinger and his coworkers observed the linear proportionality of η_{sp}/c to molecular weight M in a number of polymer solutions (Staudinger equation or Staudinger rule).[32,33]

$$\eta_{sp}/c = K_m M \tag{1.3.4}$$

Staudinger and Schulz evaluated the constant $K_m(= \eta_{sp}/c/(P_n m))$ (P_n, the number-average degree of polymerization by osmometry) of cellulose nitrate–acetone solutions using four whole polymers (nitrates from a viscose rayon and three bleached liters) and a fraction.[34] Staudinger also analyzed the literature data of η_{sp}/c and M_n on a cellulose nitrate–acetone system, reported by Bücher and Steuel, and Dobry, concluding that K_p ($\equiv K_m \cdot m$(m: molecular weight of monomer)) $= 7.0 \times 10^{-4} \sim 20 \times 10^4$) for the former and 8×10^{-4} ($M_n = 10 \times 10^4$) for the latter.[35] Staudinger and Daumiller obtained an approximately constant value of K_m for seven cellulose triacetates-m-cresol systems.[18] Thereafter, Schulz studied the relation between viscosity and the number-average degree of polymerization P_n over a wide P range for 16 cellulose-nitrate fractions.[36,37] He also noticed an approximate constancy of K_p within a group of samples.

Staudinger's equation (1.3.4) was theoretically derived for free draining random coil molecules.[38–42] (i.e. on the assumption that the flow inside the chain is the same as that on the outside).

Husemann *et al.* compiled a table of K_p for solutions of cellulose and 8 cellulose derivatives.[43] Sakurada commented on the influence of the polymolecularity of the polymer samples used, and the nature of its average on the K_p value.[44] He also pointed out the necessity of detailed further experiments using well-fractioned samples.

REFERENCES

1. See, for examples, K Hess, *Liebigs Ann, Chem.*, 1924, **435**, 1; K Hess, *Z. angew. Chem.*, 1924, **37**, 993.
2. M Bergmann, *Liebigs Ann. Chem.*, 1925, **452**, 149.
3. K Hess, *Chemie zur Zellulose*, Liepzig, 1928, p. 585.
4. RO Herzog, *Ber. Dtsch. Chem. Ges.*, 1925, **58**, 1254.
5. T Hata, *Polymer Chemistry*, In Fundamental, **Vol. 2** (eds K Kanamaru and H Sobue), Shukosha, Tokyo, 1944, p. 1031.
6. See, for example, Hermann Staudinger, Arbeitserinnerungen, Dr. Alfred Hüthig Verlag, 1961.
7. JH van't, Hoff. *Z. Physik. Chem.*, 1887, **1**, 481.
8. K Kamide and T Dobashi, *Physical Chemistry of Polymer Solutions: Theoretical Background*, Elsevier, 2000, p. 25.

9. Wo Ostwald, *Kolloid-Z.*, 1918, **23**, 68; Wo Ostwald, *Kolloid-Z.*, 1929, **69**, 60; K Kamide and T Dobashi, *Physical Chemistry of Polymer Solutions: Theoretical Background*, Elsevier, 2000, pp. 25–26, Equation (2.8.12).

10. JH van't Hoff, *Vorlesunger über Theorerische und Phzsikalische Chemie, Part 2.* 1903; K Kamide and T Dobashi, *Physical Chemistry of Polymer Solutions: Theoretical Background*, Elsevier, 2000, p. 25.

11. PJ Flory, *J. Chem. Phys.*, 1942, **10**, 51; PJ Flory, *J. Chem. Phys.*, 1945, **13**, 453.

12. ML Huggins, *J. Am. Chem. Soc.*, 1942, **64**, 1712.

13. ML Huggins, *Ann N.Y. Acad. Sci.*, 1942, **43**, 1.

14. AR Miller, *Proc. Cambridge Phil. Soc.*, 1943, **39**, 54.

15. EA Guggenheim, *Proc. Roy. Soc.(London)*, 1944, **A183**, 203.

16. H Staudinger and GV Schulz, *Ber. Dtsch. Chem. Ges.*, 1935, **68**, 2320.

17. GV Schulz, *Z. Phys. Chem.*, 1936, **A716**, 317.

18. H Staudinger and G Daumiller, *Ann. Chem.*, 1937, **529**, 219.

19. ML Huggins, in *Cellulose and Cellulose Derivatives* (eds Emil Otto and HM Spurlin), Wiley, New York, 1954, Chapter X, B3, Fig. 21.

20. A Hatano and Y Tabata, in *Cellulose Handbook* (eds H Sobue and N Migita), Asakura, Tokyo, 1958, p. 506, Fig. 14.3.

21. RM Fuoss and DJ Mead, *J. Phys. Chem.*, 1943, **47**, 59.

22. See also, GV Schulz, Das Makromolekül in Lösungen, Springer Verlag, Berlin, 1953, §59 (pp. 380–387), §60 (pp. 387–394) and §61 (pp. 394–406).

23. A Dobry, *J. chim. Phys.*, 1935, **32**, 51.

24. A Dobry, *Kolloid-Z.*, 1937, **81**, 190 (ref. 25 lacks data on acetone and cyclohexanon +5.8% ethanol).

25. I Sakurada, *Polymer Chemistry*, Publishing Association of Polymer Chemistry, Kyoto, 1955, p. 151 and 154–155, Fig. 27.

26. GV Schulz, *Z. Phys. Chem.*, VII 17, p. 399.

27. PM Doty and HM Spurlin, in *Cellulose and Cellulose Derivatives* (eds Emil Otto and HM Spurlin), Band II, D. Chapter X, Fig. 40.

28. H Staudinger, Arbeitserinrerungen, Sr. Hütig Verlag, 1961, IV.

29. I Sakurada, *Polymer Chemistry*, Publishing Association of Polymer Chemistry, Kyoto, 1955, p. 175 and ref. cited there: H Ost, *Z. ang. Chem.*, 1911, **24**, 1892; F Baker, *J. Chem. Soc., London*, 1913, **12**, 1653; WH Gibson, *J. Chem. Soc., London*, 1920, **117**, 479; J Duclaux and E Wollmann, *Bull. Soc. Chim.*, France, 1920, **27**, 414; E Heuser and N Hiemer, *Cellulose Chemie*, 1925, **6**, 127; E Berl and A Lange, *Cellulose Chemie*, 1920, **7**, 146; Esselen, *Ind. Eng. Chem.*, 1926, **18**, 1031; J Reitstöer, *Kolloid-Z.*, 1927, **41**, 362; W Biltz, *Ber. Dtsch. Chem. Ges.*, 1913, **46**, 1532; W Biltz, *Z. Physik. Chem.*, 1923, **83**, 703.

30. A Einsterin, *Ann. Phys.*, 1906, **19**, 289; A Einsterin, *Ann. Phys.*, 1911, **34**, 591; See, also, K Kamide and T Dobashi, *Physical Chemistry of Polymer Solutions: Theoretical Background*, Elsevier, 2000, p. 495, "Problem 8–27", Equation (8.27.1).

31. H Staudinger and R Nozu, *Ber. Dtsch. Chem. Ges.*, 1930, **63**, 721.

32. H Staudinger and W Heuer, *Ber. Dtsch. Chem. Ges.*, 1930, **62**, 222; H Staudinger and W Heuer, *Z. Physik. Chem. A*, 1931, **153**, 391; H Staudinger and W Heuer, *Ber. Dtsch. Chem. Ges.*, 1932, **65**, 267.

33. GV Schulz, *Z. Phys. Chem. B*, 1936, **32**, 27.

34. H Staudinger and GV Schulz, *Z. Angew. Chem.*, 1936, **49**, 804.

35. I Sakurada, *Polymer Chemistry*, p. 178.

36. GV Schulz, *J. Prakt. Chem.*, 1942, **161**, 147.

37. E Husemann and GV Schulz, *Z. Phys. Chem. B*, 1942, **52**, 1.

38. K Kamide and T Dobashi, *Physical Chemistry of Polymer Solutions: Theoretical Background*, Elsevier, 2000, Problem 8–30, p. 503.

39. W Kuhn, *Kolloid-Z*, 1934, **68**, 2.

40. ML Huggins, *J. Phys. Chem.*, 1938, **42**, 911; ML Huggins, *J. Phys. Chem.*, 1939, **43**, 439; ML Huggins, *J. Appl. Phys.*, 1939, **10**, 700.

41. HA Kraemer, *J. Chem. Phys.*, 1946, **14**, 415.

42. P Debye, *J. Chem. Phys.*, 1946, **14**, 636.
43. E Husemann, E Plötze and GV Schulz, *Natuwiss.*, 1941, **29**, 257.
44. I Sakurada, *Polymer Chemistry*, p. 180.

1.4 FURTHER DEVELOPMENT OF MOLECULAR CHARACTERIZATION OF CELLULOSE AND CELLULOSE DERIVATIVES FROM THE 1940s TO THE 1970s

1.4.1 Distribution of degree of substitution

Relationships between the nitrogen content ($N_c\%$) of cellulose nitrate and the solubility into ether-alcohol (2:1 in volume), established by Will in 1902, indicate that cellulose nitrate with $N_c = 10.5 \sim 12.6\%$ is soluble in the mixture (the solubility $= 95 \pm 5\%$).[1] Figure 1.4.1 is the plot of the solubility against $N_c\%$ of cellulose nitrate in the mixture of ether-alcohol, reproduced from Ref.1.

The first attempt to determine the average degree of substitution of the hydroxyl group attached to the C_k ($k = 2, 3, 6$) atoms in glucopyranose unit (see Section 2.1), doubly average over all glucopyranos units in all molecules (see Section 2.1) $\langle\langle f_k \rangle\rangle$ for a CD was made on cellulose xanthate (CX) to determine the selective activity of CS_2 to the C position. Lieser[2] hydrolyzed methylcellulose (MC) which was synthesized from CX using nitromethyl urethane followed by removal of unsubstituted glucose by fermentation, and tried to confirm that xanthate groups substituted mainly at the C_2 position, reacting phenylhydrazine to methylglucose. Until the mid-1960s, the distribution of substituent xanthate groups was determined by chemical methods[3-9]; that is, CX was converted to more stable derivatives (for example, methylcellulose) and then hydrolyzed to glucose and fractions followed by analysis such as paper chromatography (Table 1.4.1).

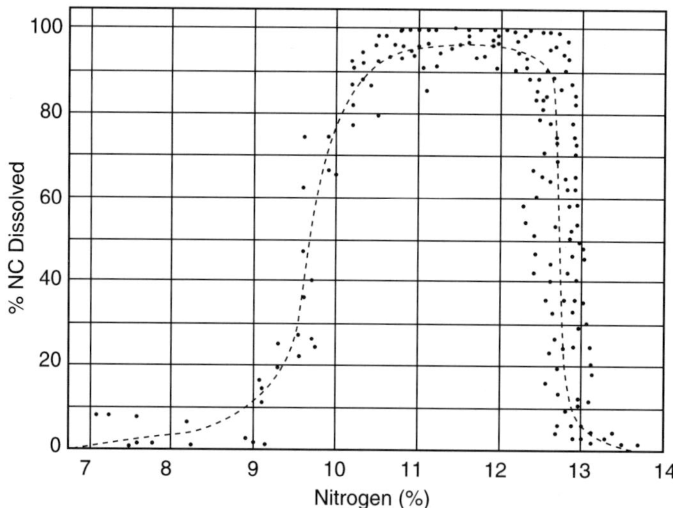

Figure 1.4.1   Nitrogen content and solubility in ether-alcohol(2-1)(Will).

Table 1.4.1

Determination of distribution of degree of substitution

Polymer	1940	1950	1960	1970	1980
Sodium cellulose xanthate (NaCX), cellulose acetate (CA), cellulose nitrate (CN)	1929: Lieser[3] (NaCX)	1942: Gardner and Purois[12] (CA)	1950: Noguchi[4] (NaCX) 1951: Lauer[5] (NaCX) 1951: Chen et al.[6] (NaCX) 1950: Malm et al.[13] (CA) 1958: Horio et al.[7] (NaCX) 1959: Willard and Pascu[8] (NaCX) 1959: Philipp and Liu[9] (NaCX)	1963: Horio et al.[10] (NaCX) 1966: Jaregg et al.[11] (NaCX)	1971: Goodlett et al.[14] (CA)	1981: Kamide and Okajima[2] (CA) 1983: Saito[15] (CA) 1984: Miyamoto et al.[16] (CA) 1986: Kowsaka et al.[17] (CA) 1981, 1982: Clark et al.[19,20] (CN) 1980: Wu[18] (CN) 1986: Kamide et al.[22] (CX)
Sodium cellulose sulfate (NaCS), sodium carboxymethylcellulose (NaCM), methylcellulose (MC),					1972: Ho et al.[23] (HPC) 1977: Parfondry and Perlin[25] (MC, NaCMC, HEC)	1981: Kamide and Okajima[21] (NaCS) 1982: Kamide et al.[24] (NaCMC)
Hydroxypropylcellulose (HPC), hydroxyethyl-cellulose (HEC)						1982: Lee and Perlin[26] (HPC) 1980: Ho and Klosiewicz[27] (NaCMC)

However, $\langle\langle f_k \rangle\rangle$ of CX cannot be estimated accurately by this method because of the possibility of a change of distribution at each process.

For cellulose acetate (CA), several attempts have been made to evaluate $\langle\langle f_k \rangle\rangle$ by chemical methods,[10,11] usually based on the differences in reactivity of these unsubstituted hydroxyl groups at tosylation with *p*-toluene-sulfonyl chloride or at tritylation with trityl chloride in the presence of pyridine. These chemical methods are very tedious and time-consuming. In addition, $\langle\langle f_k \rangle\rangle$ and $\langle\langle f_3 \rangle\rangle$ cannot be evaluated, because there are no significant differences between the reactivities of hydroxyl groups at C_2 and C_3. For this reason, since the 1970s, high-resolution ^{1}H- and ^{13}C-NMR spectroscopy has been extensively applied to structural analysis of CA and other CD polymers. Goodlett *et al.*[14] have acetylated incompletely substituted CA polymers with deuterated acetyl chloride, and determined the distribution of *O*-acetyl groups by ^{1}H-NMR measurements. However, in Goodlett *et al.*'s method, the experimental determination of the peak intensity is not sufficiently accurate because of the low separation of three peaks in the ^{1}H-NMR spectra and because $\langle\langle f_k \rangle\rangle$ values have a significant possibility of changing during the deuteration process.

1.4.2 Molecular weight fractionation

Well-characterized polymer materials with narrow molecular-weight distributions are widely used in studying the establishment of structure–properties–processability relations. The first application of fractional precipitation to cellulose derivatives was made in 1920 by Duclaux and Wollma,[28] who fractionated cellulose nitrate into three fractions with the use of acetone as a solvent, and water (or acetone–water) as a precipitant. They found that the viscosity of the fractions was different, but the nitrogen content was not. (Now, instead of water, *n*-heptane is used. The *n*-heptane has a weaker power to precipitate than water.) From the 1920s to the 1930s, many kinds of cellulose derivatives (CD) were fractionated by the precipitational method: cellulose diacetate (CDA) (in 1933, acetone/water, solvent/nonsolvent)[29]; ethylcellulose (EC) (1933, glacial acetic acid solution/water)[30]; benzylcellulose (1936, ethanol/ethanol-benzene)[31]; methylcellulose (1938, water/ sodium sulfate)[32]; CX (1934, aq. pyridine/methanol).[33] For cellulose and CDs, the successive solution fractionation (SSF) method, which is superior to the successive precipitational (SPF) method, did not come into wide use until the late 1970s. For the molecular-weight fractionation of CTA, acetic acid and chlorinated hydrocarbons, which have a low dielectric constant ε, were extensively employed as solvent. Unfortunately, the fractionation efficiency achieved by using these solvents was poor, and numerous attempts made so far have met with very limited success.[34–42]

In contrast to CD solution, the direct fractionation of cellulose has never been carried out successfully. Many alkaline metal complex solutions, which are solvents of cellulose, are not suitable as the solvent of the solution fractionation, because some of them break up the cellulose chain, especially under atmospheric oxygen, and/or precipitate themselves by non-solvent of cellulose. In alternative procedures, a derivative (usually the nitrate and acetate) is prepared and fractionated, then cellulose is regenerated. However, in addition to the probability of the breakdown of the cellulose chain during esterification, this procedure is laborious and time-consuming. For these reasons, molecular-weight fractionation of cellulose has been achieved by the hydrolysis method, using acid such as hydrochloride,[43] and consequently it is very difficult to obtain fractions with $M_w/M_n \leq 2$ (M_w, the weight-average molecular weight).

1.4.3 Light scattering

In polymer chemistry, light scattering is one of the most efficient means to determine M_w, the radius of gyration $\langle S^2 \rangle^{1/2}$, and the second virial coefficient A_2 of polystyrene solutions. Light-scattering measurements were first made on CA (2. 35)-acetone solutions as early as 1946 by Stein and Doty[44] (Table 1.4.2). However, as pointed out by many researchers,[59,60,78–85] commercially available CDA polymer forms aggregate, even in good solvents. In the mid-1970s, Kamide *et al.* studied the dissolved state of CDA in acetone and tetrahydrofuran by means of gel permeation chromatography GPC, thin-layer chromatography TLC, and infrared (IR) spectroscopy. The studies on light scattering of cellulose solutions are extremely few owing to the well-known fact that a stable and simple solvent, in which cellulose can be dissolved molecularly, has not been found. Molecular decomposition cannot be avoided if strong acids or alkalis are used. Most metal complex alkali solutions decompose cellulose by oxidation and are unstable and colored. In addition, cellulose dissolves with formation of the complex into these solutions. Hydrozine and N-morphorin-N-oxide also can dissolve cellulose, but these are sometimes explosive and highly toxic. Therefore, the above solvents are not used for molecular characterization of cellulose.

1.4.4 Molecular weight distribution

Since Schulz[86] suggested that fractional precipitation is useful in investigating the molecular weight distribution (MWD) of the original polymer, this fractionation method was applied extensively to cellulose acetate and nitrate[87] until the early 1960s (Table 1.4.3).

Table 1.4.2a

Light scattering measurements on cellulose and its derivatives solution

	1940	1950
Cellulose (Cell)		
Cellulose acetate (CA)	1946: Stein and Doty[44] CA (2.35)-acetone	1953: Experimental failure reported[45]
Cellulose nitrate (CN)		1954: Holtzer *et al.*[46] CN (2.98)-acetone 1956: Hunt *et al.*[47] CN (2.77)-ethylacetate 1958: Haque *et al.*[48] CN (2.58)-acetone, -ethylacetate
Cellulosetricarbanilate (CTC) and sodium carboxymethyl-cellulose (NaCMC)		1954: Sneider and Doty[49] NaCM (1.2)-NaCl

Table 1.4.2b

Light scattering measurements on cellulose derivatives solution

	1960	1970	1980
Cellulose (Cell)	1961: Henley[43] Cell-cadoxen 1965: Brown and Wirkström[50] Cell-cadoxen	1971: Valtassari[51] Cell-FeTNa	1986: Kamide and Saito[52] Cell-LiOH 1986: McCormick *et al.*[53] cell-DMAc-LiCl
Cellulose acetate (CA)	1967: Shakhparonov *et al.*[54] CA (2.7–2.9)- meth.chlor./methanol	1974: Tanner and Berry[57] CA (2.45)-TFE, -meth, chlor./methanol 1979–1984: Kamide *et al.*[58–61] CA (2.92), CA (2.46), CA (0.49) in various solvents	1981: Suzuki *et al.*[55] CA (2.46)-acetone 1982: Suzuki *et al.*[56] CA (2.46)-MEK 1983: Saito[15] CA (1.75)-DMAc
Cellulose nitrate (CN)	1968: Schulz and Penzel[62] CN (2.91)-, CN (2.55)- acetone		
Cellulosetricarbanilate (CTC) and sodium carboxymethyl- cellulose (NaCMC)	1961: Burchard and Husemann[63] CTC-acetone, -dioxane, -pyridine 1962: Sitarmaiah and Goring[64] NaCMC-NaCL 1963: Brown *et al.*[65] NaCMC (0.9)-cadoxed 1964: Brown and Henley[66] NaCMC (1)-NaCl 1965: Burchard[67] CTC-dioxine 1968: Shanbhag *et al.*[68,69] CTC-anisole, -cyclohexanol,[68] -acetone, -dioxane, cyclohexanon[69]		
Miscellanea	1956: Manley[70] EHEC-water 1960: Krigbaum and Sperling[71] Cu. CT-Cap. 1963: Neely[72] MC (1.8)-water 1963: Brown *et al.*[73] HEC (0.88)-water 1963: Friedberg *et al.*[74] HEC (1.67)-DMSO, -formamide 1967: Gohsh and Chaudhury[75] CX-NaOH 1969: Das *et al.*[76] CX (0.76)-, CX (0.82)-NaOH		1981: Kishino *et al.*[77] NaCS-NaCl

GPC, which has the advantages of rapidity and ease of determining the molecular weights and molecular weight distribution, is now used for many polymers. Two years later, when Moore[108] first devised GPC, Segal[99–102] measured GPC of CN samples in THF, converted from cellulose I to cellulose IV, and evaluated the MWD of cellulose

Table 1.4.3

Molecular-weight distribution

	1940	1950	1960	1970	1980
Cellulose		1955: Timell[88] F (nitration) 1958: Marx[89] F (nitration)	1961: Henley[43] L/M 1967, 1968: Erikson et al.[90,91] GPC 1968: Brown and Wirkström[50] L/M 1969: Pettersson[92] GPC	1977: Berek et al.[93] GPC	1980: Bao et al.[94] GPC
Cellulose acetate (CA)	1942: Sookne et al.[95] CA (2.35) F 1947: Badgley and Mark[96] CA (2.35) L/M		1968: Brewer et al.[97] GPC 1968: Mueller and Alexander[98] GPC	1974: Tanner and Berry[57] CA (2.45), GPC, L/M 1979–1984: Kamide et al.[58–61] CA (2.92)-CA (0.49) L/M, GPC	
Cellulose nitrate (CN)		1954: Holtzer et al.[46] CN (2.98) L/M 1956: Hunt et al.[47] CN (2.77) L/M	1966–1970: Segal[99–102] GPC 1967: Meyerhoff et al.[103] GPC 1968: Mueller and Alexander[98] GPC 1969: Huang and Jenkins[104] GPC	1971: Alexander and Mueller[105] GPC	
Cellulose tricarbanilate (CTC)			1961: Burchard and Husemann[63] L/M 1968: Shanbhag[69] L/M	1975: Guthrie et al.[106] L/M 1976: Danhelka et al.[107] GPC	
Miscellanea		1954: Schneider and Doty[49] NaCMC (1.2) L/M 1956: Manley[70] EHEC L/M	1960: Krigbaum and Sperling[71] CT-Bu, CT-Cap. L/M 1963: Neely[72] MC (1.8), L/M 1963: Brown et al.[65] NaCMC (0.9), L/M		

F, fractionation method; L/M, M_w/M_n determined by light-scattering and membrane osmometry; GPC, gel permeation chromatography.

samples, calibrating the GPC data using polystyrene. However, the weight-average degree of polymerization DP (P_w) of cellulose I, II, and III obtained by the GPC method was approximately four times larger than their viscosity-average DP (P_v). Meyerhoff and Javanovic[109] pointed out that if fractionated cellulose nitrate is used as a standard, P_w of CN, prepared from cellulose I–IV, is very close to P_v. But in the application of GPC to cellulose nitrate there still remains many unsolved problems, such as the change in calibration curve of cellulose nitrate with time[109] and the possibility of degradation of CN in THF solution.[100]

Numerous attempts at estimating the MWD of cellulose were made using cadoxen as an eluent and polyamide,[90] agarose,[91,92] or another gel[93,94] for chromatographic support.

1.4.5   Mark–Houwink–Sakurada equation

As the first step of polymer characterization, it is of paramount importance to determine the relationship between the limiting viscosity number $[\eta](\equiv \lim_{c\to 0} \eta_{sp}/c)$ and the molecular weight. The above relations have been utilized not only as a conventional method for estimating the molecular weight (the viscosity-average molecular weight M_v), but also as a useful method for obtaining information on the thermodynamic and hydrodynamic interactions between polymer and solvent molecules. Modifying Staudinger viscosity law, (eq. (1.3.4)), Mark,[110] Houwink,[111] and Sakurada[112] independently proposed an empirical equation relating $[\eta]$ and molecular weight M (MHS equation) as follows:

$$[\eta] = K_m M^a \tag{1.4.1}$$

where K_m and a are constant parameters determined by the combination of polymer, solvent, and temperature. Table 1.4.4 lists the works on the MHS equation of cellulose and CD solutions since the 1940s. The value $[\eta]$ of cellulose solution in FeTNa is the largest among the three solvents. The exponent α for aq LiOH, cadoxen, and FeTNa lies between 0.76 and 0.79. Contrary to this, α in the MHS equation established using sedimentation-diffusion average molecular weight M_{SD} for cupriammonium hydroxide (Cuoxam), cupriethylene-diamine hydroxide (Cuen), and EWNN (Eisen-Weinsäure-Natrium Komplex, discovered by Jayme and Verburg[129]) is larger than 0.9. The MHS equations, obtained using sedimentation-diffusion average molecular weight M_{SD} are theoretically expected to be rather sensitive to the polymolecularity of the polymer samples. It should be necessary in advance to conclude about the flexibility of cellulose in these solvents, to confirm that the samples can be regarded as monomolecular or that the molecular weight dependence of the polymolecularity of the sample is small enough to be neglected. Cellulose derivative solutions have situations similar to those of cellulose. For example, the α value in the MHS equation for CDA[119] and CTN[127]-acetone systems, which were established using M_n, scattered largely from 0.67 to 1 and 0.75 to 1, respectively. $[\eta]$ of the polymer solution was affected by the molecular weight distribution as well as the average molecular weight of the polymer.

In 1953, Flory referred[130] to some experimental results on cellulose derivative solutions:

(1)   Plots of intrinsic viscosity for cellulose tricaprylate in γ-phenyl propyl alcohol at 48 °C (i.e. theta solvent) against the molecular weight.[131]

Table 1.4.4a

Mark–Houwink–Sakurada equation

	1940	1950	1960	1970	1980
Cellulose (Cell)	1941: Gralen and Svedberg[113] Cell (M_{SD})-cuoxam	1959: Vink[114] Cell (M_n)-cuen	1961: Henley[43] Cell (M_W)-cadoxen 1965: Brown and Wirkstrom[50] Cell (M_W)-cadoxen	1971: Valtassari[51] Cell (M_W)-FeTNa 1977: Lyubina et al.[115] Cell (M_{SD})-cadoxen	1986: Kamide and Saito[52] Cell (M_W)-LiOH 1986: McCormic et al.[53] Cell (M_W^*)-DMAc/LiCl
Cellulose acetate (CA)	1943: Bartovics and Mark[116] CA (2.35) (M_n)-acetone 1945: Sookne and Harris[117] CA (2.36) (M_n)-acetone 1947: Badgley and Mark[96] CA (2.40) (M_n)-acetone	1951: Philipp and Bjork[118] CA (2.38) (M_n)-acetone 1958: Cumberbirch and Harland[119] CA (2.34) (M_n)-acetone 1958: Moore and Tidswell[120] CA (2.34) (M_n)-acetone 1958: Sharples and Major[121] CA (2.91) (M_n)-acetone	1964: Dymarchuk et al.[122] CA (2.91) (M_n)-TCM 1967: Golubev and Frenkel[123] CA (2.25–2.38) (M_n, M_{SD}, M_{SV})-acetone	1974: Tanner and Berry[57] CA (2.46) (M_W)-THF 1977: Nair et al.[36] CA (2.93) (M_W)-TCM 1979–1983: Kamide et al.[58–61] CA (2.92)-CA (0.49) (M_W, M_n) in various solvents	1983: Saito[15] CA (1.75) (M_W)-DMAc
Cellulose nitrate (CN)	1943: Mosimann[124] CN (2.32) (M_{SD})-acetone	1954: Holtzer et al.[46] CN (2.98) (M_W)-acetone 1956: Hunt et al.[47] CN (2.77) (M_W)-ethyl. acetate 1956: Moore and Epsten[125] CN (2.42) (M_n)-acetone 1958: Meyerhoff[126] CN (2.87) (M_{SD}, M_n)-acetone 1958: Harland[127] CN (2.95) (M_n)-acetone 1958: Huque et al.[47] CA (2.85) (M_W)-ethyl. acetate, -acetone	1968: Schulz and Penzel[62] CN (2.91), CN (2.55) (M_W)-acetone 1960: Moore and Edge[128] CN (2.91), CN (2.55) (M_n)-acetone		

Table 1.4.4b

Mark–Houwink–Sakurada equation

	1940	1950	1960	1970	1980
Miscellanea		1956: Manley[70] EHEC (M_W, M_n, M_{SD})-water	1961: Burchard and Husemann[63] CTC (M_W)-acetone-, dioxan 1961: Uda and Mayerhoff[74] MC (1.66) (M_{SD})-water 1962: Sitarmaiah and Goring[64] NaCMC (0.62–0.74)-NaCl 1963: Brown *et al.* HEC (0.88) (M_W)-cadoxen, -water[73] NaCMC (0.9-cadoxen[65] 1963: Neely[72] MC (1.8), (M_w)-water 1964: Brown and Henley[66] NaCMC (1)-NaCl 1968: Shanbhag[69] CTC (M_W)-dioxane, -acetone, -cyclohexanon	1975: Guthrie *et al.*[106] CTC (M_W)-dioxan	

M_{SD}, sedimentation-diffusion-average mol.; M_W, weight-average mol.; M_n, number-average mol; M_{SV}, sedimentation-viscosity-average mol.; M^*, apparent mol.

(2) The unperturbed end-to-end chain lengths $\langle R^2 \rangle_0^{1/2}/M^{1/2}$ of cellulose tributyrate calculated from intrinsic viscosities and molecular weight $K(= K_m$ at theta solvent) (eq. (1.4.1)) values using the values of $\Phi(2.1 \times 10^{23})$ empirically obtained.[132]

Flory wrote in the same book,[133] that cellulose tributyrate is exceptional with respect to the effect of temperature in reducing $\langle R^2 \rangle_0^{1/2}/M^{1/2}$. At high temperatures its $\langle R^2 \rangle_0^{1/2}/\langle R^2 \rangle_{0f}^{1/2}$ ratio $\langle R^2 \rangle_{0f}$ is $\langle R^2 \rangle_0$ for chains with free internal rotation is more or less normal, and compliance with viscosity relations set forth above is indicated by the values obtained for $K(\equiv [\eta]/M^{1/2})$ using fractions differing in molecular weight. He also noted[134] that "Other cellulose derivatives (than cellulose tricaprylate and tributyrate), such as the nitrate and the acetate, give evidence of abnormally low chain flexibility... It seems necessary to postulate some sort of specific interaction between successive units, including their substituent, in order to account for energies of magnitude required". In the same year, another excellent book[135] written by 14 authors, cited by far abundant data on cellulose derivatives: for example,

(1) χ^- parameter (eq. (3.3.2)) of cellulose nitrate(CN)-acetone, CN-cyclohexanon, cellulose acetate(CA)-THF.[136]

(2) Heat of dilution and entropy of CA-THF, CN-acetone, cellulose triacetate (CTA)-dioxan.[137]

(3) Solubility of methyl-, ethyl-, butyl- and benzyl-cellulose.[138]

(4) Mixture of non-solvent working as solvent of CN; ethyl alcohol + ethylether.[139]

(5) Mark–Houwink–Sakurada equations of cellulose, CN, CA, cellulose acetate-buthylate (CAB).[140]

(6) Plot of $[\eta]$ versus M (or η_{sp}/c versus P) of CN.[141]

(7) Plot of $M/[\eta]$ versus $M^{1/2}$ of CN-acetone and CA-acetone.[142]

(8) Shear rate dependence of viscosity of CN-butylacetate.[143]

Note that in reference 135 much attention was paid to describe the solution properties of CN and CA, but no word was given on the new concepts, which were developed for polymer solutions in the 1940s and 1950s, such as the unperturbed chain dimensions, theta conditions, and two-parameter concept (the short-range and the long-range interaction).

Techniques of polymer characterization, for example light scattering (LS), quickly developed in the 1950s, but could not straightforwardly be applied to cellulose and cellulose derivatives until the 1970s. Originally, LS was first attempted to apply to cellulose acetate–acetone system (Doty-Zimm-Mark), but soon the reliability of their LS data was denied, due to the difficulty of complete exclusion of gel-like materials contaminated in the solution. On the other hand, cellulose nitrate is reliably soluble in solvents (such as acetone) and research on the polymer was most advanced worldwide. Flory, Schulz, Classon, and Moore and their collaborators were familiar names in this field.

REFERENCES

1. FD Miles, *Cellulose Nitrate*, Interscience, 1955, p. 234, Fig. IV.
2. K Kamide and K Okajima, *Polym. J.*, 1981, **13**, 27.
3. T Lieser, *Ann. Chemi.*, 1930, **470**, 104; T Lieser, *Ann. Chemi.*, 1930, **483**, 135.
4. T Noguchi, *Sen-i Gakkuishi*, 1950, **6**, 153 see also pages 155, 217, 312, 314, 379, 381, 444.
5. K Lauer, *Makromol. Chem.*, 1951, **5**, 289.
6. C-Y Chen, RE Montonna and CS Grove, *Tappi*, 1951, **34**, 420.
7. M Horio, R Imamura and H Sakata, *Sen-i Gakkaishi*, 1958, **14**, 909.
8. JJ Willard and E Pacsu, *J. Am. Chem. Soc.*, 1960, **82**, 4350.
9. B Philipp and K-T Liu, *Faserforsch. Textiltech.*, 1959, **10**, 555.
10. M Horio, R Imamura, N Komatsu, H Sakata and T Kato, *Sen-i Gakkaishi*, 1963, **19**, 102.
11. PJ Garegg, B Lindberg and E Pettersson, *Acta Chemica Scandinavica*, 1966, **20**, 2763.
12. TS Gardner and CB Purves, *J. Am. Chem. Soc.*, 1942, **64**, 1539.
13. CJ Malm, JJ Tanghe and BC Laird, *J. Am. Chem. Soc.*, 1950, **72**, 2674.
14. VM Goodlett, JT Bougherty and HW Patton, *J. Polym. Sci., A-1*, 1971, **9**, 155.
15. M Saito, *Polym. J.*, 1983, **15**, 249.
16. T Miyamoto, Y Sato, T Shibata, H Inagaki and T Tanahashi, *J. Polym. Sci., Chem. Ed.*, 1984, **22**, 2363.
17. K Kowsaka, K Okajima and K Kamide, *Polym. J.*, 1986, **18**, 843.
18. TK Wu, *Macromolecules*, 1980, **13**, 74.

19. DT Clark, PJ Stephenson and F Heatley, *Polymer*, 1981, **22**, 1112.

20. DT Clark and PJ Stephenson, *Polymer*, 1982, **23**, 1295.

21. K Kamide and K Okajima, *Polym. J.*, 1981, **13**, 163.

22. K Kamide, K Kowsaka and K Okajima, *Polym. J.*, 1987, **19**, 231.

23. FF-L Ho, RR Kohler and GA Ward, *Anal. Chem.*, 1972, **44**, 178.

24. K Kamide, K Okajima, K Kowsaka, T Matsui, S Nomura and K Hikichi, *Polym. J.*, 1985, **17**, 909.

25. A Parfondry and AS Perlin, *Carbohydrate Res.*, 1977, **57**, 39.

25. A Parfondry and AS Perlin, *Carbohydrate Res.*, 1977, **57**, 39.

26. DS Lee and AS Perlin, *Carbohydrate Res.*, 1982, **106**, 1.

27. FF-L Ho and DW Klosiewicz, *Anal. Chem.*, 1980, **52**, 913.

28. J Duclaux and E Wollman, *Bull. Soc. Chim.*, 1920, **27**, 414.

29. EWJ Mardles, *J. Chem. Soc.*, 1923, **123**, 1951.

30. I Okamura, *Cellulose Chem.*, 1933, **14**, 135.

31. SA Glikman, *Kolloid-Z.*, 1936, **76**, 84.

32. R Singer and J Liechti, *J. Helv. Chim. Acta*, 1938, **21**, 530.

33. H Fink, R Stahn and A Matthes, *Z. Angew. Chem.*, 1934, **47**, 602.

34. K Kamide, Y Miyazaki and T Abe, *Makromol. Chem.*, 1979, **180**, 2801.

35. FL Strauss and RM Levy, *Paper Trade J.*, 1942, **114**, 33.

36. PRM Nair, RM Gohil, KC Patel and RD Patel, *Eur. Polym. J.*, 1977, **13**, 273.

37. H Lachs, K Kronman and J Wajs, *Kolloid Z.*, 1937, **79**, 91.

38. GR Levi and A Giera, *Gazz. Chim. Ital.*, 1937, **67**, 719.

39. FR Levi, U Villota and M Montirelli, *Gazz. Chim. Ital.*, 1938, **68**, 589.

40. S Bezzi and U Croatta, *Atti Inst. Veneto Sci.*, 1939-1940, **99**, 905.

41. A Munster, *J. Polym. Sci.*, 1950, **5**, 333.

42. H Sobue, K Matsuzaki and K Yamakawa, *Sen-i Gakkaishi*, 1956, **12**, 100.

43. D Henley, *Ark. Kemi*, 1961, **18**, 327.

44. RS Stein and P Doty, *J. Am. Chem. Soc.*, 1946, **68**, 159.

45. P Doty, NS Schneider and AM Holtzer, *J. Am. Chem. Soc.*, 1953, **75**, 754.

46. AM Holtzer, H Benoit and P Doty, *J. Phys. Chem.*, 1954, **58**, 624.

47. ML Hunt, S Newman, HA Scheraga and PJ Flory, *J. Phys. Chem.*, 1954, **58**, 624.

48. MH Haque, DAI Goring and SG Mason, *Can. J. Chem.*, 1958, **36**, 952.

49. NS Schneider and P Doty, *J. Phys. Chem.*, 1954, **58**, 762.

50. W Brown and R Wikström, *Eur. Polym. J.*, 1965, **1**, 1.

51. L Vaitassari, *Makromol. Chem.*, 1971, **150**, 117.

52. K Kamide and M Saito, *Polym. J.*, 1986, **18**, 68.

53. CL McCormic and PA Hutchinson, *Macromolecules*, 1985, **18**, 2394.

54. MI Shakhparonov, NP Zahurdayeva and YK Podgarodetshii, Vysokomol Soyed. A, 1967, **9**, 1212.

55. H Suzuki, K Ohno, K Kamide and Y Miyazaki, *Netsusokutei (Calor, Therm. Anal.)*, 1981, **8**, 67.

56. H Suzuki, Y Muraoka, M Saito and K Kamide, *Eur. Polym. J.*, 1982, **18**, 837.

57. D Tanner and G Berry, *J. Polym. Sci., Phys. Ed.*, 1974, **12**, 941.

58. K Kamide, Y Miyazaki and T Abe, *Polym. J.*, 1979, **11**, 523.

59. K Kamide, T Terakawa and Y Miyazaki, *Polym. J.*, 1979, **11**, 285.

60. K Kamide and M Saito, *Polym. J.*, 1982, **14**, 517.

61. K Kamide, M Saito and T Abe, *Polym. J.*, 1981, **13**, 421.

62. GV Schulz and E Penzel, *Makromol. Chem.*, 1968, **112**, 260.

63. W Burchard and E Husemann, *Makromol. Chem.*, 1961, **44–46**, 358.

64. G Sitarmaiah and D Goring, *J. Polym. Sci.*, 1962, **58**, 1107.

65. W Brown, D Henley and J Ohman, *Makromol. Chem.*, 1963, **62**, 164.

66. W Brown and D Henley, *Makromol. Chem.*, 1964, **79**, 68.

67. W Burchard, *Makromol. Chem.*, 1965, **88**, 11.

68. VP Shanbhag and J Ohman, *Ark. Kemi*, 1968, **29**, 163.

69. VP Shanbhag, *Ark. Kemi*, 1968, **29**, 1.

70. RS Manley, *Ark. Kemi*, 1956, **9**, 519.

71. WR Krigbaum and LH Sperling, *J. Phys. Chem.*, 1960, **64**, 99.
72. WB Neely, *J. Polym. Sci.*, 1963, **1**, 311.
73. W Brown, D Henley and J Ohman, *Makromol. Chem.*, 1963, **64**, 49.
74. F Friedberg, W Brown, D Henley and J Öhman, *Makromol. Chem.*, 1963, **66**, 168.
75. K Gohsh and P Choudhury, *Makromol. Chem.*, 1967, **102**, 217.
76. B Das, AK Ray and PK Choudhury, *J. Phys. Chem.*, 1969, **73**, 3413.
77. K Kishino, T Kawai, T Nose, M Saito and K Kamide, *Eur. Polym. J.*, 1981, **17**, 623.
78. M Waies and DL Swanson, *J. Phys. Colloid Chem.*, 1951, **55**, 203.
79. LH Sperling and M Easterwood, *J. Appl. Polym. Sci.*, 1960, **4**, 25.
80. LO Sundelöf, *Cellulose and Cellulose Derivatives, Part IV*, **Vol. V**, (eds NM Bikales and L Segal), Wiley, New York, 1971, p. 432.
81. VJ Klenin and GP Denisova, *J. Polym. Sci., Polym. Symp.*, 1973, **42**, 1563.
82. KD Goeble and GC Berry, *J. Polym. Sci., Phys. Ed.*, 1977, **15**, 555.
83. K Kamide, T Terakawa and S Manabe, *Sen-i Gakkaishi*, 1974, **30**, T-464.
84. K Kamide, T Terakawa, S Manabe and Y Miyazaki, *Sen-i Gakkaishi*, 1975, **31**, T-410.
85. K Kamide, S Manabe and T Terakawa, Japanese Patents No.885,873, 1977, and 1978, 909, 158.
86. GV Schulz and Z Phys, *Chem. B*, 1935, **30**, 379.
87. See, for example, Chap. E in Ref. 80.
88. TE Timell, *Ind. Eng. Chem.*, 1955, **47**, 2166.
89. M Marx, *J. Polym. Sci.*, 1958, **30**, 119.
90. KE Eriksson, F Johanson and B Pettersson, *Sven. Papperstidin.*, 1967, **70**, 610.
91. KE Eriksson, BA Pettersson and B Steenberg, *Sven. Papperstidin.*, 1968, **71**, 695.
92. BA Pettersson, *Sven. Papperstidin.*, 1969, **72**, 14.
93. D Berek, G Katuscakova and I Novak, *Br. Polym. J.*, 1977, **9**, 62.
94. YT Bao, A Bose, ML Ladisch and GT Tsao, *J. Appl. Polym. Sci.*, 1980, **25**, 263.
95. A Sookne, H Rutherford, H Mark and H Harris, *Res. Natl. Bur. Std.*, 1943, **30**, 1.
96. W Badgley and H Mark, *J. Phys. Colloid Chem.*, 1947, **51**, 58.
97. RJ Brewer, LJ Tanghe, S Bailey and JR Burr, *J. Polym. Sci, A-1*, 1968, **6**, 1635.
98. TE Mueller and WS Alexander, *J. Polym. Sci., C*, 1968, **21**, 283.
99. L Segal, *J. Polym. Sci., B*, 1966, **4**, l011.
100. L Segal, *J. Polym. Sci. C*, 1968, **21**, 267.
101. L Segal and JD Timpa, *Tappi*, 1969, **52**, 1666.
102. L Segal, JD Timpa and JI Wadsworth, *J. Polym. Sct., A-1*, 1970, **8**, 25.
103. G Meyerhoff and M Stitterlin, *Makromol. Chem.*, 1965, **87**, 258.
104. RYM Huang and RG Jenkins, *Tappi*, 1963, **52**, 1503.
105. WJ Alexander and TE Mueller, *Sep. Sci.*, 1971, **6**, 47.
106. JT Guthrie, MB Huglin and RW Richards, *Eur. Polym. J.*, 1947, **11**, 527.
107. J Danhelka, I Kossler and V Bohackova, *J. Polym. Sci.*, 1972, **16**, 2583.
108. JC Moore, *J. Polym. Sci., A*, 1964, **2**, 835.
109. G Meyerhoff and S Jovanovic, *J. Polym. Sci. B*, 1967, **5**, 495.
110. H Mark, *Der Feste Körper, Hirzerl, Leipzig*, 1938, p. 103.
111. R Houwink, *J. Pract. Chem.*, 1940, **155**, 241.
112. I Sakurada, *Kasen Koenshu (Japan)*, 1941, **6**, 177.
113. N Gralen and T Svedberg, *Nature*, 1943, **152**, 625; N Gralen, *Kolloid-Z*, 1941, **95**, 188; N Galen, *Dissertation, Uppsala*. 1944.
114. H Vink, *Ark. Kend*, 1956, **14**, 195.
115. SY Lyubina, SI Klenin, IA Strelina, AV Troitskaya, AK Khripunov and EU Vrinov, *Vysokomol. Soyed., A*, 1977, **19**, 244.
116. A Bartovics and H Mark, *J. Am. Chem. Soc.*, 1943, **65**, 1901.
117. A Sookne and M Harris, *Ind. Eng. Chem.*, 1945, **37**, 475.
118. H Philipp and C Bjork, *J. Polym. Sci.*, 1951, **6**, 383.
119. C Cumberbirch and W Harland, *Textile Inst.*, 1958, **49**, T664–T679.
120. W Moore and B Tidswell, *J. Appl. Chem.*, 1958, **8**, 232.
121. A Sharples and H Major, *J. Polym. Sci.*, 1958, **27**, 433.
122. N Dymarchuk, K Michenko and T Fomina, *Zhwr. Prikl. Khim (Leningrad)*, 1964, **37**, 2263.

123. VM Golubev and SY Frenkel, *Vysokomol. Soyed., A*, 1967, **9**, 1847.
124. H Mosimann, *Helv. Chem. Act*, 1943, **2661**, 369.
125. WR Moore and J Epstein, *J. Appl. Chem.*, 1956, **6**, 168.
126. G Meyerhoff, *J. Polym. Sci.*, 1958, **29**, 399.
127. WG Harland, *J. Textile Inst.*, 1955, **46**, T483.
128. WR Moore and GD Edge, *J. Polym. Sci.*, 1960, **47**, 469.
129. G Jayme and W Verburg, *Rayon Zellwolle Chemifasern*, 1954, **32**, 193, 275.
130. PJ Flory, *Principles of Polymer Chemistry*. Cornell Univ. Press, Ithaca, 1953.
131. PJ Flory, *Principles of Polymer Chemistry*, p. 613, Fig. 141. and L Mandelkern and PJ Flory, *J. Am. Chem. Soc.*, 1952, **74**, 2517.
132. PJ Flory, *Principles of Polymer Chemistry*, p. 618, Table XXXIX. and S Newman (unpublished results).
133. PJ Flory, *Principles of Polymer Chemistry*, p. 619.
134. PJ Flory, *Principles of Polymer Chemistry* 19, p. 620.
135. HA Stuard (ed), *Das Makuromolekül in Lösungen, Die Physik Der Hochpolymeren*. Zweiten Band, 1953.
136. A Münster, *Das Makuromolekül in Lösungen*, p. 152, Table II.4.
137. A Münster, *Das Makuromolekül in Lösungen*, Table II.5, p. 161; Abb. II.43, p. 168; Abb. II.44, p. 168.
138. A Münster, *Das Makuromolekül in Lösungen*, Table III.14, p. 248; Table III.16, p. 249.
139. A Münster, *Das Makuromolekül in Lösungen*, Table IV.6, p. 216.
140. A Peterlin, *Das Makuromolekül in Lösungen*, Table V.6, p. 305.
141. A Peterlin, *Das Makuromolekül in Lösungen*, Abb. V.14, p. 305; Abb. XI.10, p. 599; Abb. XI.II, p. 559.
142. A Peterlin, *Das Makuromolekül in Lösungen*, Abb. XI.12, p. 561.
143. A Peterlin, *Das Makuromolekül in Lösunge*, Abb. XI.16, p. 567.

1.5   SOLVED AND UNSOLVED PROBLEMS IN MOLECULAR CHARACTERIZATION OF CELLULOSE DERIVATIVES IN THE 1970s

Recognition of characteristic features of cellulose derivative solutions in the 1970s can be deduced from several eminent books published between 1968 and 1975.

Burchard[1] commented in 1968 "Unfortunately both suggested procedures (Stockmayer–Fixman[2] and its modified plots[3,4] (eq. (3.16.12)) fail for cellulose trinitrate ... up to now, I do not see any possibility of evaluating unperturbed dimensions of this polymer by viscosity measurements alone."

In 1971, Yamakawa[5] described "Further, we must note that some cellulose derivatives exhibit the draining effect (see, Section 3.15), i.e. the dependence of the viscosity constant Φ (eq. (3.15.1)) on the molecular weight M,[6,7] and for others $[\eta]_\theta/M^{1/2}$ is independent of M over a wide range of M.[8,9] On the other hand, if the two-parameter scheme is assumed to be valid for cellulose derivatives, the Kurata–Stockmayer(KS)[10] or Stockmayer–Fixman(SF)[2] plot gives values of the long-range interaction parameter B about twice as large as those predicted from the values of $A_2 M/[\eta]$ (A_2 is the second virial coefficient).[11] Both Florys and Kurata's version involve several weak points, as mentioned above, and seem inconclusive. Further theoretical and experimental investigations of cellulose derivatives are needed."

In 1971 Vink[12] wrote "One of the reasons for deficiency of current theories may lie in their failure to account adequately for possible draining effect present in cellulose polymers. This has been studied by Hermans,[13] Kamide and Moore,[14] and Banks and

Greenwood."[15] In the same book, Brown[16] noticed that Stockmayer–Fixman,[2] Kurata–Stockmayer[10] equations, from which K (accordingly, A) can be evaluated, requires condition of spherically systematical form and asymptotic (maximum) constant value of Φ, independent of the molecular weight (i.e. impermeability of coil to solvent flow).

Kurata stated in 1975 (in Japanese) that "for cellulose derivatives including cellulose nitrate there are two conflicting theories due to undiscovered Flory's theta solvent and other reasons."[17]

Until the 1970s the following discrepancy was found from the above researches: Cellulose derivative solutions exhibit (A) the high viscosity, the large radius of gyration, and small second virial coefficient, but they also reveal (B) small unperturbed chain dimensions and large excluded volume effect, both estimated by the viscosity plot. These experimental results contradict with each other (i.e. (A) suggests that the cellulose chain is rigid and the solvents are poor, but (B) supports that the cellulose chain is flexible and the solvents are good.) No definite conclusion was reached. Note that in the mid-1960s the experimental accuracy of light scattering (LS) was doubted and the viscosity was considered as reliable. During the 1970s it became widely understood that many advanced techniques of synthesis and analysis were needed for the LS study of cellulose and cellulose derivatives solution. A trend to study cellulose from a new point of view was brought about between the late 1970s and early 1980s, resulting in enormous progress in cellulose science. (This is a renaissance of cellulose!)

In the following four chapters molecular characterization and the commercial applications of cellulose and cellulose derivatives made in this period are described.

REFERENCES

1. W Burchard, *Solution Properties of Natural Polymers*. Special Publication No.23, Chap. 3 Polysaccharides, Hydrodynamic Properties of Cellulose, Amylose, and Their Derivatives, The Chemical Society, London, 1968, p. 135.
2. WH Stockmayer and M Fixman, *J. Polym. Sci.*, 1963, **C1**, 137.
3. W Burchard, *Makromol. Chem.*, 1961, **50**, 20.
4. W Burchard, *Solution Properties of Natural Polymers*, pp. 143–146.
5. H Yamakawa, *Modern Theory of Polymer Solutions*, Harper & Row, New York, 1971, p. 389.
6. M Hunt, S Newman, HA Sheraga and PJ Flory, *J. Phys. Chem.*, 1956, **60**, 1278.
7. E Penzel and GV Schulz, *Makromol. Chem.*, 1968, **13**, 64.
8. L Mandelkern and PJ Flory, *J. Am. Chem. Soc.*, 1952, **74**, 2517.
9. WR Krigbaum and LH Sperling, *J. Phys. Chem.*, 1960, **64**, 99.
10. M Kurata and WH Stockmayer, *Fortcher. Hochpolym.-Forsch., Bd.*, 1963, **3**, 196–312.
11. H Yamakawa, *J. Chem. Phys.*, 1966, **45**, 2606.
12. Hans Vinks, in *Cellulose and Cellulose Derivatives, Part IV* (eds CM Bikalée and L Segal), Wiley, 1971, D. Viscometry, p. 469.
13. JJ Hermans, *J. Polym. Sci.*, 1963, **C-12**, 117.
14. K Kamide and WR Moore, *J. Polym. Sci. B*, 1964, **2**, 1029.
15. W Banks and CT Greewood, *Makromol. Chem.*, 1968, **114**, 245.
16. W Brown, in *Cellulose and Cellulose Derivatives, Part IV*, G. Molecular Coformation and Dimensions, p. 557.
17. M Kurata, *Chemistry of Polymer Industry, Modern Industrial Chemistry 18*, Asakawa, Tokyo, 1975, p. 296.

– 2 –

Characterization of Molecular Structure of Cellulose Derivatives

---

2.1 DEFINITIONS OF MOLECULAR STRUCTURE PARAMETERS[1]

A large number of the industrially useful cellulose derivatives are those in which the hydroxyl groups of the anhydroglucose units are not completely substituted. The extent of substitution of these hydroxyl groups with substituent groups may differ with different carbon positions (i.e. 2, 3, 6 position) due to their different reactivities. The average degree of substitution (DS) of each hydroxyl group, as well as the total DS within a glucose residue $\langle\!\langle F \rangle\!\rangle$, have profound effects on cellulose derivative (CD) properties.

To define the molecular structure parameters, all molecules (total number of molecules, N) that were different in degree of polymerization p were arranged in order of increasing p and defined as the first, second, third,… Nth molecules from the smallest p, and the p of the jth polymer was recorded as p_j. The glucopyranose units, constituting the jth molecule from 1 to p_j were then numbered. The probability of substitution of the hydroxyl group attached to the C_2, C_3, and C_6 atoms of the ith glucopyranose unit in the jth molecule, with the functional group by $f^{j_{2,i}}$, $f^{j_{3,i}}$, and $f^{j_{6,i}}$, respectively (see Figure 2.1.1) was defined and designated the total DS of the ith ring in the jth molecule by F^{j_i}. The following relationships are readily derived.

$$f^{j_{2i}} + f^{j_{3i}} + f^{j_{6,i}} = F^{j_i} \tag{2.1.1}$$

$$\sum_{i=1}^{p_j} F^{j_i}/p^j = \langle F^j \rangle \tag{2.1.2}$$

$$\sum_{j=1}^{N} p_j \langle F^j \rangle \Big/ \sum_{j=1}^{N} p_j = \langle\!\langle F \rangle\!\rangle (= \mathrm{DS}) \tag{2.1.3}$$

$$\sum_{j=1}^{N} \sum_{i=1}^{p_j} f^{j_{k,i}} \Big/ \sum_{j=1}^{N} p_j = \langle\!\langle f_k \rangle\!\rangle, \quad k = 2, 3, \text{ and } 6 \tag{2.1.4}$$

$$\langle\!\langle f_2 \rangle\!\rangle + \langle\!\langle f_3 \rangle\!\rangle + \langle\!\langle f_6 \rangle\!\rangle = \langle\!\langle F \rangle\!\rangle \tag{2.1.5}$$

25

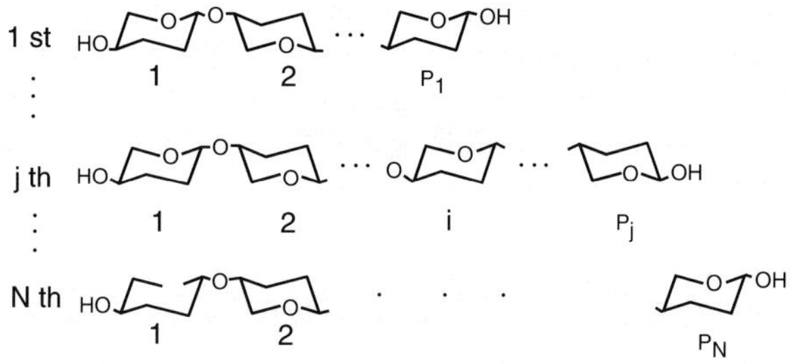

Figure 2.1.1   Schematic representation of the *j*th cellulose molecule having degree of polymerization p_j.

$$\sum_{j=1}^{N} p_j / N = P_n \tag{2.1.6}$$

$$\sum_{j=1}^{N} p_{j^2} \bigg/ \sum_{j=1}^{N} p_j N = P_w \tag{2.1.7}$$

where $\langle F^j \rangle$ is the average DS of the *j*th polymer, $\langle\langle F \rangle\rangle$ is the DS weight, averaged overall glucopyranose units of all molecules, and $\langle\langle f_k \rangle\rangle$ is the average probability of substitution of hydroxyl group attached to the C_k position ($k = 2$, 3, and 6). $\langle\langle f_k \rangle\rangle$ represents an overall distribution of, for example, the *O*-acetyl group in a trihydric alcohol unit of the cellulose acetate (CA) molecule. The distribution of $\langle F^j \rangle$ is designated as $g(\langle F^j \rangle)$. $\langle\langle F \rangle\rangle$ and $\langle\langle f_k \rangle\rangle$ are doubly averaged quantities.

REFERENCE

1. K Kamide and K Okajima, *Polym. J.*, 1981, **13**, 127.

2.2   TOTAL DEGREE OF SUBSTITUTION (DS)

The physical and chemical properties of CDs are considered primarily to clearly depend on the total DS ($= \langle\langle F \rangle\rangle$). The total DS ($= \langle\langle F \rangle\rangle$) of cellulose derivatives is estimated by (1) the chemical analysis (classical method) and (2) the nuclear magnetic resonance (NMR) from $\langle\langle f_2 \rangle\rangle$, $\langle\langle f_3 \rangle\rangle$, and $\langle\langle f_6 \rangle\rangle$ using eq. (2.1.5) and (3) thin layer chromatography (TCL).

2.2.1   Chemical analysis

(a) *Cellulose Acetate (CA)*: $\langle\langle F \rangle\rangle$ was determined, according to the procedure of ASTMD-871-61T, by the hyrolysis of the sample with sodium hydroxide, followed by neutralization titration with hydrochloric acid.[1]

(b) *Cellulose Nitrate (CN)*: $\langle\langle F \rangle\rangle$ was evaluated by elemental analysis and by the nitometer method by Lunge. Both methods gave identical values.[2]

(c) *Cellulose Sulfate (CS)*: $\langle\!\langle F\rangle\!\rangle$ was determined gravimetrically according to the method described by Schweiger.[3] One gram of NaCS was hydrolyzed with 2 N hydrochloric acid. Barium sulfate, precipitated with barium hydroxide, was gravimetrically determined,[4] and the amount of sulfuric acid liberated by boiling NaCS (sodium CS) in 1N hydrochloric acid was taken as that of barium sulfate.[5]

(d) *Sodium Carboxymethylcellulose (NaCMC)*: Before determining $\langle\!\langle F\rangle\!\rangle$ by chemical analysis, NaCMC was converted to the acid form and immersed in a 3 wt% aqueous (aq.) sodium chloride solution. The hydrochloric acid thus produced was diluted with alkali, the excess of which was back titrated with hydrochloric acid.[6]

(e) *Carboxyethyl Carbamoylethyl Cellulose (CECEC)*: Carboxyethyl content A (%) was determined by back titration of NaCECEC suspended in a given strength of aq. hydrochloric acid with aq. NaOH and carbamoylethyl content B (%) was evaluated from nitrogen content determined by elementary analysis (CHN analyzer model CHN 1-A, Shimadzu Ltd., Japan). From A and B, the total degrees of substitution by carboxyethyl group ($\langle\!\langle F\rangle\!\rangle_{CO}$), that of carbamoylethyl group ($\langle\!\langle F\rangle\!\rangle_{NH}$) and total DS $\langle\!\langle F\rangle\!\rangle_t$, were calculated using the following relationships.[7]

$$\langle\!\langle F\rangle\!\rangle_{CO} = [(44A)/(45B)]162\{4400 - 71B - (3168A^2/45)\}^{-1} \tag{2.2.1}$$

and

$$\langle\!\langle F\rangle\!\rangle_{NH} = 162\{4400 - 71B - (3168A^2/45)\}^{-1} \tag{2.2.2}$$

$$\langle\!\langle F\rangle\!\rangle_t = \langle\!\langle F\rangle\!\rangle_{CO} + \langle\!\langle F\rangle\!\rangle_{NH} \tag{2.2.3}$$

Table 2.2.1 shows $\langle\!\langle F\rangle\!\rangle_t$, $\langle\!\langle F\rangle\!\rangle_{CO}$, $\langle\!\langle F\rangle\!\rangle_{NH}$ values of CECEC samples determined by chemical analysis.

2.2.2   Nuclear magnetic resonance (NMR)

$\langle\!\langle F\rangle\!\rangle$ values can also be evaluated as a summation of $\langle\!\langle F_k\rangle\!\rangle$ (eq. (2.1.5)), if $\langle\!\langle F_k\rangle\!\rangle$ is determined by the NMR method.

Table 2.2.1

DS of CECEC determined by various methods[7]

Sample	Chem. anal.			^{1}H NMR			^{13}C NMR		
	$\langle\!\langle F\rangle\!\rangle_t$	$\langle\!\langle F\rangle\!\rangle_{CO}$	$\langle\!\langle F\rangle\!\rangle_{NH}$	$\langle\!\langle F^*\rangle\!\rangle_t$	$\langle\!\langle F^*\rangle\!\rangle_{CO}$	$\langle\!\langle F^*\rangle\!\rangle_{NH}$	$\langle\!\langle F^{\cdot}\rangle\!\rangle_t$	$\langle\!\langle F^{\cdot}\rangle\!\rangle_{CO}$	$\langle\!\langle F^{\cdot}\rangle\!\rangle_{NH}$
CECEC-1	0.28	0.27	0.01	–	–	–	–	–	–
CECEC-2	0.60	0.40	0.02	0.62	0.48	0.14	0.54	0.46	0.08
CECEC-3	0.82	0.43	0.39	0.85	0.44	0.41	0.80	0.58	0.22
CECEC-4	1.51	1.08	0.43	1.53	1.06	0.47	1.67	1.46	0.21

(a) *Cellulose Acetate*: $\langle\!\langle F\rangle\!\rangle$ values can also be evaluated to a certain extent from only the ^{13}C NMR spectra of carbons 1–6 and O-acetyl carbonyl carbon or O-acetyl methyl carbon by the equation[1]

$$\langle\!\langle F\rangle\!\rangle = 6I_2/I_1 \qquad\qquad (2.2.4)$$

$$\langle\!\langle F\rangle\!\rangle = 6I_3/I_1 \qquad\qquad (2.2.4')$$

where I_1 is the integrated peak intensity of carbon at C_1–C_6 positions at 60–104 ppm; I_2, the integrated peak intensity of the O-acetyl carbonyl carbon at 167–171 ppm; I_3, the integrated peak intensity of the O-acetyl methyl carbon (approximately 20 ppm). The results are tabulated for CA (DS $= 2.46$) and CA (DS $= 0.49$) samples in Table 2.2.2,[1] and a comparison is made with those evaluated by ^{13}C NMR/chemical analysis method. For CA (DS $= 2.46$), the $\langle\!\langle F\rangle\!\rangle$ value evaluated from the ratio $I_2{:}I_1$ by ^{13}C NMR spectrum is found to fit exactly with that from the ratio $I_3{:}I_1$ and coincides fairly well with that determined by the titration method. The corresponding values for CA (DS $= 0.49$), however, do not coincide as closely, probably due to the experimental uncertainty inherent in both methods for determining the low $\langle\!\langle F\rangle\!\rangle$. Generally, if the acetic acid contained in the sample is an impurity, the chemical analysis tends to overestimate the $\langle\!\langle F\rangle\!\rangle$ values.

Kowsaka *et al.*[9] estimated the molar fractions of eight kinds of anhydroglucopyranose (AHG) units for CA $\langle\!\langle f_{\ell mn}\rangle\!\rangle$, where ℓ, m, and n denote the existence of the acetyl group at C_2, C_3, and C_6 positions, respectively, and they can take the value of 0 (unsubstituted) or 1 (substituted) (see Table V of Ref. 8). It was confirmed that the $\langle\!\langle F\rangle\!\rangle$ values thus calculated from $\langle\!\langle f_{\ell mn}\rangle\!\rangle(\langle\!\langle F\rangle\!\rangle_{\text{calc}})$ agree well with $\langle\!\langle F\rangle\!\rangle$ values estimated from $\langle\!\langle F_k\rangle\!\rangle(\langle\!\langle F\rangle\!\rangle_{\text{conc}})$.

(b) *CS*: For low molecular weight polymers, ^{1}H NMR can be applied to determine $\langle\!\langle F\rangle\!\rangle$ as well as $\langle\!\langle F_k\rangle\!\rangle$.[9] For a CS whole sample ($M_n = 1.5 \times 10^4$), Kamide *et al.*[10] obtained $\langle\!\langle F\rangle\!\rangle = 1.95$ from $\langle\!\langle F_k\rangle\!\rangle$ data by ^{1}H NMR, which corresponds well with that of the chemical analysis.

It can been seen that $\langle\!\langle F\rangle\!\rangle_{\text{NMR}}$ corresponds well with $\langle\!\langle F\rangle\!\rangle_g \pm 0.10$ (see Table 2.2.3).[10] The correlation coefficient between these was estimated to be 0.998.

Table 2.2.2

$\langle\!\langle F\rangle\!\rangle$ values evaluated by chemical analysis and ^{13}C NMR of cellulose acetate

Sample code	$\langle\!\langle F\rangle\!\rangle$	
	^{13}C NMR/Chem. anal.	^{13}C NMR
EF 3W	2.46	2.44[a]; 2.44[b]
MAW-7	0.49	0.44; 0.46

[a]Eq. (2.2.4).
[b]Eq. (2.2.4').

Table 2.2.3

Characterization of sodium cellulose sulfate[10]

Sample	$M_n \times 10^{-4}$	Total $\langle\!\langle F \rangle\!\rangle$		Distribution		
		Gravimetry	NMR	$\langle\!\langle F_2 \rangle\!\rangle$	$\langle\!\langle F_3 \rangle\!\rangle$	$\langle\!\langle F_6 \rangle\!\rangle$
CS-1	36.8	1.66	–	–	–	–
CS-2	9.39	1.94	–	–	–	–
DSH	6.91	1.05	1.05	0.55	0.00	0.50
CS-3	6.79	1.98	–	–	–	–
CS-4	6.65	2.39	2.46	1.00	0.74	0.72
CS-5	5.32	1.36	1.36	0.75	0.42	0.19
CS-6	5.31	1.84	–	–	–	–
CS-7	5.06	1.84	–	–	–	–
CS-8	5.06	1.73	–	–	–	–
CS-9	4.98	1.81	–	–	–	–
CS-10	4.32	2.75	2.75	1.00	0.92	0.83
CS-A	3.15	1.97	1.97	0.71	0.26	1.00
CS-11	2.48	1.98	1.96	1.00	0.61	0.34
HB-1	2.26	0.95	0.95	0.52	0.00	0.43
CS-12	1.95	2.03	1.93	1.00	0.65	0.28
CS-13	1.80	1.74	–	–	–	–
CS-14	1.70	2.60	2.62	1.00	0.87	0.85
CSD	1.69	2.05	1.99	0.75	0.55	0.67
CS-15	1.57	1.96	–	–	–	–
CS-16	1.57	1.94	1.97	1.00	0.60	0.37
HB-2	1.24	0.70	–	–	–	–
HB-3	1.16	0.74	–	–	–	–
HBSD	0.99	0.95	0.95	0.29	0.33	0.33
HB-4	0.92	0.76	–	–	–	–
HBH	0.08	0.50	0.58	0.31	0.27	0.00
Heparin	–	–	–	–	–	–

(c)   *Cellulose Xanthate (CX)*: $\langle\!\langle F \rangle\!\rangle$ was determined by $^{13}\text{C}\{^1\text{H}\}$ NMR for CX in aq. alkali.[5]

(d)   *Carboxyethyl Carbamoylethyl Cellulose (CECEC)*: Based on the peak assignment, the degrees of substitution by carboxyethyl group $\langle\!\langle F \rangle\!\rangle_{\text{CO}}$, or carbamoylethyl group $\langle\!\langle F \rangle\!\rangle_{\text{NH}}$ and total DS $\langle\!\langle F \rangle\!\rangle_t (= \langle\!\langle F \rangle\!\rangle_{\text{CO}} + \langle\!\langle F \rangle\!\rangle_{\text{NH}})$ were evaluated by the ^1H NMR method and the ^{13}C NMR method. $\langle\!\langle F \rangle\!\rangle_{\text{CO}}$, $\langle\!\langle F \rangle\!\rangle_{\text{NH}}$ produced by the chemical analysis, and those by ^1H NMR, coincide reasonably well with those by the ^{13}C NMR analysis (see Table I of Ref. 7).

2.2.3   Thin layer chromatography (TLC)

The TLC method provides information on the distribution of $\langle F^j \rangle$ (eq. (2.1.2)) ($g\langle F^j \rangle$) together with $\langle\!\langle F \rangle\!\rangle$.

$$\langle\!\langle F \rangle\!\rangle = \sum_g \langle F^j \rangle \tag{2.2.6}$$

(a) *Cellulose Acetate (CA)*: Experimental relationships between the rate flow R_f and $\langle\!\langle F \rangle\!\rangle$ (in this case, the acetyl content A_c (wt%)) were established for CA (see Figure 2.5.20).[11]

(b) *Cellulose Nitrate (CN)*: Similar relationships between R_f and the nitrogen content $N\%$ were also obtained for CN (see eq. (2.5.17) and Figure 2.5.30).[12]

REFERENCES

1. K Kamide and K Okajima, *Polym. J.*, 1981, **13**, 127.
2. K Kamide, T Shiomi, H Ohkawa and K Kaneko, *Kobunshi Kagaku*, 1965, **22**, 785.
3. R Schweiger, *Carbohydr. Res.*, 1972, **21**, 219.
4. K Kishino, T Kawai, T Nose, M Saito and K Kamide, *Eur. Polym. J.*, 1981, **17**, 623.
5. K Kamide, K Kowsaka and K Okajima, *Polym. J.*, 1987, **19**, 231.
6. K Kamide, K Okajima, K Kowsaka, T Matsui, S Nomura and K Hikichi, *Polym. J.*, 1985, **17**, 909.
7. K Kamide, K Okajima, T Matsui and M Ohnishi, *Polym. J.*, 1987, **19**, 347.
8. K Kowsaka, K Okajima and K Kamide, *Polym. J.*, 1988, **20**, 827.
9. K Kamide and K Okajima, *Polym. J.*, 1981, **13**, 163.
10. K Kamide, K Okajima, T Matsui, M Ohnishi and H Kobayashi, *Polym. J.*, 1983, **15**, 309.
11. K Kamide, T Matsui, K Okajima and S Manabe, *Cell. Chem. Technol.*, 1982, **16**, 601.
12. K Kamide, T Okada, T Terakawa and K Kaneko, *Polym. J.*, 1978, **10**, 547.

2.3 DETERMINATION OF SUBSTITUENT GROUP AMONG THREE HYDROXYL GROUPS IN GLUCOPYRANOSE UNIT $\langle\!\langle f_k \rangle\!\rangle$

The extent of substitution of these hydroxyl groups with acetyl groups may differ with the different C positions to which the hydroxyl group is attached due to different reactivity. The average DS for each hydroxyl group $\langle\!\langle f_k \rangle\!\rangle$ in eq. (2.1.1) is another important parameter for characterizing cellulose derivatives.

2.3.1 Cellulose acetate

Assignment of carbonyl carbon peaks on NMR Spectra

To date, several attempts have been made to evaluate the *O*-acetyl group distribution in a trihydric alcohol unit by chemical[1,2] and ^{1}H NMR methods.[3] The former is generally based on the difference in reactivity of three unsubstituted hydroxyl groups in tosylation[1] with *p*-toluenesulfonyl chloride, followed by iodination with sodium iodide or tritylation[2] with tritylchloride in the presence of pyridine, and is very tedious and time consuming. In addition, by this chemical method, the DS at C_2 and C_3 positions cannot be principally evaluated. This method is a relative method, which requires the total DS value to be determined by another method. In contrast to this, the latter is an absolute method, and unfortunately is only applicable to completely, or at least nearly, fully substituted cellulose acetate. Otherwise, the peaks caused by the *O*-acetyl proton cannot be resolved to the components corresponding to the C_2, C_3, and C_6 positions, since a large number of glucopyranose units with magnetically nonequivalent O-acetyl groups are possible. To avoid this, Goodlett et al.[3] have acetylated incompletely substituted CA polymer with deuterated acetyl chloride before ^{1}H NMR measurements. They proposed a method for estimating $\langle\!\langle f_k \rangle\!\rangle$ of CA by analyzing the acetyl methyl proton region of their ^{1}H NMR spectra.[3] They observed that the acetyl methyl proton peak at *ca.* 2 ppm of cellulose acetate, whose unsubstituted hydroxyl groups were fully substituted with deuteroacetyl group by acetyl-d_3 chloride, was split into three peaks (see Figures 1, 4, and 5 of Ref. 3). By comparing the peak intensities with $\langle\!\langle f_k \rangle\!\rangle$, chemically evaluated according to a procedure by Malm et al.[2] and Gardner and Purves,[1] they assigned three peaks from the low magnetic field, as acetyl methyl protons at C_6, C_2, and C_3 positions. Note that in the Goodlett et al. experimental determination of the method, the peak intensity is not sufficiently accurate owing to the low separation of three peaks in ^{1}H NMR spectra, and $\langle\!\langle f_k \rangle\!\rangle$ values are likely to change during the deuteration process. Wu analyzed the distribution of substituted glucopyranose unit in CN by the ^{1}H and ^{13}C NMR methods.[4]

Kamide and Okajima[5] proposed a method using ^{13}C NMR for determining the relative amounts of acetyl substitution on each of three different hydroxyls. In other words, the distribution of O-acetyl groups in trihydric alcohol units of CA using samples, which had sharp molecular weight distributions (MWDs; the ratio of the weight-to number-average molecular weight M_w/M_n is 1.4 for (DS = 2.92), and 1.3 for (DS = 2.46 and 0.49).

Figure 2.3.1a and 2.3.1b shows the 100 MHz ^{1}H NMR and ^{13}C NMR spectra of CA (DS 2.92) sample in TCM-d_1. The O-acetyl-proton signal of CA (DS 2.92) in TCM-d_1 splits into three distinguishable peaks centered at $\delta = 2.13$, 2.02, and 1.98 ppm. These chemical shift values are nearly equal to those reported by Goodlett et al. (2.10, 1.99, and 1.94 ppm),[3] who assigned these three peaks at 2.10, 1.99, and 1.94 ppm to the O-acetyl-protons attached to the positions C_6, C_2, and C_3, respectively, by comparing the reaction rate of p-toluene sulfonylchloride with three hydroxyl groups at C_6, C_2, and C_3 positions.

Recently, Shiraishi et al.[6] confirmed the validity of Goodlett et al.'s assignment by the correlation between the order of acidity of O-acetyl methyl attached to C_6, C_2, and C_3 as well as the electron deshielding effect and a comparison of the shift of the O-acetyl-proton peaks of CA (DS 2.92) in TCM-d_1 to lower the magnetic assignment from the ratio between peak heights or the area under the peaks and the total DS $(= \langle\!\langle F \rangle\!\rangle = 2.92)$

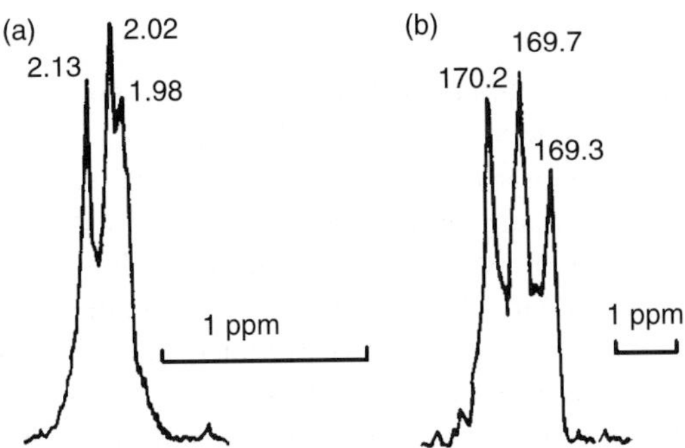

Figure 2.3.1 ^{1}H and ^{13}C NMR spectra of TCM-d solution of CA (DS 2.92) sample in the O-acetyl proton region and O-acetyl carbonyl carbon region, respectively.[5] (a) ^{1}H NMR; (b) ^{13}C NMR.

by the chemical analysis. We obtain $\langle\!\langle f_6 \rangle\!\rangle = 0.98$, or 0.99, $\langle\!\langle f_2 \rangle\!\rangle = 1.09$, or 1.01 and $\langle\!\langle f_3 \rangle\!\rangle = 0.85$ or 0.92, respectively, for the CA (DS 2.92) samples.

When peak areas are used, the accuracy for evaluating $\langle\!\langle f_k \rangle\!\rangle$ is ± 4–8%.

The $\langle\!\langle f_k \rangle\!\rangle$ value exceeding unity is rather unrealistic because, theoretically, the upper limit of $\langle\!\langle f_k \rangle\!\rangle$ is 1.00 and may be due to experimental uncertainty.

Kamide and Okajima[5] observed three peaks at 170.2, 169.7, and 169.3 ppm in carbonyl carbon region of 25 MHz ^{13}C{^{1}H} NMR spectrum of CA with the total DS $\langle\!\langle F \rangle\!\rangle = 2.92$ dissolved in trichloromethane-d (TCM-d), and their intensities were 1.00, 1.02, and 0.89, reflecting the slight deviation of $\langle\!\langle F \rangle\!\rangle$ from 3 (see Figure 2.3.1). These intensities were compared with those of three acetylmethyl proton peaks in the ^{1}H NMR spectrum of the same sample and they assigned the three carbonyl carbon peaks to C_6, C_2, and C_3 positions from the low magnetic field. Using the assignment made thus, Kamide and Okajima determined $\langle\!\langle f_k \rangle\!\rangle$ of various CA samples with $\langle\!\langle F \rangle\!\rangle = 0.49$–2.92. Note that they observed at least four peaks in the carbonyl carbon region for CA with low $\langle\!\langle F \rangle\!\rangle$, and they assigned one of these peaks as that from acetic acid contaminated in CA sample, by assuming CA with $\langle\!\langle F \rangle\!\rangle$ less than unity. Prepared homogeneous hydrolysis should have only three peaks corresponding to three types of monosubstituted glucopyranose residue.

Thereafter, Miyamoto et al.[7] approximately divided the carbonyl carbon region in the ^{13}C NMR spectrum of CA with $\langle\!\langle F \rangle\!\rangle = 1.91$ ($\langle\!\langle f_2 \rangle\!\rangle = 0.63$, $\langle\!\langle f_3 \rangle\!\rangle = 0.48$, and $\langle\!\langle f_6 \rangle\!\rangle = 0.80$ as determined by the Goodlett et al. method[3]) into three major peaks, although at least four peaks were observed for CA samples with $\langle\!\langle F \rangle\!\rangle = 0.84$ and 1.23 (Figure 3 of Ref. 7) assigning from low magnetic field as acetyl groups at C_6, C_3, and C_2 positions by comparing the peak intensities with those of ^{1}H NMR spectrum (Figure 4a of Ref. 7), which were not well resolved.

In their assignment, two peaks between those responsible for carbonyl carbons of acetyl groups attached to C_6 and C_2 positions were attributed O-acetyl carbon attached to the C_3 position. Their assignment on the C_2 and C_3 positions is simply the reverse of that first reported by Kamide and Okajima.[5] Miyamoto et al. utilized a CA ($\langle\!\langle F \rangle\!\rangle = 1.91$) unfractionated sample for assignment of carbonyl carbon peaks. CA, with $\langle\!\langle F \rangle\!\rangle$ ranging from 0.5 to 2.5, is known to consist of mono-, di-, and trisubstituted glucose residues so that a total of twelve peaks (i.e. four peaks due to C_2 position, four peaks due to C_3 position, and four peaks due to C_6 position) are theoretically expected to appear in the carbonyl carbon region in the ^{13}C NMR spectra. As far as incompletely substituted CA is concerned, any one of the roughly divided three major peaks from these less resolved spectra in Miyamoto et al.'s paper (see Figures 3 and 4 of Ref. 7) is not expected to consist of four peaks, where all belong to a specific carbon positions, and the results on the integrated intensity estimated in this manner may not be completely reliable. This is an unavoidable methodological problem in Miyamoto et al.'s procedure. In contrast to this, if we employ CA with $\langle\!\langle F \rangle\!\rangle = 3$, as Kamide and Okajima did, then trisubstituted glucose residue is a predominant component of the polymer chain and all 12 peaks theoretically expected in the ^{13}C NMR spectra due to the carbonyl carbon of insufficiently substituted CA are overwhelmingly simplified when CA with $\langle\!\langle F \rangle\!\rangle = 3$ is used, so that only three peaks remain. Therefore, it is recommended that the carbonyl carbon peaks of trisubstituted glucose residue are assigned with high accuracy. Thus, Kamide and Okajima's original method is reasonable, at least in principal, but unfortunately the resolving of acetyl methyl proton was relatively poor (see Figure 2 of Ref. 7) because they used a 60 MHz (^{1}H) CW-NMR spectrometer, and also their method has a shortcoming that $\langle\!\langle f_k \rangle\!\rangle$ estimation is restricted to CA with almost $\langle\!\langle F \rangle\!\rangle = 3$.

^{13}C and ^{1}H NMR measurements (Kamide–Okajima)[5]: ^{13}C and ^{1}H NMR spectra (100 MHz) were obtained from a JOEL FX 100 Pulse-Fourier Transform NMR spectrometer at 37 °C. In order to obtain reliable integral curves, the nuclear Overhauser effect was eliminated by the addition of ferric chloride (5 ppm) and by the selection of proper operating conditions. The chemical shifts for CA in various deuterated solvents were measured with respect to the internal tetramethylsilane (TMS) as a reference at the concentration of 5–7 g dl^{-1}. ^{13}C NMR spectra were recorded in ^{1}H-decouple mode. ^{1}H-off resonance mode for ^{13}C NMR was also carried out for CA (DS = 2.46) acetone system to ascertain that the carbonyl carbon peaks were all singlets.

Assignment of the carbonyl carbon peaks on NMR spectra of CA by applying the low-power selective spin decoupling method[8]

The assignment of the carbonyl carbon peaks on NMR spectra of CA was recorded on a 50 MHz FT-NMR spectrometer by applying the low-power selective spin decoupling method to *O*-acetyl methyl protons located at specific carbon positions with the help of well-resolved ^{1}H NMR spectra for acetyl proton region, obtained by choosing proper operating conditions.

^{13}C\{^{1}H\} NMR measurement (Kowsaka–Okajima–Kamide)[8]: 7 wt% solutions were prepared. NMR spectra were recorded on a FX-200 FT-NMR spectrometer (JEOL, Japan), at 199.5 MHz for ^{1}H and 50.15 MHz for ^{13}C. Spectral measurement conditions were utilized as follows: ^{1}H NMR: flip angle 45°, spectral width 2000 Hz (10 ppm), repetition time 10 s, data points 16,384, and accumulation 16 times. ^{13}C NMR: ^{1}H noise decoupled, ^{1}H low-power selective spin decoupled and nondecoupled mode, flip angle 45°, spectral width 10,000 and 1000 Hz (200 and 20 ppm, respectively), repetition time 2.5–5.0 s, data points 16,384 and 8192 (followed by 8192 points of zero filling), accumulation 3000–15,000 times. Frequency domain accumulation was employed if necessary. TMS was used as an internal standard. ^{13}C NMR spectra with 1000 Hz spectral width were measured only over the carbonyl carbon region and signal out of this width were cut by a Butterworth type bandpass filter of 500 Hz width to avoid folding back signals. Chemical shift (0 ppm for TMS) was determined by spectra recorded with 10,000 Hz spectral width. The zero-filling method was employed when necessary to save acquisition time. The FID signal obtained for each solution was reduced by using different broadening factors ($n = 4.0$–0.3 Hz for ^{1}H- and $n = 5.0$–0.6 Hz for ^{13}C NMR) to four or five frequency domain spectra in order to obtain a reasonable average value for the peak intensity from their integral curves. The low-power selective spin decoupling technique was applied to the acetyl methyl proton located at a specific carbon position. The other nuclei had a resonance frequency similar to that used for selective decoupling of the acetyl methyl proton in question and are more or less decoupled. In order to make this side effect minimal, the irradiation power was carefully selected.

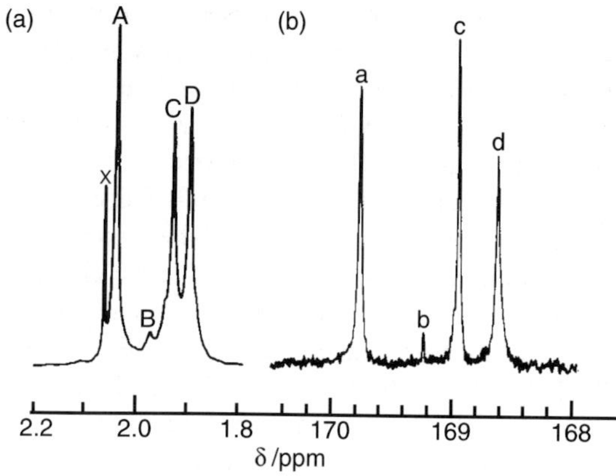

Figure 2.3.2 ^{1}H(a) and ^{13}C{^{1}H}(b) NMR spectra of CA ($\langle\!\langle F \rangle\!\rangle = 2.92$)/DMSO-$d_6$ system in the acetyl methyl proton and carbon region, respectively.[8]

Figure 2.3.2a and 2.3.2b shows the ^{1}H NMR spectrum in the acetyl methyl region and the ^{13}C{^{1}H} NMR spectrum in the carbonyl carbon region with 1000 Hz spectral width for CA ($\langle\!\langle F \rangle\!\rangle = 2.92$)/DMSO-$d_6$ system at 100 °C, respectively. In the figure, four peaks were observed in the acetyl methyl proton region of ^{1}H NMR spectrum and the same number of peaks were observed in the carbonyl carbon region of ^{13}C{^{1}H} NMR spectrum. In the latter spectrum, peak widths obviously differ from each other. A peak attached with an x mark in the former spectrum is attributed to an impurity (acetone) contaminated in the CA sample. The chemical shift δ, average integrated intensity I, and its standard deviation of these peaks (A–D in ^{1}H and a–i in ^{13}C{^{1}H} NMR spectrum), are presented in Table 2.3.1. Here, the total I for A–D or a–d peaks was adjusted to give $\langle\!\langle F \rangle\!\rangle = 2.92$. Under the operating conditions of ^{13}C{^{1}H} NMR used the nuclear Overhauser effects (NOE) for carbonyl carbon peaks (a–d), which are not very different

Table 2.3.1

Chemical shift δ, average integrated peak intensity I, and its standard deviation σ of acetyl methyl proton and carbonyl carbon peaks[8]

Nucleus	Peak code	δ (ppm)	I^a	σ^b
Acetyl methyl proton	A	2.04	1.02	0.06
	B	1.98	0.07	0.01
	C	1.93	0.94	0.01
	D	1.89	0.91	0.02
Carbonyl carbon	a	169.8	1.01	0.02
	b	169.3	0.05	0.02
	c	169.0	0.90	0.00
	d	168.6	0.95	0.01

[a]Average integrated intensity normalized as $I_{\text{total}} = 2.92$. I was determined from integral curves of frequency domain spectra reduced by Fourier transform of each FID signal using different broadening factors (0–0.3 Hz for proton and 0–0.6 Hz for ^{13}C spectra).
[b]Standard deviation of I.

from each other because these carbons do not directly bond to protons and their spin lattice relaxation time T_1 (a, 2.25 s; c, 2.33 s, and d, 2.41 s in cellulose acetate/DMSO-d_6 system at 40 °C) differs by only a few percent. Thus, I of carbon nuclei can be regarded as quantitative, although the pulse interval employed here is not sufficient to avoid the progressive saturation of the carbons. The above experimental results strongly suggest that I (shown in Table 2.3.1) quantitatively represents the existence ratio of the nuclei concerned. Considering I, shown in Table 2.3.1, a one-to-one correspondence between peaks A–D in the ^{1}H NMR spectrum and peaks and in the ^{13}C{^{1}H} NMR spectrum occurs in the following manner: A → a, B → b, C → d and D → c.

Figure 2.3.3a shows the carbonyl carbon region of the ^{13}C{^{1}H} NMR spectrum of CA($\langle\!\langle F\rangle\!\rangle$ = 2.92)/TCM-d system at 40 °C. Here, a noise modulated and high-powered RF wave was irradiated over all ^{1}H nuclei, and the spectral width was 10,000 Hz (200 ppm). Three sharp peaks are observed from the lower magnetic field, and are denoted as a, c, and d based on the notation given in Figure 2.3.2b for the CA ($\langle\!\langle F\rangle\!\rangle$ = 2.92)/DMSO-d_6 system.

Figure 2.3.3b shows the nondecoupled ^{13}C NMR spectrum of the same system as that in Figure 2.3.3a. The three peaks that appear in Figure 2.3.3a become broad and split when the irradiation is not applied on ^{1}H nuclei. The above experimental result can be explained as follows. Each of the three peaks in Figure 2.3.3a should theoretically split into four peaks, which are not well resolved in the figure, due to the existence of the long-range coupling effect on the three protons in acetyl methyl group neighboring the carbonyl carbon. Accordingly, when only the acetyl methyl proton located at a specific

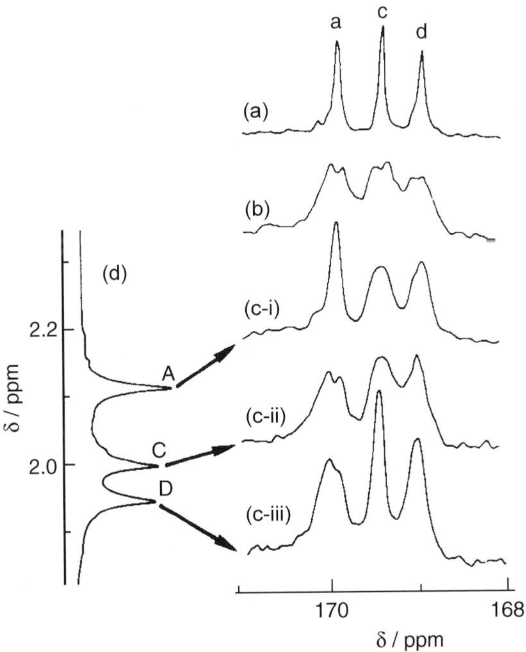

Figure 2.3.3 ^{13}C (a, b, c) and ^{1}H(d) NMR spectra of CA ($\langle\!\langle F\rangle\!\rangle$ = 2.92)/TCM-d. (a) noise decoupled; (b) nondecoupled; (c-i, c-ii, c-iii), selective decoupled.[8]

position (C_2, C_3, and C_6) and it is preferentially decoupled, the peak of carbonyl carbon bonded directly to the decoupled methyl group can be expected to be observed very sharply. This is hereafter defined as a low-power selective spin decoupling method.

Figure 2.3.3d shows the ^{1}H NMR spectrum in the acetyl methyl proton region and three peaks observed from the lower magnetic field (denoted as A, C, and D with reference to Figure 2.3.2a). We recorded the resonance frequencies as the decoupling frequency of low-power selective spin, decoupling method.

Figure 2.3.3c shows the experimental results. In spectrum c-i, c-ii, and c-iii, A, C, and D peaks of the acetyl methyl proton are selectively decoupled. Comparison of Figure 2.3.3a and 2.3.3c shows that each a, d, and c peak becomes sharp by applying the low-power selective spin decoupling method to each A, C, and D peak. Thus, we can conclude that a one-to-one correspondence exists between the acetyl methyl proton peak and the carbonyl carbon peak as follows: A → a, C → d, and D → c.

Figure 2.3.4 shows similar results for CA ($\langle\!\langle F \rangle\!\rangle = 2.92$)/DMSO-$d_6$ system at 100 °C. From Figure 2.3.4, we can also confirm the close correlation between A → a, C → d, and D → c peaks.

An inspection of Table 2.3.1 and Figures 2.3.3 and 2.3.4 leads us to the conclusion that peak A correlates to peak a, peak C to peak d, and peak D to peak c. Combining the above experimental data with the proton assignment by Goodlett *et al.* for ^{1}H NMR spectra, we can assign three major peaks, except for peak b, in the carbonyl carbon region as the acetyl groups located at C_6, C_3, and C_2 positions, respectively, from the low magnetic field as illustrated in Figure 2.3.5.

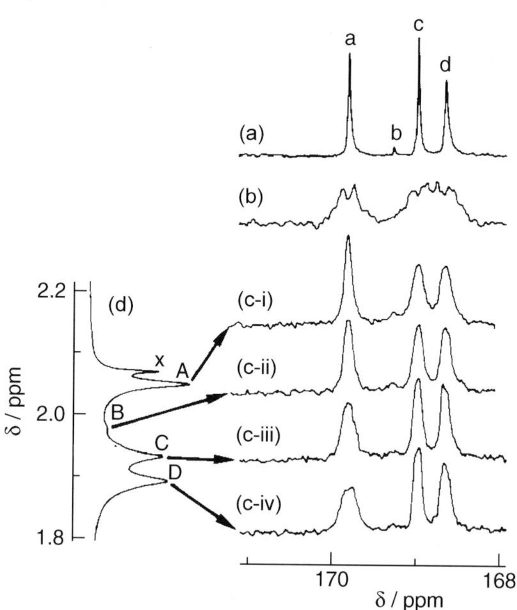

Figure 2.3.4 ^{13}C (a, b, c) and ^{1}H(d) NMR spectra decoupled; (c-i, c-ii, c-iii), selective decoupled.[8]

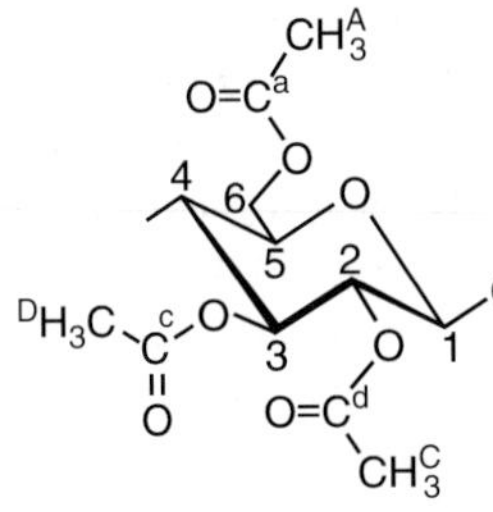

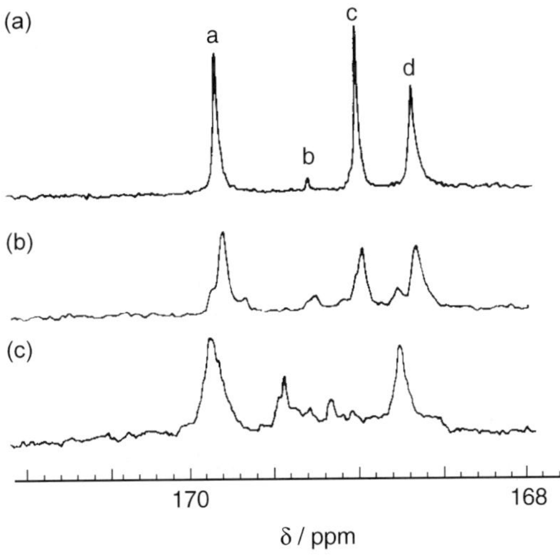

Figure 2.3.5   Schematic representation of peak assignments of CTA. Symbols on the carbonyl carbons and *O*-acetyl methyl protons denote the peaks shown in Table 2.3.1.[8]

These findings are different from those first proposed by Kamide and Okajima.[5] The former is simply the reverse of the latter in the C_2 and C_3 positions. In Kamide and Okajima's earlier work, the overlapping of the acetyl methyl proton peaks and the *O*-acetyl carbon peak are significantly large and, consequently, in both ^1H and ^{13}C NMR spectra, the peaks located at the middle position were unavoidably overestimated, leading to erroneous assignment. The conclusion reached here agrees well with that of Miyamoto *et al.*[7] but note that although Kowsaka *et al.*[8] provided a rigorous assignment of carbonyl carbon peaks of trisubstituted CA, peak b is still unassignable to any acetyl carbons attached to C_2, C_3, and C_6 positions.

Figure 2.3.6 shows $^{13}C\{^1H\}$ NMR spectra of CA with $\langle\!\langle F \rangle\!\rangle = 2.92$(a), 2.46(b), and 0.68(c) in DMSO-d_6 at 100 °C, with 1000 Hz spectral width. Compared with the spectrum for CA ($\langle\!\langle F \rangle\!\rangle = 2.92$), a decrease in $\langle\!\langle F \rangle\!\rangle$ of CA rings about peak envelopes, which are broadened as a result of overlapping of peaks originating from di- and monosubstituted glucopyranose units. In addition to three major peaks for carbonyl carbon due to trisubstituted glucopyranose units, in peak envelopes observed for CA with

Figure 2.3.6 $^{13}C\{^1H\}$ NMR spectra of CA/DMSO-d_6 system in carbonyl carbon region. (a) $\langle\!\langle F \rangle\!\rangle = 2.92$; (b) $\langle\!\langle F \rangle\!\rangle = 2.46$; (c) $\langle\!\langle F \rangle\!\rangle = 0.86$.[8]

$\langle\langle F\rangle\rangle = 2.46$, several additional peaks (including shoulder peaks), which probably originate from mainly disubstituted glucopyranose units, are observed. Inspection of the spectrum for CA with $\langle\langle F\rangle\rangle = 0.68$ reveals that there is no peak responsible for the carbonyl carbon of the trisubstituted glucopyranose unit. Since CA samples with lower $\langle\langle F\rangle\rangle$ were prepared by homogeneous acid-hydrolysis of CA ($\langle\langle F\rangle\rangle = 2.92$) in acetic acid media, CA with $\langle\langle F\rangle\rangle = 0.68$ can be regarded as constituted of mainly monosubstituted cellulose acetate. The three peaks appearing at approximately 169.8, 169.4, and 168.7 ppm seem to be derived from C_6, C_3, and C_2 monosubstituted glucopyranose units. These spectra for CA obtained here are much better resolved than those reported by Kamide and Okajima[5] and Miyamoto et al.[7] The lower resolution of the spectra might have permitted the division of the carbonyl carbon peak region into three major peaks. Unfortunately, in 1986, the complete assignment for all peaks observed in Figure 2.3.6 was impossible.

Determination of $\langle\langle f_k\rangle\rangle$

From the ^{13}C NMR spectra and the $\langle\langle F\rangle\rangle$ value (2.92) obtained by chemical analysis, we can assign the chemical shifts at 170.2, 169.7, and 169.3 ppm to the *O*-acetyl carbonyl carbons in C_6, C_2, and C_3 positions. It was found that $\langle\langle f_6\rangle\rangle = 1.00$, $\langle\langle f_2\rangle\rangle = 0.89$, and $\langle\langle f_3\rangle\rangle = 1.02$ when calculated from peak area ratios. It is clear that the hydroxyl group in the C_3 position is less reactive than those in the C_6 and C_2 positions.

For the CA (DS 0.49) sample, there are three differently substituted glucopyranose units, in which all carbonyl carbons are magnetically different from each other (see Figure 2.3.7). Note that the sample of CA (DS 0.49) was prepared from the CA (DS 2.46) sample by the hydrolysis reaction and, accordingly, acetyl groups can be evenly hydrolyzed with respect to different glucopyranose units. Hence, the hydrolysis reaction seems to accurately reflect the reactivity of acetyl groups attached to the C_6, C_2, and C_3 positions, and the chemical shifts of carbonyl carbons in CA (DS 0.49) sample are expected to split exactly into three peaks.

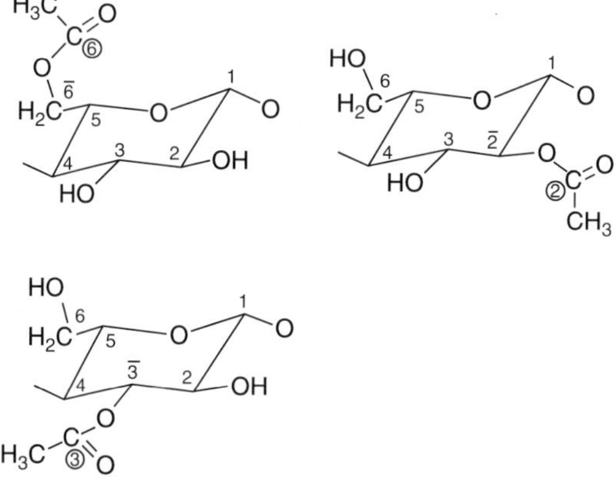

Figure 2.3.7   Possible conformations of glucopyranose units in CA (DS 0.49).[5] Numbers denote the position of carbon atoms to which a substituted hydroxyl group[5] is attached, constituting glucopyranose units. Numbers in circles denote the position of carbonyl carbon atoms. Numbers with a bar denote the position of carbon atoms, to which *O*-acetyl group is attached.

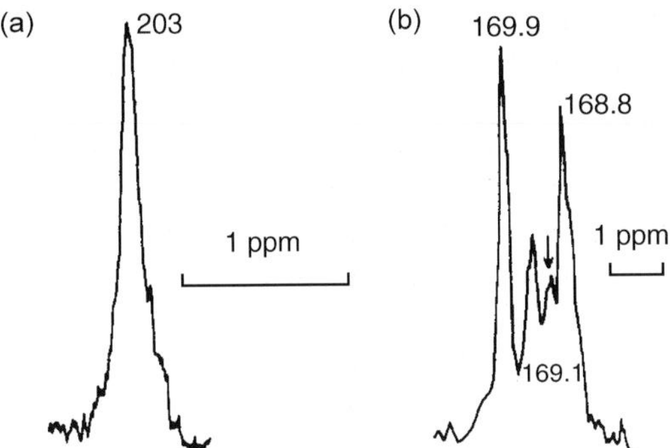

Figure 2.3.8 ^{1}H and ^{13}C NMR spectra of DMSO-d_6 solution of CA (DS 0.49) sample in the O-acetyl proton region and the O-acetyl carbon region, respectively. (a) ^{1}H NMR; (b) ^{13}C NMR.[5]

Figure 2.3.8 shows the ^{1}H and ^{13}C NMR spectra of the CA (DS 0.49) sample in DMSO-d_6. The ^{13}C NMR spectrum exhibits four resonance peaks, covering the range of *ca.* 168.8–170 ppm from TMS: $\delta = 169.9$, 169.4, 169.1, and 168.8 ppm. Among these, the peak at $\delta = 169.4$ ppm originates from contaminated acetic acid in the CA sample. In the ^{1}H NMR spectrum of CA (DS 0.49), the sample single broad peak was observed at 2.03 ppm, and corresponds to the O-acetyl-proton and the $\langle\langle f_k \rangle\rangle$ value cannot be determined.

Figure 2.3.9 shows the conformation of glucopyranose units in cellulose acetate. For CA (DS 2.46) polymer, one unsubstituted hydroxyl remains for each cellobiose unit. Therefore, there are as many as 15 magnetically nonequivalent O-acetyl groups, as illustrated in Figure 2.3.9. Figure 2.3.10a and b depicts the ^{1}H NMR and ^{13}C NMR spectra of the CA (DS 2.46) sample in acetone-d_6. Three carbonyl carbon peaks are definitely observed at 170.9, 170.1, and 169.7 ppm. The relative peak fields of these peaks are in good agreement with those of CA (DS 0.49) in DMSO-d_6 and can be clearly resolved. A small peak at 170.4 ppm is probably a reflection of contaminated acetic acid in the polymer. Thus, we may assign the above three peaks from the lower magnetic field to the C_6, C_3, and C_2 positions, as in the case of those for the CA (DS 2.92 and 0.49) polymers. From the peak area data from the ^{13}C NMR spectrum and the chemical analysis data on $\langle\langle F \rangle\rangle (= 2.46)$, we can estimate $\langle\langle f_k \rangle\rangle$ for this polymer: $\langle\langle f_6 \rangle\rangle = 0.82$, $\langle\langle f_3 \rangle\rangle = 0.75$, and $\langle\langle f_2 \rangle\rangle = 0.89$. In contrast to the ^{13}C NMR spectrum, the corresponding ^{1}H NMR spectrum shows two O-acetyl-proton peaks, from which $\langle\langle f_k \rangle\rangle$ cannot be determined.

Table 2.3.2 collects the observed proton and carbon resonance peak positions, evaluated by the NMR method. It should be noted that the absolute magnitude of $\langle\langle f_k \rangle\rangle$ values is determined with the aid of the $\langle\langle F \rangle\rangle$ value by the chemical analysis and, hereafter, this is referred to simply as the ^{13}C NMR/chemical analysis method. To a certain extent, the $\langle\langle F \rangle\rangle$ value can also be evaluated from only the ^{13}C NMR spectra of carbons 1–6 and O-acetyl carbonyl carbon or O-acetyl methyl carbon by the equation:

$$\langle\langle F \rangle\rangle = 6I_2/I_1 = 6I_3/I_1 \tag{2.3.1}$$

where I_1 is the integrated peak intensity of carbons at C_1–C_6 positions at 60–104 ppm; I_2, the integrated peak intensity of the O-acetyl carbonyl carbon

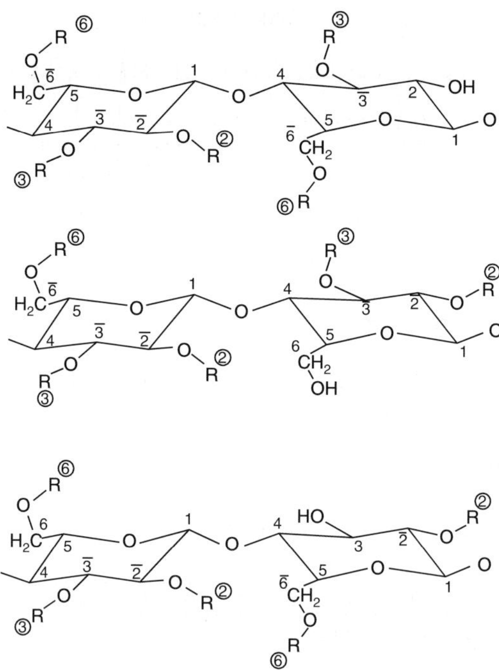

Figure 2.3.9   Possible conformations of glucopyranose units in CA (DS 2.46). Numbers and symbols are the same as described in Figure 2.3.7.[5]

at 167–171 ppm; I_3, the integrated peak intensity of the *O*-acetyl methyl carbon (*ca.* 20 ppm). Figure 2.3.11a and b illustrates the ^{13}C NMR spectra of the CA (DS 2.46) sample in acetone-d_6 and the CA (DS 0.49) sample in DMSO-d_6 over a wide range of magnetic field. A combination of ratio of peak areas of the *O*-acetyl carbons at three different positions (C$_2$, C$_3$, and C$_6$) and $\langle\!\langle F \rangle\!\rangle$ enables us to evaluate $\langle\!\langle f_k \rangle\!\rangle$.

For CA (DS 2.46), the $\langle\!\langle F \rangle\!\rangle$ value evaluated from the ratio I_2/I_1 by the ^{13}C NMR spectrum is found to fit exactly with that from the ratio I_3/I_1 and agrees reasonably well

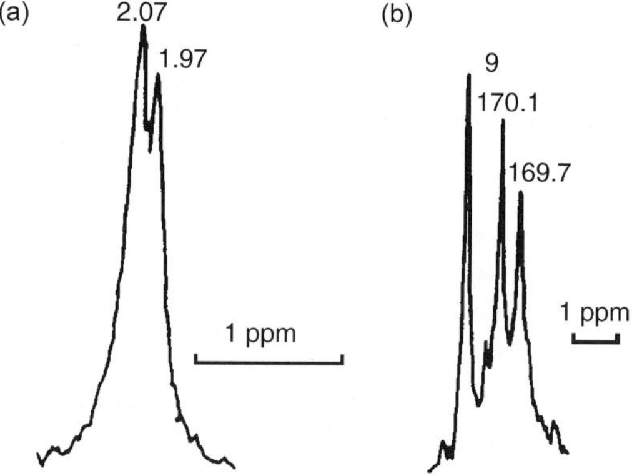

Figure 2.3.10 ^{1}H and ^{13}C NMR spectra of DMSO-d_6 solution of CA (DS 2.46) sample in the *O*-acetyl proton region and the *O*-acetyl carbonyl region, respectively. (a) ^{1}H NMR; (b) ^{13}C NMR.[5]

Table 2.3.2

Evaluation of $\langle\!\langle f_k\rangle\!\rangle$ for CA by chemical analysis, ^{1}H NMR and ^{13}C NMR measurement[9]

Polymer	Solvent	$\langle\!\langle F\rangle\!\rangle$ Chemical analysis	Position	^{1}H NMR O-acetyl proton peak (ppm)	^{13}C NMR		
					$\langle\!\langle f_k\rangle\!\rangle^a$	Carbonyl carbon peak	$\langle\!\langle f_k\rangle\!\rangle^b$
CA (2.92)	Trichloro-methane-d_1	2.92	C_6	2.13	0.99	170.2	1.00
			C_3	1.98	1.01	169.7	1.02
			C_2	2.02	1.092	169.3	0.89
CA (2.46)	Acetone-d_6	2.46	C_6	2.07	–	170.7	0.82
			C_3	1.97	–	170.1	0.75
			C_2	–	–	169.7	0.89
CA (1.75)	Dimethyl-sulphoxide-d_6	1.75	C_6	–	–	170.5	0.59
			C_3	–	–	169.6	0.53
			C_2	–	–	169.3	0.63
CA (0.49)	Dimethyl-sulphoxide-d_6	0.49	C_6	–	–	169.9	0.19
			C_3	2.03	–	169.1	0.10
			C_2	–	–	168.8	0.20

^{a}Estimated from the peak area ratios calculated from a triangle approximation.
^{b}Estimated from integrated peak intensity ratios.

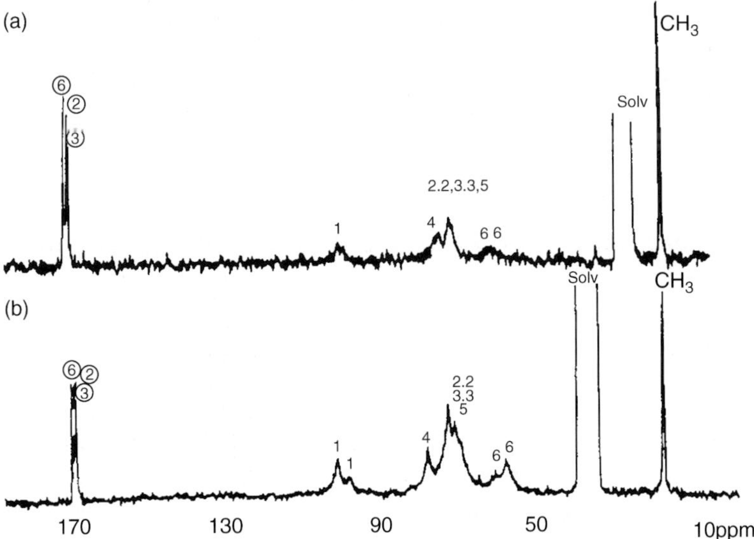

Figure 2.3.11 ^{13}C NMR spectra of CA (DS 2.46) in acetone-d_6 (a) and CA (DS 0.49) in DMSO-d_6 (b) Numbers and symbols are the same as described in Figure 2.3.7.[5]

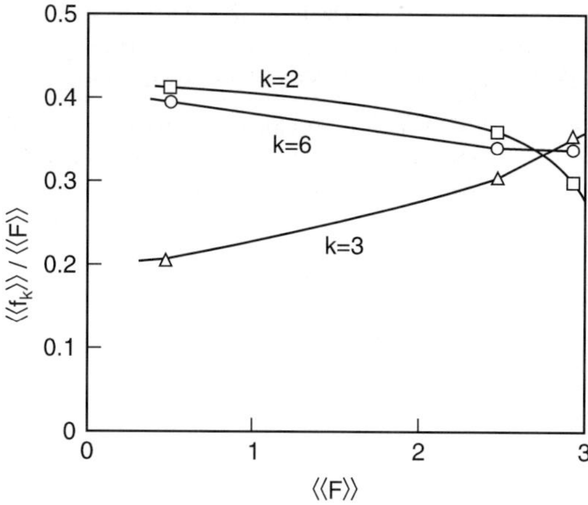

Figure 2.3.12   Plot the ratio $\langle\!\langle f_k\rangle\!\rangle / \langle\!\langle F\rangle\!\rangle$ ($k = 2$, 3, and 6) against $\langle\!\langle F\rangle\!\rangle$ for cellulose acetate.[5]

with that determined by the titration method. The agreement between the corresponding values for CA (DS = 0.49) is not very good, probably due to experimental uncertainty inherent in both methods for determining the low $\langle\!\langle F\rangle\!\rangle$. Generally, if the acetic acid is contained in the sample as an impurity, then the chemical analysis tends to overestimate the $\langle\!\langle F\rangle\!\rangle$ values. In this manner, the $\langle\!\langle f_k\rangle\!\rangle$ for CA can be determined by the ^{13}C NMR method, without the measurements of $\langle\!\langle F\rangle\!\rangle$ by the chemical analysis.

Figure 2.3.12 shows the variation in $\langle\!\langle f_k\rangle\!\rangle / \langle\!\langle F\rangle\!\rangle$ ($k = 2$, 3, and 6) of CA during the hydrolysis reaction of cellulose acetate. Note that CA (DS 2.92) and CA (DS 2.46) have no one-to-one correspondence because the latter polymer is not synthesized from exactly the same polymer as CA (DS 2.92) sample. The ratio of primary to total hydroxyl $\langle\!\langle f_6\rangle\!\rangle / \langle\!\langle F\rangle\!\rangle$ and a similar ratio $\langle\!\langle f_2\rangle\!\rangle / \langle\!\langle F\rangle\!\rangle$ increased. However, $\langle\!\langle f_3\rangle\!\rangle / \langle\!\langle F\rangle\!\rangle$ decreases with decreasing $\langle\!\langle F\rangle\!\rangle$ (i.e. as the hydrolysis proceeds), indicating that it is preferable to remove the acetyl group from the C_3 position. The reactivity of O-acetyl groups toward hydrochloric acid decreases in the order: $C_3 > C_6 > C_2$. By using the ^{13}C NMR technique, a more detailed discussion on the reactivity of the hydroxyl groups belonging to three different positions will be clarified.

2.3.2   Cellulose NaOD and CTA-TCM(C–H COSY NMR spectrum)[11]

The assignment of NMR peaks was a prerequisite for employing the NMR method for determining $\langle\!\langle f_k\rangle\!\rangle$. It is extremely difficult to obtain high resolution ^{1}H NMR spectra for cellulose and its derivatives with a high degree of polymerization (DP), enabling us to obtain information on vicinal proton coupling, even if employing selective homonuclear decoupling. Peak assignment of ^{13}C NMR spectrum is generally more difficult than that of ^{1}H NMR, as the former requires considerably more complicated and skillful techniques. For example, by applying the selective heteronuclear spin decoupling technique in measurement of ^{13}C NMR of cellulose oligomer with DP = 10 in DMSO, and also by taking into account an isotope effect of deuterated compounds, Gagnaire $et\ al.$[12] assigned, from a lower magnetic field, the peaks of cellulose oligomer, to C_1, C_4, C_5, C_3, C_2, and C_6

carbons, respectively. By applying a low-power selective spin decoupling technique,[8] Kowsaka[10] and his coworkers assigned the peaks of carbonyl carbon of acetyl group directly combined with C_2, C_3, and C_6 positions of CA and, with the help of distortionless enhancement by polarization transfer (DEPT) technique,[13] also assigned C_6 peak of CX. In the literature, except for the works described here, the peak assignments were not carried out by only the NMR method, but by an indirect method including a comparison of the peak intensity ratio in NMR spectrum of the sample with $\langle\!\langle f_k \rangle\!\rangle$, determined in advance by a non-NMR method or from analogy of the peak assignment of small molecular weight model compounds or of other derivatives. The assignment given by Gagnaire et al.[12] for cellulose can be considered reasonable in DMSO-d_6. However, mention should be made of the assignment of peaks in other solvents. This is not always possible because C_5, C_3, and C_2 peaks of cellulose oligomer in DMSO-d_6 were located very close to each other and the order of the peak positions may possibly change in other solvents. In addition, note that the positions of the ring carbon peaks are significantly influenced by substitution of hydroxyl groups in cellulose derivatives. In such cases, the method used by Gagnaire et al. cannot be straightforwardly applied in the analysis of NMR spectra of cellulose derivatives as they are too elaborate and time consuming. In addition, they are not sufficiently accurate. Advances in two-dimensional (2D) NMR techniques allow us to perform an absolute (i.e. using a pure NMR method) assignment of ^{1}H and ^{13}C NMR peaks relatively easily. Using a 2D homonuclear ^{1}H shift correlation spectrum and a 2D heteronuclear ^{13}C–^{1}H shift correlation spectrum (with 5% ^{13}C enriched cellulose), Narden and Vincendon[14] attempted to assign ^{1}H and ^{13}C peaks of cellulose in deuterated dimethylacetamide (DMAc-d_9)/lithium chloride mixture but, unfortunately, they failed to distinguish the C_3 peak from C_5 peaks. This section provides a more reliable assignment, using 2D NMR alone, of peaks in ^{1}H and ^{13}C NMR spectra of cellulose in NaOD/D$_2$O solution and cellulose triacetate (CTA) ($\langle\!\langle F \rangle\!\rangle = 2.92$) in trichloromethane-$d$ (TCM-d) and in DMSO-d_6.

NMR measurement

A JEOL FT–NMR spectrometer JNM-FX-200(C/H, 10 mm Φ probe) was used. ^{1}H and ^{13}C resonance frequencies were 199.5 and 50.18 MHz, respectively.

Homonuclear ^{1}H shift correlation spectroscopy (COSY)[15] experiments were carried out using the pulse sequence as shown in Figure 2.3.13a. The trapezoidal window was employed for Fourier transform. For cellulose/NaOD/D$_2$O system, homogate decoupling was simultaneously employed in order to reduce signals originating from remained protons in the solvent mixture.

The heteronuclear ^{13}C–^{1}H shift correlation spectrum (C–H COSY)[15] was recorded using the pulse sequences shown in Figure 2.3.13b. The mixing time Δ was set up to be 2.0 ms (= $1/(4J_{C-H})$), where J_{C-H} is the coupling constant between C and H nuclei. For CTA, long range C–H COSY was concurrently employed to determine correlations between carbonyl carbon and acetyl methyl protons. In this case, Δ was first taken as 166 ms (= $1/(4J_{C-C-H})$, J_{C-C-H} is the long-range C–H coupling constant), but the signal was extremely weak to obtain a sufficiently large signal noise (S/N) ratio due to the remarkably short ($<$50 ms) spin–spin relaxation time T_2. We then chose an adequate

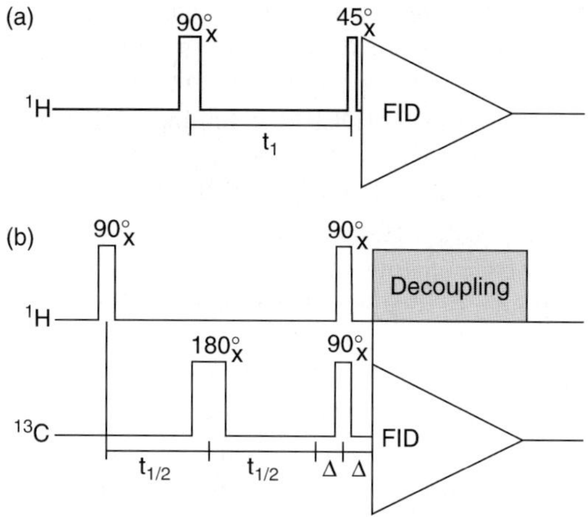

Figure 2.3.13   Pulse sequence used for 2D NMR measurements: (a) proton COSY; (b) C–H COSY.[11]

value of Δ, determined by trial and error under the conditions of $\Delta < 50$ ms. The exponential and trapezoidal windows were employed for $t_2(^{13}C)$ and $t_1(^{1}H)$, respectively.

We demonstrated 2D spectra by a stacked plot and plane figures were constructed by taking into consideration the 1D spectra of cross sections. ^{13}C and ^{1}H chemical shifts for cellulose and CTA were determined from 1D spectra, independently measured, by an internal reference of sodium trimethylsilyl propionate-d_4 (TSP; $= 0$ ppm) and TMS ($= 0$ ppm), respectively.

Figure 2.3.14a and b shows ^{1}H (homogate decoupled) and ^{13}C 1D NMR spectra of cellulose/NaOD/D_2O systems.

Figure 2.3.15 shows COSY (stacked plot, cellulose region) of the same system. Evidently, a peak at 3.57 ppm is relatively higher in intensity than the other peaks, and suggests plural proton peaks. Despite homogate decoupling, a large peak due to residual proton in the solvent was observed. A rapid exchange in proton of the hydroxyl group in glucopyranose ring with residual proton in the solvent does not enable us to distinguish, in ^{1}H NMR spectra, the hydroxyl proton peak from the residual proton peak. Over the 3.8–4.0 ppm region of the 1D spectrum, a broad peak is observed. The existence of cross peak in COSY indicates more explicitly that this is a triplet, consisting of closely overlapping two peaks (3.80 and 3.95 ppm), both combined with a single 3.57 ppm peak.

Figure 2.3.16 shows a schematic 2D spectrum of the cellulose/NaOD/D_2O system, constructed from Figure 2.3.15 and its cross section. In the figure, the projection of spectra in Figure 2.3.15 on the two axes is also illustrated. Diapeaks (open circle) in 2D spectrum correspond to each peak in the 1D spectrum and cross peaks (filled circle) reveal the existence of coupling between neighboring protons. For a doublet peak at 4.5 ppm in 1D spectrum, only a single cross peak is observed in the 2D spectrum. This means that the peak has only a single vicinal coupling proton and that the proton is obviously a proton at C_1 position (H_1). Accordingly, a triplet proton peak reveals vicinal coupling with the H_1 peak at 3.29 ppm and is assigned to H_2. In a similar manner, a peak

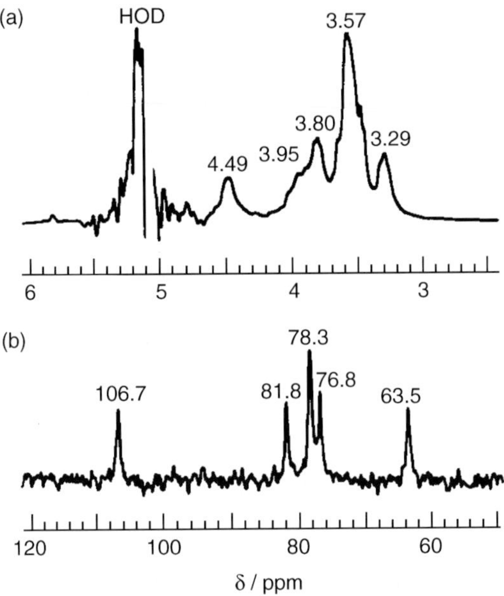

Figure **2.3.14**   1D NMR spectra of cellulose/NaOD/D_2O system: (a) ^1H NMR (homogate decoupled), (b) ^{13}C NMR (gate decoupled).[11]

at 3.57 ppm, coupled with H_2 peak is assigned to H_3 but, as described previously, this peak is heavily overlapped by other peaks and is not separable. Peaks at 3.80 and 3.95 ppm are coupled to each other and are also both coupled with a 3.57 ppm peak, strongly suggesting that the two former peaks are nonequivalent H_6 proton peaks and the latter peak is assigned to H_5 proton. Then, it can be considered that a not yet assigned H_4 peak overlaps with an H_5 peak (3.57 ppm).

Figure 2.3.17 shows the projection of C–H COSY of cellulose/NaOD/D_2O system on each axis and its cross section spectra at ^1H peak position. A schematic 2D spectrum was drawn from the figure and is shown in Figure 2.3.18. Horizontal and longitudinal axes are ^{13}C and ^1H chemical shifts, respectively, and projections on each axis are also shown. Peaks on 2D spectrum clearly show C–H scalar coupling, indicating the correlation between ^{13}C–^1H peaks. Based on the assignment on proton peaks shown in

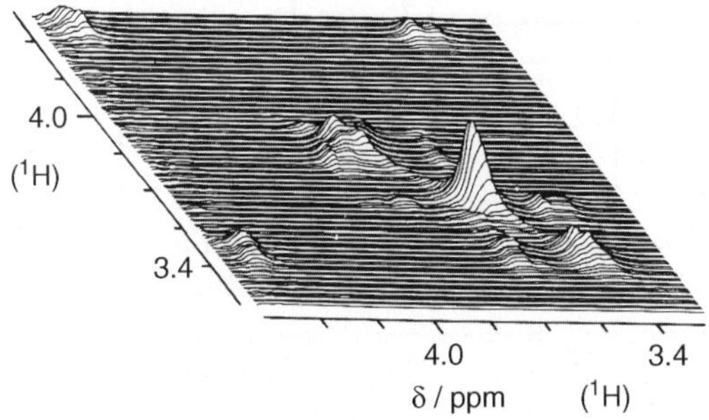

Figure **2.3.15**   Stacked plot for COSY spectrum of cellulose/NaOD/D_2O (cellulose region).[11]

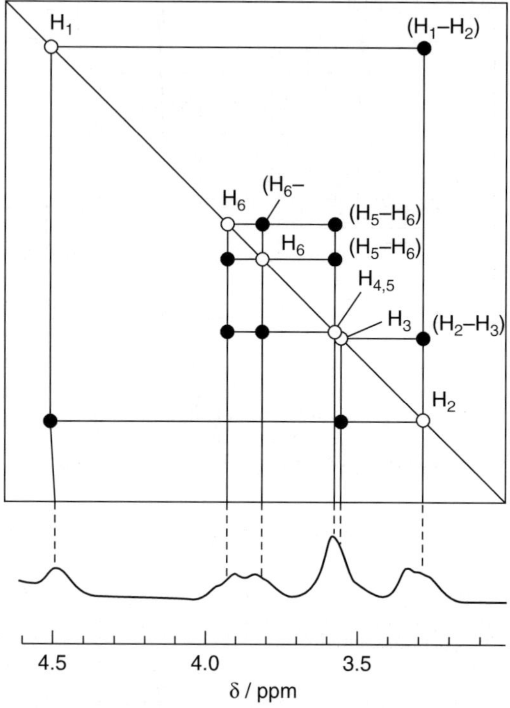

Figure 2.3.16 Schematic COSY spectrum of cellulose/NaOD/D$_2$O. Filled and unfilled marks denote cross peaks and diapeaks, respectively. In the figure, correlations between protons are shown in parenthesis.[11]

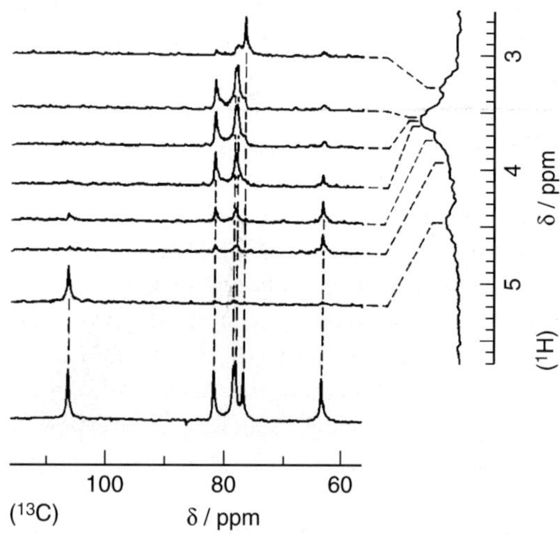

Figure 2.3.17 Cross sections of C–H COSY spectrum of cellulose/NaOD/D$_2$O.[11]

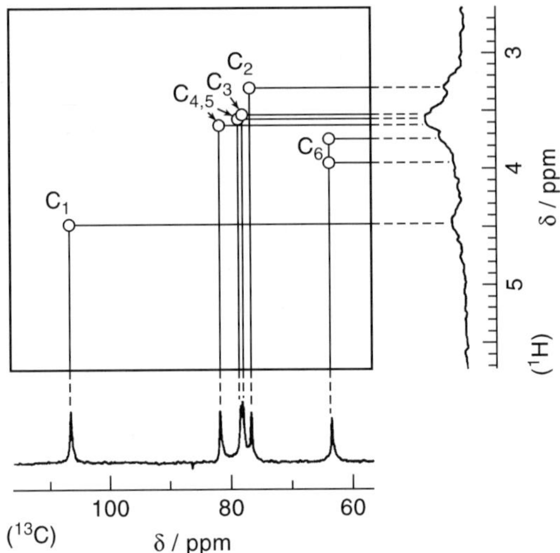

Figure 2.3.18 Schematic C–H COSY spectrum of cellulose/NaOD/D$_2$O.[11]

Figure 2.3.16, three peaks at 106.7, 76.8, and 63.5 ppm in ^{13}C NMR spectrum are exactly assigned to the C_1, C_2, and C_6 carbon peaks, respectively. A peak at 81.8 ppm and a doublet peak at 78.3 ppm (three carbon peaks in total) cannot be assigned explicitly due to the heavy overlapping of the proton peaks at 3.57 ppm, which are directly correlated with the above three carbon peaks. The assignment carried out here for ^{13}C peak is, as far as C_1, C_2, and C_6 carbon peaks are concerned, comparable to those by Gagnaire *et al.*[12] for cellulose oligomer/DMSO-d_6 system and by Nardin and Vincendon[14] for cellulose/DMAc/LiCl system.

Figure 2.3.19a shows the ^{1}H 1D NMR spectrum of CTA/TCM-d system. In the spectrum, proton peaks due to the glucopyranose skeleton are observed in 3.5–5.2 ppm, and three acetyl methyl proton peaks are observed at 2.11, 1.99, and 1.95 ppm. Figure 2.3.19b, c, and d shows acetyl, glucopyranose, and acetyl methyl regions of ^{13}C 1D NMR spectrum of the CTA/TCM-d system.

Figure 2.3.20 shows a schematic COSY 2D spectrum of the glucopyranose skeleton region of the CTA/TCM-d system, constructed from a stacked plot and cross sections of COSY spectrum, as well as its projection. A peak at 4.45 ppm reveals vicinal coupling with a peak at 4.78 ppm alone being assigned to H_1. In the same way as in the case of the cellulose/NaOD/D$_2$O system, we can assign H_2 and H_3. In addition, in the case of CTA, H_3, H_4, and H_5 peaks are observed separately and then each peak is able to be assigned completely. As a result, all peaks are assigned from the lower magnetic field to H_3, H_2, H_1, H_6, H_6', H_4, and H_5, respectively (here, H_6 is the other half of unsymmetrical H_6 peaks). This assignment is denoted on diapeaks (open circle) in Figure 2.3.20. Note that in the 1D spectrum (Figure 2.3.19a), an H_6 peak heavily overlaps an H_1 peak so that peak separation is impossible, but such a separation becomes very easy with the help of a cross peak in the 2D spectrum.

Figure 2.3.21 shows the glucopyranose region and acetyl region of C–H COSY (stacked plot) for the CTA/TCM-d system. In the 1D–^{13}C spectrum (Figure 2.3.19c),

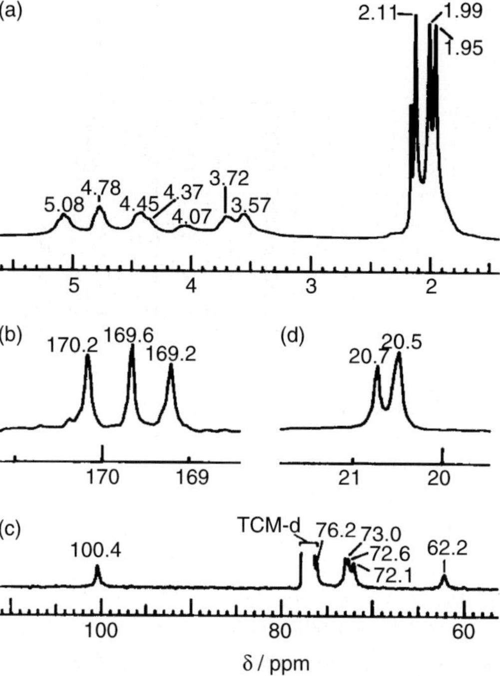

Figure 2.3.19 1D NMR spectra of CTA/TCM-*d* system: (a) ^{1}H NMR; (b, c, d) ^{1}H noise decoupled ^{13}C NMR; (b) carbonyl carbon region; (c) glucopyranose carbon region; (d) acetyl carbon region.[11]

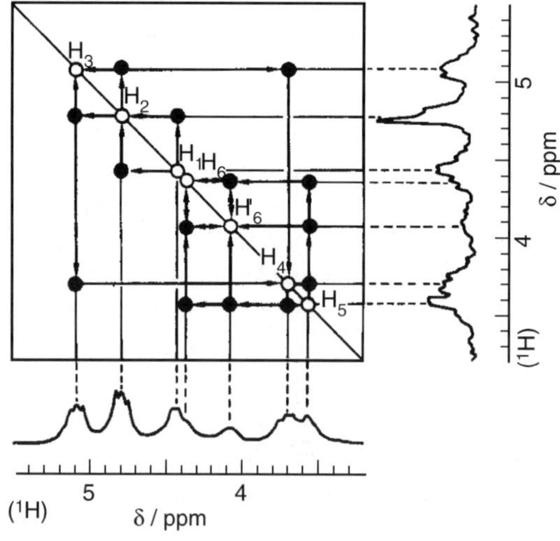

Figure 2.3.20 Schematic COSY spectrum of CTA/TCM-*d* (glucopyranose region). Filled and unfilled marks denote cross peaks and diapeaks, respectively.[11]

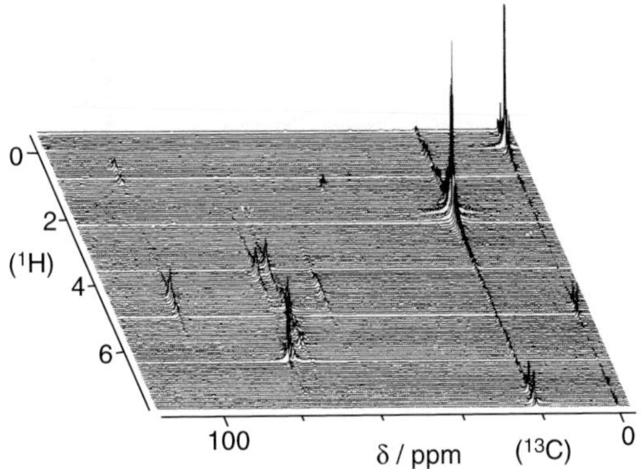

Figure 2.3.21   Stacked plot for C–H COSY spectrum of CTA/TCM-*d*.[11]

a solvent (TCM-*d*) peak at 77 ppm is extraordinarily large and overlaps some of the CTA peaks, making further analysis on CTA spectrum impossible. In contrast, in the 2D spectrum, no peak of TCM-*d*, in which C–H coupling does not exist, is observed.

Figure 2.3.22 shows a schematic C–H COSY 2D-spectrum constructed from Figure 2.3.21 and its cross section for the CTA/TCM-*d* system. In the figure, the projection of spectrum on the ^{13}C (horizontal) axis and the ^{1}H (longitudinal) axis are also shown. Carbon peaks can be assigned based on assignment of proton peaks, using the C–H correlations shown in Figure 2.3.22 from the lower magnetic field to C_1, C_4, C_3, C_5, C_2, and C_6. However, here C_3 and C_5 overlap and are inseparable.

Figure 2.3.23 shows cross section spectra of long-range C–H COSY spectrum between the acetyl methyl proton and the acetyl carbonyl carbon in CTA/TCM-*d*. Three correlation peaks in these figures are peaks, originating from long-range scalar coupling of the carbonyl carbon of acetyl group and acetyl methyl proton.

If we denote carbonyl carbon peaks from the lower magnetic field as a, b, and c, then the acetyl methyl proton peaks are assigned to a, c, and b.

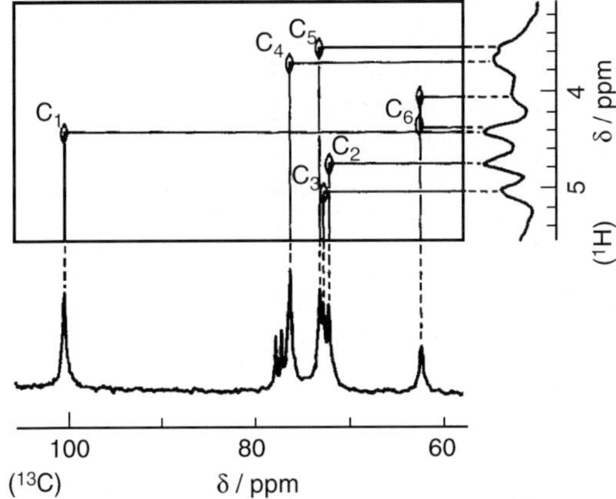

Figure 2.3.22   Schematic C–H COSY spectrum of CTA/TCM-*d* (glucopyranose region).[11]

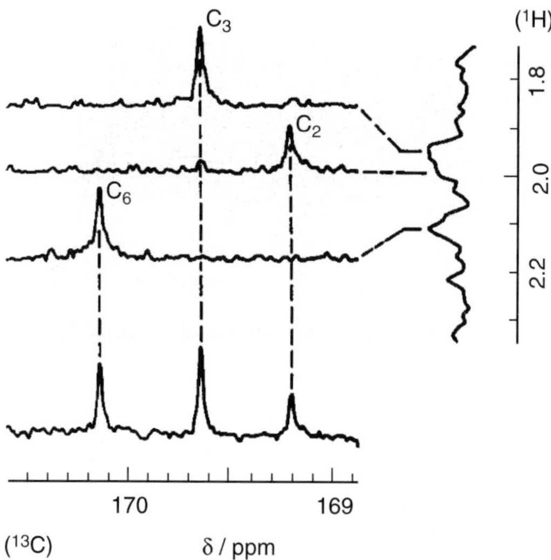

Figure 2.3.23   Cross sections of long-range C–H COSY spectrum of CTA/TCM-*d* (acetyl carbonyl carbon region).[11]

These relationships are in excellent agreement with the correlation obtained by the long-range selective spin decoupling method by Kowsaka *et al.*[11] Using these results together with the assignment of the acetyl methyl proton peak chemically speculated by Goodlett *et al.*,[3] carbonyl carbon peaks can be assigned from the lower magnetic field to C_6, C_3, and C_2, respectively. It is also expected that if long-range scalar coupling between the glucopyranose skeleton proton and carbonyl carbon could be observed, then carbonyl carbon peaks at C_2, C_3, and C_6 positions could occur. However, a possible long-range coupling constant between the proton attached to glucopyranose carbon and carbonyl carbon (i.e. H_6 proton and carbonyl carbon at C_6 position) is actually too small to detect. In addition, in long-range C–H COSY measurements, the S/N ratio is significantly low due to the rapid transverse relaxation of magnetization of nuclei in CTA molecules. For this reason, we could not observe this kind of coupling for this polymer system. Similarly, we measured long-range COSY of protons in order to determine the correlation between the ring proton and the acetyl methyl proton, but failed due to the low S/N ratio.

In the future, it may become possible to make a long-range measurement by (1) ^{13}C enrichment, (2) decrease in internal viscosity by lowering the polymer molecular weight, (3) 2D NMR by high magnetic field NMR apparatus, and (4) using other advanced 2D NMR techniques such as correlation spectroscopy *via* long-range coupling (COLOC).[15]

From COSY and C–H COSY for the CTA/DMSO-d_6 system, the almost comparable peak assignment was obtained for the CTA/DMSO-d_6 system in the same manner as was carried out for CTA/TCM-*d*.

Figure 2.3.24 shows a stacked plot of long-range C–H COSY (carbonyl carbon peak region) for CTA/DMSO-d_6 system. The figure confirms the correlation between carbonyl carbons and acetyl methyl protons obtained by the long-range selective spin decoupling method by Kowsaka *et al.*[8]

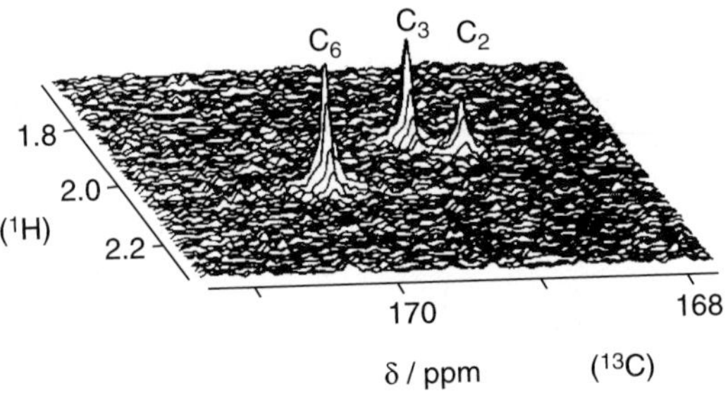

Figure 2.3.24 Stacked plot for long-range C–H COSY spectrum of CTA/TCM-*d* (acetyl carbonyl carbon region).[11]

In summary, the reliability of peak assignment of ^{1}H and ^{13}C NMR spectra for cellulose and its triacetate was undoubtedly strengthened by the use of 2D NMR. We demonstrated that overlapping peaks in 1D NMR, whose existence could only be speculated from their intensity, were experimentally verified often from the splitting of peaks on 2D spectrum.

With the use of 2D NMR, more complete and reliable calculations may be possible for almost all peaks in NMR spectra of not fully substituted cellulose derivatives with $\langle\langle F\rangle\rangle$ ranging from 0.5 to 2.5, for which peak assignment was unfortunately impossible by 1D spectrum alone due to the coexistence of various kinds of glucopyranose units (i.e. unsubstituted, mono-, di-, and trisubstituted units; see Section 2.4).

2.3.3 Sodium cellulose sulfate[16]

As an extension of the previous sections, an attempt is made to evaluate $\langle\langle f_k\rangle\rangle$ as well as $\langle\langle F\rangle\rangle$ for sulfate groups in sodium CS (NaCS) by ^{1}H and ^{13}C NMR methods.[16]

Experiments

Purified cotton lint was hydrolyzed with 1.0 mol aq. sulfuric acid at 60 °C for 6 h to give a cellulose sample having a viscosity-average molecular weight, $M_v = 0.9 \times 10^4$. NaCS was synthesized by reacting cellulose with sulfur trioxides dimethylformamide (DMF) complex at 10 °C according to the method proposed by Schweiger.[17] This was followed by the addition of sodium hydroxide. NaCS thus prepared was redissolved in water to give a solution of 5 g dl^{-1}, dialyzed with purified water until the electroconductivity of the dialysate became below 1×10^{-5} Ω^{-1} cm^{-1}, and dried *in vacuo*. The NaCS sample had a viscosity average molecular weight, $M_v = 15 \times 10^4$, which was determined by putting the limiting viscosity number $[\eta](= 544)$ in a 0.5 mol aq. NaCl solution at 25 °C into the Mark–Houwink–Sakurada equation: $[\eta] = 7.91 \times 10^{-2}$ $M_w^{0.93}$,[18] $\langle\langle F\rangle\rangle$ was determined by chemical analysis[17] (method of gravimetric analysis by converting sulfuric group into barium sulfate following decomposition of NaCS with hydrochloric acid), and was found to be 1.96.

The ^{1}H and ^{13}C NMR spectra (100 MHz) of the NaCS solution in deuterium oxide (D$_2$O) were obtained by a JOEL FX 100 Pulse-Fourier Transform NMR spectrometer at 37 °C.

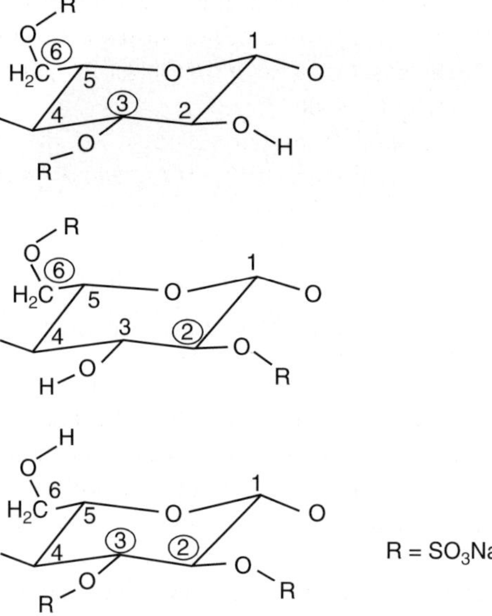

Figure 2.3.25 Possible conformation of glucopyranose units in sodium CS ($\langle\!\langle F\rangle\!\rangle = 2.0$). Numbers denote positions of carbon atoms to which a nonsubstituted hydroxyl groups are attached, constituting the glucopyranose units.[16] Numbers in circles denote positions of carbon atoms, to which substituted hydroxyl groups are attached.[16]

Figure 2.3.25 demonstrates all three differently sulfonated anhydroglucose units of NaCS with $\langle\!\langle F\rangle\!\rangle = 2.00$. Figure 2.3.26 shows ^{13}C NMR spectrum in D_2O. Carbons at the C_2–C_5 positions yield complicated peak signals at 70–80 ppm, from which neither $\langle\!\langle f_2\rangle\!\rangle$ nor $\langle\!\langle f_3\rangle\!\rangle$ could be directly estimated. C_6 carbon gave three peaks at 63.4, 61.2, and 60.8 ppm. Since the deshielding effect due to sulfate groups is expected to be almost equivalent to that of the acetyl group, the peak at 63.4 ppm could be attributed to the C_6 carbon bearing the sulfate group. The peak signal at 60.8 ppm is considered a result of the unsubstituted C_6 carbon. The peak at 61.2 ppm may be assigned to either

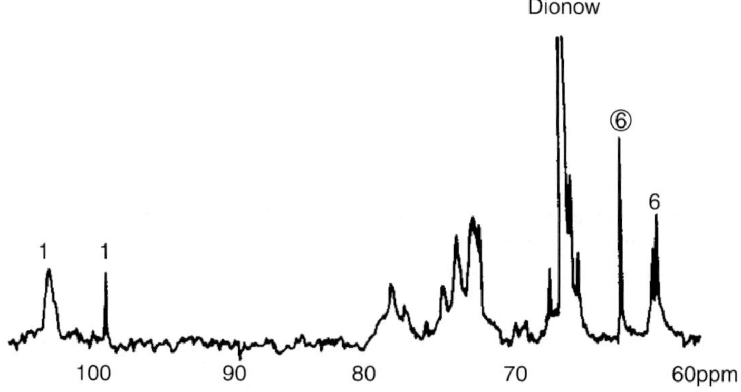

Figure 2.3.26 ^{13}C NMR spectrum of NaCS ($\langle\!\langle F\rangle\!\rangle = 1.96$) in deuterium oxide. Numbers and symbols are the same as described in Figure 2.3.25.[16]

the unsubstituted C_6 carbon (Hypothesis A) or the C_6 carbon bearing free sulfate group $(-OSO_3H)$ (Hypothesis B). The $\langle\!\langle f_6 \rangle\!\rangle$ value differs in regard to the assignment of the 61.2 ppm peak signal, which was found to be 0.34 for Hypothesis A and 0.44 for Hypothesis B from the area under peaks.

In Table 2.3.3, the chemical shift and assignment of the peaks of ^{13}C NMR spectrum are presented. Three peaks at 103.1, 102.7, and 99.4 ppm are assigned to the C_1 carbon, which was employed by Wu[4] for estimating $\langle\!\langle f_2 \rangle\!\rangle$ and $\langle\!\langle f_3 \rangle\!\rangle$ of the nitrate group. However, in this case, these peaks cannot be used for estimating $\langle\!\langle f_2 \rangle\!\rangle$ and $\langle\!\langle f_3 \rangle\!\rangle$ due to insufficient knowledge about the shielding effect of sulfate groups in the C_2 and C_3 positions. Of course, the integrated intensity of the C_1 carbon region was nearly equal to that of the C_6 carbon region.

Figure 2.3.27 shows the corresponding ^{1}H NMR spectrum of NaCS solution in D_2O. OH groups from NaCS were all converted into OD groups. Peak assignment was performed by a mutual comparison of peak areas and a comparison of the spectrum with that of cellulose acetate.[19] The proton signal at the C_3 position, which was sulfated, appears at the lowest magnetic field strength (4.97 ppm). A side band from HOD may be included in Peaks 3 and 1. In Table 2.3.3, the chemical shifts, assignments, and intensities of the peaks are shown together. Thus, $\langle\!\langle f_2 \rangle\!\rangle$, $\langle\!\langle f_3 \rangle\!\rangle$, and $\langle\!\langle f_6 \rangle\!\rangle$ can be accurately evaluated from ^{1}H NMR spectrum alone within an accuracy of ± 0.030, 0.035, and 0.012, respectively, by the following equations.

$$\langle\!\langle f_2 \rangle\!\rangle = \frac{7I_2}{\sum\limits_{i=1}^{6} I_i} \tag{2.3.2}$$

$$\langle\!\langle f_3 \rangle\!\rangle = \frac{7I_1}{\sum\limits_{i=1}^{6} I_i} \tag{2.3.3}$$

$$\langle\!\langle f_6 \rangle\!\rangle = \frac{7I_4/2}{\sum\limits_{i=1}^{6} I_i} \tag{2.3.4}$$

Table 2.3.3

Chemical shift and assignment of the peaks in ^{13}C NMR spectrum of sodium CS[16]

$\langle\!\langle F \rangle\!\rangle$ by chemical analysis	Chemical shift/ppm		
	C_1	C_2–C_5	C_6
1.96	103.1, 102.7, 99.3	79.2, 78.0 75.6, 74.7, 73.6, 73.2	63.4, 61.2, 60.8

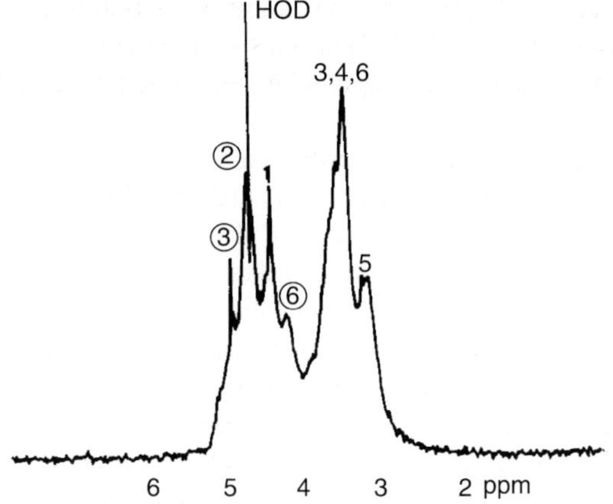

Figure 2.3.27 ^{1}H NMR spectrum of NaCS ($\langle\!\langle F \rangle\!\rangle = 1.96$) in deuterium oxide.[16] Numbers and symbols are the same as described in Figure 2.3.25.

where I_1, I_2, I_3, I_4, I_5, and I_6 are the integrated peak intensities at 4.97, 4.83, 4.56, 4.39, 4.0–3.5, and 3.42 ppm, respectively. Equation (2.3.4) was derived assuming that the proton directly attached to the substituted C_6 position is not contaminated in the integrated peak intensity I_3. The validity of this assumption is confirmed by

$$\frac{7I_3}{\sum\limits_{i=1}^{6} I_i} = 1.00 \qquad (2.3.5)$$

Putting the data from Table 2.3.4 into eqs. (2.3.2)–(2.3.4) gives $\langle\!\langle f_2 \rangle\!\rangle = 1.00$, $\langle\!\langle f_3 \rangle\!\rangle = 0.61$, and $\langle\!\langle f_6 \rangle\!\rangle = 0.34$. The latter value (0.34) is in good agreement with that (0.34) evaluated with Hypothesis A. This confirms the validity of the assignments proposed for the ^{13}C NMR spectrum. In addition, from the $\langle\!\langle f_k \rangle\!\rangle$ data by ^{1}H NMR, we obtain $\langle\!\langle F \rangle\!\rangle = 1.95$, which agrees well with that of the chemical analysis.

Table 2.3.4

Chemical shift, assignment, integrated peak intensity, and $\langle\!\langle f_k \rangle\!\rangle$ of ^{1}H NMR spectrum for sodium cellulose sulfate in deuterium oxide[16]

$\langle\!\langle F \rangle\!\rangle$		Chemical shift δ (ppm)	Assignment	Peak intensity	$\langle\!\langle f_k \rangle\!\rangle$
Chemical analysis	^{1}H NMR				
		4.97	H_3 Proton[a]	$I_1 = 16.2$	0.61
		4.83	H_2 Proton[a]	$I_2 = 26.7$	1.00
1.96	1.95	4.56	H_1 Proton	$I_3 = 26.7$	–
		4.39	H_6 Proton[a]	$I_4 = 18.0$	0.34(0.34)^{13}C NMR
		4.0–3.5	H_3, H_6, H_4 protons	$I_5 = 720$	
		3.42	H_5 Proton	$I_1 = 26.7$	–

[a]Signal of protons attached to C_k ($k = 2$, 3, and 6) position bearing a sulfate group.
[b]By ^{13}C NMR.

The C_1 proton signals in ^1H NMR spectra possibly overlap with the proton signals of the C_6 position bearing sulfate group. This makes the evaluation of $\langle\!\langle f_6 \rangle\!\rangle$ by ^1H NMR analysis difficult. Even in this case, $\langle\!\langle f_k \rangle\!\rangle$ can be determined by the following procedure.

(1) determination of $\langle\!\langle F \rangle\!\rangle$ by chemical analysis
(2) determination of $\langle\!\langle f_2 \rangle\!\rangle$ and $\langle\!\langle f_3 \rangle\!\rangle$ using eqs. (2.3.2) and (2.3.3)
(3) determination of $\langle\!\langle f_6 \rangle\!\rangle$ by

$$\langle\!\langle f_6 \rangle\!\rangle = \langle\!\langle F \rangle\!\rangle - (\langle\!\langle f_2 \rangle\!\rangle + \langle\!\langle f_3 \rangle\!\rangle) \tag{2.3.6}$$

The reactivity of hydroxyl groups at the C_2, C_3, and C_6 positions of cellulose with the SO_3–DMF complex decreased in the following order: $C_2 > C_3 > C_6$. If the hydrolysis reactions of NaCS are assumed to be the reverse of the order mentioned above, then the hydrolysis of the substituent at C_2 is the lowest. This clearly contradicts the fact that the reactivity of O-acetyl groups in CA with hydrochloric acid, previously evaluated, is of the order: $C_2 > C_3 > C_6$. This means that the reactivity of hydroxyl or substituents of cellulose or its derivatives cannot be primarily predicted according to the position alone.

2.3.4 CA whose acetyl groups are located only at C_6 position (i.e. 6-O-acetyl cellulose or CA with $\langle\!\langle f_6 \rangle\!\rangle \neq 0$ and $\langle\!\langle f_2 \rangle\!\rangle = \langle\!\langle f_3 \rangle\!\rangle = 0$)[20]

It is not sufficient to characterize CA in terms of $\langle\!\langle F \rangle\!\rangle$, $\langle\!\langle f_k \rangle\!\rangle$, and $\langle\!\langle f_{\ell mn} \rangle\!\rangle$ in order to establish the correlations between the molecular structure and their physical and physiological properties. Generally, any CA prepared by conventional methods has wide

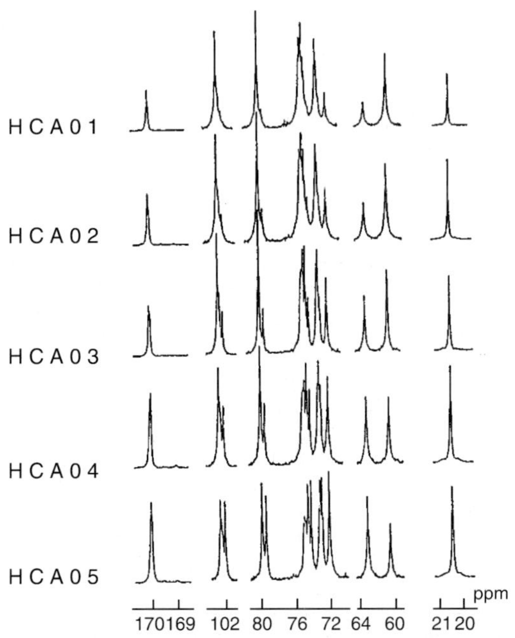

Figure 2.3.28 ^{13}C NMR spectra of 6-O-acetyl cellulose in DMSO-d_6.

variation with $\langle\!\langle f_k \rangle\!\rangle$. If CA, whose acetyl group is located at specific carbon position alone, is available, then the polymer is expected to contribute significantly to establishment of structure–properties relationships. Recently, Yasuda and Kamide[21] synthesized 6-*O*-acetyl cellulose by homogeneous acetylation and subsequent homogeneous deacetylation or their repetition.

6-*O*-acetyl cellulose

Step 1: Cellulose was acetylated at 30 °C in dimethylacetoamide (DMAc)/LiCl mixture (92 and 8 wt/wt) using pyridine and acetic anhydride[22] to give CA whose distribution of acetyl group is $\langle\!\langle f_6 \rangle\!\rangle \gg \langle\!\langle f_2 \rangle\!\rangle \approx \langle\!\langle f_3 \rangle\!\rangle$.

Step 2: CA thus prepared was deacetylated at 30 °C in DMSO using 80 wt% aq. solution of hydrazine monohydrate. The detailed reaction conditions were carefully chosen so as to yield CA whose acetyl groups are preferentially located only at the C_6 position (i.e. $\langle\!\langle f_6 \rangle\!\rangle \neq 0, \langle\!\langle f_2 \rangle\!\rangle = \langle\!\langle f_3 \rangle\!\rangle = 0$).

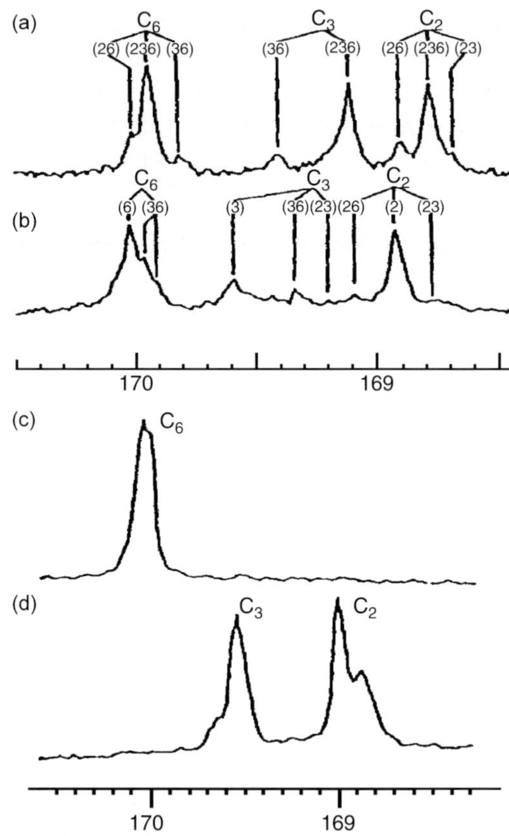

Figure 2.3.29 ^{13}C NMR spectra of the carbonyl carbon of CA (a) two-step method $\langle\!\langle F \rangle\!\rangle = 2.46$, (b) two-step method $\langle\!\langle F \rangle\!\rangle = 0.54$, (c) 6-*O*-acetyl cellulose $\langle\!\langle F \rangle\!\rangle = 0.62$, (d) 2,3-*O*-acetyl cellulose $\langle\!\langle F \rangle\!\rangle = 1.75$.

Table 2.3.5

$\langle\!\langle f_k \rangle\!\rangle$ of CA samples, prepared by homogenous acetylation and subsequent homogenous deacetylation and, if necessary, the repetition of these for adequate time (n)

Sample code	n	$\langle\!\langle f_2 \rangle\!\rangle$		$\langle\!\langle f_3 \rangle\!\rangle$		$\langle\!\langle f_6 \rangle\!\rangle$			Average
		A	B	A	B	A	B	C	
HCA01	1	0.0	0.0	0.0	0.0	0.28	0.27	0.30	0.28
HCA02	1	0.0	0.0	0.0	0.0	0.35	0.33	0.37	0.35
HCA03	2	0.0	0.0	0.0	0.0	0.42	0.42	0.42	0.42
HCA04	2	0.0	0.0	0.0	0.0	0.53	0.51	0.52	0.52
HCA05	3	0.0	0.0	0.0	0.0	0.62	0.62	0.62	0.62

A, carbonyl carbon; B, skeletal carbon; C, methyl carbon.

Steps 1 and 2 were repeated in order to prepare CA with relatively higher $\langle\!\langle F \rangle\!\rangle$.

Figure 2.3.28 shows ^{13}C NMR spectra of CA samples, whose acetyl groups are located only at the C_6 position in DMSO-d_6. The peaks at *ca.* 170, 60–105, and 20 ppm are assigned to the carbonyl carbon,[23] the skeleton carbon,[11] and the methyl carbon of acetyl group,[22] respectively.

Figure 2.3.29 shows ^{13}C NMR spectra of the carbonyl carbon of several CA samples in DMSO-d_6. Kowsaka *et al.* gave a very detailed assignment of this region as shown in

Table 2.3.6

Peak assignments in ^{13}C NMR spectra of cellulose acetate, in which all acetyl groups are located at C_6 position

Peak number	Peak position δ ppm (in DMSO)	Assignment
1	170.04	Acetylcarbonyl($f_{000}-f_{001}, f_{001}-f_{000}$)
2	170.00	Acetylcarbonyl($f_{001}-f_{001}$)
3	102.69	C_1 ($f_{000}-f_{000}, f_{000}\ f_{001}$ or $f_{001}\ f_{000}$)
4	102.47	C_1 ($f_{000}-f_{001}$ or $f_{001}-f_{000}$)
5	102.13	C_1 ($f_{001}-f_{001}$)
6	79.95	C_4 ($f_{000}-f_{000}, f_{000}-f_{001}, f_{001}-f_{000}$)
7	79.44	C_4 ($f_{001}-f_{001}$)
8	75.05	C_5 (f_{000})
9	74.98	C_3
10	74.92	C_3
11	74.82	C_3 (f_{000} or $f_{000}-f_{000}$)
12	74.56	C_3
13	74.19	C_3 ($f_{001}-f_{001}$)
14	73.20	C_2 ($f_{000}-f_{000}$)
15	73.11	C_2
16	72.86	C_2 ($f_{001}-f_{001}$)
17	72.03	C_5 (f_{001})
18	63.15	C_6 (f_{001})
19	60.56	C_6 (f_{000})
20	20.42	Acetylmethyl

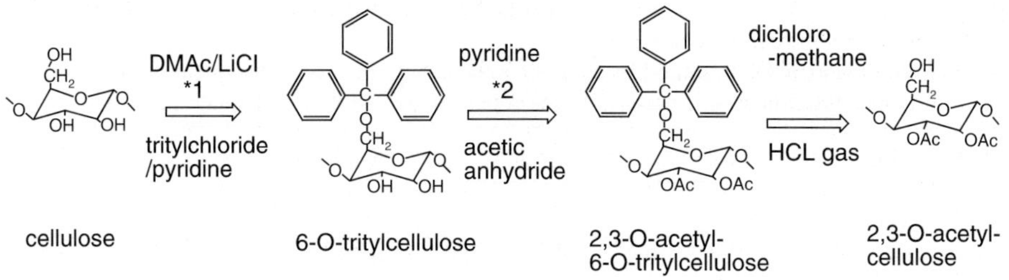

Figure 2.3.30   Schema of synthesis of 2,3-*O*-acetyl cellulose; [*]1: see Ref. 24; [*]2: see Ref. 25.

Figure 2.3.29a and b. The peak of Figure 2.3.29c is quite sharp, although it contains a shoulder (see also Figure 2.3.28). The CA sample in Figure 2.3.29d apparently has $\langle\!\langle f_6 \rangle\!\rangle = 0$.

Analysis on the carbonyl carbon peaks (peaks 1 and 2 in Figure 2.3.28) and skeleton carbon peaks (peaks 18 and 19 in Figure 2.3.28) allows us to estimate $\langle\!\langle f_k \rangle\!\rangle$ ($k = 2, 3,$ and 6). The results are summarized in Table 2.3.5. Inspection of Table 2.3.5 leads us to the conclusion that CA samples synthesized by homogeneous acetylation and subsequent homogeneous deacetylation (i.e. two homogeneous steps) have, without exception, $\langle\!\langle f_2 \rangle\!\rangle = \langle\!\langle f_3 \rangle\!\rangle = 0$. In addition to these methods, $\langle\!\langle f_6 \rangle\!\rangle$ can be determined from the acetyl methyl carbon peak (Peak 20; Yasuda and Kamide,[21] as included in Table 2.3.5). Three methods gave almost the same $\langle\!\langle f_6 \rangle\!\rangle$ within ± 0 to ± 0.02.

CA, whose acetyl groups are solely located on C_6 position, is very helpful to reinforce the assignment of ^{13}C NMR spectra of CA. Table 2.3.6 collates the peak assignments.

2,3-O-acetyl cellulose (CA whose acetyl groups are not located at C_6 position)

2,3-*O*-acetyl cellulose was synthesized *via* 6-*O*-tritylcellulose and 2,3-di-*O*-acetyl-6-*O*-tritylcellulose from cellulose. A scheme of the synthesis is shown in Figure 2.3.30. ^{13}C NMR spectra of carbonyl carbon of this sample (CA with $\langle\!\langle f_6 \rangle\!\rangle = 0$) is included in Figure 2.3.29d).

REFERENCES

1. TS Gardner and CB Purves, *J. Am. Chem. Soc.*, 1942, **64**, 1539.
2. CJ Malm, JJ Tanghe and BC Laird, *J. Am. Chem. Soc.*, 1950, **72**, 2674.
3. W Goodlett, JT Dougherty and HW Patton, *J. Polym. Sci. A-1*, 1971, **9**, 155.
4. TK Wu, *Macromolecules*, 1980, **13**, 74.
5. See, for example, K Kamide and K Okajima. *Polym. J.*, 1981, **13**, 127.
6. N Shiraishi, T Katayama and T Yokota, *Cell. Chem. Technol.*, 1978, **12**, 429.
7. T Miyamoto, Y Sato, T Shibata, H Inagaki and M Tanahashi, *J. Polym. Sci. Polym., Chem., Ed.*, 1984, **22**, 2363.
8. K Kowsaka, K Okajima and K Kamide, *Polym. J.*, 1986, **18**, 843.
9. K Kamide and M Saito, *Eur. Polym. J.*, 1984, **20**, 903.
10. K Kamide, K Okajima, K Kowsaka and T Matsui, *Polym. J.*, 1987, **19**, 1405.
11. K Kowsaka, K Okajima and K Kamide, *Polym. J.*, 1988, **20**, 1091.
12. D Gagnaire, D Mancier and M Vincendon, *J. Polym. Sci. Polym. Chem. Ed.*, 1980, **18**, 13.
13. K Kamide, K Kowsaka and K Okajima, *Polym. J.*, 1987, **19**, 231.
14. R Nardin and M Vincendon, *Macromolecules*, 1986, **19**, 2452.

15. See, for example, E Breitmeier and W Voelter, ^{13}C *NMR Spectroscopy*, 3rd Edn., Verlag Chemie, Weinheim, New York, 1986.
16. K Kamide and K Okajima, *Polym. J.*, 1981, **13**, 163.
17. FG Schweiger, *Carbohydr. Res.*, 1942, **21**, 219.
18. K Kishino, T Kawai, T Nose, M Saito and K Kamide, *Eur. Polym. J.*, 1981, **17**, 623.
19. H Friebolin, G Keilich and E Siefert, *Angew. Chem. Int. Ed.*, 1969, **8**, 766.
20. K Kamide and M Saito, *Macromol. Symp.*, 1994, **83**, 233.
21. K Yasuda and K Kamide, unpublished results.
22. See, for example, T Miyamoto, Y Sato, T Shibata, M Tanahashi and H Inagaki, *J. Polym. Sci., Polym. Chem. Ed.*, 1985, **23**, 1373.
23. K Kowsaka, K Okajima and K Kamide, *Polym. J.*, 1988, **20**, 827.
24. S Takahashi, T Fujimoto, BM Barua and T Miyamoto, *J. Polym. Sci., A. Polym. Chem.*, **36**, 952.
25. T Kondo and DG Gray, *Carbohydr. Res.*, 1991, **220**, 173.

2.4   MOLAR FRACTION OF EIGHT KINDS OF UNSUBSTITUTED AND PARTIALLY OR FULLY SUBSTITUTED GLUCOPYRANOSE UNITS ($\langle\!\langle f_{lmn}\rangle\!\rangle$)[1]

When we see the possible substituted glucopyranose units of cellulose derivatives, there are eight kinds of unsubstituted and partially or fully substituted glucopyranose units

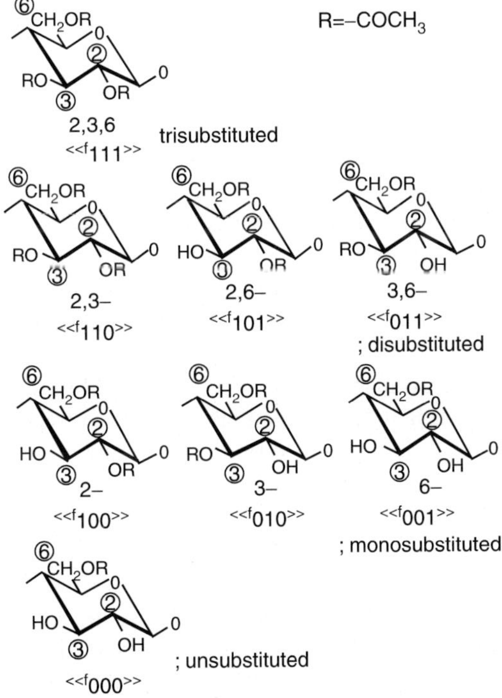

Figure 2.4.1 Substituted and unsubstituted glucopyranose units in CA molecules[5]: $\langle\!\langle f_{lmn}\rangle\!\rangle$ represents molecular fraction of the units.

as shown in Figure 2.4.1. These are single trisubstituted, three disubstituted, three monosubstituted, and one unsubstituted glucopyranose units. Molar fractions of these glucopyranose units $\langle\!\langle f_{lmn} \rangle\!\rangle$, defined in detail below, are an effective measure of the distribution of the substituent group within a glucopyranose unit, enabling us to judge whether the reaction is homogeneous or not. In addition, an accurate evaluation of these fractions is important in order to understand, on a molecular basis, the solubility of cellulose derivatives against various solvents and the physiological properties. Since the 1950s, the separation and quantitative determination of tri-, di-, monosubstituted, and unsubstituted glucopyranose units have been exclusively performed by applying distillation and chromatographic techniques to chemically decomposed cellulose derivatives, particularly sodium cellulose xanthate.[1] However, the decomposition of CD molecules into glucose units is extremely difficult without desubstituting reactions, in spite of the numerous attempts to prevent such reactions. In fact, experimental results reported about $\langle\!\langle f_{lmn} \rangle\!\rangle$ of cellulose xanthate differ depending on the researchers who carried them out. However, the conversion of the xanthate group into a more stable form involves very complicated chemical reactions.[2] Methods have been proposed by Wu,[3] and Clark and Stephenson[4] for estimating molar fractions of 2,3,6-tri-, 2,6-di-, 3,6-di-, and 6-mono substituted glucopyranose units of cellulose nitrates, whose C_6 position was fully substituted (i.e. $\langle\!\langle f_6 \rangle\!\rangle = 1$) from their ^{13}C NMR spectra. However, their methods cannot be applied to CN whose hydroxyl group at the C_6 position is not fully substituted.

2.4.1 Cellulose acetate[5]

This section assigns all peaks in the carbonyl carbon region of ^{13}C NMR spectra of CA and provides a firm basis for estimating $\langle\!\langle f_k \rangle\!\rangle$ and molar fractions of eight kinds of glucopyranose units of CA by NMR alone.

Sample preparation

A CTA whole polymer with $\langle\!\langle F \rangle\!\rangle = 2.92$ (sample code CA-0) and nine incompletely substituted CA samples, prepared by acid hydrolysis of sample CA-0 in acetic acid (sample code CA-1 to CA-9), were used. The procedures are described in detail elsewhere.[6,7]

 Table 2.4.1 collects the average molecular weight M_w of sample code CA-0 and -1, determined by light scattering in DMAc, and the viscosity average molecular weight M_v of sample code CA-2 to CA-7, determined from the limiting viscosity number in DMAc solution.[6-8]

NMR measurement

Proton noise decoupled ^{13}C NMR ($^{13}C\{^1H\}$ NMR) spectra of these CA solutions in deuterated dimethylsulfoxide (DMSO-d_6) were recorded on an FX-200 FT-NMR spectrometer (JEOL, Japan) at a resonance frequency of 50.18 MHz at 90 °C. The detailed operating conditions were almost the same as those reported in Section 2.4.2. TMS was the internal reference. Integrated peak intensity was determined from an

Table 2.4.1

Degree of substitution, weight, and viscosity average molecular weight and peak chemical shift in carbonyl region of CAs[5]

Sample code	$\langle\!\langle F\rangle\!\rangle$	M_w, $(M_v)/10^5$	Chemical shift/ppm (± 0.02 ppm)		
CA-0	2.92	2.32[a]	169.94	169.40, (169.17), 169.11	168.93, 168.78
CA-1	2.46	1.05[a]	170.02, 169.95, 169.82	169.41, 169.11	168.91, 168.78, 168.73
CA-2	1.75	0.82[b]	170.04, 169.97, 169.83	169.43, 169.17, 169.14, 169.11	168.93, 168.80
CA-3	1.23	0.80[b]	170.06, 170.00, 169.97, (169.85)	(169.61), 169.45, 169.21, 169.16	168.92, 168.79, (168.72)
CA-4	1.06	0.64[b]	170.04, 169.96, 169.87	169.58, 169.46, 169.34, 169.22, 169.10	168.92, 168.77
CA-5	0.95	0.47[b]	170.00, 168.87	169.61, 169.48, 169.36, 169.22	168.93, 168.87
CA-6	0.77	0.36[b]	170.04, (168.88)	169.59, 169.47, 169.36, 169.18	168.93, 168.75
CA-7	0.69	0.33[b]	170.04, 169.88, (168.89)	169.61, 169.46, 169.34	168.93, (168.71)
CA-8	0.54	–	170.02, 169.97, (169.90)	169.59, (169.43), 169.34, 169.09	168.92, 168.79
CA-9	0.43	–	170.04, (169.99), 169.90, 169.78	169.61, 169.34, (169.10)	168.94

[a] M_w, from light scattering.
[b] M_v, from $[\eta]$ in DMAC at 25 °C.

integral curve. $\langle\!\langle F\rangle\!\rangle$ was evaluated from the integrated intensity ratio of peaks in acetyl methyl carbon region (20–22 ppm) and peaks in C_1 carbon region (91–105 ppm).

The second column of Table 2.4.1 shows the $\langle\!\langle F\rangle\!\rangle$ of these CA samples.

Figure 2.4.2a–j shows the carbonyl carbon region of $^{13}C\{^1H\}$ NMR spectra of samples CA-0 to 9 in DMSO-d_6. These spectra were recorded at a spectral width of 1 kHz (4096 data points) in order to attain high digital resolution. The chemical shift from TMS as an internal reference was determined from the spectra obtained independently at a spectral width of 10 kHz (8192 data points). The digital resolution of these spectra was estimated to be approximately 0.01 ppm and the relative error of chemical shifts was less than 0.02 ppm. In the spectrum of sample code CA-0 ($\langle\!\langle F\rangle\!\rangle = 2.92$) in Figure 2.4.2a, three main peaks were observed, as reported in a previous paper,[9] originating from trisubstituted glucopyranose unit, and these peaks are assigned, from the lower magnetic field, to three carbonyl carbons at C_6, C_3, and C_2 positions, respectively. In the same spectrum, small peaks or shoulders, observed at 169.4, 169.2, and 168.9 ppm (as denoted by arrows in the figure), may possibly have originated from disubstituted glucopyranose units. In the spectrum of sample CA-9 ($\langle\!\langle F\rangle\!\rangle = 0.43$) in Figure 2.4.2j, three peaks (170.0, 169.6, and 168.9 ppm) due to three monosubstituted glucopyranose units, are observed. In addition, a group of small peaks (as denoted by arrows in the figure), considered to

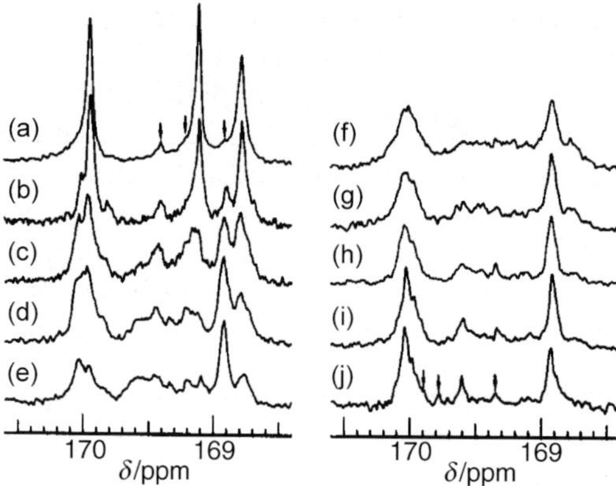

Figure 2.4.2   Carbonyl carbon region of ^{13}C{^{1}H} NMR spectra of cellulose acetates: (a) CA-0; (b) CA-1; (c) CA-2; (d) CA-3; (e) CA-4; (f) CA-5; (g) CA-6; (h) CA-7; (i) CA-8; (j) CA-9.[5]

have originated from disubstituted glucopyranose units, are detected at 169.9, 169.8, and 169.3 ppm. ^{13}C{^{1}H} NMR spectra for CA samples with intermediate $\langle\!\langle F \rangle\!\rangle$ (1.75–0.77) show very complicated patterns, possibly due to superposition of the above mentioned peak groups and of peaks from disubstituted units.

Table 2.4.1 (fourth column and onwards) summarizes the chemical shifts of all peaks (in 0.01 ppm unit) in the carbonyl carbon region of ^{13}C{^{1}H} NMR spectrum of each CA sample. In the table, chemical shifts of shoulders are shown in parentheses. The table indicates that for each CA sample CA-6 to CA-11, peaks or shoulders in total are observed in the carbonyl carbon region. Considering the relative error ($\pm$0.02 ppm) of chemical shifts, the peaks commonly observed for two or more CA samples are carefully chosen as listed in Table 2.4.2. Here, the peaks are numbered from the lower magnetic field.

Table 2.4.2.

Peak assignments in the carbonyl carbon region of CAs[5]

Peak no.	δ (ppm)	Carbon position	Glucopyranose unit	
1	170.04	6	6-Mono-	
2	170.00	6	2,6-Di-	
3	169.94	6	2,3,6-Tri-	
4	169.89	6	3,6-Di-	
5	169.83	6	3,6-Di-	
6	169.60	3	3-Mono-	
7	169.46	3	3,6-Di-	
8	169.41	3	3,6-Di-	
9	169.35	3	3,6-Di-	
10	169.22	3	2,3-Di-	
11	169.11	3(2)	2,3,6-Tri-	(2,6-Di-)
12	168.93	2	2-Mono-	2,6-Di-
13	168.71	2	2,3-Tri-	2,3-Di-
14	168.71	2	2,3-Di-	

As previously described, the total number of carbonyl carbon peaks is theoretically 12, but from actual experiments, 14 peaks are observed (Table 2.4.2).

Peak numbers 3, 11, and 13 are the main peaks in the spectrum of highly substituted CA ($\langle\!\langle F \rangle\!\rangle = 2.92$) as shown in Figure 2.4.2a of the 2,3,6-trisubstituted glucopy ranose unit and are assigned to the carbonyl carbons at C_6, C_3, and C_2 carbon positions in the 2,3,6-trisubstituted unit, respectively. This mutual positional order in the trisubstituted unit can be maintained in low substituted CA ($\langle\!\langle F \rangle\!\rangle = 0.43$); thus, the main peaks in Figure 2.4.2j No. 1, 6, and 12 can be assigned to the carbonyl carbons in 6-, 3-, and 2-monosubstituted glucopyranose units, respectively. The assignments for these six peaks are collected in Columns 3 and 4 (in part) of Table 2.4.2.

Generally, the degree of change in chemical shift due to the introduction of a specific substitution group into polymer is the maximum at the nucleus, to which the substituent group is directly combined (i.e. α-position). It is smaller at the nucleus, which is combined indirectly with (and separated through) a large number of nuclei from the substituent group. For example, the chemical shift of the carbonyl carbon at C_3 position is significantly influenced by whether the C_2 position is substituted or not, but is not so much affected by the C_6 position. Thus, it is reasonably expected that the chemical shifts of C_3 carbonyl carbons of 3-mono- and 3,6-disubstituted glucopyranose units are relatively close, and C_3 carbonyl carbons of 2,3-di- and 2,3,6-trisubstituted units give similar chemical shifts. According to this expectation, we can assign the number 10 peak in the vicinity of Peak 11 to the C_3 carbonyl carbon in the 2,3-disubstituted glucopyranose unit, and Peaks 7, 8, and 9 in the vicinity of Peak 6 to C_3 carbonyl carbons in the 3,6-disubstituted glucopyranose unit, respectively. Accordingly, Peak 14 near peak 13 is assigned to the C_2 peak in the 2,3-disubstituted glucopyranose unit. Since no peak exists near the position of Peak 12, which was assigned already to the C_2 carbonyl carbon in the 2-monosubstituted unit, then the C_2 carbonyl carbon in the 2,6-disubstituted unit is considered to be heavily superposed with that of the 2-monosubstituted unit. In a similar manner, Peaks 4 and 5, observed in the vicinity of Peak 3, is assigned to the C_6 carbonyl carbon in the 3,6-disubstituted unit and Peak 2 in the vicinity of Peak 3 to the C_6 carbonyl carbon in the 2,6-disubstituted unit. The above assignments are also shown in Columns 3, 4, and 5 of Table 2.4.2.

Table 2.4.3 collects the integrated peak intensities of carbonyl carbon Peaks 1–14, I_n (n is the peak number and $n = 1$–14), for samples CA-0, 1, 2, 3, 8, and 9. Here, I_n is normalized with total integrated intensity $\sum_{n=1}^{14} I_n (\equiv \langle\!\langle F \rangle\!\rangle)$. Several peaks overlap each other, and some I_n cannot be estimated separately.

The overlapping peaks are shown as underlined, and their total peak intensities are shown under the main peak (I_n) in the table. Inspection of Tables 2.4.2 and 2.4.3 leads us to the conclusion that in CA with $\langle\!\langle F \rangle\!\rangle > 2.4$ (CA-0 and CA-1), a large amount of the 2, 3, 6-trisubstituted (from Peaks 3, 11, and 13) unit coexists with a small amount of 2,6- and 3,6-disubstituted (from Peaks 5, 8, and 12) glucopyranose units, and CAs with $\langle\!\langle F \rangle\!\rangle < 0.6$ (CA-8 and CA-9) are mainly constituted by 2-, 3-, and 6-monosubstituted glucopyranose (from Peaks 1, 6, and 12) units with a small amount of 3,6-disubstituted glucopyranose unit (from Peak 9). We can conclude that the highly substituted CA consists mainly of trisubstituted and disubstituted glucopyranose units and low substituted CA consists predominantly of monosubstituted units. Note that all not fully substituted CA samples were prepared by acid

Table 2.4.3

Integrated peak intensity I_n of carbonyl carbon peaks of CA samples[5]

Sample code	$\langle\!\langle F \rangle\!\rangle$	I_n													
		$n=1$	2	3	4	5	6	7	8	9	10	11	12	13	14
CA-0	2.92			0.97		0.02	0.00		0.06	0.00		0.95	0.01	0.91	0.00
CA-1	2.46			0.77		0.09	0.00		0.16	0.00		0.66	0.21	0.57	
CA-2	1.75			0.56			0.00	0.23		0.00	0.40		0.23	0.33	0.00
CA-3	1.23		0.40				0.00	0.20		0.03	0.17		0.24	0.18	
CA-8	0.54	0.22			0.00	0.00	0.08	0.03		0.04		0.02	0.13	0.02	
CA-9	0.43	0.18			0.00	0.00	0.08	0.00		0.05	0.00	0.01	0.11	0.00	0.00

hydrolysis, which is expected to occur randomly (or homogeneously), and the above conclusion is consistent with our expectations for random acid hydrolysis of cellulose acetate. Kamide *et al.*[10] pointed out from TLC analysis on CTA that 'CTA' with $\langle\langle F\rangle\rangle$ of 2.92 is a mixture of CTA with $\langle\langle F\rangle\rangle = 3$ (i.e. trisubstituted unit) and not fully substituted cellulose acetate. This finding was supported in this NMR study. Although Peak 11 (C_3 carbonyl carbon in trisubstituted gluconose units) was observed for CA samples CA-8 and CA-9 (low substituted CAs), the corresponding C_6, C_2 carbonyl carbons for the trisubstituted one (Peak 3 and 13, respectively) were not found. Thus, in this case, peak 11 should be assigned to the C_2 carbonyl carbon of the 2,6-disubstituted glucopyranose unit. On the other hand, the same carbonyl carbon (i.e. the C_2 carbonyl carbon at the 2,6-disubstituted unit) peak is observed as Peak 12 for highly substituted CA samples (CA-0 and CA-1), indicating that change in $\langle\langle F\rangle\rangle$ from 2.5 to 2.9 to *ca.* 0.5 brings about a shift of the chemical shift of the C_2 carbonyl carbon peak in the 2,6-disubstituted unit by *ca.* 0.14 (± 0.04) ppm. Similar variation in the chemical shift with $\langle\langle F\rangle\rangle$ was noticed on other peaks. One or two peaks, observed over the range of 169.34–169.48 ppm, are considered due to the C_3 carbonyl carbon of the 3,6-disubstituted unit. Its location varies depending on $\langle\langle F\rangle\rangle$. We already assigned Peak 14 to the C_2 carbonyl carbon of the 2,3-disubstituted unit, but the low substituted CA Peak 13 (in place of Peak 14) should be assigned to the above carbonyl carbon. A tentative explanation of these phenomena would be the presence of a possible long-range (weak) effect from the substituent group in the nearest neighboring glucopyranose unit on the carbonyl carbon in question. That is, the peak position of the chemical shift of the C_2 carbonyl carbon in the 2,6-disubstituted unit may change depending on whether the substituent group exists or not, probably at the C_3 position, which is the nearest to the β-glucoside linkage, in the neighboring glucopyranose unit. This long-range effect will become more pronounced for the disubstituted units, existing in a wide range of $\langle\langle F\rangle\rangle$. Then, peaks due to either C_3 or C_2 carbonyl carbon are expected to be located at the position of Peak 11. Additional assignments of Peaks 11 and 13, determined from integrated peak intensity, are also shown in Table 2.4.2. Figure 2.4.3a and b demonstrate carbonyl carbon region of $^{13}C\{^1H\}$ NMR spectra of samples CA-1 and CA-8 with full assignment.

Using the assignment for all the peaks in carbonyl carbon region (as given in Table 2.4.2), we can derive equations to give $\langle\langle f_k\rangle\rangle$:

$$\langle\langle f_2\rangle\rangle = \langle\langle F\rangle\rangle\left(\sum_{j=12}^{14} I_j\right)\bigg/\left(\sum_{j=1}^{14} I_j\right) \qquad \text{(for CA with } \langle\langle F\rangle\rangle \geq 2) \qquad (2.4.1)$$

$$\langle\langle f_2\rangle\rangle = \langle\langle F\rangle\rangle\left(\sum_{j=11}^{14} I_j\right)\bigg/\left(\sum_{j=1}^{14} I_j\right) \qquad \text{(for CA with } \langle\langle F\rangle\rangle < 2) \qquad (2.4.2)$$

$$\langle\langle f_3\rangle\rangle = \langle\langle F\rangle\rangle\left(\sum_{j=6}^{11} I_j\right)\bigg/\left(\sum_{j=1}^{14} I_j\right) \qquad \text{(for CA with } \langle\langle F\rangle\rangle \geq 2) \qquad (2.4.3)$$

$$\langle\langle f_3\rangle\rangle = \langle\langle F\rangle\rangle\left(\sum_{j=6}^{10} I_j\right)\bigg/\left(\sum_{j=1}^{14} I_j\right) \qquad \text{(for CA with } \langle\langle F\rangle\rangle < 2) \qquad (2.4.4)$$

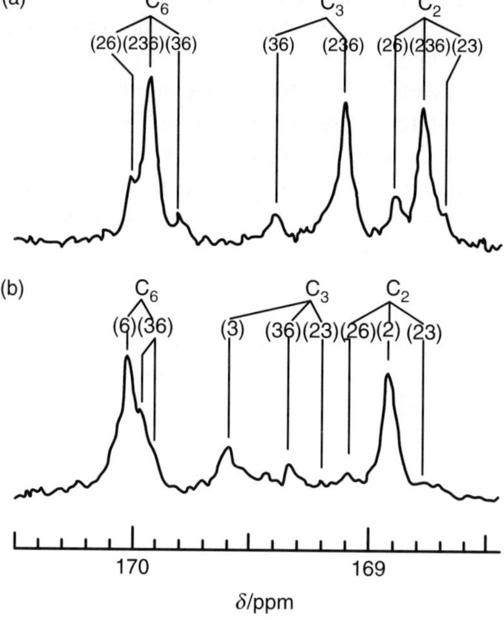

Figure 2.4.3   Peak assignments of the carbonyl carbon peaks of cellulose acetates: (a) CA-1; (b) CA-8.[5]

$$\langle\!\langle f_6 \rangle\!\rangle = \langle\!\langle F \rangle\!\rangle \left(\sum_{j=1}^{5} I_j \right) \Big/ \left(\sum_{j=1}^{14} I_j \right) \tag{2.4.5}$$

Table 2.4.4 collects $\langle\!\langle f_k \rangle\!\rangle$ of all CA samples calculated using eqs. (2.4.1)–(2.4.5) from the data in Table 2.4.3. Evidently, it is confirmed for CA with $\langle\!\langle F \rangle\!\rangle \geq 0.7$, that $\langle\!\langle f_2 \rangle\!\rangle \approx \langle\!\langle f_3 \rangle\!\rangle \approx \langle\!\langle f_6 \rangle\!\rangle$. Kamide *et al.*[11] reported a similar tendency for CA synthesized by a two-step method ($\langle\!\langle f_2 \rangle\!\rangle \approx \langle\!\langle f_3 \rangle\!\rangle \approx \langle\!\langle f_6 \rangle\!\rangle$), although in their study, determination of $\langle\!\langle f_k \rangle\!\rangle$ was

Table 2.4.4

Total degree of substitution $\langle\!\langle F \rangle\!\rangle$ and distribution of degree of substitution $\langle\!\langle F \rangle\!\rangle$ of CA samples[5]

Sample code	$\langle\!\langle F \rangle\!\rangle$	$\langle\!\langle f_2 \rangle\!\rangle$	$\langle\!\langle f_3 \rangle\!\rangle$	$\langle\!\langle f_6 \rangle\!\rangle$
CA-0	2.92	0.92	1.01	0.99
CA-1	2.46	0.79	0.82	0.85
CA-2	1.75	0.56	0.63	0.56
CA-3	1.23	0.42	0.40	0.40
CA-4	1.06	0.33	0.42	0.31
CA-5	0.95	0.29	0.33	0.33
CA-6	0.77	0.26	0.24	0.27
CA-7	0.69	0.20	0.23	0.26
CA-8	0.54	0.17	0.15	0.22
CA-9	0.43	0.12	0.13	0.18

made by rough approximation. In addition, from I_n ($n = 1$–14) data for any CA sample, we can evaluate the distribution of seven different substituted glucopyranose units, as shown in Figure 2.4.1.

We denote molar fractions of eight (seven substituted and one unsubstituted) glucopyranose units with $\langle\!\langle f_{lmn}\rangle\!\rangle$, where l, m, and n denote the existence of the acetyl group at the C_2, C_3, and C_6 positions, respectively, and they can take the value of 0 (unsubstituted) or 1 (substituted). According to this notation, the doubly averaged molar fraction of the trisubstituted unit is $\langle\!\langle f_{111}\rangle\!\rangle$, those of disubstituted units are $\langle\!\langle f_{110}\rangle\!\rangle$, $\langle\!\langle f_{101}\rangle\!\rangle$, and $\langle\!\langle f_{011}\rangle\!\rangle$, and that of unsubstituted unit is $\langle\!\langle f_{000}\rangle\!\rangle$ and so on (see Figure 2.4.1). The summation of all $\langle\!\langle f_{lmn}\rangle\!\rangle$ is unity:

$$\sum_{l=0}^{1} \sum_{m=0}^{1} \sum_{n=0}^{1} \langle\!\langle f_{lmn}\rangle\!\rangle = 1 \tag{2.4.6}$$

According to this definition, $\langle\!\langle f_k\rangle\!\rangle$ and $\langle\!\langle F\rangle\!\rangle$ can be expressed in terms of $\langle\!\langle f_{lmn}\rangle\!\rangle$ as follows:

$$\langle\!\langle f_2\rangle\!\rangle = \sum_{m=0}^{1} \sum_{n=0}^{1} \langle\!\langle f_{lmn}\rangle\!\rangle \tag{2.4.7}$$

$$\langle\!\langle f_3\rangle\!\rangle = \sum_{l=0}^{1} \sum_{n=0}^{1} \langle\!\langle f_{lmn}\rangle\!\rangle \tag{2.4.8}$$

$$\langle\!\langle f_6\rangle\!\rangle = \sum_{l=0}^{1} \sum_{m=0}^{1} \langle\!\langle f_{lmn}\rangle\!\rangle \tag{2.4.9}$$

$$\langle\!\langle F\rangle\!\rangle = \langle\!\langle f_2\rangle\!\rangle + \langle\!\langle f_3\rangle\!\rangle + \langle\!\langle f_6\rangle\!\rangle \tag{2.4.10}$$

Based on the assignment of all carbonyl carbon peaks in Table 2.4.2 and neglecting peak overlapping, we can calculate $\langle\!\langle f_{lmn}\rangle\!\rangle$ from I_n through the following equations:

$$\langle\!\langle f_{111}\rangle\!\rangle = I_3 = I_{13} + I_{14} - \langle\!\langle f_{110}\rangle\!\rangle \tag{2.4.11}$$

$$\langle\!\langle f_{111}\rangle\!\rangle = I_{11} = I_{13} \qquad \text{(for cellulose acetate with } \langle\!\langle F\rangle\!\rangle \geq 2) \tag{2.4.12}$$

$$\langle\!\langle f_{111}\rangle\!\rangle = I_{11} - \langle\!\langle f_{101}\rangle\!\rangle \qquad \text{(for cellulose acetate with } \langle\!\langle F\rangle\!\rangle < 2) \tag{2.4.13}$$

$$\langle\!\langle f_{110}\rangle\!\rangle = I_{10} = I_{13} + I_{14} - \langle\!\langle f_{111}\rangle\!\rangle \tag{2.4.14}$$

$$\langle\!\langle f_{101}\rangle\!\rangle = I_2 = I_{11}2 + I_{12} - \langle\!\langle f_{111}\rangle\!\rangle - \langle\!\langle f_{100}\rangle\!\rangle \tag{2.4.15}$$

$$\langle\!\langle f_{101}\rangle\!\rangle = I_{12} \qquad \text{(for cellulose acetate with } \langle\!\langle F\rangle\!\rangle \geq 2) \tag{2.4.16}$$

$$\langle\!\langle f_{101}\rangle\!\rangle = I_{11} \qquad \text{(for cellulose acetate with } \langle\!\langle F\rangle\!\rangle < 2) \tag{2.4.17}$$

$$\langle\!\langle f_{011}\rangle\!\rangle = I_4 + I_5 = I_7 + I_8 + I_9 \tag{2.4.18}$$

$$\langle\!\langle f_{100}\rangle\!\rangle = I_{12} \qquad \text{(for cellulose acetate with } \langle\!\langle F\rangle\!\rangle < 2) \tag{2.4.19}$$

$$\langle\!\langle f_{010}\rangle\!\rangle = I_6 \tag{2.4.20}$$

$$\langle\!\langle f_{001}\rangle\!\rangle = I_1 \tag{2.4.21}$$

The peaks in $^{13}C\{^1H\}$ NMR spectra of CA samples (except for CA-0) are quite broad indeed and overlap significantly, as shown in Figure 2.4.2. Thus, an accurate evaluation of $\langle\!\langle f_{lmn}\rangle\!\rangle$ for CA is quite difficult.

Since for the sample code CA-8 peaks 6, 11, and 12 are separately observed without overlapping with other peaks, $\langle\!\langle f_{010}\rangle\!\rangle$, $\langle\!\langle f_{100}\rangle\!\rangle$, and $\langle\!\langle f_{101}\rangle\!\rangle$ for the sample can be evaluated using the I_n data of these peaks (Table 2.4.3) from eqs. (2.4.17), (2.4.19), and (2.4.20), respectively. The fraction $\langle\!\langle f_{110}\rangle\!\rangle$ for the sample can also be determined from eq. (2.4.14) using data on $I_{13} + I_{14}$ in Table 2.4.3 and assuming $\langle\!\langle f_{111}\rangle\!\rangle = 0$, because (1) Peaks 13 and 14 overlap with each other, but not with other peaks, enabling an estimate of $I_{13} + I_{14}$ and (2) $\langle\!\langle F\rangle\!\rangle$ for this sample is low (0.54). $\langle\!\langle f_{011}\rangle\!\rangle$ for the sample can also be evaluated using data on $I_7 + I_8 + I_9$ from eq. (2.4.18), in the same manner. In addition, for this CA sample, Peak 1 significantly overlaps with Peaks 2, 3, 4, and 5. Thus, $\langle\!\langle f_{001}\rangle\!\rangle$ cannot be simply estimated from eq. (2.4.21) using I_1 data. An alternative way of estimating $\langle\!\langle f_{001}\rangle\!\rangle$ is given by the equation:

$$\langle\!\langle f_{001}\rangle\!\rangle = \sum_{j=1}^{5} I_j - \langle\!\langle f_{101}\rangle\!\rangle - \langle\!\langle f_{011}\rangle\!\rangle - \langle\!\langle f_{111}\rangle\!\rangle \qquad (2.4.22)$$

A combination of eqs. (2.4.5) and (2.4.10) leads to eq. (2.4.22). Neglecting $\langle\!\langle f_{111}\rangle\!\rangle$ in eq. (2.4.22) for sample CA-8, $\langle\!\langle f_{001}\rangle\!\rangle$ can be estimated roughly from $\langle\!\langle f_{101}\rangle\!\rangle$ and $\langle\!\langle f_{011}\rangle\!\rangle$ data previously determined by eqs. (2.4.16) and (2.4.18).

Equation (2.4.6) can be rearranged as follows:

$$\langle\!\langle f_{000}\rangle\!\rangle = 1 - (\langle\!\langle f_{111}\rangle\!\rangle + \langle\!\langle f_{110}\rangle\!\rangle + \langle\!\langle f_{101}\rangle\!\rangle + \langle\!\langle f_{011}\rangle\!\rangle + \langle\!\langle f_{100}\rangle\!\rangle + \langle\!\langle f_{010}\rangle\!\rangle + \langle\!\langle f_{001}\rangle\!\rangle) \qquad (2.4.23)$$

Then, eq. (2.4.23) enables us to estimate $\langle\!\langle f_{000}\rangle\!\rangle$ from already known $\langle\!\langle f_{lmn}\rangle\!\rangle$ data.

The values of $\langle\!\langle f_{lmn}\rangle\!\rangle$ for sample CA-8 thus determined are provided in Table 2.4.5, where the corresponding data for samples CA-1 and CA-9, both obtained in a similar manner, are also included. Here, for sample CA-1, it is assumed that the molar fractions of mono- and unsubstituted glucopyranose units are neglected ($\langle\!\langle f_{100}\rangle\!\rangle = \langle\!\langle f_{010}\rangle\!\rangle = \langle\!\langle f_{001}\rangle\!\rangle = \langle\!\langle f_{000}\rangle\!\rangle = 0$).

From $\langle\!\langle f_{lmn}\rangle\!\rangle$ data shown in Table 2.4.5, $\langle\!\langle f_k\rangle\!\rangle$ and $\langle\!\langle F\rangle\!\rangle$ can be directly evaluated using eqs. (2.4.7), (2.4.9), and (2.4.10). The results are also summarized in columns 10 and 14 in Table 2.4.5. Here, $\langle\!\langle f_k\rangle\!\rangle$ and $\langle\!\langle F\rangle\!\rangle$ are denoted by $\langle\!\langle f_k\rangle\!\rangle_{\text{calc}}$ and $\langle\!\langle F\rangle\!\rangle_{\text{calc}}$, respectively. $\langle\!\langle f_k\rangle\!\rangle$ and $\langle\!\langle F\rangle\!\rangle$, estimated previously by the conventional method for these samples through

Table 2.4.5

$\langle\!\langle f_{lmn}\rangle\!\rangle$, $\langle\!\langle f_k\rangle\!\rangle_{\text{calc}}$, $\langle\!\langle F\rangle\!\rangle_{\text{calc}}$, $\langle\!\langle f_k\rangle\!\rangle_{\text{con}}$, and $\langle\!\langle F\rangle\!\rangle_{\text{con}}$ values of CA smples

Sample code	$\langle\!\langle f_{111}\rangle\!\rangle$	$\langle\!\langle f_{110}\rangle\!\rangle$	$\langle\!\langle f_{101}\rangle\!\rangle$	$\langle\!\langle f_{011}\rangle\!\rangle$	$\langle\!\langle f_{100}\rangle\!\rangle$	$\langle\!\langle f_{010}\rangle\!\rangle$	$\langle\!\langle f_{110}\rangle\!\rangle$	$\langle\!\langle f_{000}\rangle\!\rangle$
CA-1	0.48	0.18	0.21	0.16	0[a]	0[a]	0[a]	0[a]
CA-8	0[a]	0.02	0.02	0.07	0.13	0.08	0.13	0.45
CA-9	0[a]	0.00	0.01	0.05	0.11	0.08	0.12	0.63

Sample code	$\langle\!\langle f_2\rangle\!\rangle_{\text{calc}}$	$\langle\!\langle f_3\rangle\!\rangle_{\text{calc}}$	$\langle\!\langle f_6\rangle\!\rangle_{\text{calc}}$	$\langle\!\langle F\rangle\!\rangle_{\text{calc}}$	$\langle\!\langle f_2\rangle\!\rangle_{\text{con}}$	$\langle\!\langle f_3\rangle\!\rangle_{\text{con}}$	$\langle\!\langle f_6\rangle\!\rangle_{\text{con}}$	$\langle\!\langle F\rangle\!\rangle_{\text{con}}$
CA-1	0.87	0.82	0.85	2.54	0.79	0.82	0.85	2.46
CA-8	0.17	0.17	0.22	0.56	0.17	0.15	0.22	0.54
CA-9	0.12	0.13	0.18	0.43	0.12	0.13	0.18	0.43

[a]Approximated previously.

application of eqs. (2.4.1), (2.4.5), and (2.4.10), are denoted as $\langle\!\langle f_k \rangle\!\rangle_{con}$ and $\langle\!\langle F \rangle\!\rangle_{con}$, and are also shown in columns 15 and 18 in Table 2.4.5. $\langle\!\langle f_k \rangle\!\rangle_{calc} \approx \langle\!\langle f_k \rangle\!\rangle_{con}$ ($k = 2, 3,$ and 6) and $\langle\!\langle F \rangle\!\rangle_{calc} \approx \langle\!\langle F \rangle\!\rangle_{con}$ are thus confirmed.

In summary, all peaks observed in the carbonyl carbon region in ^{13}C NMR spectra of fully and not fully substituted CA were unambiguously assigned and a method was proposed for evaluating the molar fractions of eight kinds of glucopyranose units such as 2,3,6-tri-, 2,3-di-, 2,6-di-, 3,6-di-, 2-mono-, 3-mono-, and 6-mono substituted units and an unsubstituted unit (denoted here as $\langle\!\langle f_{lmn} \rangle\!\rangle$), based on these assignments. It was confirmed for three CA samples that $\langle\!\langle f_k \rangle\!\rangle$ and $\langle\!\langle F \rangle\!\rangle$, calculated from $\langle\!\langle f_{lmn} \rangle\!\rangle$, coincide approximately with those by conventional methods, supporting the reliability of the present method of assignment.

2.4.2 Cellulose sulfate (CS)[12]

In Section 2.4.1, we used the Kowsaka *et al.*[5] definition of the molar fractions of these eight AHG units as $\langle\!\langle f_{lmn} \rangle\!\rangle$ ($l = 1$ or 0 means that a hydroxyl group attached to C_2 position is substituted or not, m and n indicate corresponding values in C_3 and C_6 positions) and succeeded in determining $\langle\!\langle f_{lmn} \rangle\!\rangle$ of CA samples by analyzing the carbonyl carbon region of their ^{13}C{^{1}H} NMR spectra.

Studies on $\langle\!\langle f_k \rangle\!\rangle$ of CS were first made in 1981 by Kamide and Okajima,[13] who measured the ^{1}H and ^{13}C{^{1}H} NMR spectra of sodium salt of CS (NaCS) with $\langle\!\langle F \rangle\!\rangle = 1.96$ dissolved in deuterium oxide (D$_2$O) and assigned six ^{1}H peaks originating from AHG units from the lower magnetic field as H_3^*, H_2^*, H_1, H_6^*, (H_3, H_4, H_6), and H_5 protons, with reference to the assignment made for cellulose acetate.[14] Here, H_k ($k = 1–6$) denotes each proton attached to the corresponding C_k carbon position in AHG unit, and * denotes the corresponding proton when the hydroxyl group at the C_k position is substituted (see Section 2.3.3). They demonstrated that blood anticoagulant activity and acute toxicity (lethal dose 50 for rat, administrated intravenously) of NaCS was mainly governed by $\langle\!\langle f_2 \rangle\!\rangle + \langle\!\langle f_3 \rangle\!\rangle$ determined by NMR spectroscopy.[15] Note that in some cases, they unavoidably estimated $\langle\!\langle f_6 \rangle\!\rangle$ as the difference between $\langle\!\langle F \rangle\!\rangle$ determined chemically and $\langle\!\langle f_2 \rangle\!\rangle + \langle\!\langle f_3 \rangle\!\rangle$ determined above.[15] However, recent advances in NMR techniques raises the following issues regarding the previous ^{1}H NMR peak assignment for NaCS carried out by Kamide and Okajima[13] (Section 2.3.3): (1) the resolution of their 100 MHz ^{1}H NMR spectra was not enough to separate peaks accurately and this situation might be enhanced when D$_2$O is used, (2) their oversimplified assumption of a one-to-one correspondence of the peak assignment between CA and NaCS cannot always be accepted, and the possible difference in magnetic influence on other protons caused by substitution at C_k position between CA and NaCS should have been carefully considered, (3) their assumption that only nine nonequivalent protons (H_1 to H_6, H_2^*, H_3^*, H_6^*) exist is crude. These difficulties lead us to the conclusion that the past determination of $\langle\!\langle f_k \rangle\!\rangle$ of CS by ^{1}H NMR analysis entails low accuracy.

The peak separation of ^{13}C{^{1}H} NMR spectrum for water soluble CD is generally better than that for ^{1}H NMR spectrum when D$_2$O is used as solvent. The peak assignment for carbons constituting glucopyranose backbone might not be so easy but recent developments in pulse techniques in Fourier transform NMR may overcome this situation. In fact, Kowsaka *et al.*[16] succeeded in assigning the carbon peaks of CA employing 2D NMR technique (Section 2.3.2).

In this section, we attempt to give complete assignments for ^{13}C, and ^{1}H NMR spectra of NaCS samples will be given with the aid of 2D NMR and the validity of the ^{1}H peak assignment previously given by Kamide and Okajima will be checked.[13]

Sample preparation

The sodium salt of CS (NaCS) was prepared according to the procedures described[15] using an acid-hydrolyzed soft wood pulp (the viscosity average molecular weight $M_v = 9.7 \times 10^4$, as measured by solution viscosity in cadoxen at 25 °C using the Mark–Houwink–Sakurada equation established by Brown and Wikström[17]).

The detailed procedures are as follows:

(1) Two grams of acid-hydrolyzed pulp were dispersed in 18 g of dimethyl formamide (DMF) at 25 °C. The DMF wetted cellulose was mixed with 9 g of DMF sulfur trioxide (SO_3) complex 2:1, mol mol^{-1}; solid) precooled at 0 °C. The mixture was stirred vigorously at 25 °C for 1 h to make a transparent solution, which was poured into 2*l* of NaOH–methanol–water (1:90:9, w/w/w) to precipitate the crude NaCS. The crude NaCS separated on a filter was washed with 1.5 *l* of methanol water (9:1 v/v at 25 °C) and then dissolved in distilled water. The polymer was reprecipitated with methanol, washed several times with a mixture of methanol–water (7:3 v/v at 25 °C), and finally subjected to drying *in vacuo*. The sample thus prepared was denoted as DS-1. Two samples (denoted as DS-2 and DS-3) were synthesized by the same procedure as that described above by using 6 or 4 g of DMF/SO_3 complex against 2 g of cellulose, respectively. In these cases, the unreacted and not dissolved cellulose in the system was first removed by centrifuging (1.5×10^4 *g* for 1 h) the reaction mixture. The sample DSH-1 was prepared by the same procedure as for DS-1 except that the reaction temperature was 70 °C.

(2) 10 g of acid-hydrolyzed pulp were mixed with 40 g of a mixture of 98% sulfuric acid butanol (110:29 v/v) at 20 °C, stood for 10 min and this reaction mixture was poured into a system (1 *l*) of diethylether containing dry ice. On standing, the system separated into a supernatant solution and brownish viscous liquid. The former phase was subjected to the same procedure as that described in (1) and the recovered polymer was coded as HBH′-1. Note that HBH series samples in Ref. 15 were recovered from the brownish viscous liquid.

NMR measurements

NMR measurements on NaCS samples dissolved in D_2O were carried out using a JNM-FX200 type FT-NMR spectrometer (JEOL, Japan) with C/H10 mm ∅ probe. [1]H and ^{13}C resonance frequencies were 199.5 and 50.18 MHz, respectively. TSP was used as the internal standard. ^{1}H and ^{13}C{^{1}H} NMR spectra were measured according to the method described elsewhere[16] (see Section 2.3.3). The water elimination Fourier transform (WEFT) method was applied for ^{1}H NMR measurements to diminish peak intensity originating from HDO. Homonuclear ^{1}H shift correlation spectroscopy (^{1}H COSY) and heteronuclear $^{13}C-^{1}H$ shift correlation spectroscopy (CH-COSY) were recorded as follows:

(1) ^{1}HCOSY experiments were carried out at 70 °C using a $90°-t_1-90°$ pulse sequence under the following operating conditions: spectral width, 500 Hz; data matrix, 512×256; pulse interval, 1.5 s; accumulation, 128. The polymer concentration was *ca.* 5 wt%, Homo-gate decoupling was employed simultaneously in order to reduce signals originating from remaining protons in the solvent. Other operating conditions are indicated in Ref. 16.

(2) The CH-COSY spectrum was recorded using the pulse sequence method described in Ref. 16. The mixing time A was set up to be 2.0 ms $(= 1/(4J_{C-H}))$, where J_{C-H} is the average coupling constant between C and H nuclei. Operating conditions are follows: ^{13}C spectral width, 2500 Hz; ^{1}H spectral width, 500 Hz; pulse interval, 1 s; accumulation, 2048; data matrix; 1024×40 (1000×128 after zero filling on t_1 axis). The polymer concentration was *ca.* 15 wt%. Other operating conditions are described in Ref. 16.

For ^{1}H and COSY measurements, sample DS-1 was once dissolved in D_2O, freeze dried and again dissolved in D_2O in order to diminish HDO peak.

Figure 2.4.4 illustrates the possible eight AHG units constituting NaCS chains, predicting the existence of magnetically nonequivalent 48 ^{13}C and 56 ^{1}H nuclei. Here, the number of ^{1}H nuclei does not include three ^{1}H nuclei for the unreacted OH protons at C_2, C_3, and C_6 positions in AHG unit because they are almost instantaneously exchanged by D nuclei of the solvent. Note that none of these expected peaks (i.e. 48 ^{13}C and 56 ^{1}H peaks) will be actually observed as separate peaks because of the overlapping of the peaks.

Figure 2.4.5 shows ^{1}H WEFT NMR spectra of the samples DS-1 (a), DSH-1 (b), and HBH′-1 (c). The negative resonance (arrowed in the figure) observed around 4.3 ppm

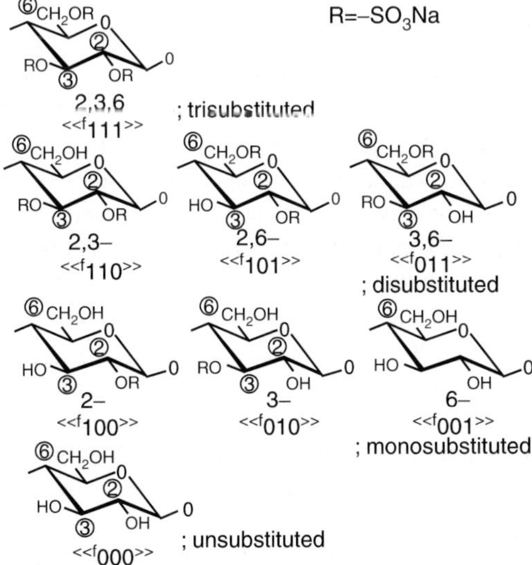

Figure 2.4.4   Substituted and unsubstituted glucopyranose units in NaCS molecules:[12] $\langle\!\langle f_{lmn} \rangle\!\rangle$ represents molar fraction of the units.

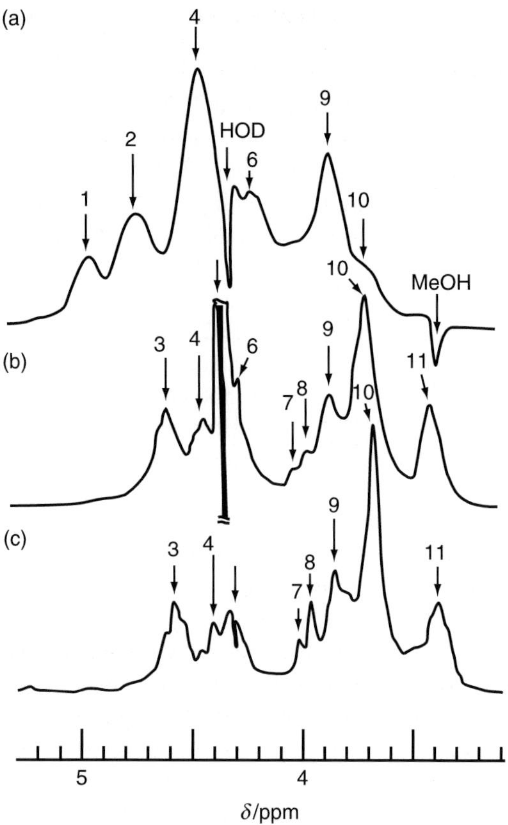

Figure 2.4.5 ^{13}C{^{1}H} NMR spectra of NaCS samples in D$_2$O solution using WEFY technique: (a) sample DS-1; (b) sample DSH-1; (c) sample HBH-1.[12]

might be due to a shorter pulse interval, applied for WEFT here, than the most adequate pulse interval ('null point') and the resonance is assigned to HDO proton. The peak appearing at 3.36 ppm in Figure 2.4.5c is responsible for methanol, which is introduced as a contaminant during the preparation of NaCS, and this peak was ignored in further analysis. For DS-1, ^{1}H peaks are observed in the chemical shift (δ) range of 3.6–5.0 ppm. On the other hand, for DSH-1 and HBH$'$-1, ^{1}H peaks are observed in the chemical shift range of 3.3–4.7 ppm. Taking the estimated reading error of δ (0.02 ppm) into account, 10 ^{1}H peaks were detected besides the HDO peak. The δ values of all the peaks detected for each NaCS sample are listed in Table 2.4.6. The peaks appearing around 4.4 and 4.6 ppm in Figure 2.4.5b and c are triplet peaks split by vicinal coupling.

Therefore, the δ values at the central peaks are employed in the table. Note that, judging from the number of peaks observed and their δ values, the sample DSH-1 is similar to that reported for NaCS with $\langle\!\langle F \rangle\!\rangle = 1.96$, synthesized by Kamide and Okajima.[13]

Figure 2.4.6 shows the ^{13}C{^{1}H} NMR spectra of DS-1 (a), DSH-1 (b), and HBH$'$-1 (c). The spectra of DSH-1 and HBH$'$-1 are similar to each other, except for nine small sharp peaks, marked by arrows in Figure 2.4.6c, which probably originated from

Table 2.4.6

Chemical shifts δ of proton peaks of NaCS sample in D_2O[12]

Peak no.	δ (ppm)		
	DS-1	DSH-1	HBH$'$-1
1	4.93		
2	4.70		
3		4.58	4.57
4	4.41	4.43	4.42
5	4.28	4.32	4.30
6	4.18		
7		4.02	4.02
8		3.96	3.96
9	3.84	3.85	3.84
10	3.66	3.68	3.67
11	3.39	3.38	

contaminants such as cellulosic oligomers produced as a by-product during synthesis of HBH$'$-1. Sixteen peaks in total were observed and their δ values are indicated in Table 2.4.7.

Table 2.4.8 gives the chemical shifts and peak assignments of ^{1}H and ^{13}C NMR peaks for cellulose/NaOD at 20 °C, which was cited from Figures 2.3.16 and 2.3.18. Here, the assignments for the ^{13}C peaks at 81.8 and 78.2 ppm are made with reference to Nardin and Vincendon's study,[18] which was carried out for the cellulose/dimethyl

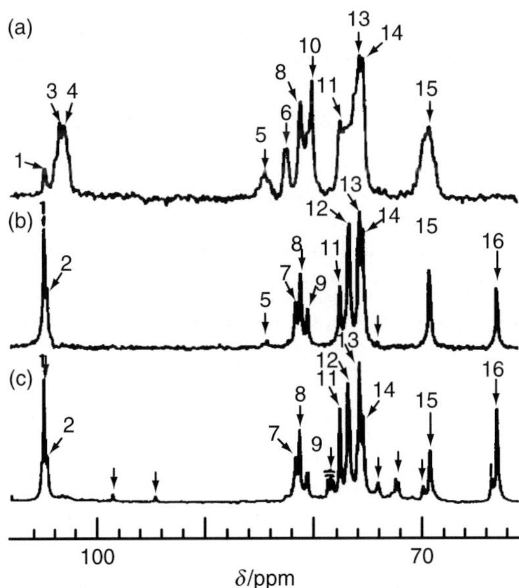

Figure 2.4.6 ^{13}C{^{1}H} NMR spectra of NaCS samples in D_2O solution:[12] (a) sample DS-1; (b) sample DSH-1; (c) sample HBH$'$-1.

Table 2.4.7

Chemical shifts δ of proton peaks of NaCS sample in D_2O[12]

Peak no.	δ (ppm)			Peak no.	δ (ppm)		
	DS-1	DSH-1	HBH$'$-1		DS-1	DSH-1	HBH$'$-1
1	104.8	104.8	104.8	9	–	80.6	80.6
2		104.6	104.6	10	80.1	–	–
3	103.4	?	?	11	77.5	77.5	77.5
4	102.9	?	?	12	(76.8*)	76.8	76.8
5	84.5	84.3	–	13	75.8	75.8	75.8
				14	75.5	75.5	75.5
6	82.4	?	?		(74.3?)	(74.3?)	(74.3?)
7	–	81.7	81.7	15	69.2	69.2	69.2
8	81.2	81.3	81.3	16	–	63.0	63.0

?, not clearly detected; –, not detected; *, included in peak 11; (74.3?), detected but not clear to arise from AHG units polymer.

acetamide/lithium chloride system. A comparison of Tables 2.4.6 and 2.4.7 with Table 2.4.8 shows that DSH-1 and HBH$'$-1 have NMR peaks at almost the same positions as those observed for cellulose, but DS-1 has numerous peaks besides those observed for cellulose. This suggests that DS-1 is highly substituted.

Figure 2.4.7 shows the contour plot of ^{1}H COSY (power spectrum) of DS-1 with a projection on the horizontal axis. The projection spectrum shown in Figure 2.4.7 apparently has a higher resolution compared with the spectrum shown in Figure 2.4.5a, but a small peak (peak 10) at 3.66 ppm is suppressed. This might be due to the application of a trapezoidal window function to the observed free induction decay signal before Fourier transformation on the COSY measurement. The peaks, observed at 4.92 (doublet), 4.68 (triplet), 4.42 (triplet?), 4.14 (triplet), and 3.88 ppm (doublet) on the projection correspond to Numbers 1, 2, 4, 6, and 9 peaks in Table 2.4.6, respectively. The peak at 4.92 ppm shows

Table 2.4.8

Proton and carbon chemical shifts δ of cellulose in aq. NaOD collected from Figures 4 and 6 in Ref. 16

Observing nucleus	δ (ppm)	Assignment
^{1}H	4.49	H_1
	3.95	H_6
	3.80	H_6
	3.57	H_6, H_4, H_5
	3.29	H_2
^{13}C	106.7	C_1
	81.8	C_4
	78.2	C_3, C_5
	76.8	C_2
	63.5	C_6

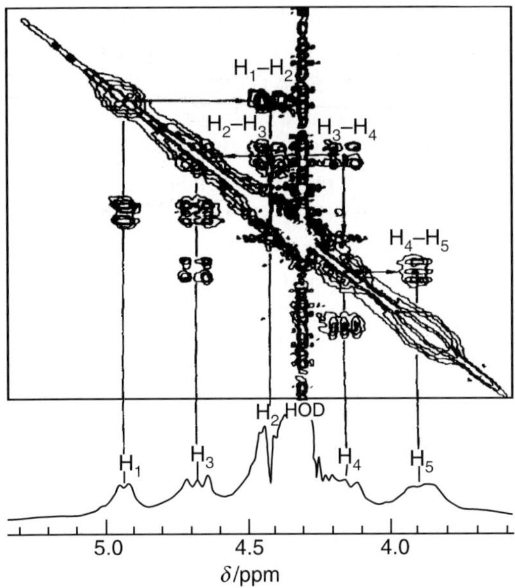

Figure 2.4.7 Homogate decouple proton COSY spectrum of NaCS (sample DA-1).[12]

only one cross peak in the contour plot in Figure 2.4.7 and is attributed to H_1 proton. The appearance of the cross peak at 4.42 ppm, shown as H_1–H_2 in Figure 2.4.7, enables us to attribute this peak to H_2 proton. In a similar manner, H_3, H_4, H_5 proton peaks can be assigned and the final assignments attained, excluding HDO proton peak region, are given on the projection spectrum, that is, H_1, H_3, H_2, H_4, and H_5 proton peaks from the lower magnetic field.

Figure 2.4.8 shows the contour plot of CH-COSY (power spectrum) for DS-1 with projection spectra on the horizontal (^{13}C) and longitudinal (^{1}H) axes. Six peaks, observed at 103.5, 81.2, 80.1, 77.5, 76.0, and 69.6 ppm on the ^{13}C projection spectrum, correspond to Numbers 3, 8, 10 (and 9), 11, 13 (and 14), and 15 peaks, respectively, in Table 2.4.7. The peaks assigned in Figure 2.4.7 are also observed in the ^{1}H projection spectrum.

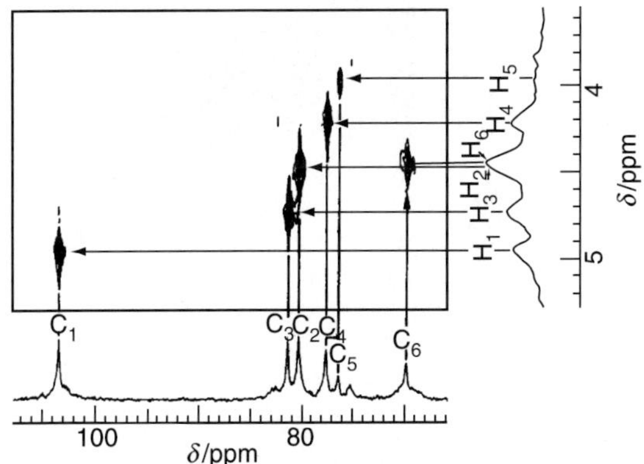

Figure 2.4.8 CH-COSY spectrum of NaCS (sample DS-1).[12]

The contour plot revealed that the ^{13}C peak at 103.5 ppm was correlated with the ^{1}H peak at 4.93 ppm, assigning this ^{13}C peak to C_1 carbon. In the same manner, three ^{13}C peaks at 81.2, 77.5, and 76.0 ppm were assigned to C_3, C_4, and C_5 carbons, respectively. The ^{1}H peak at 4.42 ppm (H_2 proton) is correlated with two ^{13}C peaks at 80.1 and 69.6 ppm and, therefore, the peak at 4.42 ppm is expected to be an overlapping peak of H_2 proton and another proton. The ^{13}C peak at 69.6 ppm is assigned to C_6 carbon because the C_6 carbon peak is, without exception, observed in the highest magnetic field in the spectra of cellulose,[16] cellulose acetate,[16] CN,[4] and CX.[9] This leads to the conclusion that the ^{1}H peak at 4.42 ppm originates from H_2 and H_6 (and H_6') protons. Note that there are two magnetically nonequivalent H_6 protons, H_6 and H_6'. The difference between δ values of ^{1}H and ^{13}C peaks for DS-1 (*cf.* Figures 2.4.7 and 2.4.8) and those for cellulose (*cf.* Table 2.4.8) are summarized as follows: (1) the peaks due to H_2, H_3, and H_6 protons for DS-1 appear in the lower magnetic field by 0.6–1.1 ppm than those of cellulose, (2) C_2, C_3, and C_6 carbon peaks for DS-1 substantially shifted to a lower magnetic field (3.0–6.0 ppm), compared with the corresponding carbon peaks for cellulose, (3) hardly any of the δ values of ^{1}H and ^{13}C peaks for DS-1 coincide with those of cellulose. These facts suggest that almost all NMR peaks for DS-1 should appear as a result of the almost complete substitution of three hydroxyl groups at C_2, C_3, and C_6 positions. Therefore, almost all peaks observed in Figures 2.4.7 and 2.4.8 originate mainly from the trisubstituted AHG unit although there are some peaks (84.0, 82.4, and 75.2 ppm) without a detectable cross peak on the CH-COSY spectra.

Figure 2.4.9 shows ^{1}H COSY of HBH'-1. The spectrum is quite similar to that for cellulose/NaOD system,[16] suggesting that $\langle\!\langle F \rangle\!\rangle$ of this polymer is very low. The following assignments are easily made H_1, 4.57 ppm; H_6, 3.99 ppm (doublet); H_6, 3.84 ppm; (H_3, H_4, H_5), 3.67 ppm; H_2, 3.38 ppm. Besides these peaks, there remain three unassigned

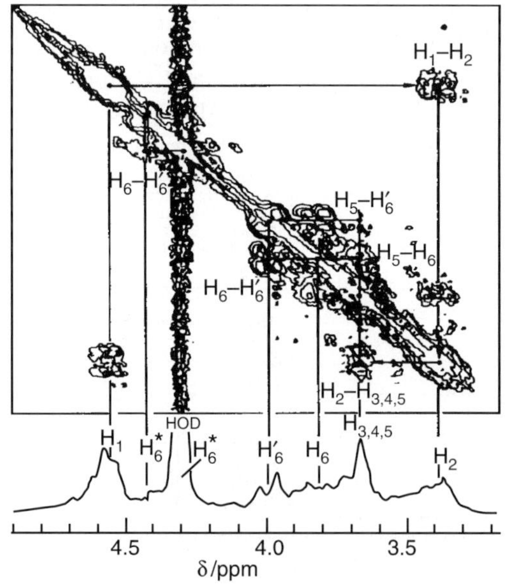

Figure 2.4.9   Homogate decoupled proton COSY spectrum of NaCS (sample HMH'-1).[12]

peaks at 5.22, 4.4, and 4.3 ppm, and the full assignment is not possible by ^{1}H COSY alone.

Figure 2.4.10 shows CH-COSY of HBH$'$-1. On the basis of ^{1}H peak assignment given in Figure 2.4.9, the following ^{13}C peak assignment is made: 104.8 ppm, C_1; 75.8 ppm, C_2; 63.0 ppm, C_6. Three ^{13}C peaks (81.3, 77.5, and 76.8 ppm) are correlated with a (H_3, H_4, H_5) proton peak (3.67 ppm) and then, these ^{13}C peaks should be assigned to either C_3, C_4 or C_5 carbons. Although these three peaks are cannot be assigned completely only from Figure 2.4.10, the peak at 81.3 ppm can be attributed to C_4 carbon from the assignment in Table 2.4.7. The remaining two peaks at 77.5 and 76.8 ppm were assigned to C_5 and C_3, respectively, with reference to Nardin and Vincendon's assignment[18] carried out for cellulose/dimethylacetamide/lithium chloride system and to Gagnaire et al.'s assignment[19] on low molecular weight cellulose in dimethyl sulfoxide. Since δ values of ^{1}H and ^{13}C peaks assigned above for HBH$'$-1 are confirmed to be almost the same as those for cellulose, almost all these peaks should originate from unsubstituted AHG units. Inspection of two cross peaks (see Figure 2.4.10) indicates that the ^{13}C peak at 69.2 ppm is due to 6-mono-substituted AHG unit because there are no cross peaks due to substituted C_2–H_2 and substituted C_3–H_3 in the dotted circle region, as shown in Figure 2.4.10, and because the total DS of HBH$'$-1 is quite low. The peaks at 81.7 and 80.6 are correlated with ^{1}H peak around 3.67 ppm and therefore attributable to either C_3, C_4 or C_5 carbon. The peak at 75.5 ppm is correlated with ^{1}H peak around 3.84 ppm.

2D NMR spectra shown in Figures 2.4.7–2.4.10 enables us to assign the following peaks: (1) all ^{13}C and ^{1}H peaks for 2,3,6-trisubstituted AHG unit (C_1, 103.4 ppm; C_2, 80.1 ppm; C_3, 8 1.3 ppm; C_4, 77.5 ppm; C_5, 75.8 (75.5) ppm; C_6, 69.2 ppm; H_1, 4.93 ppm; H_2, 4.42 ppm; H_3, 4.70 ppm; H_4, 4.18 ppm; H_5, 3.84 ppm; H_6, 4.42 ppm); (2) all ^{13}C and ^{1}H peaks for unsubstituted AHG unit (C_1, 104.8 ppm; C_2, 75.8 (75.5) ppm; C_3, 76.8 ppm; C_4, 81.3 ppm; C_5, 77.5 ppm; C_6, 63.0 ppm; H_1, 4.58 ppm; H_2, 3.39 ppm; H_3, 3.67 ppm; H_4, 3.67 ppm; H_5, 3.67 ppm; H_6, 3.96 and 3.84 ppm).

Table 2.4.9 gives the observed δ values of ^{1}H peaks for the 2,3,6-trisubstituted AHG unit and unsubstituted AHG unit. Denoting the carbon atom directly attached to the proton in question as the neighboring carbon to C_α as C_β, and the next neighboring

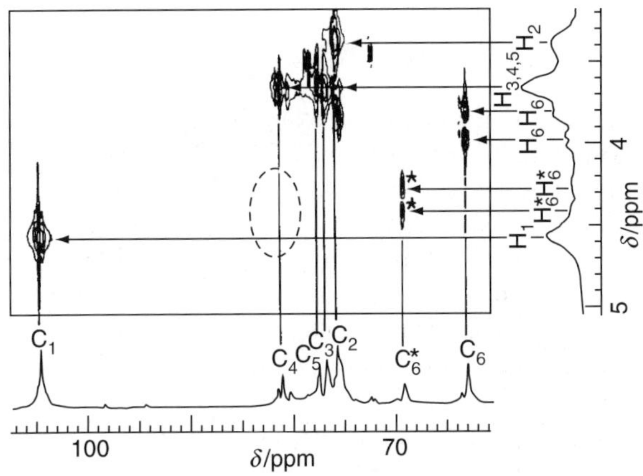

Figure 2.4.10   CH-COSY spectrum of (sample HBH-1).[12]

Table 2.4.9

Observed and calculated proton chemical shifts of AHG units in NACS[12]

AHG unit	δ (ppm)						Remarks
	H_1	H_2	H_3	H_4	H_5	H_6	
2,3,6-Trisub.	4.93	4.41	4.70	4.18	3.84	4.41	Observed
2,3-Disub.	4.93	4.41	4.70	4.18	3.44	3.81	Calculated
2,6-Disub.	4.93	4.01	4.10	3.78	3.84	4.41	Calculated
3,6-Disub.	4.53	3.81	4.30	4.18	3.84	4.41	Calculated
2-Monosub.	4.93	4.01	4.10	3.78	3.44	3.81	Calculated
3-Monosub.	4.53	3.81	4.30	4.18	3.44	3.81	Calculated
6-Monosub.	4.53	3.41	3.70	3.78	3.84	4.41	Calculated
Unsub.	4.53	3.41	3.70	3.78	3.44	3.81	Calculated
Unsub.	4.57	3.38	3.67	3.67	3.67	3.84, 3.96	Observed

Here, $\Delta H_\alpha = 0.6$ ppm, $\Delta H_\beta = 0.4$ ppm, $\Delta H_\gamma = 0.0$ ppm are assumed. $H_k = H_{k_0} + (\Delta H_\beta + \Delta H_\gamma)$; $k = 1\text{--}6$. H_{k_0} means standard value (trisubstituted AHG unit) of H_k.

carbon as C_γ, we define the magnetic influence on the proton in question as shift factor ΔH_α, ΔH_β, ΔH_γ,... when the hydroxyl groups attached to C_α, C_β, C_γ,... carbons are substituted, respectively. The shift factors can be estimated from the observed δ values in Table 2.4.9 and using these estimated shift factors we can calculate δ values for ^{1}H protons belonging to other AHG units. Since δ values of H_1 and H_4 protons for the trisubstituted AHG unit are larger than those for the unsubstituted one by 0.36 and 0.51 ppm, respectively, the following relationships are expected to hold:

$$0.36 = \Delta H_\beta + \Delta H_\gamma + \Delta H_\zeta \qquad \text{(for } H_1) \qquad\qquad (2.4.24)$$

$$0.51 = \Delta H_\beta + 2\Delta H_\gamma \qquad \text{(for } H_4) \qquad\qquad (2.4.25)$$

Similar analysis of H_2 and H_3 protons gives the following relationships:

$$1.03 = \Delta H_\alpha + \Delta H_\beta + \Delta H_\varepsilon \qquad \text{(for } H_2) \qquad\qquad (2.4.26)$$

$$1.03 = \Delta H_\alpha + \Delta H_\beta + \Delta H_\gamma \qquad \text{(for } H_3) \qquad\qquad (2.4.27)$$

Provided that the long range effects as expressed by ΔH_γ, ΔH_δ, ΔH_ε, and ΔH_ζ are assumed to be 0, $\Delta H_\beta = 0.36\text{--}0.51$ ppm and $\Delta H_\alpha = 0.52\text{--}0.67$ ppm are estimated from the above equations. If we take $\Delta H_\alpha = 0.6$ and $\Delta H_\beta = 0.4$, the δ values for H_1–H_6 protons belonging to any other AHG units can be calculated and the values thus calculated are listed in Table 2.4.9. Here, the observed δ values for 2,3,6-trisubstituted AHG unit were used as standards because the δ values were precisely determined with an unquestionable peak assignment. The calculated δ values for the unsubstituted AHG unit are in good agreement with those observed except for H_5 proton. This disagreement may be explained by conformational change around C_5–C_6 linkage, which is expected to depend on $\langle\!\langle F \rangle\!\rangle$ or $\langle\!\langle f_k \rangle\!\rangle$. The chemical shifts of the ring protons for cellulose derivatives are expected to depend on changes in the magnetic environment, conformation, and solvation state due to the introduction of substituents. Nevertheless, the relatively simple assumption introduced here proved to be adequate to predict the δ values of the protons belonging to AHG units.

Table 2.4.10 compiles the assignments of the 11 proton peaks observed for $NaCS/D_2O$ systems in Table 2.4.6 based on the results in Table 2.4.9. In the table, the symbol H_k (lmn) is employed to express the carbon position k ($= 2$, 3, and 6), to which the proton in question is attached, and to denote one of eight AHG units by (lmn). The definition of (lmn) is the same as lmn used for $\langle\!\langle f_{lmn}\rangle\!\rangle$. Table 2.4.10 indicates that the peak at the lowest magnetic field (4.93 ppm) is exclusively assigned to the H_1 proton when the hydroxyl group attached to C_2 position is substituted. Thus, the peak intensity of this proton peak ($I_{4.93}$) is proportional to $\langle\!\langle f_2\rangle\!\rangle$. That is, $\langle\!\langle f_2\rangle\!\rangle$ is given by the following equation:

$$\langle\!\langle f_2\rangle\!\rangle = 7I_{4.93}\Big/\sum I \qquad (2.4.28)$$

Here, $\sum I$ is the integrated proton peak intensity of NaCS. Kamide and Okajima[13] erroneously assigned this peak (at $\delta = 4.93$ ppm) to H_3^* proton, and employed this peak to estimate $\langle\!\langle f_3\rangle\!\rangle$. This means that the $\langle\!\langle f_3\rangle\!\rangle$ values estimated by Kamide and Okajima,[13,15] (Table I of Ref. 15, Table II of Ref. 13) should be corrected as $\langle\!\langle f_2\rangle\!\rangle$. The peak at 4.7 ppm, which was assigned as H_2^* by Kamide and Okajima,[13] now proves to correspond to two H_3 protons of 2,3,6-trisubstituted AHG unit and 2,3-disubstituted AHG unit. Therefore, the peak intensity of this peak ($I_{4.7}$) should be proportional to ($\langle\!\langle f_3\rangle\!\rangle - \langle\!\langle f_{011}\rangle\!\rangle - \langle\!\langle f_{010}\rangle\!\rangle$) but not to $\langle\!\langle f_2\rangle\!\rangle$. The value (($\langle\!\langle f_3\rangle\!\rangle - \langle\!\langle f_{011}\rangle\!\rangle - \langle\!\langle f_{010}\rangle\!\rangle$) (hereafter defined as $\langle\!\langle f_3\rangle\!\rangle'$) can be approximated to $\langle\!\langle f_3\rangle\!\rangle$ only when $\langle\!\langle F\rangle\!\rangle$ is sufficiently high. Of course, their experimentally important finding that anticoagulant activity χ, as determined by the method according to the Commentary of Japanese Pharmacopoeia,[20] is almost linearly governed by $\langle\!\langle f_2\rangle\!\rangle + \langle\!\langle f_3\rangle\!\rangle$ for the NaCS with $\langle\!\langle F\rangle\!\rangle \geq 2$ and keeps its validity even now. From Table 2.4.10, it is obvious that ^1H NMR analysis, except for aforementioned two peaks (4.93

Table 2.4.10

Assignments of proton peaks of $NaCS/D_2O$ systems[12]

Peak no.	δ (ppm)	Assignment
1	4.93	H_1 (111), H_1 (101), H_1 (110), H_1 (100)
2	4.70	H_3 (111), H_3 (110)
3	4.58	H_1 (011), H_1 (001), H_1 (010), H_1 (000)
4	4.42	H_2 (111), H_2 (110), H_6 (111), H_6 (101), H_6 (011), H_6' (001)
5	4.30	H_3 (011), H_3 (010), H_6 (001)
6	4.18	H_3 (101), H_3 (100), H_4 (111) H_4 (011), H_4 (010), H_4 (110)
7	4.02	H_2 (101), H_2 (100)
8	3.96	H_6' (110), H_6' (100), H_6' (010), H_6' (000)
9	3.84	H_2 (011), H_2 (010), H_5 (111), H_5 (101), H_5 (011), H_5 (001), H_6 (110), H_6 (100), H_6 (010), H_6 (000)
10	3.67	H_3 (001), H_3 (000), H_4 (101) H_4 (100), H_4 (001), H_4 (000), H_5 (110), H_5 (100), H_5 (010) H_5 (000)
11	3.39	H_2 (001), H_2 (000)

H_k(lmn) means H_k ($k = 1$–6) proton attached to corresponding C_k carbon position in one of eight AHG units denoted by lmn. $l = 1$ or 0 denotes that a hydroxyl group attached to C_2 position is substituted or not, m and n indicated the corresponding values at the C_3 and C_6 positions, respectively.

and 4.70 ppm), does not provide a reliable way for estimation of $\langle\!\langle f_k \rangle\!\rangle$ owing to heavy peak overlapping.

Three additional ^{13}C shift factors ΔC_α, ΔC_β, and ΔC_γ, which were defined in analogy with ΔH_α, ΔH_β, and ΔH_γ, were also estimated by trial and error to minimize the difference between the calculated and observed δ values and ΔC_α, ΔC_β, and ΔC_γ were determined as 6.1 ppm, -1.4 ppm and 0.1 ppm, respectively. Here, 12,000 combinations of ΔC_α(6.0–8.0 ppm), ΔC_β(-3.0 to 0.0 ppm), and ΔC_γ(0.0–2.0 ppm) with 0.10 ppm interval were examined, and experimental δ values observed for ^{13}C nuclei in unsubstituted AHG unit were used as standards. The values of ΔH_α and ΔH_β determined here lie well in the ranges found for CMC,[21] cellulose acetate,[16] and CN[1] ($\Delta C_\alpha = 8$–10 ppm and $\Delta C_\beta = -7$–0 ppm). The validity of this analysis can be confirmed if we note the excellent agreement between the observed and calculated values of the carbon chemical shifts of the trisubstituted AHG unit as shown in Table 2.4.11. Comparison of Table 2.4.11 with Table 2.4.7 also ascertains the wide applicability of our analytical procedure. However, the calculated C_4 value for trisubstituted AHG unit (C_4 (111)) proved to largely deviate from the observed value. Considering the conformational specificity around C_4–O–C_1 when the C_3 position is substituted, the calculated values for C_4 (111) (observed), C_4 (010), C_4 (011), and C_4 (110) AHG units are given in parenthesis in Table 2.4.11 using the observed values for trisubstituted AHG units as standard ($\Delta C_\alpha = 6.1$ ppm, $\Delta C_\beta = -1.2$ ppm, and $\Delta C_\gamma = 1.0$ ppm).

The peak assignment for the observed 16 ^{13}C peaks shown in Table 2.4.7 was made as follows: (1) adopt the peak assignment described before irrespective of the calculated δ values (Table 2.4.11), (2) for other ^{13}C peaks, find the possible ^{1}H cross peaks for each ^{13}C peak by referring the results in Figures 2.4.8 and 2.4.10 and Table 2.4.10, (3) assign the possible AHG units expressed as C_k (*lmn*) to each ^{13}C peak, (4) compare the observed with the calculated (Table 2.4.11) δ values and assign the AHG units having the lowest difference to the observed values considering the

Table 2.4.11

Observed and calculated carbon chemical shifts of AHG units in NaCS[12]

AHG unit	δ (ppm)						Remarks
	C_1	C_2	C_3	C_4	C_5	C_6	
2,3,6-Trisub.	103.4	80.1	81.3	77.5	75.8	69.2	Observed
2,3,6-Trisub.	103.3	80.2	81.5	80.1 (77.5)	76.2	69.1	Calculated
2,3-Disub.	103.3	80.2	81.5	80.0 (76.5)	77.6	63.0	Calculated
2,6-Disub.	103.2	81.6	75.4	81.5	76.1	69.1	Calculated
3,6-Disub.	104.7	74.1	82.9	80.0 (78.7)	76.2	69.1	Calculated
2-Monosub.	103.2	81.9	75.4	81.4	77.5	63.0	Calculated
3-Monosub.	104.7	74.1	82.9	79.9 (75.5)	77.6	63.0	Calculated
6-Monosub.	104.6	75.5	76.8	81.4	76.1	69.1	Calculated
Unsub.	104.6	75.5	76.8	81.3	77.5	63.0	Observed

Here, $\Delta C_\alpha = 6.1$ ppm, $\Delta C_\beta = -1.4$ ppm, and $\Delta C_\gamma = 0.1$ ppm are assumed. $C_k = C_{k_0} + (\Delta C_\alpha + \Delta C_\beta + \Delta C_\gamma)$; $k = 1$–6. C_{k_0} means standard value (unsubstituted AHG unit) of C_k. The value in parenthesis is the calculated one using the observed value for trisubstituted AHG unit as the standard.

Table 2.4.12

Assignments of carbon peaks of NaCS/D$_2$O system[12]

Peak no.	δ (ppm)	Assignment
1	104.8	C$_1$(010), C$_1$(011)
2	104.6	C$_1$(000), C$_1$(001)
3	103.4	C$_1$(110), C$_1$(111)
4	102.9	C$_1$(100), C$_1$(101)
5	84.4	C$_3$(010), C$_3$(011)
6	82.4	C$_2$(101), C$_2$(100)
7	81.7	C$_4$(001), C$_4$(100)
8	81.3	C$_3$(110), C$_3$(111), C$_4$(000)
9	80.6	C$_4$(101)?
10	80.1	C$_2$(110), C$_2$(111)
11	77.5	C$_5$(000), C$_5$(010), C$_5$(100), C$_5$(110), C$_4$(111)
12	76.8	C$_3$(000), C$_3$(001), C$_4$(011), C$_4$(110)
13	75.8	C$_5$(001), C$_5$(101), C$_5$(011), C$_5$(111)
14	75.5	C$_2$(000), C$_2$(001), C$_3$(101), C$_3$(100), C$_4$(010)
	(74.3)	C$_2$(010), C$_2$(011)
15	69.2	C$_6$(001), C$_6$(011), C$_6$(101), C$_6$(111)
16	63.0	C$_6$(000), C$_6$(010), C$_6$(100), C$_6$(110)

C$_k$(lmn) means C$_k$($k = 1$–6) carbon in one of eight AHG units denoted by lmn. $l = 1$ or 0 denotes whether a hydroxyl group attached C$_2$ position is substituted or not. m and n indicate the corresponding values at the C$_3$ and C$_6$ positions, respectively.

rough assignment made in (3) and the order of δ values. The results are summarized in Table 2.4.12. Here, for C$_4$ (010), C$_4$ (110), and C$_4$ (011) AHG units the calculated δ values in parenthesis were employed. From Table 2.4.12 it is concluded that the peak at 63.0 ppm originates from four AHG units (2,3-di-, 2-mono-, 3-mono-, and unsubstituted), of which C$_6$ position is not substituted. The NaCS samples (DS-1 and DS-2), which do not show this peak in their ^{13}C NMR spectra, can be regarded to be constituted of only four other AHG units (6-mono-, 3,6-di-, 2,6-di- and 2,3,6-tri-). The three peaks (84.4, 82.5, and 81.3 ppm) observed in DS-1, as collected in Table 2.4.7, are attributable to C$_3$ carbon in 3,6-di-substituted AHG unit, to C$_2$ carbon in 2,6-di-substituted AHG unit and to C$_3$ carbon in 2,3,6-trisubstituted AHG unit, respectively. Therefore, $\langle\!\langle f_{011}\rangle\!\rangle$, $\langle\!\langle f_{101}\rangle\!\rangle$, and $\langle\!\langle f_{111}\rangle\!\rangle$ for this polymer are readily elucidated from the ratios of the corresponding peak intensities to the total peak intensity for C$_2$–C$_5$ carbon peaks. Furthermore, for this polymer, $\langle\!\langle f_{001}\rangle\!\rangle$ can be determined through use of the following relation:

$$\langle\!\langle f_{011}\rangle\!\rangle + \langle\!\langle f_{101}\rangle\!\rangle + \langle\!\langle f_{111}\rangle\!\rangle + \langle\!\langle f_{001}\rangle\!\rangle = 1 \tag{2.4.29}$$

The results of ^{13}C NMR analysis proposed here on DS-1 and DS-2 are summarized in Table 2.4.13. Using the values of $\langle\!\langle f_{lmn}\rangle\!\rangle$ obtained thus, $\langle\!\langle f_k\rangle\!\rangle$ and $\langle\!\langle F\rangle\!\rangle$ are also calculated from the equations as follows:

$$\sum_{l=0}^{1} \sum_{m=0}^{1} \sum_{n=0}^{1} \langle\!\langle f_{lmn}\rangle\!\rangle = 1 \tag{2.4.30}$$

Table 2.4.13

$\langle\!\langle f_{lmn}\rangle\!\rangle$, estimated from ^{13}C NMR spectra using assignments in Table 2.4.12, $\langle\!\langle f_k\rangle\!\rangle$ calculated from $\langle\!\langle f_{lmn}\rangle\!\rangle$ using eqs. (2.4.31)–(2.4.33), $\langle\!\langle F\rangle\!\rangle$ calculates from $\langle\!\langle f_k\rangle\!\rangle$ using eq. (2.4.34), $\langle\!\langle f_2\rangle\!\rangle$ and $\langle\!\langle f_3\rangle\!\rangle'$ estimated from the ^{1}H NMR spectra of NaCS samples[12]

Sample code	^{13}C NMR							
	$\langle\!\langle f_{111}\rangle\!\rangle$	$\langle\!\langle f_{110}\rangle\!\rangle$	$\langle\!\langle f_{101}\rangle\!\rangle$	$\langle\!\langle f_{011}\rangle\!\rangle$	$\langle\!\langle f_{100}\rangle\!\rangle$	$\langle\!\langle f_{010}\rangle\!\rangle$	$\langle\!\langle f_{001}\rangle\!\rangle$	$\langle\!\langle f_{000}\rangle\!\rangle$
DS-1	0.48	0.00	0.26	0.20	0.00	0.00	0.06	0.00
DS-2	0.21	0.00	0.36	0.20	0.00	0.00	0.18	0.00
DS-3	0.02	0.00	0.16	0.11	0.02	0.01	0.61	0.07
DSH-1	0.01	0.01	0.09	0.04	0.08	0.04	0.39	0.34
HBH$'$-1	0.00	0.00	0.06	0.02	0.09	0.02	0.34	0.47

Sample code	^{13}C NMR				^{1}H NMR	
	$\langle\!\langle f_2\rangle\!\rangle$	$\langle\!\langle f_3\rangle\!\rangle$	$\langle\!\langle f_6\rangle\!\rangle$	$\langle\!\langle F\rangle\!\rangle$	$\langle\!\langle f_2\rangle\!\rangle$	$\langle\!\langle f_3\rangle\!\rangle'$
DS-1	0.74	0.68	1.00	2.42	0.75	0.81
DS-2	0.57	0.46	1.00	2.03	–	–
DS-3	0.20	0.14	0.90	1.24	–	–
DSH-1	0.19	0.10	0.53	0.82	–	–
HBH$'$-1	0.15	0.04	0.42	0.61	–	–

$$\langle\!\langle f_2\rangle\!\rangle = \sum_{m=0}^{1}\sum_{n=0}^{1}\langle\!\langle f_{lmn}\rangle\!\rangle \tag{2.4.31}$$

$$\langle\!\langle f_3\rangle\!\rangle = \sum_{l=0}^{1}\sum_{n=0}^{1}\langle\!\langle f_{lmn}\rangle\!\rangle \tag{2.4.32}$$

$$\langle\!\langle f_6\rangle\!\rangle = \sum_{l=0}^{1}\sum_{m=0}^{1}\langle\!\langle f_{lmn}\rangle\!\rangle \tag{2.4.33}$$

$$\langle\!\langle F\rangle\!\rangle = \langle\!\langle f_2\rangle\!\rangle + \langle\!\langle f_3\rangle\!\rangle + \langle\!\langle f_6\rangle\!\rangle \tag{2.4.34}$$

Table 2.4.13 also includes the values of $\langle\!\langle f_k\rangle\!\rangle$ and $\langle\!\langle F\rangle\!\rangle$ evaluated for DS-1 and DS-2. The values of $\langle\!\langle f_2\rangle\!\rangle$ and $\langle\!\langle f_3\rangle\!\rangle'$ (approximately $\langle\!\langle f_3\rangle\!\rangle$) for the former polymer, both determined by the above mentioned ^{1}H NMR analysis, are included in the table for comparison. ^{1}H NMR and ^{13}C NMR analyses give quite similar $\langle\!\langle f_2\rangle\!\rangle$ values, but $\langle\!\langle f_3\rangle\!\rangle$ by ^{13}C NMR method are insignificantly smaller than $\langle\!\langle f_3\rangle\!\rangle'$ by the ^{1}H NMR method. This means that ^{1}H NMR analysis tends to overestimate $\langle\!\langle f_3\rangle\!\rangle'$, owing to considerable overlapping with neighboring peaks, resulting in relatively low accuracy intrinsic to ^{1}H NMR analysis.

Assuming that the reaction rate of the hydroxyl group at the C_2, C_3, or C_6 position with sulfating reagent is independent of whether the other two hydroxyl groups are already substituted or not, the following equations are derived:

$$\langle\!\langle f_{111}\rangle\!\rangle = \langle\!\langle f_2\rangle\!\rangle\langle\!\langle f_3\rangle\!\rangle\langle\!\langle f_6\rangle\!\rangle \tag{2.4.35}$$

$$\langle\!\langle f_{110}\rangle\!\rangle = \langle\!\langle f_2\rangle\!\rangle\langle\!\langle f_3\rangle\!\rangle(1 - \langle\!\langle f_6\rangle\!\rangle) \tag{2.4.36}$$

$$\langle\!\langle f_{101}\rangle\!\rangle = \langle\!\langle f_2\rangle\!\rangle(1 - \langle\!\langle f_3\rangle\!\rangle)\langle\!\langle f_6\rangle\!\rangle \tag{2.4.37}$$

$$\langle\!\langle f_{011}\rangle\!\rangle = (1 - \langle\!\langle f_2\rangle\!\rangle)\langle\!\langle f_3\rangle\!\rangle\langle\!\langle f_6\rangle\!\rangle \tag{2.4.38}$$

$$\langle\!\langle f_{100}\rangle\!\rangle = \langle\!\langle f_2\rangle\!\rangle(1 - \langle\!\langle f_3\rangle\!\rangle)(1 - \langle\!\langle f_6\rangle\!\rangle) \tag{2.4.39}$$

$$\langle\!\langle f_{010}\rangle\!\rangle = (1 - \langle\!\langle f_2\rangle\!\rangle)\langle\!\langle f_3\rangle\!\rangle(1 - \langle\!\langle f_6\rangle\!\rangle) \tag{2.4.40}$$

$$\langle\!\langle f_{001}\rangle\!\rangle = (1 - \langle\!\langle f_2\rangle\!\rangle)(1 - \langle\!\langle f_3\rangle\!\rangle)\langle\!\langle f_6\rangle\!\rangle \tag{2.4.41}$$

$$\langle\!\langle f_{000}\rangle\!\rangle = (1 - \langle\!\langle f_2\rangle\!\rangle)(1 - \langle\!\langle f_3\rangle\!\rangle)(1 - \langle\!\langle f_6\rangle\!\rangle) \tag{2.4.42}$$

Table 2.4.14 summarizes the $\langle\!\langle f_{lmn}\rangle\!\rangle$ values indirectly calculated from $\langle\!\langle f_k\rangle\!\rangle$ using eqs. (2.4.35)–(2.4.42) for DS-1 and DS-2. The values for DS-1, shown in Table 2.4.14, are in excellent agreement with those directly determined from ^{13}C NMR spectrum for the same polymer, shown in Table 2.4.13. For DS-2, it seems probable that the experimental values for disubstituted AHG units are larger than those indirectly evaluated, but the difference is not so large. This suggests that the assumption adopted in deriving eqs. (2.4.35)–(2.3.42) is roughly acceptable to NaCS samples.

It is difficult to determine $\langle\!\langle f_{lmn}\rangle\!\rangle$ for DS-3, DSH-1, and HBH$'$-1 by ^{13}C NMR method because they have AHG units whose C_6 is not substituted, as is obvious from the existence of C_6 carbon peak at 63.0 ppm. For these polymers, $\langle\!\langle f_6\rangle\!\rangle$ can be estimated from the intensity ratio of the peak at 69 ppm to that at 63 ppm. Analysis of the peak intensities at 84.4 ppm and 80–81 ppm gives $\langle\!\langle f_{011}\rangle\!\rangle + \langle\!\langle f_{010}\rangle\!\rangle$ and $\langle\!\langle f_{101}\rangle\!\rangle + \langle\!\langle f_{100}\rangle\!\rangle$, respectively. Putting these experimental values into eqs. (2.4.35)–(2.4.42), the $\langle\!\langle f_k\rangle\!\rangle$ and $\langle\!\langle f_{lmn}\rangle\!\rangle$ values are determined and presented in Table 2.4.13. Using the $\langle\!\langle f_k\rangle\!\rangle$ values in Table 2.4.13, $\langle\!\langle f_{lmn}\rangle\!\rangle$ is recalculated by using eqs. (2.4.35)–(2.4.42) and the results are shown in Table 2.4.14. For all NaCS samples, $\langle\!\langle f_{lmn}\rangle\!\rangle$ values in Tables 2.4.13 and 2.4.14 are consistent, supporting the validity of eqs. (2.4.35)–(2.4.42), and hence the reaction of the hydroxyl group with sulfating reagent can be regarded to occur irrespective of the substitution reaction of the hydroxyl groups at different carbon positions belonging to the same AHG unit.

From Table 2.4.13, the relation $\langle\!\langle f_6\rangle\!\rangle > \langle\!\langle f_2\rangle\!\rangle > \langle\!\langle f_3\rangle\!\rangle$ holds for all NaCS samples prepared here, leading us to the conclusion that reactivity of C_6 hydroxyl is the highest in

Table 2.4.14

$\langle\!\langle f_{lmn}\rangle\!\rangle$ of NaCS samples calculated from $\langle\!\langle f_k\rangle\!\rangle$ in Table 2.4.13 using eqs. (2.4.35)–(2.4.42)[12]

Sample code	^{13}C NMR							
	$\langle\!\langle f_{111}\rangle\!\rangle$	$\langle\!\langle f_{110}\rangle\!\rangle$	$\langle\!\langle f_{101}\rangle\!\rangle$	$\langle\!\langle f_{011}\rangle\!\rangle$	$\langle\!\langle f_{100}\rangle\!\rangle$	$\langle\!\langle f_{010}\rangle\!\rangle$	$\langle\!\langle f_{001}\rangle\!\rangle$	$\langle\!\langle f_{000}\rangle\!\rangle$
DS-1	0.58	0.00	0.24	0.18	0.00	0.00	0.08	0.00
DS-2	0.26	0.00	0.31	0.20	0.00	0.00	0.23	0.00
DS-3	0.03	0.00	0.15	0.10	0.02	0.01	0.62	0.07
DSH-1	0.01	0.01	0.09	0.04	0.08	0.04	0.39	0.34
HBH$'$-1	0.00	0.00	0.06	0.01	0.08	0.02	0.34	0.47

sulfation of cellulose in DMF/SO$_3$ system. Similar results have been reported by Kamide *et al.* for carboxymethylation[21] and xanthation[9] of cellulose. The reason why $\langle\langle f_2 \rangle\rangle$ is larger than $\langle\langle f_3 \rangle\rangle$ is not clear at present and is open to further study.

In summary, peak assignments for ^{1}H and ^{13}C peaks of sodium salt of CS (NaCS) was tentatively made with the aid of 2D NMR. $\langle\langle f_{lmn} \rangle\rangle$, $\langle\langle f_k \rangle\rangle$, and $\langle\langle F \rangle\rangle$ of NaCS with $\langle\langle f_6 \rangle\rangle = 1.0$, and $\langle\langle f_k \rangle\rangle$ and $\langle\langle F \rangle\rangle$ of NaCS with $\langle\langle f_6 \rangle\rangle < 1.0$ can be determined on the basis of the ^{13}C peak assignment. ^{1}H NMR analysis is useful for estimation of $\langle\langle f_2 \rangle\rangle$ and $\langle\langle f_3 \rangle\rangle$ with relatively low accuracy, compared with ^{13}C NMR analysis. Sulfation reaction methods of cellulose employed in this study gave the highest $\langle\langle f_6 \rangle\rangle$, and the reaction at a given hydroxyl group was predicted to occur irrespective of the reaction of the other hydroxyl groups at different carbon positions.

REFERENCES

1. See, for example, T Noguchi, *J. Soc. Text. Cell. Ind. Jpn (Sen-i Gakkaishi)*, 1950, **6**, 153, see also pages 155, 217, 270, 312, 314, 379, 381, 444; K Lauer, *Makromol. Chem.*, 1951, **5**, 287.
2. K Kamide, K Okajima and K Kowsaka, *Polym. J.*, 1987, **19**, 231.
3. T-K Wu, *Macromolecules*, 1980, **13**, 74.
4. DT Clark, PJ Stephenson and F Heatley, *Polymer*, 1981, **22**, 1112.
5. K Kowsaka, K Okajima and K Kamide, *Polym. J.*, 1988, **20**, 827.
6. K Kamide, T Terakawa and Y Miyazaki, *Polym. J.*, 1979, **11**, 285.
7. K Kamide, Y Miyazaki and T Abe, *Polym. J.*, 1979, **11**, 523.
8. K Kamide, M Saito and T Abe, *Polym. J.*, 1981, **13**, 421.
9. K Kowsaka, K Okajima and K Kamide, *Polym. J.*, 1986, **18**, 843.
10. K Kamide, T Matsui, K Okajima and S Manabe, *Cell. Chem. Technol.*, 1982, **16**, 601.
11. K Kamide, K Okajima, K Kowsaka and T Matsui, *Polym. J.*, 1987, **19**, 1405.
12. K Kowsaka, K Okajima and K Kamide, *Polym. J.*, 1991, **23**, 823.
13. K Kamide and K Okajima, *Polym. J.*, 1981, **13**, 163.
14. K Kamide and K Okajima, *Polym. J.*, 1981, **13**, 127.
15. K Kamide, K Okajima, T Matsui, M Ohnishi and H Kobayashi, *Polym. J.*, 1983, **15**, 309.
16. K Kowsaka, K Okajima and K Kamide, *Polym. J.*, 1988, **20**, 1091.
17. W Brown and R Wikström, *Eur. Polym. J.*, 1966, **1**, 1.
18. R Nardin and M Vincendon, *Macromolecules*, 1986, **19**, 2452.
19. D Gagnaire, D Mancier and M Vincendon, *J. Polym. Sci. Polym. Chem. Ed.*, 1980, **18**, 13.
20. Commentary of Japanese Pharmacopoeia, No. C1235-C1242, Nankaido, Tokyo, 1969.
21. K Kamide, K Okajima, K Kowsaka, T Matsui, S Nomura and K Hikichi, *Polym. J.*, 1985, **17**, 909.

2.5 THIN LAYER CHROMATOGRAPHY (TLC)

2.5.1 Evaluation of $g(\langle F \rangle)$ and M_w from TLC

We define the distribution of the acetyl content A_c and the molecular weight M $g(A_c, M)$ as an example of $g(\langle F \rangle)$; see eq. (2.2.6) as follows: the weight fraction of the polymer molecule whose combined acetic acid content (i.e. $\langle F \rangle$) is between A_c and $A_c + dA_c$, and whose molecular weight is between M and $M + dM$, is given by $g(A_c, M)dA_c dM$. By using $g(A_c, M)$, the weight average combined acetic acid content $A_{c,w}$ (%), and the weight and number average molecular weight M_w and M_n are given by

$$A_{c,w} = \int_0^{62.5} \int_M^{M_2} A_c g(A_c, M) \mathrm{d}M \mathrm{d}A_c \tag{2.5.1}$$

$$M_w = \int_0^{62.5} \int_M^{M_2} M g(A_c, M) \mathrm{d}A_c \mathrm{d}M \tag{2.5.2}$$

$$M_n = \left[\int_0^{62.5} \int_M^{M_2} \{ g(A_c, M)/M \} \mathrm{d}A_c \mathrm{d}M \right]^{-1} \tag{2.5.3}$$

where 62.5 is the A_c value of fully substituted CA ($\langle\!\langle F \rangle\!\rangle = 3$) in percent, and M_1, and M_2 denote the minimum and maximum molecular weights of MWD of the polymer, respectively.

Kamiyama et $al.$[1] and Otacka et $al.$[2] have independently obtained an experimental equation for the TLC separation on the basis of the molecular weight by using acrylic and vinyl polymers as follows:

$$R_f = A + B \log M_w \tag{2.5.4}$$

where R_f denotes the rate of flow, and A, B are the constants independent of M_w. Kotacka and White[3] studied the TLC separation on the basis of the composition and the molecular weight by using styrene/butadiene copolymer and found that R_f can be expressed as a function of weight average styrene content $\bar{X}$ and M_w in the form:

$$R_f = A + B\bar{X} + C \log M_w \tag{2.5.5}$$

where $A, B,$ and C are constants independent of $\bar{X}$ and M_w. However, it is difficult to obtain the equations relating to the monodisperse polymer from eq. (2.5.5). On the other hand, Kamide et $al.$[4] obtained the following empirical equations for cellulose nitrate:

$$R_f = A + B\bar{X} \qquad \text{(for constant } M_w) \tag{2.5.6}$$

$$R_f - C \mid DM_w \qquad \text{(for constant } \bar{X}) \tag{2.5.7}$$

where A and B are constants independent of $\bar{X}$ (the weight average nitrogen content) and C and D are also constants independent of M_w. The combination of eqs. (2.5.6) and (2.5.7) can produce a more general equation such as eq. (2.5.8)

$$A_c = A_0 + B_0\bar{X} + C_0 M_w + D_0\bar{X}M_w \tag{2.5.8}$$

where $A_0,$ $B_0,$ $C_0,$ and D_0 are the constants independent of $\bar{X}$ and M_w. We can choose the operation conditions of TLC so that D_0 might be reduced to zero. When $D_0 = 0$, eq. (2.5.8) can be immediately transformed to the equation for monodisperse polymers. Thus, we can assume that the R_f of monodisperse polymers can be approximately expressed as a function of A_c (wt%) and molecular weight M in the form:

$$R_f = R_{f0} + a_1 A_c + b_1 M + b_{-1}/M \tag{2.5.9}$$

where $R_{f0}, a_1, b_1, b_{-1},$ are constants independent of A_c and M. a_1 and b_1 reflect the contribution of A_c and M on R_f. The consideration of effect of molecular chain end leads to the last term on the right-hand side of eq. (2.5.9). The applicability of eq. (2.5.9) for CA will be demonstrated in Section 2.5.2.

The weight average R_f value, $R_{f,w}$, is defined by

$$R_{f,w} = \int_0^{62.5} \int_M^{M_2} R_f g(A_c, M) dM dA \tag{2.5.10}$$

The substitution of eq. 2.5.9 into eq. 2.5.10 gives

$$R_{f,w} = R_{f0} + a_1 A_{c,w} + b_1 M_w + b_{-1}/M_n \tag{2.5.11}$$

The parameter a_1, b_1, and b_{-1} in eq. (2.5.11) can be evaluated from the experimental $R_{f,w}$ data for the samples with known values of $A_{c,w}$, M_w, and M_n. It is interesting to note that all of the parameters R_{f0}, a_1, b_1, and b_{-1} in eq. (2.5.11) are independent of the MWD and combined acetic acid content distribution (ACD). Here, the distribution function of R_f $g(R_f)$ is defined as being the weight fraction of polymer at the place R_f and the $R_f + dR_f$ on the thin layer given by $g(R_f)dR_f$. The third and fourth terms on the right-hand side of eq. (2.5.9) or eq. (2.5.11) are small in comparison with the first and the second terms in these equations. Moreover, the inherent spreading width of the TLC pattern of monodisperse sample can be neglected; the $g(R_f)$ can be explicitly transformed into the distribution function of combined acetic acid content $g(A_c)$ by using the following equation, which is derived directly from the definition of $g(R_f)$ and $g(A_c)$

$$g(A_c)d(A_c) = g(R_f)dR_f \tag{2.5.12}$$

Equation (2.5.12) is rearranged to give

$$g(A_c) = a_1 g(R_f) \tag{2.5.13}$$

2.5.2　Cellulose acetate[5,6]

We can evaluate the MWD and chemical composition distribution (CCD) $g(\langle\!\langle\!\langle F \rangle\!\rangle\!\rangle)$ (in this case, the distribution of the acetyl content in cellulose acetate) by use of TLC. It is currently a common practice to employ successive precipitational fractionation for elucidating MWD and CCD in CA (see Section 3.1). As is well known, commercially available CA differs considerably in molecular weight as well as acetyl content, and the choice of solvent/nonsolvent. Determining the solubility of CA molecules by either molecular weight or chemical composition alone is exceedingly difficult. Using the cross fractionation is unavoidable, even if the MWD in CA is to be evaluated. No study concerning the complete cross fractionation on CA has been published to date, but the possibility was demonstrated by Rosenthal and White.[7] According to these authors, a mixture of acetone and ethanol enables us to separate CA molecules on the basis of molecular weight, irrespective of acetyl content. Kamide *et al.* tried to fractionate CA according to Rosenthal *et al.* but the fractions obtained indicated CA dependence.[5]

In view of the above mentioned situations, it is apparently important to establish a method such as TLC for evaluating MWD and CCD for cellulose acetate, which is not elaborate and does not require a highly skilled technique.

TLC measurements

Table 2.5.1 shows the samples used. The polymer sample was dissolved in a mixture of methylene chloride and methanol (4:1 v/v at 20 °C). 1×10^{-2} cm^3 of a 1.0% solution (w/v) thus prepared was dropped using a microsyringe 1.5 cm away from the lower edge of the plate, making a spot 5 mm in diameter. The apparent density of the polymer on the silica gel was estimated to be 5.10×10^{-4} g/cm^2. Developing was carried out at 25 °C by setting the thin layer in a developing tank under the unsaturated vapor phase of the developer. The dipping level of the liquid phase (developer) was 5 mm above the lower edge of the thin layer. The end point was determined as the solvent front reached 10 cm above the original point of samples. The distance of end point of the solvent front from the starting point is defined as Z_f. After developing, the plate was dried by blowing air at room temperature. The continuous gradient developing method was employed in a similar manner to that proposed by Kamiyama and Inagaki,[8] and the stepwise development was described in the text.[5] After removal from the development chamber of TLC, the thin layer was sprayed with 10% aq. sulfuric acid fumes. The polymer on the layer was then carbonized by heating at 110 °C in an electric oven or the thin layer was directly heated by gas flame to a black trace. Thus, the polymer spread on the plate was made visible.

The phase ratio, defined by the ratio of the weight of solvent (mobile phase) retained by unit weight of silica gel (stationary phase),[8] was determined from the difference between the weight of silica gel containing solvent for the area arbitrarily cut off and that of the gel dried.

Change in composition of the solvent with the distance from the starting point (i.e. in the direction of development) was estimated by infrared (IR) spectroscopy. For example, in the case of the methylene chloride/methanol system, a given portion of the thin layer after development was removed from a glass plate and put into acetone with agitation. The content of methylene chloride in the supernatant phase was determined from the ratio of optical density of the band at 730 cm^{-1} to that of the band at 780 cm^{-1}. The content of methanol was determined by subtracting the amount of methylene chloride from the total solvent absorbed or by the optical density at the band at 1028 cm^{-1}.

Table 2.5.1

The viscosity average degree of polymerization (P_v) and the acetyl content (CA) for fractions

Sample number	(P_v)	% CA	Sample number	(P_v)	% CA
M1	234	54.0	A1	240	60.5
M2	319	52.5	A2	218	58.0
M4	264	54.2	A3	223	56.0
M5	202	54.2	A4	202	55.0
M7	138	54.2	A5	209	54.1
M9	54	52.8			

Dependence of R_f on the composition of the developer

The applicability of TLC as a method for evaluating MWD and CCD depends largely on a successful choice of proper developer. One of the necessary conditions for a useful developer is a good solvent for cellulose acetate.

Figures 2.5.1–2.5.4 illustrate the chromatogram obtained for ternary mixtures such as methylene chloride/acetone/methanol, methylene chloride/acetone/butanol, methylene chloride/tetrahydrofuran (THF)/methanol, and methylene chloride/THF/butanol, respectively. Polymers are, from the left-hand side, M2, M5, M9, and A1 (acetyl content $A_c = 60.5\%$ CTA). Table 2.5.2 summarizes the developer's composition at the numbered positions shown in Figures 2.5.1–2.5.4. The efficiency of separation with respect to the molecular weight can be judged by comparing R_f among M2, M5, and M9 and that with respect to the acetyl content by comparing R_f between M2 or M5, and A1. Evidently, R_f depends on the molecular weight as well as the acetyl content of the polymer. However, these two effects on R_f are, unfortunately, not separable with each other in Figures 2.5.1–2.5.4. As will be demonstrated later, the effective and independent evaluation of MWD and CCD by TLC becomes successful by using other compositions than those utilized in Figures 2.5.1–2.5.4 and/or by adopting the improved operating conditions.

In Figure 2.5.5, the change in R_f with composition of developer is qualitatively demonstrated for a system of methylene chloride/acetone/butanol. The arrow attached to the broken line implies the direction of increase in R_f of CA with $A_c = 54\%$. From the experiments, two dotted regions are found to be most suitable for CCD evaluation and the shadowed region is found to be profitable for MWD evaluation.

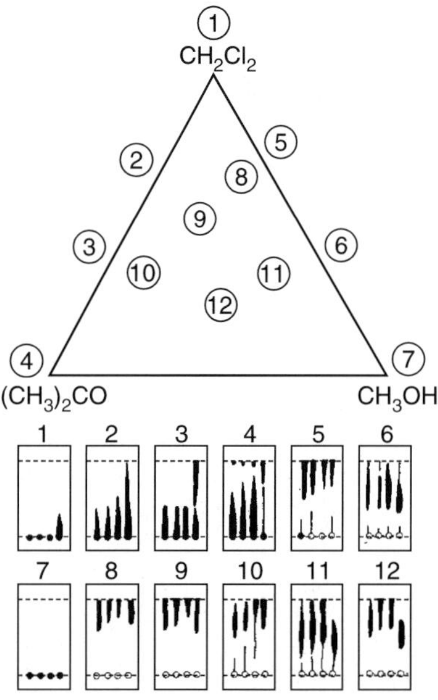

Figure 2.5.1   Thin layer chromatograms of samples (from left side M2, M5, M9, and A1), obtained by use of the system: methylene chloride, acetone, and methanol as developer.[5]

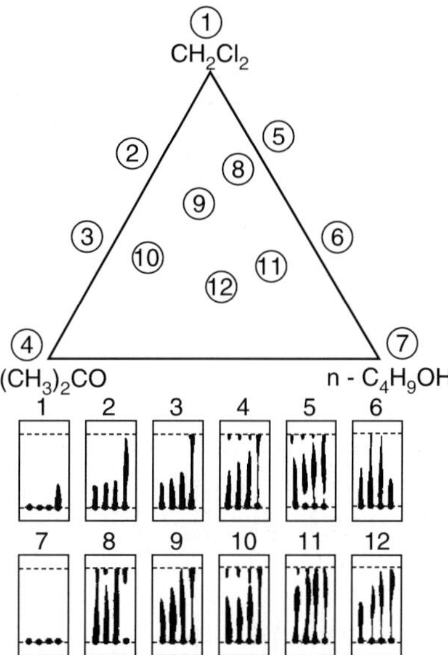

Figure 2.5.2   Thin layer chromatograms of samples (from left side M2, M5, M9, and A1), obtained by use of the system: methylene chloride, acetone, and butanol as developer.[5]

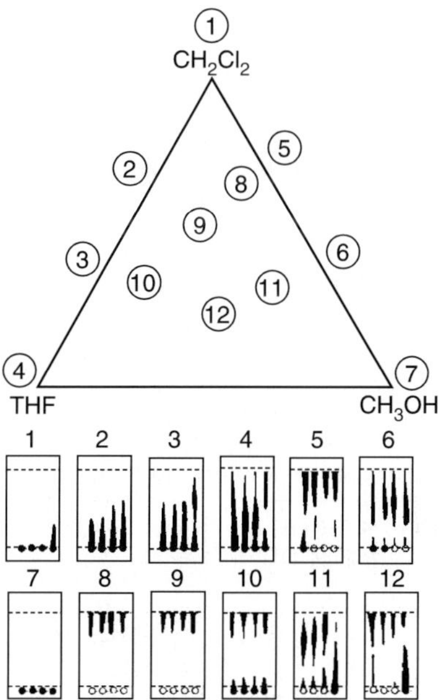

Figure 2.5.3   Thin layer chromatograms of samples (from left side M2, M5, M9, and A1), obtained by use of the system: methylene chloride, acetone, THF, and methanol as developer.[5]

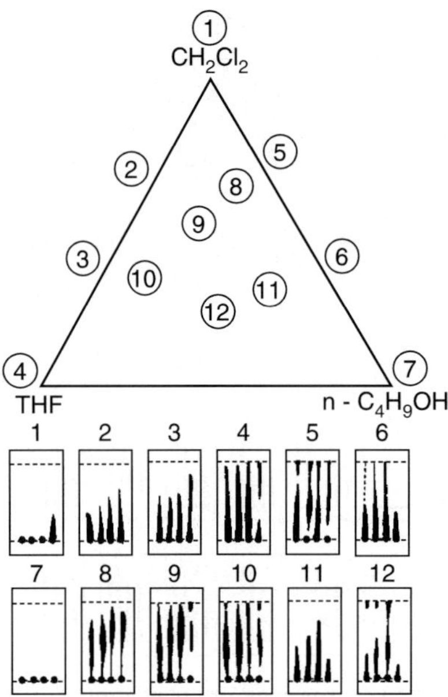

Figure 2.5.4   Thin layer chromatograms of samples (from left side M2, M5, M9, and A1), obtained by use of the system: methylene chloride, acetone, THF, and butanol as developer.[5]

Inspection of Figure 2.5.5 leads to the following conclusions:

(1) R_f obtained for systems having a constant composition of methylene chloride/ acetone or butanol/acetone shows maximum at a specific composition of methylene chloride/butanol. This value depends on CA content of polymer, i.e. 15/85 for A_c with $A_c = 60.5\%$ and 33/67 for A_c with $A_c = 54\%$. In other words, this kind of maximum line can be observed as a ridge in the three-dimensional (3D) R_f-composition diagram.

(2) A similar maximum line is found for a specific composition of acetone/butanol (0.55:1 for $A_c = 60.5\%$ and 1:1 for $A_c = 54\%$). It should be noted that the former is more prominent as compared with the latter.

Table 2.5.2

Compositions of the developer in Figures 2.5.1–2.5.4

Num-ber	%CH$_2$Cl$_2$	%(CH$_3$)$_2$CO	%CH$_3$OH or n-C$_4$H$_9$OH	Num-ber	%CH$_2$Cl$_2$	%(CH$_3$)$_2$CO	%CH$_3$OH or n-C$_4$H$_9$OH
1	100	0	0	7	0	0	100
2	65	35	0	8	64	9	27
3	40	60	0	9	51	27	22
4	0	100	0	10	34	51	15
5	70	0	30	11	33	17	50
6	40	0	60	12	25	37	38

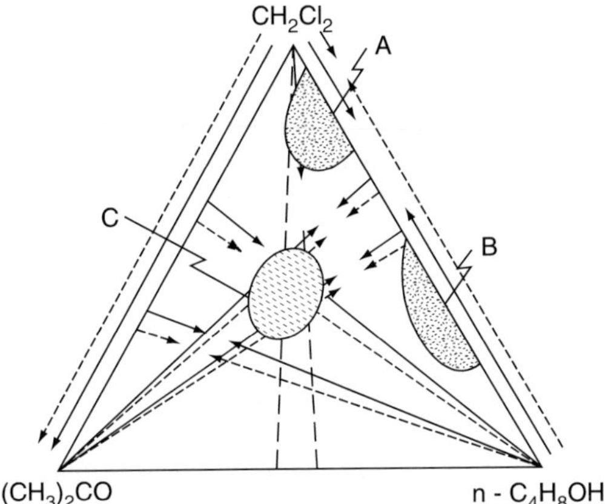

Figure 2.5.5   Change of R_f with the compositions of a ternary mixture of methylene chloride/acetone/butanol, in the case of sample A1 (broken line) and A5 (full line). R_f increases with the direction of the line. The area suitable for evaluating CA distribution is dotted and the shadowed area is for evaluating the MW distribution.[5]

(3)   As was described above, profitable regions of the mixture composition for evaluating CCD exist for this system. In Region A, which has a higher methylene chloride content, R_f increases with the increase in acetyl content. In Region B, which has a lower methylene chloride content, R_f decreases with the increase in A_c.

(4)   In Region C, suitable for estimating MWD, R_f becomes smaller as the molecular weight is larger. The above mentioned features are qualitatively observed for other solvent systems. When methanol is used in place of butanol, the maximum line is shifted to the composition rich in methylene chloride, the value of R_f becomes larger in the whole composition region, and region C splits into two regions.

Figure 2.5.6 shows the variation of R_f values obtained for three CA samples having different molecular weights (M2 (dotted line), M5 (broken line), and M9 (full line)) with

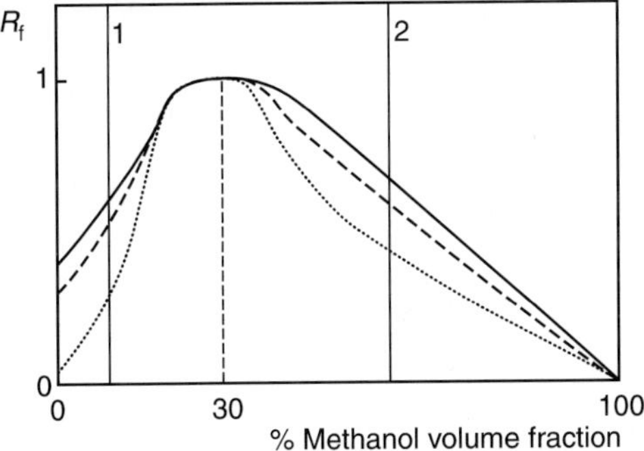

Figure 2.5.6   Schematic representation of the variation of R_f, — for small, - - - middle, and $\cdots$ large molecular weight samples, with the composition of developer (methylene chloride/ acetone /methanol). In this case, the volume ratio of methylene chloride to acetone is kept at unity.[5]

changing composition of a developer system consisting of methylene chloride/acetone/
methanol at a constant composition of methylene chloride/acetone (1:1 v/v). R_f passes
through maximum at a point of a methanol content (v_m in vol%) of 30%. The maximum
value of R_f remains almost at unity, independent of the molecular weight. Evidently,
there exist two compositions (shown by two vertical lines 1 and 2 in the figure) at which
the most effective MWD evaluation can be made.

Stepwise development

By the use of a conventional developing procedure, R_f depends on molecular weight
and chemical composition, but the spot thus obtained is always accompanied by tailing.
In order to diminish this tailing effect, Kamiyama and Inagaki[8] employed the continuous
gradient development. The mechanism of suppressing the tailing effect by this method is,
however, not clear.

In this section, at first the continuous gradient method was studied and thereafter
a stepwise development (hereafter briefly referred to as stepwise eluent method) if a
stepwise variation of composition was applied. For this purpose, the methylene chloride/
butanol/methanol system and the methylene chloride/acetone/methanol system are
utilized for evaluating CCD and MWD, respectively. Coexistence of butanol in the
former system makes the chromatogram narrower and yields better reproducibility.
In the method of stepwise development, after being developed to a proper height Z_f'
(where Z_f' is the distance from solvent front to starting point observed at a given moment)
with a developer with a given composition, the eluent composition was abruptly changed
by adding a solvent or mixture and the experiment was intercepted when the distance
between solvent front to starting point became Z_f.

Various amounts of methanol were added to a binary mixture of methylene chloride and
butanol (9:1 v/v) and the solutions thus prepared were used as developer. Relationships
between R_f for sample A1 in this system (R_{f1}) and R_f for sample A5 (R_{f5}) are illustrated
in Figure 2.5.7. In the figure, the composition of methanol (in vol%) is shown on the curve.

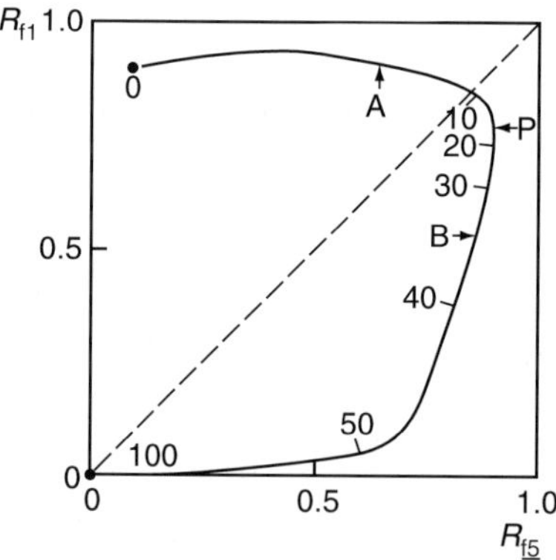

Figure 2.5.7   Relationships between R_f for A1 (R_{f1}) and R_f for A5 (R_{f5}) obtained with methylene
chloride/butanol/methanol. The figure on the curve denotes the content of methanol (vol%).[5]
In this case, the volume ratio of methylene chloride/butanol is kept constant (9:1). Meanings of A,
B, and P are described in the text.

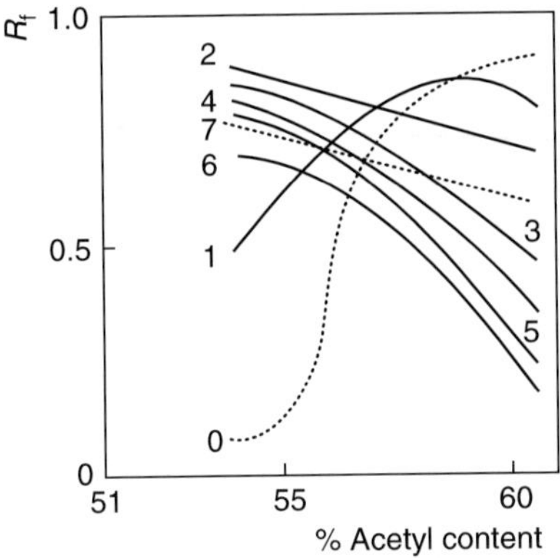

Figure 2.5.8   Dependence of R_f obtained by the continuous gradient method with the methylene chloride/butanol/methanol system, on the acetyl content of cellulose acetate. Initial and final methanol contents in the ternary mixture of methylene chloride/butanol/methanol are 0% (Curve 0), 0 and 3.8% (Curve 1), 3.8 and 16.2% (Curve 2), 16.2 and 24.2% (Curve 3), 24.2 and 36.8% (Curve 4), 36.8 and 37.9% (Curve 5), 37.9 and 41.5% (Curve 6), and 36.8% (Curve 7). The volume ratio of methylene chloride/butanol is maintained constant (9:1).[5]

Figure 2.5.8 shows the dependence of R_f on the acetyl content, which, using the continuous gradient method, is obtained for CA with different acetyl contents with the methylene chloride/butanol/methanol system. The dependence of R_f on the acetyl content changes its sign from large positive to negative with an increasing methanol content. The value of R_f obtained by the continuous gradient method can be roughly approximated with that obtained by the constant eluent composition method at a composition identical with the end point of the continuous gradient method.

The spots corresponding to curves 3 and 4 in Figure 2.5.8 are a clearly cut off form[8] at the tail. From this, at least a region of a large R_{f5} should be adopted as an initial composition of the developer. On the contrary, if the methanol content is decreased during the development process, then the tailing cannot be diminished. Similar phenomena were also observed for TLC in ternary mixtures suitable for MWD evaluation. These experimental results indicate that the method of stepwise variation of the eluent composition, as well as the continuous gradient elution method, can be expected to be useful because both methods satisfy the conditions necessary to eliminate tail. The first and second eluent compositions, A and B in Figure 2.5.7, are chosen so that the point P giving maximum R_{f5} lies between A and B. The points A and B, and the time when solvent composition at A is to be transformed via P through the curve APB to B, are experimentally determined in advance. In the gradient column method, it is very difficult to adjust the solvent front to an appropriate height at the instant at which the final eluent composition is reached.

In contrast, in the method proposed here, the second eluent composition and height of solvent front can be adjusted independently. The total amount of alcohols (methanol and butanol) at point A should be smaller than that of point B, as previously mentioned. The location of the polymer sample developed by the first eluent does not move downward even if the R_f value of the second eluent is smaller than that of the first one. By using this result, one can readily regulate the position

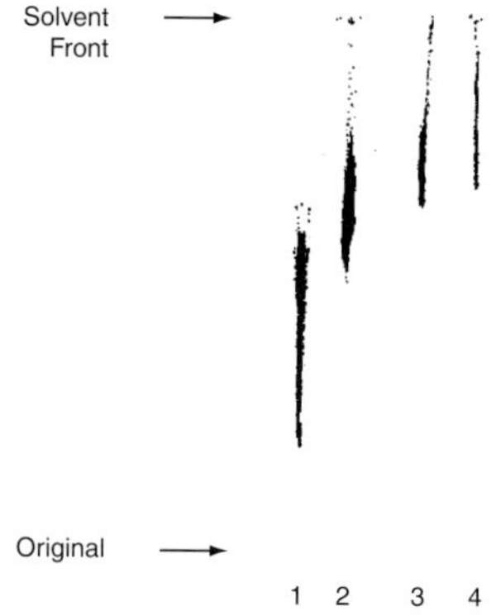

Figure 2.5.9   Thin layer chromatograms obtained by a method of stepwise variation of the eluent composition. 1. Sample A1; 2. Sample A2; 3. Sample A4; 4. Sample A5. The first eluent composition: methylene chloride/methanol (9:1 v/v) and the second eluent composition: methylene chloride/methanol (5:5 v/v); $Z'_f/Z_f = 0.2$.[5]

of the sample. For example, it is effective to increase the methanol content in the first eluent or to change the first eluent into the second eluent at a later time, in order to make the R_f value large.

Figure 2.5.9 shows chromatograms obtained for samples A1, A2, A4, and A5 by using stepwise variation of the eluent composition (in this case, the first eluent composition, methylene chloride/methanol (9:1 v/v); the second eluent composition, methylene chloride/methanol (5:5 v/v). The R_f value varies by a factor of 0.45 over the range of the acetyl content from 54 to 60%. A similar experiment shows that the R_f value varies by a factor of 0.05 if the viscosity average degree of polymerization P_v of the polymers with the same acetyl content (54%) changes from 138 to 264. That is, a variation of 0.6% in acetyl content and a change of 126 of P_v brings about the same change in R_f. Therefore, the effect of P_v on R_f for this solvent system is found to be negligible from a practical point of view.

Figure 2.5.10 shows chromatograms for samples A1, A5 and a 1:1 (by weight) mixture of A1 and A5 obtained by the stepwise eluent method. The chromatogram of the blend can be approximated as half of the summation of those of A1 and A5 (as is theoretically predicted) when a specific interaction between molecules of differing acetyl content can be neglected.

Figure 2.5.11 are shown chromatograms for samples M2, M4, M7, and M9 in methylene chloride/acetone/methanol (10:10:1 v/v/v) and methylene chloride/acetone/methanol (10:10:36 v/v/v) at $Z_f/Z'_f = 0.81$ using the stepwise developing method. Evidently, the stepwise developing method significantly eliminates the tailing effect (see Figure 2.5.1).

Figure 2.5.12 shows chromatograms for samples M2, M9 and a 1:1 (by weight) mixture of M2 and M9 obtained under the following conditions: methylene chloride/acetone/methanol (6:4:0.5 v/v/v) and methylene chloride/acetone/methanol (6:4:18 v/v/v) and $Z'_f/Z_f = 0.85$. The chromatogram of the blend can also be approximated as half of the summation of those of M2 and M9.

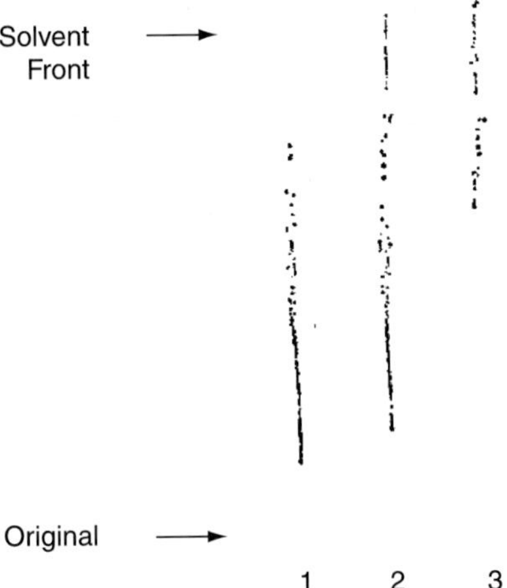

Figure 2.5.10   Thin layer chromatograms obtained by a method of stepwise variation of the eluent composition. 1. Sample A1; 2. Sample mixture of A1 and A5 (1:1 wt/wt); 3. Sample A5.[5]

Figure 2.5.11   Thin layer chromatograms obtained by a method of stepwise variation of the eluent composition.[5] 1. Sample M2; 2. Sample M4; 3. Sample M7; 4. Sample M9.[5] The first eluent composition: methylene chloride/acetone/methanol (10:10:1 v/v/v) and the second eluent composition: methylene chloride/acetone/methanol (1:1:3.6 v/v/v); $Z'_f/Z_f = 0.81$.

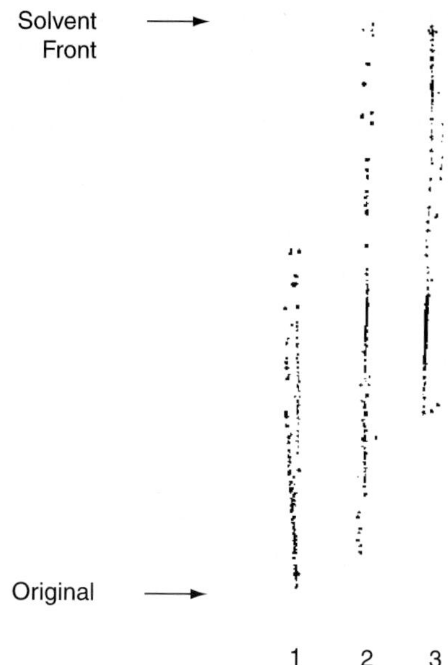

Figure 2.5.12   Thin layer chromatograms obtained by a method of stepwise variation of the eluent composition[5] 1. Sample M2; 2. mixture of Samples M2 and M9 (1:1 w/w); 3. Sample M9.[5] The first eluent composition: methylene chloride/acetone/methanol (6:4:0.5 v/v/v), and the second eluent composition: methylene chloride/acetone/methanol (3:2:9 v/v/v); $Z'_f/Z_f = 0.85$.

Fractionation mechanism of TLC

Thus far, the fractionation of TLC has been interpreted in terms of an adsorption/desorption equilibrium of the polymer on the solid phase, as a partition of macromolecules between mobile and stationary phases on the basis of solubility (phase–separation) and as a molecular sieve mechanism. For CA, R_f becomes smaller with increasing molecular weight, so that molecular sieve mechanisms should be discarded, at least in this case.

With increasing alcohol content, R_f passes through a maximum and the dependence of R_f on the acetyl content changes its sign at the peak, suggesting that fractionation is governed by an adsorption–desorption and a phase–separation mechanism. The latter mechanism should be taken into account if the following factors play an important role: (a) variation of polymer concentration due to the variation of phase ratio along the direction of development; (b) variation of solvent composition along the development direction caused by the difference of adsorption power of solvent on adsorbent; (c) variation of solvent composition due to the difference of the rate of vaporization (accordingly, vapor pressure); (d) variation of solvent composition due to the difference of moving rate (of solvent components), which depends on the surface tension and density.

The fractionation mechanism of the acetyl content and molecular weight will now be discussed in some detail. The experiment shows that in the range of an alcohol concentration (v_m) less than 10%, R_f increases with an increase in the acetyl content. Moreover, as is well known, a hydroxyl group has an adsorption power stronger than an acetyl group on silica gel or alumina. Therefore, if the fractionation with respect to the chemical composition occurs according to the adsorption–desorption mechanism, R_f is expected to increase with an increasing acetyl content ranging from 54 to 61% and it does not separate into two phases at room temperature providing that v_m is less than 30%. Thus, the fractionation of the polymer by TLC from the above mentioned solution cannot be explained by a phase separation mechanism. The R_f of the polymer increases with decreasing activity of absorbent by standing for a long period after being activated. This means that the activity of absorbent is intimately correlated with R_f. Accordingly, as the developer increases its polarity, solvent molecules are preferentially adsorbed on silica gel and polymer molecules become difficult to absorb on absorbent, resulting in a larger R_f value. These experimental results reveal that in the range of v_m less than 10%, sorption and desorption phenomena are the most predominant factors controlling the fractionation with regard to the acetyl content. Separation based on phase equilibrium occurs, but to a small extent. This is based on the various experimental evidence. For example, when a TLC experiment is carried out under saturated vapor or at a temperature low enough to suppress vaporization of the solvent (in this case, methylene chloride) from a thin layer plate, two kinds of solvent fronts are observed. The solvent front with the larger Z_f value is conventionally nominated as the first solvent front and the other solvent as the second solvent front. The solvent developing between the first and second solvent front is found to be methylene chloride. Moreover, the polymer developed between two solvent fronts has an acetyl content higher than 60%. The abrupt change in solvent composition and acetyl content of the polymer at the second solvent front indicates that, at this boundary, the difference of solubility of polymer molecules is a predominant factor.

In the concentration region larger than 30% of alcohol, the R_f value decreases with increasing acetyl content, and this does not coincide with the prediction from the adsorption–desorption theory.

Figure 2.5.13 shows the plot of the phase ratio against Z/Z_{fs} for two plates of different thickness. The phase ratio is the weight of eluent retained by unit weight of adsorbent and Z_{fs} is the distance between solvent front and solvent level. Z denotes the point located Z apart from solvent level. It is apparent that the phase ratio is larger for the thicker layer. Kamiyama et al.[8] have identified a change in the phase ratio along the direction of development using dimethylformamide as a model developer. The results obtained for dimethylformamide are not always applicable to other solvents or solvent mixtures. Taking into consideration that the polymer sample is fractionated according to the acetyl content in the range of an almost constant phase ratio ($Z/Z_f = 0.25$–0.75), R_f cannot be considered to be varied due to cause (a).

Figure 2.5.14 shows the composition of the solvent as a function of Z/Z_{ts} in the case when a mixture of methylene chloride/methanol (10:1 v/v) is developed on the plate. In Figure 2.5.14, the weight ratio of methylene chloride to silica gel is also shown as a parameter representing the phase ratio. The methanol concentration in the mixture shows a maximum at $Z/Z_f = 0.4$, approaching zero at $Z/Z_f = 1$, due to the preferential adsorption of methanol in the mixture onto the adsorbent in the range $Z/Z_f \leq 0.65$. For a mixture of methylene chloride and alcohol having an alcohol content higher than 20%, the solubility of CA decreases with the increase of alcohol concentration, irrespective of molecular weight and acetyl content. If the characteristic change in solvent composition with Z/Z_f, as shown in Figure 2.5.14, can be postulated to hold up to the mixture of

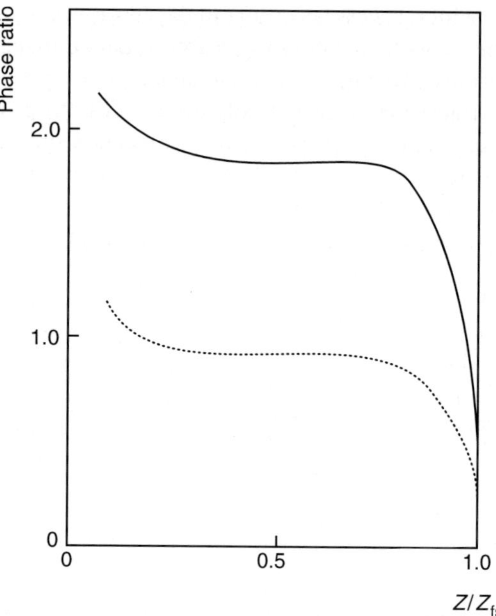

Figure 2.5.13   Plot of the phase ratio against Z/Z_{fs} for two plates of different thickness:[5] solid line, 0.73 nm; dotted line, 0.25 nm. Composition of the developer: methylene chloride/methanol (1:1 v/v).

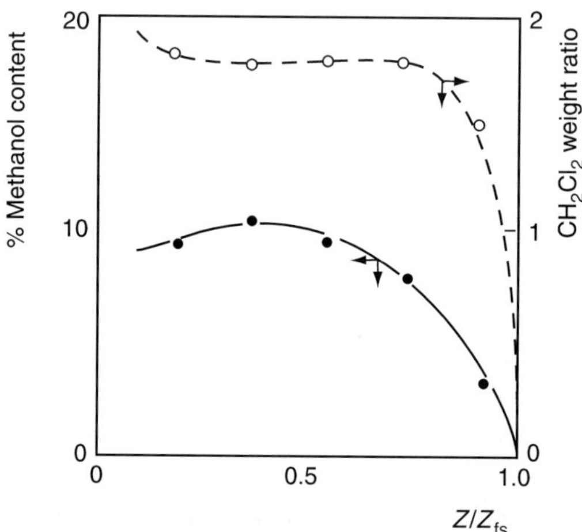

Figure 2.5.14   Variation of the eluent composition along the direction of development.[5] Composition of the developer: methylene chloride/methanol (10:1 v/v). Methanol content is expressed by the volume percent of methanol in the methylene chloride/methanol system and methylene chloride weight ratio is the weight ratio of methylene chloride to silica gel.

high alcohol content (in this case, 50% or more), in the comparatively higher Z/Z_f region, the polymer might decrease its solubility as Z/Z_f increases. In other words, R_f might increase and approach unity with an increase in alcohol concentration. This expectation is, however, inconsistent with the experimental results (see Figure 9 of Ref. 5). Therefore, it can be concluded that in the range of a lower alcohol concentration, factor (b) should not be ignored, but in the range of the higher alcohol region, this factor cannot explain the behavior of R_f.

Next, the dependence of R_f, obtained under identical conditions on the acetyl content of CA will be discussed. Figure 2.5.15 shows the plots of polymer concentration against methanol concentration at the cloud point at 20 °C for A1 and M1 in methylene chloride/methanol system are shown. In TLC, the initial and final concentrations of polymer were determined as 0.6 and 0.1%, respectively by experiment. The range of methanol concentration yielding phase separation from a solution ranging from 0.6 to 0.1% is estimated from Figure 2.5.15 as between 50 and 70%, The composition of the mixture, in which a 0.1% solution separates into two phases at 20 °C, is estimated from Figure 2.5.15, $v_m = 63\%$ for A 1. The Z/Z_f value corresponding to $v_m = 63\%$ is roughly in agreement with the observed R_f value. This experimental fact strongly supports the validity of the phase separation theory.

The value of R_f is larger for a smaller P_v regardless of the alcohol concentration in a mixture. In the lower alcohol concentration range (e.g. less than 15%), in which the molecular weight fractionation occurs, the polymer solution does not separate into two phases, suggesting that fractionation in this region is based on adsorption–desorption phenomena. On the other hand, in the higher alcohol concentration range, the fractionation obeys a phase separation mechanism. The dependence of R_f on the molecular weight can be significantly improved by an appropriate choice of the variation of the phase ratio and the solvent composition in the direction of development.

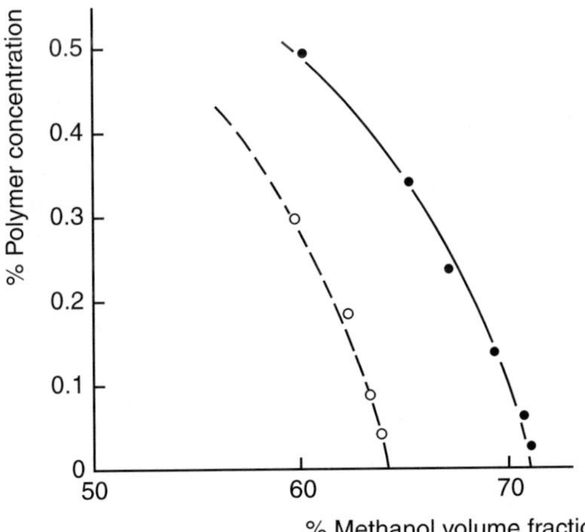

Figure 2.5.15 Plots of polymer concentration versus methanol concentration at the cloud point at 20 °C.[5] (O):A1 (CA = 61%); (●):M1 (CA = 54%) in the methylene chloride/methanol system.

Determination of average combined acetic acid content $A_{c,w}$ ($\equiv \langle\!\langle F \rangle\!\rangle \equiv DS$) for samples fractionated by TLC

There are two main reasons why the quantitative evaluation of ACD by using the TLC technique has not been well developed: a poor reproducibility of the TLC pattern and a lack of adequate technique for determining the amounts of polymer molecule absorbed on the thin layer. The latter is now solved by the rapid development of spectroscopic equipment.

For a better understanding of the solubility behavior of CA in terms of ACD, an extensive study has been conducted to establish a method for evaluating the ACD of CA having a variety of the combined acetic acid content.

Table 2.5.3 lists the molecular characteristics of CA samples used.

Using the procedure described previously, 10 spots of a CDA sample (EF3-9) ($M_w/M_n = 1.26$) were developed on a silica plate. Two spots at both sides in these 10 spots were used as reference for carbonization. After TLC run and drying the plate, fractionated samples were collected at different R_f values (width 2 mm) and were extracted with methylene chloride/methanol (9:1 v/v) and subjected to pulse Fourier transform IR analysis (Joel. JIR-40X) for evaluating $A_{c,w}$. The IR spectra were taken for the dried samples on the rock salt plate. The absorbance at $3460\ \text{cm}^{-1}$ of hydroxyl stretching was utilized as inner standard against the absorbance at $1749\ \text{cm}^{-1}$ of carbonyl stretching band for cellulose acetate.

Figure 2.5.16 shows typical chromatograms of CA whole polymer with various $A_{c,w}$, which are indicated on the curves. The pattern of chromatograms of CA has a rather simple (at most, doubly peaked) form without tailing in contrast to those of cellulose nitrate.[4] As is evident from Figure 2.5.16, the noticeable change in the shape of the chromatograms with $A_{c,w}$, which decreases in the course of hydrolysis reaction of CTA to CDA, is very complicated, suggesting the heterogeneity nature of the hydrolysis reaction.

Figure 2.5.17 shows the chromatograms of CDA whole polymer ($A_{c,w} = 55.6$ wt%, dotted curve) and its fractions (real curves) having various M_w, indicated on the curves. In the same manner, the chromatograms of CTA whole polymer ($A_{c,w} = 61.0$ wt%, dotted curve) and its fractions (real curves) are illustrated in Figure 2.5.18. For the CDA and CTA fractions, the TLC chromatogram becomes apparently sharp with an increase in M_w.

The doubly peaked form of chromatograms is characteristic of the CTA samples without exception, although these gel permeation chromatography (GPC) curves have been found to be single peaked. This fact implies that the peak at the lower field of $R_f(R_{f,l})$ corresponds to fully substituted triacetate, and that the peak at the higher field of $R_f(R_{f,h})$ is obviously due to the existence of not fully substituted acetate. In this sense, CTA is a binary mixture of ideal CTA and CDA. For example, for the chromatogram of CTA whole polymer ($A_{c,w} = 61.0$ wt%) in Figure 2.5.18, R_f at the lower field ($R_{f,l}$) peak corresponds to fully substituted triacetate ($A_{c,w} = 62.5$ wt%) and R_f at the higher field ($R_{f,h}$) peak corresponds to CA with $A_{c,w} = 58.8$ wt%. The content of ($R_{f,h}$) peak increases and that of $R_{f,l}$ gradually decreases as far as the fractions are concerned, although the $A_{c,w}$ is practically independent of M_w.

In other words, the $A_{c,w}$ increases only from 61.0 to 61.7 wt% as M_w increases from 13.7×10^4 to 50.0×10^4. This means that higher M_w fractions consist of a larger amount of completely substituted triacetate. The CTA whole polymer, from which a series of

Table 2.5.3

Molecular characteristics of various CA samples[6]

Sample code	F^a or X^b	Weight average combined acetic acid content $A_{c,w}$(%)				Average molecular weight		$R_{f,w}$ by TLC experiment
		By titration method[c]	By TLC			M_w or $M_v \times 10^4$	$M_n \times 10^4$	
			Eq. (2.5.14)	Eq. (2.5.16)	Eq. (2.5.16)			
WM-1	W	53.9	–	53.9	53.1	7.7[c]	–	0.63
WM-2	W	54.6	–	–	54.2	–	4.5[d]	0.57
WM-3	W	54.7	–	54.5	54.2	7.7[e]	–	0.57
WM-4	W	55.2	–	–	54.4	–	4.7[d]	0.56
EF3-w	W	55.6	55.6	54.9	54.8	12.0[c]	4.4[f]	0.51
WM-5	W	56.1	–	55.7	55.5	10.2[e]	–	0.50
WM-6	W	56.3	–	55.7	55.5	10.6[e]	–	0.50
WM-7	W	56.6	–	56.0	55.9	10.0[e]	–	0.48
WM-8	W	56.8	–	–	55.9	–	5.1[d]	0.48
WM-9	W	58.8	–	58.6	58.5	12.5[e]	–	0.34
WM-10	W	60.5	–	–	60.0	–	7.0[h]	0.26
TA2-W	W	61.0	60.5	60.3	60.3	23.5[g]	5.85[h]	0.24
EF3-1	F	54.8	55.0	54.5	53.1	1.5[f]	1.1[f]	0.63
EF3-3	F	54.9	54.7	54.6	54.1	3.8[f]	3.1[f]	0.58
EF3-4	F	55.0	55.1	55.0	54.6	5.1[f]	4.1[f]	0.55
EF3-5	F	55.1	55.0	54.9	54.6	5.9[f]	4.9[f]	0.55
EF3-6	F	55.2	54.7	54.7	54.4	7.6[f]	6.0[f]	0.56
EF3-7	F	55.2	55.3	55.2	55.0	7.0[f]	5.8[f]	0.53
EF3-9	F	55.3	55.0	54.9	54.8	9.8[f]	7.8[f]	0.54
EF3-10	F	55.4	55.1	55.1	55.0	11.1[f]	8.8[f]	0.54
EF3-11	F	55.4	55.9	55.8	55.7	11.9[f]	8.9[f]	0.49
EF3-12	F	55.7	56.0	56.0	55.9	14.2[f]	11.9[f]	0.48
EF3-13	F	55.7	55.8	55.7	55.7	15.9[f]	12.3[f]	0.49
EF3-14	F	55.7	56.3	56.3	56.3	17.5[f]	12.7[f]	0.46

(*Continued*)

Table 2.5.3

Continued

Sample code	F^a or X^b	Weight average combined acetic acid content $A_{c,w}$(%)				Average molecular weight		$R_{f,w}$ by TLC experiment
		By titration method[c]	By TLC			M_w or $M_v \times 10^4$	$M_n \times 10^4$	
			Eq. (2.5.14)	Eq. (2.5.16)	Eq. (2.5.16)			
TA2-5	F	60.9	61.0	60.9	60.7	8.22[g]	5.60[h]	0.22
TA2-6	F	61.3	61.4	61.3	61.2	13.7[g]	9.78[h]	0.19
TA2-7	F	61.3	61.2	61.1	61.1	14.9[g]	11.4[h]	0.20
TA2-8	F	61.4	61.6	61.6	61.6	20.0[g]	14.0[h]	0.17
TA2-9	F	61.6	61.7	61.7	61.8	26.2[g]	20.2[h]	0.16
TA2-10	F	61.5	61.7	61.7	61.8	30.8[g]	21.9[h]	0.16
TA2-11	F	61.6	61.6	61.6	61.8	44.4[g]	31.7[h]	0.16
TA2-12	F	61.7	61.5	61.5	61.8	50.0[g]	35.7[h]	0.16

[a]F means fraction.
[b]W means whole polymer.
[c]By ASTMD-871-61-T.
[d]By MO in THC at 15 °C.
[e]By viscometry.
[f]By GPC in THF.
[g]By LS in DMAc.
[h]MO in chloroform.

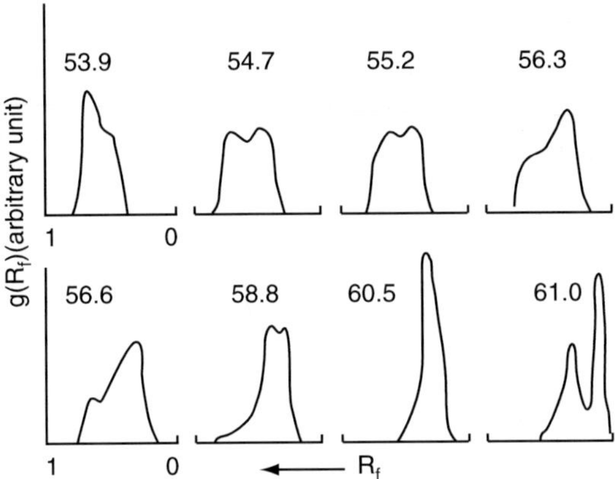

Figure 2.5.16 Typical thin layer chromatograms of CA whole polymers with various $A_{c,w}$ indicated on curves.[6]

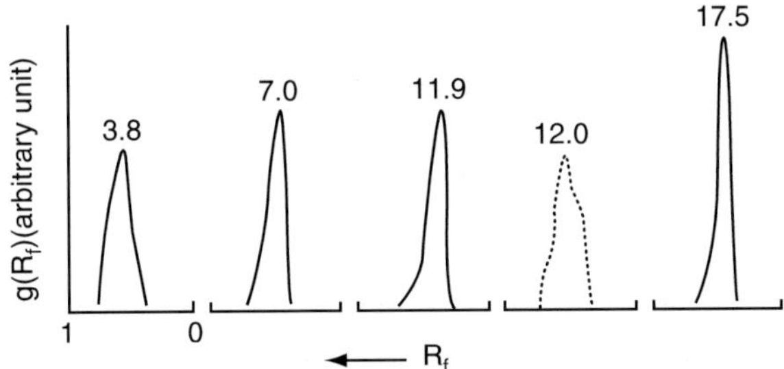

Figure 2.5.17 Typical thin layer chromatograms of cellulose diacetate fractions and whole polymer ($A_{c,w} = 55.6$ wt%) having various M_w: full line, fractions; broken line, whole polymer; numbers on curves mean $10^{-4}\, M_w$.[6]

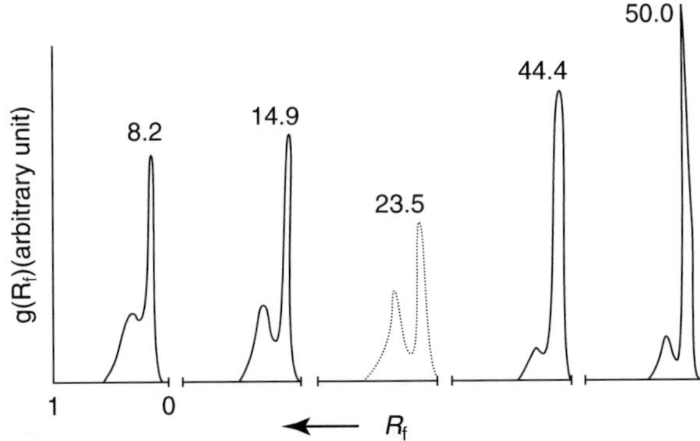

Figure 2.5.18 Typical thin layer chromatograms of cellulose diacetate fractions and whole polymer ($A_{c,w} = 61.0$ wt%) having various M_w: full line, fractions; broken line, whole polymer; numbers on curves mean $10^{-4}\, M_w$.[6]

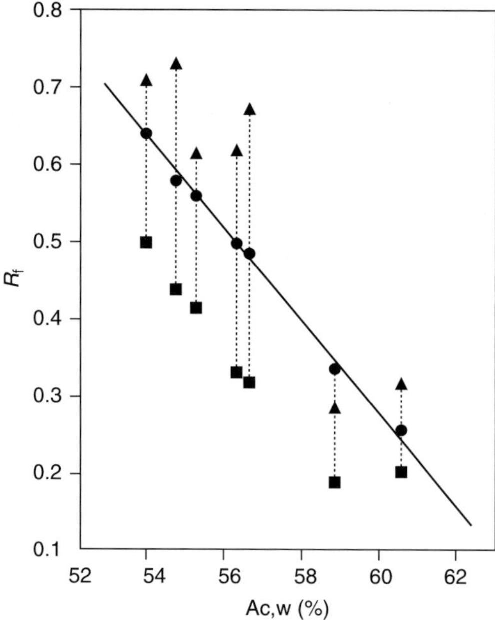

Figure 2.5.19   Correlation between the weight average combined acetic acid content $A_{c,w}$ of CA samples: ($\bullet$), $R_{f,w}$ for total chromatogram; ($\blacksquare$), $R_{f,l}$ for the peak at lower field of R_f; ($\blacktriangle$), $R_{f,h}$ for the peak at higher field of R_f.[6]

fractions were fractionated by the successive solutional fractionation method, has the largest $R_{f,h}$ peak in its TLC chromatogram.

In Figure 2.5.19, the values of $R_{f,l}$ and $R_{f,h}$ together with $R_{f,w}$ (the weight average rate of flow) are plotted as a function of $A_{c,w}$ of CA whole polymers with various $A_{c,w}$ and M_w, although neither $R_{f,l}$ nor $R_{f,h}$ reveals linear correlation with $A_{c,w}$, $R_{f,w}$ decreases linearly to a very good approximation with an increase in $A_{c,w}$. In the ninth column of Table 2.5.3, the $R_{f,w}$ values of samples are collected (Figure 2.5.20).

A careful analysis of the experimental data on $R_{f,w}$, M_n, M_w, $A_{c,w}$ for the 22 samples was made by the methods of least square using eq. (2.5.11) giving the following relation:

$$R_{f,w} = 3.51 - 5.42 \times 10^{-2}A_{c,w} - 3.66 \times 10^{-8}M_w + 1.13 \times 10^{-3}M_{n^{-1}} \qquad (2.5.14)$$

for $1.5 \leq M_w \times 10^{-4} \leq 50$, $1.0 \leq M_w/M_n \leq 4.0$, and $53.9 \leq A_{c,w} \leq 61.7$.

In other words, we obtain parameters in eq. (2.5.11) as

$$R_{f,0} = 3.51, \qquad a_1 = -5.42 \times 10^{-2}, \qquad b_1 = -3.66 \times 10^{-8}, \qquad b_{-1} = 1.13 \times 10^{3}$$
$$(2.5.15)$$

Under the conditions $2 \times 10^{4}(M_w/M_n) < M_w < 2 \times 10^{4}(M_w/M_n) + 1.36 \times 10^{6}$, $R_{f,w}$ values are almost independent of molecular weight and eq. (2.5.14) can be simplified into eq. (2.5.16):

$$R_{f,w} = 3.51 - 5.42 \times 10^{-2}A_{c,w} \qquad (2.5.16)$$

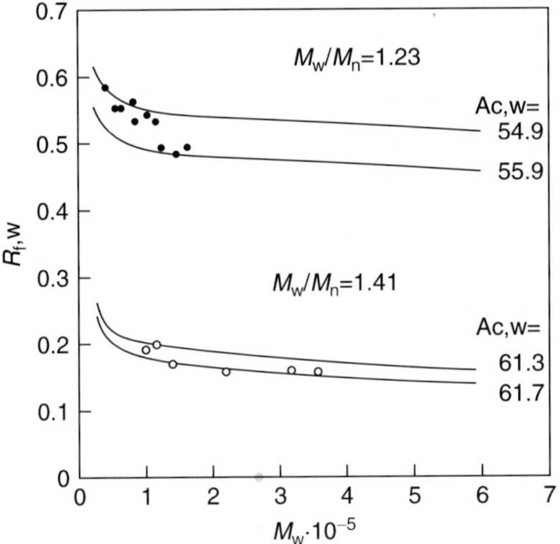

Figure 2.5.20   Comparison between calculated and experimental $R_{f,w}$ values: —, calculated $R_{f,w}$–M_w curve by using eq. (2.5.14); ($\bullet$), observed value of the sample having $A_{c,w} = 55.4 + 0.5$, $M_w/M_n = 1.23 \pm 0.04$; ($\bigcirc$), observed value of the same sample having $A_{c,w} = 61.5 \pm 0.2$, $M_w/M_n = 1.41 \pm 0.02$.[6]

The experimental data which satisfy the above conditions are graphed according to eq. (2.5.16) as shown in Figure 2.5.21. Thus, eq. (2.5.16) appears to fit the data well (the correlation coefficient $\gamma = 0.992$). We obtain a similar relation for the monodisperse polymer M in the range of $2 \times 10^4 - 1.39 \times 10^6$ with respect to A_c:

$$R_f = -3.51 - 5.42 \times 10^{-2} A_c \tag{2.5.17}$$

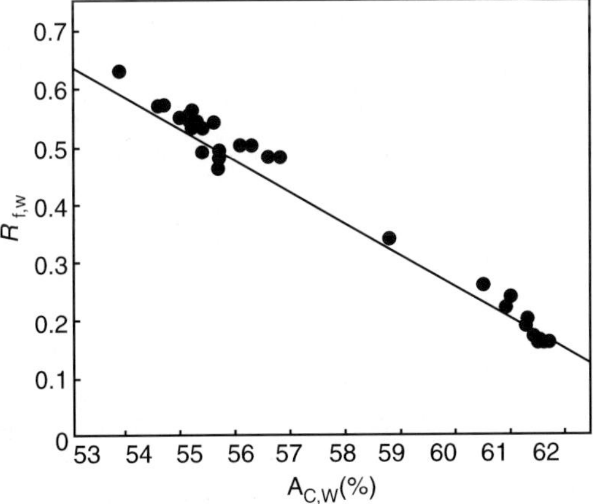

Figure 2.5.21   Comparison between the observed of $R_{f,w}$ and the calculated one using eq. (2.6.17)[6]: ($\bullet$), observed valued: — , calculated value.

Equation (2.5.17) enables us to evaluate the ACD directly from R_f distribution function $g(R_f)$ obtained by the TLC experiment. By using either eqs. (2.5.14), (2.5.16), or (2.5.17), and from the experimental data on $R_{f,w}$, M_w, and M_n, we can evaluate $A_{c,w}$. The $A_{c,w}$ values thus obtained are shown in the fourth, fifth, and sixth columns of Table 2.5.3, respectively. The uncertainty of $A_{c,w}$ by the TLC method is estimated to be $\pm 0.6\%$ (eq. (2.5.14)), $\pm 0.7\%$ (eq. (2.5.16)), and $\pm 0.8\%$ (eq. (2.5.17)), respectively, indicating that the TLC method appears to be reliable for determining the acetic acid content. With eq. (2.5.17), we can evaluate $g(A_c)$ from the observed $g(R_f)$. Some typical results are illustrated in Figures 2.5.22–2.5.24.

Figure 2.5.22 shows the combined acetic ACD for CA whole polymers. Figures 2.5.23 and 2.5.24 demonstrate ACD for the CDA and CTA fractions, together with the corresponding whole polymers, from which the fractions are fractionated. In general, the ACD of whole polymers are far broader than that of the fractions. This means that the ACD is closely associated with the MWD of the samples and the higher molecular weight components have the larger acetic acid content. The correlation is caused by the preparative method of the polymer.

Moreover, the combined acetic ACD of the CDA fractions become narrower as M_w becomes larger. All the CTA samples, including a whole polymer, have distinct doubly peaked ACD. As M_w for the fractions increases, this consequently results in the narrowing of the ACD curve.

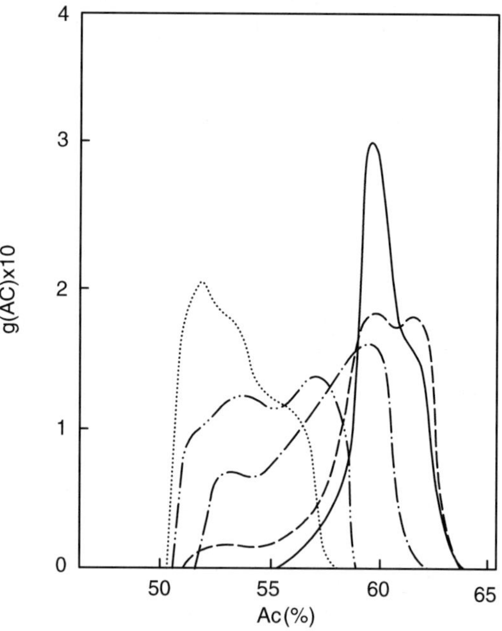

Figure 2.5.22   Combined acetic ACD evaluated by TCL method for cellulose acetate whole polymers with various $A_{c,w}$ (wt%): dotted line, $A_{c,w} = 53.9$; two dotted chain line, $A_{c,w} = 55.2$; chain line, $A_{c,w} = 56.6$; broken line, $A_{c,w} = 58.8$; solid line, $A_{c,w} = 60.5$.

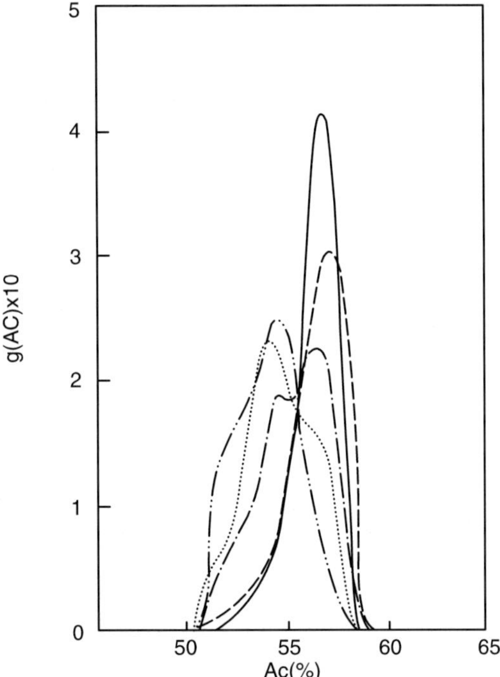

Figure 2.5.23   Combined acetic ACD evaluated by TLC method for cellulose diacetate fractions with various M_w, dotted line, $M_w = 12.0 \times 10^4$ (whole polymer); two dotted and chain line, $M_w = 3.8 \times 10^4$; chain line, $M_w = 9.8 \times 10^4$; broken line, $M_w = 11.9 \times 10$; solid line, $M_w = 17.5 \times 10^4$.

2.5.3   Cellulose nitrate (CN)[4]

CN is commercially utilized by changing its average molecular weight and the average degree of esterification (i.e. the nitrogen content by wt% as expressed by $N_c\%$). A number of attempts have been made to separate a whole CN polymer with respect to molecular weight or N_c. Unfortunately, however, no effective method of compositional fractionation has yet been proposed for CN, except by Kamiyama and Inagaki.[8,9] Using TLC, they studied the compositional heterogeneity of some commercial CN samples by a concentration gradient development using a binary acetone–ethyl acetate (20:3 v/v) and another binary chloroform–ethyl acetate (1:2 v/v) as initial and second solvents, respectively. However, they have not established a method for estimating MWD or chemical compositional distribution (CCD) of CN by TLC technique. This section shows a TLC method to fractionate CN on the basis of N_c or its molecular weight.

Experimental procedure

CN was prepared by nitrating purified cotton linter with a mixture of nitric acid and sulfuric acid, followed by a conventional thermal degradation at elevated temperatures in water (Table 2.5.4). Here, M_v was calculated from the limited viscosity number $[\eta]$

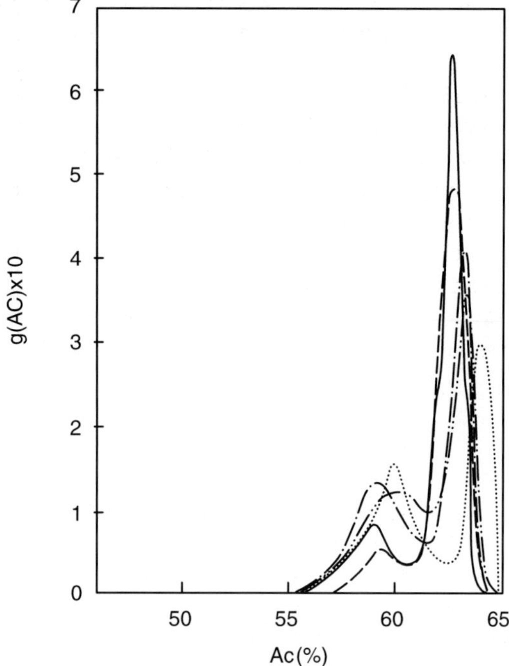

Figure 2.5.24   Combined acetic ACD evaluated by TLC method for CTA fractions with various M_w; dotted line, $M_w = 23.5 \times 10^4$ (whole polymer); two dotted and chain line, $M_w = 8.5 \times 10^4$; chain line, $M_w = 14.9 \times 10^4$; broken line, $M_w = 44.4 \times 10^4$; solid line, $M_w = 50.0 \times 10^4$.

Table 2.5.4

Characteristics of CN samples

Sample code	Fraction or whole polymer	Nitrogen content, (N_c) (wt%)	$[\eta]$ in acetone, 25 °C (cm^3 g^{-1})	$M_v \times 10^{-4}$[a]
HIG 1/2	W[b]	11.9	65.4	1.96
HIG 2	W	11.8	87.5	2.82
HIG 7	W	11.7	127	4.51
HIG 20	W	11.8	156	5.85
HIG 80	W	11.9	212	8.60
HIG 120	W	11.8	240	10.0
HIG 1000	W	12.1	370	17.3
W 115	W	11.5	431	21.0
W 119	W	11.9	436	21.3
W 121	W	12.1	512	26.0
W 127	W	12.7	531	27.3
S19-5	F[c]	11.7	831	47.9

[a] M_v was determined from $[\eta] = 2.53 \times 10^{-2} M_n^{0.795}$.
[b] Whole polymer.
[c] Fraction.

data in acetone at 25 °C using $[\eta] = 2.53 \times 10^{-2} \times M_n^{0.795}$ ($[\eta]$: cm^3 g^{-1}), which was originally established by Moore and Edge for CN fractions with $N_c = 12.2$ wt%.[10]

Preparation of thin layer. Silica gel (Kieselgel Gnach Stahl, type 60, manufactured by E. Merck, Darmstadt, Germany), aluminum oxide (aluminum oxide G, type 60/E by Merck), and kieselguhr (Kieselguhr G, Merck) in powder form were used as the substrate for TLC. A slurry composed of two parts water and one part adsorbent (by weight) was spread evenly on a (thoroughly washed) glass plate by a conventional applicator (Automatic TLC coater with adsorbent hopper manufactured by Camag, Switzerland). The plate was air dried at room temperature for 15 min and was activated for 1 h in an oven at 110 °C, immediately before use. It was then cooled to room temperature and preserved in a desiccator. The thickness of the adsorbent layer was adjusted to 0.25 mm.

TLC experiment. All solvents employed were of guaranteed grade and, to remove any trace of impurities, were mixed with granular active carbon which was subsequently filtered with porous polymeric membrane with an average pore size of 0.1 μm, distilled under reduced pressure, when necessary.

CN samples were dissolved in acetone at a concentration of 0.76 wt%. The solution thus prepared was dropped using a microsyringe onto the plate 1.5 cm away from the lower edge of the plate, making a 5 mm diameter spot. Each spot on the layer was formed by a 42 μg CN polymer.

Two kinds of TLC experiments (open and vapor programming (VP) TLC) were carried out. In the open TLC experiment, the TLC plate was placed in a rectangular box with an open end as a development vessel, and the polymer sample was developed by the natural ascending method. In this case, the development was performed in an unsaturated vapor phase. High reproducibility was attained by carefully controlling the atmosphere surrounding the development vessel. This technique was first successfully applied to the separation of cellulose acetate.[5] When a closed development vessel was employed in place of the box chamber with an open end (i.e. conventional TLC), no efficient separation occurred under the same operating conditions (developer, stationary phase, and developing method) as those used in open TLC. VP-TLC is a very precise kind of gradient elution. Its experimental procedure was briefly described by Zeeuw,[11] and was utilized with modification. A vapor programming chamber (Vario-KS chamber manufactured by Camag, Switzerland) was used. With this chamber we could accurately control the vapor pressure during the development on any portion of the plate. The cross sectional view of the chamber is illustrated in Figure 2.5.25.

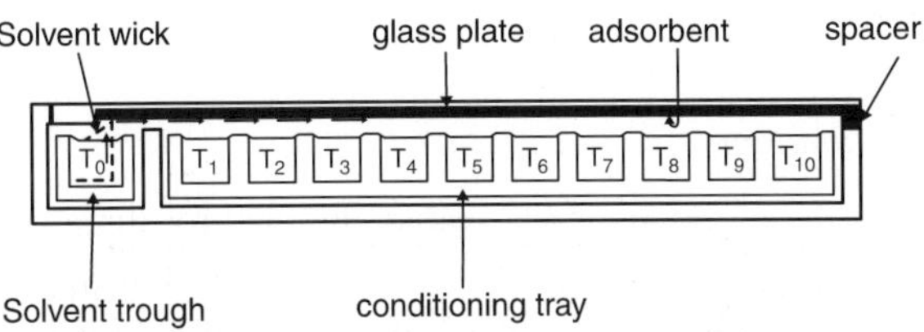

Figure 2.5.25   Cross sectional view of the chamber for vapor programming (VP) TLC.[4]

Each subdivision in the trough chamber was filled with a different solvent mixture having a given composition before development, and was carefully determined by preliminary experiment to bring about the best separation. On the basis of our experience, if the phase separation mechanism is predominantly operative in the TLC, we strongly recommend, for continuous gradients, the concentration of good solvents (e.g. chloroform in the compositional fractionation; Figure 9 of Ref. 4) and 1,4-dioxane in the molecular weight fractionation; Figure 12 of Ref. 4) in a subdivision is decreased (accordingly, that of poor solvents is increased) as the distance of the subdivision from the starting points is increased (i.e. $T_1–T_{10}$ in Figure 2.5.25). The solvent mixture as developer was then poured into the solvent trough. In this experiment, a shaped paper spacer frame of 0.2 cm thickness was inserted between the cleared rim of the TLC plate and the corresponding ridge of the conditioning tray so that the continuous vapor phase gradient along the direction of development could be built up. Without this kind of spacer, transverse bands were frequently observed in the carbonized chromatograms, corresponding to the distance of the walls of the subdivisions in the tray.

The glass plate with the polymer samples loaded and the thin layer facing downwards were pressed horizontally on the vapor programming chamber. Prior to the chromatographic development, the TLC layer was allowed to stand for 20 min in order to absorb vapor molecules evaporating from the conditioning trays. Next, the CN samples on the starting point were developed with solvents which transferred through the solvent wick, made of filter paper from the solvent trough.

The development was stopped when the solvent front reached 10 cm above the original point of samples (i.e. the starting point). After the run was completed, the plate was dried by blowing air at room temperature. To visualize polymers on the plate, the thin layer plate was sprayed with 10 wt% aq. sulfuric acid fumes. The polymer on the layer was carbonized by heating at 110 °C in an electric oven. The sensitivity of detecting the polymers spread on the plate was 0.1 μg CN mm^{-1}. The profiles were recorded in a Shimadzu dual wavelength TLC scanner at a wavelength λ of 380 nm. In this case, a zigzag scanning method was employed (beam 1.25×1.25 mm^2; width of swing stroke 30 mm). Although the spots of chromatogram have a very complicated form (as illustrated in Figure 2.5.27), the intensity curve has a single peak or at most is doubly peaked. The weight average value of $R_f(R_{f,w})$ was calculated from the intensity curve.

The phase ratio defined by the weight of solvent (mobile phase), retained by a unit weight of thin layer (stationary phase), was determined by the difference between the weight of the gel layer containing the solvent for the area arbitrarily scraped off and that of the gel dried.

The change in composition of the solvent as a function of the distance from the starting points was determined by the Shimadzu gas chromatograph model GC-4CMPF, which has a flame ionization detector.

Compositional fractionation

Figure 2.5.26 shows the variation of $R_{f,w}$ value obtained for three CN samples (W 115, W 121, and W 127) having different N_c ($N_c = 11.5$ wt% (full line), 12.1 wt% (broken line) and 12.7 wt% (dotted line), with changing composition of the developer system of nitromethane and methanol. $R_{f,w}$ passes through a maximum at a point of a nitromethane

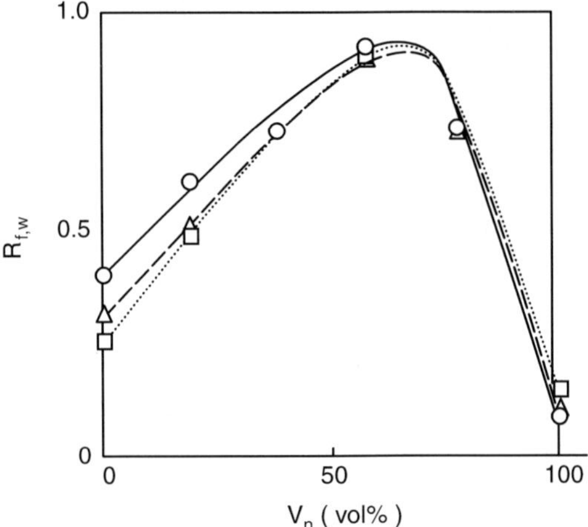

Figure 2.5.26 Dependence of $R_{f,w}$ for CN on volume percent of nitromethane[4] V_n: ($\bigcirc$), W115 ($N_c = 11.5$ wt%); ($\triangle$), W121 ($N_c = 12.1$ wt%); ($\square$), W127 ($N_c = 12.7$ wt%). R_f was determined on silica gel with nitromethane–methanol mixture (open TLC).

content (v_n in vol%) of 70%. In other words, the dependence of R_f on the nitrogen content of the samples changes its sign at this point. The maximum $R_{f,w}$ value remains almost constant, and independent of N_c. Similar $R_{f,w}-v_n$ relationships have been reported for isotactic poly(methyl methacrylate) fractions in a chloroform–methanol system[8] and CA fractions in a methylene chloride n-butanol system.[5] Clearly, there exists a composition (in this case, $v_n = 0-40$ vol%) at which the CCD evaluation can be performed most effectively. However, if the phenomenon of the downward tailing of the spot is taken into account, then $v_n = 20-40$ vol% is much more desirable. Hereafter, the system of nitromethane and methanol (20:80 v/v at 25 °C) was used to fractionate CN samples with regard to N_c by open TLC. Methanol is well-known for its displacing action.

Figure 2.5.27 shows a separation of five different N_c samples. Evidently, all the samples (except for a fraction S19-5) are developed as very complicated bands. The R_f value decreases with increasing N_c and varies by a factor of more than two over a N_c range from 11.5 to 12.7 wt%.

Figure 2.5.28 illustrates the limiting viscosity number $[\eta]$ of samples W 115 and W 127 in mixtures of nitromethane and methanol. The $[\eta]$ of both samples exhibits a minimum when the composition of nitromethane–methanol is 20:80 (v/v at 25 °C). That is, at this composition, as the most effective for compositional separation, the polymer molecules exist in the most contracted form. It was confirmed by the measurement of the phase ratio as a function of the position on layer that the polymer molecules do not precipitate over the entire range of the composition of solvent mixture examined. This indicates that the separation according to the chemical composition by open TLC with nitromethane–methanol mixture is based not on the phase separation mechanism, but probably on the adsorption–desorption mechanism.

Figure 2.5.29 shows the dependence of $R_{f,w}$ on N_c for CN samples with similar viscosity average molecular weight M_v ($21-27 \times 10^4$) as an open circle (silica gel) or

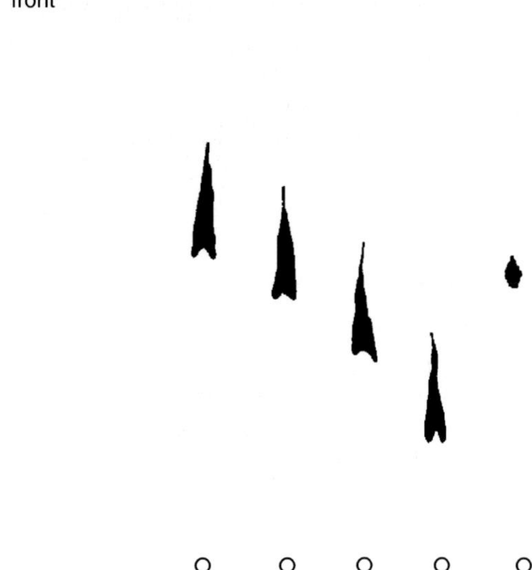

Figure 2.5.27   TLC chromatograms obtained for CN samples having different nitrogen content and almost the same molecular weight:[4] silica gel, nitromethane–methanol (open TLC); 1, W115; 2, W119; 3, W121; 4, W127; 5, S19-5.

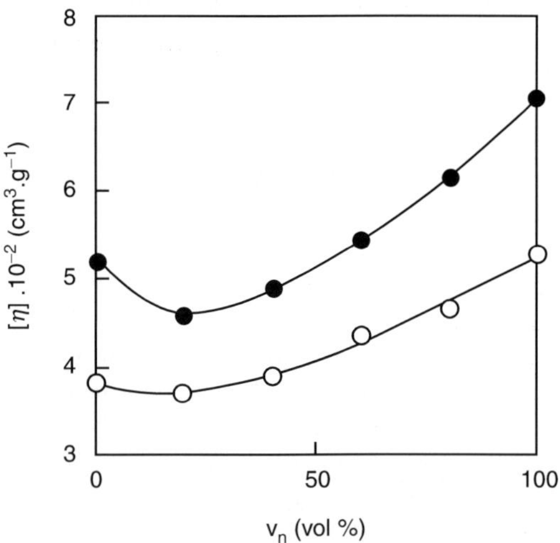

Figure 2.5.28   Plot of limiting viscosity number $[\eta]$ at 25 °C against the composition of nitromethane–methanol mixture, expressed by nitromethane content (vol%) v_n:[4] (●), W 115 ($N_\mathrm{c} = 11.5$ wt%); (○), W 127 ($N_\mathrm{c} = 12.7$ wt%).

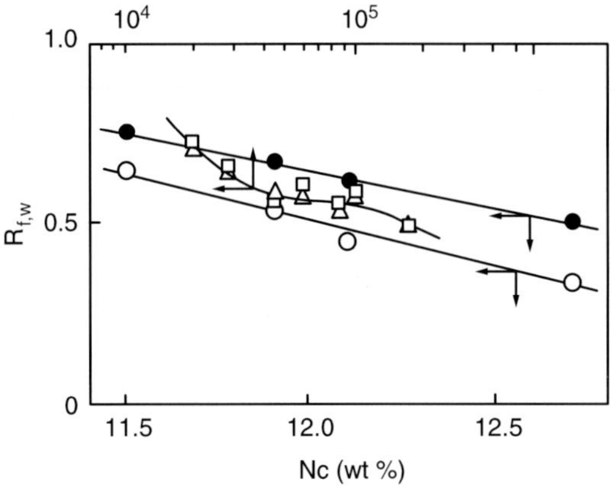

Figure 2.5.29   Dependence of $R_{f,w}$ on nitrogen content (N_c) and on the viscosity average molecular weight M_v:[4] open TLC, nitromethane–methanol, 20/80 v/v; ($\bigcirc$), silica gel; ($\bullet$), alumina; ($\square$), first run; ($\triangle$), second run.

a closed circle (aluminum oxide), which are eluted with constant composition of mixed solvent (nitromethane–methanol (20:80 v/v at 25 °C) in open TLC. CN with higher N_c exhibits a lower $R_{f,w}$ value than polymer with lower N_c. The following relationships are obtained by the least square method:

$$R_{f,w} = 3.43 - 0.244N_c(\text{wt\%}) : \text{silica gel, } 11.5\times \leq N_c(\text{wt\%}) \leq 12.7 \qquad (2.5.18)$$

and

$$R_{f,w} = 2.95 - 0.192N_c(\text{wt\%}) : \text{aluminum oxide} \qquad (2.5.19)$$

Figure 2.5.30 shows the differential weight distribution(s) of nitrogen content, $g(N_c)$, converted by the relationship between linearized absorbance of carbonized

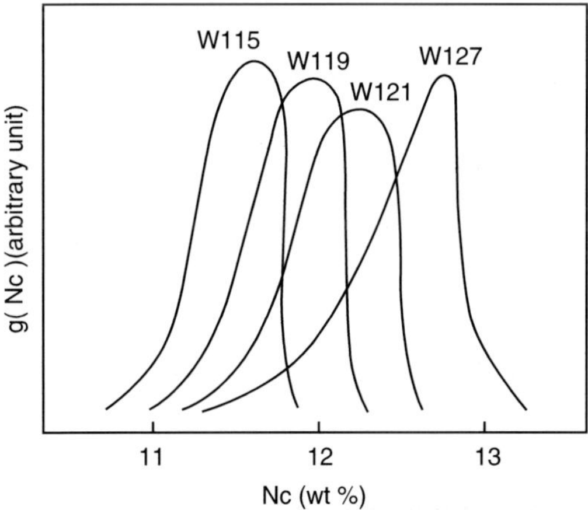

Figure 2.5.30   Differential weight distribution of nitrogen content (N_c by wt%) $g(N_c)$ for several CN samples evaluated by open TLC (silica gel, nitromethane–methanol, 20:80 v/v).[4]

chromatogram (incident beam $\lambda = 380$ nm) and the R_f for various CN samples with different N_c. The preliminary experiments confirmed a reasonable proportionality between the absorbance of the carbonized chromatogram and the amount of CN developed on the chromatogram, independent of the nitrogen content, at least over the entire experimentally accessible range. Inagaki *et al.*[12] observed for poly(methyl methacylate) that the above relationships vary markedly according to the configuration of the polymer. This variation is attributed to the large difference in affinity of the iodine molecules used for visualizing a chromatographic spot for poly(methyl methacrylate). The absorbance is proportional to the differential weight distribution $g(N_c)$.

The absolute value of $R_{f,w}$ is larger in the case of aluminum oxide, used as stationary phase, than for silica gel. This suggests that silica gel, at least as far as this experiment is concerned, has a larger absorbing power than aluminum oxide. The accuracy in evaluating N_c, expressed by $dR_{f,w}/dN_c$, seems dependent on the adsorbent nature. The polymer samples having different M_v and almost the same $N_c\,(=11.8$ wt%) were chromatogrammed under the same conditions as those employed for the open circle in Figure 2.5.29. These results are also shown in Figure 2.5.30 as an open triangle (first run) and an open rectangle (second run). $R_{f,w}$ can be regarded as approximately constant within an M_v range of $4 \times 10^4 - 15 \times 10^4$ and, in this limited range, N_c can be accurately determined from $R_{f,w}$ value independent of the average molecular weight by open TLC technique (nitromethane–methanol). However, in the range $M_v < 4 \times 10^4$ and $M_v > 1.5 \times 10^5$, $R_{f,w}$ depends considerably on both N_c and M_v.

Figure 2.5.31 shows the experimental and calculated weight distribution curves of N_c, $g(N_c)$ for 1:2 and 1:1 (by weight) mixtures of samples W 115 and W 127. The experimental curves were obtained from chromatograms for the 1:2 and 1:1 (by weight) mixtures of samples W 127 and W 115 developed under the following conditions: nitromethane–methanol (20:80 v/v), natural ascending method, silica gel, open TLC. The broken lines were calculated on the basis of chromatograms for samples W 115 and W 127, developed under the same conditions as those for the mixtures. The mixtures are separated into two upper and lower spots overlapped slightly at the intermediate region.

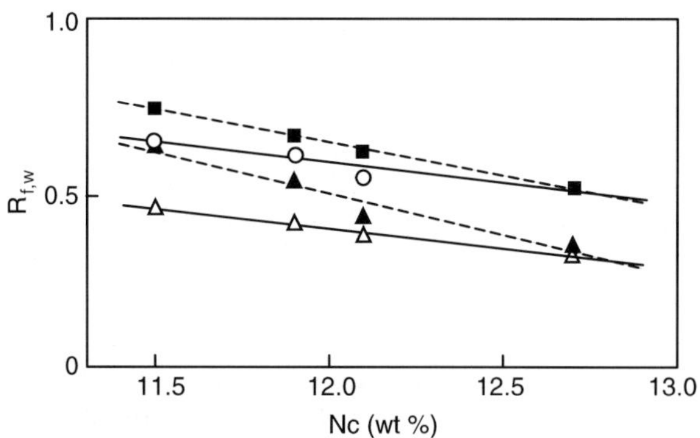

Figure 2.5.31   Dependence of $R_{f,w}$ for CN on nitrogen content (N_c):[4] nitromethane–methanol, 20:80 v/v; open TLC, silica gel; (■), open TLC, alumina; (△), VP-TLC, silica gel; (○), VP-TLC, kieselguhr.

It is important to note that the experimental curve is a fairly good reflection of the CCD curve, although N_c of the lower N_c components is slightly overestimated.

The VP-TLC technique was applied for CN samples with the nitromethane–methanol (20:80 v/v) system. The following relationships are evaluated for VP-TLC:

$$R_{f,w} = 1.82 - 0.119N_c(wt\%) : \text{silica gel at } 25\,°C, \quad 11.5 \leq N_c(wt\%) \leq 12.7$$

$$(2.5.20)$$

$$R_{f,w} = 2.04 - 0.121N_c(wt\%) : \text{kieselguhr at } 25\,°C, \quad 11.5 \leq N_c(wt\%) \leq 12.7$$

$$(2.5.21)$$

Obviously, the separation efficiency is much better in open TLC than in VP-TLC if the nitromethane–methanol system is employed as the developer. Thus, other solvent mixtures were surveyed in order to improve the efficiency of N_c evaluation by VP-TLC.

A good separation of the CN polymer with compositional heterogeneity was achieved with a 10:10:5 (v/v/v 25 °C) acetone–methanol–chloroform mixed solvent elution by the VP-TLC technique (stationary phase = kieselguhr).

The first two solvents are good solvents for CN and have high dielectric constant ($\varepsilon = 20.7$ for acetone and 32.6 for methanol), but the last solvent is nonsolvent for CN and has a low dielectric constant ($\varepsilon = 4.62$). From the plots of $R_{f,w}$ against N_c for four samples, W 115, W 119, W 121, and W 127, established under the above conditions, we obtain

$$R_{f,w} = 3.78 - 0.275N_c(wt\%) : \text{silica gel at } 25\,°C, \quad 11.5 \leq N_c(wt\%) \leq 12.7$$

$$(2.5.22)$$

and

$$R_{f,w} = 2.59 - 0.184N_c(wt\%) : \text{kieselguhr at } 25\,°C, \quad 11.5 \leq N_c(wt\%) \leq 12.7$$

$$(2.5.23)$$

Equation (2.5.22) should be compared with eq. (2.5.20) where eq. (2.5.23) corresponds to eq. (2.5.21). In the case of VP-TLC with acetone methanol chloroform (silica gel), the $R_{f,w}$ value of the fraction S 19-5 deviates slightly from the value expected from eq. (2.5.22). This is due to the large difference in the average molecular weight between the fraction and the whole polymers.

Figure 2.5.32 shows the differential weight distribution of N_c, $g(N_c)$ for the two samples W 121 and W 127 estimated by two TLC techniques. A full line is obtained by open TLC with nitromethane–methanol (20:80 v/v) mixture (silicagel) and a broken line means VP-TLC with acetone–methanol–chloroform (10:10:5 v/v/v). The agreement between the $g(N_c)$ curves evaluated by both methods is good, considering the potential for experimental uncertainty.

Figure 2.5.33 is a graph of the compositional variation of solvent mixtures on the layer in the development process of VP-TLC. The open mark denotes the experimental data points. In this figure, r is the ratio of Z to Z_f where Z is the distance of an arbitrarily chosen point from the starting point and Z_f is the distance of the end point from the starting point (in this case, $Z_f = 10$ cm). At the initial stage, an increase in r occurs as the methanol content increases. Then, r passes through a maximum, and finally decreases

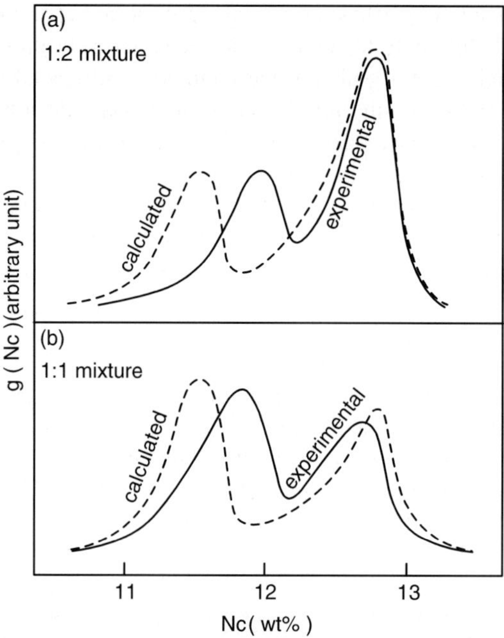

Figure 2.5.32 Experimental and calculated weight distribution curves of nitrogen content (N_c), $g(N_c)$, for 1:2 and 1:1 (by weight) mixtures of samples W 127 and W 115[4]; full line, experimental curve; broken line, line calculated from experimental chromatogram of each component, nitromethane–methanol, 20:80 v/v at 25 °C, open TLC.

to zero at $r = 1.0$ where the developer consists only of chloroform. The open triangle is the solvent composition at the $R_{f,w}$ value of sample W 127 whereas the open rectangle corresponds to the $R_{f,w}$ value of sample W 115. From a knowledge of the phase ratio, the weight of the polymer concentration deposited was found to be 0.34 g/100 g solvent

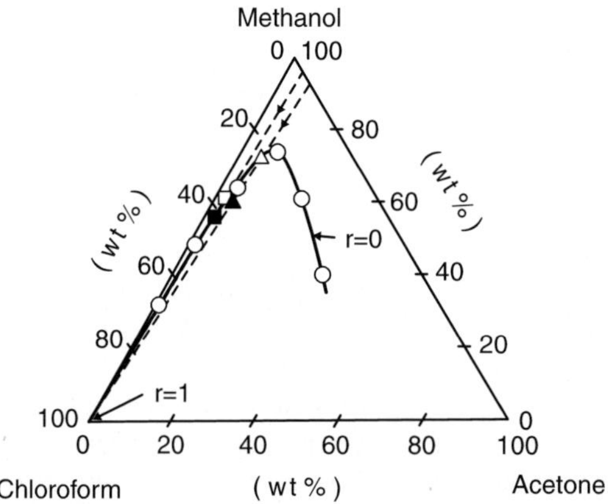

Figure 2.5.33 Variation of the composition of solvent mixtures in the development of VP-TLC:[4] open mark, experimental data point; open triangle, R_f for sample W 127; open rectangle, R_f for sample W 115; closed triangle, precipitating point for sample W 115; r, see text.

mixture for W 127 and 0.33 g/100 g solvent mixture for W 115. Solutions of the samples W 127 and W 115 in mixtures of acetone and methanol were prepared under the same conditions (weight ratio of polymer-acetone-methanol) as those of the open triangle and the open rectangle. Chloroform was added to the solutions thus prepared and the composition of each solution at turpidity was determined and is shown as closed marks in Figure 2.5.33. The solvent mixture at R_f of CN samples can be considered as roughly equal to that at the precipitating point. In other words, the fractionation with respect to N_c by these solvent mixtures (VP-TLC) is explained in terms of the polymer phase-separation mechanism.

Molecular weight fractionation

In a preliminary experiment, the combination of 1,4-dioxane-methanol-2-propanol mixture (15:10:1 v/v/v) as the developer and kieselguhr as the stationary phase in VP-TLC proved successful for molecular weight fractionation.

Figure 2.5.34 shows photometric recording curves of blackness of the chromato-graphic bands obtained under the above conditions for seven CN samples with the same N_c (11.8 wt%) and different M_v. Chromatograms shift to small R_f region as M_v increases.

Figure 2.5.35 displays the molecular weight dependence of $R_{f,w}$, evaluated from Figure 2.5.35. In this figure, open and closed circles mean the first and second runs, respectively.

From Figure 2.5.36 we obtain:

$$R_{f,w} = 0.7982 - 2.50 \times 10^{-6} M_v : 1.96 \le M_v \times 10^{-4} \le 17.3,$$
$$N_c = 11.8 \text{ wt\%, kieselguhr}$$

(2.5.24)

Under the same operating conditions as employed for establishing eq. (2.5.24), the chromatograms for the polymer samples with similar M_v ($21 \times 10^4 - 27.3 \times 10^4$) and the different N_c were obtained. The results are also shown as a rectangle in Figure 2.4.35.

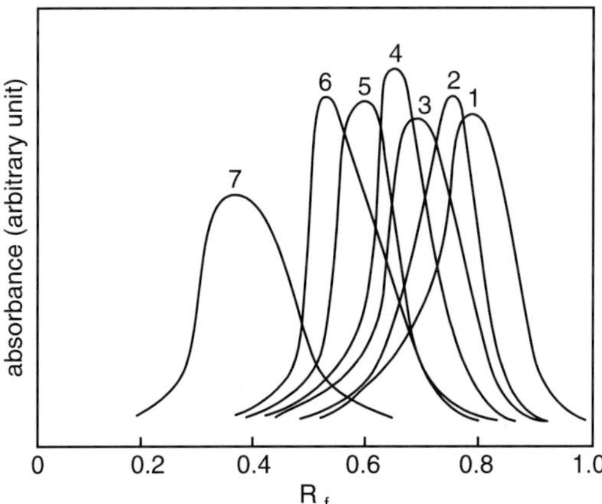

Figure 2.5.34   TLC chromatograms obtained for CN samples having different M_v and the same N_c,[4] kieselguhr, 1,4-dioxane-methanol-2-propanol (15:10:1 v/v/v at 25 °C), VP-TLC: 1, HIG 1/2; 2, HIG 2; 3, HIG 7; 4, HIG 20; 5, HIG 80; 6, HIG 120 7, HIG 1000.

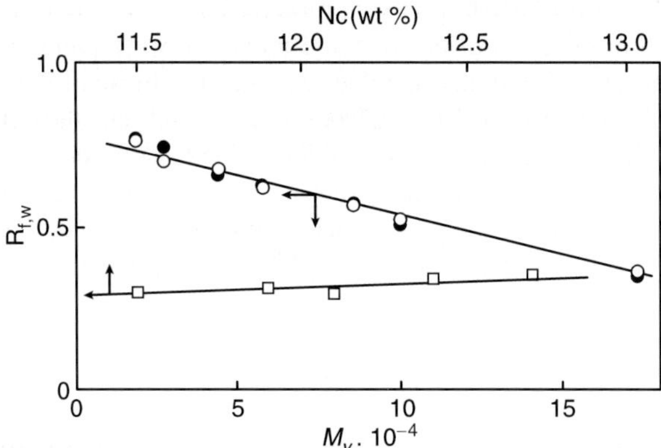

Figure 2.5.35 Molecular weight dependence of $R_{f,w}$ for CN samples having the same N_c ($= 11.8$ wt%) (circle) and nitrogen content (N_c) dependent N_c of $R_{f,w}$ for CN samples having almost the same M_v and different N_c (rectangular):[4] kieselguhr, 1,4-dioxane-methanol-2-propanol (15:10:1 v/v/v at 25 °C).

Evidently, $R_{f,w}$ depends only slightly on N_c, according to the relation:

$$R_{f,w} = -0.205 + 0.044 N_c: \quad 11.5 \leq N_c(\text{wt\%}) \leq 12.9, \quad M_v = 21-27.3 \times 10^4$$

$$(2.5.25)$$

Strictly speaking, it is impossible to fractionate CN samples according to molecular weight without incurring interference from the difference in chemical constitution. However, from a practical perspective, the R_f value can be considered independent of the difference in N_c.

Figure 2.5.36 shows the compositional variation of solvent mixtures (1,4-dioxane–methanol–2-propanol) on the layer. As r increases, the methanol content decreases gradually and the dioxane content decreases drastically, both approaching zero at $r = 1.0$. Although the difference between the solvent composition at the R_f value for polymer

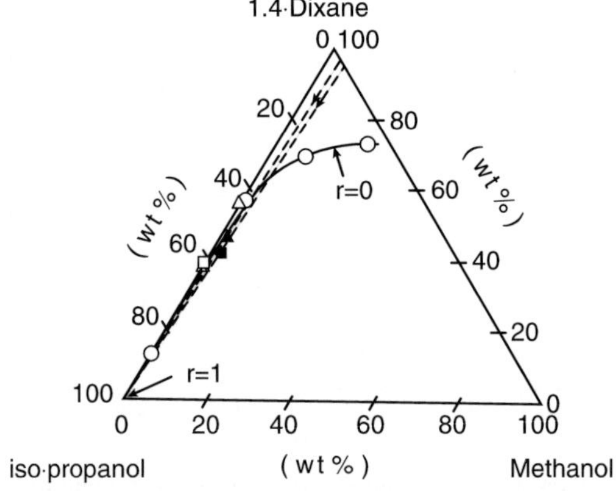

Figure 2.5.36 Variation in the composition of solvent mixture (1,4-dioxane, methanol, isopropanol) in the development of VP-TLC.[4] All symbols are the same as those in Figure 2.5.33.

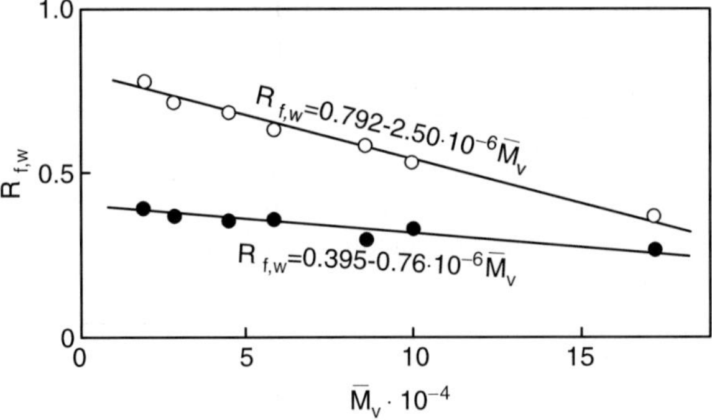

Figure 2.5.37   Influence of substrate on $R_{f,w}$ versus M_v relationships for CN($N_c = 11.8$ wt%)[4]: VP-TLC, 1,4-dioxane-methanol-2-propanol (15:10:1 v/v/v at 25 °C); (○), kieselguhr; (●), silica gel.

sample W 121 and that at the precipitating point is not small, we can consider, as a first approximation, that the molecular weight fractionation by VP-TLC of 1,4-dioxane–methanol–2-propanol is governed mainly by the phase-separation mechanism.

Figure 2.5.37 demonstrates the influence of the substrate on $R_{f,w}$ versus the M_v relationships for CN samples ($N_c = 11.8$ wt%). Data points were obtained by VP-TLC in 1,4-dioxane–methanol–isopropanol mixture (15:10:1 v/v/v at 25 °C). The open mark in the figure is the result in Figure 2.5.36. For silica gel we obtain:

$$R_{f,w} = 0.395 - 0.76 \times 10^{-6}M_v, \quad 1.96 \le M_v \times 10^{-4} \le 17.3,$$
$$N_c = 11.8 \text{ wt\%, silica gel.}$$

$$(2.5.26)$$

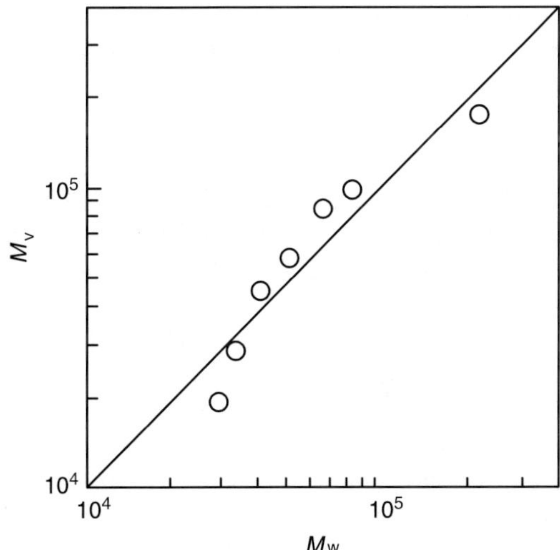

Figure 2.5.38   Relationship between the viscosity average molecular weight M_v and the weight average molecular weight M_w, as determined by VP-TLC (1,4-dioxane-methanol-2-propanol, 15:10:1 v/v/v at 25 °C, kieselguhr).[4]

The adsorbing power is stronger in silica gel than in kieselghur and the latter is evidently much more effective. Provided that the chromatograms obtained under the above conditions represent the true MWD of the polymer, we calculated the weight average molecular weight M_w from chromatograms (see Figure 2.5.35). The relationships between M_w thus determined and M_v are plotted in Figure 2.5.38. The M_w (by TLC) $\approx M_v$ was obtained over the molecular weight range investigated.

It should be noted, however, that the obtained chromatograms do not always represent the true molecular weight distribution of the samples. Surprisingly, the chromatogram of the 1:1 mixture shows only a spot lying just between their components. Similar results were obtained with various two-component mixtures. At present, we cannot explain this unexpected phenomenon explicitly. The apparent molecular weight evaluated from these chromatograms M_w (by TLC) is not far from the M_w calculated.

REFERENCES

1. F Kamiyama, M Matsuda and H Inagaki, *Polym. J.*, 1970, **1**, 518.
2. E Otacka and MY Helmann, *Macromolecules*, 1970, **3**, 362.
3. T Kotacka and JL White, *Macromolecules*, 1974, **7**, 106.
4. K Kamide, T Okada, T Terakawa and K Kaneko, *Polym. J.*, 1978, **10**, 547.
5. K Kamide, S Manabe and E Osahune, *Makromol. Chem.*, 1973, **168**, 173.
6. K Kamide, T Matsui, K Okajima and S Manabe, *Cell. Chem. Technol.*, 1982, **16**, 601.
7. AJ Rosenthal and BB White, *Ind. Eng. Chem.*, 1952, **44**, 2693.
8. T Kamiyama and H Inagaki, *Bull. Inst. Chem. Res., Kyoto Univ.*, 1971, **49**, 5; H Inagaki, F Kamiyama and T Yagi, *Macromolecules*, 1971, **4**, 133.
9. H Inagaki, Thin layer chromatography. In *Fractionation of Synthetic Polymers* (ed. HL Tung), Marcel Dekker, New York, 1977; H Inagaki, *Adv. Polym. Sci.*, 1977, **24**, p. 189.
10. WR Moore and GD Edge, *J. Polym. Sci.*, 1960, **47**, 469.
11. RA de Zeeuw, In *Progress in Separation and Purification*, **Vol. 3**, (eds ES Perry and CJ van Oss), Wiley, New York, 1970, p. 1.
12. H Inagaki, T Miyamoto and F Kamiyama, *J. Polym. Sci. Polym. Lett.*, 1969, **7**, 329.

2.6 SEQUENCE DISTRIBUTION OF SUBSTITUTED AND UNSUBSTITUTED GLUCOPYRANOSE UNITS IN WATER SOLUBLE CA CHAIN AS REVEALED BY ENZYMATIC DEGRADATION[1,2]

When we carry out the molecular characterization of cellulose derivatives, we should take into account the well-known experimented fact that a sample is a mixture of numerous chains with different $\langle F^j \rangle$ (the average total DS of the jth molecule) and that substituted β-D-anhydro-glucopyranose (AHG)s are connected through forming blocks in a given chain. Kamide and his coworkers[3-5] showed that the TLC method affords invaluable information on the distribution of $\langle F^j \rangle$, expressed as $g\langle F^j \rangle$, together with $\langle\!\langle F \rangle\!\rangle$, for cellulose derivatives such as CA[3,5] and CN[4] (Section 2.5). They also demonstrated by TLC that CTA is a binary mixture of ideal (completely substituted) CTA and not fully substituted CA,[5] and the solubility of so-called CTA in dichloromethane is significantly influenced by $g\langle F^j \rangle$ of the sample.[5] It seems reasonable to consider that the distribution of the substituent group along a molecular chain of CD (the pattern of arrangement of eight different AHG) depends strongly on the methods of sample preparation (i.e. the synthesis of derivatives), particularly on whether its synthesis is carried out through homogeneous reactions or heterogeneous reactions and also controls physical properties such as solubility. Kamide *et al.*[6] showed that in a carboxymethyl cellulose (CMC) chain with relatively low $\langle\langle F \rangle\rangle$, some consecutively connected AHG units may exist as blocks. Buchanan *et al.*[7] tried to disclose the relationship between the molecular weight, monomer composition, and water solubility of cellulose monoacetates.

As a useful method for evaluating the distribution of the substituent groups in a chain, enzymatic hydrolysis of cellulose derivatives by cellulase has been extensively studied.[8,9] Cellulase is a common name given to all enzymes having the ability to decompose cellulose. In spite of these studies, the molecular characteristics of cellulase and the mechanism of decomposition of cellulose by cellulase are not yet fully known. Isolated cellulase can be classified into the following three categories: Endo-1,4-β glucanase (hereafter referred to as endocellulase; EC 3.2.1.4), which breaks randomly β-1,4 linkages by hydrolysis reaction of cellulose chain; β-Exocellobiohydrolase (hereafter referred to Exocellulase; EC 3.2.1.9.1), which selectively acts on nonreducing end groups of cellulose chains to produce cellobiose; β-D-Glucosidase (EC 3.1.1.21), which acts on cellobiose and cello-oligosaccharide to produce glucose (see Figure 2.6.3).

In their pioneering work, Reese *et al.*[10] evaluated the relationships between the amount of glucose, produced by enzymatic hydrolysis of CMC, methyl cellulose (MC), hydroxyethylcellulose (HEC), CS, cellulose ethylsulfate and cellulose acetatephthalate, and $\langle\langle F \rangle\rangle$. They concluded that the substituted glucose residues are not hydrolyzed with enzymes because the amount of glucose produced from cellulose derivatives with higher $\langle\langle F \rangle\rangle$ was smaller. Several studies[11,12] carried out thereafter on CMC clearly indicate that enzymatic hydrolysis occurs only between two adjacent unsubstituted AHG units. Cellulase was found to split bonds between two adjacent unsubstituted AHG units in HEC.[13,14]

By applying dialysis with distilled water to the enzymatically hydrolyzed CMC, Bhattacharjee and Perlin[15] isolated the nondialyzable portion (degree of polymerization (DP) $= 10-112$) with $\langle\langle F \rangle\rangle = 1.2$ from dialyzate to determine the compositions of CMC, cellobiose, and glucose in the dialyzate. Considering the yield of large fragments as nondialyzable products and cellobiose, they concluded that there are relatively elongated regions in which the incidence of substitution is high. Moreover, they fractionated the dialyzable portion by column chromatography into eight fractions and their compositions were analyzed by gas liquid chromatography (GLC). They estimated the substitution index of the nondialyzable portion by applying GLC to acid-hydrolyzed products of the portion. However, they did not carry out quantitative evaluation of $\langle\langle F \rangle\rangle$ of oligosaccharides fractions. By combining cellulose hydrolysis techniques with solution viscosity and reducing sugar measurements, Gelman[16] evaluated the number of unsubstituted AHG units and number and average length of blocks of two or more contiguous unsubstituted residues in CMC. In studies published to date, the DP of the degraded products was estimated from the solution viscosity and both $\langle\langle F \rangle\rangle$ and the DP of the products was only indirectly speculated after further complete decomposition of products to glucopyranose units. Except for Bhattacharjee and Perlin,[15] no attempt has been made to fractionate the degraded products to evaluate the distribution of substituted or unsubstituted AHG units in a chain.

For the purpose of evaluating the distribution of DS in incompletely substituted glucopyranose unit sequences, Iijima et $al.$[1] used cellulase to water soluble CA with known $\langle\!\langle F \rangle\!\rangle$ and $\langle\!\langle f_k \rangle\!\rangle$ and fractionated, by liquid chromatography, the degraded products and determined, by ^{13}C NMR analysis, the distribution of the DS $\langle\!\langle F \rangle\!\rangle$ and $\langle\!\langle f_k \rangle\!\rangle$ for each fraction separated. They also deduced the acting mechanism of enzymatic hydrolysis of cellulosic main chains by comparing the above experimental results with computer experiments. An attempt was also made by Iijima and Kamide[17] to investigate the distribution of substituent groups along the chains of water soluble, partially substituted, CA molecules. For this purpose, two water soluble CA samples (i.e. one with the total DS $\langle\!\langle F \rangle\!\rangle = 0.88$ and the weight average molecular weight $M_w = 35,900$ (sample code CA (88)-0) and the other with $\langle\!\langle F \rangle\!\rangle = 0.60$, and $M_w = 42,000$ (sample code CA (60)-0), prepared by acid hydrolysis of cellulose diacetate (CDA) ($\langle\!\langle F \rangle\!\rangle = 2.46$) synthesized by two-step method), were subjected to enzymatic hydrolysis in water solution.

Figure 2.6.1 illustrates the scheme of experimental procedures for enzymatic degradation of CA and separation of enzymatic hydrolysis products. The degraded products were fractionated by preparative GPC into 50 and 20 fractions, analyzed by analytical GPC (Figure 2.6.2) and by ^{13}C NMR to determine peak molecular weight (M_p) and $\langle\!\langle F \rangle\!\rangle$, respectively.

Table 2.6.1 shows data on M_w, $\langle\!\langle F \rangle\!\rangle$, $\langle\!\langle f_2 \rangle\!\rangle$, $\langle\!\langle f_3 \rangle\!\rangle$, and $\langle\!\langle f_6 \rangle\!\rangle$ for sample code CA (88)-0S (dissolved portion of CA (88)-0 into deionized water). Similar results are obtained for CA (60)-0S (see Table II of Ref. [1]). $\langle\!\langle F \rangle\!\rangle$ of water soluble components in the degraded products was nearly constant ($= 1.0$) and the degree of polymerization DP was in the range between 1 and 7 in both enzymatic hydrolysis experiments. The above characteristics were quantitatively consistent with computer experiments made under the condition of the endo–exo–exo, degradation hypothesis.

2.6.1 Computer experiments of enzymatic degradation reaction

Iijima et $al.$[1] assumed as the basic tenet that enzymes, such as endo-, exocellulase, and β-glucosidase, react against CA molecules in the following manner: (a) endocellulase breaks the bond between two contiguous unsubstituted AHG units, but it does not decompose cellobiose to yield glucose; (b) exocellulase produces cellobiose by separating single cellobiose units from nonreducing ends of cello-oligosaccharide and acetylated cello-oligosaccharide; (c) β-glucosidase cleaves cellobiose, cello-oligosaccharide, and acetylated cello-oligosaccharide to remove single unsubstituted AHG units from the nonreducing end of those molecules and yields glucose as an end product. Note that the cooperative reaction between exocellulase and β-glucosidase produces glucose from the nonreducing end of those chain molecules and in computer experiments, the reaction caused by β-glucosidase and exocellulase was designated as the exo-reaction.

The above mechanism, illustrated in Figure 2.6.3, can be summarized as follows: (1) Bonding between two contiguous unsubstituted AHG units in a chain is broken by

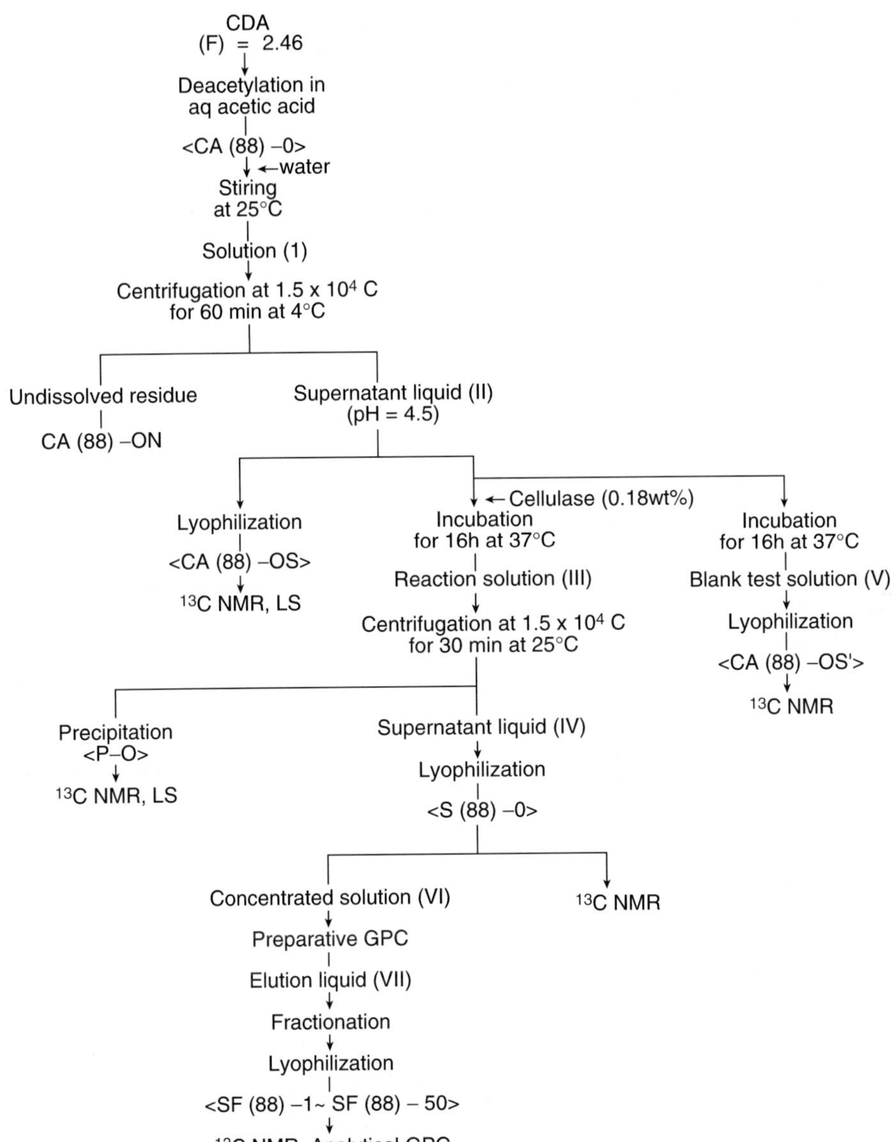

Figure 2.6.1 Experimental procedures of preparation of water soluble CA sample, its enzymatic hydrolysis and separation of degraded products.[1]

endocellulase reaction, and (2) when the nonreducing end residue is an unsubstituted AHG unit, the AHG unit is cleaved by exocellulase reaction.

In computer experiments, the following hypotheses of the cooperative reactions of cellulases on CA were employed: (i) only endocellulase reacts on CA; (ii) endo-and exocellulases react cooperatively on CA and the unsubstituted AHG unit is separated from the nonreducing end of a chain as glucose; and (iii) (ii) is satisfied and the unsubstituted

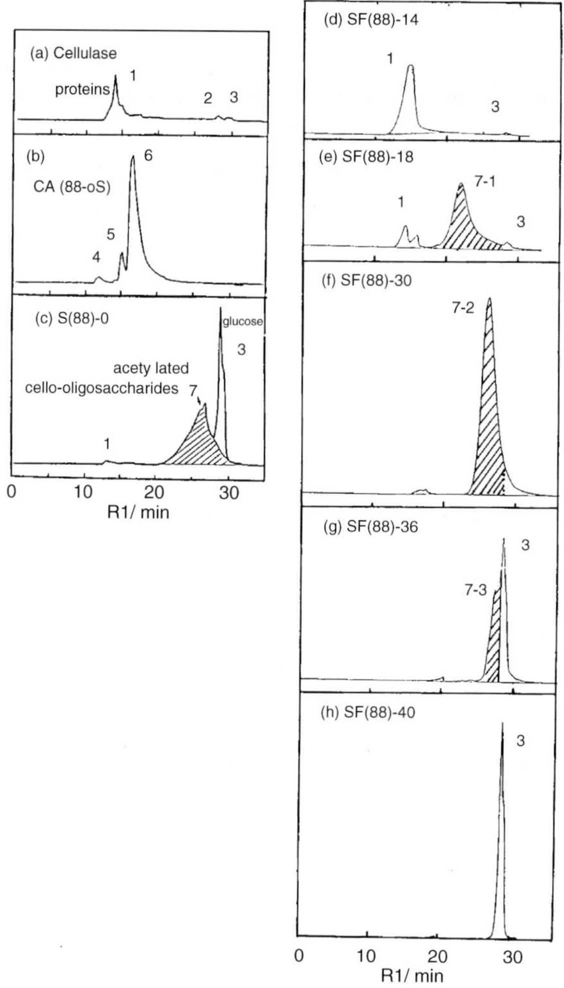

Figure 2.6.2 Gel permeation chromatograms of cellulose (a) water soluble portion of acid-hydrolyzed CA (88)-0S (b) water soluble portion of enzymatically degraded product S(88)-0 (c) and its fractions SF(88)-14 (d) SF(88)-18 (e) SF(88)-30 (f) SF(88)-36 (g) and SF (88)-40 (h); eluent, water; shadowed area, acetyl cello-oligosaccharides.[1]

AHG unit on the reducing end of a chain is cleaved as glucose. Hereafter, we call the reaction condition (i) endo degradation, (ii) endo–exo degradation, and (iii) endo–exo–exo degradation.

Computer experiments of enzymatic degradation reaction was carried out as follows: assume an assembly of 5000 chain molecules consisting of 180 AHG units each, corresponding to DP of CA (88)-0S. Choose any single chain in the assembly, any AHG unit in the chain, and any hydroxyl group in the AHG unit, and introduce a substituent group (i.e. acetyl group) to the hydroxyl group thus chosen. Repeat this procedure until the total degree of substitution $\langle\!\langle F \rangle\!\rangle$ for the assembly attains 0.88, which is the value of $\langle\!\langle F \rangle\!\rangle$ of CA (88)-0S sample. The assembly thus prepared is regarded as the starting

Table 2.6.1

Experimental results on M_w[a], M_p[b], and $\langle\!\langle F\rangle\!\rangle$[c] and $\langle\!\langle f_k\rangle\!\rangle$[c] ($k = 2,3$, and 6) for water soluble deacetylated CA (88)-OS)[d] and water soluble and soluble portions of enzymatically degraded CA (P(88)-0 and S(88)-0) and the fractions from S(88)-0

Sample code	Molecular weight		Degrees of substitution[a]			
	M_w[b]	M_p[c]	$\langle\!\langle F\rangle\!\rangle$	$\langle\!\langle f_2\rangle\!\rangle$	$\langle\!\langle f_3\rangle\!\rangle$	$\langle\!\langle f_6\rangle\!\rangle$
Deacetylated CA by acid-hydrolysis						
CA (88)-OS[d]	35,900	–	0.88	0.27	0.32	0.29
Blank test[e]						
CA (88)-OS'		–	0.86			
Enzymatically degraded CA						
P(88)-0		–	2.48	0.79	0.90	0.79
S(88)-0		–	1.12			
Fractions from S(88)-0						
SF(88)-18		1250	1.01	0.31	0.37	0.33
SF(88)-19		1200	–	–	–	–
SF(88)-20		1100	–	–	–	–
SF(88)-21		1030	1.19	0.38	0.38	0.43
SF(88)-22		–	–	–	–	–
SF(88)-23		1040	1.01	0.31	0.37	0.33
SF(88)-24		–	–	–	–	–
SF(88)-25		920	1.28	0.44	0.41	0.44
SF(88)-26		–	–	–	–	–
SF(88)-27		840	–	–	–	–
SF(88)-28		750	1.01	0.30	0.40	0.31
SF(88)-29		–	–	–	–	–
SF(88)-30		600	1.06	0.34	0.32	0.41
SF(88)-31		–	–	–	–	–
SF(88)-32		–	–	–	–	–
SF(88)-33		460	1.01	0.40	0.22	0.39
SF(88)-34		460/180	–	–	–	–
SF(88)-35		460/360/180	–	–	–	–
SF(88)-36		320/260/180	1.02	0.35	0.27	0.40
SF(88)-37		320/180	–	–	–	–
SF(88)-38		360/180	1.05	0.31	0.30	0.44
SF(88)-39		180	–	–	–	–
SF(88)-40		180	0	–	–	–

[a]Average DS by NMR.
[b]Weight average molecular weight by light scattering method.
[c]Peak molecular weight by GPC.
[d]Starting material for enzymatic degradation.
[e]CA (88)-OS was treated without enzyme under the same conditions as those used for the enzymatic degradation.

material for hypothetical enzymatic hydrolysis. Another assembly of 5000 chain molecules consisting of 224 AHG units each, corresponding to DP of CA (60)-0S, was assumed and the same procedure as above was repeated until the total DS $\langle\!\langle F\rangle\!\rangle$ for the assembly attained 0.60, which is the value of $\langle\!\langle F\rangle\!\rangle$ of CA (60)-0S sample.

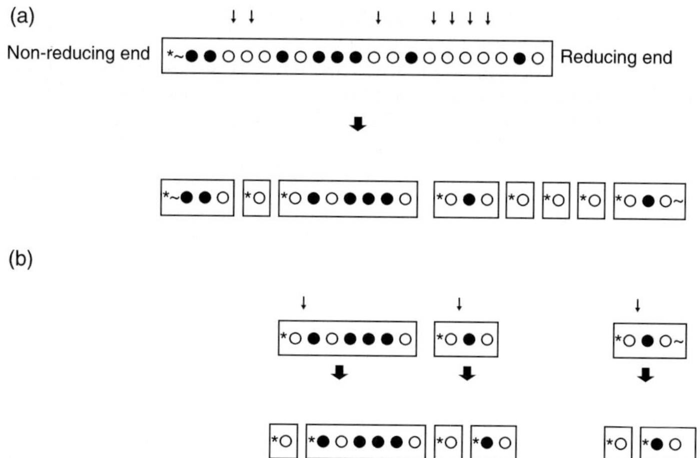

Figure 2.6.3 Schematic representation of possible reactions of cellulase on CA used in computer experiments: (●) substituted AHG; (○) unsubstituted AHG; (∗) nonreducing end (↓); linkage where enzymes can hydrolyze; (a) endocellulase reaction; (b) exocellulase reaction.[1]

Examination of the existence of the substituted group(s) in every AHG unit was conducted starting from the nonreducing end of every CA chain one by one. When the chain end is an unsubstituted AHG unit, the unit is cleaved. If the chain end is a substituted AHG unit, then all AHG units constituting the CA chains are examined from chain ends one by one until two consecutive unsubstituted AHG units are found and cut the bonding between above two units. Continue this examination and cleavage operation,

Table 2.6.2

Weight ratio of glucose and glucose acetate produced in actual enzymatic degradation or in its computer experiment to the original water soluble cellulose acetate[1]

Run number	Production ratio	Actual experiment	Computer experiment $\times 10^{-2}$		
			Endo	endo–exo	endo–exo–exo
1st	W_{g}/W_0	16.5×10^{-2}	3.80	10.86	17.91
	W_{gma}/W_0	$+^{\mathrm{a}}$	0.00	0.04	0.79
	W_{gda}/W_0	$+$	0.00	0.02	0.38
	W_{g}/W_0	$+$	0.00	0.00	0.06
2nd	W_{g}/W_0	34.0×10^{-2}	12.49	24.59	36.70
	W_{gma}/W_0	$-^{\mathrm{b}}$	0.00	0.05	3.14
	W_{gda}/W_0	$-$	0.00	0.02	0.89
	W_{g}/W_0	$-$	0.00	0.01	0.08

[a]Existence is suggested.
[b]Not examined.

Figure 2.6.4 Sequence distribution of substitution glucopyranose units along a chain cellulose acetate (degree of polymerization, 96; $\langle\langle F \rangle\rangle = 0.64$: unfilled circle, unsubstituted; filled triangle, substituted glucopyranose unit DS 1.0; filled star, DS of 0.25).[2]

starting from the nonreducing end to the other end (i.e. the CA molecule in question is completely decomposed by enzyme). This operation is repeated for all CA molecules. As a result, *ca.* 1000 and *ca.* 430 kinds of oligomers with different molecular weights were isolated as fragments from 5000 CA chains in cases of $\langle\langle F \rangle\rangle = 0.88$ and 0.60, respectively. For each oligomer, the molecular weight M and the DS $\langle F \rangle$ were calculated.

The experimental characteristics were quantitatively consistent with computer experiments made under the condition of the endo–exo–exo degradation hypothesis (Table 2.6.2).

Cleavage of water soluble CA samples yielded a small portion of water insoluble CA component with the same $\langle\langle f_k \rangle\rangle$ as that of the original CA samples, from which water soluble CA was prepared by acid-hydrolysis. Acetyl cello-oligosaccharide in the water soluble portion of enzymatically degraded products is monoacetate. Therefore, the original water soluble CA sample can be treated as a kind of a block copolymer of a 2.45 acetate block, several monoacetate blocks, and glucose. A water soluble chain can be reconstructed with good certainty using experimental data on the enzymatically degraded products. Figure 2.6.4 illustrates 20 reconstructed chains of water soluble CA with $\langle\langle F \rangle\rangle = 0.6$.

REFERENCES

1. H Iijima, K Kowsaka and K Kamide, *Polym. J.*, 1992, **24**, 1077.
2. K Kamide and M Saito, *Macromol. Symp.*, 1994, **83**, 233.
3. K Kamide, S Manabe and E Osafune, *Makromol. Chem.*, 1973, **180**, 168.
4. K Kamide, T Okada, T Terakawa and K Kaneko, *Polym. J.*, 1978, **10**, 547.
5. K Kamide, T Matsui, K Okajima and S Manabe, *Cell. Chem. Technol.*, 1982, **16**, 601.
6. K Kamide, K Okajima, K Kowsaka, T Matsui, T Nomura and K Hikichi, *Polym. J.*, 1985, **17**, 909.
7. CM Buchanan, KJ Edgar, JA Hyatt and AK Wilson, *Macromolecules*, 1991, **24**, 3060.
8. K Nishizawa, *Cellulase: Their Enzymology and Application*. Chem. Monograph Series 8, Nankodo Co. Ltd, Kyoto, Japan, 1974.
9. S Murai, M Arai and R Sakamoto, *Cellulase*. Kodansha Ltd., Tokyo, 1987.
10. ET Reese, *Ind. Eng. Chem.*, 1957, **49**, 89.

11. RGH Sui, RT Darby, PR Burkholder and ES Barghoorn, *Text. Res. J.*, 1949, **19**, 484.
12. MG Wirick, *J. Polym. Sci.*, 1968, **6**(A-1), 1965.
13. OW Klop and P Kooiman, *Biochim. Biophys. Acta*, 1965, **99**, 102.
14. PJ Gareggand and M Han, *Svensk. Papperstidning*, 1969, **21**, 6951.
15. SS Bhattacharjee and AS Perlin, *J. Polym. Sci.*, C, 1971, **36**, 509.
16. RA Gelman, *J. Appl. Polym. Sci.*, 1982, **27**, 2957.
17. H Iijima and K Kamide, unpublished results, see, for example, Ref. 1.

2.7   SODIUM CS AS BLOOD ANTICOAGULANT

2.7.1   Molecular parameters governing blood coagulation[1]

The chemical structure of CS resembles those of mucopolysaccharides, such as heparin and condroitin sulfate, which are now in wide use as naturally occurring blood anticoagulants. Figure 2.7.1 shows (a) three possible positions of the sulfate group in the glucopyranose unit of sodium CS (NaCS) and (b) the chemical structure of the uronic linkage in heparin. NaCS is linked together by 1,4-α-glucoside, but heparin by 1,4-α-glucoside. An anticoagulant activity of NaCS was first reported by Bergström[2] as early as 1935. Thereafter, the Biological Institute of Carlsberg Foundation in Copenhagen demonstrated that the coagulation time of whole blood increased by the addition of cellulose trisulfate and other polysaccharide sulfates such as amylose sulfate and amylopectin sulfate.[3–7] Felling and Wiley, and Rothschild and Castania found many interesting pharmaceutical characteristics of NaCS including (1) inhibitory action to pancreatic ribonuclease,[8] (2) kininogen depleting action,[9] and (3) endotoxin shock of dogs by the treatment of NaCS.[10] The aminosulfate group at the C_2 position of uronic units in heparin was found to play an important role in anticoagulant activity.[11] Desulfation of this aminosulfate group was found to lower the anticoagulant activity of heparin. These experimental results on heparin strongly suggest that the physiological activity, including anticoagulant activity, of polymers is closely related to their

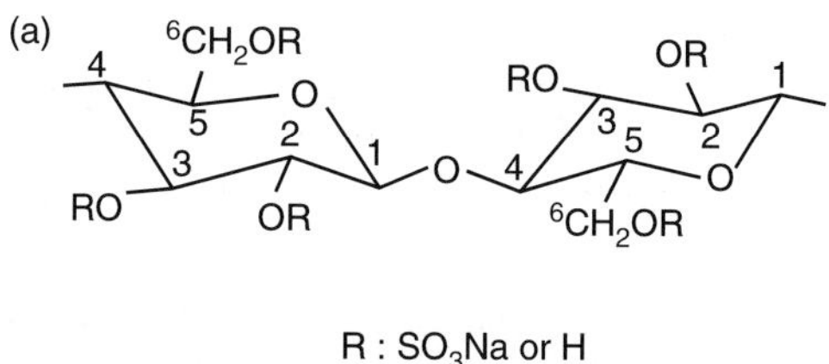

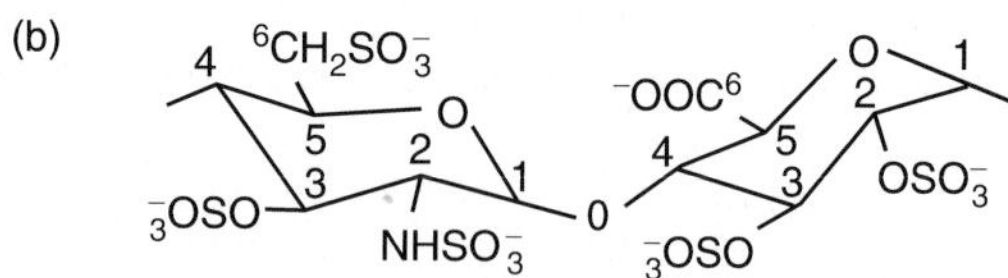

Figure 2.7.1   Chemical structure of sodium CS (a) and heparin (b).[1] Here, the C_1 chair conformation is assumed for both compounds.

molecular characteristics. Thus, for NaCS we may expect that its anticoagulant activity and other pharmaceutical activities are influenced by molecular parameters such as molecular weight, chain structure, distribution of substituent groups along the molecular chain, the probability of substitution at C_2, C_3, and C_6 positions of glucopyranose units ($\langle\!\langle f_k \rangle\!\rangle$, $k = 2$, 3, and 6), and the total DS $\langle\!\langle F \rangle\!\rangle$. Unfortunately, knowledge of the molecular characteristics of NaCS was as yet in a very primitive state because of its less practical importance and the experimental difficulty associated with its polyelectrolytic nature.

Kamide *et al.* succeeded, by NMR, in determining the distribution of sulfate groups in glucopyranose units of NaCS[12] and also evaluated, by viscometry, membrane osmometry and light scattering, the dilute solution properties of this polymer.[13] This section shows a correlation of molecular characteristics of NaCS with its anticoagulant activity and acute toxicity.

Synthesis of sodium cellulose sulfate (NaCS)

Twenty-five samples of sodium CS (NaCS), having the number average molecular weight $M_n = 800-36.8 \times 10^4$ and total degree of substituent $\langle\!\langle F \rangle\!\rangle = 0.50-2.75$, were synthesized. A commercial sample of heparin sodium salt, which had an anticoagulant activity χ of 1.52 IU mg^{-1} as measured by Japanese Pharmacopoeia,[14] was obtained from Nakarai Chemicals Co. Ltd (Kyoto). A purified heparin with χ of 188–228 was prepared by treating this commercial heparin with ethanol.

Evaluation of the anticoagulant activity of NaCS

Method of Lee–White. Anticoagulant activity (A_{LW}) was evaluated by a modification of the method of Lee–White[15] in a vessel thermostated at 37 °C.

The modification was made as follows. The inner wall of a polyethylene injector (inner diameter 1.0 cm, volume 2.5 ml) was first washed with physiological saline. 1.0 ml of physiological saline dissolving an anticoagulant at a concentration of 0.1 wt% was sucked into the injector and the anticoagulant solution was pressed out by hand. It was expected that a trace of anticoagulant would remain attached to the inner wall of the injector. We admitted 0.1 ml of fresh human whole blood into the injector and 0.1 ml of air was sucked into the injector. During this operation, the injector was maintained in an upright position. We measured the time, t_1, necessary for the test blood to start forming small coagulated blood gel particles by observing the blood flow pattern on the injector wall while the injector was gently being shaken. We also measured the time t_2 required for completion of blood coagulation (i.e. clotting or thrombi), as is usually done in measuring anticoagulant activity. These two times, t_1 and t_2, were found to be closely correlated to each other. After the completely coagulated blood gel was removed by placing the injector upside down, the amount of blood, m_1, remaining on the wall was roughly estimated. By comparing the results with those of heparin, the anticoagulant activity A_{LW} of NaCS estimated by the Lee–White method was classified into four grades: superior (S), good (G), inferior (I), and nonactive (N).

Method of Imai.[16] The amount of formaldehyde fixed thromb was determined at 37 °C. A watch glass of 11 cm in diameter was first washed with physiological saline, wiped with pure cotton gauze, and then 0.02 ml of physiological saline containing 1.0 wt% anticoagulant was dropped onto the center of the watch glass and spread by a swirling action. 0.30 ml of the fresh human whole blood was placed on this glass, made to stand for a given time, fixed with 37% formaldehyde, washed with water, dried in air, and the weight of thromb formed, m_2, was measured.

Method according to the Commentary of Japanese Pharmacopoeia. The method described in the Commentary of Japanese Pharmacopoeia[14] provides another parameter of anticoagulant activity, χ, which is expressed in international units (IU) mg^{-1} of heparin. For the evaluation of χ for NaCS, the original procedure was slightly modified as follows:

(1)   Prepare solutions of different concentrations by dissolving a heparin standard with χ of 152 IU mg^{-1} (hereafter denoted as χ_0) or NaCS in physiological saline.
(2)   Add 0.02 ml of these solutions to 1.0 ml of the fresh human whole blood (B$^-_?$ male adult) and measure the coagulation time (t_3) in seconds. Measurements are made at different concentrations ($0-75 \times 10^{-3}$ mg ml^{-1}) of the heparin standard and the polymer sample.
(3)   For each solution, plot log t_3 against the absolute amount m_3 (in mg) of the anti-coagulant used.
(4)   Determine the slopes of the plots for both the standard and the polymer sample.
(5)   Calculate χ (IU mg^{-1}) of the polymer sample from

$$\chi = \chi_0 \frac{(d \log t_3/dm_3)}{(d \log t_3/dm_3)_0^{-1}} \tag{2.7.1}$$

where the subscript 0 denotes the heparin standard. The plot in Step 4 was found to give a straight line.

Physiological action of NaCS to various coagulation factors in blood

All assays of the physiological action of NaCS to coagulation factors were made at 37 °C with heparins as the reference. We used specially treated plasma, in which one of the coagulation factors was deficient (hereafter referred to as a coagulation factor deficient substrate plasma), and evaluated the percentage of the coagulation factors existing in a given sample plasma. For this purpose, the conventional procedure was modified in advance by adding an anticoagulant to the sample plasma. The procedure scheme is given in Figure 2.7.2.
   The details are as follows:
   (a) Determination of the existence ratios Y of internal coagulation factors II, V, VII, and X.

(1)   Put 0.01 ml of physiological saline solution containing 2×10^{-4} mg of heparin or 1.6×10^{-4} mg of NaCS in a test tube having an inner diameter of 0.8 cm and maintained at 37 °C.

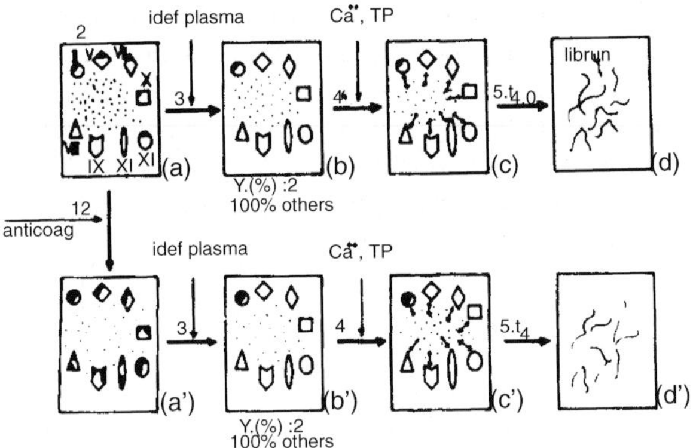

Figure 2.7.2   Scheme of the principle for measuring the existence ratio Y of coagulation factor II[1]: Numbers on the figure denote the steps described in the text. (a) Sample plasma from normal human whole blood. Shaded area indicates $(100 - Y_0)$ (%). (b) Sample plasma containing a coagulation Factor II deficient substrate plasma. Y of all factors, other than that of Factor II, gives 100% when added. (c) Fibrinogen in (b) was activated by addition of activated thromboplastin. (d) Coagulated plasma in which fibrin was formed. (a') Sample plasma containing anticoagulant NaCS. Shaded area indicates $(100 - Y_0)$ % and black area $(Y_0 - Y)$ %. (b')–(d') equals (b)–(d); dots denote fibrinogen.

(2)   Dilute one volume of the sample plasma with nine volumes of an olenveronal buffer solution, supplied by Diagnostic Inc., USA, and put 0.1 ml of this solution in the above mentioned tube and leave the system at 37 °C.

(3)   Prepare four kinds of coagulation factor deficient plasmas (deficient in II, V, VII, and X, respectively; Diagnostic Inc., USA). Add 0.1 ml of the plasma to 0.1 ml of the solution in Step 1 and allow the tube to stand for 2 min.

(4)   Blow 0.1 ml of an activated thromboplastin/calcium chloride mixture (Diagnostic Inc., USA) into the tube in Step 3.

(5)   Measure the time (t_4) necessary for fibrin to start growing.

(6)   Calculate the existence ratios Y (%) of individual coagulation factors using the following equations, derived from the experimental relationships supplied in the form of graphs by Diagnostic Inc., USA.[17]

$$\log t_4 = -0.19 \log Y + 1.475 \text{ for factor II} \tag{2.7.2}$$

$$\log t_4 = -0.30 \log Y + 1.879 \text{ for factor V} \tag{2.7.3}$$

$$\log t_4 = -1.445 \log Y + 1.742 \text{ for factor VII} \tag{2.7.4}$$

$$\log t_4 = -0.2824 \log Y + 1.799 \text{ for factor X} \tag{2.7.5}$$

(b) Determination of the existence ratios Y of external coagulation factors VIII, IX, XI, and XII.

For this purpose, we slightly modified the steps for the determination of internal coagulation factors as follows. In Step 2, one volume of the sample plasma was diluted with four volumes of the Olenveronal buffer, and in Step 4, a cepharo-plastincalcium chloride mixture was used in place of the thromboplastin/calcium chloride mixture. The values of Y of the external coagulation factors were calculated from the experimental relationships:

$$\log t_4 = -0.1709 \log Y + 2.0325 \text{ for factor VIII} \qquad (2.7.6)$$

$$\log t_4 = -0.1776 \log Y + 2.0363 \text{ for factor IX} \qquad (2.7.7)$$

$$\log t_4 = -0.250 \log Y + 2.1236 \text{ for factor XI} \qquad (2.7.8)$$

$$\log t_4 = -0.2189 \log Y + 2.22147 \text{ for factor XII} \qquad (2.7.9)$$

In the procedures described in (a) and (b) above, some loss may occur in the activity of the coagulation factors when the sample plasma and the anticoagulant are mixed. However, it is assumed that when coagulation factor deficient substrate plasma is added to the mixture of normal plasma and an anticoagulant, the activities of all coagulation factors other than the deficient coagulation factor recovered to 100%. The fundamental concept underlying the method for determining Y is illustrated in Figure 2.7.2. Since the values of Y for various coagulation factors in the sample plasma were determined after adding the anticoagulant, the ability of an anticoagulant to inactivate the coagulation factor under consideration can be defined by $100(Y_0 - Y)/Y_0$, where Y_0 is the value of Y when no anticoagulant is added. It should be noted that Y was obtained only at one dose level (2×10^{-4} mg). Linear relationships between the logarithmic coagulation time t_4 (expressed in seconds) and the absolute amount m_4 of anticoagulant were confirmed at least within the concentration range of the anticoagulant studied. Thus, the slope of the plot of $\log t_4$ versus m_4 could be taken as another measure of the inhibitory action of the anticoagulant toward the coagulation factor.

LD_{50} when anticoagulant is injected into the vein of a rat

In this experiment, 4-week-old male rats weighing 21–23 g were used. The rats were confined to shaving beds in a plastic cage and allowed to eat and drink at will. The cage was cleaned every 3–4 days. An NaCS solution (concentration 0.25–0.8 wt%) in physiological saline was injected into a vein at a rate of 0.1 ml per 10 s, and general symptoms and changes in body weight were observed during the following 3 weeks, using between 10 and 20 rats for a given dose level. LD_{50} was determined from the death ratio 1 week after the injection, according to the Canola-SX-50 (Probit) method.[18] During the test both dead and live rats were subjected to anatomy and intestinal abnormalities were visually examined.

Figure 2.7.3 shows some typical ^{1}H NMR spectra of NaCS in deuterium oxide. Figure 2.7.3(a) and (b) refers to two NaCS samples having nearly the same $\langle\!\langle F \rangle\!\rangle$ (1.96–1.97), but different $\langle\!\langle f_k \rangle\!\rangle$. $\langle\!\langle F \rangle\!\rangle$ decreased in the order: a $\cong$ b > c > d. The NMR signal for the proton at the C_1 position shifts slightly to a higher magnetic field as $\langle\!\langle F \rangle\!\rangle$ decreases. The structure of the spectrum in the range from 3.4 to 4.0 ppm becomes complicated with a decrease in $\langle\!\langle F \rangle\!\rangle$, approaching that for pure cellulose.

Table 2.7.1 lists the values of M_n, $\langle\!\langle F \rangle\!\rangle$, and the anticoagulant parameters A_{LW}, m_2, and χ, and LD_{50} of 25 NaCS samples and heparin. For the selected 13 NaCS samples,

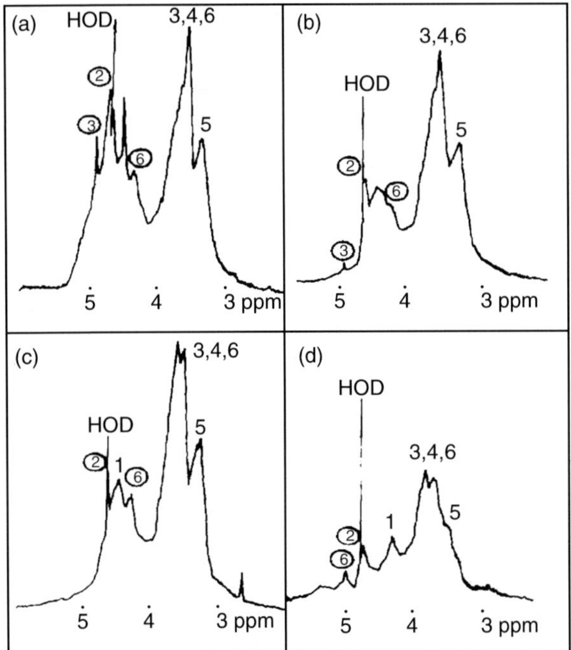

Figure 2.7.3   Typical ^{1}H NMR spectra of NaCS samples in deuterium oxide: (a) CS-A; (b) CS-11; (c) HBSD; (d) HBH.[1] Numbers indicate the position of carbon atoms (see Figure 2.7.1) and those in the circles denote the position of carbon atoms linking the hydroxyl residues.

the distribution of sulfate groups over the three possible positions in a glucopyranose unit ($\langle\!\langle f_k \rangle\!\rangle$ ($k = 2$, 3, and 6)) is also given in this table. It can be seen that $\langle\!\langle F \rangle\!\rangle_{\mathrm{NMR}}$ ($\langle\!\langle F \rangle\!\rangle$ by the NMR method) agrees with $\langle\!\langle F \rangle\!\rangle_{\mathrm{g}}$ ($\langle\!\langle F \rangle\!\rangle$ by the chemical method) within an uncertainty of ± 0.10. The correlation coefficient between these was estimated to be 0.998. In a previous study,[13] it was found that the ratio of weight-to-number average molecular weight $M_{\mathrm{w}}/M_{\mathrm{n}}$ of CS samples ranged from 3.1 to 4.1. The procedures employed here for synthesis of NaCS afforded samples with a great variety of M_{n} and $\langle\!\langle F \rangle\!\rangle$. Thus, the M_{n} of these samples varied from 800 to 36.8×10^4 and their $\langle\!\langle F \rangle\!\rangle$ ranged from 0.5 to 2.75.

Figure 2.7.4 shows the relation between A_{LW} and m_2 or χ. A_{LW} of the S grade corresponds to m_2 of 0.4–2.8 and χ of 188–246 and A_{LW} of the grade G to m_2 of 4.8–5.3 and χ of 152–206. A_{LW} of the I grade ranges in m_2 from 12.6 to 26.3, and in χ from 58 to 124 and A_{LW} of the N grade corresponds to m_2 larger than 42 and χ lower than 15. Some overlaps and gaps in m_2 and χ for different grades exist because of the qualitative nature of A_{LW} and also because of the insufficient number of polymer samples. It is of interest to see that NaCS has a wide range of anticoagulant activity.

Figure 2.7.5 shows the relation between A_{LW} and $\langle\!\langle F \rangle\!\rangle$ of NaCS. Qualitatively, A_{LW} has a tendency to approach the G or S grade with an increase in $\langle\!\langle F \rangle\!\rangle$. In particular, for samples with $\langle\!\langle F \rangle\!\rangle$ less than 1.94, A_{LW} was of neither G nor S grade, and all HB samples that had $\langle\!\langle F \rangle\!\rangle$ less than unity, showed very low anticoagulant activity.

Table 2.7.1

Characterization of sodium CS[1]

Sample	M_n ($\times 10^{-4}$)	Total $\langle\!\langle F \rangle\!\rangle$		Distribution of sulfate			Anticoagulant activity			LD_{50} (mg kg^{-1}-rat)
		Gravimetry	NMR	$\langle\!\langle f_2 \rangle\!\rangle$	$\langle\!\langle f_3 \rangle\!\rangle$	$\langle\!\langle f_6 \rangle\!\rangle$	A_{WL}	m_2 (mg)	χ (IU mg^{-1})	
CS-1	36.8	1.66	–	–	–	–	I	26.3	–	–
CS-2	9.39	1.94	–	–	–	–	G	–	–	–
DSH	6.91	1.05	1.05	0.55	0.00	0.50	N	–	0	730
CS-3	6.79	1.98	–	–	–	–	G	–	–	–
CS-4	6.65	2.39	2.46	1.00	0.74	0.72	G	4.8	184–228	536
CS-5	5.32	1.36	1.36	0.75	0.42	0.19	I	–	55–61	513
CS-6	5.31	1.84	–	–	–	–	I	12.6	–	–
CS-7	5.06	1.84	–	–	–	–	I	–	–	–
CS-8	5.06	1.73	–	–	–	–	I	–	–	–
CS-9	4.98	1.81	–	–	–	–	I	–	–	–
CS-10	4.32	2.75	2.75	1.00	0.92	0.83	S	0.8	230–242	25.2
CS-A	3.15	1.97	1.97	0.71	0.26	1.00	N	42.0	7–8	5000
CS-11	2.48	1.98	1.96	1.00	0.61	0.34	S	2.8	175–219	43
HB-1	2.26	0.95	0.95	0.52	0.00	0.43	N	46.3	0	4634–5326
CS-12	1.95	2.03	1.93	1.00	0.65	0.28	S	1.2	200–222	63.8
CS-13	1.80	1.74	–	–	–	–	I	–	–	–
CS-14	1.70	2.60	2.62	1.00	0.87	0.85	S	0.4	242–251	38.3
CSD	1.69	2.05	1.99	0.75	0.55	0.67	I	–	117–130	726
CS-15	1.57	1.96	–	–	–	–	S	0.6	–	–
CS-16	1.57	1.94	1.97	1.00	0.60	0.37	S	0.6	167–209	184.5
HB-2	1.24	0.70	–	–	–	–	I	18.5	–	–
HB-3	1.16	0.74	–	–	–	–	I	13.7	–	–
HBSD	0.99	0.95	0.95	0.29	0.33	0.33	N	–	10–11	1004
HB-4	0.92	0.76	–	–	–	–	I	12.7	–	–
HBH	0.08	0.50	0.58	0.31	0.27	0.00	N	–	12–15	712
Heparin	–	–	–	–	–	–	G	5.3	152	1200

I denotes inferior; G denotes good equivalent; N denotes nonactive; S denotes superior.

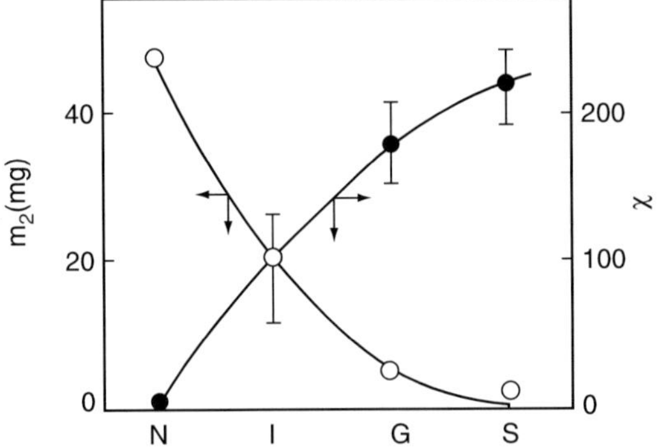

Figure 2.7.4   Correlation between anticoagulant activity A_{LW}, measured by the Lee–White method, and weight of thromb m_2 determined by the Imai method, or anticoagulant activity χ (IU mg^{-1}) measured according to the Commentary of Japanese Phamacopoeia.[1]

$\langle\!\langle F \rangle\!\rangle$ is not the only factor controlling A_{LW}, because when $\langle\!\langle F \rangle\!\rangle$ is in the range of 1.88–2.05, A_{LW} covers all grades varying from the N to the S. All CS samples with $\langle\!\langle F \rangle\!\rangle$ larger than 1.88 showed A_{LW} of the S or G grade but sample CS-A, which had $\langle\!\langle F \rangle\!\rangle$ of 1.97, showed A_{LW} of the N grade. It should be noted that there is a great difference in $\langle\!\langle f_2 \rangle\!\rangle + \langle\!\langle f_3 \rangle\!\rangle$ between the CS samples (> 1.6) and sample CS-A (0.97). Sample CSD had almost the same $\langle\!\langle F \rangle\!\rangle$ ($= 2.00$) as sample CS-11. The most significant difference between these two polymers is that CSD had a lower $\langle\!\langle f_2 \rangle\!\rangle + \langle\!\langle f_3 \rangle\!\rangle$ (or $\langle\!\langle f_2 \rangle\!\rangle$ and $\langle\!\langle f_3 \rangle\!\rangle$) and its A_{LW} was one grade below that of CS-11. Attempts to increase $\langle\!\langle F \rangle\!\rangle$ by resulfation of sample HB-1 with the DMF/SO$_3$ complex failed, and no improvement of A_{LW} was obtained in spite of the resulting marked changes in $\langle\!\langle f_k \rangle\!\rangle$ (e.g. $\langle\!\langle f_3 \rangle\!\rangle$ changed from 0 to 0.34). These facts suggest that A_{LW} is related not only to $\langle\!\langle F \rangle\!\rangle$ but also to the average distribution of the substituent groups in glucopyranose units ($\langle\!\langle f_k \rangle\!\rangle$, $k = 2$, 3, and 6).

Figure 2.7.6 shows A_{LW} of NaCS as a function of $\langle\!\langle f_2 \rangle\!\rangle$, $\langle\!\langle f_3 \rangle\!\rangle$, $\langle\!\langle f_6 \rangle\!\rangle$, and $\langle\!\langle f_2 \rangle\!\rangle + \langle\!\langle f_3 \rangle\!\rangle$. $\langle\!\langle f_2 \rangle\!\rangle$, $\langle\!\langle f_3 \rangle\!\rangle$, and that $\langle\!\langle f_2 \rangle\!\rangle + \langle\!\langle f_3 \rangle\!\rangle$ show good correlation with A_{LW} whereas $\langle\!\langle f_6 \rangle\!\rangle$ does not. All NaCS samples with $\langle\!\langle f_2 \rangle\!\rangle < 0.75$ or $\langle\!\langle f_3 \rangle\!\rangle < 0.48$ have A_{LW} of the N grade (in this case, $\langle\!\langle f_2 \rangle\!\rangle$ and $\langle\!\langle f_3 \rangle\!\rangle$ are closely correlated with one another). $\langle\!\langle f_2 \rangle\!\rangle + \langle\!\langle f_3 \rangle\!\rangle$ has the best correlation with A_{LW}.

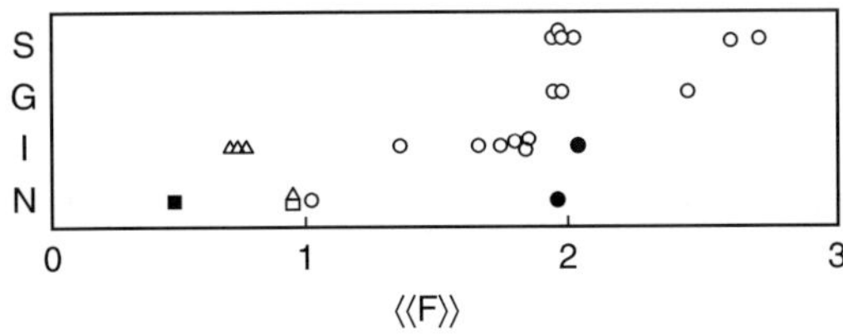

Figure 2.7.5   Effect of total degree of substitution $\langle\!\langle F \rangle\!\rangle$ of NaCS on anticoagulant activity A_{LW}:[1] (○), CS series; united half circle, CSD; half black circle, DSH; (△), HB series; (●), CS-A; (□), HBSD; (■), HBH.

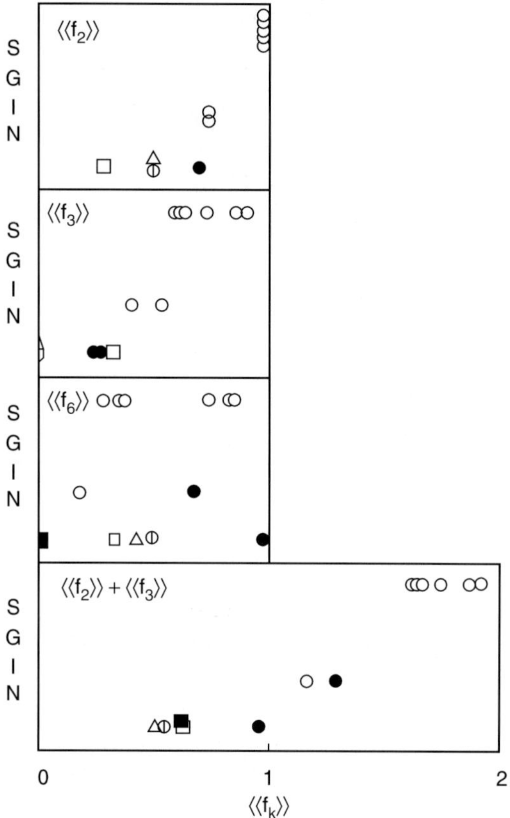

Figure 2.7.6 Effect of $\langle\!\langle f_2\rangle\!\rangle$, $\langle\!\langle f_3\rangle\!\rangle$, $\langle\!\langle f_6\rangle\!\rangle$ and $\langle\!\langle f_2\rangle\!\rangle + \langle\!\langle f_3\rangle\!\rangle$ of NaCS on anticoagulant activity A_{LW}.[1] Symbols have the same meaning as those in Figure 2.7.5.

Table 2.7.1 suggests that the average molecular weight M_{n} has a significant effect on A_{LW}. As far as CS samples are concerned, A_{LW} tends to increase with a decrease in M_{n}.

Figure 2.7.7 shows the relation between χ and the distribution of the sulfate groups, represented by $(\langle\!\langle f_2\rangle\!\rangle + \langle\!\langle f_3\rangle\!\rangle)/3$, $\langle\!\langle f_6\rangle\!\rangle/3$, and $(3 - \langle\!\langle F\rangle\!\rangle)/3$. The theoretical maxima of $(\langle\!\langle f_2\rangle\!\rangle + \langle\!\langle f_3\rangle\!\rangle)/3$, $\langle\!\langle f_6\rangle\!\rangle/3$, and $\langle\!\langle F\rangle\!\rangle$ are 0.67, 0.33, and 3, respectively. The full lines in the figure indicate the contours of the same χ. χ is predominantly governed by $(\langle\!\langle f_2\rangle\!\rangle + \langle\!\langle f_3\rangle\!\rangle)/3$, being larger for larger $(\langle\!\langle f_2\rangle\!\rangle + \langle\!\langle f_3\rangle\!\rangle)/3$ at constant $\langle\!\langle f_6\rangle\!\rangle/3$ or $\langle\!\langle F\rangle\!\rangle/3$. If compared at a fixed $(\langle\!\langle f_2\rangle\!\rangle + \langle\!\langle f_3\rangle\!\rangle)/3$, then χ is larger for smaller $\langle\!\langle f_6\rangle\!\rangle/3$ and $\langle\!\langle F\rangle\!\rangle/3$. However, $\langle\!\langle f_6\rangle\!\rangle/3$ and $\langle\!\langle F\rangle\!\rangle/3$ are not very sensitive to χ. Figure 2.7.8 indicates that χ corresponding to the maximum value of $(\langle\!\langle f_2\rangle\!\rangle + \langle\!\langle f_3\rangle\!\rangle)(= 2)$ is 285.

These results seem to be the first demonstration of a close correlation between the physiological activity and the chemical structure of macromolecules. To obtain a polymer drug with desirable physiological activity, it is important to make a precise design of the molecular structure. For many years, it has been known that the anticoagulant activity

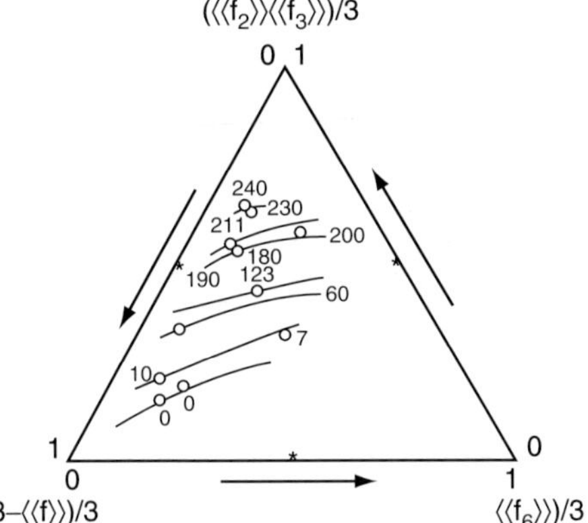

Figure 2.7.7   Dependence of anticoagulant activity χ of NaCS on $(\langle\!\langle f_2\rangle\!\rangle + \langle\!\langle f_3\rangle\!\rangle)/3$, $\langle\!\langle f_6\rangle\!\rangle/3$ and $(3 - \langle\!\langle F\rangle\!\rangle)/3$:[1] $\langle\!\langle f_k\rangle\!\rangle$ is the distribution of sulfate groups on the carbon position k in glucopyranose units. χ is given on each curve. Open marks, experimental data.

of naturally occurring heparin varies significantly depending on its source and preparation method.

The cause of this phenomenon can be understood if the distribution of chemical structure in heparin is fully elucidated, as in the case of NaCS.

Figures 2.7.9 and 2.7.10 illustrate χ and LD_{50} as functions of $\langle\!\langle f_2\rangle\!\rangle + \langle\!\langle f_3\rangle\!\rangle$ and M_n. The full lines represent the contours of the same χ (Figure 2.7.8) and LD_{50} (Figure 2.7.9). Larger $\langle\!\langle f_2\rangle\!\rangle + \langle\!\langle f_3\rangle\!\rangle$ gives larger χ when compared at the same M_n. For fixed $\langle\!\langle f_2\rangle\!\rangle + \langle\!\langle f_3\rangle\!\rangle$, χ tends to increase with a decrease in M_n. From the anticoagulant activity perspective, it is highly desirable to prepare NaCS with higher $\langle\!\langle f_2\rangle\!\rangle + \langle\!\langle f_3\rangle\!\rangle$ and lower M_n.

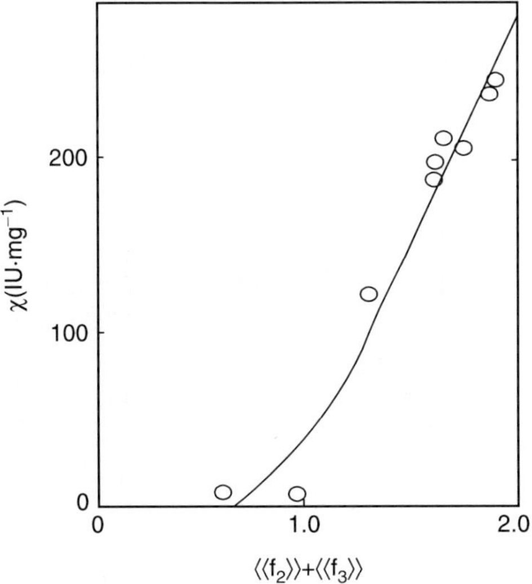

Figure 2.7.8   Effect of the sum of the substitution degree at carbon position 2 and 3, $\langle\!\langle f_2\rangle\!\rangle + \langle\!\langle f_3\rangle\!\rangle$, on anticoagulant activity χ.[1]

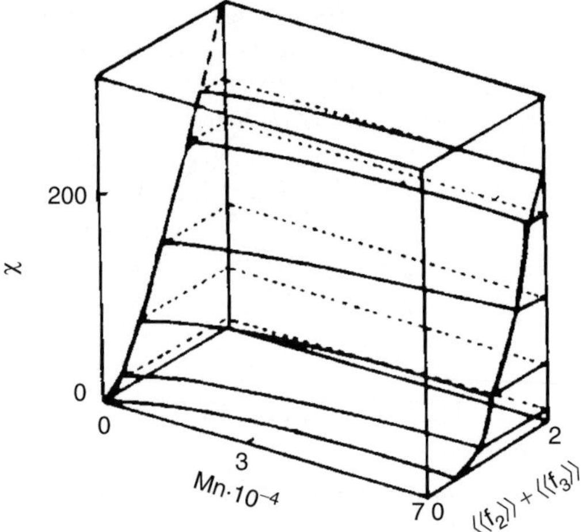

Figure 2.7.9   Anticoagulant activity χ plotted against number average molecular weight M_n and the sum of the substitution degree at carbon positions 2 and 3, $\langle\!\langle f_2 \rangle\!\rangle + \langle\!\langle f_3 \rangle\!\rangle$.[1]

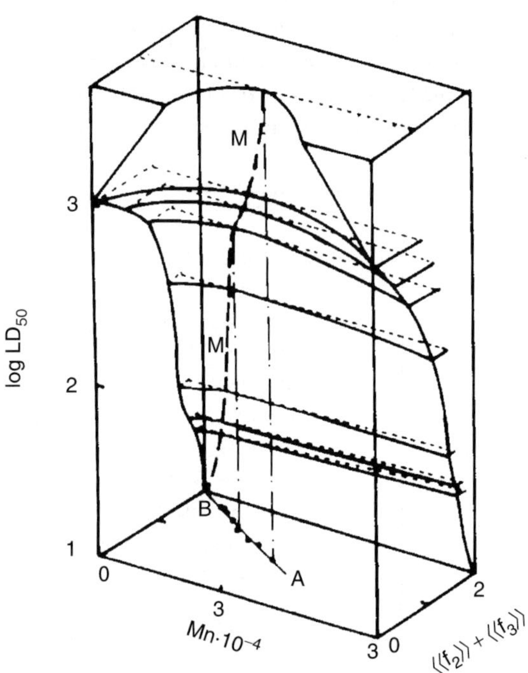

Figure 2.7.10   Plots of acute toxicity LS_{50} of NaCS against M_n and $\langle\!\langle f_2 \rangle\!\rangle + \langle\!\langle f_3 \rangle\!\rangle$.[1] The line M represents maximum values of $\langle\!\langle f_2 \rangle\!\rangle + \langle\!\langle f_3 \rangle\!\rangle$ for a given LD_{50} (hereafter denoted by $(\langle\!\langle f_2 \rangle\!\rangle + \langle\!\langle f_3 \rangle\!\rangle)_m$. M_n on the line M is denoted by $(M_n)_m$. AB denotes the position of the line M on the plane $LD_{50} = 0$.

This suggests that the mobility of NaCS in solution is a minor factor in its anti-coagulant activity. In Figure 2.7.10, the contours of the same LD_{50} are convex against the M_n axis, and the broken line M_n, connecting the point for maximum $\langle\!\langle f_2 \rangle\!\rangle + \langle\!\langle f_3 \rangle\!\rangle$ at a given LD_{50}, shifts to lower M_n, with a decrease in LD_{50}, as can be seen from the projected curve AB on the plane of $LD_{50} = 0$. $\langle\!\langle f_2 \rangle\!\rangle + \langle\!\langle f_3 \rangle\!\rangle$ M_n at a point on the line M_n are denoted hereafter by $(\langle\!\langle f_2 \rangle\!\rangle + \langle\!\langle f_3 \rangle\!\rangle)_m$ and $(M_n)_m$, respectively. When compared at the same M_n, LD_{50} decreases as $\langle\!\langle f_2 \rangle\!\rangle + \langle\!\langle f_3 \rangle\!\rangle$ increases. This condition causes trouble in obtaining NaCS with a high anticoagulant activity and a desirable LD_{50}. The line M_n represents maximum $\langle\!\langle f_2 \rangle\!\rangle + \langle\!\langle f_3 \rangle\!\rangle$ for an NaCS sample, hence the maximum anticoagulant activity for a given LD_{50}. This line gives $\langle\!\langle f_2 \rangle\!\rangle + \langle\!\langle f_3 \rangle\!\rangle = 2$ at $M_n = 0$ (i.e. the fully substituted glucose sulfate).

Figures 2.7.11 and 2.7.12 show the effects of $(\langle\!\langle f_2 \rangle\!\rangle + \langle\!\langle f_3 \rangle\!\rangle)_m$ and $(M_n)_m$ on χ and LD_{50}. From these figures, χ and LD_{50} at $\langle\!\langle f_2 \rangle\!\rangle + \langle\!\langle f_3 \rangle\!\rangle = 2$ and at $M_n = 0$ are estimated to be 305 and 8, respectively. The former is in very good agreement with the value 285 estimated from Figure 2.7.8. At $\langle\!\langle f_2 \rangle\!\rangle + \langle\!\langle f_3 \rangle\!\rangle = 0$ and $M_n = 3.7 \times 10^4$, LD_{50} is expected to be 10,000, implying that such a NaCS sample is practically nontoxic but its A_{LW} is of the N grade. No NaCS sample having an anticoagulant activity and acute toxicity equivalent to heparin ($\chi = 152$ and $LD_{50} = 1200$) can be anticipated, since these figures indicate that NaCS has an acute toxicity higher than heparin, which exists in the artery wall of animals and human beings. The next to the best molecular and structural parameters of NaCS as anticoagulants can be estimated from Figures 2.7.11 and 2.7.12.

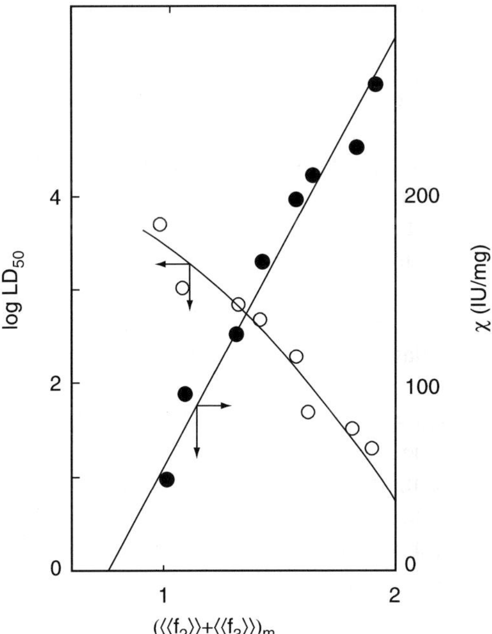

Figure 2.7.11   Relation between $(\langle\!\langle f_2 \rangle\!\rangle + \langle\!\langle f_3 \rangle\!\rangle)_m$ and acute toxicity LD_{50} or anticoagulant activity χ.[1]

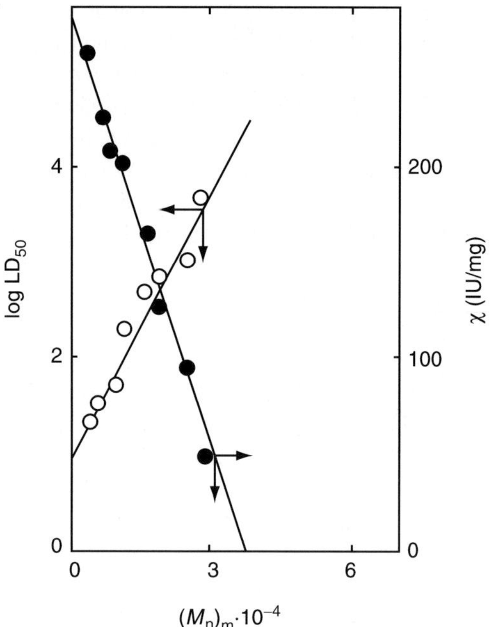

Figure 2.7.12   Relation between $(M_n)_m$ and acute toxicity LD_{50} or anticoagulant activity χ.[1]

Given LD_{50} at the level of heparin, NaCS with $M_n = 1.65 \times 10^4$ and $\langle\!\langle f_2 \rangle\!\rangle + \langle\!\langle f_3 \rangle\!\rangle = 1.23$ (so that χ is 120) is acceptable. If χ is desired to be 152, $M_n = 2.4 \times 10^4$ and $\langle\!\langle f_2 \rangle\!\rangle + \langle\!\langle f_3 \rangle\!\rangle = 1.40$ are relevant and give $LD_{50} = 320-160$. It should be noted here that the oral LD_{50} of NaCS for a rat was found to be larger than 15,000 mg kg^{-1}, and hence presents no toxic problem.

The above discussion is concerned entirely with whole blood. Since 12 blood-coagulation factors are known, it should be of interest to determine how (and to what extent) NaCS acts on these factors. Table 2.7.2 shows the existence ratio Y (%) of each coagulation factor and the anticoagulating ability $100(Y_0 - Y)/Y_0$ as well as χ of the anticoagulants.

Figure 2.7.13 shows $(Y_0 - Y)/Y_0$ for various coagulation factors plotted against χ. From this figure, the coagulation factors may be classified into two categories:

(1) $100(Y_0 - Y)/Y_0$ varies with χ: factors V, VIII, IX, XI, and XII. Factors VIII, XI, and XII increase in linear proportion to χ and factor IX increases rapidly with χ, approaching some asymptotic value. In contrast, factor V has a tendency to decrease gradually with increasing χ.
(2) $100(Y_0 - Y)/Y_0$ is almost independent of χ and decreases in the following order: factor VII > factor X > factor II.

We cannot explain why some factors are independent of χ and why others decrease with χ and, in particular, why factor VII vanished by addition of NaCS, even if the polymer has no anticoagulant activity ($\chi = 0$), as in the case of sample HB-1.

Table 2.7.2

Existence ratios of coagulation factors and degree of inhibitory action $100(Y_0 - Y)/Y_0$ of anticoagulants[1]

Anticoagu-lants	χ (IU mg^{-1})	Y (%)								$100(Y_0-Y)/Y_0$ (%)							
		II	V	VII	X	VIII	IX	XI	XII	II	V	VII	X	VIII	IX	XI	XII
Heparin[a]	188–228	41.5	102.5	2.4	42.2	90.6	0.0	86.6	3.5	57.5	24.2	97.6	64.8	30.1	100	27.7	99.1
CS-12	200–222	31.6	112.2	2.2	38.9	26.3	20.8	78.4	51.5	68.4	16.9	97.8	67.5	80.0	76.0	34.6	85.6
CSD	117–130	–	112.2	2.4	34.1	63.0	24.4	103.4	157.0	–	16.9	97.6	71.5	51.9	71.9	13.7	56.0
CS-5	51–61	30.1	94.7	2.4	47.3	115.5	78.3	119.8	483.9	69.2	29.8	97.6	60.5	11.9	9.8	8.0	35.6
HBH	12–15	–	85.1	2.3	47.3	115.5	62.8	–	–	–	37.0	97.7	60.5	11.9	27.7	–	–
HBSD	10–11	–	78.7	–	–	–	–	–	–	–	41.6	–	–	–	–	–	–
CS-A	7–8	–	107.1	2.4	–	120.8	68.8	111.3	–	–	20.6	97.6	–	7.9	20.8	7.1	–
HB-1	0	44.1	89.1	2.3	50.0	–	–	–	–	54.9	34.0	97.7	57.8	–	–	–	–
None[b]	–	97.7	134.2	100	119.8	131.1	86.8	119.8	356.7	–	–	–	–	–	–	–	–

[a]Purified heparin.
[b]Denotes Y_0 for each coagulation factor in sample plasma.

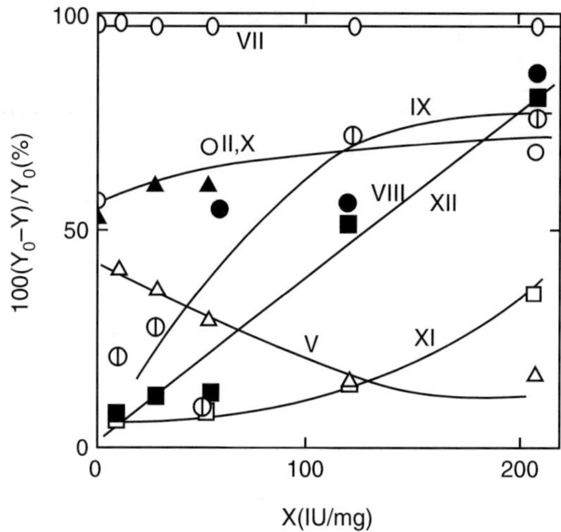

Figure 2.7.13 Correlation between $100(Y_0 - Y)/Y_0$ of various coagulation factors and anticoagulant activity of NaCS:[1] factor II, (O); V, (△); VII, (O); VIII, (■); IX, united half circle; X, (▲); XI, (●).

The inhibitory action of NaCS is compared with that of heparin in Figure 2.7.14, where $100(Y_0 = Y)/Y_0$ for the two polymers having almost the same χ value (190 ± 10) are plotted on the horizontal and vertical axes. Except for factor VIII, the action of NaCS on all coagulation factors is the same as that of heparin. NaCS inhibits the action of factor VIII (antithermophilic globulin) much more effectively than heparin. Therefore, it may be concluded that NaCS acts mainly on factor VIII and partly on factor IX (plasma thromboplastin component); the latter was found to be suppressed by heparin.[19]

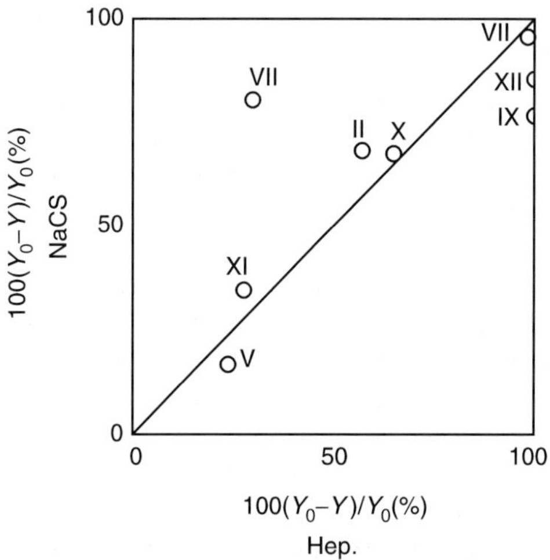

Figure 2.7.14 Comparison of $100(Y_0 - Y)/Y_0$ between NaCS(CS-12) and heparin:[2] roman numbers indicate the order of coagulation factors. Symbols are the same as those in Figure 2.7.13.

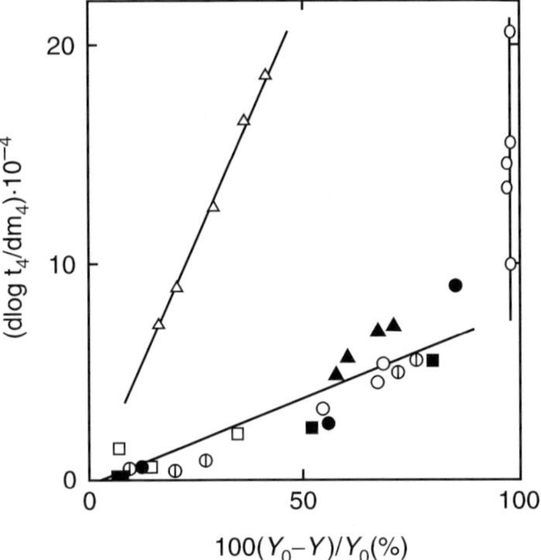

Figure 2.7.15 Plots of dlog t_4/dm_4 against $100(Y_0 - Y)/Y_0$ for various coagulation factor.[1] Symbols are the same as those in Figure 2.7.13.

Figure 2.7.15 illustrates the relation between d log t_4/dm_4 and $(Y_0 - Y)/Y_0$. Except for factor VII, a straight line passing the origin is obtained, although factor V is expected to depend more strongly on the concentration than the other factors.

Figures 2.7.16(a)–(d) illustrates time changes in the body weight of rats after saline solutions of NaCS samples with four different LD_{50} (43, 53.6, 63.8, and 184.5) were injected. In obtaining these data, we prepared polymer solutions of different concentrations and injected the same volume of each solution into a rat. The numbers on the lines denote the injected NaCS expressed in mg kg^{-1}. For all NaCS samples, when the dose amount was less than the LD_{50} of a given NaCS, there was progressive body weight increase. When the dose amount was more than the LD_{50}, body weight decreased at least initially, passed a minimum, and then increased monotonically. If the body weight loss ratio (B_{WL}) is defined as the maximum body weight loss M (g) (in this case, this is not molecular weight) divided by the dose amount m (mg kg^{-1}) and the LD_{50} of a given NaCS; that is, $B_{WL} = M/(m \times LD_{50})$, then B_{WL} for all samples comes close to 0.08. This indicates that body weight loss in rats is determined primarily by the dose amount and the LD_{50} of the NaCS sample used.

General symptoms displayed by rats after the injection of NaCS were occasional staggering motion and depression of breathing. Some rats lay down and often fell into a fit of convulsions. However, these symptoms disappeared between 1 and 4 days after the injection. The dead rats showed bleeding from the nose and mouth. The bleeding seemed greater for larger volumes of injected NaCS solution. In live rats, anemia was found.

Inspection of the intestinal organs of live and dead rats after injection of NaCS indicated that in the dead rats, the liver showed anemia and the lung showed congestion and bloodspots. These abnormal symptoms were conspicuous in rats to which larger amounts of NaCS had been administered. In the live rats, no anomalous symptoms were found.

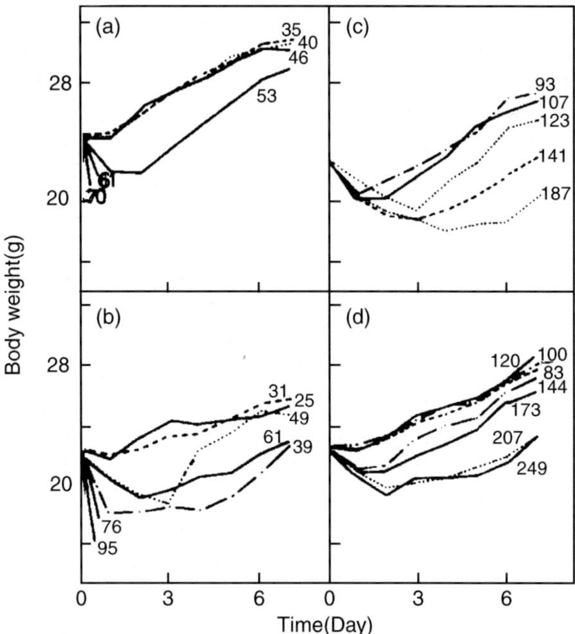

Figure 2.7.16   Change in body weight of rats with time following injection of NaCS solution:[1] (a) sample CS-11, (b) sample CS-4, (c) sample CS-12, and (d) sample CS-6. Numbers on the curves indicate the injected amounts of NaCS in mg kg^{-1}.

In this section, it is shown that the most important factor governing the anticoagulant activity of NaCS is not the average molecular weight but the average chemical structure instead.

2.7.2   Pharmacodynamic properties[20]

The pharmacodynamic properties of NaCS were first studied by Rothschild and his coworkers.[9,21,22] Their findings were as follows: (1) NaCS depletes bradykinin from blood plasmas of rat, guinea pig, and human; (2) administration of NaCS to rats results in a depression of blood pressure with tachyphylaxis, the reduction of kininogen in plasma, and moderate leucocytosis; (3) the blood pressure depressing activity of NaCS in the rats can be inhibited by pre-treatment with trypsin from soybean, but not with a histamine antagonist (mepyramine). In their papers, the method of synthesis (i.e. pyridine/ chlorosulfonic acid method) and these pharmacodynamic properties were described very briefly but the work was restricted to rats and no data on the molecular characteristics of NaCS were given. Therefore, it is of prime importance to confirm whether their findings are transferable to other animals, using samples of NaCS whose molecular characteristics are definitely known. An attempt was made by Kamide *et al.*[20] to evaluate certain pharmacodynamic properties of sodium CS (NaCS) using rats, rabbits, dogs, and cats. For this purpose, two NaCS samples, whose molecular characteristics, anticoagulant activity and acute toxicity were determined in eq. (2.7.1)[1] (sample code CS-4; $M_n = 6.65 \times 10^4$, $\langle\!\langle F \rangle\!\rangle = 2.46$, $\langle\!\langle f_2 \rangle\!\rangle = 1.00$, $\langle\!\langle f_3 \rangle\!\rangle = 0.74$, and $\langle\!\langle f_6 \rangle\!\rangle = 0.72$, $\chi = 188 - 228$ IU mg^{-1}, LD$_{50} = 53.6$ mg kg^{-1}, and CS-16; $M_n = 1.57 \times 10^4$, $\langle\!\langle F \rangle\!\rangle = 1.97$, $\langle\!\langle f_2 \rangle\!\rangle = 1.00$, $\langle\!\langle f_3 \rangle\!\rangle = 0.60$, $\langle\!\langle f_6 \rangle\!\rangle = 0.37$, $\chi = 167 - 209$, LD$_{50} = 184.5$) were used. A pyrogen test was made, and the changes in blood pressure, heart rate, respiration, and the effect of histamine antagonists pretreatment were examined. The following was confirmed: (1) An intravenous administration of 1 mg kg^{-1} of the two NaCS to dogs and cats had no effect on the cardiac function of these animals, although Rothschild[9] reported that injection of 0.3 mg kg^{-1} of NaCS into a rat depressed the average blood pressure by 20–68 mmHg. The minimum value of dosage level for inducing the blood pressure depression was different from that in Rothschild's experiments. This may be due to differences in animal species or in molecular characteristics and purity of NaCS samples. Although the results from animal tests do not apply to human beings, the NaCS samples used here may be useful as anticoagulant in blood dialysis for chronical renal insufficiency. Because 10,000 IU of heparin is usually used for this therapy, only about 1 mg kg^{-1} human (average body weight: 60 kg) of NaCS ($= 185-220$ IU mg^{-1}) is sufficient as an alternative for heparin. The dose value of 1 mg kg^{-1}, which was found to have no effect on the cardiac function of dogs, may have no influence on that of human beings. (2) At a dose level of 10 mg kg^{-1} of the two NaCS to dogs, an abrupt blood pressure depression amounting to 40–100 mmHg in the average blood pressure was observed, and found to be proportional to the reciprocal of the acute toxicity (LD$_{50}$). This finding indicated that the blood pressure depression by NaCS administration is dependent on the molecular weight and the distribution of the substituent ($\langle\!\langle f_2 \rangle\!\rangle + \langle\!\langle f_3 \rangle\!\rangle$) in NaCS. (3) NaCS administration to a cat induced an abrupt rise in blood pressure in contrast to the case of the dog. (4) NaCS administration was accompanied by tachyphylaxis; (5) The liberation of histamine and acetylcholine as the cause of these cardiac changes by NaCS administration was ruled out experimentally, and the liberation of bradykinin was considered the most probable cause.

REFERENCES

1. K Kamide, K Okajima, T Matsui, M Ohnishi and H Kobayashi, *Polym. J.*, 1983, **15**, 309.
2. S Bergström, *Naturwissenshaften*, 1935, **25**, 706.
3. T Astrup, I Galsmar and M Volkert, *Acta Physiol. Scand.*, 1944, **8**, 215.
4. P Karrer, H Köenig and E Usteri, *Helv. Chim. Acta*, 1943, **26**, 1296.
5. J Astrup and J Piper, *Acta Physiol. Scand.*, 1945, **9**, 351.
6. J Piper, *Acta Physiol. Scand.*, 1945, **9**, 28.
7. J Piper, Farmakologiske Undersögelser över Syntetiske Heparin–lignande Stoff, (Disp.) Copenhagen, 1945.
8. J Felling and CE Wiley, *Arch. Biochem. Biophys.*, 1959, **85**, 313.
9. AM Rothschild, *J. Pharmacol. Chemother.*, 1968, **33**, 501.
10. AM Rothschild and A Castania, *J. Pharm. Pharmacol.*, 1968, **20**, 77.
11. J Kiss, Heparin. In *Chemical Structure of Heparin*. (ed. K Thomas), Academic Press, London, 1976, p. 9.
12. K Kamide, K Okajima. *Polym. J.*, 1981, **13**, 163.
13. K Kishino, T Kawai, T Nose, M Saito and K Kamide, *Eur. Polym. J.*, 1981, **17**, 623.
14. Commentary of Japanese Pharmacopoeia, No. C1235–C1242, Nankodo, Tokyo, 1965.

15. I Kanai and M Kanai (eds), *Rinsho Kensaho Teiyo*, Chapter VI, Kinbara Publishing Company, Tokyo, 1975, p. 85.
16. Y Imai, Y Nose, *J. Biomed. Mater. Res.*, 1972, **6**, 165.
17. Technical Sheet, No. CO 15-4313, CO 16-4828E, CO 17-4379, CO 18-4378E, Diagnostic Incorporated, USA, 1978.
18. B Probit, *Ann. Appl. Biol.*, 1934, **22**, 134.
19. MA Lyapina, *Frigiologia Chelveka*, 1978, **4**, 295.
20. K Kamide, K Okajima, T Matsui and H Kobayashi, *Polym. J.*, 1984, **16**, 259.
21. AM Rothschild and LA Gascon, *Nature* (London), 1966, **212**, 1364.
22. AM Rothschild, International Symposium on Vaso-active Polypeptides: Bradykinin and Related Kinins, **Vol. 197**, Sao Paulo, 1967.

2.8   SODIUM SALT OF CARBOXYMETHYLCELLULOSE AS ABSORBENT TOWARD AQUEOUS LIQUID[1]

As early as 1918, Jansen synthesized CMC using the reaction of alkali cellulose with sodium monochloroacetate.[2] Since then, CMC with a total DS $\langle\!\langle F \rangle\!\rangle$ of 0.5–1.0 has been commercialized worldwide and found numerous applications as stabilizing, thickening and absorbing agents in printing, detergent, foodstuffs, medicine, toilet, sanitary, and petroleum industries. At present, the industrial manufacturing of CMC is carried out either in aqueous or organic media.[3,4] Natural cellulose with the crystal form of cellulose I (hereafter simply referred to as cellulose I) is exclusively used in the present commercial process, but no regenerated cellulose with the crystal form of cellulose II (hereafter referred to as cellulose II) is used.

Characteristic features, including solubility in water and aqueous salt solutions and interaction with cationic compounds, have been discussed so far only in a nonsystematic manner.[5,6]

This section deals with the high degree of the absorbency of the sodium salt of CMC (NaCMC), prepared from regenerated cellulose having the crystal form of cellulose II, toward various liquids and explain the absorbency on the basis of $\langle\!\langle f_k \rangle\!\rangle$ and $\langle\!\langle F \rangle\!\rangle$.

2.8.1   Experimental procedure

Synthesis of carboxymethylcellulose

As starting materials, four kinds of cellulose were used. Cellulose II (the viscosity average molecular weight $M_v = 7.3 \times 10^4$ and the degree of crystallinity by the X-ray diffraction method $\chi_c = 46\%$ as determined by the Segal method[7]), regenerated from a cuprammonium cotton linter cellulose solution, was used and is referred to as sample BLC. Natural wood pulp (cellulose I, $M_v = 21 \times 10^4$ and $\chi_c = 76\%$) and two acid-hydrolyzed wood pulps, prepared in 6 N sulfuric acid at 60 °C for 15 and 65 min ($M_v = 9.4 \times 10^4$ and 7.4×10^4, $\chi_c = 77\%$), were used and are referred to as NC-1, NC-2, and NC-3, respectively.

Twenty-six samples of NaCMC having $\langle\!\langle F \rangle\!\rangle_{chem} = 0.01–0.64$, chemically determined, were synthesized in the following manner: 20 g of cellulose were dipped in 80 ml of a system containing 3.3 g of sodium hydroxide and a mixture of 2-propanol–methanol–water (56:28:16 v/v/v) at 25 °C and allowed to stand for 30 min. At 60 °C, 2-propanol solution saturated with monochloroacetic acid was added to the system. The reaction appeared heterogeneous and was processed at 60 °C for 120 min without mechanical agitation. In this case, 0.07–7.15 g of monochloroacetic acid per 10 g of cellulose were added so as to obtain CMC with the desired $\langle\!\langle F \rangle\!\rangle$. Immediately on termination of the reaction, the resultant system was neutralized with a mixture of methanol and 30 wt% hydrochloric acid (9:1, w/w). The CMC in acid form was separated by filtration and washed with excess aq. methanol, immersed in a mixture of methanol to 2 wt% aq. sodium chloride (9:1, w/w) for 24 h for conversion into the salt form, washed again with aq. methanol, dried in air and then *in vacuo* at 70 °C for 8 h. By this method, nine NaCMC samples, coded the BL series, were synthesized from BLC, nine samples from NC-1 (N-1 series), five samples from NC-2 (N-2 series), and three samples from NC-3 (N-3 series).

Viscosity average molecular weight

The number average and weight average molecular weights of the NaCMC samples were not determined owing to experimental difficulty. We determined only M_v of starting cellulose from the limiting viscosity number $[\eta]$ in cadoxen (cadomium oxide–sodium hydroxide–ethylenediamine–water (4:1:12:83, w/w/w/w)) using the Brown–Wikström relation:[8]

$$[\eta] = 3.85 \times 10^{-2} M_w^{0.76} \qquad \text{(at 25 °C)} \qquad (2.8.1)$$

^{13}C NMR measurement

Each of the four samples of the BL and N-1 series in a mixture of sodium hydroxide–deuterium oxide (1:9, w/w) was measured for the ^{13}C NMR spectrum on a JEOL FX-400 and FX-500 pulse Fourier Transform NMR spectrometer (100.7 and 125.9 MHz for ^{13}C nuclei, respectively). The proton decoupled ^{13}C NMR method in the NNE mode was applied using dioxane (67.8 ppm) as the internal standard. For detection of CH_2 carbon peaks, the insensitive nuclei enhanced by the polarization transfer/complete decoupling (INEP/TCOM) method[9] (delay time, $\Delta = 3(4J)^{-1}$, J being the scalar coupling constant between ^{13}C and ^{1}H), was carried out.

NMR peak assignment

Using the reported results[10,11] for β-glucose and its C_2-, C_3-, and C_6-mono-carboxymethylated β-glucoses, the shielding or deshielding effect on ring carbons, induced by monosubstitution, was estimated (see Table 2.8.1). By this result, along with the ring carbon peaks of unsubstituted cellulose (i.e. C_1, 104.7; C_2, 75.0; C_3, 76.4; C_4, 80.0; C_5, 76.4; and C_6, 61.9 ppm), the ring carbon peaks for all possible substituted forms of cellulose was calculated. For example, the peak position of the C_1 carbon of NaCMC with $\langle\!\langle f_6 \rangle\!\rangle = 1.00$, $\langle\!\langle f_2 \rangle\!\rangle = 1.00$, and $\langle\!\langle f_3 \rangle\!\rangle = 0.00$, is $104.7 - 0.4 - 0.3 = 104.0$ ppm. The results are shown in Table 2.8.1. The calculated peaks were compared with those observed.

Table 2.8.1

Shielding and deshielding effects of the *o*-carboxymethyl group in 2-, 3-, and 6-substituted carboxymethyl glucose

Type of substitution	Chemical shift from unsubstituted group					
	C_1	C_2	C_3	C_4	C_5	C_6
2-substituted	-0.4	-8.1	-1.2	-0.4	-0.3	-0.5 to -1.1
3-substituted	-0.4	-0.9	9.1	-0.8	-0.7	-0.9 to -1.0
6-substituted	-0.3	-0.7	-0.8	-1.1	-1.8	8.1

Determination of the distribution of substitution by the NMR method

Total DS $\langle\!\langle F\rangle\!\rangle_{\mathrm{NMR}}$, $\langle\!\langle f_6\rangle\!\rangle$, $\langle\!\langle f_2\rangle\!\rangle + \langle\!\langle f_3\rangle\!\rangle$ were calculated from ^{13}C NMR spectra using the following relationships:

$$\langle\!\langle F\rangle\!\rangle_{\mathrm{chem}} = I_{c0}/I_{c1} \tag{2.8.2}$$

$$\langle\!\langle f_6\rangle\!\rangle = I_{sc6}/(I_{c6} + I_{sc6}) \tag{2.8.3}$$

$$\langle\!\langle f_2\rangle\!\rangle + \langle\!\langle f_3\rangle\!\rangle = (I_{sc2} + I_{sc3})/I_{c1} \tag{2.8.4}$$

where I_{c1}, I_{c0}, I_{c6}, I_{sc6}, I_{sc2}, and I_{sc3}, are the integrated peak intensities of C_1 carbon (103–105 ppm), C_0 carbon (180 ppm), unsubstituted C_6 carbon (61 ppm), substituted C_6 carbon (72.3 ppm), and substituted C_2 and C_3 carbons (81–86 ppm), respectively. Note that $\langle\!\langle f_6\rangle\!\rangle$ is independently determined from $\langle\!\langle F\rangle\!\rangle_{\mathrm{NMR}}$ or $\langle\!\langle f_2\rangle\!\rangle + \langle\!\langle f_3\rangle\!\rangle$ since the Overhauser effect in the C_6 carbon was found somewhat smaller than in other carbons.

Total DS determined by chemical analysis

Before determining $\langle\!\langle F\rangle\!\rangle_{\mathrm{chem}}$ by chemical analysis, NaCMC was converted to the acid form and immersed in a 3 wt% aq. sodium chloride solution. The hydrochloric acid thus produced was diluted with alkali, the excess of which was back titrated with hydrochloric acid.

Absorbency

W_0 gram (approximately 0.5 g) was placed in a nonwoven fabric bag having a weight of a gram. The bag was immersed for 10 min at 37 °C in pure water, 0.9 wt% of NaCl, CaCl$_2$, and AlCl$_3$ aq. solutions, respectively. It was then suspended for 20 min in air to remove water adhering to the bag and sample and weighed ($(W_1 + a)$ gram). The bag containing the NaCMC sample was dried and weighed ($(W_2 + a)$ gram). Absorbency A and solubility S were then defined by the following relationships:

For water,

$$A = (W_1/W_2) \times 100(\%) \tag{2.8.5}$$

$$S = \{(W_0 - W_2/W_0\} \times 100(\%) \tag{2.8.6}$$

and for aq. salt solutions,

$$A = W_1 - \{W_2 - 0.009(W_0 - W_1)\} \times 100(\%) \qquad (2.8.7)$$

$$S = \{W_0 - W_2 + 0.009(W_0 - W_1)\} W_0^{-1} \times 100(\%) \qquad (2.8.8)$$

Here, W_0, W_2, and a were determined for the sample which was conditioned at 20 °C at 65% relative humidity for 24 h.

Figure 2.8.1 shows the change in X-ray diffraction patterns of cellulose I (A) and II (B) by alkali treatment and carboxymethylation.

The characteristic diffraction peaks for both cellulose I ($2\theta = 9.0$, 14.7, 16.4, and 22.6°) and cellulose II ($2\theta = 9.5$, 12.0, 20.0, and 21.5°) remain distinct despite any treatment, although the X-ray diffraction pattern of NaCMC with relatively high $\langle\!\langle F \rangle\!\rangle$ (0.5–0.6) shows only the main peaks of the original celluloses. The original celluloses are thus not converted to alkali celluloses and the reaction is principally heterogeneous. Carboxymethylation first occurs in the amorphous region and proceeds to the crystalline part. Thus, the reaction from start to finish is influenced by the structure of the original celluloses.

Figure 2.8.2 shows $\chi_c(X)$, estimated by the Segal method,[7] plotted against $\langle\!\langle F \rangle\!\rangle_{\text{chem}}$. $\chi_c(X)$ of the NaCMC sample prepared from cellulose I decreases gradually as $\langle\!\langle F \rangle\!\rangle$ increases and abruptly drops to 10% at $\langle\!\langle F \rangle\!\rangle \cong 0.4 - 0.5$. The hydroxyl group present in the crystalline region does not react with monochloroacetic acid in a random manner. $\chi_c(X)$ of the sample, prepared from cellulose II, decreases linearly with increasing $\langle\!\langle F \rangle\!\rangle$.

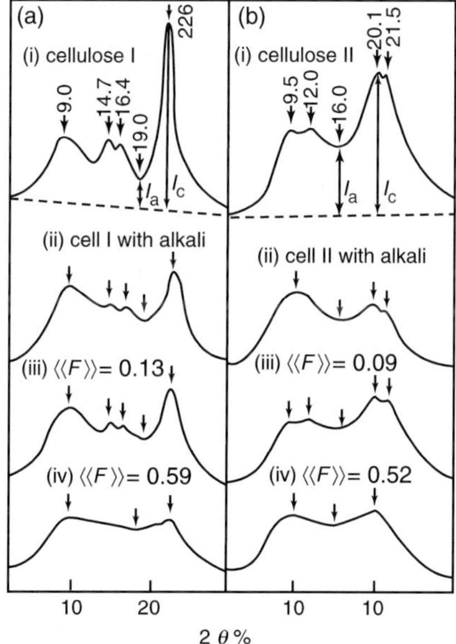

Figure 2.8.1 Change in X-ray diffraction curves of cellulose solid of cellulose I (A) and II (B) with alkali treatment and carboxymethylation:[1] (A) (i) cellulose I; (ii) cellulose I with alkali; (iii) CMC ($\langle\!\langle F \rangle\!\rangle_{\text{chem}} = 0.13$) from cellulose I; (iv) CMC ($\langle\!\langle F \rangle\!\rangle_{\text{chem}} = 0.59$) from cellulose I; (B) (i) cellulose II; (ii) cellulose II with alkali; (iii) CMC ($\langle\!\langle F \rangle\!\rangle_{\text{chem}} = 0.09$) from cellulose II; (iv) CMC ($\langle\!\langle F \rangle\!\rangle_{\text{chem}} = 0.52$) from cellulose II, I_a and I_b denote the relative intensity from amorphous and crystalline regions.

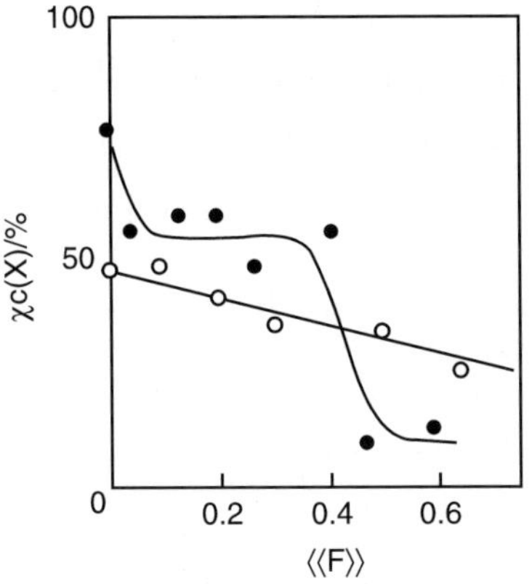

Figure 2.8.2   Relationships between crystallinity $\chi_c(X)$ and total DS $\langle\!\langle F\rangle\!\rangle_{chem}$:[1] (O), CMC from cellulose II (BL series); (●), CMC from cellulose I (N-1 series).

The cellulose II sample is regenerated cellulose with low $\chi_c(X)$ and is very reactive with chemical reagents (Figure 2.8.3) (Table 2.8.2).

The solubility S in water and aq. NaCl is plotted as a function of $\langle\!\langle F\rangle\!\rangle_{chem}$ in Figure 2.8.4. NaCMC did not dissolve significantly in either aq. $CaCl_2$ or aq. $AlCl_3$. S in water and in aq. NaCl was always less than 5% for $\langle\!\langle F\rangle\!\rangle_{chem} < 0.55$ and increased remarkably when $\langle\!\langle F\rangle\!\rangle_{chem}$ exceeded 0.55.

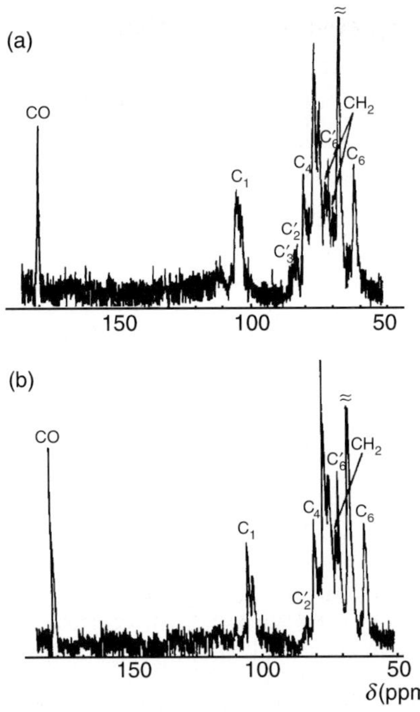

Figure 2.8.3   Typical ^{13}C NMR spectra of CMC (N-1-8) from cellulose I (a) and CMC (BL-9) from cellulose II (b).[1]

Table 2.8.2

Calculated chemical shift of CMC

Type of substitution	Chemical shift δ (ppm)					
	C_1	C_2	C_3	C_4	C_5	C_6
Unsubstituted	104.7	75.0	76.4	80.0	76.4	61.9
2-substituted	104.3	85.9	75.2	79.6	76.1	61.4
3-substituted	104.3	74.1	84.3	79.2	75.7	61.1
6-substituted	104.4	74.3	75.6	78.9	74.6	70.0
2,3-substituted	103.9	83.0	85.3	78.8	75.5	60.6
2,6-substituted	104.0	83.1	74.4	78.5	74.3	69.5
3,6-substituted	104.0	73.4	84.7	78.1	73.9	69.2
2,3,6-substituted	103.6	82.3	83.5	77.7	73.6	68.7

CH_2 for substituent group is expected to be around 71.2 ppm. COONa for substituent group was seen at 178.4 and 178.8 ppm.

Figure 2.8.5 shows a plot of the ratio $\langle\!\langle f_6 \rangle\!\rangle / \langle\!\langle F \rangle\!\rangle_{NMR}$ versus $\langle\!\langle F \rangle\!\rangle_{NMR}$ for the BL (open mark) and the N-1 (closed mark) series CMC. $\langle\!\langle f_6 \rangle\!\rangle / \langle\!\langle F \rangle\!\rangle_{NMR}$ for the BL series CMC was 0.9 ± 0.1, when $\langle\!\langle F \rangle\!\rangle_{NMR} < 0.7$, indicating carboxymethylation to occur almost preferentially at the hydroxyl group attached to the C_6 carbon on using cellulose II. $\langle\!\langle f_6 \rangle\!\rangle / \langle\!\langle F \rangle\!\rangle_{NMR}$ for N-1 series was 0.43 ± 0.1 within $\langle\!\langle F \rangle\!\rangle_{NMR} < 0.7$. The reactivity of the hydroxyl groups is thus seen to depend very significantly on the cellulose sample, particularly its crystalline form (Table 2.8.3).

2.8.2   Factors influencing absorbency of NaCMC

Figure 2.8.6 shows the relationships between absorbency A against pure water (a), aq. NaCl (b), aq. $CaCl_2$ (c), and aq. $AlCl_3$ (d), and $\langle\!\langle F \rangle\!\rangle_{chem}$ for N-1 ($M_v = 21.0 \times 10^4$, closed

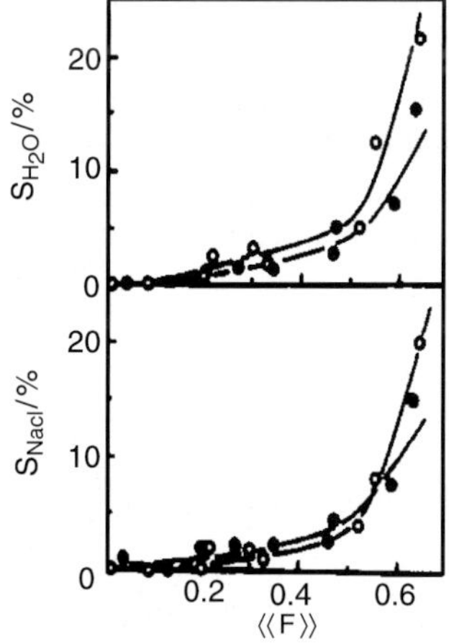

Figure 2.8.4   Solubility of NaCMC in H_2O and 0.9 wt% aq. NaCl as a function of total DS $\langle\!\langle F \rangle\!\rangle_{chem}$: (○) CMC from cellulose II (BL series), (●) CMC from cellulose I (N-1 series).[1]

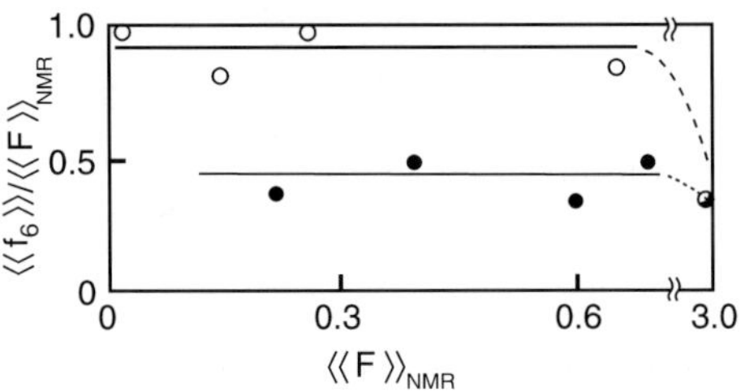

Figure 2.8.5   Plot of $\langle\!\langle f_6 \rangle\!\rangle / \langle\!\langle F \rangle\!\rangle_{NMR}$ versus $\langle\!\langle F \rangle\!\rangle_{NMR}$.[1] Symbols are the same as in Figure 2.8.4.

circle) and BL ($M_v = 7.3 \times 10^4$, open circle) series. Figure 2.8.6(a) and (b) include the data for N-2 ($M_v = 9.4 \times 10^4$, closed triangle) and N-3 ($M_v = 7.4 \times 10^4$, closed rectangle) series in order to elucidate the molecular weight dependence of A. For the N series, A increases somewhat with an increase in M_v at a given $\langle\!\langle F \rangle\!\rangle_{chem}$. Compared with BL series at the same M_v level, the absolute A value for the N-3 series in water and aq. NaCl is much lower than that of the BL series, indicating molecular weight not to be the main factor controlling A. On comparing the BL series with the N-1 series, maximum A in aq. NaCl of the former is 1.5 times that of the latter, so that NaCMC synthesized from cellulose II is particularly suitable as an absorbing agent, especially for aq. NaCl and similar solutions.

Neither $\langle\!\langle F \rangle\!\rangle_{chem}$ is the main factor since its value giving maximum A against water and aq. NaCl differ according to the sample. A in aq. CaCl$_2$ increased gradually with an increase in $\langle\!\langle F \rangle\!\rangle_{chem}$ for both samples. The BL series samples had a somewhat larger A than that of the N-1 series. However, A in aq. AlCl$_3$ decreased slightly with an increase in $\langle\!\langle F \rangle\!\rangle_{chem}$ due, of course, to cross-linking between carboxyl groups and Al$_3^+$ cations.

Figure 2.8.7 shows the effects of the crystallinity $\chi_c(X)$ on A in pure water (a) and in aq. NaCl (b). As $\chi_c(X)$ increased A tended slightly to decrease, and the correlation coefficients for A and χ_c were as follows: $A_{H_2O} - \chi_c(X)$, -0.34; $A_{NaCl} - \chi_c(X)$, -0.88 for the BL series and $A_{H_2O} - \chi_c(X)$, -0.79 and $A_{NaCl} - \chi_c(X)$, -0.82 for the N-1 series. Thus, there is no significant correlation between A and $\chi_c(X)$ and $\chi_c(X)$ is concluded not to be an additional factor controlling A.

Figure 2.8.8 shows a plot of the absorbency A against $\langle\!\langle f_6 \rangle\!\rangle$, giving a single master curve for each liquid, irrespective of the sample series. As long as NaCMC does not dissolve in the liquid being used, the absorbency can be accurately determined by $\langle\!\langle f_6 \rangle\!\rangle$. The following relationships were obtained using the least squares method:

$$A = 400\langle\!\langle f_6 \rangle\!\rangle + 7.0 \text{ in pure water,} \qquad \langle\!\langle f_6 \rangle\!\rangle < 0.17 \qquad (2.8.9)$$

$$A = 79\langle\!\langle f_6 \rangle\!\rangle + 14.2 \text{ in 0.9 wt\% aq. NaCl,} \qquad \langle\!\langle f_6 \rangle\!\rangle < 0.30 \qquad (2.8.10)$$

$$A = 19.5\langle\!\langle f_6 \rangle\!\rangle + 14.2 \text{ in 0.9 wt\% aq. CaCl}_2, \qquad \langle\!\langle f_6 \rangle\!\rangle < 0.57 \qquad (2.8.11)$$

$$A = -3.25\langle\!\langle f_6 \rangle\!\rangle + 13.7 \text{ in 0.9 wt\% aq. AlCl}_3, \qquad \langle\!\langle f_6 \rangle\!\rangle < 0.57 \qquad (2.8.12)$$

Table 2.8.3

Characterization and properties of CMC samples[1]

Sample	M_v ($\times 10^4$)	Total $\langle\!\langle F \rangle\!\rangle$		Distribution		S (%)		χ_c (%)	A (100%)			
		Chem. anal.	NMR	$\langle\!\langle f_2 \rangle\!\rangle + \langle\!\langle f_3 \rangle\!\rangle$	$\langle\!\langle f_6 \rangle\!\rangle$	H_2O	NaCl		H_2O	NaCl	$CaCl_2$	Al_2Cl_3
BL-1	7.3	0.01	–	–	–	0.0	0.0	–	16.2	18.0	–	–
BL-2	7.3	0.09	0.03	0.00	0.03	0.0	0.0	46.9	17.5	17.5	13.5	15.2
BL-3	7.3	0.20	0.14	–	0.11	0.5	0.0	40.2	49.6	22.3	13.0	40.2
BL-4	7.3	0.22	–	–	–	2.5	2.0	–	51.9	25.0	–	–
BL-5	7.3	0.30	0.27	0.01	0.27	3.0	1.5	34.7	52.0	32.0	18.0	13.0
BL-6	7.3	0.33	–	–	–	2.0	1.0	–	60.0	38.5	–	–
BL-7	7.3	0.52	–	–	–	5.0	4.0	32.8	22.5	48.5	–	–
BL-8	7.3	0.55	–	–	–	12.5	8.0	–	20.0	45.0	–	–
BL-9	7.3	0.64	0.65	0.09	0.54	21.5	20.0	25.0	15.4	29.1	24.0	12.0
N-1-1	21.0	0.04	–	–	–	0.0	1.0	54.3	14.5	16.0	–	–
N-1-2	21.0	0.13	–	–	–	0.0	0.0	57.8	19.3	20.0	–	–
N-1-3	21.0	0.20	0.24	0.16	0.08	1.0	2.0	57.8	29.1	22.0	13.5	13.7
N-1-4	21.0	0.27	–	–	–	1.5	2.5	47.0	42.6	21.0	–	–
N-1-5	21.0	0.35	0.41	0.20	0.21	1.5	2.5	54.3	44.0	19.6	17.0	17.8
N-1-6	21.0	0.46	–	–	–	3.0	2.5	–	61.8	29.3	–	–
N-1-7	21.0	0.47	–	–	–	5.0	4.5	9.1	61.0	27.4	–	–
N-1-8	21.0	0.59	0.60	0.39	0.20	7.0	7.5	12.8	49.0	28.6	16.5	11.0
N-1-9	21.0	0.63	0.71	0.34	0.37	15.0	15.0	–	30.7	26.5	21.0	11.8
N-2-1	9.4	0.14	–	–	–	–	–	–	19.4	16.0	–	–
N-2-2	9.4	0.24	–	–	–	–	–	–	29.0	15.5	–	–
N-2-3	9.4	0.31	–	–	–	–	–	–	37.2	18.0	–	–
N-2-4	9.4	0.38	–	–	–	–	–	–	23.0	16.6	–	–
N-2-5	9.4	0.51	–	–	–	–	–	–	22.6	13.0	–	–
N-3-1	7.4	0.16	–	–	–	–	–	–	12.5	7.5	–	–
N-3-2	7.4	0.23	–	–	–	–	–	–	22.5	12.6	–	–
N-3-3	7.4	0.39	–	–	–	–	–	–	23.5	12.5	–	–

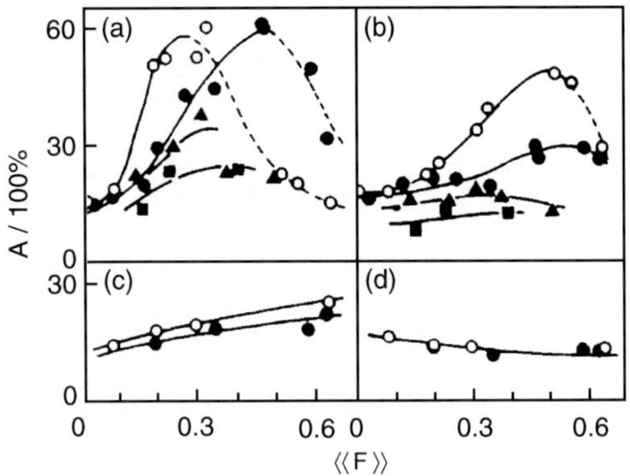

Figure 2.8.6   Absorbency A of NaCMC toward various liquids as a function of total DS $\langle\!\langle F\rangle\!\rangle_{\text{chem}}$.[1] (a) H_2O; (b) 0.9 wt% aq. NaCl; (c) 0.9 wt% aq. $CaCl_2$; (d) 0.9 wt% aq. $AlCl_3$; (O), CMC from cellulose II (BL series, $M_v = 7.3 \times 10^4$); (●) CMC from cellulose I (N-1 series, $M_v = 21.0 \times 10^4$); (▲) CMC from cellulose I (N-2 series, $M_v = 9.4 \times 10^4$); (■) CMC from cellulose I (N-3 series, $M_v = 7.4 \times 10^4$).

2.8.3   Mechanisms of carboxymethylation and water absorption by NaCMC

Figure 2.8.9 schematically illustrates the mechanism of carboxymethylation of cellulose (a–c) and that of water absorption by NaCMC (a → a′, b → b′, c → c′). In this figure, the full line is the cellulose chain and the shadowed area, the crystalline region. The filled circle is carboxymethylated glucose and open circle, water molecules. On immersing cellulose (a) in water, only a part of the amorphous region absorbs water (a′). Carboxymethylation of the cellulose occurs first in the amorphous region, the crystalline

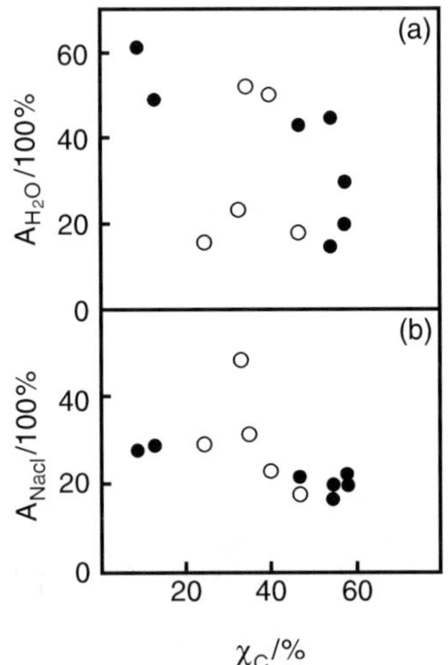

Figure 2.8.7   Absorbency A of NaCMC towards H_2O (a) and 0.9 wt% aq. NaCl (b) as function of the degree of crystallinity $\chi_c(X)$.[1] Symbols are the same as those in Figure 2.8.4.

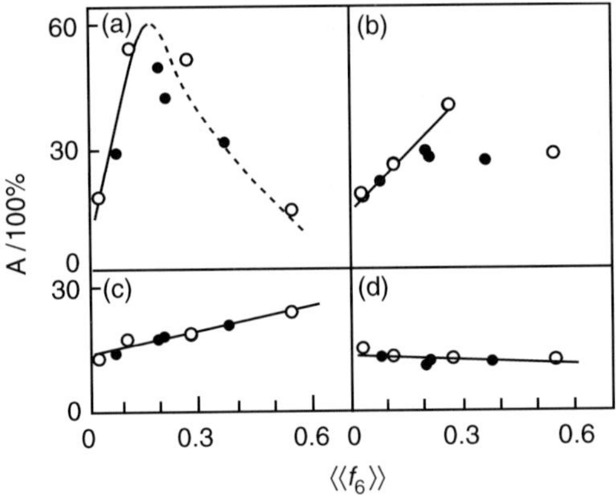

Figure 2.8.8   Absorbency A of NaCMC towards various liquids as a function of $\langle\langle f_6 \rangle\rangle$:[1] (a) H_2O; (b) 0.9 wt% aq. NaCl; (c) 0.9 wt% aq. $CaCl_2$; (d) 0.9 wt% aq. $AlCl_3$. Symbols are the same as those in Figure 2.8.4.

region collapses from the surface with further carboxymethylation (b and c). CMC with relatively low $\langle\langle F \rangle\rangle$ (b) dissociates, resulting in strong solvation with the surrounding water molecules and the consequent swelling (b′). CMC with relatively high $\langle\langle F \rangle\rangle$ (c) remains its smaller crystalline region and becomes hydrogel with its crystalline region as a kind of cross-linking point when immersed in water (c′). This mechanism is considered to be applicable to both BL and N series samples.

However, the reason for preferential carboxymethylation at the C_6 position of cellulose having the crystalline form of cellulose II and the paramount role of $\langle\langle f_6 \rangle\rangle$ in the absorbency

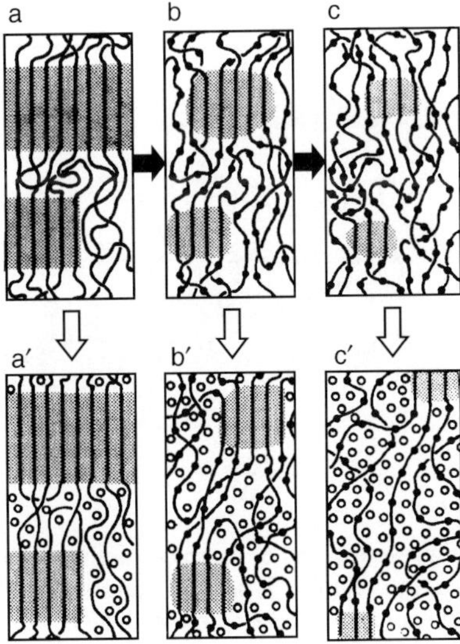

Figure 2.8.9   Schematic representation of carboxymethylation and swelling by water:[1] —, cellulose chain; (●) carboxymethylated glucose unit; (○) H_2O molecule; shadowed area denotes crystalline part.

A should be clarified in great detail. In considering the present carboxymethylation, it should be noted that: (1) cellulose is not converted to alkali cellulose, (2) mono-chloroacetic acid reacts as an electrophilic reagent,[12] and (3) carboxymethylation proceeds from an amorphous to crystalline region. A hydroxyl group having strong acidity in cellulose molecules thus reacts much more easily with monochloroacetic acid. This acidity is proportional, for the most part, to the electron density on its adjacent carbon atom. This density in turn governs the ^{13}C NMR resonance position, resonating at a higher magnetic field when the carbon has a higher electron density.

A recent NMR analysis on cellulose showed the C_6 carbon NMR peak to appear at approximately 65–62 ppm, while both C_2 and C_3 carbons resonate at 77–72 ppm, regardless of the crystalline form. The hydroxyl group attached to the C_6 carbon thus reacts more easily with monochloroacetic acid than those at the C_2 and C_3 carbons. This

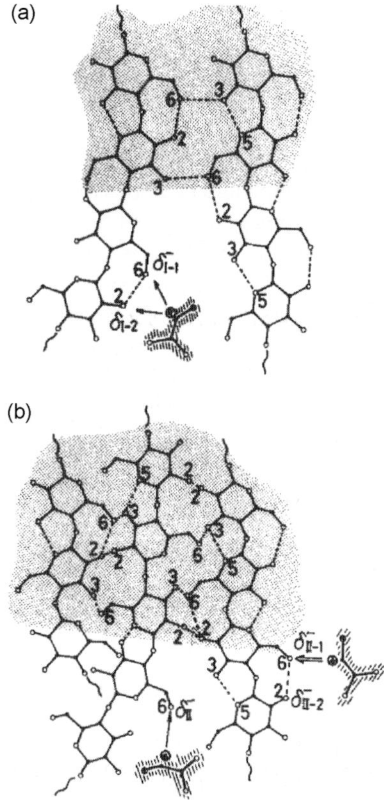

Figure 2.8.10 Schematic representation of carboxymethylation toward O_2–H$\cdots O_6'$ type intramolecular hydrogen bonds in cellulose I (a) and cellulose II (b).[1] Shadowed area denotes the cationized monochloroacetic acid. δ_{I-1}^- and δ_{I-2}^-, mean the electronegativity (EN) for cellulose I and δ_{II}^-, δ_{II-1}^-, and δ_{II-2}^-, EN for hydroxyl groups at the C_6 and C_2 positions of cellulose II. EN has the following order; $\delta_{II-1}^- > \delta_{II}^- > \delta_{I-1}^- > \delta_{I-2}^- > \delta_{II-2}^-$. The O_6–H$\cdots O_2'$ type intramolecular hydrogen bonds for cellulose II are not shown. NE on the hydroxyl group at the C_6 position is much stronger than δ_{I-1}^-.

agrees with the experimental facts that $\langle\!\langle f_6 \rangle\!\rangle / \langle\!\langle F \rangle\!\rangle_{NMR}$ for celluloses I and II were found to be 0.43 ± 0.1 and 0.9 ± 0.1, respectively.

A more detailed investigation of the NMR spectrum of the C_6 carbon region, carried out in Kamide $et\ al.$'s work[13] indicated that: (1) hydroxyl groups at the C_6 position for cellulose II participate in two types of intramolecular hydrogen bonds (O_2–H$\cdots O_6'$, and O_6'–H$\cdots O_2'$, the latter being much more acidic), (2) hydroxyl groups of cellulose I does not form the O_6'–H$\cdots O_2'$ type intramolecular hydrogen bond, and (3) the electron density on C_6 carbon for cellulose II in both ordered (crystalline) and amorphous regions is higher than that for cellulose I and the reverse is true for the electron density on the C_2 carbon. Thus, the reactivity of the hydroxyl group at the C_6 position of cellulose II may possibly be relatively higher than that of cellulose I.

The reverse situation applies at the C_2 position. The preferential substitution of the hydroxyl group at the C_6 position does not occur in the case of cellulose I under the same conditions of carboxymethylation, as evident from Figure 2.8.10. The introduction of a bulky substituent at the C_6 position destroys the intermolecular hydrogen bonds and widens the distance between molecular chains, creating a wider space to readily receive the absorbed liquids. This is because the hydroxyl groups at the C_6 position in the original cellulose mainly govern the intermolecular hydrogen bonds.

REFERENCES

1. K Kamide, K Okajima, K Kowsaka, T Matsui, S Nomura and K Hikichi, $Polym.\ J.$, 1985, **17**, 909.
2. E Jansen, Ger. Patent, No. 332,203, 1918.
3. See, for example, WF Waldeck and FW Smith, $Ind.\ Eng.\ Chem.$, 1952, **44**, 2803.
4. See, for example, RH Charles, US Patent, No. 2,607,772, 1952.
5. CD Callihan, in $Cellulose\ Technology\ Research.$ (ed. AF Turbak), ACS Series 10, 1975, p. 33.
6. E Otto and HM Spurlin, in $Cellulose\ and\ Cellulose\ Derivative,\ II$, Wiley, New York, 1954, p. 944.
7. L Segal, $Text.\ Res.\ J.$, 1959, **29**, 786.
8. W Brown and R Wikström, $Eur.\ Polym.\ J.$, 1965, **1**, 1.
9. DM Doddrell and DT Pegg, $J.\ Am.\ Chem.\ Soc.$, 1980, **102**, 6388.
10. AS Perlin, B Casu and HJ Koch, $Can.\ J.\ Chem.$, 1970, **48**, 2596.
11. A Parfondry and AS Perlin, $Carbohydr.\ Res.$, 1977, **57**, 39.
12. See, for example, L Fieser and M Fieser, $Advanced\ Organic\ Chemistry.$ Reinhold, New York, 1961.
13. K Kamide, K Okajima, K Kowsaka and T Matsui, $Polym.\ J.$, 1985, **17**, 701.

2.9 RIPENING OF VISCOSE[1]

The alkali solution of cellulose xanthate (hereafter referred to as CX) was first made by Cross and Bevan[2] as early as 1892. Since then, the solution, known commercially as viscose solution, has been utilized worldwide for the production of cellophane, viscose rayon, etc. for almost 100 years. CX is one of the most popular cellulose derivatives. CX is prepared by various methods:[3] reaction of solid alkali cellulose with gaseous carbon disulfide (CS_2; hereafter, denoted simply as the gas–solid reaction), and reaction of fibrous cellulose in emulsified CS_2/aqueous alkali mixture (emulsion method) and the reaction of CX solution in alkali with CS_2 (after xanthating method). Note that no homogeneous xanthation of cellulose has been carried out. The total DS $\langle\langle F\rangle\rangle$ of CX lies between 0.4 and 0.6 for commercial use[3] although CX having $\langle\langle F\rangle\rangle = 3$ has been obtained experimentally.[3] Generally, CX is known to be very unstable in water, rapidly decomposing to cellulose and in alkali $\langle\langle F\rangle\rangle$ of CX gradually decreases, accompanying a significant change in viscosity as well as other properties. The latter phenomenon is widely known as 'ripening'. Variations in preparation and ripening conditions have a significant influence on the properties of resultant viscose solution and this has encouraged many investigators to make efforts to establish a method for determining the probability of substitution of xanthate groups to hydroxyl groups located at C_2, C_3, and C_6 positions in trihydric glucopyranose unit ($\langle\langle f_k\rangle\rangle$, $k = 2$, 3, and 6). In 1928, Lieser[4] first reported the method for evaluating $\langle\langle f_k\rangle\rangle$ and thereafter many improved methods appeared. The typical procedures proposed so far are briefly illustrated in Figure 2.9.1. These methods are principally based on the quantitative conversion of CX to methylated glucose derivatives, which are chemically and gravimetrically analyzed. As is obvious from Figure 2.9.1, these methods are very time consuming, and not reliable in accuracy of $\langle\langle f_k\rangle\rangle$ because of the inclusion of severe destructive analysis and the possibility of change in $\langle\langle f_k\rangle\rangle$ during many steps of analysis, and are evidently not the correct way to study $\langle\langle f_k\rangle\rangle$ of CX systematically. On the basis of the methods illustrated in Figure 2.9.1, many authors reported $\langle\langle f_k\rangle\rangle$ of CX. For example, Lieser[4] insisted that $\langle\langle f_2\rangle\rangle$ was predominant for CX with $\langle\langle F\rangle\rangle = 0.5$. Noguchi[5] pointed out that CX with $\langle\langle F\rangle\rangle = 0.5$ and 0.7, both prepared by the gas–solid reaction, produced only 2-methyl glucose and, 2- and 2,6-methyl glucoses, respectively, when reacted with diazomethane. By using CX prepared in the same manner as Noguchi did, Lauer[6] obtained only 2- and 2,3-methyl glucoses from CX with $\langle\langle F\rangle\rangle = 0.75 \sim 0.80$. Chen et al.,[7] by applying Lauer's method for evaluation of CX, concluded that $\langle\langle f_2\rangle\rangle > \langle\langle f_6\rangle\rangle > \langle\langle f_3\rangle\rangle$ was generally obtained for CX prepared by the gas–solid reaction. For ripened CX with $\langle\langle F\rangle\rangle = 0.31$, $\langle\langle f_2\rangle\rangle \approx \langle\langle f_6\rangle\rangle > \langle\langle f_3\rangle\rangle$ was reported by Willard and Pacsu[8] and contrary to this, Phillip and Liu[9] insisted

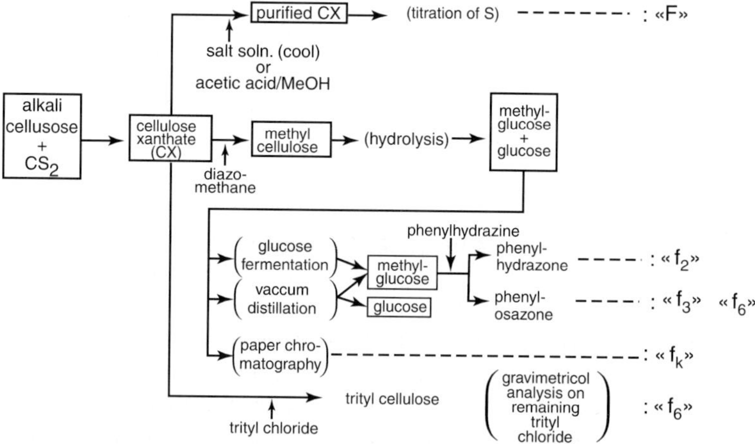

Figure 2.9.1 Typical procedures for determination of $\langle\langle F\rangle\rangle$ and $\langle\langle f_k\rangle\rangle$ by chemical analysis.[1]

that only CX substituted at C_6 position existed in ripened viscose solution. Horio *et al.*[10] extensively studied the difference in $\langle\langle f_k \rangle\rangle$ of CX prepared by different methods (the gas–solid reaction and the emulsion methods), and concluded that $\langle\langle f_2 \rangle\rangle > \langle\langle f_6 \rangle\rangle > \langle\langle f_3 \rangle\rangle$ held in both cases and the ripening did not bring about the drastic change in the order of the magnitude of $\langle\langle f_k \rangle\rangle$ discovered above. These results indicate that $\langle\langle f_2 \rangle\rangle$ is the largest in CX, but an apparent disagreement of conclusion on $\langle\langle f_k \rangle\rangle$ values, especially of the ripened CX led by many researchers, is obviously large, requiring a more rigorous and accurate evaluation method other than the conventional methods used so far, which have several commonly fatal short points described before. Accordingly, it is still one of the unsolved problems to determine $\langle\langle f_k \rangle\rangle$ of CX and their change during ripening despite 60 year's voluminous studies.

There has been no study on $\langle\langle f_k \rangle\rangle$ of CX in viscose solution by NMR analysis, probably due to an inevitable decomposition of xanthate group (i.e. gradual decrease in $\langle\langle F \rangle\rangle$ and $\langle\langle f_k \rangle\rangle$) when stored and during NMR measurement. ^{1}H NMR spectrum can be recorded in a very short time (3–100 s), compared with chemical analysis, but the resolution of NMR peaks is not sufficiently high and also peaks overlap heavily with large water proton signals. In contrast, ^{13}C NMR is a very convenient tool for analyzing the change in $\langle\langle f_k \rangle\rangle$ of CX in viscose solution as a function of ripening time.

In this section, we first intend to select the optimum operating conditions of ^{13}C$\{^1$H$\}$ NMR spectra for CX in aq. alkali, and to assign NMR peaks of CX by applying DEPT method, and by inspecting the change in NMR spectra obtained as a function of ripening time. In addition, we also intend to clarify the difference in $\langle\langle f_k \rangle\rangle$ between CXs prepared by the gas–solid (heterogeneous) and the liquid–liquid (homogeneous) reactions. For the latter case, we employed an alkali soluble cellulose, reported by Kamide *et al.*[11,12]

2.9.1   Experimental procedure

CX samples

A cellulose with the viscosity average molecular weight $M_v = 3 \times 10^4$ (evaluated by (Mark–Houwink–Sakurada equation established by Brown and Wikström[13] in cadoxen at 25 °C) and with the low degree of intramolecular hydrogen bonding[12] (crystal form: cellulose I) was used as a starting material for the preparation of CX in the liquid–liquid reaction. The liquid–liquid reaction was carried out as follows: A 0.5 g of the cellulose was immersed into 5 g of 2.5 N sodium hydroxide (NaOH)/deuterium oxide (D_2O) solution at 0 °C and stirred vigorously. Into the resultant cellulose solution, 0.3 g of carbon disulfide (CS_2) was added and the reaction was carried out for 48 h at 0 °C. The resultant xanthate solution was subjected to the apparent ripening at 20 °C and, after given ripening times, the solutions were subjected to the NMR measurement.

A hydrolyzed wood pulp ($M_v = 6 \times 10^4$, cellulose *I*) was also used as a starting material of the gas–solid reaction. Alkali cellulose was prepared from the cellulose sample as described in Ref. 11. 1.4 g of the alkali cellulose (0.5 g of cellulose, 0.9 g of 17.5 wt% aq. NaOH in D_2O) and 0.5 g of CS_2 were reacted in a sealed tube for 2 h at 30 °C and the reaction mixture was diluted with NaOH/D_2O solution at 20 °C for 2 h to produce viscose solution with 10 wt% cellulose content and *ca.* 2.5 N NaOH concentration. The resultant xanthate solution was subjected to ripening at 20 °C for given times and subjected to NMR measurement.

The cellulose solution obtained in the preparation process for CX in the liquid–liquid reaction was employed as reference solution for NMR measurement in order to aid the peak assignment of CX.

$^{13}C\{^{1}H\}$ NMR

NMR spectra were recorded on a 200 MHz JEOL FT NMR spectrometer JNM-FX-200. Measuring conditions are as follows: Frequency 50.15 NHz, irradiation 199.5 MHz (^{1}H), gated decoupling mode (NNE), pulse width 10 µs (*ca.* 40°), data point 16384, sampling data points 8192 (followed by 8192 points zero filling), temperature 20 °C. These conditions were carefully selected so as to minimize the temperature rise induced by the ^{1}H irradiation in alkaline solution. Accumulation was more than 3600 times.

Distorsionless enhancement by polarization transfer (DEPT)[14] method was applied to assign the peaks. In this NMR technique, the flip angle was set at 135°. Chemical shifts were determined using methylene carbon peak (67.4 ppm) of dioxane as an internal reference, after confirming the peak for dioxane experimentally. Peak intensity was estimated from integral curve and, when peaks are overlapped, the integral curve was visually separated.

2.9.2   Peak assignment in $^{13}C\{^{1}H\}$ NMR spectra of cellulose xanthate

Figure 2.9.2 shows typical $^{13}C\{^{1}H\}$ NMR spectrum of whole region (a) and magnified spectra of substituent peak region (b) and of glucose ring carbon region (c) of the 5-h ripened xanthate solution obtained by the liquid–liquid reaction. Twenty peaks, numbered as 1–20 from the lower magnetic field, are detected in the range 270–60 ppm. Chemical shifts of these peaks are summarized in Table 2.9.1. Peak 19 will be proved later as having originated from methylene carbon of dioxane. Peaks 1 and 4 may be

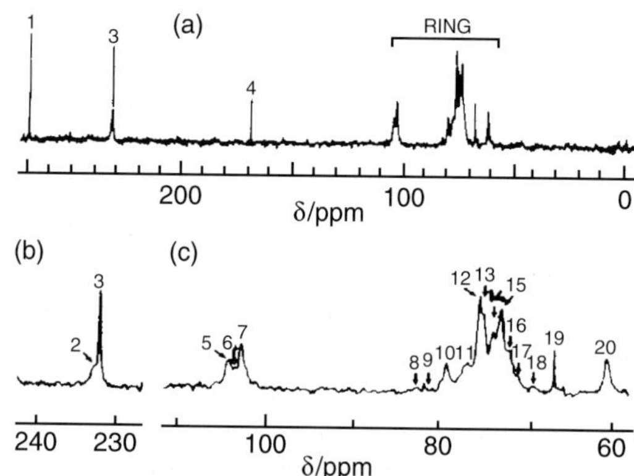

Figure 2.9.2 $^{13}C\{^{1}H\}$ NMR spectra of CX obtained by the liquid–liquid reaction and ripened for 5 h:[1] (a), entire spectrum; (b), substituent region; (c), ring carbon region.

Table 2.9.1

Chemical shift (δ), species of atomic group and assignment of carbon peaks in ^{13}C NMR spectrum of CX obtained from liquid–liquid reaction and ripened at 4 °C for 48 h and 20 °C for 10 h[1]

Code	δ (ppm)[a]	Species[b]	Assignment
1	269.9	q	$NaCS_3$ (byproduct from CS_2 and NaOH)
2	233.2	q	$-CS_2Na$ (substituent group at C_2, C_3)
3	232.5	q	$-CS_2Na$ (substituent group at C_6)
4	169.1	q	$NaCO_3$ (byproduct from CS_2 and NaOH)
5	104.4	t	C_1 in unsubstituted glucose unit
6	103.5	t	C_1 in substituted glucose unit
7	102.9	t	C_1 in substituted glucose unit
8	83.2	t	C_3 in 3-substituted glucose unit
9	82.0	t	C_2 in 2-substituted glucose unit
10	79.7	t	C_4 in ubsubstituted glucose unit
11	77.3	t	? (ring carbon in substituted unit)
12	76.0	t	C_3 and C_5 in unsubstituted glucose unit
13	75.5	t	? (ring carbon in substituted unit)
14	74.4	t	C_2 in unsubstituted glucose unit
15	73.5	t	? (ring carbon in substituted unit)
16	72.7	s	C_6 in 6-substituted glucose unit
17	71.6	t?	? (ring carbon in substituted unit)
18	70.0	t?	? (ring carbon in substituted unit)
19	67.4	s	Dioxane (reference)
20	61.5	s	C_6 in 6-substituted glucose unit

[a]67.4 ppm for dioxane.

[b]s, secondary (methylene) carbon; t, tertiary (methyne) carbon; q, quaternary carbon.

attributable to impurities produced by reaction of CS_2 with alkali because of their narrow peak width, and were discarded in further analyses.

Figure 2.9.3 shows ^{13}C{^{1}H} NMR spectra of cellulose/aq. NaOD (a), ripened CX solution for 5 h (b), ripened CX for 120 h (c), and DEPT spectrum for ripened CX for 75 h (d). Peak assignment for cellulose/NaOH/D_2O system is indicated on the spectrum, according to Gagnaire *et al.*[15]: 104.4 ppm, C_1 carbon; 79.7 ppm, C_4 carbon; 76.0 ppm, C_3 and C_5 carbons; 74.4 ppm, C_2 carbon and 61.5 ppm, C_6 carbon. These peaks appeared in the NMR spectrum of CX and can be assigned to unsubstituted ring carbons of glucopyranose unit. Other peaks in the region 110–60 ppm are derived from ring carbons magnetically influenced by derivatization of cellulose and standard peak (dioxane).

From the DEPT spectrum shown in Figure 2.9.3(d), together with a well-known experimental fact that phase of DEPT spectrum is positive for methyl and methine carbons and negative for methylene carbon and nonobservable for quaternary carbon, we can easily assign the peaks based on carbon species (number of coupling protons). The results are listed in the third column of Table 2.9.1. Here, note that methyl carbon does not exist in the system and any positive carbon peaks must be attributed to methine carbons. The number of coupling protons for carbons corresponding to peaks 17 and 18 could not be evaluated due to their very weak intensities and a heavy overlapping with a neighboring strong negative peak located at lower magnetic field. Of methylene carbon peaks (16, 19, and 20), peak 19 is assignable to that for dioxane because of its narrow

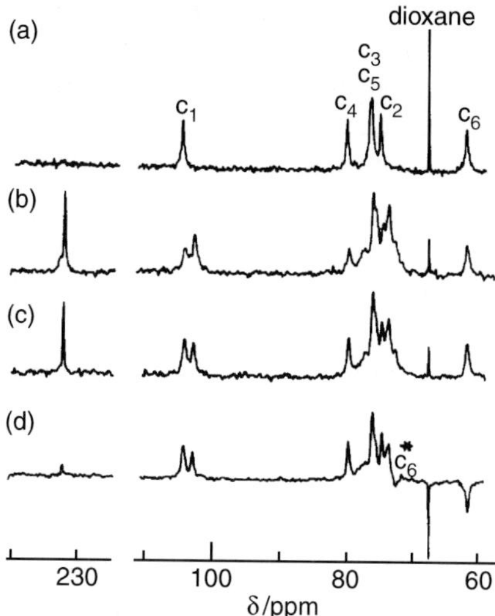

Figure 2.9.3 ^{13}C{^{1}H} NMR spectra of cellulose and CXs (obtained by the liquid–liquid reaction):[1] (a) cellulose in 9.1 wt% aq. NaOH solution; (b) CX ripened for 5 h; (c) CX ripened for 120 h; (d) DEPT specttrum of CX ripened at 75 h.

peak width and strong intensity and this peak was defined as 67.4 ppm. Peak 20 appeared at the same position as that for C_6 carbon of cellulose and is therefore assigned as unsubstituted C_6 carbon. Methylene carbon is available only for C_6 carbon and peak 16 can be attributed to substituted C_6 carbon. Peaks 1 and 4 are derived from carbons having no proton of inorganic impurities produced by the side reaction of CS_2 and alkali. Peaks 2 and 3 at 233.2 and 232.5 ppm, respectively, were also proved to be quaternary carbons and their intensities decreased after ripening as shown in Figure 2.9.3(b) and (c). These peaks are attributable to carbons for xanthate substituent and, in fact, these chemical shifts are reasonable by taking into consideration that the chemical shift of CS_2 carbon is approximately 200 ppm.

The existence of two xanthate carbon peaks may reflect the different OH positions substituted and will be discussed later. Figure 2.9.3(b) and (c) can help to roughly assign ring carbons by considering that intensity of the peak which newly appeared by xanthation, compared with those for cellulose, should decrease after dexanthation occurred during ripening.

Thus, peaks 8 (83.2 ppm) and 9 (82.0 ppm) are assigned to substituted C_3 and C_2 carbons, respectively. These chemical shifts are always found for other cellulose ethers.[16] Since α-substitution of C_2 and C_3 may induce the same order of deshielding effect, the lower field peak (C_3) for cellulose might appear at the lower magnetic field even for xanthate. Peaks 6 (103.5 ppm) and 7 (l02.9 ppm) proved to decrease in their intensities by ripening and, in contrast, peak 5 (104.4 ppm) increased by ripening. Increased peaks by dexanthation should be regarded as those for unsubstituted ring carbons. Peaks at 104.4 ppm are attributable to unsubstituted C_1 carbon. As reported by Kamide et al.[16] and others,[17] ether-type substituents (such as carboxymethyl and methyl groups) give rise to higher magnetic field shifts of 0.5–1.8 ppm for the β and γ carbons. Two peaks at 103.5 and 102.9 ppm are responsible for C_1 carbons influences by xanthate substitution at either C_2, C_3, or C_6. As mentioned before, peaks 10 (79.7 ppm), 12 (76.0 ppm), and 14 (74.4 ppm) are attributed to unsubstituted C_4, $C_3 + C_5$, and C_2 carbons, respectively. Other unassignable peaks (11, 13, 15, 17, and 18) are responsible for either C_2, C_3, C_5, and C_6 carbons influenced by xanthation. Peak assignment is given in Table 2.9.1.

2.9.3   Determination of $\langle\!\langle F \rangle\!\rangle$ and $\langle\!\langle f_k \rangle\!\rangle$

Based on the peak assignment shown in Table 2.9.1, $\langle\!\langle F \rangle\!\rangle$ and $\langle\!\langle f_k \rangle\!\rangle$ can be estimated as follows:

$$\langle\!\langle F \rangle\!\rangle = (I_2 + I_3)/(I_5 + I_6 + I_7) \tag{2.9.1}$$

$$\langle\!\langle f_6 \rangle\!\rangle = 1 - \{I_{20}/(I_5 + I_6 + I_7)\} \tag{2.9.2}$$

$$\langle\!\langle f_2 \rangle\!\rangle = I_9/(I_5 + I_6 + I_7) = 1 - \{I_{14}/(I_5 + I_6 + I_7)\} \tag{2.9.3}$$

$$\langle\!\langle f_3 \rangle\!\rangle = I_8/(I_5 + I_6 + I_7) = 2 - \{I_{12}/(I_5 + I_6 + I_7)\} \tag{2.9.4}$$

$$\langle\!\langle f_6 \rangle\!\rangle = I_{16}/(I_5 + I_6 + I_7) \tag{2.9.5}$$

It should be noted that under the NMR measuring conditions employed here, neither NOE nor progressive saturation exists because we adopted the gated decoupling mode and used pulse interval (10 s) sufficiently larger than the anticipated spin lattice relaxation time T_1 value (0.1–2 s) for xanthate. I_k in eqs. (2.9.1)–(2.9.5) denotes the peak intensity for the corresponding peak k. Equations (2.9.3)–(2.9.5) cannot be practically used for estimation of $\langle\!\langle f_2 \rangle\!\rangle$, $\langle\!\langle f_3 \rangle\!\rangle$, and $\langle\!\langle f_6 \rangle\!\rangle$ due to experimental uncertainty in I_8, I_9, I_{12}, I_{14}, and I_{16}. The former two are too small to determine accurately and the peaks for others heavily overlap. The most reasonable value for $\langle\!\langle f_2 \rangle\!\rangle + \langle\!\langle f_3 \rangle\!\rangle$ can be obtained using the following equation:

$$\langle\!\langle f_2 \rangle\!\rangle + \langle\!\langle f_3 \rangle\!\rangle = \langle\!\langle F \rangle\!\rangle - \langle\!\langle f_6 \rangle\!\rangle \tag{2.9.6}$$

Here, $\langle\!\langle F \rangle\!\rangle$ and $\langle\!\langle f_6 \rangle\!\rangle$ are determined from eq. (2.9.1) and (2.9.2), respectively. The values estimated from eqs. (2.9.1), (2.9.2), and (2.9.6) have an uncertainty of ± 0.03.

2.9.4 $\langle\!\langle f_k \rangle\!\rangle$ of CXs prepared under different conditions

Figure 2.9.4 shows $^{13}C\{^1H\}$ NMR spectra of a 7-h ripened CX solution obtained by the gas–solid reaction (a. entire spectrum; b. substituent group region; c. glucopyranose ring region). Compared with the spectra of CX solution obtained by the liquid–liquid reaction shown in Figure 2.9.1, the spectra in Figure 2.9.4 are broader and less resolved, except for the peaks responsible for impurities (low molecular weight substance). This is because of the higher molecular weight and the higher viscosity of the CX sample in the latter case. However, in Figure 2.9.4, peaks 8 and 9 responsible for substituted C_2 and C_3 carbons are much more intense than those in Figure 2.9.2, and peak 2, which appeared as a shoulder peak in Figure 2.9.2, is distinctively observable. This strongly suggests that peak 2 is correlated to peaks 8 and 9, and that $\langle\!\langle f_2 \rangle\!\rangle + \langle\!\langle f_3 \rangle\!\rangle$ is higher in the CX prepared

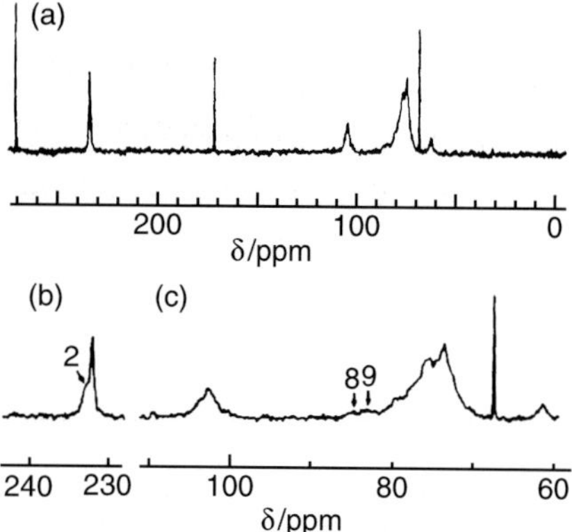

Figure 2.9.4 $^{13}C\{^1H\}$ NMR spectra of CX obtained by the gas–solid reaction and ripened for 7 h;[1] (a) entire spectrum; (b) substituent region; (c) ring carbon region.

by the gas–solid reaction than in the CX by the liquid–liquid reaction. The experimental fact mentioned above points to the fact that the peak 2 is attributable to the substituent groups attached to C_2 and C_3 positions, and consequently we can also assign peak 3 to the substituent groups at C_6 position.

2.9.5   Change in $\langle\!\langle F \rangle\!\rangle$ and $\langle\!\langle f_k \rangle\!\rangle$ of CXs during ripening

Variation of $\langle\!\langle F \rangle\!\rangle$ and $\langle\!\langle f_k \rangle\!\rangle$ of CX, prepared by the liquid–liquid reaction is shown as a function of ripening time in Figure 2.9.5. Here, NMR measurement usually required a 10-h pulse accumulation at 20 °C. Thus, ripening of CX during NMR measurement could not be avoided. The apparent ripening time is often defined as the time between the end of the reaction and the initiation of the NMR measurement. However, strictly speaking, the reasonable ripening time, shown on the horizontal axis in Figure 2.9.5, is the value in which 5 h (average of 10 h pulse accumulation) is added to the apparent ripening time. From the figure, the following relationships for CX, prepared by the liquid–liquid

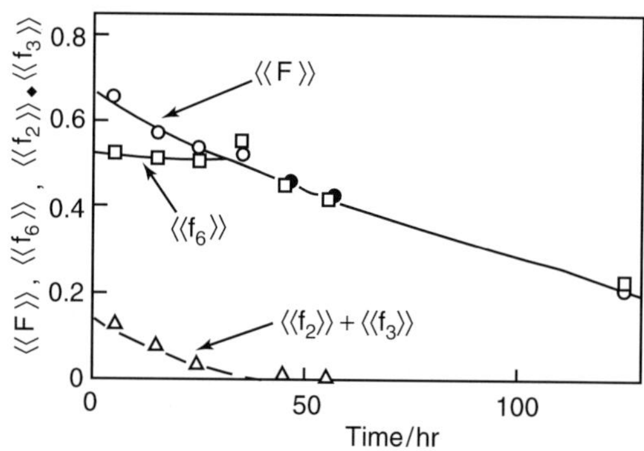

Figure 2.9.5   Change in $\langle\!\langle F \rangle\!\rangle$ ($\bigcirc$), $\langle\!\langle f_2 \rangle\!\rangle + \langle\!\langle f_3 \rangle\!\rangle$ ($\triangle$), and $\langle\!\langle f_6 \rangle\!\rangle$ ($\square$) of CX (obtained by the liquid–liquid reaction) as a function of ripening time.[1]

reaction, are observed.

$$\langle\!\langle f_6 \rangle\!\rangle > \langle\!\langle f_2 \rangle\!\rangle + \langle\!\langle f_3 \rangle\!\rangle \qquad \text{(CX in initial stage of ripening)} \qquad (2.9.7)$$

$$\langle\!\langle f_6 \rangle\!\rangle \gg \langle\!\langle f_2 \rangle\!\rangle + \langle\!\langle f_3 \rangle\!\rangle = 0 \qquad \text{(CX ripened for more than 50 h)} \qquad (2.9.8)$$

$\langle\!\langle F \rangle\!\rangle$ and $\langle\!\langle f_k \rangle\!\rangle$ of the CX at zero ripening time was estimated from extrapolating dexanthation curve in Figure 2.9.5 as follows:

$$\langle\!\langle F \rangle\!\rangle = 0.67, \qquad \langle\!\langle f_6 \rangle\!\rangle = 0.53 \qquad (2.9.9)$$

and

$$\langle\!\langle f_2 \rangle\!\rangle + \langle\!\langle f_3 \rangle\!\rangle = 0.14 \qquad (2.9.10)$$

Figure 2.9.6 shows similar results for CX prepared by the gas–solid reaction, as in Figure 2.9.5. From the figure, the following relationships are obtained:

$$\langle\!\langle f_6 \rangle\!\rangle > \langle\!\langle f_2 \rangle\!\rangle + \langle\!\langle f_3 \rangle\!\rangle \qquad \text{(CX at the initial stage of ripening)} \qquad (2.9.11)$$

$$\langle\!\langle f_6 \rangle\!\rangle > \langle\!\langle f_2 \rangle\!\rangle + \langle\!\langle f_3 \rangle\!\rangle = 0 \qquad \text{(CX ripened for more than 50 h)} \qquad (2.9.12)$$

$\langle\!\langle F \rangle\!\rangle$ and $\langle\!\langle f_k \rangle\!\rangle$ values of the CX extrapolated to zero ripening time is:

$$\langle\!\langle F \rangle\!\rangle = 1.24, \qquad \langle\!\langle f_6 \rangle\!\rangle = 0.68 \qquad (2.9.13)$$

$$\langle\!\langle f_2 \rangle\!\rangle + \langle\!\langle f_3 \rangle\!\rangle = 0.56 \qquad (2.9.14)$$

Inspection of Figures 2.9.5 and 2.9.6 clearly shows that at the initial stage of ripening the xanthate groups substituted at C_2 and C_3 decompose much faster than those substituted at C_6, independent of the preparation methods of CX. In other words, at the initial stage the xanthate groups substituted at C_6 is relatively stable, although the xanthate groups at C_2 and C_3 decompose rapidly.

The xanthate group at only C_6 position was proven to exist after a long ripening time. The results obtained in this study, especially for CX prepared by the gas–solid reaction, can be compared with those reported so far, and are listed in Table 2.9.2. The CXs with $\langle\!\langle F \rangle\!\rangle = 0.7 \sim 0.8$ obtained by the gas–solid reaction in Table 2.9.2 revealed that $\langle\!\langle f_2 \rangle\!\rangle$ was the highest among $\langle\!\langle f_k \rangle\!\rangle$.

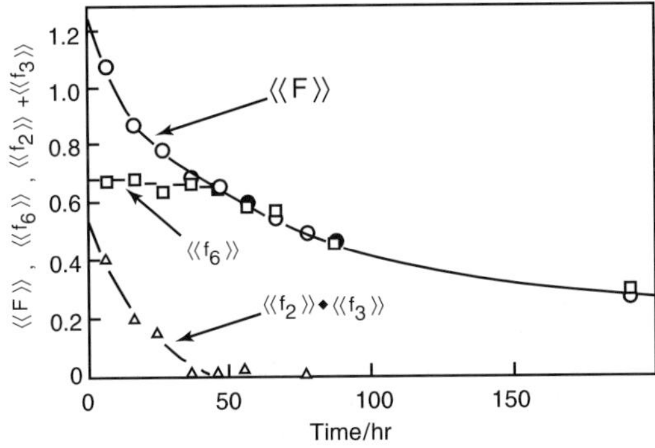

Figure 2.9.6   Change in $\langle\!\langle F \rangle\!\rangle$ (O), $\langle\!\langle f_2 \rangle\!\rangle + \langle\!\langle f_3 \rangle\!\rangle$ ($\triangle$), and $\langle\!\langle f_6 \rangle\!\rangle$ ($\square$) of CX (obtained by the gas–solid reaction) as a function of ripening time.[1]

Table 2.9.2

Results for distribution of xanthate groups[1]

Researchers	Year	Technique[a]	Sample[b]	$\langle\langle F\rangle\rangle$	Result	Reference
Lieser	1929	Ch(MC)	G–S, Un	–	At least C_2	4
Noguchi	1950	Ch(MC, TC)	G–S, Un	0.7	2-Monosub. and 2,6-disub. only	5
Lauer	1950	Ch(MC, VD)	G–S, Un	0.75	2-Monosub. and 2,3-disub. only	6
Chen et al.	1951	Ch(MC, VD)	G–S, Un	0.72	$\langle\langle f_2\rangle\rangle = 0.27, \langle\langle f_3\rangle\rangle = 0.21, \langle\langle f_6\rangle\rangle = 0.24,$	7
Swan and Purves	1957	Ch(SMX)	G–S, Un	–	6-Sub. (53.5%) and 3-sub. (46.5%)	19
			G–S, Ri	–	81% 6-Sub.	19
Philipp and Liu	1959	Ch	G–S, Un	–	41% 6-Sub.	9
			G–S, Ri	–	100% 6-Sub.	9
Willard and Pacsu	1960	Ch(MC', PC)	G–S, Ri	0.31	$\langle\langle f_2\rangle\rangle = 0.13, \langle\langle f_3\rangle\rangle = 0.06, \langle\langle f_6\rangle\rangle = 0.11$	8
Horio et al.	1963	Ch(MC, PC)	G–S, Un	0.62	$\langle\langle f_2\rangle\rangle = 0.37, \langle\langle f_3\rangle\rangle = 0.09, \langle\langle f_6\rangle\rangle = 0.13$	10
			G–S, Ri	0.50	$\langle\langle f_2\rangle\rangle = 0.32, \langle\langle f_3\rangle\rangle = 0.08, \langle\langle f_6\rangle\rangle = 0.10$	10
			L–S, Un	0.78	$\langle\langle f_2\rangle\rangle = 0.47, \langle\langle f_3\rangle\rangle = 0.13, \langle\langle f_6\rangle\rangle = 0.18$	10
Kamide et al.	1986	NMR	G–S, Un	1.24	$\langle\langle f_2\rangle\rangle + \langle\langle f_3\rangle\rangle = 0.56, \langle\langle f_6\rangle\rangle = 0.68$	1
			G–S, Ri	0.30	$\langle\langle f_2\rangle\rangle + \langle\langle f_3\rangle\rangle = 0.0, \langle\langle f_6\rangle\rangle = 0.30$	1
			L–L, Un	0.67	$\langle\langle f_2\rangle\rangle + \langle\langle f_3\rangle\rangle = 0.14, \langle\langle f_6\rangle\rangle = 0.53$	1
			L–L, Ri	0.22	$\langle\langle f_2\rangle\rangle + \langle\langle f_3\rangle\rangle = 0.0, \langle\langle f_6\rangle\rangle = 0.22$	1

[a]Ch, chemical: MC, conversion to methylcellulose; MC', multistep conversion to methylcellulose; SMX, conversion to cellulose S = methylxanthate; TC, conversion to tritylcellulose; VD, vacuum distillation; PC, paper chromatography.
[b]G–S, gas–solid reaction; L–S, liquid–solid (emulsion) reaction; L–L, liquid–liquid reaction; Un, unripened; Ri, ripened.

From Figure 2.9.6 it is obvious that xanthate group exists almost only at C_6 position in the CX having $\langle\!\langle F \rangle\!\rangle = 0.7 \sim 0.8$ prepared by the gas–solid reaction in this study, being quite in contrast to the literature data. However, note that the CXs described in the existing literature were prepared according to the commercial process and the amount of CS_2 reacted to alkali cellulose was around 30–40% against original cellulose used, and are thus far lower than that (60%) employed here. It is also anticipated from Figure 2.9.6 that substitution at C_2 and C_3 is much larger than that at C_6 and $\langle\!\langle f_2 \rangle\!\rangle$ may become highest among $\langle\!\langle f_k \rangle\!\rangle$ for CX immediately after the start of reaction. For the ripened CX obtained by the gas–solid reaction shown in Table 2.9.2, the order of magnitude of $\langle\!\langle f_k \rangle\!\rangle$ is quite different among researchers. Philip and Liu[9] suggested the substitution only at C_6 position in the ripened CX by the combination of chemical and paper chromatographic analyses although $\langle\!\langle F \rangle\!\rangle$ of the ripened CX is not clear. Our results showed that the ripened CXs having $\langle\!\langle F \rangle\!\rangle$ less than 0.6 remained as xanthate groups only at C_6 position, as seen from Figure 2.9.6. Unfortunately, literature data for the CX prepared by the liquid–liquid reaction are not available. Our results show that the preparation method influences significantly $\langle\!\langle F \rangle\!\rangle$ as well as $\langle\!\langle f_k \rangle\!\rangle$ of CX: The liquid–liquid reaction exhibits the preferential substitution of xanthate group at C_6 and in contrast to this, substitution at C_2 and C_3 are greatly enhanced by the gas–solid reaction, as described before. The enhanced substitution at C_2 in the gas–solid reaction can be reasonably explained by an experimental fact that, in solid alkali cellulose, NaOH preferentially coordinates to the hydroxyl group at C_2 position.[18] In contrast, no preferential coordination of NaOH to any hydroxyl groups for cellulose dissolved in 9.1 wt% aqueous NaOH solution (confirmed by Kamide et al.[16]) leads to the conclusion that the most acidic hydroxyl group (OH at C_6 position in this case) is most reactive. Note that even if the gas–solid reaction started when alkali cellulose contacted with gaseous CS_2, the nature of the xanthation changes as the reaction proceeds. In other words, the xanthate formed on the skin of cellulose microfibrils is soluble in alkali and then, the further reaction becomes similar to the liquid–liquid reaction. Lower $\langle\!\langle F \rangle\!\rangle$ of CX attained by the liquid–liquid reaction, compared with that obtained by the gas–liquid reaction, is mainly due to the lower amount of CS_2 used. The amounts of CS_2 used for the liquid–liquid and the gas–solid reactions were 60 and 100% against original cellulose, respectively. $\langle\!\langle F \rangle\!\rangle$ values attained at zero ripening time for both reactions were found to be proportional to the amount of CS_2 employed.

In summary, a method for determining $\langle\!\langle F \rangle\!\rangle$ and $\langle\!\langle f_k \rangle\!\rangle$ was established by ^{13}C NMR analysis, based on the peak assignment of the NMR spectra. The difference in $\langle\!\langle F \rangle\!\rangle$ and $\langle\!\langle f_k \rangle\!\rangle$ between cellulose xanthates obtained under different preparation conditions (the liquid–liquid and the gas–solid reactions) was clarified. It was proven that (1) substitution at C_2 and C_3 positions was greatly enhanced by the gas–solid reaction, (2) the substitution at C_6 was preferential for the liquid–liquid reaction and (3) in both preparation methods, after long ripening time, the xanthate group remained at only C_6 position.

REFERENCES

1. K Kamide, K Kowsaka and K Okajima, *Polym. J.*, 1981, **19**, 231.
2. C Cross, E Bevan and C Beadle, Br. Patent No.8700, 1883; C Cross, E Bevan and C Beadle, *Ber. Dtsch. Chem. Ges.*, 1901, **34**, 1513.

3. See, for example, E Ott and H Spurlin (eds), *Cellulose and Cellulose Derivatives*, 2nd Edn., Interscience Publishers, Inc., New York, 1954.
4. T Lieser, *Ann. der Chem.*, 1928, **43**, 464; T Lieser, *Ann. der Chem.*, 1929, **470**, 140; T Lieser, *Ann. der Chem.*, 1930, **483**, 132.
5. T Noguchi, *Sen-i Gakkaishi*, 1950, **6**, 153, see also: 155, 217, 270, 312, 314, 379, 381 444.
6. K Lauer, *Makromol. Chem.*, 1951, **5**, 287.
7. CY Chen, R Montana and C Grove, *Tappi*, 1951, **34**, 420.
8. J Willard and E Pascu, *J. Am. Chem. Soc.*, 1960, **82**, 4350.
9. B Philipp and KT Liu, *Faserforsch. Textiltech.*, 1959, **10**, 555.
10. M Horio, R Imamura, N Komatsu, A Sakata and T Kako, *Sen-i Gakkaishi*, 1963, **19**, 102.
11. K Kamide, K Okajima, T Matsui and K Kowsaka, *Polym. J.*, 1984, **16**, 857.
12. K Kamide, K Okajima, K Kowsaka and T Matsui, *Polym. J.*, 1985, **17**, 701.
13. W Brown and R Wikström, *Eur. Polym. J.*, 1966, **1**, 1.
14. W Randall, D Pegg, D Doddrell and J Field, *J. Am. Chem. Soc.*, 1981, **103**, 934.
15. D Gagnaire, D Mancier and M Vincendon, *J. Polym. Sci., Polym., Chem. Ed.*, 1980, **18**, 13.
16. K Kamide, K Okajima, K Kowsaka and T Matsui, *Polym. J.*, 1985, **17**, 701.
17. A Parfondry and A Perlin, *Carbohydr Res.*, 1977, **57**, 39.
18. K Kamide, K Kowsaka and K Okajima, *Polym. J.*, 1985, **17**, 707.
19. E Swan and C Purves, *Can. J. Chem.*, 1957, **35**, 1522.

2.10   THERMAL PROPERTIES OF CA SOLIDS[1]

Since Ueberreiter[2] attempted to determine the second order transition temperature of cellulose diacetate and triacetate (CTA) by dilatometry, the thermal analysis of CA solids has been energetically carried out using various methods, such as differential scanning analysis (DSC),[3-5] differential thermal analysis (DTA),[6,7] the dilatometry,[8-14] and mechanical methods.[12,13] However, the CA polymers used, except for those measured by Cowie and Ranson,[3] were unfractionated and their average molecular weights were not determined. Consequently, the molecular weight dependence of the glass transition temperature (T_g), the crystallization temperature (T_c), the melting point (T_m), and the onset of decomposition (T_d) have not yet been clarified. In addition, the dependence of T_g, T_c, T_m, and T_d on the total DS ($\langle\!\langle F \rangle\!\rangle$) has not been reported since research work has been directed primarily to cellulose acetate, with a very limited range of $\langle\!\langle F \rangle\!\rangle$ (i.e. $\langle\!\langle F \rangle\!\rangle > 2$). The relationship between thermal characteristic temperature and $\langle\!\langle F \rangle\!\rangle$ is paramount to the molecular design of CA fibers with high thermal stability that would render them suitable as flame proofing materials. Recently, successive solution fractionation (SSF) has made it possible to prepare CA with relatively narrow MWD ($M_w/M_v = 1.2$–1.5, M_w, and M_v, the weight average and number average molecular weights).[15-18]

In this section, we determine the T_g, T_c, T_m, and T_d of fractionated and unfractionated CA having different $\langle\!\langle F \rangle\!\rangle$ using DSC, thermal gravimetry (TG), and the X-ray diffraction method. The effect of $\langle\!\langle F \rangle\!\rangle$ and average molecular weight on T_g, T_m, and T_d of CA are discussed.

2.10.1   Cellulose acetate (DS 2.92)

The DSC curves of a CA (DS 2.92) whole polymer and fractions are shown in Figure 2.10.1. In all samples, a baseline shift toward the endothermic side corresponding

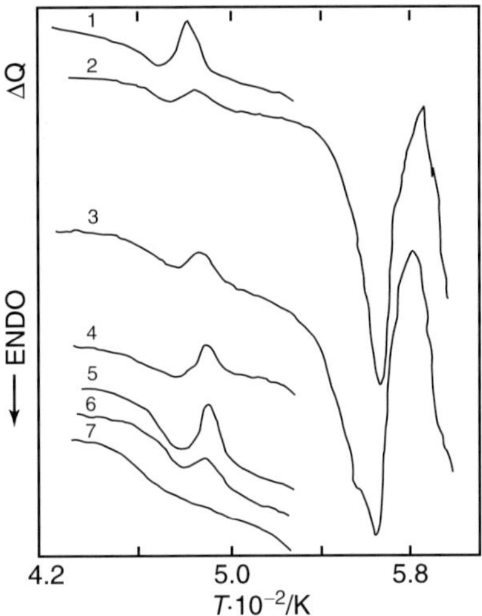

Figure 2.10.1   Differenctial scanning calorimetry (DSC) curves of CA (DS 2.92) fractions (1–6) and a whole polymer (7) at the heating rate 10 K min^{-1}. 1, Viscosity average molecular weight $M_v = 4.7 \times 10^4$;   2, $M_v = 1.97 \times 10^5$;   3, $M_v = 2.22 \times 10^5$;   4, $M_v = 3.59 \times 10^5$;   5, $M_v = 4.56 \times 10^5$; 6, $M_v = 5.83 \times 10^5$; 7, weight average molecular weight $M_w = 2.35 \times 10^5$.[1]

to the glass transition was observed between 440 and 480 K. The inflection point in the baseline shift was defined as T_g. In each DSC thermogram of the fractionated samples, an exothermic peak appeared approximately 30 K higher than T_g. The fractionated samples investigated here (sample TA3-3 with $M_v = 1.97 \times 10^5$ and sample TA3-4 with $M_v = 2.22 \times 10^5$) had very sharp endothermic peaks at 546 and 567 K, respectively. The T_g of whole CA polymers with $\langle\langle F \rangle\rangle \geq 2.9$, determined by the DSC, was reported to be 451–454 K (heating rate HR = 20 K min^{-1})[3] and these values are similar to the T_g of CA (DS 2.92) whole polymers (= 460 K) observed here.

Figure 2.10.2 shows the relation between T_g by DSC and M_v for the CA (DS 2.92) fractions. T_g increased with an increase in M_v in the range of $M_v \leq 3.5 \times 10^5$ and became constant (467 K) at $M_v \geq 3.5 \times 10^5$ within experimental error. Similar molecular weight dependence of T_g was reported for cellulose tricarbanilate, amylose triacetate, and amylose tripropionate.[19]

Figure 2.10.3 shows the effects of heating rate on the DSC curve of the sample TA3-4. In this case, HR varied from 5 to 40 K min^{-1}. T_g and the exothermic peak shifted to

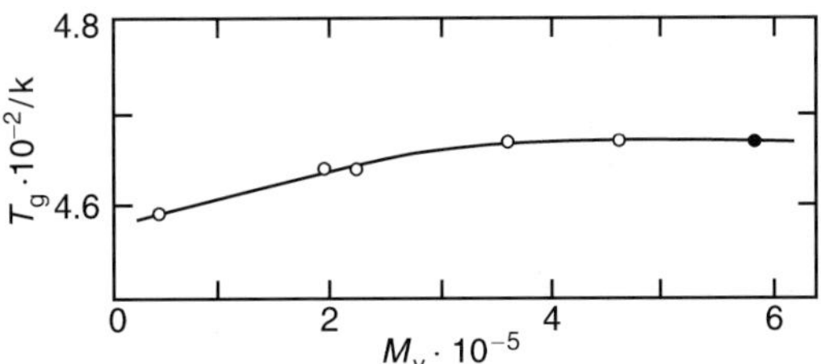

Figure 2.10.2   Glass transition temperature T_g of CA (DS 2.92) fractions as a function of the viscosity average molecular weight M_v.[1]

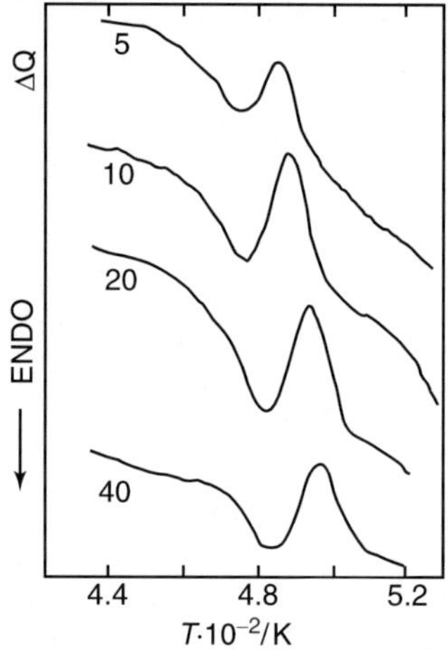

Figure 2.10.3   DSC curves CA (DS 2.92) fraction (TA3-4) with $M_v = 2.22 \times 10^5$ at a heating rate from 5 to 40 K min^{-1}. Number on each curve denotes heating rate.[1]

the higher temperature side at higher HR. Figure 2.10.4 depicts the DSC curves of fraction TA3-4 heated from room temperature to 510 K at a rate of 10 K min^{-1} (curve 1 in the figure), followed immediately by cooling to room temperature at the same HR as recorded in the curve. The exothermic peak at 483 K in the curve disappeared in the course of cooling and never appeared again in subsequent cycles of heating and cooling. Curves 2–8 denote the DSC curves during heating at a given repeating order of cycles.

The X-ray diffraction patterns of the fraction TA3-4 at five temperatures are shown in Figure 2.10.5. The diffractogram at 293 K has relatively sharp peaks at $2\theta = 7.1, 8.4, 17,$ and 20.6°, indicating the crystal CTA II type[20] structure. The diffractogram at 448 K

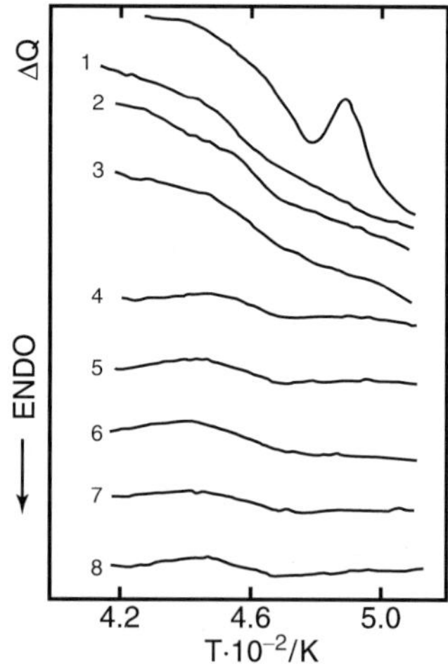

Figure 2.10.4   DSC curves of the CA (DS 2.92) fraction (TA3-4) with $M_v = 2.22 \times 10^5$ at a heating rate 10 K min^{-1}. 1, Curves in heating process; 1′, cooling immediately after heating (curve 1); 2–8, repeating order of cycle and curves during heating.[1]

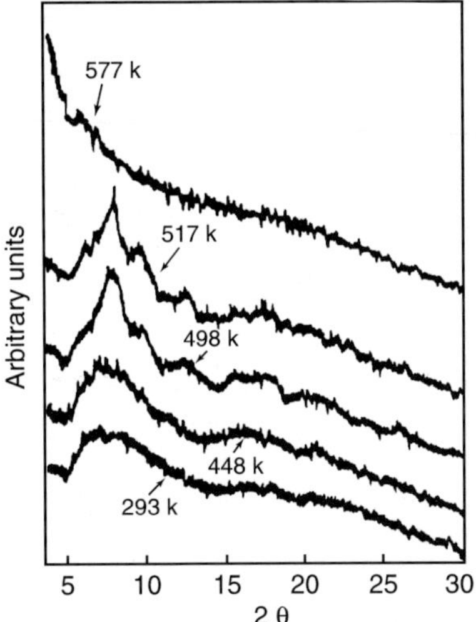

Figure 2.10.5   X-ray diffractograms of the CA (DS 2.92) fraction (TA3-4) at temperatures from 293 to 577 K; 2θ, diffraction angle.[1] Number to each curve denotes temperature.

was almost the same as that at 293 K. At 498 and 517 K, the intensity at $2\theta = 8.4$, 10.0, and 12.2° increased, and two new peaks appeared at $2\theta = 15.8$ and 17.8°. We evaluated the diffraction intensity, a measure of crystallinity, from the area under the peaks of the diffractogram between $2\theta = 5$ and 30°. Here, the X-ray diffractograms in the figure were redrawn with smooth lines to eliminate the noise and the baseline was conventionally drawn to pass through the point at $2\theta = 5$ and 14° and $2\theta = 14$ and 30°.

Figure 2.10.6 shows the relationship between the ratio of diffraction intensity at various temperatures to that at 293 K (we denote this ratio as DIR) and the temperature for cellulose acetate (CA) (DS 2.92) (solid line). The diffraction intensity ratio (DIR) of

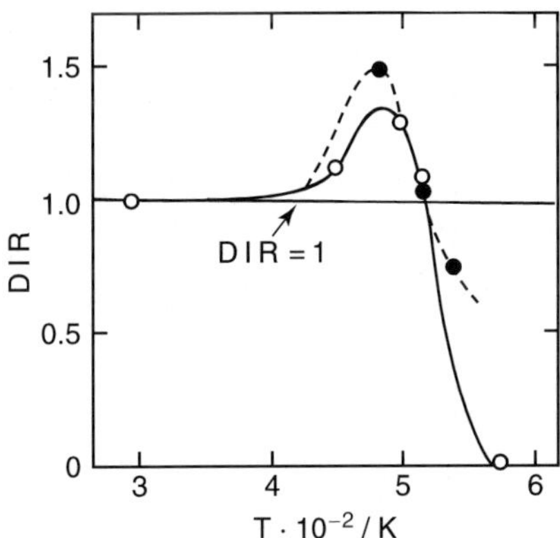

Figure 2.10.6   Relationship between diffraction intensity ratio (DIR) estimated from the X-ray diffractograms of the CA (DS 2.92) and CA (DS 2.46) and temperature.[1] —○—, CA (DS 2.92) fraction; – – •– –, CA (DS 2.46) fraction.

CA (DS 2.92) fraction at 498 K was about 30% higher than that at 293 K and reached a maximum near 490 K. The increase in X-ray diffraction intensity of the CTA solid with annealing was also reported by Sprague *et al.*[21] and Creely and Conrad.[22]

These results indicate the exothermic peak at 478 K in DSC curve of the fraction TA3-4 is caused by crystallization. The temperature of crystallization of CA (DS 2.92) was about 30° higher than T_g, corresponding to the 'cold crystallization' first observed for undrawn poly (ethylene terephthalate) fibers.[23] Above about 490 K, the crystallinity decreased monotonously and in the diffractogram at 577 K, all peaks disappeared and only an amorphous hallow remained.

We may thus conclude the endothermic peak of DSC at approximately 564 K to be T_m. This T_m value is very close to those of two fractions (fraction code W.366 ($\langle\!\langle F \rangle\!\rangle = 2.86$, $M_v = 1.63 \times 10^5$) and W.325 ($\langle\!\langle F \rangle\!\rangle = 2.88$, $M_v = 5.3 \times 10^4$)) (566 and 576 K, respectively) measured by Cowie and Ranson[3] using DSC at HR = 20 K min^{-1}. Patel *et al.*[7] regarded the endothermic peaks at 543 and 533 K in the DTA curves (HR = 10 K min^{-1}) of CA (DS 3.0) and CA (DS 2.94) whole polymer films as T_m. These are approximately 25–5 K lower than those from our results. The CA films used by Patel *et al.* were prepared by evaporating the solvent (CHCl$_3$) from the solution. There is a possibility that the crystallinity and the perfectness of the crystals of their sample are lower than those in this section.

2.10.2   CA (DS 2.46), CA (DS 1.75), and CA (DS 0.49)

Figures 2.10.7–2.10.9 show the DSC curves of some fractions and whole polymers of CA (DS 2.46), CA (DS 1.75), and CA (DS 0.49). The DSC curves of the CA (DS 2.46) fractions have the baseline shift to the endothermic side in the range of 450–480 K, due to the glass transition. However, in the case of the whole polymers, no remarkable shift in the DSC curve was detected. The fractions with $M_w = 1.08 \times 10^5$ and 1.85×10^5 have a

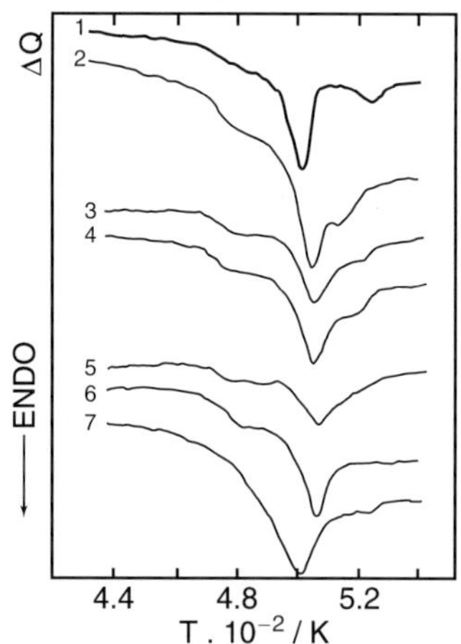

Figure 2.10.7   DSC curves of the CA (2.46) fractions (1–6) and a whole polymer (7) at heating rate of 10 K min^{-1}. 1, $M_w = 5.3 \times 10^5$; 2, $M_w = 7.4 \times 10^4$; 3, $M_w = 10.8 \times 10^5$; 4, $M_w = 1.41 \times 10^5$; 5, $M_w = 1.85 \times 10^5$; 6, $M_w = 2.65 \times 10^5$; 7, $M_w = 1.2 \times 10^5$.

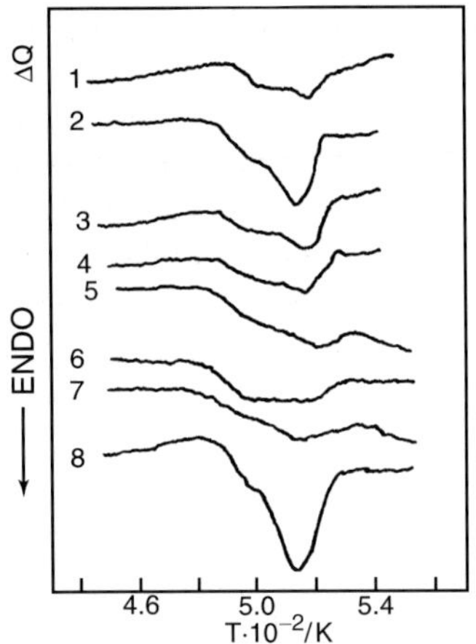

Figure 2.10.8   DSC curves of the CA (1.75) fractions (1–7) and a whole polymer (8) at heating rate of 10 K min^{-1}.[1]   1, $M_v = 2.1 \times 10^4$; 2, $M_w = 3.75 \times 10^4$; 3, $M_v = 5.53 \times 10^4$; 4, $M_w = 5.54 \times 10^4$; 5, $M_v = 7.72 \times 10^4$; 6, $M_w = 1.00 \times 10^5$; 7, $M_w = 1.31 \times 10^5$.

small exothermic peak at a temperature approximately 20 K higher than T_g. All of the samples of CA (DS 2.46) have the sharp endothermic peak in the range 500–510 K, and the fractions with $M_w \geq 1.41 \times 10^5$ have another endothermic peak (in the case of $M_w = 1.08 \times 10^5$ and 1.41×10^5, a shoulder) at approximately 520 K. The DSC curves of the CA (DS 1.75) fractions and whole polymers also have T_g in the temperature range from 480 to 500 K and an endothermic peak at approximately 510 K. The CA (DS 0.49) fractions and their whole polymers have T_g at 500–530 K, but no endothermic peak which appeared for the CA (DS 2.46) and CA (DS 1.75) polymers at approximately 500–510 K, was detected.

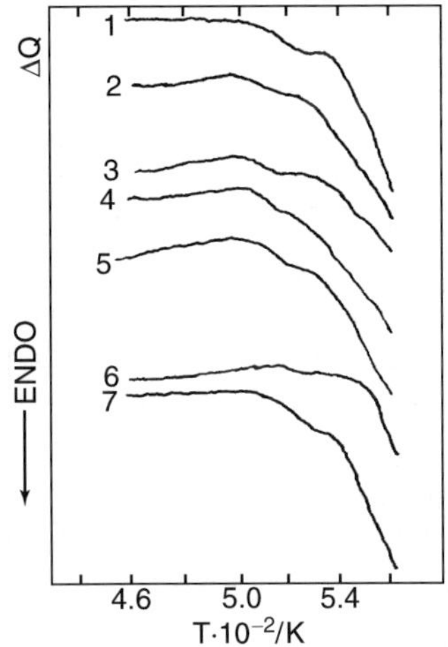

Figure 2.10.9   DSC curves of the CA (0.49) fractions (1–6) and a whole polymer (7) at a heating rate of 10 K min^{-1}.   1, $M_w = 4.55 \times 10^4$; 2, $M_w = 6.76 \times 10^4$; 3, $M_w = 7.94 \times 10^4$; 4, $M_w = 1.08 \times 10^5$; 5, $M_w = 1.11 \times 10^5$; 6, $M_w = 1.45 \times 10^5$.[1]

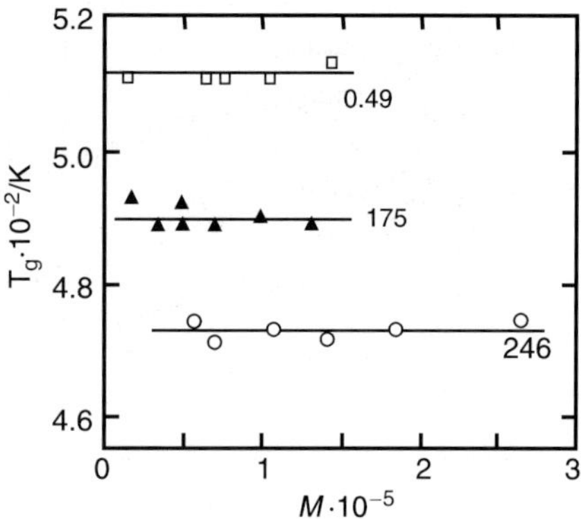

Figure 2.10.10 T_g of CA (2.46), CA (1.75) and CA (0.49) fractions determined by DSC as a function of molecular weight[2] (M_w or M_n; see, Table I of reference [1]). (○), CA (2.46); (▲), CA (1.75); (□), CA (0.49). Numbers on the lines denote total degrees of substitution.

T_g of CA with $\langle\!\langle F \rangle\!\rangle$ from 2.2 to 2.8 reported in the literature scattered a little, depending on the measuring method. For example, T_g of the cellulose acetate (DS 2.47) whole polymer was reported at 491 K by DSC[6] and that of the CA whole polymer with $\langle\!\langle F \rangle\!\rangle$ from 2.23 to 2.75 at 453–476 K by mechanical methods.[11,12] By dilatometry, a second order transition was observed at three temperature ranges: 288,[14] 310–340,[10–12,14] and 360–390 K.[10–12,14] Unfortunately, no experiment using dilatometry could be carried out at a temperature higher than 450 K, at which T_g was reported by the previous two methods. DSC and DTA data in the literature on CA with $\langle\!\langle F \rangle\!\rangle$ smaller than two are very few. Patel *et al.*[7] carried out DTA analysis on CA (DS 1.60) and CA (DS 0.9) whole polymers, but failed to detect the T_g of these polymers.

Figure 2.10.10 shows the relation between T_g of CA (DS 2.46), CA (DS 1.75), and CA (DS 0.49) fractions and their molecular weights. T_g of CA polymer was mainly determined by $\langle\!\langle F \rangle\!\rangle$ and was independent of the average molecular weight. Averaged T_g values CA (DS 2.46), CA (DS 1.75), and CA (DS 0.49) were 472.5, 489.5, and 511.5 K, respectively.

Figure 2.10.11 shows the X-ray diffractograms of a CA (2.46) fraction (EF3-14) at various temperatures. The diffraction curve at 293 K is almost the same as that of the CA (DS 2.92) polymer at 293 K, indicating the crystal form of CA (DS 2.46) to be CTA-II also. In the temperature range from 293 to 473 K, the diffraction intensity between $2\theta = 7.1$ and $17°$ increased with an increase in temperature. The relationship between the diffraction intensity ratio DIR and temperature is illustrated by the broken line in Figure 2.10.6. DIR reached a maximum at approximately 490 K. A small exothermic peak in DSC curve of CA (DS 2.46) fractions with $M_w = 1.08 \times 10^5$ and 1.85×10^5 at near 490 K resulted from crystallization. Above 490 K, DIR decreased remarkably and in this temperature region, there was an endothermic peak in DSC curve of CA (DS 2.46) fractions. In the diffractogram at 540 K, there remained peaks at 7.1 and $16°$, and DIR at this temperature was 0.75. The CA (DS 2.46) fractions may have multiple melting points as in the case of some synthetic polymers, such as heat treated polypropylene[24] and polyethylene.[25] One melting point of the CA (DS 2.46) fractions was approximately 500 K (we denote this point as T'_m). Other melting points were higher than 550 K.

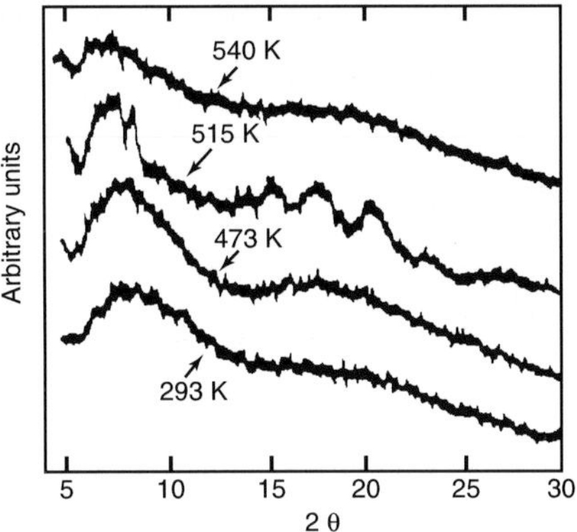

Figure 2.10.11   X-ray diffractograms of the CA (2.46) fraction (EF3-14) at temperatures from 293 to 540 K.[1] Curve numbers denote temperature.

Following the same line of the above discussion on CA (DS 2.92) and CA (DS 2.46), we can speculate that the endothermic peak observed at 510–520 K in the DSC curve of CA (DS 1.75) (Figure 2.10.8) corresponds to the melting point. Figures 2.10.7 and 2.10.8 suggest that T_m of the CA (2.46) and CA (DS 1.75) fractions is almost independent of molecular weight.

Figure 2.10.12 depicts the TG and differential thermogravimetry (DTG) curves of the CA (DS 2.92), CA (DS 2.46), and CA (DS 0.49) fractions.

T_d of the CA (DS 2.92), CA (DS 2.46), and CA (DS 0.49) fractions determined from these curves was 524, 508, and 483 K, respectively. T_d very slightly increased with increasing molecular weight and Figure 2.10.13 illustrates the case of the CA (DS 2.46) fractions, where $d\,T_d/d\,M_w$ was 8×10^{-5} K.

2.10.3   Effects of $\langle\!\langle F \rangle\!\rangle$ on T_g, T_m, and T_d

Figure 2.10.14 shows the plots of T_g and T_m determined by DSC, and T_d by TG as a function of $\langle\!\langle F \rangle\!\rangle$.

Here, if the molecular weight dependence cannot be neglected, as in the case of T_g of CA (DS 2.92), the value at $M_v = 1 \times 10^5$ can be employed. T'_m, was plotted as T_m, of CA (DS 2.46). T_d of regenerated cellulose (i.e. T_d at $\langle\!\langle F \rangle\!\rangle = 0$) was measured by Kamide et al.[26] by TG under the same conditions as those used here. T_g of CA increased almost linearly with a decrease in $\langle\!\langle F \rangle\!\rangle$ as follows:

$$T_g(\text{K}) = 523 - 20.3\langle\!\langle F \rangle\!\rangle \qquad (2.10.1)$$

Kamide et al.[27] succeeded, by the IR absorption method, in separately estimating the intra- and intermolecular interactions between O-acetyl and OH groups or between O-acetyl or between OH groups for CA (DS 2.92) and CA (DS 2.46). $\langle\!\langle F \rangle\!\rangle$ dependence of T_g for CA fractions suggests the strength of the interaction between O-acetyl and OH groups or between OH groups to be stronger than that between O-acetyl groups. T_g of cellulose, as an asymptotic value obtained by extrapolating T_g versus $\langle\!\langle F \rangle\!\rangle$ curve to $\langle\!\langle F \rangle\!\rangle = 0$, was 523 K. Manabe et al.[28] measured some isochronal dynamic

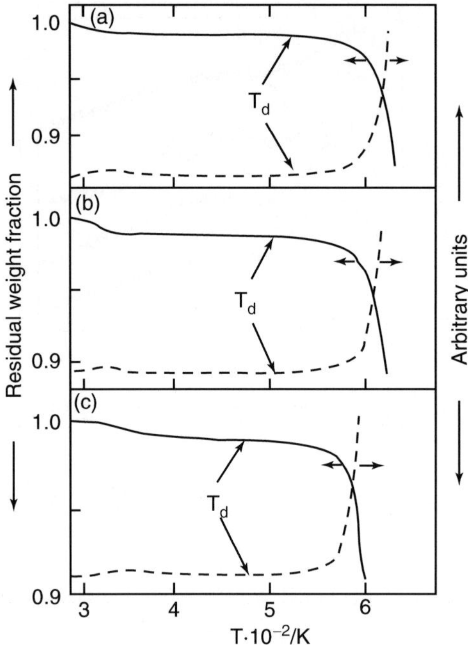

Figure 2.10.12   Thermogravimetry (TG) and differential thermogravimetry (DTG) curves of the CA (DS 2.92) ($M_w = 1.49 \times 10^5$) CA (DS 2.46) ($M_w = 1.41 \times 10^5$) and CA (DS 0.49) ($M_w = 1.45 \times 10^5$) fractions.[1] Solid line: TG curve; broken line: DTG curve. T_d, temperature at start of decomposition.

viscoelasticities of regenerated cellulose fiber at frequency of 110 Hz, observing three absorption peaks at 298 K (denoted as αH_2O), 513 K (α_2), 573 K (α_1), and one shoulder at 393 K (α_{sh}). The α_2 peak was due to the relaxation caused by microbrownian motion of the polymer molecules except for the tautly tied molecules in the amorphous region.[28] The asymptotic T_g value at $\langle\!\langle F \rangle\!\rangle = 0$ was close to 513 K of α_2 peak. T_m and T_d of CA became minimum at $\langle\!\langle F \rangle\!\rangle \approx 2.5$ and ≈ 1, respectively. $\langle\!\langle F \rangle\!\rangle$ dependence of T_d suggests that the molecular interaction in CA solids influences the thermal degradation temperature.

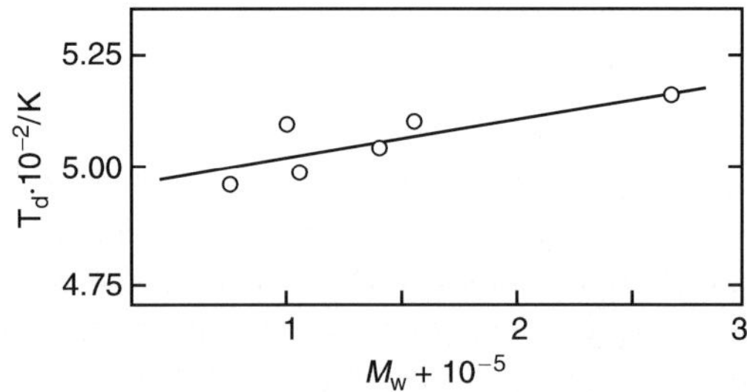

Figure 2.10.13 T_d of CA (2.46) fractions determined by TG as a function of weight average molecular weight M_w.[1]

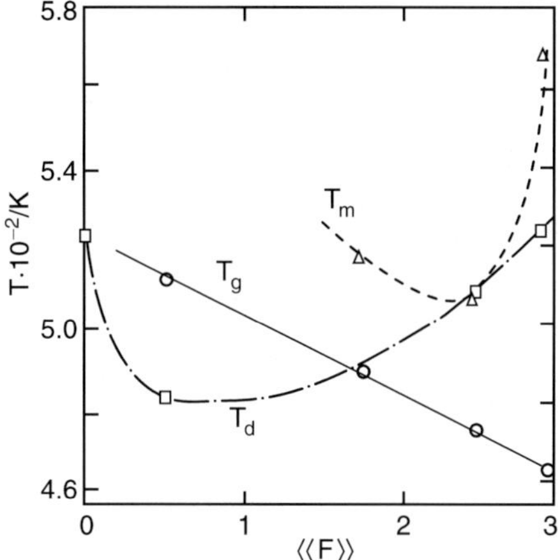

Figure 2.10.14 Total DS $\langle\langle F\rangle\rangle$ dependence of T_g, T_m, determined by DSC and T_d by TG.[1] (◯): T_g; (△), T_m; (□): T_d. Solid line, T_g; broken line, T_m; chain line, T_d.

In summary, the T_g, T_c, T_m, and T_d values of the CA fractions and whole polymers with $\langle\langle F\rangle\rangle$ ranging from 0.49 to 2.92 are listed in Table 2.10.1.

The CA (DS 2.92) and CA (DS 2.46) fractions have T_g, T_c (except some CA (DS 2.46) fractions), T_m, and T_d. In the case of the CA (DS 2.92), CA (DS 2.46), CA (DS 1.75), and CA (DS 0.49) whole polymers and all of CA (DS 1.75) and CA (DS 0.49) fractions, T_c did not appear. The CA (DS 0.49) polymer did not have T_m, since T_d is far below its melting point. T_g, T_m, and T_d depend on $\langle\langle F\rangle\rangle$ significantly, but are almost independent of average molecular weight, except for T_g of the CA (2.92) fractions.

Table 2.10.1

T_g, T_c, T_m, and T_d of cellulose acetates[1]

$\langle\langle F\rangle\rangle^a$	Fraction or whole polymer	Thermal characteristic temperature (K^{-1})			
		T_g	T_c	T_m	T_d
2.92	F^b	459–467	482–492	564, 567	525
	W^c	460	–	nm^d	nm
2.46	F	473^e	490	502–509	497–516
	W	480	–	500	nm
1.75	F	490^e	–	513–519	nm
	W	489	–	514	nm
0.49	F	512^e	–	–	472
	W	520	–	–	nm

^{a}Total degree of substitution.
^{b}Fraction.
^{c}Whole polymer.
^{d}Not measured.
^{e}Average value of all fractions.

REFERENCES

1. K Kamide and M Saito, *Polym. J.*, 1985, **17**, 919.
2. K Ueberreiter, *Z. Phys. Chem.*, 1941, **B48**, 197.
3. JM Cowie and RJ Ranson, *Makromol. Chem.*, 1971, **143**, 105.
4. S Nakamura, S Shindo and K Matuzaki, *J. Polym. Sci.*, Polym. Lett. Ed., 1971, 9, 591.
5. A Arneri and JA Sauer, *Thermochim. Acta*, 1976, **15**, 29.
6. H Trivedi, KC Patel and RD Patel, *J. Macromol. Sci.*, Chem. Ed., 1983, **A19**, 851.
7. KS Patel, KC Patel and RD Patel, *Makromol. Chem.*, 1970, **132**, 7.
8. FE Willey, *Ind. Eng. Chem.*, 1942, **34**, 1052.
9. RF Clash and LM Rynkiewicz, *Ind. Eng. Chem.*, 1944, **36**, 279.
10. L Manderkern and PJ Flory, *J. Am. Chem. Soc.*, 1951, **73**, 3206.
11. K Nakamura, *Kobunshi Kagaku*, 1956, **13**, 47.
12. J Russel and RG Kerpel, *J. Polym. Sci.*, 1957, **25**, 77.
13. A Sharple and FL Swinton, *J. Polym. Sci.*, 1961, **50**, 53.
14. JH Daane and RE Barker, *J. Polym. Sci. Polym. Lett. Ed.*, 1964, **2**, 343.
15. K Kamide, Y Miyazaki and T Abe, *Polym. J.*, 1979, **11**, 523.
16. K Kamide, T Terakawa and Y Miyazaki, *Polym. J.*, 1979, **11**, 285.
17. K Kamide, M Saito and T Abe, *Polym. J.*, 1981, **13**, 421.
18. K Kamide and M Saito, *Polym. J.*, 1982, **14**, 517.
19. JM Cowie and SAE Henshall, *Eur. Polym. J.*, 1976, **12**, 215.
20. See, for example, J Blackwell and RH Marchessault, in *Cellulose and Cellulose Derivatives* Part IV, (eds. NM Bikales and L Segel), Chapter 13, Wiley-Interscience, New York, 1971, p. 19.
21. BS Sprague, JL Riley and HD Noether, *Text. Res. J.*, 1958, **28**, 275.
22. JJ Creely and CM Conrad, *Text. Res. J.*, 1966, **32**, 184.
23. RF Schwenker and LR Beck, *Text. Res. J.*, 1960, **30**, 624.
24. K Kamide and M Sanada, *Kobunshi Kagaku*, 1967, **24**, 662.
25. T Kawai, M Hosoi and K Kamide, *Makromol. Chem.*, 1971, **146**, 55.
26. K Kamide, K Okajima and K Uchida, unpublished results.
27. K Kamide, K Okajima and M Saito, *Polym. J.*, 1981, **13**, 115.
28. S Manabe, Y Komatsu, M Iwata and K Kamide, *Polym. J.*, 1986, **18**, 1.

2.11   SURFACE ACTIVITY OF AQUEOUS SOLUTION OF CELLULOSE ACETATE[1]

The correlation between water solubility of CA and the substituent group distribution towards the three hydroxyl groups directly attached to the C_2, C_3, and C_6 positions in a glucopyranose unit of CA $\langle f_k \rangle$ ($k = 2, 3$, and 6) was intensively investigated mainly using high resolution ^{1}H NMR and ^{13}C NMR by Miyamoto *et al.*[2] and Kamide *et al.*[3] The conclusions reached by these two groups are as follows: (1) CA samples with $\langle\langle F \rangle\rangle = 0.38-1.24$, prepared by acetylation of cellulose dissolved in *N,N*-dimethylacetamide (DMAc)-lithium chloride mixture (one-step method), showed no water solubility, while these CA samples completely dissolve in DMAc and *N,N*-dimethylsulphoxide (DMSO);[2] (2) the water soluble CA obtained by the two-step method satisfies the relation $\langle\langle f_6 \rangle\rangle \approx \langle\langle f_2 \rangle\rangle \approx \langle\langle f_3 \rangle\rangle^{2,4}$ or $\langle\langle f_6 \rangle\rangle \approx (\langle\langle f_2 \rangle\rangle + \langle\langle f_3 \rangle\rangle)/2,^3$ whereas the relation $\langle\langle f_6 \rangle\rangle \approx (\langle\langle f_2 \rangle\rangle + \langle\langle f_3 \rangle\rangle)/2,^{3,4}$ holds for the CA prepared by the one-step method. Based on these facts, Kamide *et al.*[3] suggested that the water solubility of the CA is brought about by cleavage of intramolecular hydrogen bonds between the hydroxyl group at the C_3 position and a heterocyclic oxygen atom in neighboring glucopyranose units of the cellulose chain.

Cellulose ethers have been commercialized as emulsion stabilizers and polymer flocculators[5–7] using their surfactant properties. It is well known that the surface activity of aq. solutions is determined by the mole balance of hydrophilic and hydrophobic groups (HLB) in the surfactant molecule adsorbed on the surface of the solution.[8] No work relating the surface activity of aq. solutions of cellulose derivatives to their $\langle\langle f_k \rangle\rangle$ and chain conformation has hitherto been carried out.

In this section, an attempt will be made to clarify the effects of the polymer concentration, the molecular weight, and $\langle\langle F \rangle\rangle$ on the surface tension of aq. solutions of CA samples with $\langle\langle F \rangle\rangle = 0.58$–$0.80$, and to explain the surface activity of the solutions in terms of the dissolved state of the CA polymeric chain in water.

Samples

A whole polymer of CA (DS 2.46) was hydrolyzed using hydrochloric acid as catalyst at 30 °C. Three CA samples with different $\langle\langle F \rangle\rangle$ were prepared by varying the hydrolysis period. The $\langle\langle F \rangle\rangle$ of each hydrolyzed cellulose acetate, determined by neutralization titration, was 0.80 (sample code W8), 0.68 (W7), and 0.58 (W6).

Sample W8 was dissolved in distilled water at a polymer concentration (c, g cm^{-3}) of 2% at 60 °C. The solution obtained was used for successive solution fractionation (SSF).[9,10] SSF was carried out under the following conditions: fractionation steps, 10; solvent, water; nonsolvent, methanol; temperature, 30 °C. We used fractions 1, 5, and 10 (sample code F8-1, F8-5, and F8-10, respectively) for further measurements.

Membrane osmometry

Osmotic pressure π of CA in DMAc was measured using a Wescan Osmometer Model 231 (Wescan Inc., USA) at 20 ± 0.1 °C. Regenerated cellulose membrane RC 51 (Schraicher & Schnell, USA) was employed as a semipermeable membrane (see also Section 3.3).

Static light scattering

Sample W8 was dissolved in freshly distilled water or DMAc at 60 °C giving 0.5 wt% solutions. The specific refractive index increment ($\partial n/\partial c$) of these solutions was measured using a differential refractometer model DR-4 (Shimadzu, Kyoto) with incident wave length $\lambda_0 = 633$ nm at 20 °C. The values found were 0.141 cm^3 g^{-1} for CA (DS 0.8) in water and 0.078 cm^3 g^{-1} for CA (DS 0.8) in DMAc.

The solutions were centrifuged at 1.03×10^5 g for 1 h in a Hitachi model 55p-7 automatic preparative ultracentrifuge, followed by dilution with the solvents. The dilute solutions were carefully filtered through a membrane, Fluorophore type FP-010 (Sumitomo Denko, Osaka), with mean pore size 0.1 µm, were directly transferred into a light scattering cylindrical cell. The light scattering intensity was measured over an angle θ range from 30 to 150° using unpolarized light with $\lambda_0 = 633$ nm at 25 °C in

Table 2.11.1

Molecular characteristics of water soluble CA in DMAc at 20 °C[1]

Sample code	$\langle\langle F\rangle\rangle$[a]	Membrane osmometry		Viscometry $M_v \times 10^{-4}$[d]	Static light scattering	
		$M_v \times 10^{-4}$[b]	$A_2 \times 10^{3}$[c]		$A_2 \times 10^{3}$[c]	$\langle\langle S^2\rangle\rangle_z^{1/2} \times 10^{6}$[e]
W8[f]	0.80	2.4	3.1	4.1	1.2	1.90
F8-1[g]	0.79	1.5	1.5	–	–	–
F8-5[g]	0.81	3.4	1.7	–	–	–
F8-10[g]	0.81	5.5	1.5	–	–	–
W7′	0.68	2.5	3.6	–	–	–
W6′	0.58	2.0	3.5	–	–	–

[a]Total degree of substitution.
[b]Number average molecular weight.
[c]Second virial coefficient (cm^3 mol g^{-2}).
[d]Weight average molecular weight.
[e]z average radius of gyration (cm).
[f]Whole polymer.
[g]Fractionated polymer.

a photogenic diffusiometer type DLS 700 (Otsuka Electronics, Osaka). The data obtained were analyzed according to Zimm's procedure.[11]

In Table 2.11.1, the molecular characteristics of CA in DMAc, as determined by membrane osmometry and light scattering, are summarized. The ratio M_w/M_n of sample W8 is *ca.* 1.7, which is significantly smaller than that of CA (DS 0.49) whole polymer $(M_w/M_n = 2.2)^4$ prepared by the same method as that employed in this study.

Surface tension

Surface tension of CA solutions was measured using a Whilhelmy type surface tensiometer, model CBVP-A3 (Kyowa Kaimenkagaku, Tokyo) over a temperature range of $5-75 \pm 0.5$ °C. In order to confirm the cleanness of the Whilhelmy plate in the apparatus, the surface tension of distilled water was measured before each measurement on the CA solutions.

Dynamic light scattering

The first-order normalized time correlation function of the scattered light from CA (0.8) whole polymer–water system, $N(q, \tau)$ (q: the propagation vector length, τ: the correlation time; see, eq. (3.5.12)) were obtained by dynamic light scattering (DLS). DLS measurements were carried out on an Otsuka Electronics model DLS 700 under the following conditions: $\lambda_0 = 633$ nm; $\theta = 90°$; gate time = 320 ms; clock rate = 20 µs; correlation channel = 512; temperature = 20 °C (see Section 3.5.5). The hydrodynamic diameter d of the CA polymer chains in water was determined from the diffusion coefficient $\langle D\rangle$ using eq. (3.5.16), which is calculated from $\ln N(q, \tau)$.

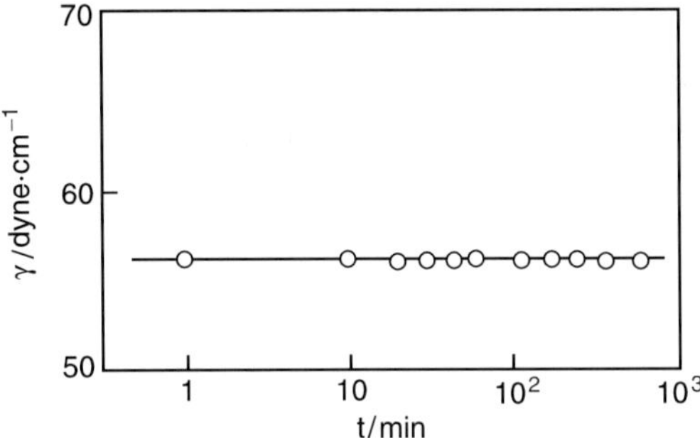

Figure 2.11.1   Time dependence of the surface tension (γ) of 0.1% CA (DS 0.8) whole polymer (sample W8) aq. solution at 20 °C.[1] The time at 5 h after the completion of dissolution is taken as zero.

Figure 2.11.1 shows the time (t) dependence of surface tension (γ) of 0.1% aq. cellulose acetate (DS 0.8) (Sample W8) solution at 20 °C. Measurements of surface tension were started at 5 h after dissolution was completed. Surface tension of the aq. CA solution is almost independent of time, indicating that, under the conditions of preparation, surface phase (where surfactant molecules are adsorbed) and bulk phase are in equilibrium. Surface tension of aq. cellulose acetate (DS 0.8) solution with $c = 0.1\%$ ($\gamma = 56$ dyn cm^{-1}) is *ca.* 22% less than that of pure water (72 dyn cm^{-1}) at 20 °C. The γ value of the aq. CA (DS 0.8) solution can be compared with the values of other CD solutions, for example, 64–66 dyn cm^{-1} for 0.1% aq. solution of the commercial hydroxyethyl cellulose (HEC) with $\langle\!\langle F \rangle\!\rangle$ ranging from 1.0 to 1.3 at 20 °C, and 71 dyn cm^{-1} for 2% aq. solution of sodium carboxymethyl cellulose with $\langle\!\langle F \rangle\!\rangle$ ranging from 0.6 to 0.8 at 25 °C.[7] It is recognized that water soluble CA is highly effective in lowering the surface tension of water, and may be useful as a polymer surfactant.

Figure 2.11.2 depicts the concentration dependence of γ of aq. solutions of three unfractionated samples, W8 (Figure 2.11.2a), W7 (Figure 2.11.2c), and W6

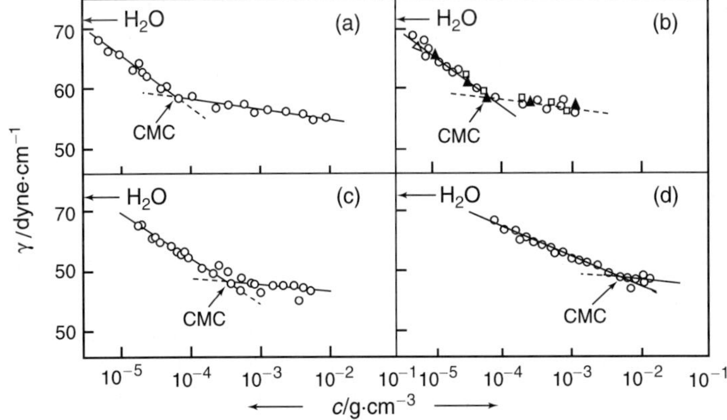

Figure 2.11.2   Concentration (c) dependence of surface tension (γ) of aq. solutions of (a) CA (DS 0.8) whole polymer; (b) CA (DS 0.8) fractions (○): F8-1, (△): F8-5; (□): F8-10; (c) CA (DS 0.68) whole polymer; (d) CA (DS 0.58) whole polymer, at 20 °C.[1] The γ value of water 20 °C is shown.

Table 2.11.2

Critical micelle concentration (c^*), surface tension (γ), surface excess concentration (Γ), and number of molecules in a micelle (n') of CA at 20 °C[1]

Sample code	$\langle\!\langle F \rangle\!\rangle$	$c^* \times 10^5$ (g cm^{-3}) ($m^* \times 10^9$ (mol cm^{-3}))	γ (dyne cm^{-1}) at CMC	Γ (mol cm^{-2}) above CMC	n'
W8	0.80	7.0 (3.0)	59.0	1.53	5.5
F8-1	0.79	7.0 (3.0)	59.0	1.53	–
F8-5	0.81	7.0 (3.0)	59.0	1.53	–
F8-10	0.81	7.0 (3.0)	59.0	1.53	–
W7	0.68	29 (14)	59.3	1.27	5.6
W6	0.58	520 (260)	59.2	0.77	5.9

(Figure 2.11.2d), and of three fractions, F8-1, F8-5, and F8-10 (Figure 2.11.2b), at 20 °C. The relationship between γ and log c for each aq. CA solution can be approximately represented by kinked lines. Here, the critical micelle concentration (CMC) c^* is defined as the polymer concentration at the kinked point. In the range above c^*, the concentration dependence of γ is significantly small. γ of the solutions of whole CA polymers at the same polymer concentration decreases monotonically with an increase in $\langle\!\langle F \rangle\!\rangle$. That is, the surface activity is higher in aq. solutions of CA with larger $\langle\!\langle F \rangle\!\rangle$.

Table 2.11.2 lists values of c^* and γ at CMC (γ^*) for aq. CA solutions. Very few data on CMC (c^* and γ^*) of nonionic polymer surfactants, including cellulose acetate, have *hitherto* been reported. For aq. solutions of HEC ($M_w = 8 \times 10^4$) chemically modified with 1,2-epoxy alkane, Landoll[12] investigated the polymer concentration dependence of interfacial tension between aq. HEC solutions and toluene, and determined c^* to be 5×10^{-4} g cm^{-3}, which is close to that of the aq. solution of sample W7. Sample W8 has almost the same c^* as typical low molecular surface active compounds (0.2×10^{-4} to 1.43×10^{-4} g cm^{-3}), such as polyethylene oxide with a degree of polymerization ranging from 8 to 12.[13] Examination of c^* for the CA (DS 0.8) fractions with different molecular weight leads to the conclusion that c^* of CA (DS 0.8) has practically insignificant molecular weight dependence. Contrary to this, Figure 2.11.3 shows that c^* of CA solution remarkably depends on the total DS $\langle\!\langle F \rangle\!\rangle$; an increase in $\langle\!\langle F \rangle\!\rangle$ of 0.22 gives *ca.* 75 times higher c^*.

Figure 2.11.4 shows the relationship between γ and c for sample W8 in DMAc at 20 °C. Evidently, γ of this system is independent of c, being almost the same as that of DMAc liquid.

This means that the CA polymer shows no surface activity in DMAc solution. DMSO and formamide as well as DMAc are known[4] as good solvents of CA polymer with $\langle\!\langle F \rangle\!\rangle$ ranging from 0.8 to 0.45, and CA dissolved in these organic solvents does not behave as a surfactant.

From analysis of the chemical shifts of O-acetyl methyl protons and hydroxyl protons by ^{1}H NMR, Kamide *et al.*[14] demonstrated the existence of specific interactions between polar groups in CA (DS 0.49) molecules, such as the O-acetyl methyl and the hydroxyl groups, and the solvent molecules, including water and DMAc. They proved that the DMAc molecules interact with both polar groups and behave as electron donors to

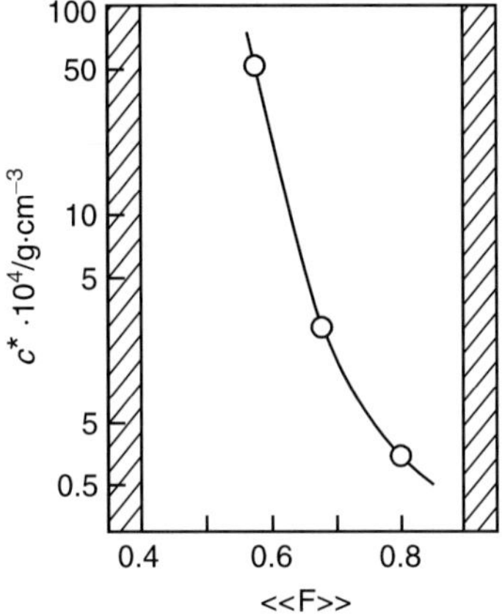

Figure 2.11.3   Total DS ($\langle\!\langle F \rangle\!\rangle$) dependence of the critical micelle concentration (c^*) of CA whole polymers in water at 20 °C.[1] Shadowed area denotes water insoluble region.

the O-acetyl methyl groups and as electron acceptors to the hydroxyl groups, while water molecules interact mainly with the hydroxyl groups. From these experimental findings, it can be considered that CA with low total DS shows surface activity in water owing to the hydrophobic nature of the O-acetyl methyl groups and the hydrophilic nature of the hydroxyl groups. On the other hand, CA does not act as a surfactant in DMAc because the CA molecule has no polar group acting hydrophobically against the DMAc molecules. CA molecules with higher DS have large numbers of hydrophobic groups and, accordingly, readily tend to aggregate, forming micelles in water at lower polymer concentrations than CA with lower degree of substitution. However, inspection of Table 2.11.1 shows that γ of aq. CA solutions is practically independent of $\langle\!\langle F \rangle\!\rangle$, and hence the molar balance of hydrophobic and hydrophilic groups plays a less effective role in the surface activity of aq. solutions of CA with $\langle\!\langle F \rangle\!\rangle$ in the range from *ca.* 0.5 to 0.80.

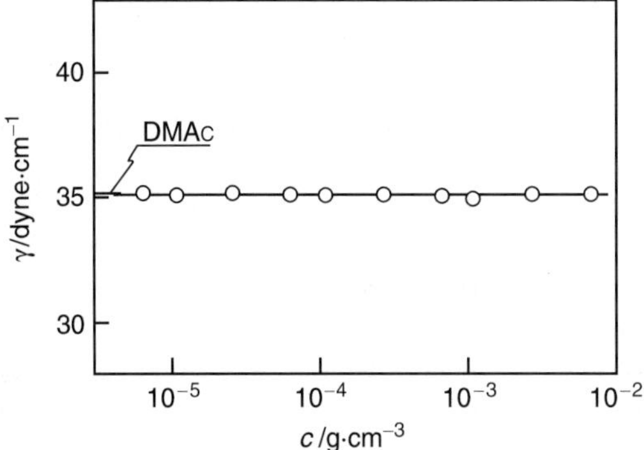

Figure 2.11.4   Relation between the surface tension (γ) of CA (0.8) whole polymer-DMAc system and the polymer concentration c at 20 °C.[1]

According to Gibbs' adsorption isotherm, γ related to the surface excess concentration γ by the equation:[15]

$$d\gamma = -\gamma\Gamma\,d\Delta\mu_1 \qquad (2.11.1)$$

where $\Delta\mu_1$ is the chemical potential of the solute.

Equation (2.11.1) is readily rewritten as:

$$\Gamma = -(d\gamma/d\ln c)/(d\Delta\mu_1/d\ln c) \qquad (2.11.1')$$

Based on the Flory–Huggins theory of polymer solutions,[16] $\Delta\mu_1$ is expressed as a function of the polymer volume fraction v_p by the equation

$$\Delta\mu_1 = RT\{\ln v_p - (X_n - 1)(1 - v_p) + \chi X_n(1 - v_p)^2\} \qquad (2.11.2)$$

where X_n is the number average degree of polymerization, v_p, the polymer volume fraction, and χ is the thermodynamic interaction parameter.

Differentiating eq. (2.11.2) with respect to $\ln v_p$ and neglecting the terms higher than v_p^2 we obtain

$$d\Delta\mu_1/d\ln v_p = RT(1 - \delta vc) \qquad (2.11.3)$$

where

$$\delta = [1 - 2X_n(0.5 - \chi)] \qquad (2.11.4)$$

and v is the specific volume of the polymer.

Combination of eqs. (2.11.1') and (2.11.3) gives

$$\Gamma = -(d\gamma/d\ln c)/\{RT(1 - \delta v^* c^*)\} \qquad (2.11.5)$$

where v^* is the specific volume of the polymer at c^*. δ can be calculated by eq. (2.11.4) from X_n for a given χ value. Putting the experimental data of c^*, cd $\gamma/d\ln c$) in the range $c \leq c^*$, and δ, calculated thus, into eq. (2.11.5), we can estimate Γ as a function of χ. Note that χ ranges from 0 to 0.5 for polymer solutions. Γ of each sample solution tends to increase slightly with an increase in χ. For example, Γ varies from 1.525×10^{-10} to 1.534×10^{-10} mol cm^{-2} corresponding to a variation of χ from 0 to 0.5 for the whole and fractionated CA (DS 0.80) polymer solutions. In Table 2.11.2, the average value, in this case $(1.525 + 1.534) \times 10^{-10}/2$ ($\equiv \langle\Gamma\rangle$), of Γ is listed. Evidently, Γ is larger in aq. solutions of CA with higher $\langle\langle F\rangle\rangle$.

Figure 3.5.7 shows the hydrodynamic diameter d, as determined by DLS, of CA (0.80) whole polymer chains in water. d increases stepwise at $c = 1 \times 10^{-4}$ g cm^{-3}, which is roughly equal to c^* estimated from the concentration dependence of γ. This indicates that just immediately above c^*, polymer chains associate together, forming micelles whose average size is almost independent of the polymer concentration. In Table 2.11.3, static light scattering data on M_w, A_2, and the z average radius of the gyration, $\langle S_2\rangle_z^{1/2}$ of CA (DS 0.8) whole polymer in water and DMAc are summarized. M_w of CA (DS 0.80) in water, determined from the Zimm plot at the polymer concentration region higher than c^* (in this case 1.25×10^{-3} g cm^{-3}), is four times larger than that in DMAc, in which CA molecules were proved to dissolve molecularly. Hence we can consider that a micelle in water consists of *ca.* four CA (DS 0.80) molecules.

Table 2.11.3

Results of static light scattering measurements on unfractionated CA
(sample W8) water system above critical micelle concentration at 20 °C

Solvent	$M_w \times 10^{-4}$	$A_2 \times 10^{-10}$ $(cm^3 \ mol \ g^{-2})$	$\langle S^2 \rangle_z^{1/2} \times 108$ (cm)
H_2O	*ca.* 16	*ca.* 0	4.0
DMAc[a,b]	4.05	1.17	1.9

[a]N,N-Dimethylacetamide.
[b]These data are listed again for comparison.

Alternatively, the number of molecules in a micelle can be estimated using the mass
action model:[17] Assume that n' CA molecules associate to form one micelle and that the
ratio (i.e. the mole fraction) of CA molecules involved in micelles to those existing in the
solution is x. If the masses of molecularly dispersed CA molecules and associated
molecules are well balanced, then the following relationship holds:

$$n'CA = (CA)n' \tag{2.11.6}$$

$$c(1 - x)cx/n'$$

The equilibrium constant K in the equilibrium (eq. (2.11.6)) is given by

$$K = (cx/n')/[c(1 - x)]^{n'} \tag{2.11.7}$$

Equation (2.11.7) can be rewritten as

$$c = c(1 - x) + n'K[c(1 - x)]^{n'} \tag{2.11.8}$$

Since the experimental results by DLS measurements (Figure 3.5.7) indicate that below
the CMC no micelles are formed (i.e. $x = 0$, $K = 0$), we obtain the relationship:

$$c = c(1 - x)(c < c^*) \tag{2.11.9}$$

d γ/d ln c in eqs. (2.11.1') and (2.11.5) is rewritten using eq. (2.11.9) in the form

$$d\gamma/d\ln c = d\gamma/d\ln [c(1 - x)]^{n'}(c < c^*) \tag{2.11.10}$$

When we assume that above CMC, most of the CA molecules are involved in micelles
(i.e. $x \approx 1$), eq. (2.11.8) can be roughly approximated to

$$c = n'K[c(1 - x)]^{n'} \quad (c > c^*) \tag{2.11.11}$$

Accordingly, dγ/d ln c is given by

$$d\gamma/d\ln c = d\gamma/(n' \ d\ln[c(1 - \chi)]) \quad (c > c^*) \tag{2.11.12}$$

Combining eqs. (2.11.10) and (2.11.12), we finally obtain the relationship:

$$n' = (d\gamma/d\ln c)_{c<c^*}/(d\gamma/d\ln c)_{c<c^*} \tag{2.11.13}$$

Equation (2.11.13) indicates that n' can be evaluated from the experimental data of
γ/d ln c at $c < c^*$ and at $c > c^*$ of aq. solutions of CA polymers, and the results for

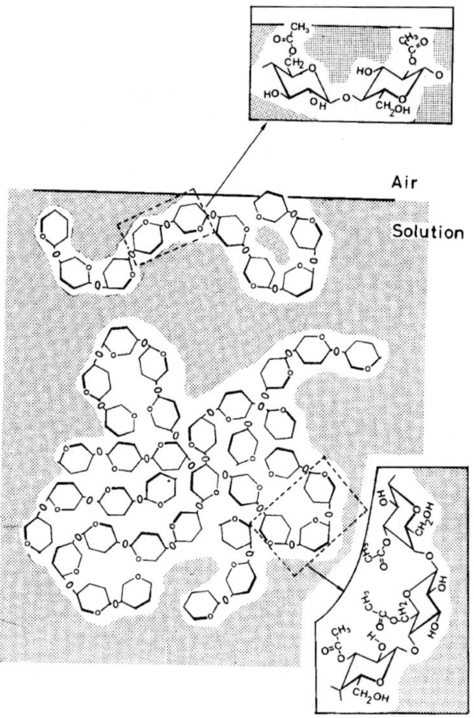

Figure 2.11.5   Schematic representation of dissolved state of CA (0.8) in water at 20 °C.[1]

the three whole CA polymers in water at 20 °C are given in Table 2.11.2. Values of n' lie
between 5 and 6, and slightly depend on $\langle\langle F \rangle\rangle$. The value of n' for CA (DS 0.80) whole
polymer (5.5) roughly coincides with the association number, defined as the ratio of M_w
in water to that in DMAc (i.e. 3.95 for CA (DS 0.8) whole polymer).

On the basis of the above experimental results, we can speculate on the dissolved state
of CA (DS 0.8) whole polymer in water above CMC at 20 °C; a schematic representation
is illustrated in Figure 2.11.5. At the air–water interface, the hydrophobic O-acetyl
methyl groups in the partially substituted glucopyranose units preferentially exist and the
hydrophilic hydroxyl groups attached to the same glucopyranose units are found only in
the bulk phase (i.e. nonsurface region) of the aq. solution. In the bulk phase the
unsubstituted glucopyranose rings, which have only hydrophilic groups, can exist. A
micelle consists of approximately four CA molecules, owing mainly to the hydrophobic
interaction between the O-acetyl methyl groups or to the hydrogen bonding between the
O-acetyl methyl and the hydroxyl groups. Most hydroxyl groups in CA polymeric chains
forming the micelle interact with water molecules surrounding the micelle. Likewise, CA
(DS 0.58) and CA (DS 0.68) molecules should dissolve in water, even if the surface
excess concentration and the polymer concentration in the bulk phase in these solutions
are different to those of CA (DS 0.80).

Figure 2.11.6 shows the concentration dependence of γ for CA (DS 0.80) solution at
20, 30, and 40 °C. $d\gamma/d\ln c$ above CMC is almost independent of T, while in the range of

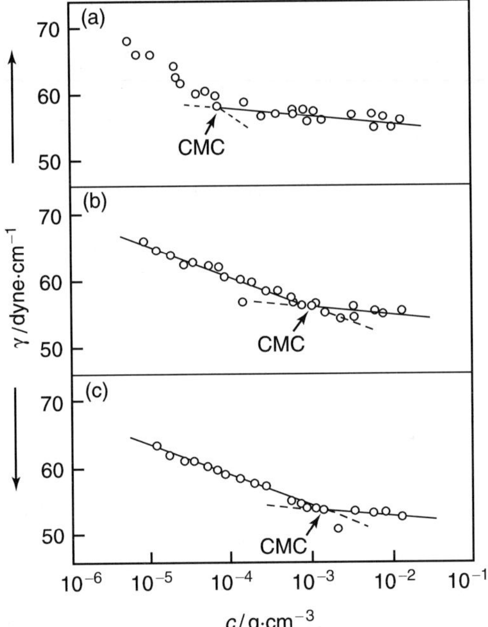

Figure 2.11.6   Concentration (c) dependence of the surface tension (γ) of aq. CA (DS 0.8) whole polymer solution at (a) 20 °C; (b) 30 °C; and (c) 40 °C.[1]

$c < c^*$, dγ/dln c is dependent on the temperature. c^* in the range of $c < c^*$ of CA (DS 0.80) solutions monotonically increases with T. γ for aq. CA (DS 0.80) solution and for pure water shows concomitant decrease with an increase in temperature at least up to 80 °C, and the ratio of the surface tension of the polymer solution to water remains almost constant (*ca.* 0.72) over the whole temperature range investigated. This implies that the effect of CA (DS 0.8) polymer as surfactant is definitely independent of the solution temperature.

REFERENCES

1. K Kamide, M Saito and T Akedo, *Polym. Int.*, 1992, **27**, 35.
2. T Miyamoto, Y Sato, T Shibata, M Tanahashi and H Inagaki, *J. Polym. Sci., Poly. Chem. Ed.*, 1985, **23**, 1373.
3. K Kamide, K Okajima, K Kowsaka and T Matsui, *Polym. J.*, 1987, **19**, 1405.
4. K Kamide, M Saito and T Abe, *Polym. J.*, 1981, **13**, 421.
5. See, for example, E Oomori, *Polymer Flocculator*, Kohbunshi Kankokai, Tokyo, 1973, p. 90.
6. N Sarkar, *Polymer*, 1984, **25**, 451.
7. See, for example, Dicell Co., Commercial brochure (1988).
8. WC Griffin, *J. Soc. Cosm. Chem.*, 1949, **1**, 311.
9. K Kamide, Y Miyazaki and T Abe, *Brit. Polym. J.*, 1981, **14**, 168.
10. K Kamide and M Saito, *Adv. Polym. Sci.*, 1987, **83**, 1.
11. BH Zimm, *J. Chem. Phys.*, 1948, **16**, 1093–1099.
12. LM Landoll, *J. Polym. Sci. Polym. Chem. Ed.*, 1982, **20**, 443.

13. See, for example, M Tanaka, in *Applied Colloid Science*, (eds. R Matsuura and T Kondo), Chapter 8, Hirokawa Shoten, Tokyo, 1963.
14. K Kamide, K Okajima and M Saito, *Polym. J.*, 1981, **13**, 115.
15. See, for example, (ed. K Durham), *Surface Activity and Detergency*, Chapter 1, Macmillan Co. Ltd., New York, 1961.
16. See, for example, PJ Flory, *Principles of Polymer Chemistry*, Chapter 12, Cornell University Press, Ithaca, NY, 1953.
17. DC Thomas and SD Christian, *J. Colloid Interface Sci.*, 1980, **78**, 466.

− 3 −

Molecular Properties of Cellulose and Cellulose Derivatives

3.1 SAMPLE PREPARATION

3.1.1 Water soluble cellulose acetate (DS 0.49)

Preparation[1]

Cellular acetate (CA) flakes (DS = 2.46, weight-average molecular weight $M_w = 1.2 \times 10^5$) synthesized in Section 3.1.2[2] was crushed into a fine powder in a ball mill and hydrolyzed using hydrochloric acid as catalyst, in a 77% aq. acetic acid solution under nitrogen atmosphere at 40 °C for different times of period. The incompletely substituted CA thus hydrolyzed was recovered by precipitation with acetone, filtered, washed with acetone to remove all traces of acetic acid in the polymer, dried *in vacuo* at room temperature, and finally stocked in a dark, cold place. Ten CA whole polymers of different DS (polymer code MAW-1 to MAW-10) were prepared in this manner. The combined acetic acid content (AC) of the unfractionated and fractionated cellulose acetates was determined by back titration, using sodium hydroxide and sulfuric acid. From ^{13}C NMR analysis of the CA whole polymer (DS = 0.49) dissolved in deuterated dimethyl sulfoxide (DMSO-d_6), the probability of finding *O*-acetyl groups attached to the C_2, C_3, and C_6 atoms was estimated to be 0.10, 0.20, and 0.19, respectively.[3]

Solubility behavior[1]

Ten CA samples with DS ranging from 0.10 to 1.96 in the water were subjected to solubility testing. Table 3.1.1 shows the results of this test. CA is water soluble in a DS range of 0.39–0.81. It has long been believed that CA is soluble in water in a DS range from 0.6 to 0.8 at room temperature.[4] However, CA with a DS lower than 0.6 (in this case, 0.4) is evidently soluble in water at room temperature. The CA sample (polymer code MAW-5, DS = 0.49) was used for further study.

In order to confirm the stability of aq. CA solution, a 0.6 wt% solution of the CA whole polymer (DS = 0.49) in water was prepared and stocked in a dark place at 25 °C. Figure 3.1.1 seems to indicate that although water is a poor solvent in light scattering measurements (see Table 3.5.1) and CA molecules have a rather high tendency to

Table 3.1.1

DS dependence on water solubility of cellulose acetate[1]

Polymer	MAW-1	MAW-2	MAW-3	MAW-4	MAW-5
DS	1.96	0.98	0.81	0.65	0.49
Solubility	×	△	○	○	○
	MAW-6	MAW-7	MAW-8	MAW-9	MAW-10
DS	0.39	0.34	0.33	0.16	0.10
Solubility	○	△	△	×	×

○, soluble; △, only swell; ×, insoluble.

crystallize when coagulated or precipitated as gels on heat treatment, an aqueous solution of CA (DS = 0.49) is stable at least for several months at room temperature, with no indication of crystallization.

The solubility behavior of the CA whole polymer (DS = 0.49) in different organic solvents was examined. The polymer is found to be insoluble in less polar solvents such as hydrocarbons, chlorinated hydrocarbons, amines, esters, but soluble in very polar organic solvents such as formic acid (dielectric constant $\varepsilon = 58.5$),[5] trifluoroacetic acid ($\varepsilon = 39.5$),[5] dimethylacetamide ($\varepsilon = 37.8$),[5] dimethyl sulfoxide ($\varepsilon = 46.6$),[5] water ($\varepsilon = 78.3$),[5] and formamide ($\varepsilon = 111$).[5] It should be noted that the solubility of CA (DS = 0.49) is controlled primarily by the dielectric constant of the solvent and there exists the critical value of this constant above which the polymer dissolves.

3.1.2   Cellulose acetate (DS 1.75)[6]

A CA sample (DS = 2.46 and $M_w = 1.2 \times 10^5$) synthesized in a previous study[2] (see Section 3.1.3) was hydrolyzed with hydrochloric acid as catalyst in a 77% aq. acetic acid solution for about 1 week. The hydrolyzed CA was recovered by precipitation with methanol and dried *in vacuo*.

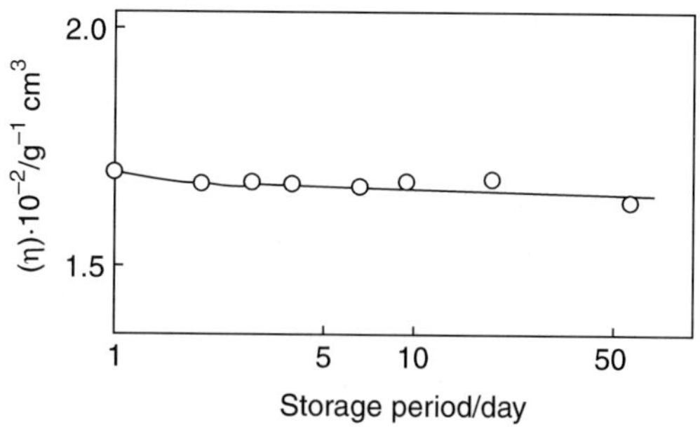

Figure 3.1.1   Limiting viscosity number (η) of CA MAW-5 in water as a function of the storage period of the original solution (0.6 wt%) at 25 °C.[7]

The DS of this unfractionated CA was determined by back titration to be 1.75. From its ^{13}C NMR spectra in deuterated dimethylsulfoxide (DMSO-d_6) the average probabilities that the hydroxyl groups attached to the C_2, C_3, and C_6 atoms were substituted by the *O*-acetyl groups (i.e. $\langle\!\langle f_2 \rangle\!\rangle$, $\langle\!\langle f_3 \rangle\!\rangle$, $\langle\!\langle f_6 \rangle\!\rangle$) were estimated to be 0.45, 0.54, and 0.76, respectively. The total DS determined from $\Sigma_{k=2,3,6} \langle\!\langle f_k \rangle\!\rangle (= \langle\!\langle F \rangle\!\rangle)$ was 1.75, which is in excellent agreement with that evaluated by the neutralizing method.

3.1.3 CA (DS 2.46): Commercially referred to as cellulose diacetate (CDA)[2]

High α-cellulose pulp, trade name Raycrod XG-LD, manufactured by ITT Rayonier Inc. (USA) (α-cellulose 95.4 wt%, β-cellulose 3.8 wt%, pentosan 1.0 wt%, ash 0.03 wt%, the viscosity-average degree of polymerization $P_v = 1189$) was employed as a starting material. The cellulose pulp was converted into CTA by using the acetic anhydride acetic acid/sulfuric acid system. CTA thus prepared was hydrolyzed with magnesium carbonate to yield CA of 55.6 wt% combined acetic acid content (AAC). Based on more than 40 years of experience in the study of cellulose derivatives, the authors believe that in order to prepare CDA solution that is absolutely free from the gel-like materials, it is of prime importance to choose mild conditions of acetylation reaction, followed by a very careful hydrolysis reaction, and to treat whole CDA polymers thus obtained with a weak acid. This decomposes the CA having a large weight-average molecular weight M_w, which contains a sulfuric acid group, sodium and calcium (particle size in acetone, 80–300 nm as determined by prehump II in GPC curve), and then to extract the component of cellulose diacetate with higher AAC (57–58 wt%; particle size: 150–450 nm in acetone, as detected by prehump I in GPC curve) with dichloromethane. It should be remembered that almost all previous investigators have used commercial products, which were not satisfactory for light scattering measurements.

3.1.4 CA CTA (DS 2.92)[7]

Purified cotton linter was acetylated by the standard method with an acetic anhydride acetic acid sulfuric acid mixture.[8,9] The product was subjected to additional stabilization by an excess magnesium acetate. A satisfactory amount of 35 wt% acetic acid was added to produce the precipitate completely, which was filtered through filter paper and then washed with water. CTA thus prepared was dried *in vacuo* and had the number-average molecular weight $M_n = 5.85 \times 10^4$ by membrane osmometry and the weight-average molecular weight $M_w = 2.35 \times 10^5$ by light scattering. The combined acetic acid content (AAC) determination of the unfractionated CTA by the back titration method using sodium hydroxide and sulfuric acid gave a value of 61.0 wt% (DS = 2.89).

The $\langle\!\langle f_k \rangle\!\rangle$ was evaluated by the ^{1}H NMR and ^{13}C NMR method. Here, the absolute magnitude of $\langle\!\langle f_k \rangle\!\rangle$ values are evaluated with the aid of the $\langle\!\langle F \rangle\!\rangle$ values by chemical analysis. The $\langle\!\langle f_k \rangle\!\rangle$ values of the CA (2.92) whole polymer determined by ^{1}H NMR and ^{13}C NMR agree fairly well.

Table 3.1.2 collects the total DS $\langle\!\langle F \rangle\!\rangle$ and the distribution of DS $\langle\!\langle f_k \rangle\!\rangle$ ($k = 2, 3$, and 6) of CA samples, prepared here.[10]

Table 3.1.2

Total DS $\langle\!\langle F \rangle\!\rangle$ and the distribution of DS $\langle\!\langle f_k \rangle\!\rangle$ of CA samples synthesized here

Polymer	$\langle\!\langle F \rangle\!\rangle^a$	$\langle\!\langle f_6 \rangle\!\rangle^b$	$\langle\!\langle f_2 \rangle\!\rangle^b$	$\langle\!\langle f_3 \rangle\!\rangle^b$
CA (2.92)	2.92	1.00 (0.99)^c	1.02 (1.01)^c	0.89 (0.92)^c
CA (2.46)	2.46	0.82	0.89	0.75
CA (1.75)	1.75	0.76	0.45	0.54
CA (0.49)	0.49	0.19	0.10	0.20

^{a}By chemical analysis.
^{b13}C NMR (carbonyl carbon).
^{c1}H NMR (O-acetyl proton).

3.1.5   Cellulose aq. LiOH[11]

Cellulose samples

Regenerated cellulose samples with different molecular weights can be obtained from a cuprammonium solution (Cu: 11; NH_3: 200; water: 1000; gram unit) of cellulose by altering the storage time of the solution in the dark in air. The viscosity (of the solution decreases gradually during storage due to acid hydrolysis as shown in Figure 3.1.2, where it η was measured at 25 °C with a cone-and-plate type viscometer Rotovisco (Haake, Germany) at an angular velocity of 1.26 and 3.14 rad s^{-1}. Thus, by adding 5 wt% aq. H_2SO_4 to six cuprammonium solutions with different storage times, six regenerated cellulose samples were prepared as precipitates, and they were washed with acetone, and dried *in vacuo*.

Preparation of cellulose aq. LiOH solution

In order to determine the optimum range of LiOH concentration Φ_L (wt%) for dissolving cellulose solid in aq. LiOH, the solubility of cellulose was investigated in the range of Φ_L from 3 to 10 wt%. For this purpose, aq. LiOH solution was prepared in the following

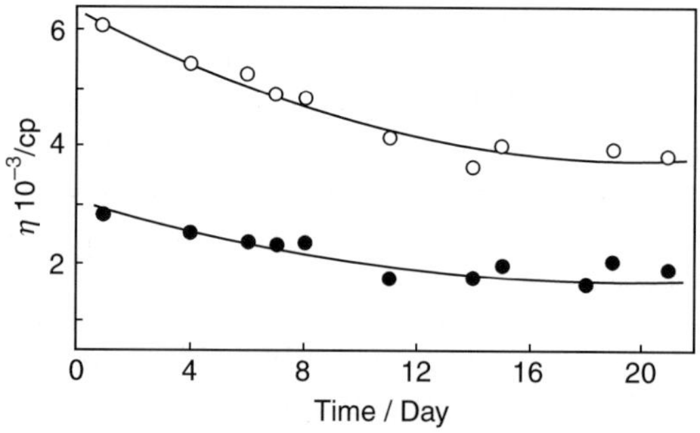

Figure 3.1.2   Change in the viscosity η of cellulose cuprammonium solution (cellulose concentration, 6 wt%) at 25 °C during the storage.[11] Angular velocity: 1.26 rad s^{-1} (○) and 3.14 rad s^{-1} (●).

Table 3.1.3

Dependence of solubility of cellulose (1 wt%) to aq. LiOH solution on the concentration of lithium hydroxide (wt%) in water at 25 °C[11]

Concentration of LiOH (wt%)	3	4	5	6	8	10
Solubility of cellulose	×	△	△–O	O	△–O	×

O, soluble; △, only swell; ×, insoluble.

manner: LiOH (purity 98.0%) anhydride (manufactured by Kishida Chemicals Co., Osaka) was dissolved at 10 °C in distilled water of liquid chromatogram grade (made by Wako Jun-Yaku Co., Osaka) and agitated. The solution was boiled at room temperature under reduced atmosphere (*ca.* 1.3 Pa) to exclude the dissolved gas. Dried regenerated cellulose flakes (sample code SA-1) were dipped in aq. LiOH, thus prepared, at 4 °C for 24 h to give a 1 wt% cellulose slurry. The slurry was agitated for 1 min by a home mixer followed by quenching at -70 °C in liquid nitrogen, stored for 1 h at -70 °C, and heated gradually up to room temperature.

Table 3.1.3 shows the experimental relationships between the solubility behavior of cellulose in aq. LiOH and Φ_L at room temperature. Obviously, cellulose II is insoluble in aq. LiOH with Φ_L less than 3 wt% or larger than 10 wt%, and is dissolved or swollen in aq. LiOH with Φ_L ranging 5–8 wt%. At $\Phi_L = 6$ wt%, cellulose gave a transparent and clean solution. On the basis of these preliminary experimental results, the aq. LiOH solution with $\Phi_L = 6$ wt% was chosen as a solvent for the light scattering and viscosity measurements. A series of 0.6 wt% solutions of cellulose in aq. LiOH ($\Phi_L = 6$ wt%) were prepared.

3.1.6 Cellulose aq. NaOH[12]

The regenerated cellulose sample used in Section 3.1.5[11] (sample code SA-3, $M_w = 8.0 \times 10^4$) was utilized. One gram of sample was dried at room temperature *in vacuo* for 1 day and immersed in 10 g of water, stood at a temperature below -10 °C for 1 day to freeze the water. Then, the solid was thawed and the cellulose suspended in water was obtained. This pretreatment promotes the swelling of cellulose and was very helpful to make a gel-free cellulose solution. On the other hand, 15.92 g of NaOH (guaranteed reagent) was mixed with 173.08 g of distilled water to give aq. NaOH solution, maintained at -6 °C, to which the cellulose water suspension was added. The mixture was agitated mechanically for 5 min and finally a clear and transparent 0.5 wt% (5.45×10^{-3} g cm^{-3}) cellulose solution in 8 wt% aq. NaOH solution was prepared. The solution was filtered through a sintered glass filter (grade 1) and used for the viscosity and light scattering measurements.

REFERENCES

1. K Kamide, M Saito and T Abe, *Polym. J.*, 1981, **13**, 421.
2. K Kamide, T Terakawa and Y Miyazaki, *Polym. J.*, 1979, **11**, 285.

3. K Kamide and K Okajima, *Polym. J.*, 1981, **13**, 127.
4. CH Malm, CF Ordyce and HA Tanner, *Ind. Eng. Chem.*, 1942, **34**, 430.
5. JA Dean (ed), *Lange's Handbook of Chemistry*, 12th Edn., **Vol. 10**, McGraw-Hill, New York, 1979, p. 103.
6. M Saito, *Polym. J.*, 1983, **15**, 249.
7. K Kamide, Y Miyazaki and T Abe, *Polym. J.*, 1979, **11**, 523.
8. See, for example, K Sobue and N Migita, *Cellulose Handbook*. Asakura, Tokyo, 1958.
9. K Kamide, S Manabe and E Osahune, *Makromol. Chem.*, 1973, **168**, 173.
10. K Kamide and M Saito, *Eur. Polym. J.*, 1984, **20**, 903.
11. K Kamide and M Saito, *Polym. J.*, 1986, **18**, 569.
12. K Kamide, M Saito and K Kowsaka, *Polym. J.*, 1987, **19**, 1173.

3.2   FRACTIONATION

3.2.1   Theoretical background and experimental procedure

Since the late 1960s, Kamide and coworkers and Koningsveld *et al.* have studied, theoretically and experimentally, molecular weight fractionation by solubility differences, such as successive precipitational fractionation (SPF) and successive solutional fractionation (SSF), for quasibinary and -ternary solutions, consisting of multicomponent polymer/single solvent or solvent mixture systems.

Kamide strongly recommended SSF, rather than SPF, for isolation of fractions with narrower molecular weight distribution. SPF or SSF (accompanied by changing the temperature of multicomponent polymer dissolved in a single solvent) should be the best in the sense that they have a sound theoretical basis. Unfortunately, SPF and SSF using quasibinary solutions have not widely been applied mainly due to the experimental difficulty of finding a suitable solvent. Accordingly, in practice, the fractionations have been carried out generally by adding nonsolvent to polymer/solvent system.[1–5]

The principal difference between SPF and SSF is schematically demonstrated in Figure 3.2.1. In the former, the polymer-rich phase is separated as the fraction and in the latter, the polmer lean phase is isolated.

Comparison of SSF with SPF is summarized in Table 3.2.1.[6–14]

Figure 3.2.2 shows a large-scale SSF apparatus constructed by Kamide *et al.* for preparation of cellulose acetates.[15]

CA (DS 0.49)[16]

The successive solutional fractionation (SSF) technique was applied. The original polymer (polymer code MAW-5; acetyl content (AC) = 16.1 wt%, DS = 0.49) in water was fractionated using methanol at 25 °C into 15 approximately equal fractions. The polymer volume fraction v_p^0 in the first fractionation step, from which a phase separation occurred, was 0.56%. In order to minimize the time for the precipitate to settle in the bottom of the fractionation vessel, a small amount of sodium chloride was added to the aqueous solution. The supernatant phase was concentrated by evaporating the water and methanol at 60 °C, leaving a dry powder which was dissolved in water to make up

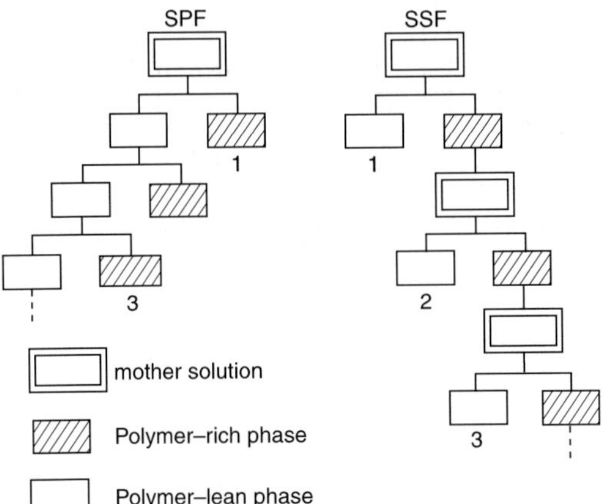

Figure 3.2.1 Schematic representation of SPF and SSF: Numerical figure denotes fraction number.

Table 3.2.1

Comparison of SSF with SPF[3]

Parameter	Comparison	Advantage of SSF	Reference
1. Partition coefficient σ	SSF(SPF	yes	6
2. Volume ratio R	SSF $\ll$ SPF	yes	6
3. Breadth in MWD of the fractions	SSF $<$ SPF (except for extremely low X_w range)	yes	7
4. Operation conditions for controlling fractionation	SSF: ρ SPF: v_p^0	—	8
5. Effect of X_w^0/X_n^0	SSF $<$ SPF	yes	6, 9
6. Effect of X_w^0	SSF $<$ SPF	yes	7, 10
7. Reverse order fractionation	only for SPF	yes	7, 11
8. Double peaked MWD in fraction	only for SPF	yes	6, 9
9. Upper limit of initial polymer conc. v_p^0	SSF $>$ SPF	yes	12
10. Ease of phase separation	SSF $\gg$ SPF	yes	13
11. Accuracy of controlling fraction size ρ	SSF $\gg$ SPF	yes	13
12. Total amount of solvent in a given run	SSF $\gg$ SPF	no	12
13. Ratio of X_w^0/X_n^0 of the first fraction to X_w^0/X_n^0 of the original polymer	SSF: always less than 1 SPF: always greater than 1	yes	7, 8
14. Effect of p_1[a]	SSF $<$ SPF	—	8, 13, 14

[a]For definition of p_1, see eq. (3.21.5); $p_2 = \cdots = p_n = 0$.

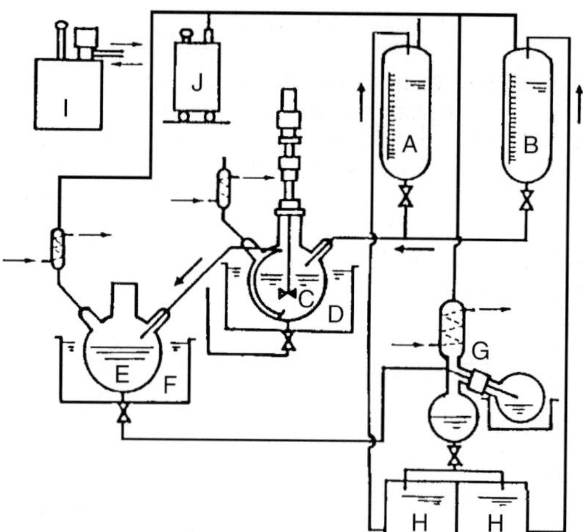

Figure 3.2.2 Outline of a large-scale successive solutional fractionation (SSF) apparatus. A and B: solvent and nonsolvent measuring vessel; C: fractionation vessel; D: thermostated bath; E: storage vessel; F: bath; G: evaporator; H: solvent and nonsolvent recovery vessel; I: thermostated bath; J: vacuum pump.[15]

a solution of 5 wt%. A large quantity of acetone was then added to completely precipitate the fraction, which was then filtered, washed with acetone, and dried *in vacuo*.

CA (DS 1.75)[17]

The CA (DS = 1.75) sample was dissolved in acetone water (7:3 vol/vol) and fractionated by the SSF method with water as the precipitant into 10 almost equal fractions.

CA (DS 2.46)[18]

A solution of the original polymer in acetone was fractionated at 30 °C by using ethanol as the precipitant, according to SSF into 16–21 fractions. The polymer volume fraction v_p^0 of the solution at the first fractionation step, from which the phase separation occurs, was 2.1%. The SSF run was duplicated to obtain sufficient fractions for further study (the first run was designated as EF-2 and the second run, EF-3). The combined acetic acid content of fractions was determined by IR spectroscopic method.[19] Of the resulting 37 fractions, 11 were chosen for further study, excluding GPC.

CA (DS 2.92)[20]

Up to now, acetic acid and chlorinated hydrocarbons, which have the low dielectric constant ε, have been extensively employed as solvents for the molecular weight fractionation of CTA. The fractionation efficiency achieved by using the above-mentioned solvents was poor, unfortunately, and the numerous attempts made so far have been met with very limited success. In order to overcome the above-mentioned experimental

Table 3.2.2

Preparative molecular weight fractionation of cellulose acetates by successive solutional
fractionation (SSF) method

Degree of substitution $\langle\langle F \rangle\rangle$	Solvent	Nonsolvent	Initial concentration	Number of fractions	M_w/M_n	Reference
0.49 (MA series)	Water	methanol	0.0056	14	1.1–1.5	16
1.75	Acetone/water 7:3 v/v	water	–	10	1.2–1.5	17
2.46 (run 1)	Acetone	ethanol	0.021	21	1.1–1.5	18
(run 2)	Acetone	ethanol	0.021	16	1.1–1.5	
2.92	epoxy-propane	n-hexane	0.005	13	1.3–1.5	20

problem, preliminary experiments on phase separation of CA solution were performed for
many solvent/nonsolvent combinations, including those employed in the literature.[21–36]
Judging from the ease of separation of the two liquid phases and of the solvent recovery, we
chose 1-chloro-2,3-epoxypropane (epichlorohydrine) as a solvent and hexane as a
precipitant. Successive solutional fractionation technique (SSF), originally advocated for
use by Kamide and coworkers[2,3] was applied:

Sixty grams of TA 2 sample was dissolved in 1-chloro-2,3-epoxypropane (6000 cm^3)
and thermostated at 35 °C. The amount of hexane predetermined by a pilot fractionation
was added to the solution, resulting in phase separation. The supenatant phase was
isolated by a vacuum line from the vessel and hexane and 1-chloro-2,3-epoxypropane in
the phase were separated by stepwise evaporation in a rotary evaporator and reused
for further fractionation. The fractionation was carried out in a totally closed system.
The fractionation apparatus was specially designed and is described in Figure 3.2.2.
Finally, 13 fractions were separated in a SSF run, in which the composition of the
hexane at each step varied from 44.5 to 33.4 vol% at 35 °C. Equilibrium between the
two phases was not difficult to attain so that the fractionation is efficient (see Table
3.3.5). The polymer fractions prepared in this way were vacuum dried at 60 °C for 1
day. No hydrolysis of the acetyl group was detected.

Table 3.2.2 collects solvent/nonsolvent combinations and the conditions employed
for preparative fractionation of cellulose acetates.

Table 3.2.3 summarizes the method of fractionation, solvent, and nonsolvent of cellulose derivatives reported in the literature.[37]

3.2.2 Fractionation results

CA (DS 0.49)[16]

The second and third columns of Table 3.2.4 give the methanol composition of the
water–methanol mixture, v_m, for each fractionation step and the fraction size ρ_s. The v_m
varies from 0.80 to 0.66. The total recovery of the polymer was 85%. Columns 4–11 in the
same table show the limiting viscosity numbers $[\eta]$ and the 8th to the 11th columns
compile Huggins coefficients k' for CA (DS 0.49) fractions in DMAc, DMSO, water,
and FA at 25 °C. $[\eta]$ of the fractions in water covers a range from 40 to 259, indicating that

Table 3.2.3

Molecular weight fractionation of cellulose derivatives

Polymer	$\langle\langle F\rangle\rangle$	Method	Solvent	Nonsolvent	Temperature (°C)	Number of fractions	Molecular weight range ($\times 10^{-4}$)	M_w/M_n	Reference
Cellulose acetate	2.99	SPF	TCM-acetate (1:1 vol)	Petroleum ether	40–45	6	1.56–3.09 (M_n)	–	38
	2.92	SSF	1-Chloro-2,3-epoxypropane	n-Heptane	35	13	6.36–69.0 (M_w)	1.3–1.5	51
	2.91	SPF	Methylene chloride	Acetone	–	13	1.36–13.0 (M_n)	–	33
	2.89	SPF	Methylene chloride	Methanol-$CaCl_2$	25	12	2.17–20.5 (M_v)	–	39
	2.81	SPF	Glacial acetic acid	n-Heptane	Room temperature	10	–	–	
	2.46	SSF	Acetone	Ethanol	30	21	6.1–26.5 (M_w)	1.25–1.38	18
	2.45	SPF	Acetone	Ethanol	23	11	4.32–80.5 (M_w)	1.15–2.04	40
	2.38	SPF	Acetone	Ethanol	–	14	1.16–18.4 (M_n)	–	41
	2.35	SPF	Acetone	Ethanol	20–25	28	3.1–3.6 (M_n)		42
	2.34	SPF	MEK-acetone (85:15)	–	20–50	15			39
			MEK-acetone (55:45)	–	20–50	15	2.92–30.5 (M_n)	–	
	1.75	SSF	Acetone-water (7:3 vol.)	Water	25	10	3.75–13.1 (M_w)	1.28–1.31	17
	0.49	SSF	Water	Methanol	25	14	4.55–14.5 (M_w)	1.28–1.31	16
Cellulose nitrate	2.98	SPF	Acetone	n-Heptane	25	11	7.7–264 (M_w)	1.7($n = 2$)	43
	2.95	SPF	Acetone	n-Heptane	25	41	3.13–52.8 (M_n)	–	44
	2.85	SPF	Acetone	Water	25	9	68–250 (M_w)	–	45

(Continued)

	2.77	SPF	Acetone–ethanol (5:3)	n-Heptane	–	14	41.3–57.3 (M_w)	1.19–2.23	46
			Acetone	n-Heptane	–	11			
	2.42	SPF	Acetone–ethanol	n-Hexane	Room temperature	6	6.5–26 (M_n)	–	47
	2.28	SPF	Acetone–ethanol	n-Heptane	–	17	–	–	48
Cellulose tributylate	3	SPF	Acetone	Water	30	7	7.72–22.2 (M_n)	–	49
Cellulose tricarbanilate	3	SPF	Acetone	Acetone-1% NaCl (5:5)	20	22	–	1.3–1.6	50
Cellulose tricaprylate	3	SPF	MEK	Ethanol	30	6	8.1–35.5 (M_n)	–	49
Sodium carboxymethyl cellulose	1.66	SPF	Water	Acetone	–	3	7.1–30.2 (M_SD)	–	51
	0.9	SPF	Water	Acetone	–	5	6.7–105.5 (M_w)	–	52
	0.62	SPF	Water–ethanol	Ethanol	25	5	19.5–158.7 (M_w)	3.0–8.8 (M_w/M_SD)	53
	0.74								
Ethylcellulose	2.73 2.77	SPF	Acetone	Water	–	5	1.15–7.88 (M_SD)	–	54
Cellulose xanthate	0.76	SPF	10 wt% NaOH	Methanol	–	3	14–125 (M_w)	–	55
	0.82	SPF	10 wt% NaOH	Methanol	–	3	39–91 (M_w)		55

Table 3.2.4

Results of successive fractionation run (water methanol) and viscosity measurements
on CA (DS = 0.49) in various solvents[16]

Sample code	v_m	ρ_s	$[\eta]$ (cm^3 g^{-1}) (at 25 °C)				Huggins coefficient k'			
			DMAc	DMSO	H$_2$O	FA	DMAc	DMSO	H$_2$O	FA
MA-1	0.800	0.102	42	38	40	–	0.45	–	0.52	–
MA-2	0.755	0.085	–	79	80	87	–	0.39	0.62	0.46
MA-3	0.730	0.120	108	106	118	124	0.43	0.53	0.52	0.49
MA-4	0.720	0.111	142	140	157	–	0.43	0.42	0.52	–
MA-5	0.715	0.044	133	132	151	156	0.38	0.45	0.47	0.57
MA-6	0.710	0.056	154	153	172	180	0.40	0.41	0.52	0.42
MA-7	0.707	0.027	167	167	172	180	0.34	0.39	0.52	0.49
MA-8	0.703	0.032	164	162	182	188	0.43	0.47	0.60	0.49
MA-9	0.699	0.020	185	178	193	198	0.35	0.40	0.59	0.60
MA-10	0.695	0.034	189	201	203	220	0.44	0.44	0.67	0.62
MA-11	0.691	0.018	200	–	215	230	0.48	–	0.69	0.68
MA-12	0.682	0.033	203	204	232	232	0.48	0.49	0.64	0.74
MA-13	0.671	0.054	235	235	259	257	0.57	0.56	0.78	0.74
MA-14	0.660	0.111	–	–	234	–	–	–	–	–
MAW-5	–	0.847	170	170	170	–	0.35	0.35	0.65	–
CMA-WI	–	–	452	–	–	–	0.46	–	–	–

Table 3.2.5

Successive solution fractionation data on CA with total DS 2.46[37]

Fractionation step	Volume fraction of ethanol	Fraction size	$M_n \times 10^{-4}$	$M_w \times 10^{-4}$	M_w/M_n
1	0.777	0.050	0.65	1.16	1.78
2	0.770	0.035	1.24	1.86	1.45
3	0.761	0.016	1.54	2.04	1.33
4	0.750	0.032	2.17	2.88	1.32
5	0.714	0.021	2.34	3.01	1.29
6	0.701	0.017	2.70	3.35	1.24
7	0.677	0.042	3.28	4.04	1.23
8	0.661	0.067	4.03	5.04	1.25
9	0.649	0.056	4.34	5.53	1.27
10	0.636	0.087	5.62	7.11	1.27
11	0.618	0.073	6.14	8.05	1.31
12	0.616	0.044	7.25	9.23	1.27
13	0.614	0.081	7.56	9.79	1.28
14	0.610	0.038	8.43	11.0	1.31
15	0.610	0.060	8.96	11.4	1.27
16	0.603	0.083	11.5	14.8	1.29
17	0.598	0.075	13.4	18.0	1.34
18	0.580	0.029	15.2	22.7	1.49
19	0.580	0.029	15.4	21.8	1.39
20	0.580	0.025	18.9	25.8	1.37
21	–	0.110	17.6	26.8	1.52

fractionation according to molecular weight was performed successfully. The DS values for all the fractions are within 0.49 $\pm$ 0.01 and thus can be regarded as independent of M_w.

The polydispersity parameter M_w/M_n was determined by LS and MO, to be 1.30 $\pm$ 0.02 for the series of CA (DS = 0.49).

CA (DS 1.75)[17]

In column 8 of Table 3.3.2, the values of the ratio of M_w (by the LS method) and M_n (by the MO method) for a series of CA (DS 1.75) fractions are presented.

CA (DS 2.46)[18]

Table 3.2.5 provides an example of successive solutional fractionation (SSF) run on CA (DS 2.46).[37] The acetic acid content for series of EF-2 and EF-3 fractions lies between 55.5 $\pm$ 1.0 wt%, independent of M_w, indicating that the fractionation with respect to the DS does not occur during the SSF run.

Figure 3.2.3 shows the MWD curve $g(M)$ as determined by gel permeation chromatography (GPC) for CA fractions in THF at 25 °C.[56]

In the figure, $g(M)$ of monodisperse atactic polystyrene in THF at 25 °C is also shown. The GPC traces of CA fractions were highly systematic, and the broadening of the chromatograms was properly corrected according to Tung *et al.*'s procedure (special solution for log normal distribution method).[57]

The prehump of the GPC curve,[58,59] which was readily observed for CDA, prepared by the conventional manner, was not detected significantly in the present case. This strongly suggests that all the CDA molecules exist in a molecularly dispersed state in the solution, and this is consistent with the experimental results that M_w (as determined by LS in acetone) is in good agreement with M_w by LS in THF and M_w by GPC in THF.

In the evaluation of the MWD of the CDA fractions, the plot of M_w (by LS in acetone) versus eluent volume V_e (in GPC) was extrapolated linearly for both M_w sides.

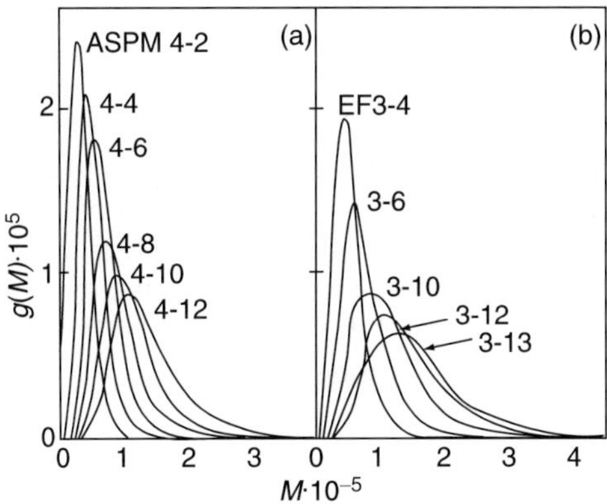

Figure 3.2.3 The MWD curves, determined by gel permeation chromatography (GPC) of (a) atactic polystyrene and (b) CA fractions.[56]

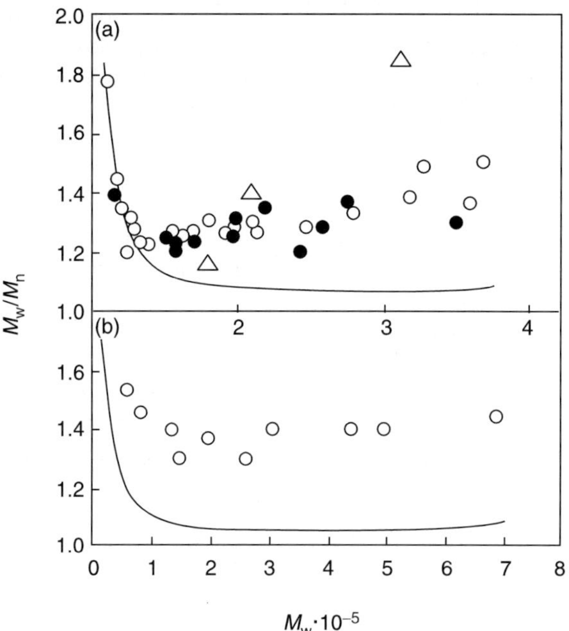

Figure 3.2.4   Plot of M_w/M_n against M_w for CA fractions obtained in CDA Run 1 and 2 (a), in CTA Run 1 (b): full line; theoretical curve for quasibinary mixture ($p = 0$): open circle; experimental data (CDA Run 1): closed circle; experimental data (CDA Run 2): open triangle; data from Tanner and Berry[40] (SPI) (a) full line; theoretical curve for quasibinary mixture ($p = 0$): open circle; experimental data (CTA Run 1) (b).[15]

The number-and weight-average molecular weights, M_n and M_w, as calculated from MWD curve thus obtained by GPC, are tabulated in Table 3.2.5 (see also Table 3.5.2).

The polydispersity parameter M_w/M_n of the fractions, as determined by GPC, is shown in Figure 3.2.4 as open mark as a function of their M_w.[15] The open circle and open triangle correspond to the first (EF-2) and second (EF-3) runs, respectively. Good reproducibility of SSF experiment is confirmed experimentally. Except for a few initial fractions, M_w/M_n of the fractions can be considered as constant (1.2–1.3), independent of M_w. Figure 3.2.4 includes in closed circles the data of Tanner and Berry,[40] who fractionated precipitationally commercial CDA (DS 2.45) by using the same solvent/ nonsolvent pair as that employed in this section. In spite of the very severe operating conditions used by them (i.e. $v_p^0 = 0.3\%$ and total number of fractions in run 1, $n_t = 19$), the polydispersity of the fractions isolated by SPF is larger than that of the fractions by SSF, showing a remarkable M_w dependence, as the theory predicts. As a result, the fractions employed here have perhaps the narrowest MWD among CD samples, which have been used previously for the solution study.

CA (DS 2.92)[20]

Table 3.3.5 lists the ratio of the number-average molecular weight M_n by the membrane osmometry and the weight-average molecular weight M_w by light scattering for a series of CTA fractions, isolated by a single SSF run, in DMAc, all evaluated at 25 °C. The values of M_w/M_n in the run was plotted against M_w in Figure 3.2.4b.

The CTA fractions separated in this work cover a wide molecular weight (M_w) range from 2.9×10^4 to 69×10^4 (i.e. the ratio of the upper to the lower limited $M_w = 23.8$). The polydispersity, expressed in terms of M_w/M_n, of the CTA fractions lies between 1.30 and 1.54, and there is a reasonably constant M_w/M_n ratio of about 1.4 over the M_w range. Theoretical calculations[7,13,60,61] as well as the actual experiments on polyethylene,[62] polystyrene,[63–66] poly(methyl styrene),[67] and cellulose diacetates,[18] carried out at our laboratory, show that SSF always yields a series of fractions with nearly the same M_w/M_n, except for a few initial fractions. The results on CTA are in excellent agreement with previous conclusions we have reached. The data on CTA shown in Table 3.3.5 do not seem to advance the preparation of CTA with very narrow MWD ($M_w/M_n \approx 1.1$) without a new single solvent suitable for SSF.

Molecules of cellulose derivatives have a peculiar tendency to associate in poor solvents. CDA in TCE below 70 °C is an example[18] (see Figure 3.3.1) of this trend. For the purpose of clarifying this point to the case of CTA solutions in the solvents employed here, osmotic pressures of solutions of two CTA fractions, differing M_w (TA2-5 and TA2-10), in various solvents were measured. The results of these determinations will be described in Section 3.3.

3.2.3 MWD of the fractions utilized for further study

It became evident that except for a few fractions of CA (0.49), CA (1.75), CA (2.46), and CA (2.92), M_w/M_n values of CA fractions, prepared by the SSF method lie between 1.2 and 1.5, independent of M_w/M_n (see Tables 3.2.5, 3.3.1–3.3.5). Note that most of the fractions of cellulose derivatives obtained hitherto were prepared by SPF (see Table 3.2.3) and values of their M_w/M_n range roughly from 1.2 to 3.7 (most of them from 1.5 to 2.0) and depend remarkably on M_w. This result indicates clearly that the SSF method is superior to SPF for cellulose derivatives, as in the case of synthetic polymers such as polystyrene.[65] The computer simulation for quasiternary system done by Kamide and Matsuda confirmed the superiority of SSF method.[68–70]

The polydispersity of the CA fractions utilized for further study is shown in Table 3.5.2 for CA (DS 0.49), Table 3.5.3 for CA (DS 2.46), and Table 3.5.5 for CA (DS 2.92). Form these tables, the average M_w/M_n is estimated to be 1.30 ± 0.01 for CA (DS 0.49), 1.33 ± 0.05 for CA (DS 2.46), and 1.40 ± 0.03 for CA (DS 2.92). It was again confirmed that M_w/M_n does not depend on M_w. Here, M_w was obtained by averaging those by LS and GPC, and M_n was averaged from those by MO and GPC, if available.

REFERENCES

1. K Kamide, *Macromolecular Chemistry – 8 (Helsinski 1972)*, 1973, **8**.
2. K Kamide, Batch Fractionation in *Fractionation of Synthetic Polymers* (ed. LH Tung), Marcel Dekker, 1977, Chapter 2.

3. K Kamide, *Thermodynamics of Polymer Solutions; Phase Equlibria and Critical Phenomena*, 3.6, Elsevier, 1990.

4. K Kamide and S Matsuda, Fractionation method for the determination of molecular weight distribution. In *Determination of Molecular Weight* (ed. AR Cooper), John Wiley, New York, 1989, Chapter 9, pp. 201–261.

5. K Kamide and T Dobashi, *Physical Chemistry of Polymer Solutions; Theoretical Background*, Problem 4-31 to 4-41, Elsevier, Amsterdam, 2000.

6. K Kamide and Y Miyazaki, *Makromol. Chem.*, 1975, **176**, 1051.

7. K Kamide and Y Miyazaki, *Makromol. Chem.*, 1975, **176**, 1029.

8. K Kamide, Y Miyazaki and K Yamaguchi, *Makromol. Chem.*, 1973, **173**, 175.

9. K Kamide, K Yamaguchi and Y Miyazaki, *Makromol. Chem.*, 1973, **173**, 133.

10. K Kamide, Y Miyazaki and K Sugamiya, *Makromol. Chem.*, 1973, **173**, 113.

11. K Kamide and Y Miyazaki, *Makromol. Chem.*, 1975, **176**, 2393.

12. K Kamide and Y Miyazaki, *Makromol. Chem.*, 1975, **176**, 3453.

13. K Kamide and Y Miyazaki, *Makromol. Chem.*, 1975, **176**, 1447.

14. K Kamide and K Sugamiya, *Makromol. Chem.*, 1970, **139**, 197.

15. K Kamide, Y Miyazaki and T Abe, *Brit. Polym. J.*, 1981, **13**, 168 and Figure 2.28 of Ref. 3.

16. K Kamide, M Saito and T Abe, *Polym. J.*, 1981, **13**, 421.

17. M Saito, *Polym. J.*, 1983, **15**, 249.

18. K Kamide, T Terakawa and Y Miyazaki, *Polym. J.*, 1979, **11**, 285.

19. K Kamide, T Terakawa, S Manabe and Y Miyazaki, *Sen-i Gakkaishi*, 1975, **31**, T-410.

20. K Kamide, Y Miyazaki and T Abe, *Polym. J.*, 1979, **11**, 523.

21. PRM Nair, RM Gohil, KC Patel and RD Patel, *Eur. Polm. J.*, 1977, **13**, 273.

22. H Lachs, J Kronman and J Wajs, *Kolloid Z.*, 1937, **79**, 91.

23. GR Levi and A Giera, *Gazz. Chim. Ital.*, 1937, **67**, 719.

24. GR Levi, U Vilota and M Montirelli, *Gazz. Chim. Ital.*, 1938, **68**, 589.

25. S Bezzi and U Croatta, *Atti Inst. Veneto Sci.*, 1939–1940, **99**, 905.

26. A Munster, *J. Polym. Sci.*, 1950, **5**, 533.

27. H Sobue, K Matsuzaki and K Yamada, *Sen-i Gakkaishi*, 1956, **12**, 100.

28. G Langhammer, *Makromol. Chem.*, 1956, **24**, 74.

29. RJE Cumberbrich and WG Harland, *J. Text. Ind.*, 1958, **28**, T679.

30. K Thirius and L Dinter, *Plaste Kautsch.*, 1959, **6**, 547.

31. I Kido and K Suzuki, *Sen-i Gakkaishi*, 1960, **16**, 83.

32. PA Okunev and OG Terakanov, *Khim. Volokna*, 1963, **6**, 44.

33. NP Dymarchuk, KP Mischenko and TV Formina, *Zh. Prikl. Khim (Leningrad)*, 1964, **37**, 2263.

34. P Howard and RS Parikh, *J. Polym. Sci.*, 1966, **4**A-1, 407.

35. KH Bischoff and B Phillipp, *Faserforsch. Textiltech.*, 1966, **17**, 395.

36. JMG Cowie and RS Ranson, *Makromol. Chem.*, 1971, **143**, 105.

37. K Kamide, Characterization of Chemically Modified Cellulose. In *Wood and Cellulose Chemistry* (eds DN-S Hon and N Shiraishi), Marcel Dekker, 1991, **17**.

38. WR Moore and B Tidswell, *J. Apply. Chem.*, 1958, **8**, 232.

39. C Camberbirch and W Harland, *Textile Inst.*, 1958, **49**, T664.

40. D Tanner and G Berry, *J. Polym. Sci., Phys. Ed.*, 1974, **12**, 941.

41. H Philipp and C Bjork, *J. Polym. Sci.*, 1951, **6**, 549.

42. A Sookne, H Rutherford, H Mark and H Harris, *Res. Natl. Bur. Std.*, 1943, **30**, 1.

43. AM Holtzer, H Benoit and P Doty, *J. Phys. Chem.*, 1954, **58**, 624.

44. WG Harland, *J. Textile Inst.*, 1955, **46**, T483.

45. MM Huque, DAI Goring and SG Mason, *Can. J. Chem.*, 1958, **36**, 952.

46. ML Hunt, S Newman, HA Scheraga and PJ Flory, *J. Phys. Chem.*, 1954, **58**, 624.

47. WR Moore and J Epstein, *J. Apply. Chem.*, 1956, **6**, 168.

48. H Spurlin, *Ind. Eng. Chem.*, 1938, **30**, 538.

49. L Manderkern and PJ Flory, *J. Am. Chem. Soc.*, 1952, **74**, 2517.

50. RJ Brewer, JJ Tanghe, JT Burr and S Bailey, *J. Polym. Sci.*, 1968, **6**A-1, 1697.

51. K Uda and G Mayerhoff, *Makromol. Chem.*, 1961, **47**, 168.

52. W Brown, D Henley and J Öhman, *Makromol. Chem.*, 1965, **87**, 258.

53. S Sitarmaiah and D Goring, *J. Polym. Sci.*, 1962, **58**, 1107.
54. G Meyerhoff and M Sütterlin, *Makromol. Chem.*, 1963, **62**, 164.
55. B Das, AK Raz and PK Choudhury, *J. Phys. Chem.*, 1969, **73**, 3413.
56. K Kamide, T Terakawa and S Matsuda, *Br. Polym. J.*, 1983, **15**, 91.
57. LH Tung, JC Moore and GW Knight, *J. Apply. Polym. Sci.*, 1966, **10**, 1261.
58. K Kamide, T Terakawa and S Manabe, *Sen-i Gakkaishi*, 1974, **30**, T-464.
59. K Kamide, T Terakawa, S Manabe and Y Miyazaki, *Sen-i Gakkaishi*, 1975, **31**, T-410.
60. K Kamide and Y Miyazaki, *Makromol. Chem.*, 1975, **176**, 1051.
61. K Kamide and Y Miyazaki, *Makromol. Chem.*, 1975, **176**, 1427.
62. K Kamide, Y Miyazaki and T Abe, *Polym. Prepr. Jpn.*, 1977, **26**, 1164.
63. K Kamide, K Sugamiya, T Ogawa, C Nakayama and N Baba, *Makromol. Chem.*, 1972, **135**, 23.
64. K Kamide, K Sugamiya, T Terakawa and H Hara, *Makromol. Chem.*, 1972, **156**, 287.
65. K Kamide, Y Miyazaki and T Abe, *Makromol. Chem.*, 1976, **177**, 485.
66. K Kamide, Y Miyazaki and T Abe, *Polym. J.*, 1977, **9**, 395.
67. I Noda, K Kamide, Y Miyazaki and H Ishikawa, *Polym. Prepr. Jpn.*, 1977, **26**, 1164.
68. K Kamide, S Matsuda, H Shirataki and Y Miyazaki, *Eur. Polym. J.*, 1989, **25**, 1153.
69. K Kamide, in Thermodynamics of Polymer Solutions; Phase Equilibria and Critical Phenomen, **3.6.6**, Elsevier, Amsterdam, 1990, pp. 333–338.
70. See, also, K Kamide, S Matsuda and Y Miyazaki, *Polym. J.*, 1984, **16**, 479.; K Kamide and S Matsuda, *Polym. J.*, 1984, **16**, 515–591; K Kamide, S Matsuda and Y Miyazaki, *Polym. J.*, 1984, **16**, 479; K Kamide and S Matsuda, *Polym. J.*, 1986, **18**, 347; S Matsuda, *Polym. J.*, 1986, **18**, 993; S Matsuda and K Kamide, *Polym. J.*, 1987, **19**, 203–211.

3.3   MEMBRANE OSMOMETRY

3.3.1   Theoretical background

When the solvent will not permit the passage of solute molecules (i.e. semipermeable) across a permeable membrane, separating the two chambers containing a pure solvent and polymer solution, the chemical potential difference (of the solvent) causes the solvent to pass into the solution. This causes the level to rise in a capillary until the hydrostatic pressure equals the osmotic pressure π. The following relationship holds between π and the concentration c (g cm^{-3}).[1,2]

$$\pi/c = RT(1/M_n + A_{2,o}c + \cdots) \tag{3.3.1}$$

where R is the gas constant, T is temperature (K), M_n is the number-average molecular weight, and $A_{2,o}$ is the second virial coefficient obtained by membrane osmometry. M_n and $A_{2,o}$ were obtained from the linear part of a plot of π/c versus c as intercept and slope, respectively.

The polymer–solvent thermodynamic interaction parameter χ is calculated from the osmotic data using the relationship:

$$\chi = -\{\pi V_0/RT + \ln(1 - v_p) + (1 - 1/X_n^*)v_p\}/v_p^2 \tag{3.3.2}$$

where V_0 is the solvent molar volume, v_p is the volume fraction of polymer ($= V_0 c/M$), is the number-average ratio of polymer to solvent ($= M_n V_m/m$), V_m is the molar volume of the monomer unit, and m is the molecular weight of the monomer unit.

Table 3.3.1

Results of light scattering, membrane osmometry, and viscosity measurements on cellulose acetate (DS = 0.49) in dimethylacetamide at 25 °C[3]

Sample code	LS				Osmotic pressure	M_w/M_n	$\Phi \times 10^{-23}$	α_s^a	X^b
	$M_w \times 10^{-4}$	$\langle S^2 \rangle_z^{1/2} \times 10^8$ cm	$\langle S^2 \rangle_w^{1/2} \times 10^8$ cm	$A_2 \times 10^4$ cm^3 mol g^{-2}	$M_n \times 10^{-4}$				
MA-1	–	–	–	–	0.97	–	–	–	–
MA-2	–	–	–	–	1.67	–	–	–	–
MA-3	4.55	137	123	4.80	–	–	1.85	1.03	4.3
MA-4	6.25	160	144	8.90	–	–	2.08	1.07	8.0
MA-5	6.32	164	148	7.50	–	–	1.82	1.05	4.2
MA-6	6.76	175	158	14.9	5.20	1.30	1.86	1.18	6.2
MA-7	7.25	184	166	13.8	–	–	1.86	1.18	6.2
MA-8	7.94	189	170	15.1	6.10	1.31	1.85	1.26	7.8
MA-9	8.00	191	172	15.0	–	–	2.01	1.24	10.5
MA-10	10.8	215	194	12.1	–	–	1.96	1.26	10.1
MA-11	11.1	223	201	10.2	–	–	1.92	1.13	6.6
MA-12	11.3	–	–	–	8.70	1.30	–	–	–
MA-13	14.5	253	223	3.1	11.3	1.28	2.18	1.03	9.0
MAW-5^c	8.00	247	198	13.8	3.90	2.29	–	–	–

[a]Via penetration function.
[b]By method 1A (3.15.2.A).
[c]Whole polymer.

Table 3.3.2

Results of light scattering, membrane osmometer, and viscosity measurements on cellulose acetate (DS 1.75) in dimethylacetamide at 25 °C[7]

Sample	$M_w \times 10^{-4}$	LS			Osmometry		M_w/M_n	Viscosity	$\Phi \times 10^{-23}$	α_s [a]	X [b]
		$\langle S \rangle_z^{1/2}(10^{-6}\,\mathrm{cm})$	$\langle S^2 \rangle_w^{1/2}$	$A_2\,(10^{-4}\,\mathrm{cm}^3\,\mathrm{mol}\,\mathrm{g}^{-2})$	$M_n \times 10^{-4}$	$A_{2,o}$		$[\eta]\,(\mathrm{cm}^3\,\mathrm{g}^{-1})$			
KS2-1	–	–	–	–	–	–	–	62	–	–	–
KS2-2	3.75	1.51	1.38	9.8	2.76	9.2	1.36	86	0.94	1.028	0.8
KS2-3	4.38	1.67	1.47	9.6	3.43	1.3	1.28	96	1.00	1.029	1.0
KS2-4	–	–	–	–	–	–	–	116	–	–	–
KS2-5	5.53	1.96	1.75	9.5	–	–	–	122	1.00	1.02	1.0
KS2-6	6.15	2.06	1.84	9.5	4.67	10.9	1.31	132	1.03	1.031	1.1
KS2-7	–	–	–	–	–	–	–	144	–	–	–
KS2-8	8.23	2.37	2.12	8.3	–	–	–	157	1.07	1.032	1.2
KS2-9	10.0	2.55	2.28	6.8	7.2	9.8	1.39	165	1.10	1.031	1.2
KS2-10	13.1	2.95	2.63	6.0	–	–	–	196	1.10	1.030	1.3

[a]Estimated using penetration function ψ.
[b]Estimated by method 1A (see Section 3.15.2).

Table 3.3.3

Results from light scattering, membrane osmometer, gel permeation chromatograph, and viscosity measurements on cellulose diacetate fractions in dimethylacetamide, acetone, and tetrahydrofuran at 25 °C[7]

Sample code	LS									Osmotic pressure[a]		GPC[a]		Viscosity		
	DMAc			Acetone			THF			THF		THF		DMAc	Acetone[a]	THF[b]
	M_w $\times 10^{-4}$	$A_2 \times 10^4$ (cm^3 mol g^{-2})	$\langle S^2 \rangle_z^{1/2}$ (10^{-8} cm)	M_w $\times 10^{-4}$	A_2 $\times 10^4$	$\langle S^2 \rangle_z^{1/2}$	M_w $\times 10^{-4}$	A_2 $\times 10^4$	$\langle S^2 \rangle_z^{1/2}$	M_n $\times 10^{-4}$	A_2 $\times 10^4$	M_w $\times 10^{-4}$	M_n $\times 10^{-4}$	$[\eta]$ (cm^3 g^{-1})	$[\eta]$	$[\eta]$
EF 3-4	5.3	10.4	212	–	–	–	–	–	–	–	–	–	–	100	–	–
EF 2-10	–	–	–	6.1	5.9	218	–	–	–	4.7	5.9	7.1	5.6	(140)[c]	117	–
EF 3-6	7.3	9.5	263	–	–	–	7.4	4.4	160	4.8	5.2	7.5	6.0	132	–	113
EF 2-11-1	–	–	–	9.6	5.3	252	–	–	–	7.6	6.8	9.2	7.3	170[c]	151	–
EF 3-8	–	–	–	9.4	3.8	243	10.0	4.9	181	–	–	9.8	7.5	178	–	138
EF 3-10	10.8	10.2	305	10.6	3.7	252	10.9	3.3	191	8.7	7.8	11.1	8.8	194 (205)[c]	160	158
EF 2-14	–	–	–	11.5	3.7	280	–	–	–	10.8	6.0	11.4	9.0	–	187	–
EF 3-12	–	–	–	14.1	3.0	282	15.5	4.0	235	12.2	7.7	14.2	11.9	243[c]	193	182
EF 2-15	–	–	–	15.4	5.2	310	–	–	–	12.0	5.4	14.9	11.5	–	218	–
EF 3-13	15.6	9.6	377	15.6	5.2	290	18.2	4.6	257	12.6	7.1	15.9	12.3	252	217	212
EF 3-14	–	–	–	18.5	4.3	310	20.0	2.5	277	14.0	5.0	17.5	12.7	315[c]	241	230
EF 3-15	27.0	9.9	512	26.5	5.0	338	–	–	–	–	–	24.9	19.1	370	277	–

[a]Ref. 5.
[b]Ref. 14.
[c]Ref. 6.

3.3.2 Experimental procedure

CA (DS 0.49)/dimethylacetamide (DMAc):[3] CA (DS 1.75)/DMAc:[4] CA (DS 2.46)/ tetrahydrofuran (THF),[5] tetrachloroethane (TCE):[5] CA (DS 2.92)/DMAc, acetone, TCE, trichloromethane (TCM) (Figure 3.3.1).[6]

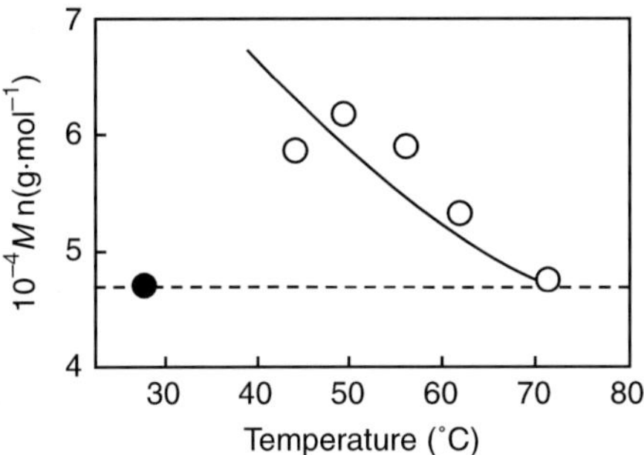

Figure 3.3.1 Plot of apparent number-average molecular weight M_n, as determined by membrane osmometry, for a cellulose diacetate fraction in tetrachloroethane against the temperature of measurement:[5] (○), M_n in tetrachloroethane; (●), M_n in tetrahydrofuran.[13]

Osmotic measurements were carried out on the solutions at 25 °C with a Hewlett–Packard high-speed membrane osmometer model 502. The membrane was the Ultracella filter type SH-1 1 539 (regenerated cellulose, allerfeinst) manufactured by Sartorius (Germany) and a Selectron membrane filter type 0-8, manufactured by Schleicher and Schuell Inc. (Germany). The diffusion of solute was not detected experimentally by comparing solvent reference values before and after the measurement.

Table 3.3.4

Weight and number averages molecular weights of cellulose acetate (DS 2.46) samples

Sample M_w/M_n code	$M_w \times 10^{-4}$				$M_n \times 10^{-4}$			
	by LS			by GPC	by MO	by GPC		by VPO
	DMAc	Acetone	THF	THF	THF	THF	THF	
EF3-4	5.3[a]	–	–	5.1	2.7	4.1	2.6	1.66
EF3-6	7.3[a]	7.2[a]	7.4[b]	7.5[b]	4.8[a]	5.0[a]	4.1[c]	1.59
EF3-7	–	–		7.55	5.5	5.8	5.4	1.36
EF3-10	10.86[a]	10.6[b]	10.9[b]	11.1[a]	8.7[a]	8.8[a]	8.1	1.25
EF3-12	–	14.1[b]	15.5[b]	14.2[a]	12.2[a]	11.9[a]	11.1	1.25
EF3-13	15.6[b]	15.6[b]	18.2[b]	15.9[a]	12.6[a]	12.3[a]	12.9	1.28
EF3-14	–	18.5	20.0[b]					
EF3-15	27.0[a]	26.5						

[a]Ref. 7.
[b]Ref. 5.
[c]Average of three determinations.

Table 3.3.5

Results of membrane osmometer and light scattering measurements with cellulose triacetate in dimethylacetamide at 25 °C[6]

Sample code	Osmotic pressure			LS			M_w/M_n	χ	$\Phi \times 10^{-23}$	α_s^a	X^b
	$M_n \times 10^{-4}$	$A_2 \times 10^4$ (cm^3 mol g^{-2})	$M_w \times 10^{-4}$	$\langle S^2 \rangle_z^{1/2} \times 10^8$ (cm^2)	$\langle S^2 \rangle_w^{1/2} \times 10^8$	$A_2 \times 10^4$					
TA2-1	–	–	–	–	–	–	–	–	–	–	–
TA2-2	1.96	13.2	–	–	–	–	–	0.29	–	–	–
TA2-3	2.71	13.1	–	–	–	–	–	0.30	–	–	–
TA2-4	4.11	12.2	6.36	180	159	13.6	1.54	0.31	1.15	1.10	1.4
TA2-5	5.60	11.8	8.22	214	186	13.4	1.46	0.32	1.19	1.10	1.5
TA2-6	9.78	11.4	13.7	287	253	15.1	1.40	0.32	1.09	1.16	1.4
TA2-7	11.4	11.4	14.9	277	249	12.5	1.30	0.33	1.32	1.17	2.1
TA2-8	14.6	11.7	20.0	342	303	12.1	1.37	0.32	1.29	1.16	2.0
TA2-9	20.2	11.4	26.2	395	356	11.3	1.30	0.33	1.33	1.15	2.2
TA2-10	21.9	10.6	30.8	421	371	9.5	1.40	0.34	1.49	1.16	2.9
TA2-11	31.7	10.8	44.4	531	468	9.9	1.40	0.32	1.38	1.19	2.5
TA2-12	35.7	10.7	50.0	565	503	10.8	1.40	0.34	1.36	1.12	2.3
TA2-13	47.8	10.4	69.0	671	587	10.4	1.44	0.33	1.47	1.15	2.6
TA2^c	5.85	9.12	23.5	467	353	7.8	4.02	0.36	1.12	–	–

[a]Via penetration function ψ.
[b]By method 1A.
[c]Whole polymer.

Table 3.3.6

Results of osmotic pressure measurements with cellulose triacetate fractions in various solvents at 25 °C[6]

Sample code	$M_n \times 10^{-1}$				$A_2 \times 10^4 / \mathrm{mol\ cm^3\ g^{-2}}$				χ from eq. (3.3.2)			
	DMAc[a]	Acetone	TCE[b]	TCM[c]	DMAc[a]	Acetone	TCE[b]	TCM[c]	DMAc[a]	Acetone	TCE[b]	TCM[c]
TA2-5	5.60	5.45	5.80	5.60	11.8	11.7	12.0	12.4	0.32	0.36	0.29	0.33
TA2-10	21.9	21.1	21.2	21.7	10.6	11.7	11.5	8.7	0.34	0.36	0.30	0.38

[a]Dimethylacetamide.
[b]Tetrachloroethane.
[c]Trichloromethane.

Table 3.3.7

Results of light scattering, membrane osmometry, and viscosity measurements with NaCS (DS = 1.9) in 0.5 M aq. NaCl at 25 °C[12]

Polymer code	LS				Osmotic pressure		M_w/M_n	$[\eta]$ ($cm^3 g^{-1}$)	k'
	$M_w \times 10^{-4}$	$\langle S^2 \rangle_z^{1/2} \times 10^8$ (cm)	$\langle S^2 \rangle_w^{1/2} \times 10^8$ (cm)	$A_2 \times 10^4$ (cm^3 mol g^{-2})	$M_n \times 10^{-4}$	$A_2 \times 10^4$ (cm^3 mol g^{-2})			
CSK-1	150	857	640	0.58	31	1.29	5.03	384	0.51
CSK-2	116	754	560	1.41	22	1.27	5.27	311	0.53
CSK-3	74	590	445	1.37	18	1.91	4.11	229	0.48
CSK-4	53	491	375	1.63	15	2.55	3.53	150	0.46
CSK-5	40	427	325	2.00	11	2.55	3.64	118	0.49
CSO-1	–	–	–	–	9.9	2.45	–	126	0.48
CSO-2	24.4	330	250	2.60	6.3	2.72	3.87	79	0.47
CSO-3	–	–	–	–	5.8	2.82	–	84	0.47
CSO-4	–	–	–	–	4.6	2.85	–	63	0.45
CSO-5	–	–	–	–	2.1	3.05	–	22	0.47
CSK-6	8.0	191	143	3.00	1.8	3.22	4.44	26	0.44
CSK-7	7.2	178	136	3.50	2.0	5.19	3.60	22	0.40
CSO-6	3.3	(190)	(173)	(2.25)	(0.94)	(6.22)	(3.51)	12	0.40

3.3.3   Number-average molecular weight and second virial coefficient

Cellulose acetates in solution

Values of M_n of six cellulose fractions and a whole polymer of CA (DS 0.49), determined in DMAc at 25 °C by membrane osmometry (MO), is collected in the sixth column of Table 3.3.1.[3]

Values of M_n and $A_{2,o}$ of CA (1.75) fractions in DMAc by MO are shown in column 6 in Table 3.3.2.[4] Results (M_n) of MO and GPC for 9–11 CA (DS 2.46) fractions in THF are shown in columns 11 and 13 of Table 3.3.3,[7] respectively, and values of $A_{2,o}$ of the samples in THF by MO are shown in column 12 in Table 3.3.3.[7] M_n values of six CA (DS 2.46) fractions in THF by MO, GPC and VPO, are shown in columns 7–9 in Table 3.3.4.[8]

Table 3.3.5 shows M_n and $A_{2,o}$ of 12 fractions and a whole polymer of CA (DS 2.92) in DMAc in the second and third columns.[6] Results of MO measurements with two CA (DS 2.92) fractions in DMAc, acetone, TCE, and TCM at 25 °C are shown in Table 3.3.6.[6] M_n of CA (2.92) fractions determined in four solvents by membrane osmometry also agree well within experimental error ($\pm 3.4\%$). This is direct experimental evidence supporting the idea that cellulose triacetate (CTA) molecules are molecularly dispersed in dilute solutions, including acetone. As early as 1953, Staudinger and Eicher[9] found by membrane osmometry a fairly good coincidence of M_n of a CA (DS 2.37) sample in acetone with that in glacial acetic acid. After the measurement of M_n for a CA (DS 2.88) whole polymer in nitroethane, TCM, and acetone, Cowie and Ranson[10] concluded that neither degradation nor aggregation is apparent. It has generally been believed that CTA is soluble only in acidic solvents.[11] However, it became clear that nonacidic solvents, such as DMAc, trifluoroacetic acid (TFA), and acetone are also adequate solvents for CTA.

Cellulose sulfate in aq. NaCl[12]

Columns 6 and 7 of Table 3.3.7 show M_n and $A_{2,o}$, both by MO, of sodium cellulose sulfate (DS 1.9) in aq. NaCl solution.[12]

REFERENCES

1. K Kamide and T Dobashi, *Physical Chemistry of Polymer Solutions; Theoretical Background,* Problem 2-7, Problem 2-8, Problem 2-9, and Problem 2-10, Elsevier, Amsterdam, 2000; K Kamide, M Saito and Y Miyazaki, Molecular weight determination. In *Polymer Characterization* (eds BJ Hunt and MI James), Blackie Academic & Professional, London, 1953, Chapter 5, pp. 115–144; K Kamide, Colligative properties. In *Polymer Characterization,* **Vol. 1,** (eds C Booth and C Price), Pergamon Press, Oxford, 1989, Chapter 4, pp. 75–102.
2. RV Bonner, M Dimbat and FH Stross, *Osmotic Pressure, Number-Average Molecular Weights.* Interscience, New York, 1958.
3. K Kamide, M Saito and T Abe, *Polym. J.,* 1981, **13,** 421.
4. M Saito, *Polym. J.,* 1983, **15,** 249.
5. K Kamide, T Terakawa and Y Miyazaki, *Polym. J.,* 1979, **11,** 285.
6. K Kamide, Y Miyazaki and T Abe, *Polym. J.,* 1979, **11,** 523.

7. K Kamide and M Saito, *Polym. J.*, 1982, **14**, 517.
8. K Kamide, T Terakawa and S Matsuda, *Br. Polym. J.*, 1983, **15**, 91.
9. H Studinger and T Eicher, *Makromol. Chem.*, 1953, **10**, 261.
10. JMG Cowie and RS Ranson, *Makromol. Chem.*, 1971, **143**, 105.
11. See, for example, WR Moore and J Russell, *J. Colloid. Sci.*, 1954, **9**, 338.
12. K Kishino, T Kawai, T Nose, M Saitoh and K Kamide, *Eur. Polym. J.*, 1981, **17**, 623.
13. T Ikeda and H Kawaguchi, *Rep. Prog. Polym. Phys. Jpn.*, 1966, **9**, 23.
14. H Suzuki, Y Miyazaki and K Kamide, *Eur. Polym. J.*, 1980, **16**, 703.

3.4 VAPOR PRESSURE OSMOMETRY (VPO)

3.4.1 Theoretical background

When the polymer solution is placed in an atmosphere of the saturated vapor phase, drops of solution and solvent on the tips of thermistors are held in a space filled with saturated solvent vapor. As the solvent condenses on the drop of solution, the drop of solution is heated from T_0 to T until a steady state is reached. In the steady state, this heating is balanced by heat losses through conduction. The temperature difference between the two drops in the steady state, $\Delta T_S(\equiv (T - T_0)_S)$ can be related to the solvent activity and may be expressed as a power series of the concentration c as[1-7]

$$\Delta T_S/c = K_S[(1/M_n) + A_{2,v}c + \cdots]$$
(3.4.1)

$$K_S = \frac{K_e}{1 + \left[\dfrac{k_1A_1 + k_2A_2}{k_3A_1}\right]\left[\dfrac{RT_0^2}{\Delta H_2 P_0(T_0)}\right]}$$
(3.4.2)

$$K_e = RV_0 T_0^2/\Delta H$$
(3.4.3)

M_n = the number-average molecular weight.
k_1 and k_2 = the coefficients of surface heat transfer (cal cm^{-2} s^{-1} K).
k_3 = the mass transfer coefficient (mol dyn^{-1} s^{-1}).
A_1 = surface area of a solution drop (cm^2).
A_2 = area of contact between the solution drop and thermistor (cm).
R = the gas constant (cal K^{-1} mol^{-1}).
T_0 = measuring temperature (K).
P_0 = saturated vapor pressure (dyn cm^{-2}).
V_0 = molar volume of the solvent (cm^3 mol^{-1}).
ΔH = heat of condensation (cal mol^{-1}).

Here, K_S is a calibration parameter (cm^3 K mol^{-1}), $A_{2,v}$ is the second virial coefficient by VPO. Note that $A_{2,v}$ does not generally coincide with $A_{2,o}$.

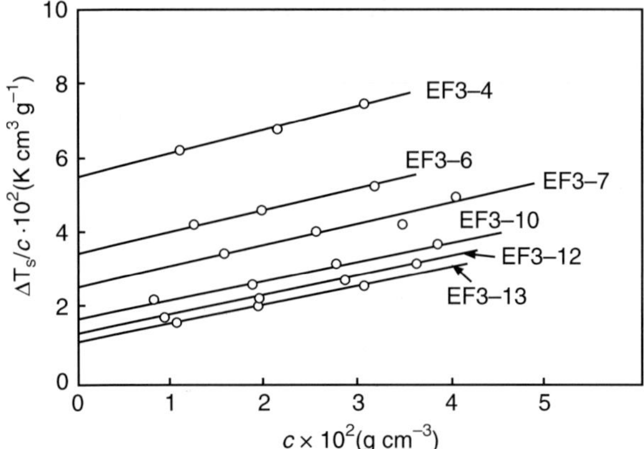

Figure 3.4.1 Plot of $\Delta T_S/c$ against c for several cellulose acetate (DS 2.46) fractions in tetrahydrofuran at 25 °C.[9]

3.4.2 Apparatus

In 1976, Kamide *et al.* constructed a vapor pressure osmometer apparatus with an appreciable sensitivity, by which the number-average molecular weights M_n up to *ca.* 1×10^5 of monodisperse polystyrene in benzene were demonstrated to be measurable with a precision less than $\pm 10\%$.[8]

3.4.3 M_n of CA by VPO

Figure 3.4.1 shows the plot of $\Delta T_S/c$ against c for six CA (DS 2.46) fractions in THF at 25 °C.[9] Values of M_n thus determined by VPO for CA (DS 2.46) are collected in column 6 of Table 3.3.4.

REFERENCES

1. K Kamide and M Sanada, *Kobunshi Kagaku*, 1960, **24**, 751.
2. K Kamide, *Kobunshi Kagaku*, 1968, **25**, 648.
3. K Kamide, K Sugamiya and C Nakayama, *Makromol. Chem.*, 1970, **132**, 15.
4. K Kamide and R Fujishiro, *Makromol. Chem.*, 1971, **147**, 261.
5. K Kamide, *Netsusokutei*, 1980, **7**, 88.
6. See K Kamide and T Dobashi, *Physical Chemistry of Polymer Solutions: Theoretical Background*, Problem 2–14 and Problem 2–20, Elsevier, Amsterdam, 2000, pp. 34–49.
7. See K Kamide, Colligative properties. In *Modern Technique for Polymer Characterization* (eds RA Pathrick and JV Dawkins), Wiley, New York, 1999, Chapter 13; K Kamide, Colligative properties In *Polymer Characterization*, **Vol. 1**, (eds C Booth and C Price), Pergamon Press, Oxford, 1989, Chapter 4.
8. K Kamide, T Terakawa and H Uchiki, *Makromol. Chem.*, 1976, **177**, 1447.
9. K Kamide, T Terakawa and S Matsuda, *Br. Polym. J.*, 1983, **15**, 91.

3.5 LIGHT SCATTERING

3.5.1 Theory of static light scattering

The scattered light intensity of natural light from polymer solution is given by the following relationships:[1–3]

$$Kc/R_\theta = \{M_w^{-1} + 2A_{2,L}c\}\{1 + (16\pi^2 n^2\langle S^2\rangle_z/3\lambda_0^2)\sin^2\theta/2\} \qquad (3.5.1)$$

$$K \equiv \{(2\pi^2 n^2)/(N_A\lambda_0^4)\}(\partial n/\partial c)^2 \qquad (3.5.2)$$

$$R_\theta \equiv I_s r^2/\{I_0(1 + \cos^2\theta)\} \qquad (3.5.3)$$

R_θ = reduced scattering light intensity.
K = optical constant.
I_0 = incident light intensity.
I_s = scattered light intensity.
n = refractive index of solution.
$\partial n/\partial c$ = refractive index increment.
λ_0 = wavelength of incident light *in vacuo*.
θ = scattering angle.

From eq. (3.5.1), the z-average radius of gyration $\langle S^2\rangle_z^{1/2}$, the second virial coefficient $A_{2,L}$, and the weight-average molecular weight M_w can be estimated from the initial slope at $c \to 0$, $\theta \to 0$, and the intercept in the plot of Kc/R_θ versus $\sin^2\theta + kc$, when k is a constant (Zimm plot).[3]

The weight-average radius of gyration $\langle S^2\rangle_w^{1/2}$, can be converted from $\langle S^2\rangle_z^{1/2}$ and the polydispersity (M_w/M_n) data through use of relationship:[4]

$$\langle S^2\rangle_w^{1/2} = \{(h + 1)/(h + 2)\}^{1/2}\langle S^2\rangle_z^{1/2} \qquad (3.5.4)$$

with

$$h^{-1} = (M_w/M_n) - 1 \qquad (3.5.5)$$

Here, the Schulz–Zimm-type molecular weight distribution (MWD) is assumed for the polymer samples.

3.5.2 Experimental procedure

CA (DS 0.49)[5]

DMAc water and formamide (FA) were chosen as the solvents for the LS measurements. CA (DS = 0.49) fractions were dissolved in freshly distilled solvents at 60 °C to make 0.6 wt% solutions. The solutions were left standing for more than 16 h and then filtered with a sintered glass filter (second grade). The specific refractive index increments $\partial n/\partial c$ were determined by a Shimadzu differential refractometer, model DR-4 at 25 °C at a wavelength λ_0 of 436 nm to be 0.068, 0.131, and 0.069 (cm^3 g^{-1}) for CA fractions in DMAc, water, and FA, respectively. The stock solutions were centrifuged at

1.03×10^5 g for 2 h in a Hitachi model 55p-7 automatic preparative ultracentrifuge. The upper two-thirds of the supernatant phase was carefully filtered through a Sartorius membrane filter 0.15 μm and directly transferred into the LS cylindrical cell. The LS intensity was measured within an angle range from 30 to 150° using unpolarized incident light (λ_0) = 436 nm at 25 °C in a FICA model 42000 photogonio diffusiometer. The data obtained were analyzed according to Zimm's procedure.[3]

Cellulose acetate (DS1.75)[6]

We centrifuged 1 wt% stock solution of the fractions in DMAc at 3.0×10^4 g for 1 h in a Hitachi model 55p-7 automatic preparative ultracentrifuge to remove gel-like material. The average of the specific refractive index increments ($\partial n/\partial c$) of three fractions in DMAc was 0.046 (cm^3 g^{-1}) at 25 °C.

LS measurements were made at 25 °C with a FICA model 42000 photogoniometer, using unpolarized light (wavelength, $\lambda = 436$ nm) in a scattering angle range from 30 to 150°.

CA (DS 2.46)[7,8]

Preparation of the solutions. The stock solution of CDA fractions was made by mechanical agitation at a concentration of approximately 5×10^{-3} g/cm^3 with acetone or THF as the solvent. The solutions of high molecular weight CA (DS 2.46) fractions were cooled down below -70 °C and were subsequently warmed at 30 °C for complete dissolution. After these solutions were centrifuged at 29,000 g for 1 h, the middle layer portions were collected by a microsyringe. The concentrations of the solutions were adjusted by adding pure solvent to the collected solutions which were then made to pass through a 200 nm Alpha metricel filter (Alpha-8), manufactured by Gelman Instrument Co. (Germany), directly into the cylindrical light scattering (LS) cell. The solutions were allowed to stand at room temperature overnight. The solutions and solvents were thus free from dust and gels.

LS measurements (1).[7] LS measurements on CA (DS 2.46) in acetone and in THF were carried out in a Shimadzu–Brice type LS photometer model PG-21 (for EF-2 series fractions) and in a FICA model 42000 photogonio diffusiometer (for EF-3 series fractions). Unpolarized light with wavelength $\lambda_0 = 546$ and 436 nm were utilized in acetone and THF, respectively. Cylindrical cells capped with a close fitting stainless stopper were employed to minimize evaporation of the solvent during measurements and were immersed in a bath containing benzene at 25 ± 0.2 °C. No scatter dissymmetry was observed. In all cases, the dissymmetry factor was always less than 1.02. The concentration range was 1.25–0.5×10^{-3} g/cm^3. The apparatus was calibrated with benzene, using unpolarized 90° scattering light and taking the Rayleigh ratio $R_{90} = 16.3 \times 10^{-6}$ cm^{-1} at $\lambda = 546$ nm and 48.5×10^{-6} cm^{-1} at 436 nm. Scattered intensities were measured at various angles from 37.5 to 142.5°.

LS measurements (2).[8] A Fica 50 LS photometer (Sofica, France) in the Institute for Chemical Research, Kyoto University, was also used. The lights scattered from the sample in acetone was measured at 11 scattering angles between 30 and 150° and at five temperatures between 12.6 ± 0.05 and 50.3 ± 0.15 °C with two lights of 436 and 365 nm in wavelength. Vertically polarized components were used both for radiation and detection. Under these conditions, correction for the anisotropy effect was easy.[9] The photometer was calibrated at 25° against benzene (spectroscopy reagent), the literature values[10] 46.4×10^{-6} cm^{-1} and 0.42 being adopted, respectively, for the Rayleigh ratio and the depolarization ratio of unpolarized light of 436 nm.

The specific refractive index increment $\partial n/\partial c$ on solutions of cellulose acetate (DS 2.46) in acetone and in THF was determined by a Shimadzu differential refractometer model DR-4 at 25 ± 0.2 °C at wavelengths of 546 and 436 nm. With this instrument, the refractive index increment measurements are accurate to $\pm 1.3 \times 10^{-5}$ cm^3 g^{-1}. The $\partial n/\partial c$ value obtained for CA (DS 2.46) solutions was 0.109 cm^3 g^{-1} in acetone at $\lambda_0 = 546$ nm and 0.068 cm^3 g^{-1} in THF at $\lambda_0 = 436$ nm. These values were found for three different fractions and two whole polymers with combined acetic acid content (AAC) = 54.2–55.6 wt%. The $\partial n/\partial c$ values for these systems are in fairly good agreement with the values found by Stein and Doty[11] (0.104–0.116 in acetone) and by Tanner and Berry[12] (*ca.* 0.11 (from Figure 2 of Ref. 10) in acetone and 0.0710 in THF). The refractive index of the solvents was measured on a Hitachi Abbe refractometer.

The specific refractive index increment $\partial n/\partial c$ of this system was found to be fairly well represented by the relationship:

$$(\partial n/\partial c) = 0.112 + 3.3 \times 10^{-4}(T - 25)^2 \qquad (3.5.6)$$

where T is temperature in Celsius. The value of 0.112 at 25° is compared with the corresponding value, 0.100, previously reported for green (546 nm) light (measurement (1)). Zimm plots were made of Kc/R_θ against $\sin^2(\theta/2) + kc$ where symbols have their usual meaning. The second virial coefficient A_2, the z-average radius of gyration $\langle S^2 \rangle_z^{1/2}$ and M_w were determined from the plots in the usual way.

CA (DS 2.92)[13]

In preference to acetone, DMAc was chosen as the solvent for the LS measurement, since it affords easy exclusion of the gel-like materials in solution. Two vacuum-dried CA (DS 2.92) fractions (TA2-6 and TA2-13) and a CA (DS 2.46) fraction (EF 3-11), obtained in a previous work[7] were dissolved in freshly distilled DMAc at 60 °C. The specific refractive index increment $\partial n/\partial c$ of these solutions was determined in a Shimadzu differential refractometer model DR-4 at 25 °C at a wavelength λ_0 of 436 nm (4360 Å). Great care was taken to ensure that DMAc was absolutely dry since it is highly hygroscopic and will yield a high $\partial n/\partial c$ value when wet. The $\partial n/\partial c$ was found to be 0.0398 (cm^3 g^{-1}) for two CA (DS 2.92) fractions, irrespective of their M_w values, and 0.0418 (cm^3 g^{-1}) for a CA (DS 2.46) fraction.

The optical clarification procedure of the solution is similar to that used in the previous study.[7] After a 0.5% stock solution was centrifuged at 3×10^4 g for 90 min, the fresh solvent was added to the isolated upper layer, yielding the solutions with four different concentrations. The solutions prepared thus far were immediately followed again by centrifugation at 3×10^4 g for 90 min. The upper two-thirds of the supernatant phase were carefully filtered through a Sartorious membrane filter 0.2 μm, directly into a LS cylindrical cell.

The LS measurements in the angle range of 30–150° were made with vertically polarized incident light ($\lambda_0 = 436$ nm (4360 Å)) at 25 °C in the Shimadzu–Brice-type LS photometer model PG-21. The vertical component of the scattered light was measured. M_w and $A_{2,L}$ (the suffix L means LS) were determined from the Zimm plot. No distortion was observed in the Zimm plot for CA (DS 2.92) solutions after ultracentrifugation for removing the gel particles attributable to non-CA (DS 2.92) materials and in part to the associated CA (DS 2.92) molecules.

Cellulose/aq. LiOH,[14] cadoxen[14]

A preliminary attempt to estimate the specific refractive index increment under constant chemical potential of LiOH, $(\partial n/\partial c)_\mu$, failed because of non-availability of a suitable membrane, not swollen in 6 wt% aq. LiOH. As a next best method, we determined $(\partial n/\partial c)$ at a constant LiOH fraction in aq. LiOH, designated as $(\partial n/\partial c)_{\Phi L}$. Using this value, the LS data were analyzed to give the apparent weight-average molecular weight M_w^*, the apparent second virial coefficient A_2^*, and the z-average radius of gyration $\langle S^2 \rangle_z^{1/2}$. In order to evaluate the absolute molecular weight of cellulose samples, cellulose (sample code SA-1 and SA-5) solutions in cadoxen were dialyzed using a commercially available cellulose tube from Union Carbide Co. (USA) in an apparatus specially designed and constructed by us. The attainment of Donnan membrane equilibrium after 5 days dialysis was confirmed by measuring the electrical conductivity of the solution. $(\partial n/\partial c)_\mu$ of cellulose cadoxen and $(\partial n/\partial c)_{\Phi L}$ of cellulose/aq. LiOH were determined by a Shimadzu differential refractometer type DR-4 at 25 °C with the incident light of wavelength λ_0 of 436 nm. $(\partial n/\partial c)_\mu$ for sample SA-1 and 5 in cadoxen were found to be 0.194 and 0.198 cm^3 g^{-1}, respectively. These values are larger than those by Vink and Dahlström (0.193)[15] and Henley (0.183–0.193).[16] $(\partial n/\partial c)_{\Phi L}$ for sample SA-1 ~ SA-6 in aq. LiOH was 0.134 $\pm$ 0.04 cm^3 g^{-1}.

The LS experiments were carried out for cellulose/cadoxen and cellulose/aq. LiOH in the following manner: the solution was centrifuged by a Hitachi model 55p-7 automatic preparative centrifuge at 1×10^5 g for 1 h and then filtered through a poly(tetrafluoroethylene) (PTFE) membrane filter FP series (pore diameters 0.1, 0.22, 0.3, and 0.45 μm), manufactured by Sumitomo Denko Co. (Osaka). In this case, the membrane with a larger pore diameter was used for more concentrated solutions. The intensity of scattered light was measured at 25 °C by a FICA photogonio diffusiometer 42000 with unpolarized incident light of $\lambda_0 = 436$ nm. Using the ratio of M_w in nondialyzed aq. LiOH to M_w in dialyzed cadoxen for the samples SA-1 and 5, M_w of other samples was converted to M_w by the linear interpolation method.

Cellulose/aq. NaOH[17]

The refractive index of 8% aq. NaOH solution n_0 was determined on a Hitachi refractometer PRA-B with an unpolarized light of wavelength 633 nm at temperatures from 3.5 to 45 °C. For the same reason as described in a previous study[14] for cellulose 6 wt% aq. LiOH solution, the specific refractive index increment under constant chemical potential μ of NaOH in the solutions $(\partial n/\partial c)_\mu$ (c, concentration of polymer, g cm^{-3})

could not be evaluated. Then, $(\partial n/\partial c)$ at constant NaOH fraction in aq. NaOH, designated as $(\partial n/\partial c)_{cN}$, was determined by a Shimadzu differential refractometer type DR-4 with an incident light of wavelength λ_0 of 633 nm at temperatures in the range of 10–45 °C. We filtered 8% aq. NaOH solution (simply referred to as aq. NaOH) through a PTFE membrane FP series (pore diameter = 0.1 μm), manufactured by Sumitomo Denko Co. (Osaka), in a circulating filtration apparatus for 24 h. The solution was then filtered three times through a PTFE membrane FP series (pore diameter = 0.1 μm) installed in a Millipore pressure holder under atmospheric pressure. The cellulose solution with $c = 0.545$ g cm^{-3} was centrifuged at $1 \times 10^5 g$ for 60 min at 10 °C in a Hitachi model 55p-7 automatic preparative ultracentrifuge. The upper two-thirds of the supernatant phase were carefully sucked up with a microsyringe and filtered through a PTFE membrane (pore diameter = 0.45 μm) installed in a Millipore pressure holder. Then, the aq. NaOH, purified as described above, was added to give cellulose solutions of five different cellulose concentrations. Each solution was filtered under atmospheric pressure of 20–27 °C through PTFE membranes (pore diameters 0.1, 0.22, 0.3, 0.45 μm), and the last filtrate was directly transferred into a cylindrical LS cell, which was then sealed with Parafilm 'M' (manufactured by American Can Co., USA) in order to avoid absorption of carbon dioxide from the air. The LS measurements for each solution were carried out successively at six different temperatures of 3.5, 10, 26, 35, 41, and 45 °C (controlled to ± 0.1 °C). After the experiments at each given temperature, and when measurements had to be intercepted temporarily, the solution was once cooled to 4 °C and stocked. In order to avoid gelation, all measurements were designed in the direction of rising temperature. An LS apparatus, DLS-700, manufactured by Otsuka Electronics (Osaka), was utilized with an incident light (vertically polarized light) of $\lambda_0 = 633$ nm. The calibration of the apparatus was done with the literature value[18] $(1.184 \times 10^{-6}$ cm$^{-1})$ of reduced Rayleigh ratio of benzene. The following operating conditions were employed: gate time 160 s; accumulation 100 times; and measuring angle 20–150° (10° each). The data were analyzed according to Zimm's procedure to evaluate the apparent molecular weight $M_{\mathrm{w}}^*, \langle S^2 \rangle_z^{1/2}$ and the apparent second virial coefficient A_2^*.

The temperature dependences of n_0 and $(\partial n/\partial c)_{\Phi N}$ of cellulose in aq. NaOH can be empirically represented by eqs. (3.5.6) and (3.5.7), respectively.

$$n_0 = 1.3535 - 2.17 \times 10^{-4}(T - 25) \qquad (3.5.7)$$

$$(\partial n/\partial c)_{\Phi N} = 0.154 - 3.0 \times 10^{-4}(T - 25)(\mathrm{cm}^3\mathrm{g}^{-1}) \qquad (3.5.8)$$

Here, T is expressed by the °C unit. The value of $(\partial n/\partial c)_{\Phi N}$ at 25 °C (0.154) is *ca.* 13% larger than that for the undialyzed cellulose (SA3)-6 wt% aq. LiOH solution system[13] $((\partial n/\partial c)_{\Phi L} = 0.134)$ at the same temperature. The CA (DS 2.46)/acetone system has a positive temperature dependence of $(\partial n/\partial c)$,[8] whereas cellulose/aq. NaOH system shows a negative temperature coefficient of $(\partial n/\partial c)_{\Phi N}$.

3.5.3   Zimm plots

Some typical Zimm plots are demonstrated in Figures 3.5.1–3.5.5 for CA (DS 0.49) in FA at 25 °C,[5] CA (DS 2.46) in acetone at 25 °C,[7] cellulose nitrate (CN; DS 2.3) in

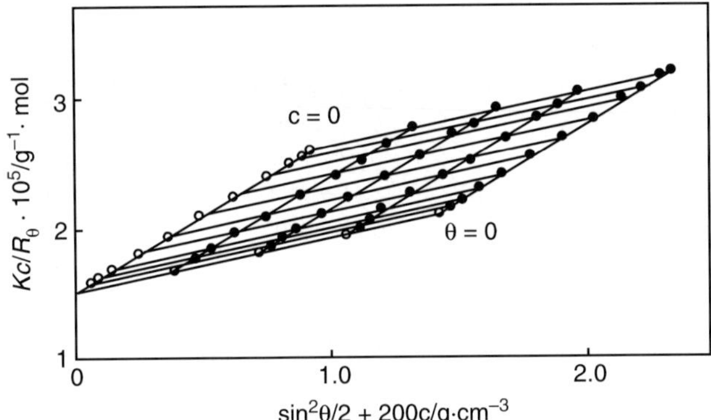

Figure 3.5.1   Zimm plot of cellulose (DS 0.49) fraction MA-5 in formamide at 25 °C: unpolarized incident light wavelength $\lambda_0 = 436$ nm.[5]

acetone at 25 °C,[19] cellulose in cadoxen, aq. LiOH,[12] and in 8% aq. NaOH at 10 and 45 °C, respectively.[15]

All of the plots can be adequately represented by a nondistorted diamond shape, exhibiting no downward curvature. The double extrapolations $\lim_{\theta\to 0} Kc/R_\theta$ and $\lim_{c\to 0} Kc/R_\theta$ cut the ordinate axis at the same point. The experimental error involved in the LS measurement of M_w was estimated to be less than a few percent.

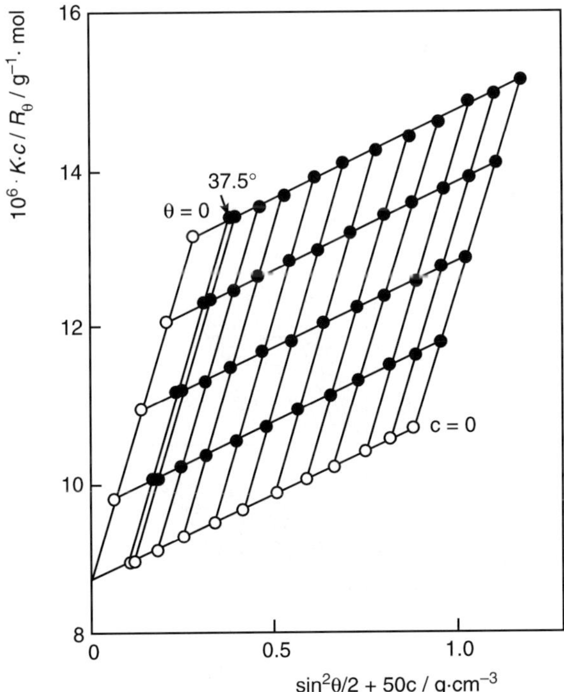

Figure 3.5.2   Zimm plot of cellulose diacetate fraction EF2-14 in acetone at 25 °C: unpolarized incident light wavelength $\lambda_0 = 546$ nm. The symbols have their usual meaning.[7]

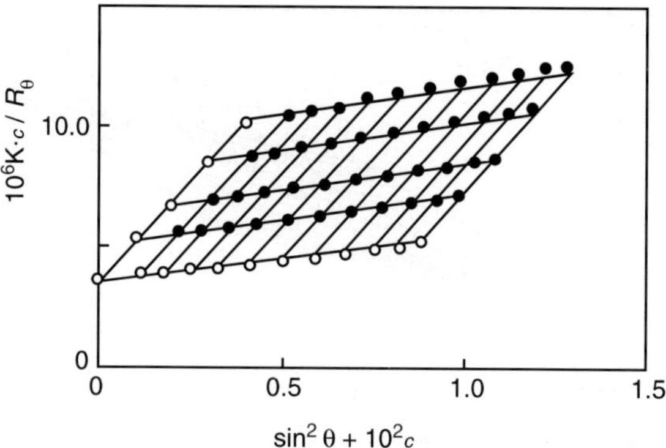

Figure 3.5.3   Zimm plot of cellulose nitrate ($\langle\!\langle F \rangle\!\rangle = 2.3$) whole polymer in acetone at 25 °C.[19]

In CDA, although the highest fractions (EF-2-16 and EF-3-21) were completely soluble in acetone and THF, their Zimm plots revealed a downward curvature at lower angle and they were thus discarded.

3.5.4   Determination of weight average molecular weight, *z*-average radius of gyration, and second virial coefficient

CA (DS 0.49)[5]

Analysis of Zimm plots of 13 fractions and a whole polymer of CA (DS 0.49) in DMAc and that of a single fraction of CA (DS 0.49) in FA and water was made. The results are shown in Tables 3.3.1 and 3.5.1.

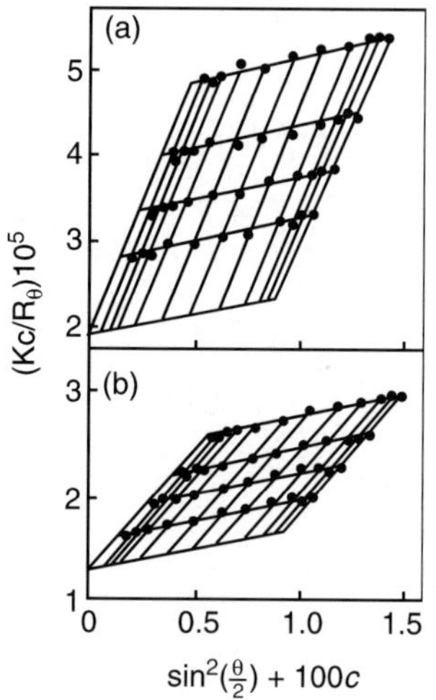

Figure 3.5.4   Typical Zimm plots of cellulose (sample code SA-5) in cadoxen (a) and in aq. LiOH (b).[12]

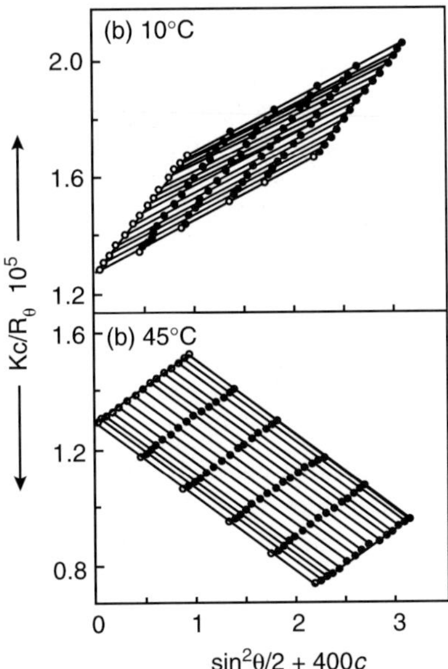

Figure 3.5.5   Zimm plots of cellulose/8% NaOH aq. solution system at temperature of 10 °C (a) and 45 °C (b). ($\bullet$), Experimental data; ($\circ$), value extrapolated from the experimental data at scattering angle $\theta = 0°$ and a concentration $c = 0\%$.[15]

In Table 3.3.1, the maximum value of M_w for these fractions is a little below 15×10^4, which is an upper limit as long as the hydrolysis method is employed and at present there is no efficient method for preparing CA (DS ≤ 0.6) with higher M_w. Excellent agreement of the M_w values of the fraction MA-5 in three different solvents (Table 3.5.1) confirms the experimental reliability.

Table 3.5.1

Results of solution viscosity and light scattering measurements on cellulose acetate (DS 0.49) fraction MA-5 in formamide, water, and dimethylacetamide at 25 °C[5]

Solvent	Dielectric constant ε	$[\eta]$ $(cm^3\ g^{-1})$	$M_w \times 10^{-4}$	$\langle S^2 \rangle_z^{1/2} \times 10^8$ (cm)
FA	111	156	6.30	349
Water	78.3	151	6.42	291
DMAc	37.8	133	6.32	164

Solvent	$A_2 \times 10^4$ $(cm^3\ mol\ g^{-2})$	$\Phi \times 10^{-23}$	$\Psi \times 10^3$	α_s	χ
FA	8.4	0.22	8.7	1.01	0.14
Water	3.1	0.38	6.7	1.01	0.26
DMAc	7.5	1.82	4.9	1.05	4.2

CA (DS 1.75)

In the second column of Table 3.3.2 M_w, values of seven fractions of CA (DS 1.75) evaluated from the Zimm plot are shown. The weight-average radius of gyration $\langle S^2 \rangle_w^{1/2}$ was calculated from the z-average radius of gyration $\langle S^2 \rangle_z^{1/2}$ and the M_w/M_n value, assuming the Schulz-Zimm MWD (eq. (3.5.4)). Columns 3 and 4 of the same table show $\langle S^2 \rangle_w^{1/2}$ of the sample.

CA (DS 2.46)

Table 3.3.3 shows the results of LS measurements on 12 fractions of CA (DS 2.46) in DMAc, acetone, and THF.[20] We see that the M_w values in the three solvents (acetone, THF, and DMAc) are almost the same. Table 3.5.2 shows some molecular parameters of a CA (DS 2.46) fraction in acetone and in THF, estimated by LS, viscometry, and the GPC method.[8] Inspection of Tables 3.3.1, 3.3.3, 3.5.1, and 3.5.2 shows the excellent agreement of the M_w values of the fractions in different solvents and confirms the experimental reliability of LS measurements. These data also indicate that the CA (DS 0.49) and CA (DS 2.46) molecules dissolve molecularly in these solvents without association, and that the LS technique established for the cellulose acetates is very reliable, at least where M_w is concerned.

We note that most of the data in acetone and THF given in Table 3.3.3 were obtained by other research personnel at our laboratory, and the M_w and $\langle S^2 \rangle_z^{1/2}$ data on sample EF3-8 (in acetone) were obtained by research personnel at the Institute for Chemical Research, Kyoto University, Kyoto.

The weight-average molecular weight M_w of two CA (DS 2.46) fractions (see Table 3.3.3) in acetone and THF were determined at 25, 25.4, and 50.3 °C by LS measurements by two different individuals at different laboratories (Asahi Chemical Industry and Kyoto University) using different apparatus under different operating conditions. The results are summarized in Table 3.5.3, which confirms the reliability of LS method established for CA solutions.

Inspection of Tables 3.5.1–3.5.3 illustrates that the CA (DS 0.49) and (DS 2.46) polymers dissolve molecularly in those solvents without association.

Table 3.5.2

Molecular parameters of a cellulose diacetate fraction EF3-8 in acetone

Molecular parameter	Light scattering		Viscosity[a]	GPC
	Acetone 25.4 °C	THF[b] 25.0 °C	Acetone 25.0 °C	THF *ca.* 25 °C
$M_w \times 10^{-4}$	9.4	10.0	9.5	9.8
$M_n \times 10^{-4}$	–	–	–	7.5
$A_2 \times 10^4\,\mathrm{cm^3\,mol/g^2}$	3.8	4.9	–	–
$\langle S^2 \rangle_z^{1/2}$, Å	234	181	–	–

[a]Calculated from viscosity data, $[\eta] = 154\,\mathrm{cm^3\,g^{-1}}$ using eq. (3.11.9).
[b]Taken from Ref. 7.

Table 3.5.3

Reliability of light scattering method for cellulose acetate (DS 2.46) solutions

Fraction no.	Solvent	M_w $(\times 10^{-4})$	Temperature (°C)	Measuring condition	Reference
EF3-12	Acetone	14.1	25	$\lambda = 546$ nm unpolarized light	Asahi Chem.[7]
		14.1	25	$\lambda = 436$ nm vertically polarized component, $\theta = 22.5\text{--}150°$ (12 analyses)	Kyoto Univ.[21]
EF3-8	Acetone	9.4	25	$\lambda = 546$ nm	Asahi Chem.[7]
	THF	10.0	25	Unpolarized component	
	THF	9.8	25	GPC	
	Acetone	9.4	25.4	$\lambda = 436$ and 365 nm	Kyoto Univ.[8]
		9.5	50.3	Vertically polarized component	

CA (DS 2.92)[11]

Table 3.3.5 displays the results on 12 fractions and a whole polymer of CA (DS 2.92) in DMAc. Table 3.5.4 compares CA (DS 2.46) and CA (DS 2.92) with the same M_w, both dissolved in common solvent (DMAc).

Cellulose/aq. LiOH and cadoxen[14]

Table 3.5.5 summarizes results of LS measurements of six fractions of cellulose/aq. LiOH at 25 °C.[14]

Table 3.5.6 shows M_w, A_2, $\langle S^2 \rangle_z^{1/2}$ of samples SA-1 and SA-5 in cadoxen and M_w^* of the samples in aq. LiOH. M_w^* of SA-1 and SA-5 in aq. LiOH is some 40–60% larger than M_w in dialyzed cadoxen solutions of SA-1 and SA-5, respectively.

A parameter representing selective adsorption, α_α is related to M_w and the apparent weight-average molecular weight M_w^* through the relationship:[22]

$$\alpha_\alpha = \{(M_w/M_w)^{1/2} - 1\}(\partial n/\partial c)_{\Phi L}/(\partial n_0/\partial \Phi_L) \tag{3.5.9}$$

where $(\partial n_0/\partial \Phi_L)$ is an increment of the refractive index of aq. LiOH, and was 0.31 cm^3 g^{-1} near 6 wt% LiOH concentration. From eq. (3.5.9), the α_α for SA-1 and SA-5 in aq. LiOH was determined as 0.117 and 0.088, respectively. These comparatively large α_α values imply the possibility of preferential adsorption of lithium cation or the hydroxyl (OH) group of LiOH on cellulose, rather than water molecules.

Cellulose/aq. NaOH[15]

At any temperature of measurements ranging from 3.5 to 45 °C, no significant distortion was observed in the Zimm plots. This means that the solutions prepared here had no

Table 3.5.4

Light scattering data on cellulose diacetate in dimethylacetamide (DMAc) as compared with those of cellulose diacetate in acetone and tetrahydroforan (THF) and of cellulose triacetate in DMAc[11]

Polymer	Sample	Solvent	Temperature (°C)	$M_w \times 10^{-4}$	$\langle S^2 \rangle_z^{1/2} \times 10^{-8}$ (cm)	$\langle S^2 \rangle_w^{1/2} \times 10^{-8}$ (cm)	$A_2 \times 10^4$ (mol cm^3 g^{-2})	$[\eta]$ (cm^{3}g^{-1})	$\Phi \times 10^{-23}$
Cellulose diacetate (AC = 55.6%)	EF3-11	DMAc	25	10.8	344	310	0.77$_5$	227	0.58
		Acetone	25	10.8	(262)[a]	(234)	–	(168)	(1.00)
		THF	25	10.8	(192)	(173)	–	(149)	(2.16)
Cellulose triacetate (AC = 61.0%)	–	DMAc	25	10.8	(272)	(240)	(1.25)	(157)	(0.85)

[a]The values in parentheses are calculated from the molecular weight dependence of $\langle S^2 \rangle_z^{1/2}$ (or $\langle S^2 \rangle_w^{1/2}$, or $[\eta]$ or Φ), experimentally determined.

Table 3.5.5

Results of light scattering measurements at 25 °C and various parameters of cellulose/6 wt% aq. LiOH solution system[14]

Sample code	$M_w^* \times 10^{-4}$	$\langle S^2 \rangle_z^{1/2} \times 10^6$ (cm)[a]	$A_2^* \times 10^3$ (cm^3 mol g^{-2})	$M_w \times 10^{-4}$	$A_2 \times 10^3$ (cm^3 mol g^{-2})[b]	$M_v \times 10^{-4c}$	$\Phi \times 10^{-22}$	α_s[d]
SA-1	19.1	3.94	1.09	12.0	1.73	11.5	7.6_9	1.04
SA-2	15.1	3.61	1.11	$(9.6)^a$	1.74	9.9	6.9_6	1.03
SA-3	12.1	3.25	1.10	$(8.0)^a$	1.67	7.8	6.5_3	1.03
SA-4	11.5	2.97	1.22	$(7.6)^a$	1.84	7.0	7.4_6	1.04
SA-5	7.6	2.46	1.12	5.2_6	1.62	5.1	6.8_5	1.03
SA-6	3.8	1.94	1.41	$(2.8)^a$	1.94	3.1	4.6_9	1.02

[a]Estimated value.
[b]Corrected value with use of the equation $A_2 = (M_w^*/M_w)A_2^*$ (M_w and A_2 are apparent M_w and A_2.
[c]Calculated from $[\eta]$ in cadoxen at 25 °C using Mark–Houwink–Sakurada equation by Brown and Wikström.[29]
[d]Calculated from using eqs. 3.13.4–3.13.6, and 3.13.10.

undissolved material or gel over a wide range of temperatures. Table 3.5.7 summarizes M_w^*, $\langle S^2 \rangle_z^{1/2}$ and A_2^* of cellulose in aq. NaOH, each evaluated from the Zimm plots. Here, the above quantities at 3.5 °C were determined using the value of $(\partial n/\partial c)_{\Phi N}$ calculated by eq. 3.5.8.

Now, we can evaluate a parameter representing selective adsorption, Γ', quantatively from eq. (3.5.10).[23]

$$\Gamma' = \{(M_w^*/M_w)^{1/2} - 1\}d_N(\partial n/\partial c)_{\Phi N}/(\partial n_0/\partial \Phi_N) \tag{3.5.10}$$

to be 0.007 at 3.5 °C, where d_N is the density of 8 wt% aq. NaOH solution (*ca.* 1.09) and $(\partial n_0/\partial_{\Phi N})$ is an increment of refractive index of aq. NaOH (0.31 cm^3 g^{-1} at 3.5 °C near 8 wt% NaOH concentration). Γ' of cellulose in aq. NaOH is almost 10% of those of cellulose in aq. LiOH at 25 °C (i.e. $\Gamma' = 0.095$ for $M_w = 5.26 \times 10^4$ and 0.126 for $M_w = 1.2 \times 10^5$). Note that the Γ' value calculated for 6 wt% aq. LiOH in Ref. 14 should be multiplied by the density of 6 wt% aq. LiOH (1.08; see also Ref. 28 in Ref. 14).

Table 3.5.6

Light scattering data of cellulose sample code SA-1 and SA-5 in cadoxen and in 6 wt% aq. LiOH and the viscosity-average molecular weight M_v determined using the Mark–Houwink–Sakurada equation in cadoxen established by Brown and Wikström at 25 °C

Sample code	Light scattering				Viscosity
	Cadoxen (dialyzed)			6 wt% aq. LiOH	Cadoxen
	$M_w \times 10^{-4}$	$\langle S^2 \rangle_z^{1/2} \times 10^6$ (cm)	$A_2 \times 10^{-4}$ (cm^3 mol g^{-2})	$M_w^* \times 10^{-4a}$	$M_v \times 10^{-4b}$
SA-1	12.0	3.51	2.29	19.1	11.5
SA-5	5.2_6	2.29	2.95	7.6	5.1_2

[a]The apparent weight average molecular weight determined for nondialyzed 6 wt% aq. LiOH solution.
[b]Calculated using Mark–Houwink–Sakurada equation established by Brown and Wikström.[29]

Table 3.5.7

Results of light scattering measurements on cellulose/8% NaOH aq. solution and hydro- and thermodynamic parameters in 8% NaOH aq. solution at temperatures ranging from 3.5 to 45 °C[15]

Tempera-ture (°C)	Light scattering							Hydro- and thermo-dynamic parameters	
	8% NaOH aq. solution			6% LiOH aq. solution[a]			Cadoxen[a]	8% NaOH aq. solution	
	$M_w^* \times 10^{-4}$	$\langle S^2 \rangle_z^{1/2} \times 10^6$ (cm)	$A_2^* \times 10^4$ $(cm^3 mol\ g^{-2})$[b]	$M_w^* \times 10^{-4}$	$\langle S^2 \rangle_z^{1/2} \times 10^6$ (cm)	$A_2^* \times 10^4$	$M_w \times 10^{-4}$	$\Phi \times 10^{-22}$	α_s
3.5	8.10	3.66	4.79 (4.91)[c]	–	–	–	–	4.14	1.005
10	7.89	3.61	3.75	–	–	–	–	3.95	1.004
25	–	–	–	12.1	3.25	11.0	(8.0)[d]	–	–
26	7.89	3.14	3.33	–	–	–	–	5.46	1.006
35	7.98	3.01	2.75	–	–	–	–	5.55	1.005
41	7.98	2.69	−0.58	–	–	–	–	6.10	0.999
45	7.81	2.64	−3.06	–	–	–	–	6.28	0.992

[a]Ref. 14.

[b]$A_2^* \simeq A_2$, except 3.5 °C.

[c]True A_2.

[d]Estimated value using light scattering data on two dialyzed cellulose cadoxen solutions.

Sodium cellulose sulfate/aq. NaCl[24]

Table 3.3.7 shows results of light scattering measurements with sodium cellulose sulfate (DS 1.9) in aq. NaCl. The values for the lowest molecular weight sample (CSO − 6) are less accurate except for $[\eta]$ and M_w by LS, shown in brackets, and were thus discarded in the further analyses.

3.5.5   Dynamic LS

When monochromatic light (laser light) is irradiated on polymer solutions, which do not absorb the light, a part of irradiated light is scattered to various directions. If the polymer solute is in a stationary state, then the angular frequency of the scattered light coincides strictly with that of incident light (elastic scattering). If the polymer solute obeys a kind of motion, then the frequency of the scattered light varies depending on its moving velocity. This is an optical Doppler effect. Dynamic light scattering (DLS) aims to obtain information of the moving state of the scattering body from analysis on the frequency variation (*ca.* 10^6 Hz) of the scattered light. Each piece of information is described in terms of frequency dependence of the frequency breath of the scattered light (i.e. autocorrelation function).

Theoretical background[25,26]

By DLS measurements, the relationship between the second-order time correlation function of the electric field of the scattered light, $I_2(q, \tau)$ (where $q = (4\pi \sin(\theta/2)/\lambda)$: τ correlation time) and τ were obtained. The hydrodynamic diameter of the solute d was determined by a cumulative method as follows.[27]

$I_2(q, \tau)$ is related to the first correlation function, $I_1(q, \tau)$ through the equation

$$I_2(q, \tau) = \{1 + |I_1(q, \tau)/I_1(q, 0)|^2\} \tag{3.5.11}$$

When the solutes diffuse at random in the continuous media, $I_1(q, \tau)$ is related to the diffusion coefficient, D_i, of the solutes with the same diameter, d, by[25]

$$I_1(q, \tau) = \sum g_i \, \exp(-q^2 D_i \tau) \tag{3.5.12}$$

with $g_i = \alpha_i^2(N_i M_i^2)$, where α_i is polarizability per unit mass, and N_i and M_i are the number and mass of the particles with D_i.

Using eq. (3.5.12), the normalized first correlation function, $N(q, \tau)$, can be expressed by the relationship

$$N(q, \tau) \equiv I_1(q, \tau)/I_1(q, 0) = \left(\sum g_i \, \exp(-q^2 D_i \tau)\right)\Big/\sum g_i \tag{3.5.13}$$

Take the logarithm of both sides of eq. (3.5.13) and expand the right-hand side in a power series in τ, giving[27] τ

$$\ln N(q, \tau) = 1 - K_1\tau + (1/2)K_2\tau^2 - (1/3!)K_3\tau^3 + (1/4!)K_4\tau^4 - \cdots \tag{3.5.14}$$

where

$$K_1 = q^2 \left\{ \sum N_i m_i^2 D_i \right\} \Big/ \left\{ \sum N_i m_i \right\} = q^2 \langle D \rangle_z$$

$$K_2 = q^4 \langle (\delta D)^2 \rangle_z$$

$$K_3 = q^6 \langle (\delta D)^3 \rangle_z \qquad\qquad (3.5.15)$$

$$K_4 = q^8 \langle (\delta D)^4 \rangle_z - 3 K_2^2$$

Here, $\langle D \rangle_z$ is the z-average diffusion coefficient and $\langle (\delta D)^2 \rangle_z$ defined by

$$\langle (\delta D)^2 \rangle_z \equiv \langle D^2 \rangle_z - \langle D \rangle_z^2 \qquad\qquad (3.5.16)$$

as a variance of $\langle D \rangle_z$.

The most probable $\langle D \rangle_z$ was estimated from K_1 by choosing the most probable combination of values of K_1, K_2, K_3, K_4 in eq. (3.5.15) to minimize the difference between experimentally obtained $\ln(I_2 - 1)$ $(= \ln|N(q, \tau)|^2)$ and the calculated ones. When the polymer chains are assumed as a rigid sphere, $\langle D \rangle_z$ is transformed to the hydrodynamic diameter d through the Stokes–Einstein equation[28]

$$d = kT / \{ 3\pi \, \eta_0 \langle D \rangle_z \} \qquad\qquad (3.5.17)$$

where k is Boltzmann's constant, T the absolute temperature, and η_0 the solvent viscosity. Note that in deriving eq. (3.5.17), the solutes are assumed to be rigid spheres (i.e. completely undraining).

Experimental procedure[25,26]

DLS measurements were carried out on an Otsuka Electronics model DLS 700 based on a homodyne method under the following conditions: $\lambda = 633$ nm (λ, wavelength of incident beam in the solution); $\theta = 45$–$90°$ (θ, scattered angle); time domain method; temperature, $20 \pm 0.2\,°C$. The refractive index and the viscosity of cadoxen at $20\,°C$ were measured by an Abbé refractometer and a capillary-type viscometer, and were found to be 1.3530 and 3.76 cP, respectively.

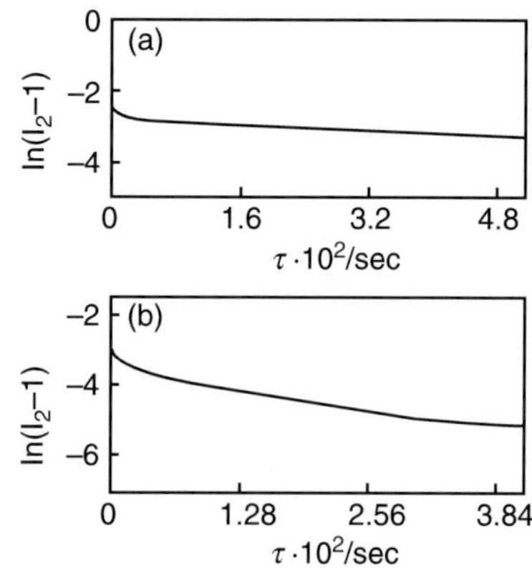

Figure 3.5.6   Correlation time τ dependence of $\ln(I_2 - 1)$ for cellulose solution diluted without centrifugation: (a) cellulose I; (b) cellulose II. Solid line, experimental; broken line, calculated.[25]

Table 3.5.8

Results of analysis on data of homodyne dynamic light scattering measurement at 20 °C[25]

| | Cellulose I solution | | | | | | Cellulose II solution | | | | | |
| | Undiluted 0.5% | Diluted 0.5% solution | | | | | Undiluted 0.5% | Diluted 0.5% solution | | | | |
	$CG^a = 0$	0	1775	7100	2830	63,600	0	0	1775	7100	2830	63,600
d (nm)	350	1415^b	863	223	140	145	752	702	752	713	672	738
		327^c	323	228	161	128		640	614	672	547	772
$\langle(\delta D)^2\rangle_z/\langle D_2\rangle^2$	1.32	4.06	3.46	1.65	2.24	1.63	1.20	1.56	1.69	1.29	1.50	1.76
		2.19	1.94	1.59	1.90	1.62		1.46	1.55	1.66	1.84	1.87
$\langle(\delta D)^3\rangle_z/\langle D_2\rangle^3$	1.77	15.7	11.5	22.6	3.81	2.02	1.72	2.97	3.31	1.78	2.40	3.25
		4.34	3.53	2.13	2.82	1.94		2.26	2.56	2.81	3.13	3.97
$\langle(\delta D)^4\rangle_z/\langle D_2\rangle^4$	1.28	31.5	20.1	1.44	3.17	1.15	1.43	2.85	3.30	1.25	1.97	2.88
		3.98	2.89	1.32	1.92	1.06		1.82	9.29	2.31	2.55	4.23

[a]Centrifugal acceleration in unit of g.
[b]Value at 1 day after dilution.
[c]Value at 5 days after dilution.

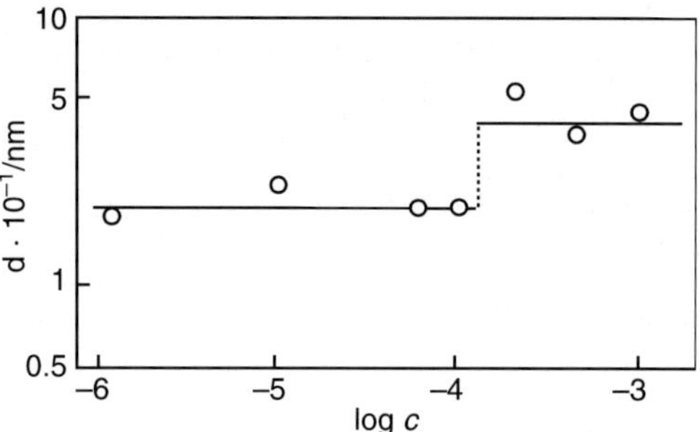

Figure 3.5.7 Concentration (c) dependence of the hydrodynamic diameter (d) as determined by the dynamic light scattering method for cellulose acetate (0.8) whole polymer in water at 20 °C.[26]

Semidilute cellulose solution in cadoxen

Figure 3.5.6(a) and (b) illustrates the experimental relationships (solid line) between $\ln(I_2 - 1)$, and τ determined by the homodyne DLS method, for cellulose I and II cadoxen solutions ($c = 0.5\%$), respectively, which were diluted from 5% solutions and not centrifuged.[25] In these figures, $\ln(I_2 - 1)$, decreases monotonically with an increase in τ. Dotted lines in the figures are the best fit curves calculated by eq. (3.5.14) from the combination of K_1 to K_4, which were determined so as to minimize the difference in $\ln(I_2 - 1)$ between the experiments and the calculation. Here, the ratios $\langle(\delta D)^2\rangle_z/\langle D^2\rangle_z^{-2}$ (i.e. square of coefficient of variation (SCV)), $\langle(\delta D)^3\rangle_z \equiv \langle D\rangle_z^3$ and $\langle(\delta D)^4\rangle_z \equiv \langle D\rangle_z^4$ are found to be 4.06, 15.73, and 31.48 for cellulose I solution (Figure 3.5.6(a)), and 1.58, 2.97 and 2.85 for cellulose II solution (Figure 3.5.6(b)), respectively. Excellent coincidence is confirmed between the experimental and theoretical $\ln(I_2 - 1)$ versus t curves, at least for $\tau > 3.2 \times 10^{-3}\,\text{s}^{-1}$.

The experimental fact of $K_3 \neq 0$ and $K_4 \neq 0$ indicates the nonGaussian nature of the solute particle size distribution. Comparison of SCV between cellulose I and II solutions suggests that the solute particles of cellulose I are larger than those of cellulose II and the former have a broader size distribution. From K_1, d was found to be 1415 nm for the cellulose I cadoxen system and 702 nm for the cellulose II cadoxen system.

Table 3.5.8 summarizes all of the experimental results on the d of cellulose in cadoxen solutions.[25] In the table, the d data of the solutions with different storage time after dilution are shown by the upper and lower lines, respectively.

CA (DS 0.80)/water[26]

Figure 3.5.7 shows the hydrodynamic diameter d as determined by DLS, of CA (0.80) whole polymer chains in water (see Section 2.11).

REFERENCES

1. See, for example, K Kamide and T Dobashi, *Physical Chemistry of Polymer Solution*, (Problem 7-10-b)–(Problem 7-11-b), Elsevier, Amsterdam, 2000.
2. MB Huglin (ed.), *Light Scattering from Polymer Solutions*, Academic Press, London, 1972.

3. BH Zimm, *J. Chem. Phys.*, 1948, **16**, 1099.
4. See, for example, K Kamide and T Dobashi, *Physical Chemistry of Polymer Solution*, (Problem 9-31) and (Problem 9-32), eq 9.31.7, Elsevier, Amsterdam, 2000.
5. K Kamide, M Saito and T Abe, *Polym. J.*, 1981, **13**, 421.
6. M Saito, *Polym. J.*, 1983, **15**, 249.
7. K Kamide, T Terakawa and Y Miyazaki, *Polym. J.*, 1979, **11**, 285.
8. H Suzuki, Y Miyazaki and K Kamide, *Eur. Polym. J.*, 1946, **16**, 703.
9. H Uchiyama, M Kurata. *Bull. Inst. Chem. Res. Kyoto Univ.*, 1962, **42**, 128.
10. S Claesson and J Ohman, *Ark. Kemi*, 1964, **23**, 69.
11. RS Stein, P Doty. *J. Am. Chem. Soc.*, 1946, **68**, 159.
12. DW Tanner and GC Berry, *J. Polym. Sci. Polym. Phys.*, 1974, **12**, 941.
13. K Kamide, Y Miyazaki and T Abe, *Polym. J.*, 1979, **11**, 523.
14. K Kamide and M Saito, *Polym. J.*, 1986, **18**, 569.
15. H Vink and G Dahlström, *Makromol. Chem.*, 1967, **109**, 249.
16. D Henley, *Ark. Kemi*, 1961, **18**, 327.
17. K Kamide, M Saito and K Kowsaka, *Polym. J.*, 1987, **19**, 1173.
18. ER Pike, WRM Pomercy and JM Vaugham, *J. Chem. Phys.*, 1975, **62**, 3188.
19. K Kamide and M Saito, *Macromol. Symp.*, 1994, **83**, 233.
20. K Kamide and M Saito, *Polym. J.*, 1982, **14**, 517.
21. H Suzuki, Y Muraoka, M Saito and K Kamide, *Eur. Polym. J.*, 1982, **18**, 831.
22. C Strazielle, in *Light Scattering from Polymer Solutions* (ed. MB Huglin), Academic Press, New York, 1972, Chapter 15, p. 652.
23. MB Huglin, In *Light Scattering from Polymer Solutions* (ed. MB Huglin), 1972, Chapter 6, p. 192.
24. K Kishino, T Kawai, T Nose, M Saito and K Kamide, *Eur. Polym. J.*, 1981, **17**, 623.
25. K Yasuda, M Saito and K Kamide, *Polym. Int.*, 1993, **30**, 393.
26. K Kamide, M Saito and T Akedo, *Polym. Int.*, 1992, **27**, 35.
27. BJ Berne and R Pecora, *Dynamic Light Scattering with Application to Chemistry, Biology and Physics*, Wiley, New York, 1976, Chapter 8, p. 195.
28. RB Bird, WE Stewart and EN Lightfoot, *Transport Phenomena*, Wiley, New York, 1960, Chapter 16, p. 514.
29. W Brown and R Wikström, *Eur. Polym. J.*, 1965, **1**, 1.

3.6   VISCOMETRY

3.6.1   Theoretical background and experimental procedure

Pure shear flow of viscous fluids with shear velocity g induces the work heat ηg^2 per unit volume and unit time. The viscosity η of solution is related to the viscosity of pure solvent η_0 as[1]

$$\eta g^2 = \eta_0 g^2 + \Delta q \tag{3.6.1}$$

This equation means that the solution generates heat more than the solvent by Δq.

The viscosity of polymer solutions indicates significant viscosity gradient dependence and, in most cases, η decreases with an increase of velocity gradient, especially in the higher gradient region. η of a polymer solution also depends on the concentration c. In physical chemistry of polymer solutions, relative increase in viscosity of dilute solution is more important than the viscosity itself.

Limiting viscosity number $[\eta]$ is defined by[2]

$$[\eta] = \lim_{c \to 0} \eta_{sp}/c \tag{3.6.2}$$

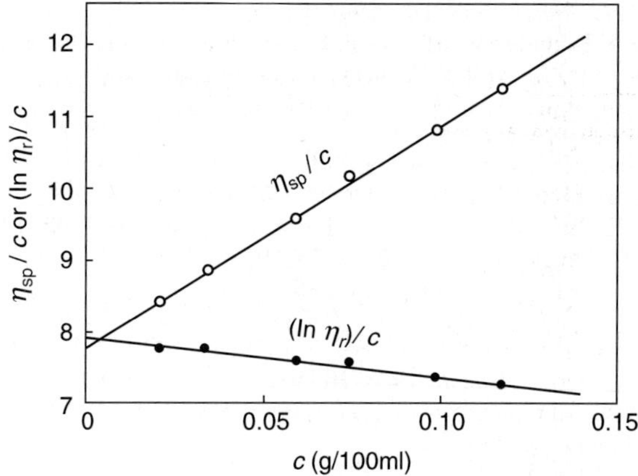

Figure 3.6.1   Plot of η_{sp}/c (or $\ln \eta_r/c$) against c for a cellulose nitrate ($N_c = 12\%$) fraction FZ4-1 acetone solution at 20 °C.[9]

or

$$[\eta] = \lim_{c \to 0} (\ln\eta_r)/c \qquad (3.6.3)$$

with

$$\eta_{sp} = (\eta/\eta_0) - 1 = \eta_r - 1 \qquad (3.6.4)$$

($\eta_{sp} =$ specific viscosity, $\eta_r =$ relative viscosity)
  η of dilute polymer solutions are expressed in the form:

$$\eta = \eta_0\{1 + [\eta]c + kc^2 + \text{higher term of } c^3\} \qquad (3.6.5)$$

$$\eta_{sp}/c = [\eta] + kc + \cdots \qquad (3.6.6)$$

Table 3.6.1

Limiting viscosity number of cellulose diacetate in various solvents[10]

Sample code	$M_w \times 10^{-4a}$	$[\eta]$ at 25 °C (cm^3 g^{-1})		
		DMAc[b]	TFA[c]	Acetone[d]
EF 2-10	6.1	140	113	117
EF 2-11-1	9.6	170	145	151
EF 3-10	10.6	205	178	160
EF 3-12	14.1	243	193	193
EF 3-14	18.5	315	247	241

[a]By light scattering in acetone solution.
[b]Dimethylacetamide.
[c]Trifluoroacetic acid.
[d]Ref. 16.

Table 3.6.2

Results of viscosity measurements with cellulose triacetate in various solvents[10]

Sample code	$M_{\mathrm{w}} \times 10^{-4}$	$\mathrm{cm^3\,g^{-1}}$ (at 25 °C)/$[\eta]$					
		DMAc[a]	TFA[b]	Acetone	DCM[c,d]	TCE[e]	TCM[f]
TA2-1	–	–	–	–	–	26	–
TA2-2	–	59	–	–	–	38	–
TA2-3	–	70	–	–	–	47	–
TA2-4	6.36	102	96	–	59	60	(47)[g]
TA2-5	8.22	131	118	110	71	73	69
TA2-6	13.7	183	164	143	100	95	97
TA2-7	14.9	195	–	154	–	98	–
TA2-8	20.0	256	228	211	136	129	129
TA2-9	26.2	327	–	265	–	163	–
TA2-10	30.8	350	286	282	184	168	171
TA2-11	44.4	451	385	357	–	214	210
TA2-12	50.0	490	434	398	260	243	228
TA2-13	69.0	607	508	480	306	270	272

[a]Dimethylacetamide.
[b]Tricluoroacetic acid.
[c]Dichloromethane.
[d]At 20 °C.
[e]Tetrachloroethane.
[f]Trichloromethane.
[g]Partially dissolved.

When higher order terms in eq. (3.6.5) are negligible, we have

$$\eta_{\mathrm{sp}}/c = [\eta]\{1 + k'[\eta]c\} \qquad (k = k'[\eta]^2) \tag{3.6.7}$$

Eq. (3.6.7) is called a Huggins equation.[3,4] $[\eta]$ and k' are determined from the intercept at $c \rightarrow 0$ and the slope, respectively, of the plot of η_{sp}/c versus c (Huggins plot). $[\eta]$ is also determined from the intercept of the plot for ln η_{r}/c versus c (Kraemer plot).[3]

methods for experimental determination and analysis of viscosity are described in the literature.[6–8] An Ubbelohde type or a modified Ubbelohde suspension type viscometer is widely utilized in a bath controlled at ± 0.01 °C. In this case, the kinetic energy correction was very small or specially designed so that no kinetic energy correction was necessary. $[\eta]$ was determined from reduced viscosities η_{sp}/c at five concentrations by extrapolation to infinite dilution.

3.6.2 Determination of limiting viscosity number, Huggins coefficient of cellulose acetates, and cellulose solutions

(a) *Huggins plot and Kraemer plot.* A Huggins plot and a Kraemer plot are exemplified in Figure 3.6.1[9] for CN ($N = 12.0\%$) in acetone (see also Table 3.7.1).

(b) *Preparation of CTA acetone solution.* The CTA solution in acetone was prepared by the cooling method.[10] It is known that CTA dissolves in acetone by cooling a slurry

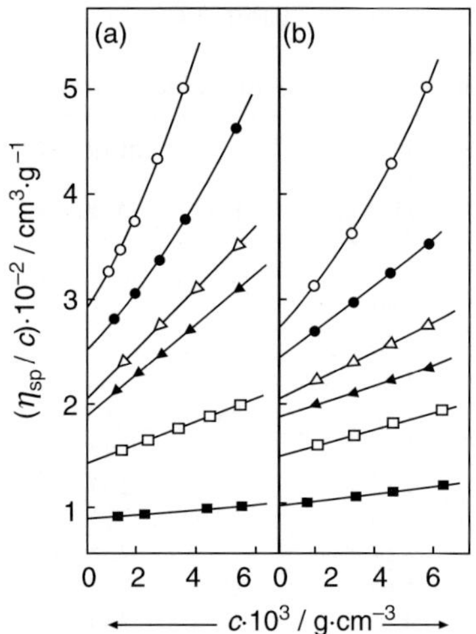

Figure 3.6.2   Plots of the ratio of specific viscosity η_{sp} to polymer concentration c, η_{sp}/c against c for cellulose in 6 wt% aq. LiOH solution (a) and in cadoxen (b) at 25 °C.[17] (O), SA-1; (●), SA-2; (△), SA-3; (▲), SA-4; (□), SA-5;(■), SA-6.

of CTA and acetone down to $-40\,°C$ or below, followed by warming it to room temperature.[11,12] As early as the late 1920s, Kita *et al.*[13] first pointed out that the acetone solubility of CTA is mainly attributable not to the acetyl content, but to the 'physical state' of polymer.

(c)   *Cellulose acetate.* Results of viscosity measurements on CA are summarized for CA (0.49) in columns 4–11 of Table 3.2.4,[14] CA (1.75) in the column 9 of Table 3.3.2,[15] CA (2.45) in the columns 15–17 of Table 3.3.3[16] and in Table 3.6.1,[10] and CA (2.9.2) in columns 3–8 in Table 3.6.2.[10]

(d)   *Cellulose/aq. LiOH and cadoxen.* Figure 3.6.2 shows the plots of η_{sp}/c versus c for cellulose in 6% aq. LiOH and in cadoxen at 25 °C.[17] For cellulose samples with

Table 3.6.3

Limiting viscosity number $[\eta]$ and Huggins constant k' of regenerated cellulose in 6 wt% aq. LiOH and in cadoxen at 25 °C.[17]

Sample	6 wt% LiOH		Cadoxen	
	$[\eta] \times 10^{-2}$ $(\text{cm}^3\,\text{g}^{-1})$	k'	$[\eta] \times 10^{-2}$ $(\text{cm}^3\,\text{g}^{-1})$	k'
SA-1	285	0.58	270	0.43
SA-2	248	0.51	241	0.32
SA-3	204	0.57	201	0.30
SA-4	187	0.70	185	0.23
SA-5	141	0.51	146	0.32
SA-6	89	0.36	99	0.27

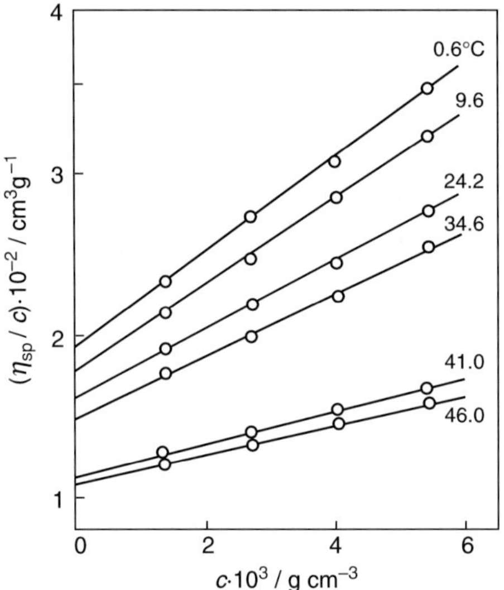

Figure 3.6.3 Plots of the ratio η_{sp}/c (η_{sp} specific viscosity; c, concentration) against c of cellulose/aq. NaOH at various temperatures ranging from 0.6 to 46 °C.[18]

a larger molecular weight, the plots became nonlinear at larger c region. $[\eta]$ and the Huggins coefficient, k', estimated from the intercept and initial slope of the plots, are summarized in Table 3.6.3. $[\eta]$ in aq. LiOH (except sample SA-6) is almost the same or very slightly (*ca.* 1%) larger than that in cadoxen, but k', in the former is some 1.7 times larger than that in the latter.

(e) *Cellulose/aq. 8% NaOH solution.* Figure 3.6.3 shows the plots of η_{sp}/c against c for cellulose sample code SA3 aq. 8% NaOH solution at different temperatures.

REFERENCES

1. K Kamide and T Dobashi, *Physical Chemistry of Polymer Solution*, Problem 8–16, Elsevier, Amsterdam, 2000.
2. K Kamide and T Dobashi, *Physical Chemistry of Polymer Solution*, Problem 8–22, Eqs. 8.22.1–8.22.6, Elsevier, Amsterdam, 2000.
3. ML Huggins, *J. Am. Chem. Soc.*, 1942, **64**, 2716.
4. K Kamide and T Dobashi, *Physical Chemistry of Polymer Solution*, Problem 8–25, Eqs. 8.25.1 and 8.25.2, Elsevier, Amsterdam, 2000.
5. EO Kraemer, *Ind. Eng. Chem.*, 1938, **30**, 1200.
6. K Kamide and M Saito, Determination of molecular weight. In *Viscometric Determination of Molecular Weight* (ed. A Cooper), Eqs. 91–104 and Figure 16, Wiley, New York, 1989.
7. M Bohdanecky and J Kovar, *Viscosity of Polymer Solutions*. Elsevier, Amsterdam, 1982.
8. K Kamide, M Saito and Y Miyazaki, 5 Molecular weight determination, 5.6 Viscometric methods. In *Polymer Characterization* (eds BJ Hunt and MI James), Blackie Academic and Professional, London, 1993, pp. 135–144.
9. K Kamide, T Shiomi, H Ohkawa and K Kaneko, *Kobunshi Kagaku*, 1965, **22**, 785.
10. K Kamide, Y Miyazaki and T Abe, *Polym. J.*, 1979, **11**, 523.

11. JMG Cowie, RS Ranson. *Macromol. Chem.*, 1971, **143**, 105.
12. CL Smart and CN Zellner, in *Cellulose and Cellulose Derivatives* (eds NM Bikales and L Segal), Wiley, New York, 1971, Chapter XIXC, p. 1152.
13. G Kita, I Sakurada and T Nakajima, *Kogyo Kagaku Zasshi*, 1927, **30**, 484.
14. K Kamide, M Saito and T Abe, *Polym. J.*, 1981, **13**, 421.
15. M Saito, *Polym. J.*, 1983, **15**, 2249.
16. K Kamide, T Terakawa and Y Miyazaki, *Polym. J.*, 1979, **11**, 285.
17. K Kamide and M Saito, *Polym. J.*, 1986, **18**, 569.
18. K Kamide, M Saito and K Kowsaka, *Polym. J.*, 1987, **19**, 1173.

3.7   SEDIMENTATION VELOCITY

3.7.1   Principle

The sedimentation coefficient s is calculated from the transport velocity of the position of the maximum of the refractive index gradient curve by the relationship:[1]

$$s = [1/(\omega^2 r_n)](\mathrm{d}r_n/\mathrm{d}t) \tag{3.7.1}$$

where r_n is the position of the maximum of the refractive index gradient, $\mathrm{d}r_n/\mathrm{d}t$ the transport velocity of the peak, ω the angular velocity of the rotor, and t the time. The relationship between $\ln r_n$ and t for all the samples was linear. The pressure effect on the sedimentation behavior was neglected.

The sedimentation coefficient s is linearly extrapolated to infinite dilution according to eq. (3.7.2):[2]

$$1/s = (1/s_0)(1 + k_s c) \tag{3.7.2}$$

where s_0 is s at infinite dilution, k_s the concentration dependent coefficient, and c, the concentration.

3.7.2   Experimental procedure

Experiment 1: CN (N_c 12%)/acetone, ethyl acetate[3]

An air-driven ultracentrifuging machine, manufactured by Phywe Ltd, Germany, was used. Measurements were taken in a standard cell at 35,000 or 40,000 rpm at room temperature. Six photographs were taken at 5 (or 7) min intervals (Figure 3.7.1)

Experiment 2: CA (DS 2.46)/acetone and CA (DS 2.92)/DMAc[4]

Ultracentrifugation was carried out with a single sector duralumin cell using a Hitachi Analytical ultracentrifuge Model UCA-1A. All runs were made at two different rotation speeds: 51,200 and 60,000 rpm. Each fraction was measured at six concentrations over the range $8 \times 10^{-4} - 1.5 \times 10^{-2}$ g cm^{-3}. Measurements were started at the time of

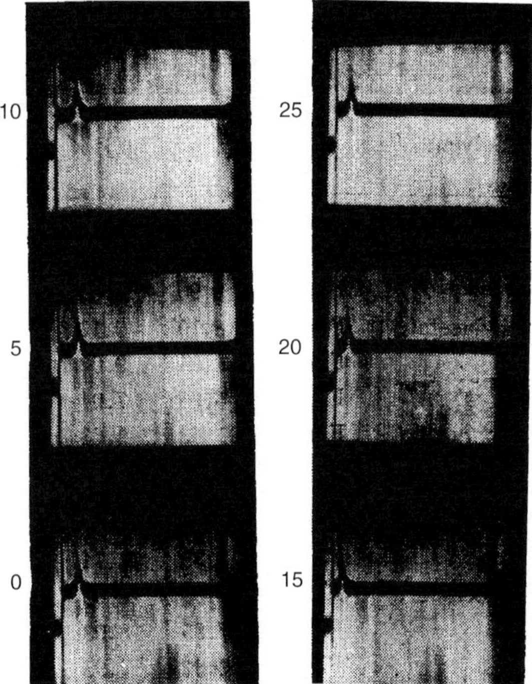

Figure 3.7.1   Schieren sedimentation diagrams of cellulose nitrate (fraction FZ4-1, $N_c = 13.3\%$) in ethyl acetate. Initial concentration: 0.049 g 100 ml^{-1}. The numerals denote the time in minutes after the initial exposure.[3]

two-thirds of the final speed. The course of the sedimentation was recorded by a Schlieren diagonal system.

Experiment 3: CA (DS 2.46)/2-butanone[5]

Solution of EF 3-5 was centrifuged on a Beckman-Spinco Model E ultracentrifuge at 52,640 rpm at temperatures of 25 and 35 °C. Sedimentation velocities were measured and sedimentation was examined in order to check possible change in the heterogeneity due to association of CDA molecules.

3.7.3   Application to experimental data

CN (N_c13.3%)[3]

Figure 3.7.2 illustrates the plot of $1/s$ against c for CN ($N = 13.3\%$) fractions in acetone and in ethyl acetate at 20 °C. The value for k_s in eq. (3.7.2) for seven fractions, determined from the above plots, are shown in Table 3.7.1. In the table, the viscosity data $[\eta]$ and k' are included. $[\eta]_H$ and $[\eta]_K$ mean $[\eta]$ determined by Huggins plot and Kraemer plot, respectively (see Figure 3.6.1).

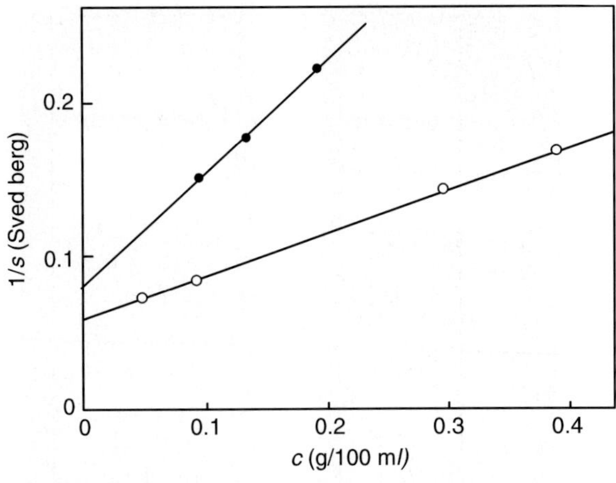

Figure 3.7.2   Plots of $1/s$ against c for CN ($N_c = 13.3\%$) $\sim$ acetone solution and for ethyl acetate solution of fraction FZ2-2 at 20 °C.[3]

CA (DS 2.46, 2.92)[4]

Figures 3.7.3 and 3.7.4 show the plots of $1/s$ versus c for CA (DS 2,46)/acetone and CA (DS 2.92)/DMAc systems, respectively.

Table 3.7.2 summarizes the $[\eta]$, s_0, and k_s, values, both evaluated from Figures 3.7.3 and 3.7.4.

Table 3.7.1

Evaluation of the ratio of concentration coefficient of sedimentation constant k_s and limiting viscosity number $[\eta]$ for the solutions of cellulose nitrate ($N_c = 13.3\%$)[3]

Solvent	Fraction number	$[\eta]_H$	$[\eta]_K$	k'	k_s	$k_s/[\eta]_K$
Ethyl acetate	FZ 1-1	12.40	13.96	1.02	16.49	1.181
	FZ 1-3	10.00	10.32	0.83	11.81	1.144
	FZ 2-1	8.90	9.35	0.70	8.69	0.929
	FZ 3-1	8.10	9.12	1.03	8.04	0.881
	FZ 3-2	8.34	9.02	0.81	6.64	0.736
	FZ 4-1	6.75	7.04	0.71	5.82	0.825
	FZ 5-1	5.63	5.63	0.59	2.58	0.458
Acetone	FZ 1-1	16.95	17.00	0.40	12.06	0.709
	FZ 1-3	14.25	14.41	0.43	7.80	0.541
	FZ 2-2	11.39	11.74	0.64	6.38	0.543
	FZ 3-1	10.00	10.06	0.52	5.51	0.547
	Z 3-2	7.85	8.03	0.65	4.80	0.598
	FZ 4-1	7.82	7.93	0.51	2.61	0.329
	FZ 5-1	6.85	6.90	0.52	1.98	0.287

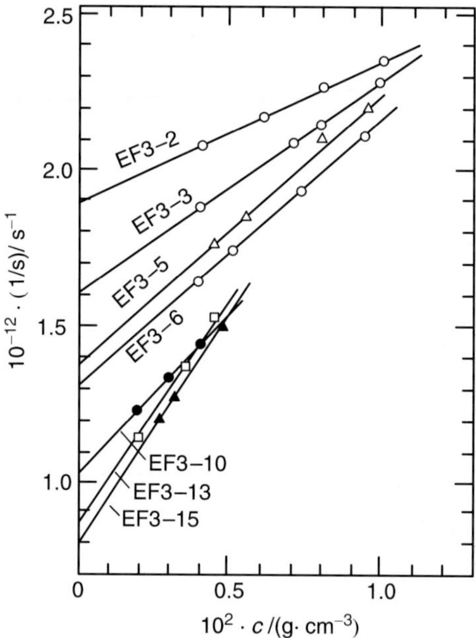

Figure 3.7.3   Reciprocal sedimentation coefficient $1/s$ versus concentration c for cellulose acetate fractions (*cf.* Table 3.7.2) with DS 2.46 in acetate at 25 °C.[4]

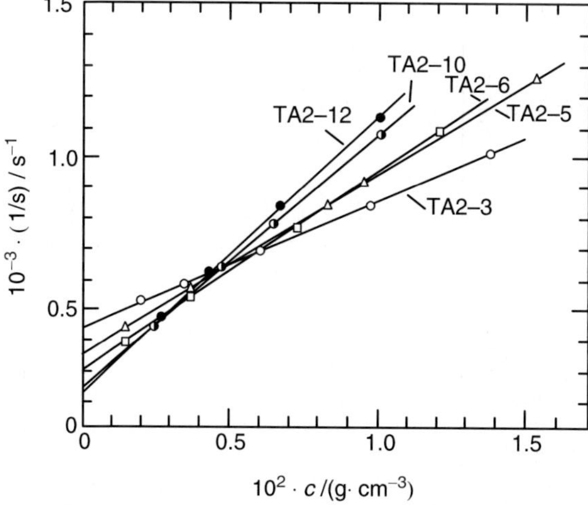

Figure 3.7.4   Reciprocal sedimentation coefficient $1/s$ versus concentration c for cellulose acetate fractions (*cf.* Table 3.7.2) with DS 2.92 in N,N-dimethylacetamide at 25 °C.[4]

Table 3.7.2

Weight-average molecular weight M_w, z-average radius of gyration $\langle S^2 \rangle_z^{1/2}$, intrinsic viscosity $[\eta]$, sedimentation coefficient at infinite dilution s_0, concentration dependent constant k_s, ratio $k_s/[\eta]$, molecular friction coefficient $\xi_{w,w}$, polydispersity correction factor $q'_{w,z}$, and Flory parameter P from viscosity and sedimentation measurements with solutions of cellulose acetate with DS = 2.46 and 2.92[4]

Polymer	Solvent[a]	Polymer code	$M_w \times 10^{-4}$	$\langle S^2 \rangle_z^{1/2}$ $(10^{-8}$ cm$)$	$[\eta]$ $(\text{cm}^3\,\text{g}^{-1})$	s_0 $(10^{-13}$ s$)$	k_s $(\text{cm}^3\,\text{g}^{-1})$	$k_s/[\eta]$	$\xi_{w,w}$ $(10^{-8}$ g s$^{-1})$	$q'_{w,z}$	P
CA (D S 2.46)	Acetone	EF3-2	2.42	166	29	5.28	24.3	0.84	3.41	–	2.77
		EF3-3	3.76	190	46	6.26	43.8	0.95	4.47	–	3.18
		EF3-5	5.90	218	85	7.31	65.1	0.77	6.01	–	3.72
		EF3-6	7.50	234	104	7.67	68.3	0.66	7.33	1.08	4.24
		EF3-10	11.1	252	160	9.80	104	0.65	8.43	1.08	4.32
		EF3-13	15.9	290	217	11.6	169	0.78	10.2	1.09	4.64
		EF3-15	24.9	338	277	12.5	188	0.68	14.8	1.09	6.31
CA (D S 2.92)	DMAc	TA2-3	3.67	150	70	2.27	71.7	1.02	7.95	1.10	2.55
		TA2-5	8.22	214	131	2.83	101	0.77	14.3	1.13	3.31
		TA2-6	13.7	287	183	3.32	131	0.72	20.3	1.12	3.47
		TA2-10	30.8	421	350	4.04	205	0.59	37.5	1.12	4.37
		TA2-12	50.0	565	490	4.17	228	0.47	58.9	1.12	5.11

[a]DMAc = N,N-dimethylacetamide.

REFERENCES

1. See, for example, HA Stuart (ed.), Die Physik Der Hochpolymeren, Zweiter Band. *Das Macromolekül in Lösungen.* Springer, Berlin, 1953, pp. 413–418; T Svedberg, *Kolloid Z. Erg. Bd.*, 1925, **36**, 53 *Z. Phys. Chem.*, 1927, **127**, 51; T Svedberg and KO Pedersen, *Die Ultrazentrifuge, Zusammerfassende Monogra phie.* Th. Steinkopff, 1940; H Fujita, *Foundation of Ultrafiltration Analysis.* Wiley-Interscience, New York, 1975; NC Billingham, *Molecular Mass Measurements in Polymer Science.* Transport measurements I ultracentrifuge, Kogan Page Ltd, London, 1977, Chapter 6.
2. See, for example, JM Burgers, *Proc. Akad. Amsterdam*, 1945, **44**, 1045–1177 1942, **45**, 9, 126, for dilute solution of sphere molecule; CO Beckmann and J Rosenberg, *Ann. NY. Acad. Sci.*, 1945, **46**, 209 for dilute solution of chain molecule; S Newman and F Eirich, *J. Colloid. Sci.*, 1940, **5**, 544 for dilute solution of chain molecule.
3. K Kamide, T Shiomi, H Ohkawa and K Kaneko, *Kobunshi Kagaku*, 1965, **22**, 785.
4. S Ishida, H Komatsu, H Katoh, M Saito, Y Miyazaki and K Kamide, *Makromol. Chem.*, 1982, **183**, 3075.
5. H Suzuki, Y Muraoka, M Saito and K Kamide, *Eur. Polym. J.*, 1982, **18**, 831.

3.8 ADIABATIC COMPRESSIBILITY

The earliest study of the interaction between the cellulose derivative and the solvent molecule (i.e. the solvation) seems to have been that of Moore and coworkers,[1–3] who, in 1965, measured the adiabatic compressibility of solutions of CA, CN, ethyl cellulose, and so on, estimating the number of the solvent molecules solvated to a glucopyranose ring. Unfortunately, as Kamide and Saito[27] later pointed out, the accuracy and preciseness of their data on the sound velocity are unsatisfactory.

3.8.1 Theoretical background

Solvation is believed to play an important part in the interaction of cellulose derivatives and solvents. Specific interactions between unlike molecules in binary liquid mixtures lead to reduction in adiabatic compressibility. The pressure in the immediate vicinity of the dipole should be very high so that the liquid in this region is effectively rendered incompressible, thereby reducing the compressibility of the liquid as a whole. Since the pressure p is inversely proportional to r^6 (r, the distance), it will fall off very rapidly with an increase in the distance of dipole and a molecule of liquid r so that the incompressible liquid will be restricted to that in close proximity to the dipole. The incompressible part of the liquid has been identified with that concerned in the solution of the dipole.

The adiabatic compressibility (β) of the solution is calculated by the Laplace equation:

$$\beta = 1/\rho V^2 \tag{3.8.1}$$

Here, ρ and V represent the density and the sound velocity of the solution, respectively. Passynsky has regarded the incompressible part of the solution as the polymer and the solvent involved in solvation, deriving an equation relating β to

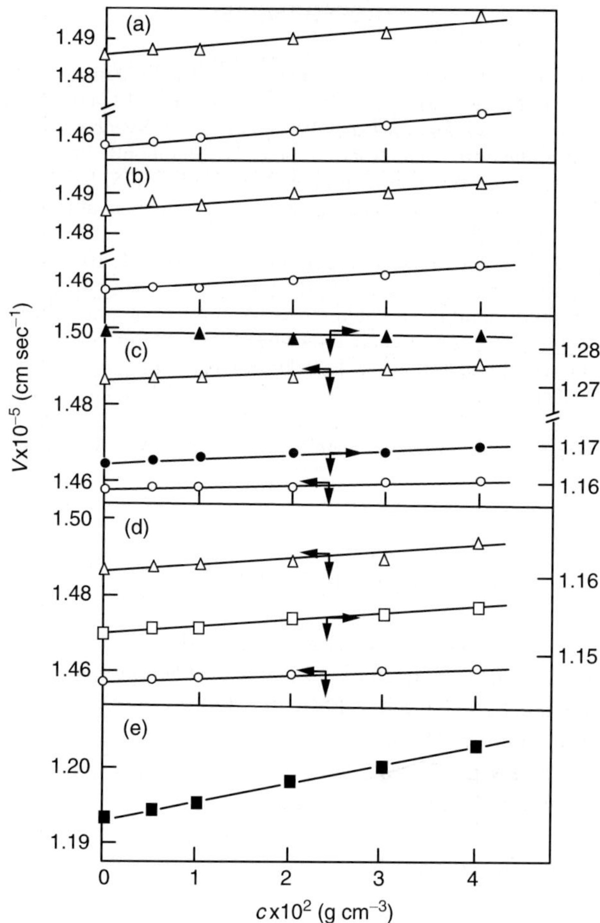

Figure 3.8.1   Plots of the sound velocity of cellulose acetate whole polymer/solvent system and polystyrene whole polymer-methylethylketone system against the concentration at 25 °C. (a) Cellulose acetate (0.49); (b) (1.75); (c) cellulose acetate (2.46); (d) cellulose acetate (2.92); (e) polystyrene. The lines are determined by the least square method. (○) dimethylacetamide; (△) dimethyl sulfoxide; (●) acetone; (▲) tetrahydrofuran; (□) tetrachloroethane; (■) methylethylketone.[5]

the number of the solvated solvent molecules per g of polymer (n) as follows:[4]

$$n = \{1 - (\beta/\beta_s)\}(100\rho - c)/c \tag{3.8.2}$$

where β_s denotes the adiabatic compressibility of the solvent.

3.8.2   Experimental procedure[4]

The ultrasonic velocity was measured with a Pierce-type ultrasonic interferometer constructed by Nomura and Miyahara,[6] operating at 4.99985 MHz. The temperature of

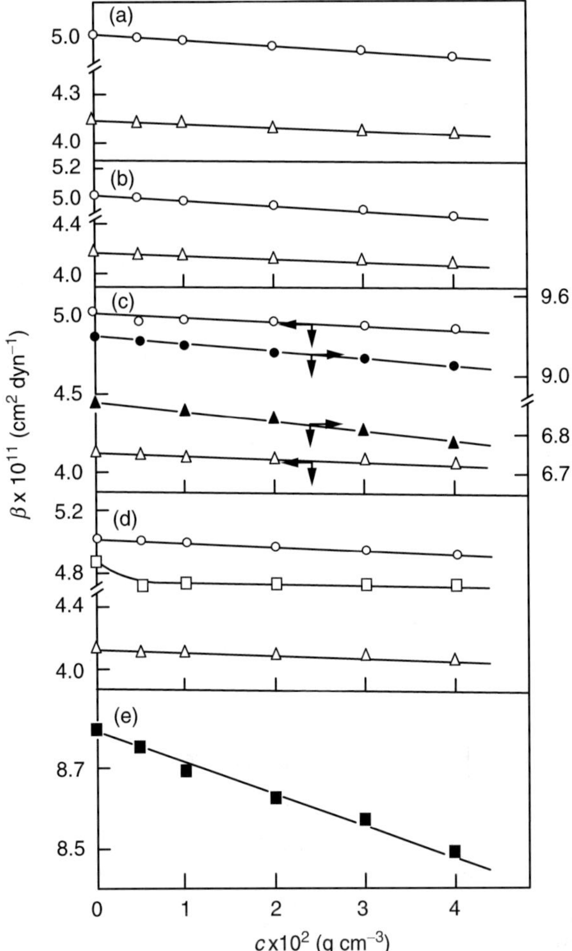

Figure 3.8.2 Plot of adiabatic compressibility of cellulose acetate whole polymer-solvent system and polystyrene whole polymer-methylethylketone system and against the concentration at 25 °C. (a) cellulose acetate (0.49); (b) cellulose acetate (1.75); (c) cellulose acetate (2.46); (d) cellulose acetate (2.92); (e) polystyrene. The lines are determined by the least square method. (○) dimethylacetamide; (△) dimethyl sulfoxide; (●) acetone; (▲) tetrahydrofuran; (□) tetrachloroethane; (■) methylethylketone.[5]

the solution was controlled to ± 0.005 °C by circulating water around the cell of the interferometer. The sound velocity of distilled water at 25.00 °C at 1 atm was found to be 1497.5 ± 0.5 m s⁻¹ with our interferometer. This value is only some 0.033% larger than that measured by a sing around method.[7] ρ of the solution was measured with an Ostwald-type pycnometer. Polymer samples were dissolved in freshly distilled solvents at 60 °C to make 4.0 wt% solutions. Then, the solutions were filtered through a sintered glass (first grade) and stored. By adding solvent to the stock solution, solutions of five different concentrations of CA (DS 2.92), CA (DS 2.46), CA (DS 1.75), and CA (DS 0.49) whole polymers, CA (DS 2.46) fractions, and an atactic polystyrene (PS) whole polymer were made. In order to clarify the effect of temperature on solvation, β of CA (DS 2.46) whole polymer/acetone system was measured in the temperature range of 12.8–45.2 °C.

3.8.3   CA[5]

Figure 3.8.1(a–e) depicts the concentration (c) dependence of the sound velocity V. Reported V values for any solvent vary to some extent. For example, for acetone at 25 °C, V was given as 1150 m s^{-1} by Moore,[2] 1174 m s^{-1} by Willard,[8] and 1168 m s^{-1}, calculated here using the experimental relationship between V and temperature T(0 °C $\leq T \leq$ 50 °C) obtained by Freyer $et\ al.$[9] We found $V = 1163.3$ m s^{-1} for acetone at 25 °C, which was very near the calculated value from the Freyer $et\ al.$ relationship. The difference is not significant when calculating adiabatic compressibility β, because β calculated using the V values of Moore is only about 3% smaller than that using the V value of Freyer $et\ al.$ Figure 3.8.1 indicates that V of CA and PS solutions increases with increasing concentration, except for the CA (2.46)/THF system in which V is almost independent of concentration. β, as shown in Figure 3.8.2(a–e), decreases with increasing concentration in all cases.

REFERENCES

1. WR Moore and B Tidswell, *Makromol. Chem.*, 1965, **81**, 1.
2. WR Moore, *J. Polym. Sci.*, 1967, **C16**, 571.
3. WR Moore, *Solution Properties of Natural Polymers: An International Symposium in Edinburgh*, The Chemical Society, Burlington House, London, 1968, pp. 185–194.
4. A Passynsky, *Acta Phys. Chim. USSR*, 1947, **22**, 137.
5. K Kamide and M Saito, *Eur. Polym. J.*, 1984, **20**, 903.
6. Y Miyahara and Y Matsuda, *J. Chem. Soc. Jpn., Pure Chem. Sect.*, 1960, **81**, 692.
7. M Greenspan and CE Tschiegg, *J. Acous. Soc. Am.*, 1959, **31**, 751.
8. GW Willard, *J. Acous. Soc. Am.*, 1947, **19**, 235.
9. EB Freyer, JC Hubbard and DH Andrews, *J. Am. Chem. Soc.*, 1929, **51**, 759.

3.9   NUCLEAR MAGNETIC RESONANCE STUDY OF THERMODYNAMIC INTERACTION

3.9.1   Cellulose acetates by ^{1}H NMR method[1]

Recently, Kamide and coworkers[2–8] have systematically studied the dilute solution properties of CA (DS = 0.49, 2.46, and 2.92). They found that the unperturbed chain dimensions, A, defined as the radius of gyration of a polymeric chain at the unperturbed state $\langle S^2 \rangle_0^{1/2}$, divided by the square root of the molecular weight M, estimated from the solution data, depends strongly on the solvent nature, especially polarity. This phenomenon is highly noticeable in polar polymers. The unusually large solvent dependence of the unperturbed chain dimensions has already been reported for cellulose, amylose, and their derivatives by Kamide *et al.*,[9–13] who analyzed available data (see Sections 3.16 and 3.18). On the other hand, the existence of a hydrogen bond between the *O*-acetyl (*O*-Ac) group in the CA molecule and halogenated hydrocarbons,[14,15] and the interaction between the *O*-Ac group and aniline or acidic solvents has been proposed[16] on the basis of infrared spectroscopic observations. These results indicate that there are specific interactions other than the van der Waals force between CA and solvent molecules, influencing the unperturbed chain dimensions. Up until now, ^{1}H NMR spectroscopy has been applied to the CA molecule for (1) the determination of its configuration,[17] (2) the identification of the chemical shifts of ring proton, (3) the evaluation of the average combined acetic acid content,[18] and (4) the determination of the distribution of acetyl group over C_2, C_3, and C_6 positions.[19,20] Nuclear magnetic resonance (NMR) spectroscopy is also expected to provide a powerful tool for studying the interaction of functional groups in cellulose derivatives and solvents.

This section illustrates a study on the thermodynamic interaction between the CA molecule and the solvent molecule by ^{1}H NMR spectroscopy.

H NMR spectroscopy[1]

^{1}H NMR spectra of CA in a wide variety of solvents were recorded on a ^{1}H NMR spectrometer model PMX60 (60 MHz, resolving power 0.04 Hz, JOEL, Japan) and the chemical shifts of the methyl protons of *O*-Ac and OH protons relative to the internal tetramethylsilane (TMS) were determined. All spectra were obtained at a probe temperature of 37 °C. Preliminary measurements showed no variation in the methyl- and OH proton magnetic resonance of CA at a concentration ranging from 3 to 10 vol%. Hence, the chemical shifts were determined at a concentration of 5–7 vol%. When the peak due to the methyl proton in the *O*-Ac group splits into more than two parts, the weight average of the characteristic chemical shifts of the methyl proton was calculated from these split speaks. The OH proton signal was detected through a comparison of the NMR spectra obtained in the absence and presence of deuterium oxide or trifluoroacetic acid. All chemical shifts are expressed in parts per million downfield from the internal TMS.

Infrared spectroscopy[1]

The infrared (IR) spectra of CA (DS 2.92, 2.46, and 0.49) in the film form and CA (DS 2.92 and 2.46) in the solvents were obtained with a Shimadzu model 430 IR spectrophotometer. A 1–5 vol% polymer solution was introduced by the capillary method between two rock salts, and the solvent compensation method was used for making measurements.

Figure 3.9.1 shows the ^{1}H NMR spectra of CA (DS 2.92), CA (DS 2.46), and CA (DS 0.49) in various solvents together with their assignments. The chemical analysis of CA (DS 2.92) indicated that there remained one OH group per 10 glucopyranoside residues. Three separate signals between 1.9 and 2.1 ppm were detected individually for CA (DS 2.92) in chlorinated hydrocarbons and assigned to the methyl proton groups of *O*-Ac substituents in three different positions C_2, C_3, and C_6 in the glucopyranose units, to which the acetyl group is attached. For example, in DCM, the three methyl

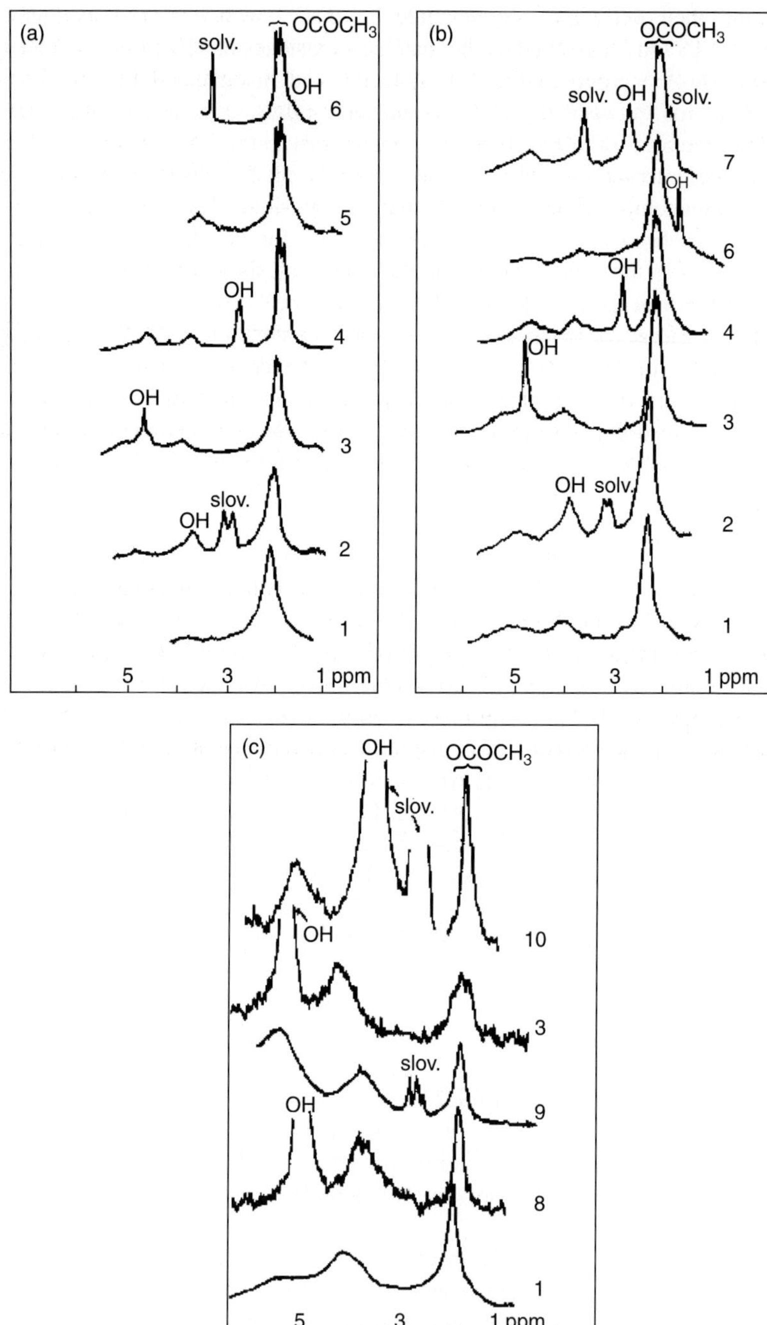

Figure 3.9.1 ^{1}H NMR spectra of (a) cellulose acetate (DS 2.92), (b) cellulose acetate (DS 2.46), and (c) cellulose acetate (DS 0.49) in various solvents.[1] (a) 1, TFA; 2, DMAc; 3, pyridine; 4, acetone; 5, TCE; 6, DCM. (b) 1, TFA; 2, DMAc; 3, pyridine 4, acetone; 6, DCM; 7, THF. (c) 1, THF; 3, pyridine; 8, D$_2$O; 9, FA; 10, DMF.

proton peaks were at 2.07, 1.98, and 1.94 ppm. These values are in reasonably good accordance with those reported by Goodlett *et al.*[19] for fully substituted CA in DCM. These authors assigned the signals at 2.09, 1.99, and 1.94 ppm to the methyl groups of O-Ac attached to the C_6, C_2, and C_3 positions, respectively. In solvents other than halogenated hydrocarbons, the signals due to the methyl proton were often superposed with each other, giving one or two peaks. This implies the existence of a specific interaction between the O-Ac groups in the CA molecule and the solvent molecule in question.

In CA (DS 2.46), it is expected from chemical analysis that there is approximately one OH group per two glucopyranose units. Therefore, the possibility of seven types of acetylated glucopyranose units yields 12 nonequivalent methyl protons with different magnetic environment. The chemical shift of the methyl proton signals of CA (DS 2.46) lies in a relatively narrow range, giving the peaks a rather complicated nature. This is very obvious in THF, DCM, and acetone, but the signals due to the methyl group attached to O-Ac enveloped two peaks in pyridine (Py) and a single peak in both DMAC and in TFA. The value, 2.24 ppm, obtained for TFA agrees well with data in the existing literature.[21] It should be noted that the hydroxylproton signals of the CA molecule occasionally overlap and are inseparable from water or HOD in solvents.

Figures 3.9.2(a–d) and 3.9.3(a–c) show the chemical shifts of the methyl proton in the O-Ac group and the OH protons of CA (DS 2.92) and CA (DS 2.46) in various solvents as a function of the weight-average molecular weight M_w of cellulose acetate. In these figures, the circles are for CA (DS 2.46) and the triangles are for CA (DS 2.92), and the open and closed marks indicate the fractions and the whole polymers, respectively. In the CA (DS 2.92)/TFA and CA (DS 2.92)/Py systems, the chemical shift of the methyl proton

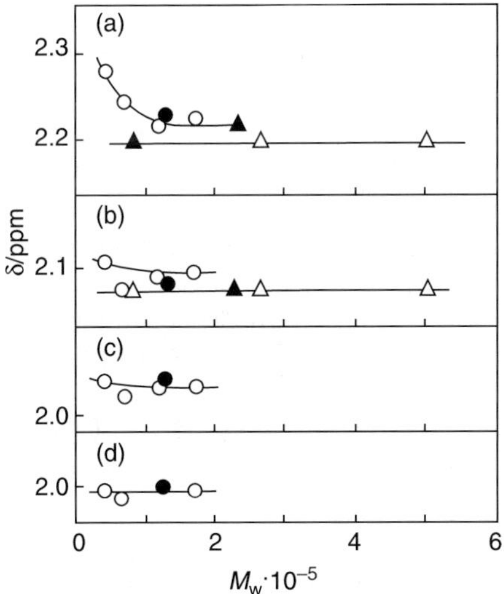

Figure 3.9.2 The molecular weight dependence of the chemical shift of O-acetyl-proton of cellulose acetate (DS 2.92 and 2.46) in trifluoroacetic acid (a), pyridine (b), and of cellulose acetate (DS 2.46) in acetone (c), and tetrahydrofuran (d); open mark, fractions; closed mark, whole polymer; triangle, cellulose acetate (DS 2.92); circle, cellulose acetate (DS 2.46).[1]

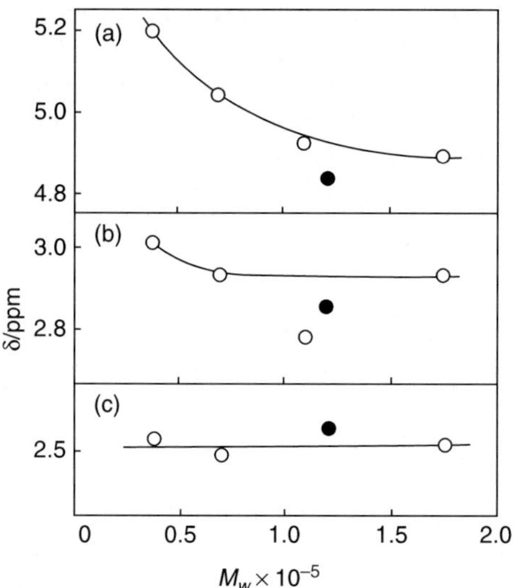

Figure 3.9.3   The molecular weight dependence of the hydroxyl proton of cellulose acetate (DS 2.46) in pyridine (a); acetone (b); and tetrahydrofuran (c).[1]

shows no variation with the molecular weights. In contrast to this, the methyl proton signal of CA (DS 2.46) in TFA shifts toward a lower magnetic field as M_w decreases.

The large disparity between CA (DS 2.92) and CA (DS 2.46) may originate from the difference in the molecular weight range studied; that is, the molecular weight range employed for CA (DS 2.46) is lower than that for CA (DS 2.92). The possibility of an interaction between solvent and polymer is higher when the molecular weight of the polymer is lower. The molecular weight dependence of the methyl proton chemical shift for CA (DS 2.46) in Py, acetone, and THF is far smaller than that in TFA. In the former two systems, the chemical shift of the OH proton shifts toward a lower field as M decreases, as can be anticipated from the behavior of the methyl proton chemical shift. In these systems, the decrease in M_w strengthens the OH solvent interaction. The molecular weight dependence of these chemical shifts in two solvents becomes remarkable with an increase in the electron donating power (the electronegativity). The OH proton chemical shift depends more markedly on the molecular weight than on the chemical shift of the O-Ac proton for CA with the same DS and M_w dissolved in a given solvent. The polydispersity of the polymer samples has no significant effect on the chemical shifts of the O-Ac and OH protons. The use of CA with M_w approximately equal to 1×10^5 allows us to neglect the molecular weight dependence of the chemical shifts. Hence, further experiments were done with the whole polymer samples.

The acetyl group in CA is known to be nucleophilic. Thus, the interaction between the O-Ac group and the solvent is expected to depend on the properties of the solvent, such as dielectric constant and electronegativity. Figure 3.9.4(a) and (b) show the plot of the weight-average methyl proton chemical shifts of CA against the dielectric constant, ε (a) and the electronegativity $\Delta\nu$ (b) of the solvent, respectively.

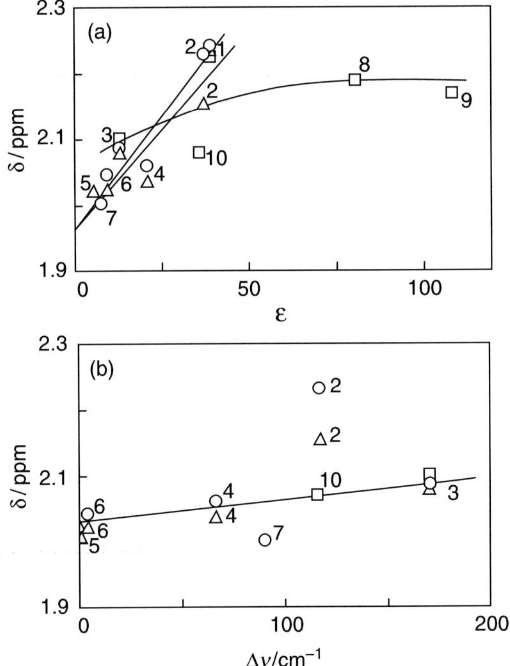

Figure 3.9.4   The plot of the weight-average methyl proton chemical shifts of cellulose acetate molecular against the dielectric constant, ε (a) and the electronegativity, $\Delta\nu$ (b) of the solvents: ($\square$) cellulose acetate (DS 0.49); ($\bigcirc$) cellulose acetate (DS 2.46); ($\triangle$) cellulose acetate (DS 2.92).[1]

In these figures, the squares stand for CA (DS 0.49), the circles, for CA (DS 2.46), and the triangles, for CA (DS 2.92). The methyl proton signals shift almost exclusively toward a lower magnetic field as the dielectric constant of the solvent becomes larger. The dependence of the chemical shift on the dielectric constant becomes stronger, depending on the DS of the CA polymers and is highest for CA (DS 2.46). For example, the methyl proton signal shifts from 2.00 to 2.24 ppm as the dielectric constant of solvent changes from 7.3 to 39.5. This suggests that solvents with high dielectric constants, such as TFA, strongly interact with the O-Ac group of the CA molecule, and that the interaction is strongest for CA (DS 2.46) in the solvent with $\varepsilon \geq 20$. This tendency is particularly notable in DMAc as an aprotic reagent. The O-Ac proton chemical shift δ_{CH_3} exhibits a linear correlation with the dielectric constant ε of the solvent as follows:

$$\delta_{CH_3} = 1.97_2 + 5.68 \times 10^{-3}\varepsilon \qquad DS = 2.92, \qquad 7.7 \leq \varepsilon \leq 39.5 \qquad (3.9.1)$$

$$\delta_{CH_3} = 1.97_2 + 6.60 \times 10^{-3}\varepsilon \qquad DS = 2.46, \qquad 7.3 \leq \varepsilon \leq 39.5 \qquad (3.9.2)$$

$$\delta_{CH_3} = 2.11_2 + 7.07 \times 10^{-3}\varepsilon \qquad DS = 0.49, \qquad 12.3 \leq \varepsilon \leq 110 \qquad (3.9.3)$$

$$\delta_{CH_3} = 2.06 + 2.93 \times 10^{-3}\varepsilon - 1.78 \times 10^{-5}\varepsilon^2 \qquad (3.9.4)$$

Eqs. (3.9.1–3.9.4) were established by the least squares method. The correlation coefficients γ for these relationships are 0.972, 0.956, 0.457, and 0.964, respectively. By using eqs. (3.9.1), (3.9.2), and (3.9.4), we can estimate the expected downfield chemical shift of the methyl proton from ε of the solvent.

The methyl proton signal linearly shifts from 2.00 to 2.10 ppm as the electronegativity of solvent increases from 0 to 170 cm^{-1}, except for all CA in DMAc ($\Delta\nu = 138$ cm^{-1}) and THF ($\Delta\nu = 92$ cm^{-1}) (2 and 7 in the figure). The O-Ac proton chemical shift, δ_{CH_3}, is expressed, except in DMAc and THF, as a function of the electronegativity $\Delta\nu$ in the form:

$$\delta_{CH_3} = 2.02 + 3.43 \times 10^{-4} \Delta\nu \qquad 0.49 \leq DS \leq 2.92, \qquad 0 \leq \Delta\nu \leq 170 \qquad (3.9.5)$$

The correlation coefficient γ for eq. (3.9.5) is 0.872. A better correlation of δ_{CH_3} is found with the dielectric constant (eqs. (3.9.1), (3.9.2), and (3.9.4)), rather than with the electronegativity (eq. (3.9.5)).

Figure 3.9.5 illustrates the change in the limiting viscosity number $[\eta]$ of CA with $M_w = 1.08 \times 10^5$ in the solvent with $\varepsilon = 40$ at 25 °C and the unperturbed chain dimensions, A (see eq. (3.13.9)), in the same solvent with DS, both interpolated from the experimental results in the figure, the O-Ac proton chemical shift δ_{CH_3} of CA in a solvent with $\varepsilon = 40$, calculated from eqs. (3.9.1–3.9.3), is also shown for comparison. There is a good correlation between $[\eta]$, A, and the chemical shift of the O-Ac proton, suggesting that the former two parameters are closely correlated with the latter. It should be noted that the expansion factor, expressing the long-range interaction parameter, is almost constant and very near unity, irrespective of DS.

Figure 3.9.6(a) and (b) show the relationships between the chemical shift of the OH proton in CA and the dielectric constant (a) or the electronegativity (b) of various solvents. In the figure, the rectangles stand for CA (DS 0.49), the circles, for CA (DS

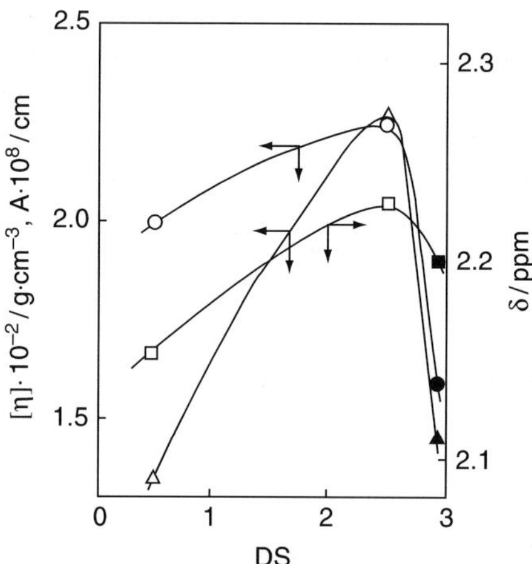

Figure 3.9.5   The effect of DS on the limiting viscosity number $[\eta]$ of cellulose acetate polymer with $M_w = 1.08 \times 10^5$ in the solvent with $\varepsilon = 40$ at 25 °C, the unpertubed chain dimensions, A in the same solvent and on the O-acetyl-proton chemical shifts, δ_{CH_3} of cellulose acetate polymer dissolved in a solvent with $\varepsilon = 40$: circle, $[\eta]$; triangle, A; rectangle, δ_{CH_3}.[1]

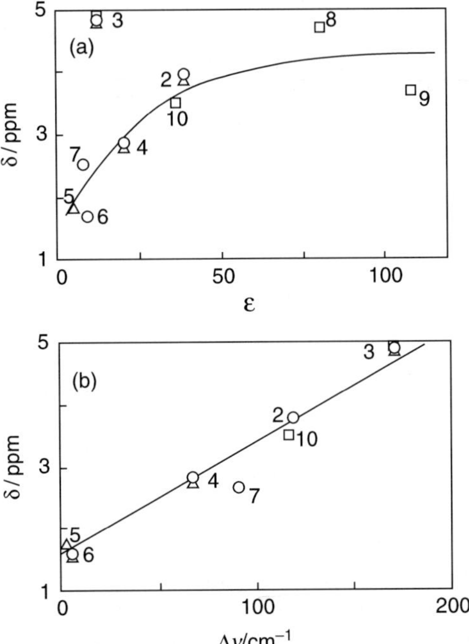

Figure 3.9.6   The plot of the weight-average hydroxylproton chemical shifts of cellulose acetate molecular against the dielectric constant, ε (a) and the electronegativity, $\Delta\nu$ (b) of the solvents: rectangle, cellulose acetate (DS 0.49); circle, cellulose acetate (DS 2.46); triangle, cellulose acetate (DS 2.92).

2.46), and the triangle, for CA (DS 2.92). Evidently, the chemical shift of the OH proton is a unique function of the dielectric constant or the electronegativity irrespective of DS. The chemical shift increases with an increase in the dielectric constant or the electronegativity of the solvent over the entire range of ε and $\Delta\nu$ investigated.

The chemical shift of the OH proton was found to be 4.8 ppm in Py ($\Delta\nu = 170\ \mathrm{cm}^{-1}$) and 1.6–2.0 ppm in halogenated hydrocarbons ($\Delta\nu = 0$), irrespective of the DS of cellulose acetate. The electronegativity of the solvent is an index for forming a hydrogen bond against the OH group in various alcohols.[22] In this sense, the OH group in CA has a hydrogen bond-forming capacity similar to low molecular weight alcohols.

The following relationships were experimentally obtained by the least squares method, between the hydroxyl-proton chemical shift, δ_{OH} and the dielectric constant, ε or the electronegativity, $\Delta\nu$ of the solvent.

$$\delta_{OH} = 4.04 - 17.1/\varepsilon + 1.68 \times 10^{-5}\varepsilon^2 \qquad 0.49 \leq DS \leq 2.92, \qquad 7.3 \leq \varepsilon \leq 110$$

$$(3.9.6)$$

and

$$\delta_{OH} = 1.61 + 1.81 \times 10^{-2}\Delta\nu \qquad 0.49 \leq DS \leq 2.92, \qquad 0 \leq \Delta\nu \leq 170 \qquad (3.9.7)$$

The correlation coefficients γ of these equations are 0.990 and 0.980, respectively. For CA (DS 2.92) and CA (DS 2.46), δ_{OH} relates linearly to ε as follows:

$$\delta_{OH} = 1.12 + 7.32 \times 10^{-2}\varepsilon \qquad DS = 2.92, \qquad 7.7 \leq \varepsilon \leq 39.5 \qquad \gamma = 0.997$$

$$(3.9.8)$$

and

$$\delta_{OH} = 1.61 + 6.02 \times 10^{-2}\varepsilon \qquad DS = 2.46, \qquad 7.3 \leq \varepsilon \leq 39.5 \qquad \gamma = 0.918$$

$$(3.9.9)$$

Moore and Russell[23] explained the viscosity behavior of CA in various solvents on the basis of acidity and basicity of the solvents. However, as is evident from Figure 3.9.4, the interaction between CA and the solvent is not as simple Moore and Russel's explanation suggests. In fact, Kamide *et al.* demonstrated that the limiting viscosity number $[\eta]$ of cellulose acetate, with a given DS and a given M_w, is a unique function of the dielectric constant of the solvent ε (see Figure 3.18.1).

Figures 3.9.4 and 3.9.6 reveal that DMAc and TFA, both with large ε and large $\Delta\nu$, strongly interact with the *O*-Ac and OH groups in cellulose acetate. In order to obtain more detailed information on the interaction between the CA molecule and solvents, the IR spectra of the carbonyl groups in the films of CA (DS 2.92), CA (DS 2.46), and CA (DS 2.92) in solvents were measured. The results are shown in Figure 3.9.7. In most cases, the maximum absorption band of the carbonyl group was found at 1741 cm^{-1} and in the dissolved state especially in DMAc the maximum band shifts to a higher wave number region (1750 cm^{-1}). However, no shift was observed in DCM. In 1943, Clermont[15] observed a significant shift of the band at between 1760 and 1740 cm^{-1} due to the presence of DCM as solvent. Clermont's experimental results have long been cited in the literature without being subjected to detailed examination, and they are in sharp

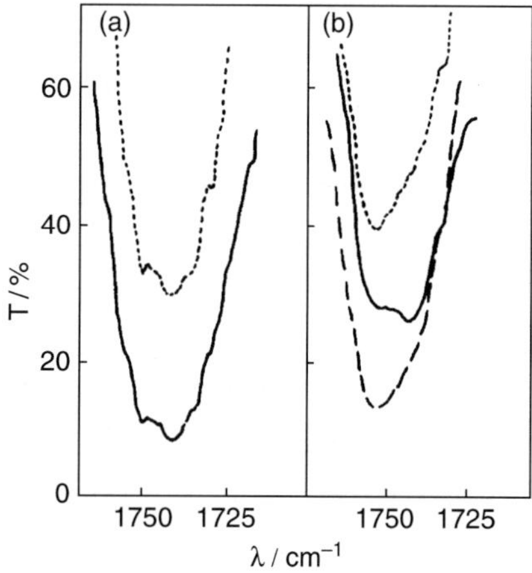

Figure 3.9.7   IR spectra of solid cellulose acetate (DA 2.46 and 2.92) (a) and that in solvents (b): (a) dotted line, (DA 2.46); solid line, cellulose acetate (DS 2.92); (b), solid line, dichloromethane; chain line, dimethylacetamide; dotted line, epichlorohydrine.[1]

contradiction with the results in Figure 3.9.7. If Clermont's experimental results are correct (as is generally accepted), then the O-Ac groups in solid CA should exist in a free state, strongly interacting with DCM. However, contrary to this, the ring conformational analysis of cellulose by Hermans[24] and Robinson and Conmar[25] show that an interaction between the O-Ac groups in the solid state is possible, even if the interacting force is weak. Figure 3.9.7 illustrates that in the solid state, the interaction between the O-Ac groups of CA (DS 2.46) and also those of CA (DS 2.92) is strong, and that the addition of a solvent weakens this interaction.

Based on C1 chain conformation by the molecular structure model, such an examination indicates that conformations, in which the O-Ac and OH groups belonging to the same cellobiose unit approach each other, are possible for CA (DS 2.46) where an OH group exists in each cellobiose unit. The above expectation is consistent with the appearance of a band due to a carbonyl group at 1732 cm^{-1} found for CA (DS 0.49). This value is significantly lower than 1741 cm^{-1} found for CA (DS 2.46) and CA (DS 2.92). In CA (DS 0.49), an O-Ac group exists in each of three glucopyranose units and the effect of the O-Ac$\cdots O$-Ac interaction on the solution behavior is negligible, compared with that of CA (DS 2.46 and 2.92), as far as a cellobiose unit is concerned. However, the O-Ac$\cdots$OH interaction may shift the band, due to a carbonyl, to a lower wave number (1732 cm^{-1}).

Figure 3.9.8 illustrates the possible conformation patterns of solid CA (DS 2.92) and CA (DS 2.46). For the former, the interaction between nearest neighboring O-Ac groups of contiguous glucopyranose units is shown and for the latter, the possible interaction between the O-Ac group, OH group, and ring oxygen atoms of contiguous glucopyranose units.

Figures 3.9.9 and 3.9.10 show the possible molecular models of the interactions of CA (DS 2.46) and CA (DS 2.92) in various solvents.

The models in these figures were constructed by taking into consideration the conclusions deduced from NMR spectra; that is, that the interaction of the solvent molecule with the O-Ac group in the CA molecule decreases in the following order: TFA > DMAc > Py > acetone > halogenated hydrocarbons > THF. The interaction with the OH group is in the order: TFA > Py > DMAc > acetone > THF > halogenated hydrocarbons. The interaction of the OH group with TFA will be discussed in detail later.

Figure 3.9.11(a) and (b) show the effect of DS on the O-Ac and OH proton chemical shifts of CA in TFA ($\varepsilon = 39.5$), DMAc ($\varepsilon = 38.6$), and Py ($\varepsilon = 12.3$). The data points of CA (DS 0.49) in DMAc were interpolated from those of CA with larger DS. The spectrum of the OH proton of CA in TFA was overlapped considerably and could not be separated from that of the carboxyl proton. In Figure 3.9.11(c), the difference, $\Delta\delta$ between the chemical shift of the caryboxyl proton of TFA and that of pure TFA represents the extent of the interaction of the carboxyl group in TFA and the OH group in cellulose acetate. The chemical shift of the O-Ac proton reveals a maximum at DS $= 2.46$ in TFA and DMAc, but is almost independent of DS in Py. These results are expected from Figure 3.9.4. The absolute magnitude of the shift is larger in TFA than in Py and that in DMAc lies intermediate between these. The DS dependence of the O-Ac proton chemical shift is much smaller in TFA than in DMAc. The large difference in DS dependence may be attributed to the differences in the solvent.

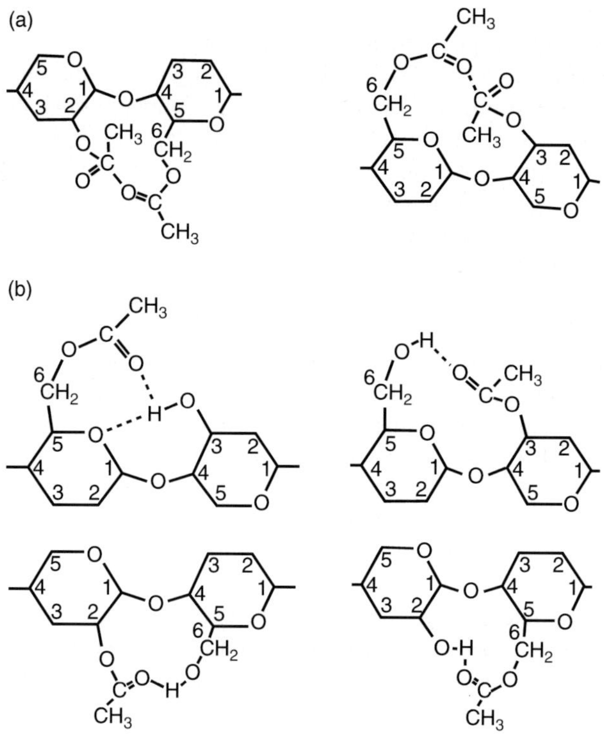

Figure 3.9.8   Possible conformation patterns of solid cellulose acetate (DS 2.92) (a) and cellulose acetate (DS 2.46) (b).[1]

TFA is a highly proton-donating solvent and Py is a typical basic reagent. Other solvents employed in this work, including DMAc, are almost neutral or amphoteric.

The interaction of the OH group with Py and TFA decreases gradually with an increase in DS, indicating that the above interaction becomes weak with increasing DS. From this, we can conclude that the interaction of the O-Ac group in CA with the solvent plays an important role in the dissolution process of CA (DS 2.92) in solvent and that, in addition to this kind of interaction, the interaction between the OH group with solvent becomes large as DS decreases.

Kamide *et al.*[2,4] demonstrated that the limiting viscosity numbers $[\eta]$ of CA (DS 2.92) fractions in DMAc are larger than those in halogenated hydrocarbons, and that these $[\eta]$ values for acetone lie between those of these two solvents: DMAc > acetone > halogenated hydrocarbon. The solvent dependence of the perturbed and unperturbed radii of gyration, $\langle S^2 \rangle_z^{1/2}$ and $\langle S^2 \rangle_{0,z}^{1/2}$, is parallel to that of $[\eta]$. Contrary to this, the χ parameter and the expansion factor α_s (see (eq. (3.13.4)) are almost independent of the solvent employed. In short, the short-range interaction is remarkably influenced by the solvent nature. This may be explained in terms of the O-Ac (or hydroxyl)$\cdots$solvent interaction.

Figure 3.9.12 shows a plot of the unperturbed chain dimensions A (eq. (3.13.9)) as a function of the chemical shifts of the O-Ac group (a) and the OH group. (b) In this

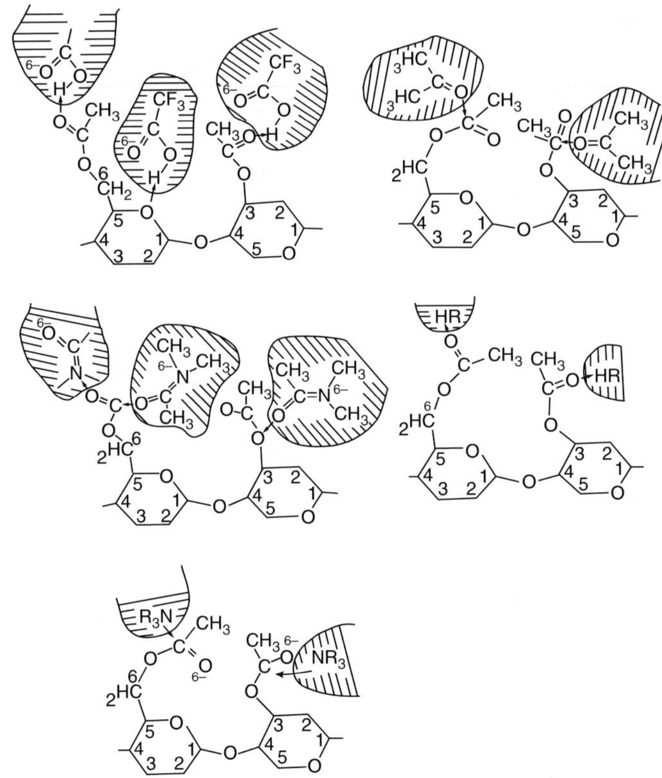

Figure 3.9.9 Possible molecular model of interaction of cellulose acetate (DS 2.92) with various solvents.[1]

figure, the rectangles stand for CA (DS 0.49), the circles for CA (DS 2.46), and the triangles, for CA (DS 2.92). The A values were determined in Section 3.16. For CA (DS 2.46), the unperturbed chain dimension becomes large as the signals of methyl and OH groups shifts to a lower magnetic field. Such a tendency, though significant, is much less remarkable for CA (DS 2.92) than for CA (DS 2.46). The strong intermolecular interaction between OH and acetyl groups formed in solid CDA (see Figure 3.9.7) is partially or fully destroyed by the interaction between the OH group and solvent and that between the acetyl group and solvent. The degree of destruction depends on the solvent nature, especially its polarity. On the other hand, the intermolecular or intra-molecular interactions in CA (DS 2.92) solid are weak enough to be fully broken by the addition of a solvent if it can dissolve the CA (DS 2.92) polymer. Although rather speculative, the above mechanism can qualitatively explain the noticeable differences in the A value when there is chemical shift of the methyl and OH protons. For CA (DS 2.46) in DMAc, the O-acetyl···solvent and the hydroxyl···solvent interactions are stronger than those in acetone. In addition, the high polarity of DMAc, through interaction with CA (DS 2.46), induces the mutual repulsion of the polymer chains, resulting in a large radius of gyration $\langle S^2 \rangle_z^{1/2}$ and a large $[\eta]$, compared with the CA (DS 2.46) acetone system.

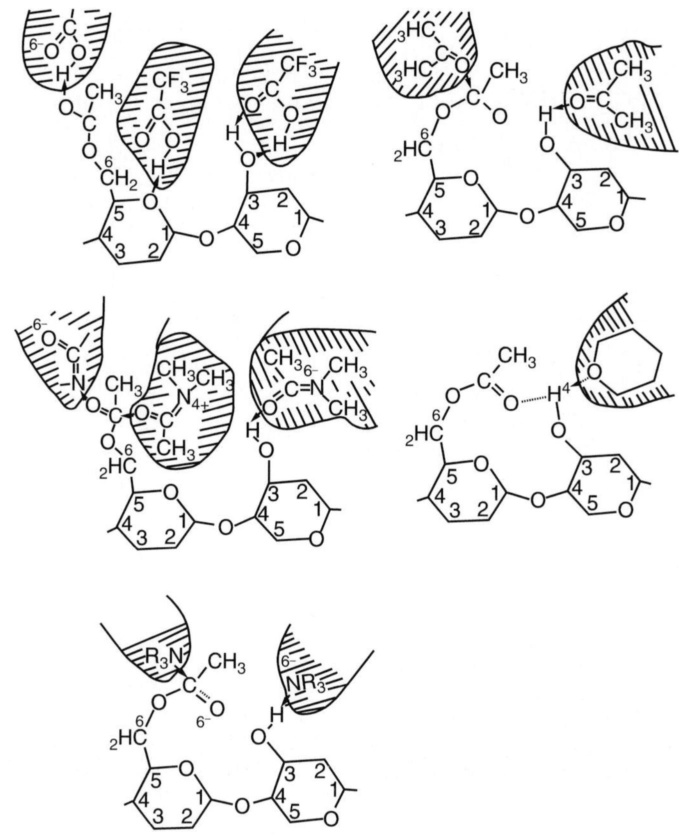

Figure 3.9.10   Possible molecular model of interaction of cellulose acetate (DS 2.46) with various solvents.[1]

Hence, we can interpret the solvation of CA in polar solvents as involving the *O*-Ac or OH···solvent bond formation with numerous arrangements (see Figure 3.9.9) and these doubtlessly strong interactions may give rise to the solvation of cellulose acetate. It seems plausible to assume that the CA molecule along with the solvated solvent dissolve into nonsolvated solvent molecules and behave as a single solute molecule. Consequently, the solvated solvent–nonsolvated solvent contact may predominate over the CA molecule···nonsolvated solvent contact. This hypothesis can explain the well-known experimental facts, such as small second virial coefficients, small expansion coefficients ($\alpha_s \sim 1$), and large unpertubed chain dimensions of cellulose derivatives in dilute solutions. The last fact may be due to the way that the solvated solvent molecules interfere with the contraction of the space occupied by the CA molecule in solution. It is noteworthy that the intensity of the interaction can be estimated by the chemical shift in NMR spectra, but the number of moles of the solvent involved in the solvation of 1 mole of a substituted glucopyranose unit cannot be counted. Therefore, unless an accurate knowledge of the degree of solvation is available, we cannot discuss the solvent dependence of A value at the molecular level (see Section 3.17).

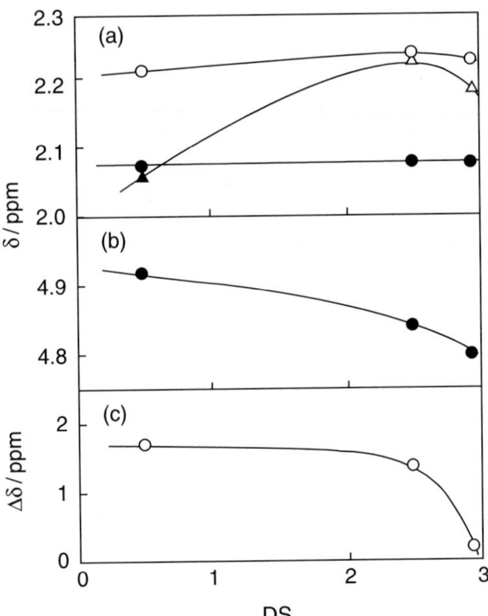

Figure 3.9.11   (a): The effect of the degree of substitution on the *O*-acetyl-proton chemical shift of cellulose acetate in various solvents: open circle, trifluoroacetic acid; closed circle, pyridine; open triangle, dimethylacetamide. Closed triangle is the interpolated value. (b): The effect of the degree of substitution on the hydroxyl proton chemical shift of cellulose acetate in pyridine. (c): The effect of the degree of substitution on the difference of the chemical shift of the carboxyl proton of trifluoroacetic acid of cellulose acetate solution and that of pure trifluoroacetic acid.[1]

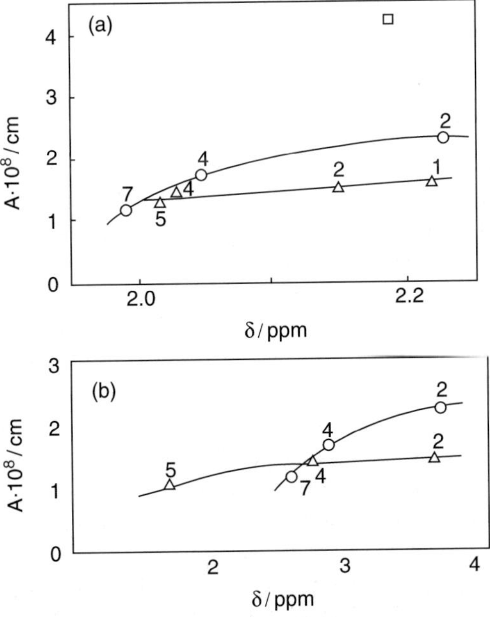

Figure 3.9.12   The plot of the unperturbed chain dimensions, *A* as a function of the chemical shift of *O*-acetyl methyl (a) and hydroxyl group (b).

As discussed in Section 3.9.2, Kamide et $al.$[3] found from the measurements of ^{1}H NMR chemical shifts[5] of the OH and O-Ac methyl protons of CA with the total degree of substitution $\langle\!\langle F \rangle\!\rangle$ 0.49–2.92 that (1) there exists the solvation of solvent molecules to O-Ac and OH groups, (2) it becomes larger in polar solvents, and (3) δ of OH proton attains maximum at $\langle\!\langle F \rangle\!\rangle = 2.5$ in N,N-dimethylacetamide (DMAc).

3.9.2   CTA by pulse-Fourier ^{1}H and ^{13}C NMR method[26]

Kamide and Saito[27] measured the adiabatic compressibility of solutions of CA samples and concluded that s at infinite dilution, $s_0 (\equiv \lim_{c \to 0} s)$ depends on the solvent nature in the same manner as δ. The strong correlation of s_0 with the unperturbed chain dimension A and the partially draining parameter $X^{2-6,8}$ (eq (3.15.7)) were experimentally demonstrated for CA with $\langle\!\langle F \rangle\!\rangle = 0.49–2.92$ in various solvents. These experimental facts illustrate that the solvation plays a predominant role in the flexibility of the unperturbed chain together with the hydrodynamic interaction in CA solutions. In other words, the solvated solvent molecule yields a significant steric hindrance and an interaction between the solvated and free solvent molecules results in partial free drainage. The solvent molecules possibly solvate to different extents with the O-Ac or OH groups at three different carbons (i.e. C_2, C_3, and C_6) in a glucopyranose unit.

If the above reasoning is acceptable, then the difference in the position of the O-Ac and or the OH group, to which the solvent molecules are solvated, is expected to cause a different effect on A and χ. A recent commercialization of Fourier transform (FT) NMR makes it possible to evaluate the intensity of the solvation.

As an extension of Section 3.9.1, this section measures the chemical shift δ and the relaxation time T_1 by FT-^{1}H and -^{13}C NMR for CTA in various solvents to disclose the solution structure of CTA solvent system.

Polymer sample and solvents

A CTA whole polymer with $\langle\!\langle F \rangle\!\rangle = 2.92$ ($M_w = 2.32 \times 10^5$ and $M_n = 5.85 \times 10^4$),[1,26,27] was used for NMR measurements. Deuterated TCM-d$_1$, Py-d$_5$, acetone-d$_6$, N,N-dimethyl-sulfoxide (DMSO)-d$_6$, and DMAc-d$_9$ were used as received from the manufacturer (E Merck, Darmstadt, Germany).

^{1}H and ^{13}C NMR measurements

The CTA sample, dried in $vacuo$ at 40 °C for 24 h, was mixed with solvents, except acetone, to give 7 wt% solution in the NMR sample tube. For TCM and DMAC, the dissolution was carried out at room temperature with 20 min mixing and for Py and DMSO the tube was placed in an NMR cell and the solution was prepared at 120 °C by rotating the tube at a rate of 20 rpm. The mixture of the CTA sample flake and acetone was quenched to − 170 °C, maintained at that temperature for 1 h, and the clean and transparent solution of 3 wt% CTA was obtained by raising the temperature of the mixture very gradually to room temperature.

δ values of the *O*-Ac methyl proton (AMH) and *O*-Ac carbonyl carbon (ACC) were determined on a FX-200 JEOL FT NMR spectrometer (Japan Electron Optics Laboratory Co., Tokyo) using TMS as a reference. ^{1}H NMR spectra were obtained under the following conditions: frequency, 199.5 MHz; pulse width, 45° (7 μs); repetition time, 8 s; data points, 16,384; and accumulation, 16 times. Only for the CTA/acetone system, $180 - \tau - 90°$ (τ, pulse interval), pulse was used to eliminate peak of acetone with long spin lattice relaxation time T_1 (in this case, $\tau = 5$ s, repetition time 30 s). ^{13}C NMR spectra were obtained under the following conditions: frequency, 50.15 μMHz; pulse width 45° (6.5 μs); repetition time, 3.2 s; and complete decoupling of ^{1}H nuclei. For acetone, DMSO, and Py, the solvent peaks overflowed by frequency domain accumulation. The repetition was *ca.* 1000–4000 times. T_1 was measured by the inversion recovery method using $180 - \tau - 90°$ pulse sequence.

Figure 3.9.13 shows the ^{1}H NMR spectra of the *O*-AMH region of CTA in five deuterated solvents. Generally, three peaks were observed for the *O*-Ac methyl proton. δ of these peaks varied over a range of between 1.8 and 2.2 ppm. The intensity ratio in TCM, calculated from the peak area, is 0.99:1.04:1.00 for three peaks at 1.95, 2.00, and 2.12 ppm. Up until now, ^{1}H NMR measurements on CTA with $\langle\!\langle F \rangle\!\rangle$ 2.9 were, without exception, carried out on 100 or less MHz NMR spectrometers, which have less resolved power for this purpose. From a comparison of the reactivity of three OH groups at C_2, C_3, and C_6 positions with toluene sulfonyl chloride, Goodlett *et al.*[19] assigned three peaks at 1.94, 1.99, and 2.09 ppm for CTA in TCM to be the *O*-Ac methyl protons at the C_3, C_2, and C_6 position, respectively. Shiraishi and coworkers[20] arrived at the same conclusion

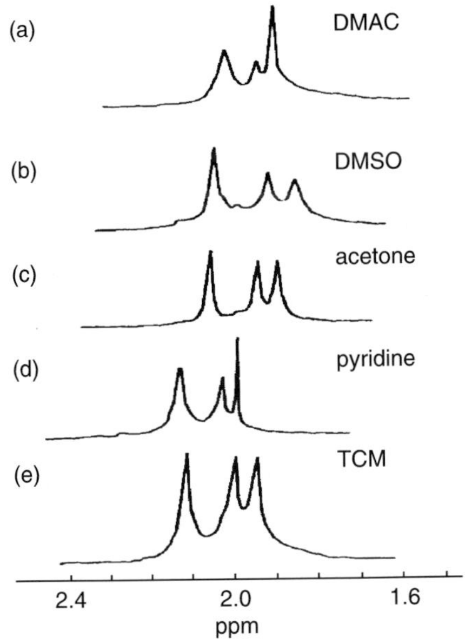

Figure 3.9.13 NMR spectra of cellulose triacetate ($\langle\!\langle F \rangle\!\rangle = 2.92$) in various deuterated solvents in *O*-acetyl methyl proton region.[26] (a) dimethylacetamide-d_9; (b) dimethyl-sulfoxide-d_6; (c) acetone-d_6; (d) pyridine-d_5; (e) trichloromethane-d_1.

by studying the correlation between the acidity of O-Ac methyl groups at C_2, C_3, and C_6 positions and their electron deshielding effect, and by estimating the degree of shift of the methyl proton peak of CTA in TCM to a lower magnetic field. The intensity ratio of these three peaks corresponds accurately to the degree of substitution of three OH groups at C_3, C_2, and C_6 positions with the acetyl group (i.e. $\langle\langle f_3 \rangle\rangle$, $\langle\langle f_2 \rangle\rangle$, and $\langle\langle f_6 \rangle\rangle$). $\langle\langle f_3 \rangle\rangle$, $\langle\langle f_2 \rangle\rangle$, and $\langle\langle f_6 \rangle\rangle$, of the CTA sample employed here were determined using $\langle\langle F \rangle\rangle(= 2.92)$, as 0.92, 1.01, and 0.99, respectively. These values are approximately the same as those (0.89, 1.02, 1.00) evaluated by Kamide and Okajima[28] in a similar CTA sample.

The peak at the highest magnetic field ($\delta = 2.03$ ppm) Py is extremely sharp compared with the other two peaks. This suggests that the peak at $\delta = 2.03$ ppm has a long spin–spin relaxation time, T_2, and may be due to low molecular weight compounds contaminated in the solution. The peak intensity at $\delta = 2.15$ ppm is almost twice that of $\delta = 2.04$ ppm, and we may assign the peak at $\delta = 2.15$ ppm to the O-AMH at C_2 and C_6 positions (AMH$_2$ and AMH$_6$) and the peak at $\delta = 2.04$ ppm to the O-AMH at C_3 position (AMH$_3$). The intensity ratios of three peaks observed in TCM, acetone, and DMSO over the range of 1.6–2.4 ppm were almost independent of the solvents. The peaks from higher magnetic field can be attributed to the O-Ac methyl protons at C_3, C_2, and C_6 positions. An exception is DMAc, in which the peak intensity at $\delta = 1.97$ ppm (probably assigned for AMH$_2$) is about 10–15% smaller than the other two peak intensities, suggesting a small probability of resonance absorption by AMH$_2$ at 1.94 and 2.07 ppm, in addition to at 1.97 ppm. Here, we analyzed the spectrum in DMAc on the assumption that each peak attributes to the O-AMH at C_3, C_2, and C_6 positions (from higher magnetic field), respectively.

Table 3.9.1 shows the chemical shifts of the O-Ac methyl protons of CTA in various solvents.

Table 3.9.1

NMR peaks of O-acetyl and hydroxyl protons of cellulose acetate in various solvents

Solvent	Dielectric constant ε	CA (0.49)[a]		CA (1.75)		CA (2.46)		CA (2.92)	
		O-Ac	OH	O-Ac	OH	O-Ac	OH	O-Ac	OH
Formamide	111	2.17	3.72	–	–	–	–	–	–
D$_2$O[b]	78	2.19	4.70	–	–	–	–	–	–
TFA	39.5	2.22	–	–	–	2.24	–	2.23	–
DMAc	37.8	2.21	3.87	2.23	3.93	2.23	3.90	2.15	3.84
Acetone	20.7	–	–	–	–	2.06	2.85	2.04	2.75
THF	8.2	–	–	–	–	2.00	2.51	–	–
Pyridine		2.10	4.92	–	–	2.09	4.83	2.08	4.80
DCM	7.77	–	–	–	–	2.05	1.62	2.03	1.64
TCE	7.29	–	–	–	–	–	–	2.00	1.70
DMF[c]		–	–	–	–	–	–	2.07	3.50

[a]Numbers in parentheses denotes total degree of substitution.
[b]Deuterium oxide.
[c]Dimethylformamide.

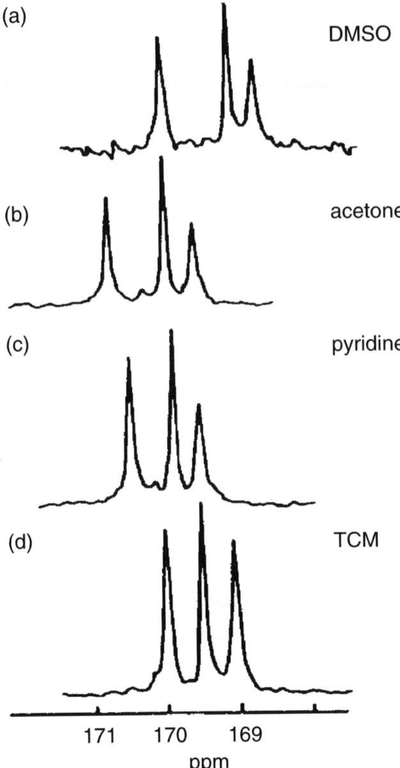

Figure 3.9.14 Nuclear magnetic resonance spectra of cellulose triacetate ($\langle\!\langle F \rangle\!\rangle = 2.92$) in various deuterated solvents in O-acetyl carbonyl carbon region.[26] (a) dimethyl-sulfoxide-d_6; (b) acetone-d_6; (c) pyridine-d_5; (d) TCM-d_1.

Figure 3.9.14 shows the ^{13}C NMR spectra of the O-ACC region of CTA. Until recently, the peak assignment of the carbonyl carbon region in ^{13}C NMR spectra was an unsolved problem. Applying a low power selective decoupling method to O-AMH located at specific carbon position, Kowsaka $et\ al.$[29] made the most reasonable assignments to date (see Section 2.3). Three large peaks were assigned to the acetyl groups at C_6, C_3, and C_2 positions, from the low magnetic field side.

Table 3.9.2 summarizes T_1 of the O-AMH and O-ACC in various solvents. Here, T_1 for O-AMH at C_6 in TCM and DMSO and for O-AMH at C_2 in Py system could not be determined because of the overlapping of the peaks, which were due to impurities in these solutions. T_1 of the O-AMH at C_3 position decreases in the following order: acetone $>$ DMSO $=$ TCM $\geq$ Py. No significant solvent dependence of T_1 of the O-AMH at C_2 position was detected experimentally. T_1 of the carbonyl carbon at C_k ($k = 2, 3, 6$) position varies depending on the carbon positions to which the O-Ac group is attached. T_1 of the carbonyl carbon at C_6 position attains a maximum in acetone and minimum in DMSO.

For the carbonyl carbon at C_2, T_1 decreases in the following order: acetone $>$ TCM $>$ Py $>$ DMSO and for the carbonyl carbon at C_3, acetone $>$ DMSO $>$ TCM $>$ Py.

Table 3.9.2

Spin lattice relaxation time (T_1) of the O-acetyl methyl proton and O-acetyl carbonyl carbon in cellulose triacetate solution[26]

Solvent	T_1/s					
	O-Ac methyl proton			O-Ac carbonyl oxide		
	$C_6{}^a$	C_3	C_2	C_6	C_3	C_2
DMSO[b]-d$_6$	$-^c$	0.80	0.77	2.25	2.23	2.41
Pyridine-d$_5$	0.77^d	0.77^d	$-^c$	2.56	2.26	2.46
TCM[e]-d$_1$	$-^c$	0.80	0.77	2.83	2.28	2.49
Acetone-d$_6$	$-^f$	0.85	0.77	3.27	3.36	2.61

^{a}Carbon position substituted by O-acetyl group.
^{b}N, N-Dimethylsulfoxide.
^{c}Overlapping with the park of impurity.
^{d}Overlapping with neighboring cellulose triacetate peaks.
^{e}Trichloromethane.
^{f}Overlapping with solvent peak.

It is generally considered that ^{13}C nucleus relaxation is most attributable to the relaxation due to dipole–dipole interaction between ^{13}C nucleus and ^1H nucleus (its relaxation time hereafter referred to as T_{1DD}).

In the case of CTA, there is no ^1H nucleus directly combined with the O-ACC. Moreover, a possible interaction of the carbonyl carbon with ^1H nucleus attached to glucopyranose ring (skeleton) carbon and ^1H nucleus in the O-Ac methyl group may be negligible because a reciprocal of T_{1DD} decreases rapidly in reverse proportion to the sixth power of the distance of ^{13}C and ^1H nuclei. Therefore, we can conclude that the difference in T_1 of the carbonyl carbon reflects a difference of the interaction between the O-Ac group and the solvent; shorter T_1 for stronger interaction. T_1 was the longest in acetone for all carbon positions. This means that the interaction of acetone molecule with the O-ACC is the weakest among the four solvents examined, regardless of the carbon position. T_1 of the carbonyl carbons at three positions decreases in the following order in each solvent: in DMSO, $C_2 > C_3 > C_6$; in Py, $C_6 > C_2 > C_3$; in TCM, $C_6 > C_2 > C_3$; in acetone, $C_3 \geq C_6 > C_2$.

Figures 3.9.15 and 3.9.16 show the NMR spectra of the O-acetylmethyl proton and the O-ACC of the CTA in five solvents. The number in the figures means the difference (ppm) from δ in acetone. When acetone is taken as the standard, the peaks of the methyl protons at three carbon positions are shifted downfield in Py and TCM and to upfield in DMSO. In DMAc, the proton peak at C_6 position slightly shifts upfield, but the proton peaks at C_2 and C_3 positions downfield. On the other hand, all the peaks of the carbonyl carbons (except for those at C_2 and C_3 positions in Py) shift more than 0.2 ppm upfield.

Based on the above experimental results, we can conclude that in CTA solvent systems, there are interactions between the O-Ac group and the solvent molecules.

TCM and Py. The chlorine atom in TCM molecule has a strong electron donating property allowing it to act with O-Ac carbonyl carbons located at C_2, C_3, and C_6

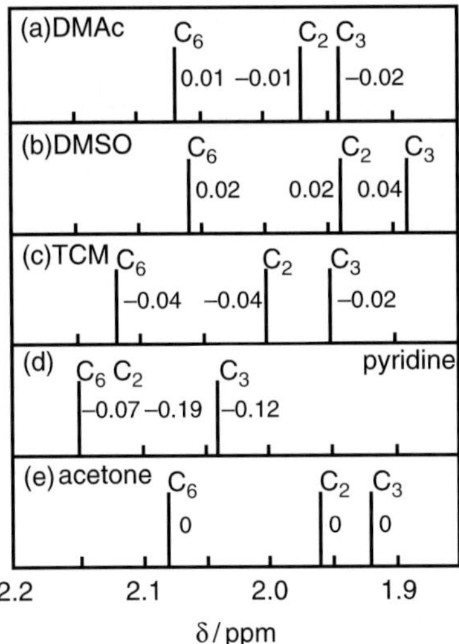

Figure 3.9.15 ^{1}H NMR chemical shift δ (in ppm) of *O*-acetyl methyl proton of cellulose triacetate molecules in various deuterated solvents.[26] Number of each bar denotes the difference of the δ value between that of acetone and other solvents.

(Figure 3.9.16(c)). This results in the shielding of the *O*-Ac carbonyl carbons, whose ^{13}C NMR peaks shift to a higher magnetic field than those of the carbonyl carbon of acetone. For this reason, weakening of the double bond nature of the carbonyl group brings about deshielding of the *O*-Ac methyl carbon, shielding before a downfield

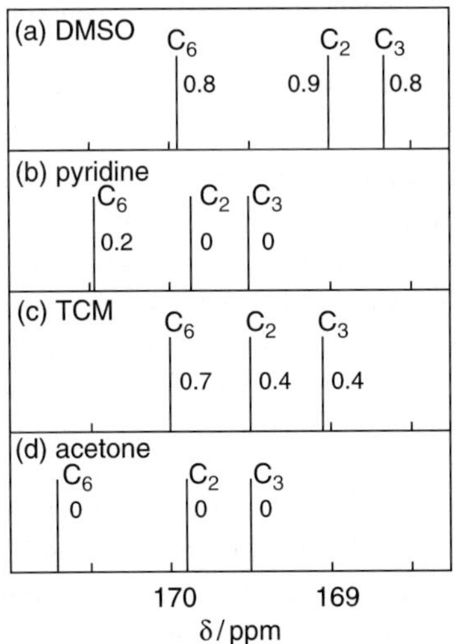

Figure 3.9.16 ^{1}H NMR chemical shift δ (in ppm) of *O*-acetyl carbonyl carbon of cellulose triacetate molecules in various deuterated solvents.[26] Number of each bar denotes the difference of the δ value between that in acetone and in other solvents.

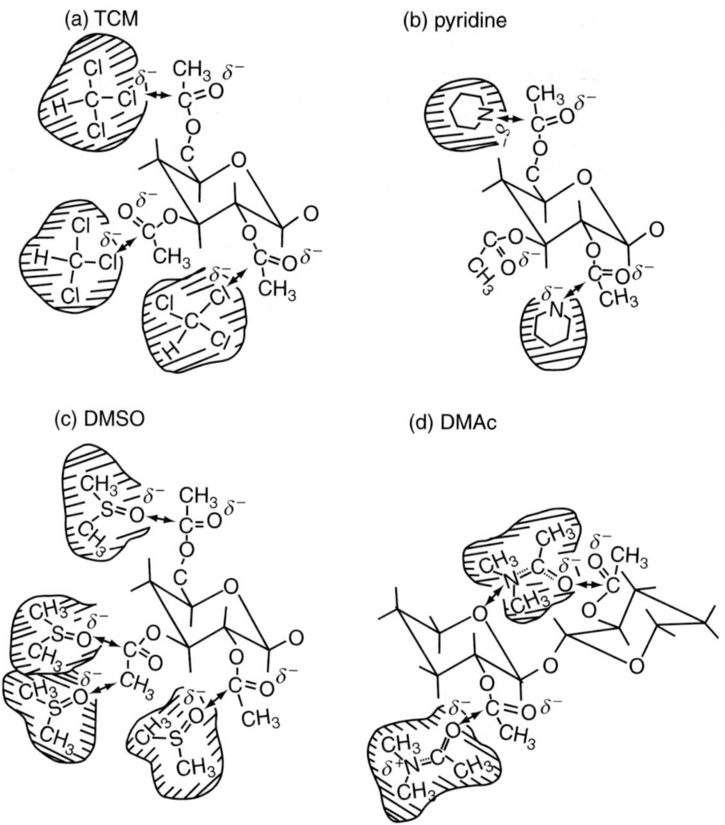

Figure 3.9.17   Schematic representation of the interaction between a cellulose triacetate molecule and various solvents.[26]

shift of the O-AMH peak. Figure 3.9.17(a) is a schematic representation of the interaction between the CTA and TCM molecules. Tertiary amine in Py also acts as an electron donor and, judging from Figure 3.9.16(b), seems to interact only with the O-ACC at C_6 position, as demonstrated in Figure 3.9.17(b). Unfortunately, it is not understood why the large shift of δ of O-Ac methyl protons at C_2 and C_3 positions toward lower magnetic field side occurs and why the tertiary amine has an interaction only to O-ACC at C_6 position.

DMSO. The oxygen atom in the DMSO molecule donates electrons to O-Ac carbonyl carbons at C_2, C_3, and C_6 positions, resulting in a significant shift of the ACC peak to upfield. This also brings about weakening of the shielding effect, originally existing between the O-ACC and carbonyl oxygen. As a result, the peaks of O-Ac methyl protons are expected to shift to a lower magnetic field. However, the experimental data show the shift of the peaks to a higher magnetic field, suggesting electrostatic interaction between the oxygen atom in DMSO and O-Ac methyl protons. Figure 3.9.17(c) illustrates the solvent effect of DMSO on the CTA molecule. Note that although DMSO has a similar chemical structure to acetone, the former interacts with the O-Ac group more strongly than the latter. This conclusion was deduced from the chemical shifts of O-AMH and carbonyl carbon and from T_1 of the O-Ac carbonyl carbon, and is quite consistent with the higher polarity of sulfonyl group than that of the carbonyl group.

DMAc. Figure 3.9.15(a) shows that an interaction between O-Ac group at C_6 position, that DMAc is expected to be small, and that there occurs an interaction between O-Ac group at C_2 and C_3 and DMAc. It also illustrates that an interaction at C_3 position may be a little stronger than that at C_2 position. This can be explained by the amphoteric nature of the DMAc molecule. Figure 3.9.17(d) shows the interaction of DMAc with the CTA molecule.

In summary, it can be concluded that the solvent molecule interacts significantly with the O-Ac groups of CTA, and that the extent of solvation varies depending on the positions at which the O-Ac group is located as well as on the solvent nature.

REFERENCES

1. K Kamide, K Okajima and M Saito, *Polym. J.*, 1981, **13**, 115.
2. K Kamide, T Terakawa and Y Miyazaki, *Polym. J.*, 1979, **11**, 285.
3. S Ishida, H Komatsu, T Terakawa, Y Miyazaki and K Kamide, *Mem. Fac. Eng. Kanazawa Univ.*, 1979, **12**, 103.
4. K Kamide, Y Miyazaki and T Abe, *Polym. J.*, 1979, **11**, 523.
5. K Kamide, M Saito and T Abe, *Polym. J.*, 1981, **13**, 421.
6. K Kamide and M Saito, *Polym. J.*, 1982, **14**, 517.
7. S Ishida, H Komatsu, H Katoh, M Saito, Y Miyazaki and K Kamide, *Makromol. Chem.*, 1982, **183**, 3075.
8. M Saito, *Polym. J.*, 1983, **15**, 249.
9. K Kamide and Y Miyazaki, *Polym. J.*, 1978, **10**, 409.
10. K Kamide and T Terakawa, *Polym. J.*, 1978, **10**, 559.
11. K Kamide and Y Miyazaki, *Polym. J.*, 1978, **10**, 539.
12. H Suzuki, K Kamide and Y Miyazaki, *Netsusokutei*, 1980, **7**, 37.
13. H Suzuki, Y Miyazaki and K Kamide, *Eur. Polym. J.*, 1980, **16**, 703.
14. HM Spurlin, in *Cellulose and Cellulose Derivatives* (ed. E Otto), Interscience, New York, 1943, p. 868.
15. P Clermont, *Ann. Chim.*, 1943, **12**, 2420.
16. RJB Marsden and AR Urquhart, *J. Text. Inst.*, 1942, **33**, T-105.
17. RU Lemiux and JD Stevens, *Can. J. Chem.*, 1965, **43**, 1059.
18. D Gagnaire and M Vincendon, *Bull. Soc. Chim. Fr.*, 1966, 204.
19. VW Goodlett, JT Dougherty and HW Patton, *J. Polym. Sci.*, 1971, A-1, **9**, 155.
20. N Shiraishi, T Katayama and T Yokota, *Cell. Chem. Technol.*, 1978, **12**, 429.
21. WW Simons and M Zanger (eds), *The Sadtler Guide to the NMR Spectra of Polymers*. Sadtler Laboratory, New York, 1973, p. 171.
22. See, for example, T Kagiya, Y Sumida and T Inoue, *Bull. Chem. Soc. Jpn.*, 1968, **41**, 767.
23. WR Moore and J Russel, *J. Colloid. Sci. Jpn.*, 1954, **9**, 338.
24. PH Hermans, *Physics and Chemistry of Cellulose Fibers*, Elsevier, New York, 1949, p. 13.
25. J Robinson, E Conmar, *Discuss Faraday Soc.*, 1954, **16**, 125.
26. K Kamide, M Saito, K Kowsaka and K Okajima, *Polym. J.*, 1987, **19**, 1377.
27. K Kamide and M Saito, *Eur. Polym. J.*, 1984, **20**, 903.
28. K Kamide and K Okajima, *Polym. J.*, 1981, **13**, 127.
29. K Kowsaka, K Okajima and K Kamide, *Polym. J.*, 1986, **18**, 843.

3.10 SMALL ANGLE X-RAY SCATTERING (SAXS)

When the X-ray scattering intensity I of polymer solutions is measured over a wide range of scattering angle (usually, the scattering vector k is used), the plot of Ik^2 versus k (Kratky plot) provides valuable information on the overall shape of the molecules and the size of segments.[1] In the plot, the small k region corresponds to the scattering Gaussian chain (i.e. overall shape of CA molecules) and the larger k region, to the rod (i.e. segment). The transition point of the region, where I is proportional to $1/k^2$ to the region and where I is proportional to $1/k$ (rod), was conventionally determined to be as point of the interception k^* of two lines drawn on the plot.[1]

3.10.1 CA

Figure 3.10.1 shows the Kratky plot of four 1-wt% CA solutions in DMAc at 25 °C.[2] Here, CA samples were prepared by acetylation of wood pulp in a mixture of acetic acid, acetic anhydride, and sulfuric acid and subsequent hydrolysis in acetic acid with hydrochloric acid. From the figures, k^* was found to be 0.35, 0.24, 0.46, and 0.6113 nm^{-1} for CA with $\langle\!\langle F \rangle\!\rangle$ = 2.9, 2.5, 1.75 and 0.9, respectively.

k^* was converted to the persistence length, q, using the equation derived on the basis of Monte Carlo simulation:

$$q = \mu^* / k^* \tag{3.10.1}$$

where μ^* is constant and 2.30[1] was employed.

3.10.2 CN

Figure 3.10.2 shows the Kratky plot of CN ($\langle\!\langle F \rangle\!\rangle$ (by ^{13}CNMR analysis) = 2.3) whole polymer in acetone at 25 °C.[2] Here, commercially available CN, manufactured by Asahi

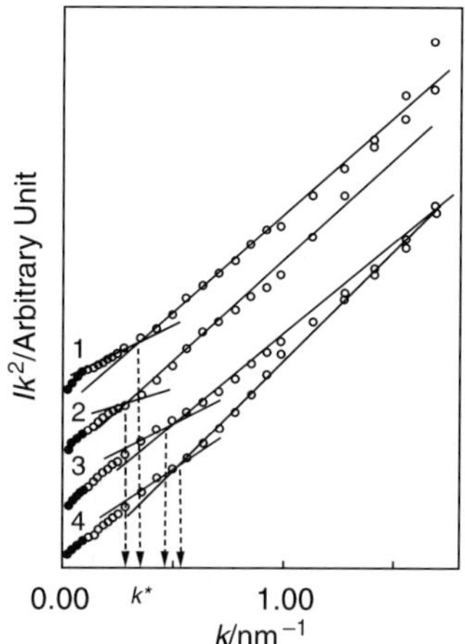

Figure 3.10.1   Kratky plot of 1 wt% CA whole polymer solutions in DMAc at 25 °C.[2] Total degree of substitution $\langle\!\langle F \rangle\!\rangle$ is 2.9 (curve 1), 2.5 (curve 2), 1.75 (curve 3), and 0.9 (curve 4).

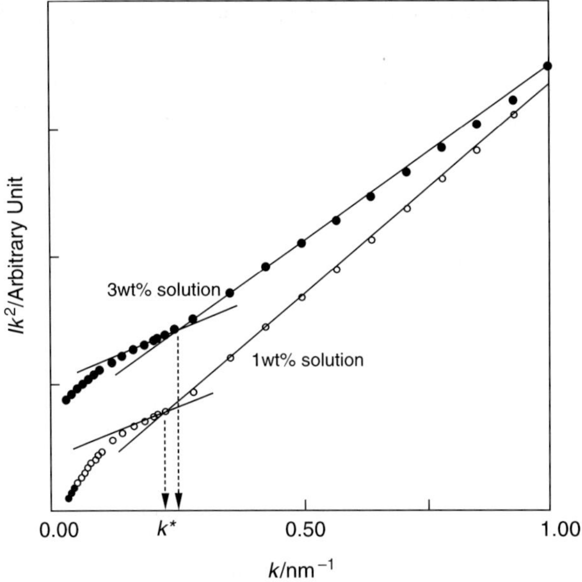

Figure 3.10.2   Kratky plot of CN($\langle\!\langle F \rangle\!\rangle = 2.3$ whole polymer) in acetone at 25 °C.[2]

Chemical Ind. Co., (Nobeoka, Japan) was used to prepare 1- and 3-wt% solution. From the plot, $k^* = 0.24$ for 1 wt% and 0.27 nm^{-1} for 3 wt% solution are estimated. Slight concentration dependence of k^* is observed. On average, we obtained $q = 9.6$ nm for CN with $\langle\!\langle F \rangle\!\rangle = 2.3$ by the SAXS method.

REFERENCES

1. O Kratky, in *Small Angle X-ray Scattering* (eds O Glatter and O Kratky), Academic Press Inc, New York, 1982, Chapter 11; O Kratky and G Porod, *Das Makromolekül in Lösungen*, Lehntes Kapitel, Röntgenkleinwinkelstreung von makro-molekularen Lösungen, §79 Klein-wirkstreung von Lösungen fadenförmiger Mole-küle, Springer, 1953.
2. K Kamide and M Saito, Advanced polymer materials. *Macromol. Sym.*, 1994, **83**, 233.

3.11   MARK–HOUWINK–SAKURADA (MHS) EQUATIONS

For most polymer solutions, the relationship between [η] and the molecular weight M is expressed as:[1]

$$[\eta] = K_{\mathrm{m}} M^a \tag{3.11.1}$$

where K_{m} and a are parameters characteristic of the polymer/solvent combination at given temperature. The relationship was theoretically derived by Kuhn (1934)[2] and was empirically proposed by Mark (1934),[3] Houwink (1941),[4] and Sakurada (1941)[5] independently. Eq. (3.11.1) is called as the MHS equations.

3.11.1 CA (DS 0.49)

Figure 3.11.1 shows the log–log plots of $[\eta]$ versus M_w for CA (DS 0.49) in DMAc, DMSO, water, and FA at 25 °C (open circles).[6]

From this figure, the parameters K_m and a in the MHS equation (3.11.1) can be established by using the least square method. The results are:

$$[\eta] = 0.191 M_w^{0.60} \qquad \text{in DMAc} \qquad (3.11.2)$$

$$[\eta] = 0.171 M_w^{0.61} \qquad \text{in DMSO} \qquad (3.11.3)$$

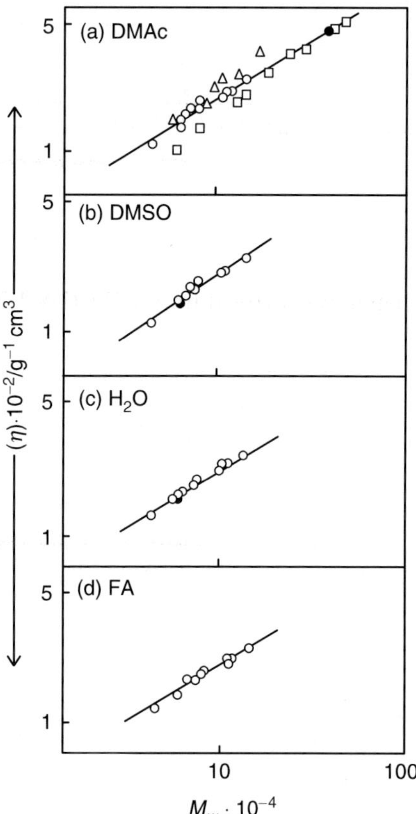

Figure 3.11.1 Log–log plot of limiting viscosity number $[\eta]$ against the weight-average molecular weight M_w for cellulose acetate (DS 0.49) in dimethylacetamide, dimethyl sulfoxide, water, and formamide at 25 °C.[6] Straight lines represent the MHS equations (a–d) determined by the least squares method: (○), cellulose acetate (DS 0.49);[6] (△), cellulose acetate (DS 2.46);[10] (□), cellulose acetate (DS 2.92).[7]

$$[\eta] = 0.209\, M_{\mathrm{w}}^{0.60} \qquad \text{in water} \tag{3.11.4}$$

$$[\eta] = 0.209\, M_{\mathrm{w}}^{0.60} \qquad \text{in FA} \tag{3.11.5}$$

The molecular weight range covered by the MHS equations is $(4.55–14.5) \times 10^4$ and the ratio of the maximum to minimum M_{w} is 3.2. The MHS equations in water and in FA coincide, and the values for exponent a in the above equations range from 0.60 to 0.61 in these different solvents ($\varepsilon = 37.8 - 1.11$). Thus, a is very insensitive to the solvent. Eq. (3.11.2) for DMAc solution should be compared with those established for CA (DS 2.92)[7] and CA (DS 2.46)[8] in the same solvent, at the same temperature.

$$[\eta] = 0.0264 M_{\mathrm{w}}^{0.75} \qquad \text{for CA (DS 2.92)} \tag{3.11.6}$$

and

$$[\eta] = 0.0395 M_{\mathrm{w}}^{0.74} \qquad \text{for CA (DS 2.46)} \tag{3.11.7}$$

In Figure 3.11.1(a), the open rectangles represent CA (DS 2.92) and the open triangles stand for CA (DS 2.46), as reported elsewhere.[6] In the lower M_{w} range, $[\eta]$ of CA (DS 0.49) is similar to that of CA (DS 2.46). However, in the higher M_{w} range, $[\eta]$ for CA (DS 0.49) approaches that for CA (DS 2.92). The exponent a for CA (DS 0.49) is substantially smaller than the values obtained for CA (DS 2.46) and CA (DS 2.92) in DMAc.

3.11.2 CA (DS 1.75)[9]

The following MHS equation was derived for the CA (DS 1.75) fractions in DMAc:

$$[\eta] = 9.58 \times 10^{-2} M_{\mathrm{w}}^{0.65} \tag{3.11.8}$$

The exponent a in the MHS equation decreases in the following order: CA (DS 2.46) $\gg$ CA (DS 2.92) $>$ CA (DS 1.75) $>$ CA (DS 0.49).

3.11.3 CA (DS 2.46)[8,10]

The Huggins constant, k', increases continuously with M_{w} from 0.4 to 0.6 for CA (DS 2.46) in acetone and THF. Log–log plots of $[\eta]$ versus M_{w} for CA (DS 2.45) in both solvents at 25 °C are shown in Figure 3.11.2. M_{w} by LS in acetone is used for $[\eta]$ in acetone and M_{w} by LS in THF is employed for $[\eta]$ in THF. From this figure, the following MHS equations are established.

$$[\eta] = 0.133 M_{\mathrm{w}}^{0.616} \qquad \text{in acetone at 25 °C} \tag{3.11.9}$$

$(6.1 \times 10^4 \le M_{\mathrm{w}} \le 26.5 \times 10^4$, sample number 9)

$$[\eta] = 0.0513 M_{\mathrm{w}}^{0.688} \qquad \text{in THF at 25 °C} \tag{3.11.10}$$

$(7.4 \times 10^4 \le M_{\mathrm{w}}\ 30.0 \times 10^4$, sample number 6)

The value of a for acetone solution, as calculated from M_{n}, is 0.643. Among the five fractions employed for Stein and Doty's LS study,[11] two fractions (23B and 31B) were

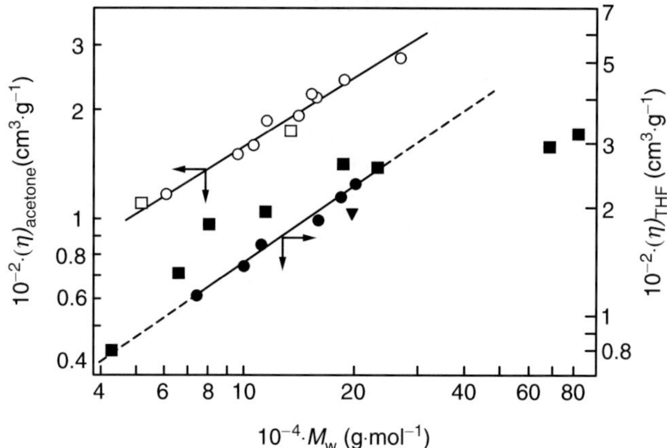

Figure 3.11.2 Log–log plot of limiting viscosity number $[\eta]$ against the weight-average molecular weight M_w for cellulose diacetate in acetone at 25 °C (open mark) and in tetrahydrofuran at 25 °C (closed mark). (○), Kamide *et al.*;[10] (□), data from Stein and Doty's work;[11] (■) and (▼), data from Tanner and Berry.[13] M_w is determined by light scattering method in acetone (○) and (□) in tetrahydrofuran (●) and (▼) and in a mixed solvent (■).

used for the viscosity measurements in acetone at 25 °C by Badgley and Mar.[12] The data ($[\eta]$ and M_w) for these two fractions are plotted as open rectangles in Figure 3.11.2 for comparison. These data fall reasonably on the experimental line (eq. (3.11.9)). Also shown are the data points obtained by Tanner and Berry for CDA (DS 2.45) in THF.[13] Here, M_w as determined by LS in a mixed solvent (closed rectangle) or in THF (closed triangle) is employed. Their data points scatter widely around eq. (3.11.10), indicating a large experimental error, particularly in the higher M_w range.

The above equations are obtained for the samples with $M_\mathrm{w}/M_\mathrm{n} = 1.2 - 1.38$. Hence, equations can be readily converted into those for monodisperse samples. The results are

$$[\eta] = 0.136M^{0.616} \qquad \text{in acetone at 25 °C} \qquad (3.11.9')$$

and

$$[\eta] = 0.0524M^{0.688} \qquad \text{in THF at 25 °C} \qquad (3.11.10')$$

In deriving eqs. (3.11.9′) and (3.11.10′), the Schulz–Zimm distribution is assumed for the CDA fractions.

Up until now, numerous different MHS equations for CDA in acetone have been published by Kraemer,[14] Bartovics and Mark,[15] Sookne and Harris,[16] Badgley and Mark,[12] Philipp and Bjork,[17] Cumberbirch and Harland,[18] Moore and Tidswell,[19] and lkeda and Kawaguchi,[20] who employed M_w except for Kraemer. All experimental equations lie to the left of eq. (3.11.9′), when $[\eta]$ and M are double logarithmically plotted, and the wide variation of K_m and a of those equations can be adequately explained by the polydispersity effect of the sample (see also (e)). It is noted that the molecular weight dependence of the polydispersity of the polymer samples utilized in the literature made the exponent a large, as compared with 0.616 obtained here.

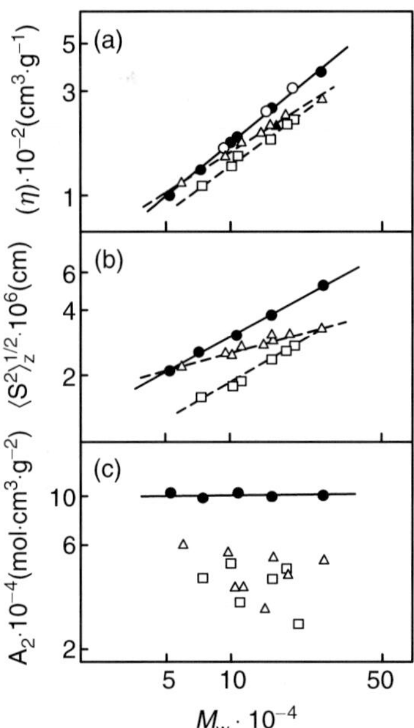

Figure 3.11.3   Molecular-weight dependence of limiting viscosity number $[\eta]$, (a) z-average radius of gyration $\langle S^2 \rangle_z^{1/2}$ (b) and the second virial coefficient A_2 (c) for cellulose acetate (DS 2.46) in dimethylacetamide (circle),[8] acetone (triangle),[10] and tetrahydrofuran (rectangle);[10] open circle.[7] (M_w was determined in acetone and tetrahydrofuran); closed circle, present work.

Figure 3.11.3 shows the molecular weight dependence of $[\eta]$, z-average radius of gyration $\langle S^2 \rangle_z^{1/2}$, and second virial coefficient A_2 of CA (DS 2.46) in DMAc, acetone, and THF at 25 °C.[8] In our previous study,[10] $[\eta]$ was measured in DMAc for several fractions, whose M_w were determined by LS in acetone and THF (Table IX of Ref. 10). In Figure 3.11.3(a), some of these data are represented by unfilled circles. Because of unexpectedly large scatter, the data for fractions EF2-10 and 3-10 shown in parentheses in Table 3.3.3 are omitted from this figure. $[\eta]$ for fraction EF3-10 was determined as 194 cm^3 g^{-1}. In the same figure, the $[\eta]$–M_w data for acetone and THF solutions, obtained in our previous study,[8] are shown by triangles and rectangles, respectively. The following MHS equation is established for DMAc solutions:

$$[\eta] = 1.34 \times 10^{-2} M_w^{0.82} \ (\text{cm}^3\text{g}^{-1}) \qquad \text{in DMAc at 25 °C} \qquad (3.11.11)$$

Obviously, $[\eta]$ is highest in DMAc and lowest in THF. The exponent a in the MHS equation increases in the order: acetone $<$ THF $<$ DMAc.

Table 3.6.1 summarizes the viscosity data on solutions of five CDA (DS 2.46) fractions in DMAc, TFA, and acetone at 25 °C.

The following MHS equations were established from these $[\eta]$ and M_w in the table for CDA in DMAc and TFA, respectively,

$$[\eta] = 3.95 \times 10^{-2} \ M_w^{0.738} \qquad \text{in DMAc} \qquad (3.11.11')$$

and

$$[\eta] = 5.27 \times 10^{-2} \, M_{\mathrm{w}}^{0.696} \qquad \text{in TFA} \qquad (3.11.12)$$

As far as DMAc, TFA, and acetone are concerned, in a given solvent $[\eta]$ for CTA is about 20–30% smaller for CTA than that for CDA of the same M_{w}.

3.11.4   CA (DS 2.92)[7]

Only the solubility behavior of CTA in various solvents and their MHS equations (eq. (3.11.1)) have been the targets for studies in the past. In addition, all MHS equations proposed hitherto are less accurate and are still not completely established. For instance, the *Polymer Handbook* compiles nine MHS equations for 'CTA' solutions.[21] However, it is particularly noteworthy that all these equations (except for Flory *et al.*'s equation) are for CDA and not for CTA. During the course of our investigation, Nair *et al.*[22] carried out LS measurements on CTA fractions in a mixed solvent (DCM/methanol = 1:1 v/v) and viscosity measurements in TCM over a limited range of molecular weight (the ratio of the maximum to minimum weight average molecular weight, M_{w}, is only 1.72). Their results are far beyond the scope of the present discussion and, in particular, the values of the radius of gyration $\langle S^2 \rangle_z^{1/2}$ are extraordinarily great. Although advances have made in the elucidation of the molecular parameters for other cellulose derivatives, the main reason why the research of CTA

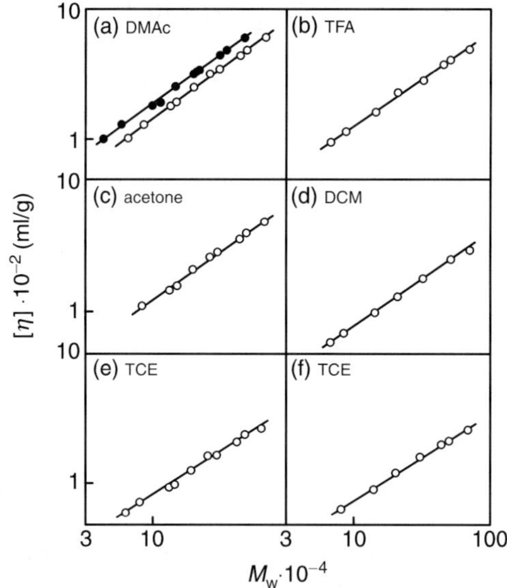

Figure 3.11.4   Log–log plot of the limiting viscosity number $[\eta]$ against the weight-average molecular weight M_{w} (open mark) or the number-average molecular weight M_{n} (closed mark) for cellulose triacetate in dimethylacetamide, trifluoroacetic acid, acetone, dichloromethane, tetrachloroethane, and trichloromethane.[7] Straight lines represent the MHS equations determined by the least squares method.

dilute solution has been delayed is due to the difficulties encountered in experiments on molecular weight fractionation. In fact, from a large number of studies of the molecular weight fractionation of CTA published since the mid-1930s, very few successful examples can be found.

Figure 3.11.4(a) shows the log–log plot of $[\eta]$ versus M_n or M_w for CTA solution in DMAc at 25 °C. In this figure, the closed circle denotes M_n and the open circle is M_w. Both plots can be accurately represented by straight lines, suggesting that the MHS equation,

$$[\eta] = K_m M_n^a \tag{3.11.1'}$$

or

$$[\eta] = K_m M_w^a \tag{3.11.1''}$$

may be well established over the entire M_w range investigated.

The parameters K_m and a of MHS equations for six solvents were evaluated using the least square method, and the MHS equations are represented in Figure 3.11.4(b–f) by straight lines through the observed points. The MHS equations of Kamide $et\ al.$ are the most comprehensive correlations made to date between $[\eta]$ and the molecular weight.

Table 3.11.1

Mark–Houwink–Sakurada equations for cellulose triacetate in various solvents

Solvent	Temp °C	K_m $\times 10^2$	a	Number of sample		Molecular weight range $M \times 10^{-4}$	Method	Acetyl content %	Reference
				Fr.	W.P.				
DCM[a]/	25	1.41	0.834	24	–	2.15–20.4	vis[f]	60.9	24
ethanol	25	0.45	0.90	–	5	3.06–18.0	vis[g]	62	25, 26
(8:2 v/v)									
TCM[b]	25	2.51	1.02	–	8		MO[h]	61.0	27
	25	4.54	0.649	7	–	8.22–69.0	LS[i]	61.0	23
	20	0.22	0.95	13	–	1.36–13.0	MP	61.0	22
DCM	20	2.47	0.704	7	–	6.36–69.0	LS	61.0	23
DMAc[c]	25	2.64	0.750	10	–	6.36–69.0	LS	61.0	23
TFA[d]	25	3.96	0.706	8	–	6.36–69.0	LS	61.0	23
Acetone	25	2.89	0.725	9	–	8.22–69.0	LS	61.0	23
TCE[e]	25	3.93	0.662	10	–	6.36–69.0	LS	61.0	23

[a]Dichloromethane.

[b]Trichloromethane.

[c]Dimethylacetamide.

[d]Trifluoroacetic acid.

[e]Tetrachloroethane.

[f]Degree of acetylation was calculated from the cuprammonium limiting viscosity number of diacetylated fractions: $[\eta] = 0.0319 \times P_n^{0.657}$ (P_n, the number-average degree of polymerization determined for cellulose diacetate by osmometry).

[g]M was calculated from the relationship for cellulose diacetate in acetone at 25 °C: $[\eta] = 8.97 \times 10^{-3} M_n^{0.90}$, which is given by Phillips and Bjork.[26]

[h]Membrane osmometry.

[i]Light scattering.

In general, $[\eta]$ values in CTA solution in chlorinated hydrocarbons are 50% smaller than those of the same molecular weight in other solvents.

The MHS equation corresponding to monodisperse CTA in DMAc can be readily derived from $[\eta]$ and M_w of CTA fractions by taking into account the MWD of the fractions as

$$[\eta] = 2.71 \times 10^{-2} M^{0.750} \qquad \text{in DMAc at 25 °C} \qquad (3.11.13)$$

In deriving eq. (3.11.13), we assumed the Schulz–Zimm distribution with $M_n/M_w = 1.4$ for the samples. Almost the same equation as eq. (3.11.13) can also be derived from $[\eta]$ and M_n of the fractions.

Table 3.11.1[11,23] lists the MHS equations for CTA established in this work, together with those proposed hitherto. The exponent a values reported in the literature for the systems studied here are generally in the range of 0.8–1.0, and are markedly larger than our values (0.65–0.75). In particular, the systematic molecular weight dependence of MWD of the samples will yield erroneously high a values. Detailed discussions of the effect of MWD on MHS equation are presented below.[23]

3.11.5 Effect of polymolecularity of samples on MHS parameters[23]

No reliable MHS equation for a CA polymer has yet been established.[7] In this section, we intend to determine the MHS equations for CTA (combined acetic acid content = 61.0 wt%) in various solvents, using fractions with reasonably narrow MWD.

The parameters K_m and a in eq. (3.11.1) were evaluated from the log–log plot of $[\eta]$ and M_w. Figure 3.11.5a and b shows similar plots ($[\eta]$ versus M_n) for DCM and TCM as open circles. It should be noted that the MHS parameters, reported in the literature for this polymer (see Table 3.11.1), were determined for polymer fractions whose MWD is wide, on the basis of M_n or the viscosity average molecular weight M_v. As shown in Table 3.11.1, serious differences in parameters in the MHS equations are not all clear. However, these differences probably originate from the significant molecular weight dependence of the polydispersity of the samples used in the literature, as was demonstrated before with CDA solutions.[10] We now turn our attention to exploring this point in more detail.

In Figures 3.11.5(a) and (b), the straight lines shown are the MHS equations corresponding to the samples with M_w/M_n values as denoted on the lines. The equations for the various M_w/M_n values were calculated from the MHS equations for the monodisperse samples using polymolecularity correction factors according to Schulz–Zimm MWD.[21] In Figure 3.11.5(a), the closed circles denote the data of Dymarchuk et al.[28] and the open triangles are the data of Shakhparonov et al.[29] Dymarchuk et al. employed fractions isolated by successive precipitational fractionation (SPF; total number of fractionations 13, but the initial concentration is not given), using the DCM/methanol system. As judged from the frequent occurrence of reverse order fractionations in their SPF run,[30,31] the MWD of the fractions might be relatively broad and they increase with M_w. Dymarchuk et al. calculated the molecular weight M by substituting the experimental $[\eta]$ value, which was utilized for establishing the MHS equation, into the MHS equation. Moreover, they regarded the M value calculated in this manner as M_w and insisted that since the ratio of M_w to the experimental M_n value is very near to unity, the polymer fractions

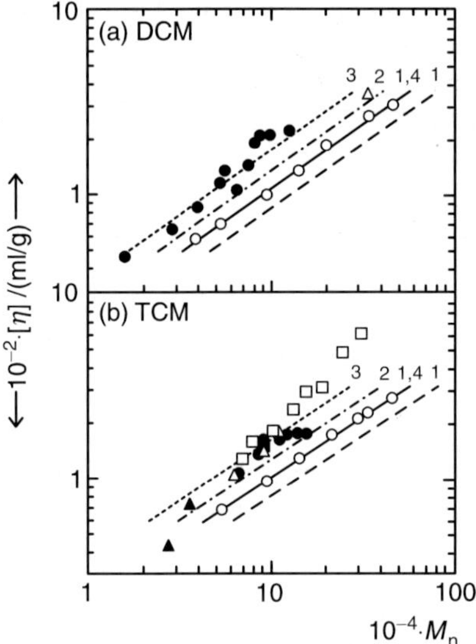

Figure 3.11.5 Double logarithmic plot of the limiting viscosity number $[\eta]$ and the number-average molecular weight M_n for cellulose triacetate solution.[23] For the meaning of the series of straight lines, see the text. (a) dichloromethane: ($\bigcirc$), this work;[23] ($\bullet$), data from Dymarchuk *et al.*;[28] ($\triangle$), data from Shakhpararonov *et al.*[29] (b) trichlorometane; ($\bigcirc$), this work; ($\triangle$), data from Staudinger and Eicher;[32] ($\square$), data from Sharples and Major;[27] ($\blacktriangle$), data from Howard and Parih;[33] ($\bullet$), data from Nair *et al.*[22]

used should be almost monodisperse. Clearly, the above calculation only exemplifies the coincidence of the MHS equation in question and the experimental data (i.e. the linearity of the data points) and M calculated is M_n, and no information pertaining to the samples polydispersity can be obtained. If we assume $M_w/M_n \approx 3$ for the samples of Dymarchuk *et al.*, then their data points are consistent with those in this work. In Figure 3.11.5b, the open triangles are from Staudinger and Eicher's work,[32] the open rectangles are the data of Sharples and Major (combined acetic acid content (AC) = 61 wt%),[27] the closed triangles are based on Howard and Parikh's (AC = 62.2 wt%),[33] and the closed circles are based on the work of Nair *et al.*[22] All experimental points lie to the left of Kamide *et al.*'s plot. Staudinger and Eicher prepared CTA by acetylation after saponification of CDA fractions, isolated by SPF with an acetone/water system. Since the MWD of these samples is not considered to be sharp, a reasonable assumption of $M_w/M_n = 2$–3 for these fractions makes their data comparable with ours. Sharples and Major used CTA by acetylation of cotton, which was prepared by hydrolysis for various times in 2 N hydrochloric acid at 60 °C.[27] Therefore, it is probable that the breadth of MWD will change systematically with M_w from $M_w/M_n = 4$–6 (e.g. unfractionated CTA (TA2) has $M_w/M_n = 4.02$).[7] The data points of Howard and Parikh[33] are similar to those in the work of Sharples and Major.[27] The two highest fractions from the work of Howard and Parikh are the first fractions, whose M_n values are similar to that of the whole polymer (i.e. the ratios of M_n of the first fraction to that of the whole polymer are only 1.71 and 1.31, respectively).

This indicates that these first fractions have a relatively large MWD. Nair *et al.* prepared CTA fractions with $M_w/M_n \approx 1.1$ by employing the selective absorption method.[22] Unexpectedly, their viscosity data points, which were based on very narrow MWD fractions, deviate remarkably from the present results. As yet, there is no adequate answer to the divergences in the viscosity behavior between the work of Nair *et al.* and Kamide *et al.* However, extremely small values of the Flory viscosity parameter Φ, calculated from the Nair *et al.* data (as compared with those for other macromolecules including a wide variety of cellulose derivatives) is a clear indication that their $[\eta]$, M_w and the weight-average radius of gyration $\langle S^2 \rangle_w^{1/2}$ data are not consistent with each other.

In short, all existing data on $[\eta]$ and M_n indicate a significant departure from the present data and the difference can be explained (with the exception of the results of Nair *et al.*[22]) by the wide MWD of samples employed. In particular, the systematic molecular weight dependence of the MWD of the samples will yield incorrectly high a values.

The MHS equation corresponding to monodisperse CTA in solvents can be readily derived from the equations obtained here (see Table 3.11.1) by acknowledging the MWD of the fractions as

$$[\eta] = 2.71 \times 10^{-2} M^{0.750} \qquad \text{in DMAc at 25 °C} \tag{3.11.13}$$

$$[\eta] = 3.19 \times 10^{-2} M^{0.706} \qquad \text{in TFA at 25 °C} \tag{3.11.14}$$

Table 3.11.2

MHS parameters for cellulose acetate solutions

DS	Solvent	Temp/°C	$K_m \times 10^2$	a	Number of fractions	Molecular weight range (10^5)	Method	Reference
0.49	DMAc[a]	25	19.1	0.60	10	4.55–14.5	LS	6
	DMSO[b]	25	17.1	0.61	10	4.55–14.5	LS	6
	water	25	20.9	0.61	10	4.55–14.5	LS	6
	FA[c]	25	20.9	0.60	10	4.55–14.5	LS	6
1.75	DMAc	25	9.58	0.60	7	3.75–13.1	LS	9
2.46	Acetone	25	13.3	0.61_6	9	6.1–26.5	LS	10
	THF[d]	25	5.13	0.68_8	6	7.4–30.0	LS	10
	TFA[e]	25	5.27	0.69_6	5	6.1–18.5	LS	10
	DMAc	25	3.95	0.73_8	5	6.1–18.5	LS	8
			1.34	0.82	10	5.3–27.0	LS	8
2.92	DCM[f]	20	2.47	0.704	7	6.36–69.0	LS	7
	DMAc	25	2.64	0.750	10	6.36–69.0	LS	7
	TFA	25	3.96	0.706	8	6.36–69.0	LS	7
	Acetone	25	2.89	0.725	9	8.22–69.0	LS	7
	TCE[g]	25	3.93	0.662	10	6.36–69.0	LS	7

[a]Dimethylacetamide.
[b]Dimethylsulfoxide.
[c]Formamide.
[d]Tetrahydrofuran.
[e]Trifluoroacetic acid.
[f]Dichloromethane.
[g]Tetrachloroethane.

$$[\eta] = 2.31 \times 10^{-2} M^{0.725} \qquad \text{in acetone at 25 °C} \qquad (3.11.15)$$

$$[\eta] = 2.02 \times 10^{-2} M^{0.704} \qquad \text{in DCM at 25 °C} \qquad (3.11.16)$$

$$[\eta] = 3.25 \times 10^{-2} M^{0.662} \qquad \text{in TCE at 25 °C} \qquad (3.11.17)$$

and

$$[\eta] = 3.76 \times 10^{-2} M^{0.649} \qquad \text{in TCM at 25 °C} \qquad (3.11.18)$$

In deriving eqs. (3.11.13–3.11.18), we assumed the Schulz–Zimm distribution with $M_w/M_n = 1.4$ for the samples. From eqs. (3.11.13–3.11.18), we can calculate the viscosity average molecular weight M_v of a CTA sample from $[\eta]$ data, irrespective of its MWD.

Table 3.11.2 displays the MHS parameters for CA solutions.

3.11.6 Cellulose/aq. LiOH[34]

Brown and Wikström[35] obtained the parameters K_m and a in the MHS relationship for cellulose in cadoxen, analyzing their own and Henley's[36] data, to be 3.85×10^{-2} cm^3 g^{-1} and 0.76, respectively. Using these data, we estimated the viscosity-average molecular weight M_v of samples SA-1 and -5 from $[\eta]$ in cadoxen as compiled in column 6 of Table 3.5.6. M_v of the two samples thus estimated was in good agreement with M_w directly determined in our laboratory by the LS method in cadoxen within experimental uncertainty. Henley observed that two cellulose samples ($M_w = 2.9 \times 10^5$ and 2.25×10^5), both prepared by hydrolysis of cotton linter ($M_w = 9.45 \times 10^5$) with 2 N HCl, had M_w/M_n of approximately 2. Brown and Wikström[35] reported that M_w/M_n of acid (2–2.5 N H$_2$SO$_4$) hydrolyzed cotton linters with M_w of 3.36×10^4–1.65×10^4 was about 2. These experimental facts indicate that acid hydrolysis of cotton linter proceeds according to random scission of the polymer chains

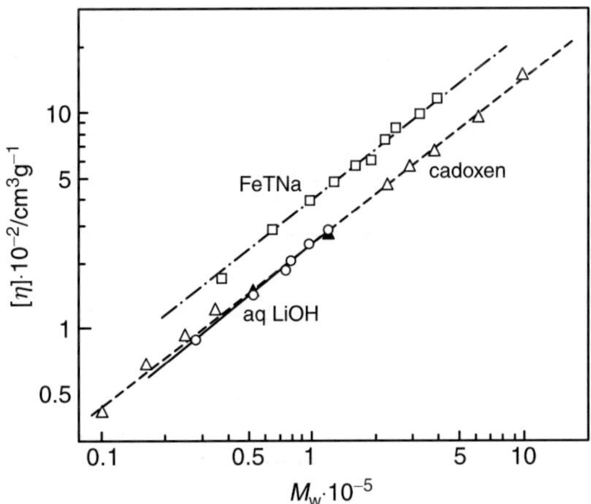

Figure 3.11.6 Log–log plots of the limiting viscosity number versus the weight-average molecular weight M_w of cellulose in 6 wt% aq. LiOH (○),[33] cadoxen (△)[36] and iron sodium tartrate (□)[36] at 25 °C; (▲) Kamide–Saito;[33] (△) Henley, Brown, and Wikström.[35]

Table 3.11.3

Parameters for the Mark–Houwink–Sakurada equation, $[\eta] = K_m M^a$, as determined for cellulose in several alkaline solvents

Solvent	$K_m \times 10^2$ cm^3 g^{-1}	a	Method for determination of M	Reference
6-wt% aq. LiOH	2.7_8	0.79	LS[a]	34
Cardoxen	3.8^5	0.76	LS	35
	5.5_1	0.75	SD[b]	36
FeTNa[c]	5.3_1	0.78	LS	37
Cuoxam[d]	0.70	0.9	SD	36
Cuen[e]	1.0_1	0.9	SD	36
EWNN[f]	0.38_7	1.0_1	SD	36

[a]Light scattering.
[b]Sedimentation diffusion.
[c]Iron sodium tartrate.
[d]Cuprammonium hydroxide.
[e]Cupriethylenediamine hydroxide.
[f]Eisen–Weisäure–Natrium Komlex (iron tartaric acid sodium complex solution).

and if it proceeds to a greater extent, then the products have $M_w/M_n \approx 2$. Hence, all the samples employed here can be regarded to have the same polymolecularity as those of Henley and Brown–Wikström.

Figure 3.11.6 shows the plot of $[\eta]$ against M_w for cellulose in aq. LiOH. The figure contains also the literature data on cellulose in cadoxen[35,36] and in iron sodium tartrate (FeTNa).[37] $[\eta]$ in FeTNa is the largest among those in three solvents.

Table 3.11.3 lists the parameters K_m and a in eq. (3.11.1) established for cellulose/aq. LiOH systems at 25 °C and the corresponding literature data for other solvents.[35–37]

The exponent a for aq. LiOH, cadoxen, and FeTNa lies between 0.76 and 0.79 and a for cuprammonium hydroxide (Cuoxam), cupriethylene diamine hydroxide (Cuen), and EWNN (Eisen–Weinsäure–Natrium Komplex, discovered by Jayme and Verburg[38]) is larger than 0.9. For the latter solvent groups, the sedimentation-diffusion average molecular weight M_{SD} were employed in place of M_w. Theoretically, the MHS equations obtained using M_{SD} are expected to be sensitive to the polymolecularity of the polymer samples. As such, it should be necessary to draw conclusions about the flexibility of cellulose in these solvents in advance in order to confirm that the samples can be regarded as monomolecular or that the molecular weight dependence of the polymolecularity of the sample is small enough to be neglected. As is well known, a exceeds 0.9 for CN (total degree of substitution $\langle\langle F \rangle\rangle$, 2.91 and 2.55) acetone, CTA, and dioxane systems.[39] These cellulose derivatives are semiflexible polymers.

REFERENCES

1. K Kamide and T Dobashi, *Physical Chemistry of Polymer Solutions*, Problems 8–30, eq. 8.30.19, 2000.
2. W Kuhn, *Kolloid-Z.*, 1934, **68**, 9; *Angew. Chem.*, 1936, **49**, 860.

3. H Mark, *Der feste Korper*, Leipzig, 1938, 103.
4. R Howink, *J. prakt. Chem.*, 1941, **157**, 15.
5. I Sakurada, N Kagakusenni Kenkyusho Kenshu. *Proc. Symp. Jpn Text. Res. Lab.*, 1940, **5**, 33.
6. K Kamide, M Saito and T Abe, *Polym. J.*, 1981, **13**, 421.
7. K Kamide, Y Miyazaki and T Abe, *Polym. J.*, 1979, **11**, 523.
8. K Kamide and M Saito, *Polym. J.*, 1982, **14**, 517.
9. M Saito, *Polym. J.*, 1983, **15**, 249.
10. K Kamide, T Terakawa and Y Miyazaki, *Polym. J.*, 1979, **11**, 285.
11. RS Stein and P Doty, *J. Am. Chem. Soc.*, 1946, **68**, 159.
12. WJ Badgley and H Mark, *J. Phys. Chem.*, 1947, **51**, 58.
13. DW Tanner and GC Berry, *J. Polym. Sci., Polym. Phys.*, 1974, **12**, 941.
14. EO Kraemer, *Ind. Eng. Chem.*, 1938, **30**, 1200.
15. A Bartovics and H Mark, *J. Am. Chem. Soc.*, 1943, **65**, 1901.
16. AM Sookne and M Harris, *Ind. Eng. Chem.*, 1945, **37**, 475.
17. HJ Phillip and CF Bjork, *J. Polym. Sci.*, 1951, **6**, 549.
18. RJE Cumberbirch and WG Harland, *J. Text. Inst.*, 1958, **49**, T664.
19. WR Moore and BM Tidswell, *J. Appl. Chem.*, 1958, **8**, 232.
20. T Ikeda and H Kawaguchi, *Rep. Prog. Polym. Sci. Jpn.*, 1966, **9**, 23.
21. M Kurata, Y Tsunashima, M Iwata and K Kamata, in *Polymer Handbook*, 2nd Edn., (eds J Brandrup and EH Immergut), Wiley, New York, 1975.
22. PRM Nair, RM Gohil, KC Patal and RD Patel, *Eur. Polym. J.*, 1977, **13**, 273.
23. K Kamide, Y Miyazaki and T Abe, *Makromol. Chem.*, 1979, **180**, 2801.
24. RJE Cumberbirch and WG Harland, *J. Text. Inst.*, 1958, **49**, T679.
25. PJ Flory, OK Spurr Jr., and DK Carpenter, *J. Polym. Sci.*, 1958, **27**, 231.
26. HJ Phillip and CF Bjork, *J. Polym. Sci.*, 1951, **6**, 549.
27. A Sharples and HM Major, *J. Polym. Sci.*, 1958, **27**, 433.
28. NP Dymarchuk, KP Mishchenko and TV Fomia, *Zhur. Prikl. Khim (Leningrad)*, 1964, **37**, 2263.
29. MI Shakhparonov, NP Zahurdayeva and YeK Podgarodetskii, *Vysokomol. Soedin., Ser. A.*, 1967, **9**, 1212.
30. K Kamide and Y Miyazaki, *Makromol. Chem.*, 1975, **176**, 2393.
31. K Kamide, In Reverse-Order Fractionation, *Thermodynamics of Polymer Solutions: Phase Equilibria and Critical Phenomena*, Elsevier, Amsterdam, 1990, **2.43**, p. 79.
32. H Staudinger and T Eicher, *Makromol. Chem.*, 1953, **10**, 261.
33. P Howard and SS Parikh, *J. Polym. Sci., Part A-1*, 1966, **4**, 407.
34. K Kamide and M Saito, *Polym. J.*, 1986, **18**, 569.
35. W Brown and R Wikström, *Eur. Polym. J.*, 1965, **1**, 1.
36. D Henley, *Ark. Kemi.*, 1961, **18**, 327.
37. L Valtassari, *Makromol. Chem.*, 1971, **150**, 117.
38. G Jayme and W Verburg, *Rayon, Zellwolle, Chemifasern*, 1954, **32**, 193 See also p. 275.
39. K Kamide and Y Miyazaki, *Polym. J.*, 1978, **10**, 409.

3.12 MOLECULAR WEIGHT DEPENDENCE OF RADIUS OF GYRATION

The molecular weight dependence of the radius of gyration $\langle S^2 \rangle_z$ is empirically expressed as

$$\langle S^2 \rangle_z^{1/2} = K_\lambda M_w^{(\lambda+1)/2} \tag{3.12.1}$$

where K_λ and λ are parameters characteristic of a polymer solvent combinations.

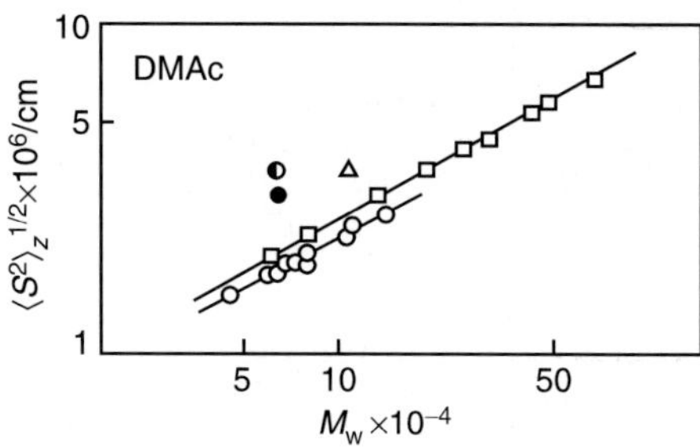

Figure 3.12.1   Molecular weight dependence of the radius of gyration $\langle S^2 \rangle_z^{1/2}$ for cellulose acetate in dimethylacetamide at 25 °C: (○), cellulose acetate (DS 0.49);[1] (△), cellulose acetate (DS 2.46);[3] (□), cellulose acetate (DS 2.92);[2] (●), cellulose acetate (DS 0.49) in water;[1] (◑), cellulose acetate (DS 0.49) in formamide.[1]

3.12.1   CA (DS 0.49)/DMAc[1]

The molecular weight dependence of the z-average radius of gyration $\langle S^2 \rangle_z^{1/2}$ for CA (DS 0.49) in DMAc is shown by open circles in Figure 3.12.1. In this figure, the data points for CA (DS 2.92)[2] and CA (DS 2.46)[3] are also included as open rectangles and an open triangles for comparison. Among these three kinds of CA with different DS, CA (DS 0.49) has the most compact form in DMAc.

$\langle S^2 \rangle_z^{1/2}$ is related to M_w by

$$\langle S^2 \rangle_z^{1/2} = 0.50 \times 10^{-8} M_w^{0.52} (\text{cm}) \quad \text{for CA(DS 0.49) in DMAc at 25 °C} \qquad (3.12.2)$$

It is interesting that $\langle S^2 \rangle_z^{1/2}$ of the CA (DS 0.49) fraction in water solution (closed circle in Figure 3.12.1) is approximately twice the value in DMAc. The CA (DS 0.49) polymer expands in the following order: DMAc < water < formamide.

3.12.2   CA (DS 1.75)/DMAc[4]

The relationship $\langle S^2 \rangle_z^{1/2}$ and M_w for the CA (DS 1.75) fractions in DMAc is represented by

$$\langle S^2 \rangle_z^{1/2} = 0.38 \times 10^{-8} M_w^{0.52} (\text{cm}) \qquad (3.12.3)$$

Compared at the same M_w, the $\langle S^2 \rangle_z^{1/2}$ value for the CA (2.46)/DMAc system is larger than those of other CA/DMAc system in the range of $M_w \leq 10^6$. This suggests that CA molecule in DMAc becomes most rigid at DS = 2.46.

3.12.3   CA (DS 2.46)/acetone, THF, DMAc[3,5]

Figure 3.12.2 displays the molecular weight dependence of $\langle S^2 \rangle_z^{1/2}$ for CDA in acetone and THF. In the figure, the data of Stein and Doty[6] on CDA in acetone and of Tanner and

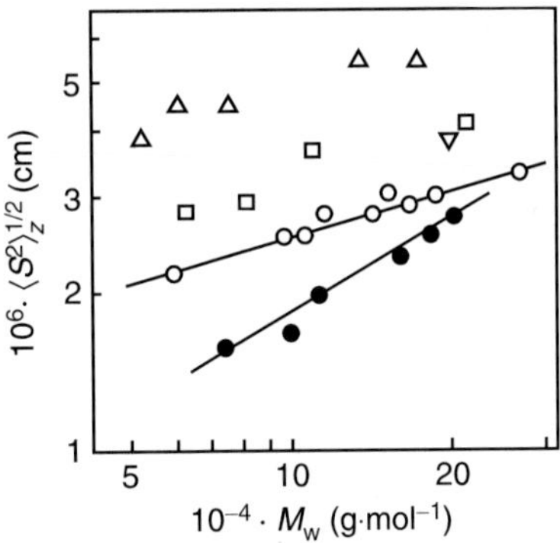

Figure 3.12.2   Molecular weight dependence of the radius of gyration $\langle S^2 \rangle_z^{1/2}$ for cellulose diacetate solution:[3] ($\bigcirc$), acetone;[5] ($\triangle$), acetone (Stein and Doty);[6] ($\square$), trifluoroethanol (Tanner and Berry);[7] ($\bullet$), tetrahydrofuran;[5] ($\blacktriangledown$), tetrahydrofuran (Tanner and Berry).[7]

Berry[7] on CDA in THF at 25 °C and TFE at 20 °C are also included for the sake of comparison. The molecular weight dependence of $\langle S^2 \rangle_z^{1/2}$ of CDA in acetone from Stein and Doty's work is similar to that obtained by Kamide *et al.* and the absolute magnitude of $[\eta]$ in acetone is almost the same as that in Kamide *et al.*'s paper[3] (see Figure 3.11.2) compared at the same molecular weight. However, the absolute magnitude of $\langle S^2 \rangle_z^{1/2}$ by Stein and Doty is about two times larger than Kamide *et al.*'s. The accuracy of Stein and Doty's work was pointed out by Doty *et al.*[8] as dubious due to the colloidal contaminations.

The value of $\langle S^2 \rangle_z^{1/2}$ of CDA in THF from Tanner and Berry's work[7] (fraction 3;2) is about 50% larger than that obtained here of the polymer with the same M_w, but $[\eta]$ value of the fraction is about 20% smaller than $[\eta]$ calculated from eq. (3.11.9) $\langle S^2 \rangle_z^{1/2}$ of CA (DS 2.46) is related to M_w by the relationship:

$$\langle S^2 \rangle_z^{1/2} = 7.39 \times 10^{-8} M_w^{0.308} \text{(cm)} \quad \text{in acetone at 25 °C} \tag{3.12.4}$$

$$\langle S^2 \rangle_z^{1/2} = 2.99 \times 10^{-9} M_w^{0.558} \text{(cm)} \quad \text{in THF at 25 °C} \tag{3.12.5}$$

and

$$\langle S^2 \rangle_z^{1/2} = 0.68 \times 10^{-8} M_w^{0.53} \text{(cm)} \quad \text{in DMAc at 25 °C} \tag{3.12.6}$$

3.12.4   CA (2.92)/DMAc[2]

Figure 3.12.3 shows the radius of gyration $\langle S^2 \rangle_z^{1/2}$ plotted against the molecular weight for CTA solution in DMAc as open circle. From this figure, we obtain the following empirical relationship.

$$\langle S^2 \rangle_z^{1/2} = 0.46 \times 10^{-8} M_w^{0.55} \text{(cm)} \quad \text{in DMAc at 25 °C} \tag{3.12.7}$$

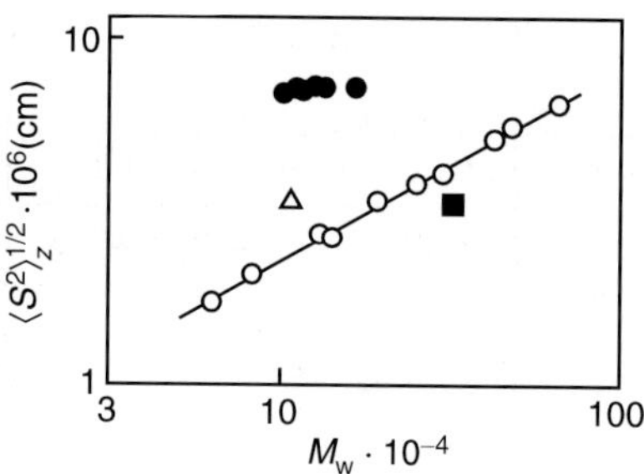

Figure 3.12.3   Molecular weight dependence of the radius of gyration $\langle S^2 \rangle_z^{1/2}$ for cellulose triacetate and cellulose diacetate solutions: (O), cellulose triacetate in dimethylacetamide (Kamide et al.[2]); (●), cellulose triacetate in dichloromethane/methanol (1:1 v/v; Nair et al.[9]); (■), cellulose triacetate in dichloromethane (Shakhparonov et al.[10]); (△), cellulose diacetate in dimethylacetamide (Kamide et al.[2]).

Figure 3.12.3 also includes the data from Nair et al.'s work[9] on CTA in a mixed solvent (DCM/methanol, 1:1 v/v) as a closed circle, together with the point by Shakhparonov et al.[10] on CTA in DCM (temperature was not described) as a closed rectangle. Indeed, Nair et al.'s value of $\langle S^2 \rangle_z^{1/2}$ is larger than the value obtained by Kamide et al. for DMAc by a factor of about 3.5, compared at the same M_w. Thus, it is obvious that Nair et al.'s data of $\langle S^2 \rangle_z^{1/2}$ are the largest among all data available for cellulose derivatives.[11] The data points obtained by Shakhparonov et al.[10] for CTA in TCM are slightly smaller than those by Kamide et al. in DMAc. It is expected from the $[\eta]$ data in Table 3.6.3 that $\langle S^2 \rangle_z^{1/2}$ is smaller in TCM than in DMAc. An open triangle in the figure corresponds to CDA in DMAc.

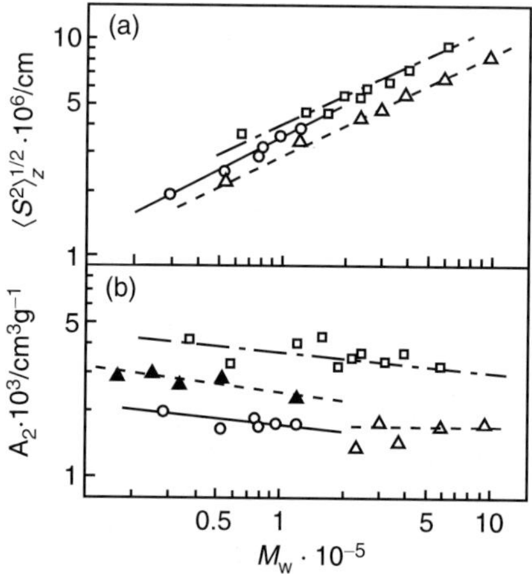

Figure 3.12.4   Molecular weight dependence of the z-average radius of gyration $\langle S^2 \rangle_z^{1/2}$ and the second virial coefficient A_2 of cellulose in 6 wt% aq. LiOH (O);[2] cadoxen (△;[13] half black triangle,[14] ▲[2]); and iron sodium tartrate (□).[15]

Table 3.12.1

Molecular weight dependence parameters K_λ and λ of the radius of gyration of cellulose acetate and cellulose in solutions at 25 °C

DS	Solvent	$K_\lambda(\times 10^8)$	$(\lambda + 1)/2$	λ	Reference
0.49	DMAc	0.50	0.52	0.04	1
1.75	DMAc	0.38	0.52	0.04	4
	Acetone	7.39	0.308	-0.384	3
2.46	THF	0.2999	0.558	0.116	3
	DMAc	0.68	0.53	0.06	5
2.92	DMAc	0.46	0.55	0.10	2
0.00	Aq. LiOH	0.91	0.52	0.04	12

3.12.5 Cellulose/aq. LiOH[12]

Figure 3.12.4 shows the molecular weight dependence of $\langle S^2 \rangle_z^{1/2}$ and A_2 of cellulose in aq. LiOH and cadoxen. The existing data[13–15] for cadoxen and FeTNa are also shown in the figure. Here, it should be noted that Brown and Wikström[14] did not determine $\langle S^2 \rangle_z^{1/2}$ because no angular dependence of the reduced scattering intensity was observed within experimental uncertainty. $\langle S^2 \rangle_z^{1/2}$ of cellulose in aq. LiOH can be expressed as

$$\langle S^2 \rangle_z^{1/2} = 9.1 \times 10^{-9} M_{\mathrm{w}}^{0.52}(\mathrm{cm}) \quad \text{in DMAc at 25 °C} \tag{3.12.8}$$

$\langle S^2 \rangle_z^{1/2}$ in cadoxen and FeTNa is proportional to $M^{0.5}$. Kamide *et al.*'s $\langle S^2 \rangle_z^{1/2}$ data for cadoxen are slightly larger than the values interpolated from the Henley's data.[13]

Table 3.12.1 summarizes the molecular dependence parameters of the radius of gyration $\langle S^2 \rangle_z^{1/2}$ of CA (DA = 0.49, 1.75, 2.46, and 2.92) and cellulose in solutions.

REFERENCES

1. K Kamide, M Saito and T Abe, *Polym. J.*, 1981, **13**, 421.
2. K Kamide, Y Miyazaki and T Abe, *Polym. J.*, 1979, **11**, 523.
3. K Kamide and M Saito, *Polym. J.*, 1982, **14**, 517.
4. M Saito, *Polym. J.*, 1983, **15**, 249.
5. K Kamide, T Terakawa and Y Miyazaki, *Polym. J.*, 1979, **11**, 285.
6. RS Stein and P Doty, *J. Am. Chem. Soc.*, 1946, **68**, 159.
7. DW Tanner and GC Berry, *J. Polym. Sci. Polym. Phys.*, 1974, **12**, 941.
8. P Doty, N Schneider and A Holtzer, *J. Am. Chem. Soc.*, 1953, **75**, 754.
9. PRM Nair, RM Gohil, KC Patal and RD Patel, *Eur. Polym. J.*, 1977, **13**, 273.
10. MI Shakhparonov, NP Zahurdayeva and YeK Podgarodetskii, *Vysokomol. Soedin. Ser. A.*, 1967, **9**, 1212.
11. K Kamide and Y Miyazaki, *Polym. J.*, 1978, **10**, 409.
12. K Kamide and M Saito, *Polym. J.*, 1986, **18**, 569.
13. D Henley, *Ark. Kemi*, 1961, **18**, 327.
14. W Brown and R Wikström, *Eur. Polym. J.*, 1965, **1**, 1.
15. L Valtassari, *Makromol. Chem.*, 1971, **150**, 117.

3.13   SECOND VIRIAL COEFFICIENT AND EXCLUDED VOLUME EFFECT

3.13.1   Second virial coefficient

CA (DS 0.49)[1]

The second virial coefficient A_2, determined for CA (DS 0.49) in DMAc at 25 °C by LS is collected in the fifth column of Table 3.3.1. There is too much scatter around 1.1×10^{-3} (cm^3 g^{-2} mol) to allow a definite conclusion as to its molecular weight dependence. The A_2 values of the CA (DS 0.49) fraction for the three solvents, shown in Table 3.5.1, are of the order of 10^{-4} cm^3 mol g^{-2}, thus showing no systematic variation in the solvent.

CA (DS 1.75)[2]

The data A_2 values in DMAc solution by LS and MO methods are collected in columns 7 and 5 of Table 3.3.2.

CA (DS 2.46)[3,4]

Data on A_2 are shown in columns 3 (DMAc; LS), 6 (acetone; LS), 9 (THF; LS), and 11 (THF; MO) in Table 3.3.3. The molecular weight dependence of A_2, as determined by LS,

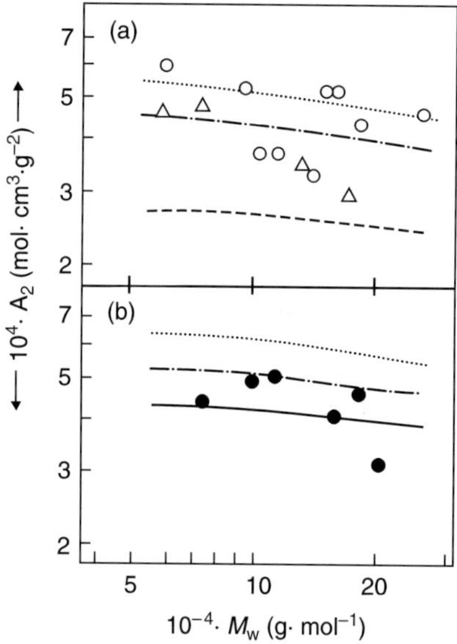

Figure 3.13.1   Molecular weight dependence of the second virial coefficient A_2 for cellulose diacetate solutions in acetone (a) and in tetrahydofuran (b) at 25 °C:[4] (O), cellulose diacetate (DS 2.45) in acetone;[4] (Δ), cellulose diacetone (DS 2.45) in acetone;[5] ($\bullet$), cellulose diacetate (DS 2.45) in tetrahydrofuran.[4] Lines are calculated by eqs. (3.13.1) and (3.13.5) from A (or K_0) and B, which are estimated by methods 2C (full line), 2D (broken line), 2E (dotted line), and 2G (chain line), and the experimental $\langle S^2 \rangle_w$ data.

is plotted in Figure 3.13.1, where data of CA (DS 2.43) in acetone from Stein and Doty's work[5] are also included as an open triangle. A_2 for CDA in acetone and in THF is of a magnitude in the order of 10^{-4} cm^3 mol g^{-2}, practically independent of M within a rather large experimental uncertainty, as found for various cellulose, amylose, and their derivatives. The value of A_2 as determined by LS in THF is somewhat (*ca.* 40–50%) larger than that by MO. Lines are calculated by eq. (3.13.16)[6–8] and from the short-range and long-range parameters, A (or K_0) and B, which are estimated by methods 2C (full line), 2D (broken line), 2E (dotted line), and 2G (chain line; see Section 3.16), and the experimental $\langle S^2 \rangle_w$ data.

$$A_2 = (N_A/2)Bh_0(z) \qquad (3.13.1)$$

(Here, N_A is Avogadoro Number, $h_0(z)$ is a function of the excluded volume parameter z, for which various theories have been presented). When $h_0(z)$ is given by eq. (3.13.5), A is given by eq. (3.13.9), and B is given by eq. (3.13.10). However, the results do not change significantly, even if other solution theories are used, because for cellulose and amylase, derivative z is in vicinity of zero.

Figure 3.13.2 illustrates also the plot of A_2 against M_w for CDA in acetone. In this figure, the open circles represent experimental values[4] and the broken and full lines represent the A_2 values calculated from A and B, which are evaluated by methods 2I and 2J (see Section 3.16), respectively. The calculated line based on method 2G is also included. Obviously, methods 2J and 2G can follow the experimental values of A_2 quite reasonably.[9]

CA (DS 2.92)/DMAc[10]

The values of A_2 by MO and LS are collected in columns 3 and 6 of Table 3.3.5.

Figure 3.13.3 shows the log–log plot of the second virial coefficient A_2 against M_w for CTA solution in DMAc at 25 °C. A_2 shows a small molecular weight dependence.

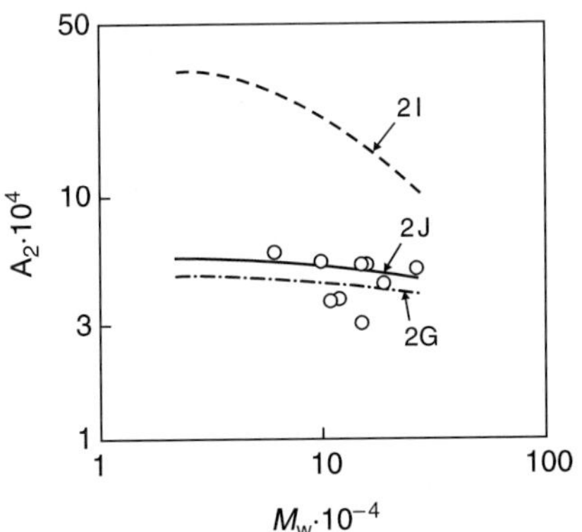

Figure 3.13.2   Molecular weight dependence of the second virial coefficient A_2 for cellulose diacetate solutions in acetone:[41] (O), experimental data point[4] by light scattering; chain, broken, and full lines; A_2 values calculated from A, B (by methods 2G, 2I, and 2J), and the experimental weight-average radius of gyration $\langle S^2 \rangle_w^{1/2}$.

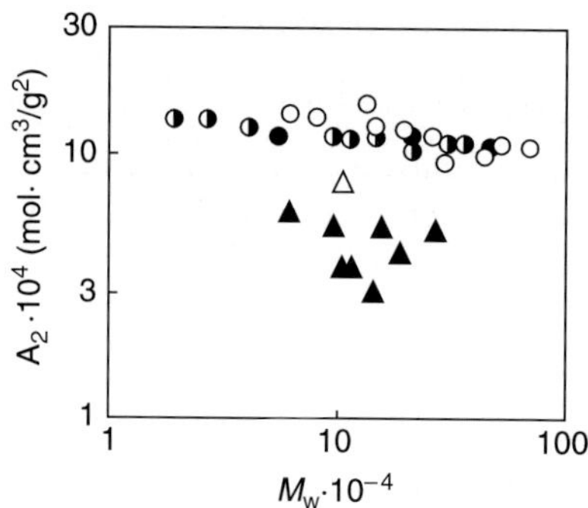

Figure 3.13.3   Molecular weight dependence of the second virial coefficient A_2 for cellulose triacetate and cellulose diacetate solutions:[10] ($\bigcirc$), cellulose triacetate in dimethylacetamide by light scattering; half black circle, cellulose triacetate in dimethylacetamide by membrane osmometry; ($\bullet$), cellulose triacetate in acetone by membrane osmometry; ($\triangle$), cellulose diacetate in dimethylacetamide by light scattering; ($\blacktriangle$), cellulose diacetate in acetone by light scattering.

In the lower M_w range, the second virial coefficient by LS $A_{2,L}$ is slightly larger than the second virial coefficient by membrane osmometry (MO) $A_{2,o}$. However, if we take into consideration the effect of MWD on A_2 and the experimental error involved, there is essentially no difference between $A_{2,L}$ and $A_{2,o}$. According to Casassa,[11] the polydispersity correction of the CTA samples reduces the observed $A_{2,L}$ only by about 1% (and $A_{2,o}$ by about 9%). $A_{2,o}$ is related experimentally to its M_w through the equation:

$$A_{2,o} = 2.64 \times 10^{-3} M_w^{-0.071} \qquad (\text{cm}^3 \text{ mol g}^{-2}) \qquad (3.13.2)$$

for a CTA solution in DMAc at 25 °C. In Figure 3.13.3, the corresponding data for CDA solutions in DMAc and acetone are plotted for the sake of comparison. The A_2 values for CTA in DMAc are more than two times the values for CDA in acetone.

The $A_{2,o}$ values of CTA in DMAc and acetone are similar in magnitude, and slightly larger than those in the chlorinated solvents. The polymer–solvent interaction parameter χ, calculated using eq. (3.3.2) (see column 10 of Table 3.3.5 and columns 10–13 of Table 3.3.6), for CTA is in the range of 0.29–0.38 in DMAc, acetone, TCE, and TCM (Tables 3.3.5 and 3.3.6). Howard and Parish obtained $\chi = 0.38$ for an unfractionated CTA sample ($M_w = 3.82 \times 10^4$) solution in TCM.[12] This value agrees fairly well with those evaluated here, if the MWD of the sample is taken into account. There is no obvious relationship between $[\eta]$ and χ (or A_2) of a given sample dissolved in different solvents.

Table 3.13.1 shows the magnitude of the MWD of A_2 of CA solutions.

Cellulose trinitrate (CTN)/CA[9]

The second virial coefficient A_2 has a large experimental uncertainty for cellulose, amylase, and their derivatives in solutions, for which A_2 is very small (in the order of 10^{-4} cm^3 mol g^{-2}), and does not exhibit any significant molecular weight dependence (except for CTN,[13,14] CN,[13,14] and CA[4]). Here, we examine the reliability of A_2.

Table 3.13.1

Molecular weight dependence of the second virial coefficient of cellulose
acetate and cellulose solutions at 25 °C

DS	Solvent	$\nu \equiv d \ln A_2 / d \ln M$	Reference
0.49	DMAc	–(too much scatter)	1
1.75	DMAc	<0	4
2.46	DMAc	-0	5
	Acetone	-0	3
	THF	-0	3
2.92	DMAc	-0.071	2
0.0 (cellulose)	Aq. LiOH	-0.08	12

When the values of the short- and long-range interaction parameters, A (eq. (3.13.9)) and
B (eq. (3.13.10)), are indirectly obtainable from plots of Baumann (method 2C), BKM
(method 2D), SF (method 2E), and KM (method 2G; see Section 3.16), we can calculate
A_2 from A, B, and the experimental $\langle S^2 \rangle_z^{1/2}$ value with the aid of eqs. (3.13.1) and (3.13.5).
The experimental value of A_2 can be compared with the calculated one thus obtained. As
is evident from Table 3.15.1, Penzel and Schulz's data[13,15] on CTN ($N = 13.9\%$) and CN
($N = 12.9\%$) are most reliable. The parameters A (or K_0) and B are determined from an
intercept at $M_w^{1/2} = 0$ and its slope of the above plots for CTN and CN in acetone (Figures
3.16.2–3.16.4, and 3.16.7). The A values thus obtained are listed in Table 3.16.1. From
these values, together with the experimental $\langle S^2 \rangle_z^{1/2}$ data, A_2 was calculated.

Figure 3.13.4 shows the smooth curves of the molecular weight dependence of A_2
calculated in this manner as full (method 2C), dotted (method 2E), broken (method 2D),
and chain (method 2G) lines.

The experimental data points are also included in this figure. There is a considerable
disagreement between the experimental points and the theoretical curves by method 2E.
However, the above disagreement should be remarkably improved by adopting methods
2C, 2D, and 2G in the case of CTN ($N = 13.9\%$). Since in the BKM and KM plots, the value
of a_2 (eq. (3.15.10)) (which was determined using A_2 value) was employed, the agreement
between the experimental A_2 and those by methods 2D and 2G does not strictly afford
direct verification of the reliability of the experimental A_2. However, it seems sufficient
to point out that not the absolute value, but the molecular weight dependence of A_2
contributes the a_2 value and the effect of a_2 on A_2 is very minor (see, Figure 10 in Ref. 9).

Cellulose/aq. LiOH[16]

Table 3.5.4 shows the experimental results of M_w and $\langle S^2 \rangle_z^{1/2}$ on cellulose/aq. LiOH
system. A_2 can be evaluated from apparent second virial coefficient A_2^* by[17]

$$A_2 = (M_w^* / M_w) A_2^* \tag{3.13.3}$$

as summarized in the sixth column of the table (M_w^* is apparent M_w).

In Figure 3.12.4, A_2 of cellulose in aq. LiOH, cadoxen and FeTNa lies roughly between
1.5 and 5×10^{-3} cm^3 g^{-1}. The log–log plots of A_2 and M_w of the data for the cellulose

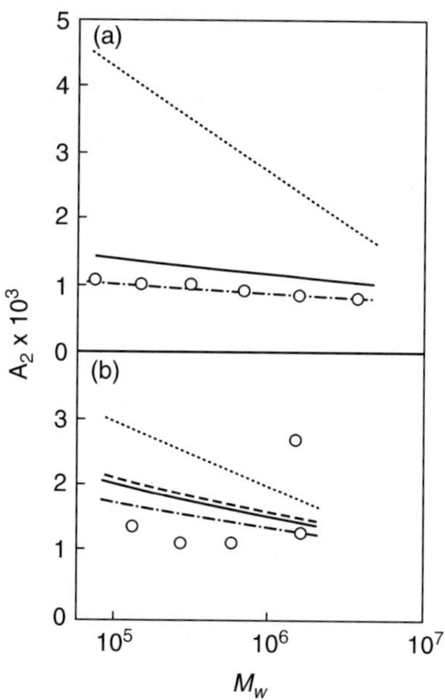

Figure 3.13.4   Second virial coefficient A_2 of cellulose trinitrate ($N = 13.9\%$) and cellulose nitrate ($N = 12.9\%$) in acetone:[13,14] open mark, experimental data. Lines are calculated by using eqs. (3.13.5) and (3.13.1) from A (or K_0) and B, which are estimated by methods 2C (full line), 2D (broken line), 3E (dotted line) band 2G (chain line), and the experimental $\langle S^2 \rangle_w$ data.

cadoxen system by Brown and Wikström ($M_w \leq 4 \times 10^4$)[18] and those by Henley ($M_w > 2 \times 10^5$)[19] make two different lines. The present data for the cellulose/cadoxen system are near the extrapolated line of Brown and Wikström data.[18] Except for Henley's data[19] (for which large experimental error was pointed out by the author himself), A_2 reveals a slightly negative molecular weight dependence; $\mathrm{d}\ln A_2/\mathrm{d}\ln M_w (\equiv \nu) = -0.08$ for aq. LiOH and FeTNa and -0.12 for cadoxen.

3.13.2   Excluded volume effect (Pearl necklace model)

The linear expansion factor a_s is defined by the relationship

$$\alpha_s = \langle S^2 \rangle^{1/2}/\langle S^2 \rangle_0^{1/2} \tag{3.13.4}$$

Here, $\langle S^2 \rangle_0^{1/2}$ is the radius of gyration at unperturbed state. Direct evaluation of $\langle S^2 \rangle_0^{1/2}$ by the LS measurement at the unperturbed state is often impossible due to the tendency for aggregation or crystallization of solute and thus indirect evaluation of α_s is necessary.

The excluded volume parameter α_s is estimated in the following manner. First, the penetration function ψ is defined by[20,21]

$$\psi \equiv \bar{z}h_0(\bar{z}) = A_2 M_w^2 / (\langle S^2 \rangle_w^{3/2} 4\pi^{3/2} N_A) = 0.746 \times 10^{-25} A_2 M_w^2 / \langle S^2 \rangle_w^{3/2} \tag{3.13.5}$$

ψ can be calculated using eq. (3.13.5) from the LS data (M_w, A_2, $\langle S^2 \rangle_w$). Generally, $\psi(z) \approx 0$ for $z \approx 0$ and converges to finite value for large z.[21]

According to Kurata–Fukatsu–Sotobayashi–Yamakawa,[22] ψ (accordingly, $h_0(\bar{z})$) is related to the excluded volume parameter $\bar{z}$ through the following equations:

$$\psi = \bar{z} h_0(\bar{z}) = (1/5.047)\{1 - (1 + 0.683\bar{z})^{-7.39}\} \qquad (3.13.6)^{[22,23]}$$

$$\psi = \bar{z} h_0(\bar{z}) = \frac{1 - \{1 + 3.903\bar{Z}\}^{-0.468}}{1.828} \qquad (3.13.7)^{[20]}$$

$\bar{z}$ is the excluded volume parameters given by[20,24]

$$\bar{z} = z/\alpha_s^3 = (3/2\pi)^{3/2}(\alpha_s^{-3})BA^{-3}M^{1/2} \qquad (3.13.8)$$

with

$$A = (6\langle S^2 \rangle_{w,0}/M_w)^{1/2} \qquad (3.13.9)$$

$$A = (\langle R^2 \rangle_{w,0}/M_w)^{1/2} \qquad (3.13.9')$$

and

$$B = \beta/m_0 \qquad (3.13.10)$$

Here, $\langle R^2 \rangle$ is the end-to-end distance of a polymer chain at unperturbed state, A and B are the short- and long-range interaction parameters, respectively, the binary cluster integral represents the interaction between the nonbonded segment of polymer chains, and m_0 is the molecular weight of a segment.

Fixman derived a relationship between α_s and z in the form[25]

$$\alpha_s^3 = 1 + 1.78z \qquad (3.13.11)$$

Eq. (3.13.6) (simply referred to as KFSY-I) was applied for cellulose, amylose, and their derivative solutions, except for the CTA/DMAc system, for which system KFSY-I was not applicable, since the ψ values for this system frequently exceed the upper limit (0.198) in KFSY-I, and then, eq. (3.13.7) (simply referred to as KFSY-II) was employed.

The coefficient (1.78) in eq. (3.13.11) differs slightly depending on the theory, but this variation does not afford a significant change in the range $\alpha_s^3 < 2$. The method for estimating α_s from ψ using eqs. (3.13.5), (3.13.6) (or 3.13.7), and (3.13.11) is very accurate, if α_s does not exceeds 1.

(a) *CA (DS 0.49)/DMAc.*[1] In column 9 of Table 3.3.1, α_s values of CA (DS 0.49) fractions in DMAc are shown.

(b) *CA (DS 1.75)/DMAc.*[2] Column 10 of Table 3.3.2 shows the α_s values of CA (DS 1.75) fractions in DMAc.

(c) *CA (DS 2.46)/THF, DMAc.*[3,4] The values of α_s of CA (DS 2.46) in acetone and THF were collected in the third and sixth columns of Table 3.15.1, respectively. α_s for CA (DS 2.46) in DMAc is shown in the fourth column of Table 3.15.2.

(d) *CA (DS 2.92)/DMAc.*[10] In column 11 of Table 3.3.5, the α_s values of CA (2.92) in DMAc are depicted.

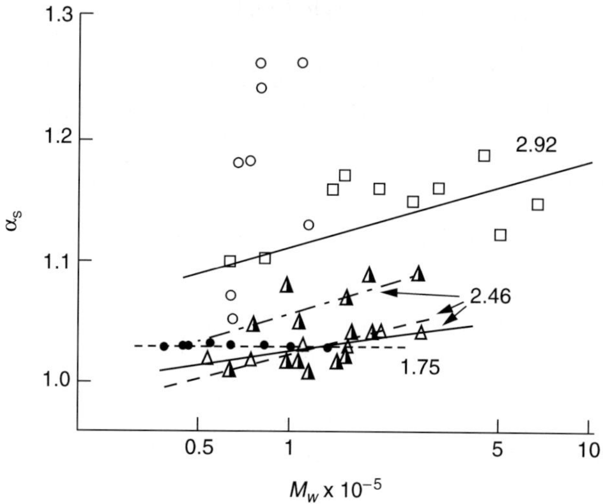

Figure 3.13.5 Linear expansion factor α_s of cellulose acetate-solvent systems plotted as a function of M_w. The lines are determined by the least square method. Numbers on the lines denote the total degree of substitution $\langle\!\langle F \rangle\!\rangle$ of cellulose acetate. (O), cellulose acetate (0.49)-dimethylacetamide; (●), cellulose acetone (1.75)-dimethylacetamide; (Δ), cellulose acetate (2.46)-dimethylacetamide; (half filled (right) triangle), cellulose acetate (2.46)-acetone; (half filled (left) triangle), cellulose acetate (2.46)-tetrahydrofuran; (□), cellulose acetate (2.92)-dimethylacetamide.

Figure 3.13.5 shows the α_s values of CA solvent systems as a function of M_w. α_s values are smaller than 1.18 for $M_w \leq 7 \times 10^5$, except for those of a few CA (0.49) fractions with molecular weights from 8×10^4 to 1×10^5.

Table 3.13.2 summarizes the range of α_s for CA solutions.

(e) *Cellulose/aq. 6 wt% LiOH solution.*[16] The values of α_s of the above system were collected in the column 9 of Table 3.5.4

Table 3.13.2

Linear expansion factor α_s for cellulose acetate solutions at 25 °C

DS	Solvent	α_s	Reference
	DMAc	1.03–1.26	Column 9 of Table 3.3.1
	FA	1.01 (one sample)	
0.49	Water	1.01 (one sample)	Column 9 of Table 3.5.1
	DMAc	1.01 (one sample)	
1.75	DMAc	1.02_8–1.03_1	Column 10 of Table 3.3.2
	Acetone	1.01–1.04 (1.09)[a]	Column 3 of Table 3.15.1
2.46	THF	1.05–1.09	
	DMAc	1.02–1.09	Column 6 of Table 3.15.2
2.92	DMAc	1.10–1.19	Column 9 of Table 3.3.5

[a]Sample EF3-15.

Table 3.13.3

Linear expansion factor α_s for cellulose alkali solutions

Solution	Temperature (°C)	α_s	Reference
6 wt% LiOH	25	1.02–1.04	Table 3.5.5 (Column 9)
8 wt% NaOH	26 3.5–35	1.004–1.005	Table 3.5.7 (Column 9)

(f) *Cellulose/aq. 8 wt% NaOH solution.*[26] In column 10 of Table 3.5.5, α_s values for cellulose in 8 wt% NaOH solution was compiled.

The range of α_s values thus calculated for cellulose alkali solutions are shown in Table 3.13.3. Evidently α_s is in the vicinity of unity.

(g) *Miscellaneous.*[9] Cellulose in cadoxen and FeTNa, amylose in DMSO, and solutions of 12 cellulose and amylose derivatives were analyzed. The α_s, estimated from ψ are summarized in the column 5 of Table 3.15.1. It can be said α_s is usually less than 1.3–1.4, even in good solvents.

3.13.3   Excluded volume effect (Worm-like chain model[27])

Kamide and coworkers showed that (1) the thermodynamic approach is preferable to the conventional hydrodynamic approach, which should be modified by introducing the concepts of the partial free draining effect and the nonGaussian nature of the polymer chain even in the unperturbed state (see Section 3.16); and that (2) solvation phenomena play a very important role in these solutions (see Section 3.17). The two approaches, both based on the pearl necklace chain model, indicate that cellulose, amylase, and their derivatives are not flexible, but are semi- or inflexible polymers. The expansion factor α_s, evaluated through use of the penetration function, is only slightly larger than unity (see Section 'A Pearl necklace model') and the solvents involved are not configurationally good, except for some cellulose derivative/solvent systems, for example, methyl cellulose/NaCl aq. solution,[28] CA (DS 2.46)/2-butanone.[29]

The semi- or inflexible polymer chain can also be analyzed by using the worm-like chain model. In this section, we attempt to evaluate the excluded volume effect in solutions of cellulose derivatives through use of the penetration function method based on the worm-like chain model and to compare it with that obtained by the pearl necklace chain model.

Theoretical background

By considering the excluded volume effect on worm-like touching beads, Yamakawa and Stockmayer[30] derived an equation relating the interpenetration function ψ to an excluded volume effect parameter $\bar{z}$. Their result can be generalized for a polydisperse polymer sample dissolved in a solvent as follows:

$$\psi (\equiv A_2 M_w^2 / (4\pi^{3/2} N_A \langle S^2 \rangle_w^{3/2})) = (\langle S^2 \rangle_{0,w} / \langle S^2 \rangle_{0,\infty,w})^{-3/2} \bar{z} h(\bar{z}) \qquad (3.13.12)$$

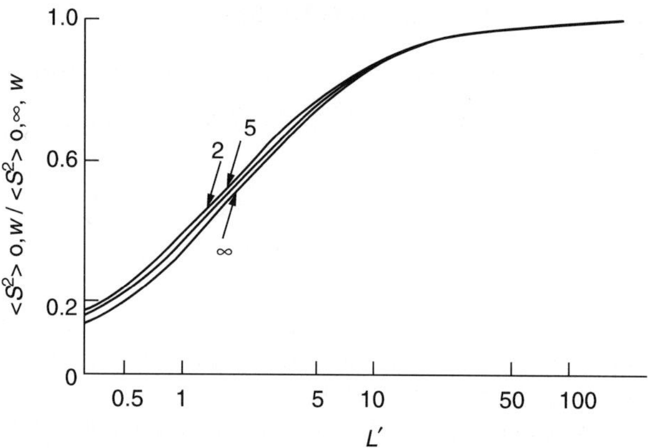

Figure 3.13.6   Plot of the ratio $\langle S^2\rangle_{0,w}/\langle S^2\rangle_{0,\infty,w}$ as a function of L' (eq. (3.13.14)). Here, $\langle S^2\rangle_{0,w}$ is the weight-average unperturbed mean square radius of gyration and $\langle S^2\rangle_{0,\infty,w}$ is that at the limit of infinite molecular weight. Number on curve denotes $h(\equiv \{M_w/M_n) - 1\}^{-1})$.

with

$$\bar{z} = z/\alpha_s^3 \tag{3.13.13}$$

$$z = (3/(2\pi))^{3/2} B'/(2_{qBD}^0/M_L)^{3/2} M_w^{1/2} \tag{3.13.14}$$

Here, A_2 is the second virial coefficient, M_w is the weight-average molecular weight, N_A is the Avogadro number, $\langle S^2\rangle_w^{1/2}$ is the weight-average radius of gyration, and the suffix 0 means the unperturbed state. $\langle S^2\rangle_{0,\infty,w}^{1/2}$ corresponds to $\langle S^2\rangle_w^{1/2}$ of the polymer with infinite molecular weight in the unperturbed state, $h(\bar{z})$ is a function of $\bar{z}$, B' is a long-range interaction parameter, which reduces to the long-range interaction parameter B in the pearl necklace chain model (see eq. (3.13.9)), q_{BD}^0 is the persistence length in the unperturbed state, and M_L is a shift factor.

The ratio $\langle S^2\rangle_w/\langle S^2\rangle_{0,\infty,w}$ was first given by Yamakawa and Stockmayer on the basis of the Benoit–Doty (BD) theory[31] for monodisperse polymers and is easily generalized for polymers with the Schulz–Zimm MWD function:

$$\langle S^2\rangle_{0,w}/\langle S^2\rangle_{0,\infty,w} = 1 - (3/2L') + 3(h+1)/(2hL'2) - \{3(h+1)^2\Gamma(h-1)/$$
$$4L'3\Gamma(h+1)\} \times [1 - \{(h+1)/(2L'+h+1)\}^{h-1}] \tag{3.13.15}$$

where

$$L' = L/2q_{BD}^0 \tag{3.13.16}$$

L is the contour length, $h^{-1} = (M_w/M_n) - 1$ (eq. (3.5.5)), $\Gamma(x)$ is the gamma function of x. q_{BD}^0 is the persistence length at unperturbed state derived by BD theory (Section 3.19).

Figure 3.13.6 shows the relationships between $\langle S^2\rangle_{0,w}/\langle S^2\rangle_{0,\infty,w}$ and L', calculated by eq. (3.13.14), for the cases of $h = 2, 5$ and ∞.

For $L' > 5$, the ratio $\langle S^2\rangle_{0,w}/\langle S^2\rangle_{0,\infty,w}$, is practically independent of the polydispersity of the polymer samples.

α_s and $h(\bar{z})$ in eqs. (3.13.2) and (3.13.13) can be expressed as infinite series functions of $\bar{z}$ respectively.

$$\alpha_s = 1 + (67/70)\, Kz + \cdots \tag{3.13.17}$$

and

$$h(\bar{z}) = 1 - Q\bar{z} + \cdots \tag{3.13.18}$$

where Q is a function of reduced contour length L' and bead diameter d, and K' is a function of L', both given in the Yamakawa–Stockmayer (YS) theory.

Yamakawa and Stockmayer approximated for $\alpha_s \approx 1$ (eqs. (3.13.17) and (3.13.18)) with the mutually corresponding closed form equations such as the Flory fifth power law-type[32] and Flory–Krigbaum–Orofino-type[33] equations.

On the other hand, Kamide and coworkers[1,3,4,9,10,34] estimated the α_s value using a third power law-type equation of α_s and the corresponding $h(\bar{z})$ equation for the pearl necklace chain model, derived by Fixman[25] and KFSY-II,[22] respectively. The latter equation was derived using the third power law-type equation of α_s for a cruciform molecule. Therefore, it is highly desirable to calculate α_s using the equation of third power law-type for the worm-like chain model. For this purpose, eqs. (3.13.17) and (3.13.18) can be converted to the third power law-type equations in the closed form thus

$$\alpha_s^3 - 1 = (3/2)(67K/70)z \tag{3.13.19}$$

$$\bar{z}h(\bar{z}) = 1 - [1 + (3.908/2.865)Q\bar{z}^{-0.468}(1.828/2.865)^{-1}Q^{-1} \tag{3.13.20}$$

At the limit of infinity of L' (hereafter, simply referred to as the coil limit), a polymeric worm-like chain behaves as a Gaussian chain (i.e. coil-like behavior) and eqs. (3.13.19) and (3.13.20) reduce to the corresponding equations of Fixman and KFSY-II, respectively. The detailed derivation of eqs. (3.13.19) and (3.13.20) was given in Ref. 27.

Figure 3.13.7 shows some theoretical relationships between ψ and α_s for monodisperse worm-like chain polymer with various reduced contour lengths L' for reduced bead diameter d' (defined by $d' = d/(2q_{BD}^0)$) of 0.1. Here, the worm-like chain polymer with $L' = \infty$ is equivalent to the pearl necklace chain polymer. In this case, h is taken as 10^5 for convenience of calculation.

The full line in the figure denotes the curve calculated from the third power law-type equation (eqs. (3.13.19) and (3.13.20)), and the broken line from the fifth power law-type equation the equation of α_s (eq. 128 of Ref. 30) and the Flory–Krigbaum–Orofino–type equation of $\bar{z}h(\bar{z})$ (eq. 129 of Ref. 30), derived by Yamakawa and Stockmayer. $\alpha_{s,3}^w$ (α_s derived by a third power law-type equation for the worm-like chain model) is always larger than $\alpha_{s,5}^w$ (α_s derived by a fifth power law-type equation for the worm-like chain model) compared at the same ψ value, and the difference between $\alpha_{s,3}^w$ and $\alpha_{s,5}^w$ becomes larger for larger ψ and larger L' (i.e. larger M_w or smaller q_{BD}^0). Generally, at a given value of ψ, the $\alpha_{s,3}^w$ value for the worm-like chain is always slightly smaller than $\alpha_{s,3}^w$ (α_s, derived by a third power law-type equation of α_s, and KFSY-II equation of $\bar{z}h(\bar{z})$ for the pearl necklace chain model).

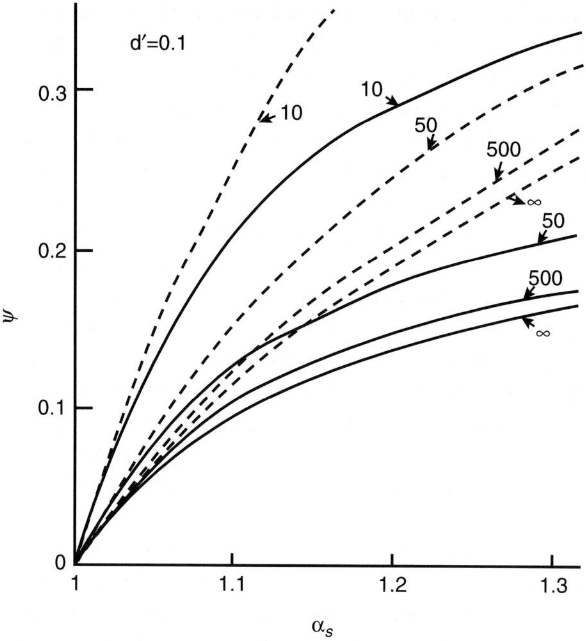

Figure 3.13.7 The interpenetration function ψ plotted against the expansion factor α_s, based on third power (full line) or fifth power (broken line) law-type equations for monodisperse worm-like chain polymers with reduced bead diameter d' of 0.1.[27] Numbers on curves refer to the reduced contour length L'. The worm-like chain with $L' = \infty$ coincides with the pearl necklace model.

Using the penetration function ψ, we can evaluate $\alpha_{s,3}^w$ (i.e. the penetration method). Putting the experimental data on $\langle S^2 \rangle_z^{1/2}$, A_2', M_w, h' and the value of the hydrodynamic diameter of a worm-like chain obtained using Ullman's theory[35] as d into eqs. (3.13.11–3.13.16), (3.13.19), and (3.13.20), we solved these equations by a numerical calculation method. q_{BD}^0 can be also evaluated concurrently. Of course, the limit of applicability of this method is primarily determined by that of eq. (3.13.19). In the higher z region ($z \geq 0.2$), α_s values estimated by the penetration method differ significantly depending on the relationships connecting α_s with z. For solutions of cellulose and its derivative z is fortunately not far from zero, supporting the high confidence in the α_s values obtained.

Application to experimental data

In order to evaluate $\alpha_{s,3}^w$, the experimental data were analyzed for cellulose and its derivatives as follows: cellulose/cadoxen,[36] cellulose/FeTNa,[37] CA (DS 2.92)/DMAc,[10] CA (DS 2.46)/DMAc,[3] CA (DS 2.46)/acetone,[4] CA (DS 2.46)/THF,[4] CA (DS 0.49)/DMAc,[1] CN (DS 2.91)/acetone,[13] CN (DS 2.55)/acetone,[13] hydroxyethyl cellulose (HEC)/water,[38] ethylhydroxyethyl cellulose (EHEC)/water.[39] Although all the LS data were treated by Zimm's procedure, we analyzed such experimental data without any reservation because the Zimm plots constructed are, at least as far as these plots were available for analysis, rectangular without distortion. Furthermore, the unperturbed persistence length by BD theory, q_{BD}^0, is without exception, in good agreement with the coil limit persistence length in the unperturbed state, q_{CL}^0, which was directly calculated from the unperturbed chain dimension A in a pearl necklace model.[40] The Gaussian nature of the chain is a concept that does not conflict with the worm-like chain.

L' values of the polymer/solvent systems analyzed here usually exceed 10, except for some low molecular weight samples of CN (DS 2.91) in acetone, and of CA (DS 2.46) in DMAc, and so the effect of sample polydispersity can be neglected for $\langle S^2 \rangle_{0,w}/\langle S^2 \rangle_{0,\infty,w}$ for most solutions of cellulose and its derivatives. The ratio $\langle S^2 \rangle_{0,w}/\langle S^2 \rangle_{0,\infty,w}$ for these systems is then above 0.9.

For a few systems such as cellulose/FeTNa, CA (2.91)/DMAc, and CA (0.49)/DMAc, ψ lies between 0.12 and 0.18. However, for most systems, $\psi = 0.1$. Taking into account these findings, together with the range of L' for the polymers examined here, we can consider that the difference between $\alpha_{s,3}^w$ and $\alpha_{s,5}^w$ is very small. For example, values of the ratio $\alpha_{s,3}^w/\alpha_{s,5}^w$ corresponding to $\psi = 0.13$ and 0.05, are 1.012 and 1.000, both at $L' = 10$, and 1.042 and 1.003, both at $L' = 500$, respectively.

Figure 3.13.8 demonstrates the relationships between $\alpha_{s,3}^w$ and M_w for cellulose derivatives. Most of the $\alpha_{s,3}^w$ data points lie below 1.1, and the significant molecular weight dependence of $\alpha_{s,3}^w$ was observed especially for cellulose/FeTNa, CA (2.92)/DMAc, and CA (0.49)/DMAc. The slope of the plots in Figure 3.13.7 tends to increase with M_w. This means that the contribution of an excluded volume effect to the limiting viscosity number, α_1, defined by $\alpha_1 = \mathrm{d} \ln \alpha_s/\mathrm{d} \ln M_w$ cannot be neglected. Neglect of this term will lead to an erroneous conclusion. As demonstrated for a pearl necklace chain model.[9] $\alpha_{s,3}^w$ values are dependent on DS as for CA and CN where the polymers with different DS are dissolved in a solvent.

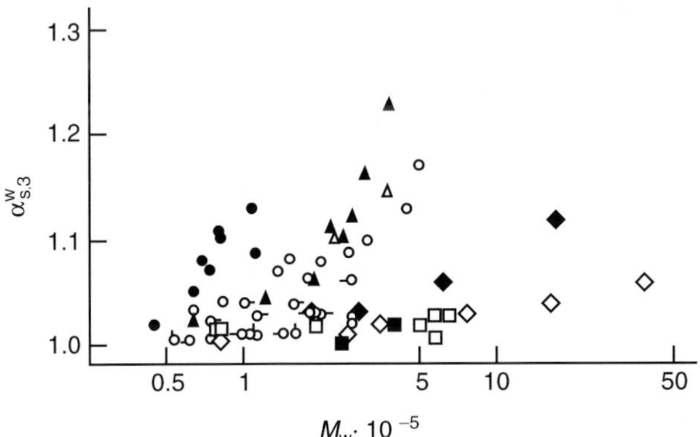

Figure 3.13.8 Relationships between the expansion factor $\alpha_{s,3}^w$ and the weight-average molecular weight M_w.[27] ($\triangle$), cellulose/cadoxen; ($\blacktriangle$), cellulose/iron sodium tartrate; ($\bigcirc$), cellulose acetate (2.92)/dimethylacetamide; (circle with upward tail), cellulose acetate (92.46)/dimethylacetamide; (circle with left tail), cellulose acetate (2.46)/acetone; (circle with right tail), cellulose acetate (2.46)/tetrahydrofuran; ($\bullet$), cellulose acetate (90.49)/dimethylacetamide; ($\diamond$), cellulose nitrate (2.55)/acetone; ($\square$), hydroxyethyl cellulose/water; ($\blacksquare$), ethylhydroxyethyl cellulose/water.

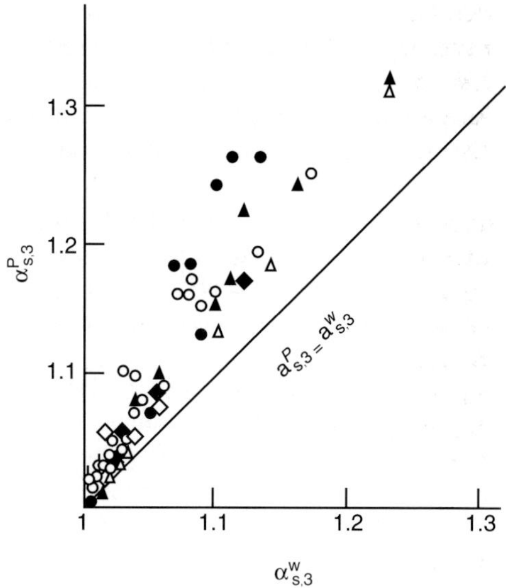

Figure 3.13.9   The expansion factor, evaluated through use of the third power law-type equations, of the pearl necklace chain model $\alpha^{p}_{s,3}$ plotted against the corresponding expansion factor of the worm-like chain model $\alpha^{w}_{s,3}$.[27] The symbols are the same meaning as in Figure 3.13.8.

Figure 3.13.9 shows the plot of $\alpha^{p}_{s,3}$ as a function of $\alpha^{w}_{s,3}$. In the figure, the full line is the line of $\alpha^{p}_{s,3} = \alpha^{w}_{s,3}$. For all the polymer samples investigated, the following relationship is obtained by the method of least squares: $\alpha^{w}_{s,3} = (0.94 \pm 0.06) \times \alpha^{p}_{s,3}$. (Here, $\alpha^{p}_{s,3}$ is α_s derived by a third power law-type equation of the pearl necklace model and coincides with α_s in other sections, unless otherwise noted.) We can conclude that the solvents employed for cellulose derivatives are configurationally poor irrespective of the molecular models with which the excluded volume effect is estimated. This indicates that the conclusions about the thermodynamic properties of cellulose derivative solutions deduced by Kamide *et al.* on the basis of the pearl necklace chain model[3,4,9,10,34] are acceptable without any restrictions.

At coil limit (i.e. at the limit of infinity of L'), the perturbed and unperturbed persistence lengths, q_{CL} and q^{0}_{CL} are directly related to $\alpha^{p}_{s,3}$ by the relationship:

$$(q_{CL}/q^{0}_{CL})^{1/2} = \alpha^{p}_{s,3} \qquad (3.13.21)$$

where

$$q_{CL} = 3M_{L}(\langle S^2 \rangle_{w}/M_{w}) \qquad (3.13.22)$$

and

$$q^{0}_{CL} = 3M_{L}(\langle S^2 \rangle_{0,w}/M_{w}) \qquad (3.13.23)$$

For solutions of cellulose and its derivatives

$$q_{BD} \approx q^{0}_{CL} \qquad (3.13.24)$$

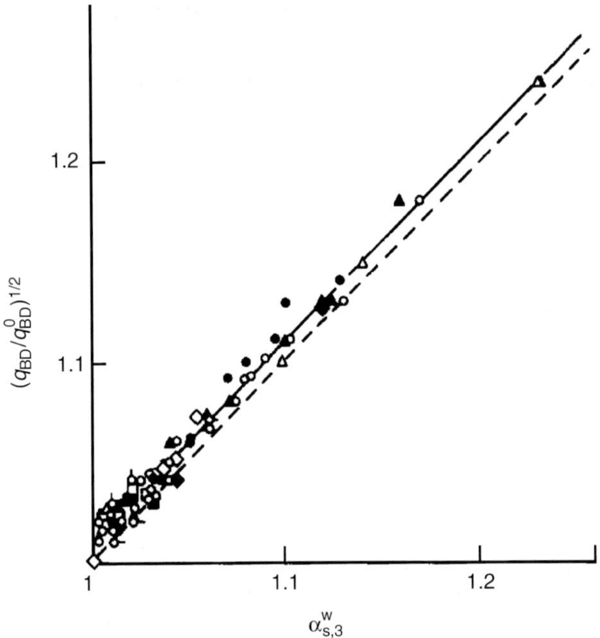

Figure 3.13.10   Plot of $(q_{BD}/q_{BD}^0)^{1/2}$ versus $\alpha_{s,3}^w$ for cellulose and its derivative in solutions.[27] Broken line is eq. (3.13.25) and full lines are eq. (3.13.26). The symbols have the same meanings as in Figure 3.13.8.

is found as described before. For these solutions, the following relationship, analogous to eq. (3.13.21), is expected to hold.

$$(q_{BD}/q_{BD}^0)^{1/2} \approx \alpha_{s,3}^w \tag{3.13.25}$$

Figure 3.13.10 depicts the experimental verification of eq. (3.13.25). Here, q_{BD} was calculated by generalizing the equation of Benoit and Doty (eq. 17 in Ref. 31) to polydisperse polymers. The broken line in the figure refers to eq. (3.13.25), and the full line is the regression line given by

$$(q_{BD}/q_{BD}^0)^{1/2} = 1.00_2\,\alpha_{s,3}^w + 8 \times 10^{-3} \qquad (1.00 \leq \alpha_{s,3}^w \leq 1.25) \tag{3.13.26}$$

The maximum difference in $(q_{BD}/q_{BD}^0)^{1/2}$ estimated by eqs. (3.13.25) and (3.13.26) is only 1.0% and therefore negligible. Eq. (3.13.25), in combination with eq. (3.13.15), provides a method for calculating $\alpha_{s,3}^w$ concurrently without use of the penetration function.

In summary, most values of expansion factor $\alpha_{s,3}^w$ estimated by third power law-type equations (eqs. (3.13.19) and (3.13.20)) on the basis of worm-like chain model do not exceed 1.1 for cellulose and cellulose derivative solutions, and $\alpha_{s,3}^w$ is about 95% of the expansion factor $\alpha_{s,3}^p$ estimated for the same system on the basis of pearl necklace model. These results strongly suggest that the excluded volume effect is small for cellulose and its derivative solutions, and that the solvents adequate for these polymers are configurationally poor. For the cellulose and its derivatives analyzed here, the root of the ratio q_{BD}/q_{BD}^0 is approximately expressed by $\alpha_{s,3}^w$ in eq. (3.13.25).

REFERENCES

1. K Kamide, M Saito and T Abe, *Polym. J.*, 1981, **13**, 421.
2. M Saito, *Polym. J.*, 1983, **15**, 249.
3. K Kamide and M Saito, *Polym. J.*, 1982, **14**, 517.
4. K Kamide, T Terakawa and Y Miyazaki, *Polym. J.*, 1979, **11**, 285.
5. RS Stein and P Doty, *J. Am. Chem. Soc.*, 1946, **68**, 159.
6. BH Zimm, *J. Chem. Phys.*, 1946, **14**, 164.
7. PJ Flory, *Principles of Polymer Chemistry*, Chapter XIV, Cornell University Press, Ithaca, NY, 1953.
8. WH Stockmayer, *Makromol. Chem.*, 1960, **35**, 54.
9. K Kamide and Y Miyazaki, *Polym. J.*, 1978, **10**, 409.
10. K Kamide, Y Miyazaki and T Abe, *Polym. J.*, 1979, **11**, 523.
11. EF Casassa, *Polymer*, 1962, **3**, 625.
12. P Howard and RS Parish, *J. Polym. Sci.*, 1966, **4**, 407.
13. GV Schulz and E Penzel, *Makromol. Chem.*, 1968, **112**, 260.
14. GV Meyerhoff, *J. Polym. Sci.*, 1958, **29**, 399.
15. E Penzel and GV Schulz, *Makromol. Chem.*, 1968, **113**, 64.
16. K Kamide and M Saito, *Polym. J.*, 1986, **18**, 569.
17. C Strazielle, in *LS from Polymer Solutions* (ed. MB Huglin), Academic Press, New York, 1972, Chapter 15, p. 652.
18. W Brown and R Wikström, *Eur. Polym. J.*, 1965, **1**, 1.
19. D Henley, *Ark. Kemi.*, 1961, **18**, 327.
20. M Kurata, *Industrial Chemistry of High Polymers*. Modern Industrial Chemistry Series No.18 **Vol. III**, Asakura, 1975, Chapter 4, p. 258.
21. See, K Kamide and T Dobashi, *Physical Chemistry of Polymer Solutions*, Problems 6-20, Elsevier, Amsterdam, 2000, Eq. 6.20.5.
22. M Kurata, M Fukatsu, H Sotobayashi and H Yamakawa, *J. Chem. Phys.*, 1964, **41**, 139.
23. See, K Kamide and T Dobashi, *Physical Chemistry of Polymer Solutions*, Elsevier, Amsterdam, 2000, Eq. 6.20.9.
24. See, K Kamide and T Dobashi, *Physical Chemistry of Polymer Solutions*, Problems 6–18, Elsevier, Amsterdam, 2000, Eq. 6.18.6.
25. M Fixman, *J. Chem. Phys.*, 1962, **36**, 3123.
26. K Kamide, M Saito and K Kowsaka, *Polym. J.*, 1987, **19**, 1173.
27. K Kamide and M Saito, *Eur. Polym. J.*, 1983, **19**, 507.
28. Y Baba and K Kagemoto, *Kobunshi Ronbubshu*, 1974, **31**, 528.
29. H Suzuki, Y Muraoka, M Saito and K Kamide, *Eur. Polym. J.*, 1982, **18**, 831.
30. H Yamakawa and WH Stockmayer, *J. Chem. Phys.*, 1972, **57**, 2843.
31. H Benoit and PM Doty, *J. Chem. Phys.*, 1953, **57**, 958.
32. PJ Flory, *J. Chem. Phys.*, 1949, **17**, 303.
33. PJ Flory and WR Kirgbaum, *J. Chem. Phys.*, 1950, **18**, 1086.
34. K Kishino, T Kawai, T Nose, M Saito and K Kamide, *Eur. Polym. J.*, 1981, **17**, 623.
35. R Ullman, *J. Chem. Phys.*, 1968, **49**, 5486.
36. D Henley, *Ark. Kemi.*, 1961, **18**, 372.
37. L Valtasaari, *Makromol. Chem.*, 1971, **150**, 117.
38. W Brown and D Henley, *Makromol. Chem.*, 1967, **108**, 153.
39. RS Manley, *Ark. Kemi.*, 1956, **9**, 519.
40. K Kamide and M Saito, *Makromol. Chem. Rapid Commun.*, 1983, **4**, 33.
41. K Kamide and Y Miyazaki, *Polym. J.*, 1978, **10**, 539.

3.13A APPENDIX

3.13A.1 Derivation of eqs. (3.13.19) and (3.13.20)

We assume the third power law-type equation, as a closed form equation for eq. (3.13.17), is given by

$$\alpha_s^3 = 1 + k(67/70)K'(L')z \tag{3.13.a1}$$

where k is a constant.

Eq. (3.13.a1) can be expanded in terms of z thus

$$\alpha_s^3 = 1 + (2k/3)(67/70)K'z + \cdots \tag{3.13.a2}$$

As in the case of YS theory, at the limit of $L' \rightarrow \infty$ the equation relating α_s and z should reduce to the corresponding equation for a Gaussian chain. According to the well-known first-order perturbation theory of the excluded volume effect,[1] the second term on the right hand side of eq. (3.13.a2) is given by

$$\lim_{L' \rightarrow \infty} (2k/3)(67K'L')/70 = 134/105 \tag{3.13.a3}$$

where

$$K'(\infty) = 4/3 \tag{3.13.a4}$$

Comparison of eq. (3.13.a3) with eq. (3.13.a4) yields $k = 3/2$: Therefore, eq. (3.13.a2) can be rewritten as

$$\alpha_{s,3}^w - 1 = (3/2)(67K'/70)z + \cdots \tag{3.13.a5}$$

Equation (3.13.a5) reduces to eq. (3.13.19) when $\alpha_{s,3}^w$ is near unity.

Next, we assume KFSY-II-type equation as a closed form equation for eq. (3.13.18)

$$\bar{z}h(\bar{z}) = 1 - [1 + \{(3.903/a)Q\bar{z}\}^{-0.468}](1.828/a)^{-1}Q^{-1} \tag{3.13.a6}$$

where a is constant.

Equation (3.13.a6) can be expanded around $\bar{z}$ in the form:

$$h(\bar{z}) = [1 - \{1 - (3.903 \times 0.468/a)Q\bar{z} + (1/2(3.903/a)^2 Q^2(-0.468)$$

$$\times (-1.468)\bar{z}^{-2} + \cdots\}]/(1.82/a)Q\bar{z} \tag{3.13.a7}$$

By comparing Eq. (3.13.a7) with eq. (3.13.18), we obtain $a = 2.865$, leading directly to eq. (3.13.20).

REFERENCE

1. H Yamakawa, *Modern Theory of Polymer Solutions*. Harper and Row, New York, 1971.

3.14   MOLECULAR WEIGHT DEPENDENCE OF SEDIMENTATION AND DIFFUSION COEFFICIENTS

In general, the sedimentation coefficient at infinite dilution s_0 is semi-empirically related to M_w by the relationship:

$$s_0 = K_s M_w^{a_s} \qquad (3.14.1)$$

Here, K_s and a_s are parameters characteristic to polymer solvent combination at a given temperature.

3.14.1   CA (DS 2.46 and 2.92)[1,2]

Figure 3.14.1(a) and (b) show the plots of s_0 as a function of M_w with open circles for CA (DS 2.46)/acetone and CA (DS 2.92)/DMAc, respectively. From the figures, the following relationships are obtained by using the least squares method.

CA (D.S. 2.46)/acetone:[2]

$$s_0 = 1.02 \times 10^{-14} M_w^{0.390} \qquad \text{(sec)} \qquad (3.14.2)$$

CA (DS 2.92)/DMAc:[2]

$$s_0 = 1.83 \times 10^{-14} M_w^{0.242} \qquad \text{(sec)} \qquad (3.14.3)$$

Table 3.14.1 lists the parameters K_s and a_s of eq. (3.14.1) for CA (D.S 2.46)/acetone, along with those proposed hitherto.[3-5] It should be noted that eq. (3.14.1), as reported in the literature for this system, was determined by using M_{SD} (the molecular weight from sedimentation-diffusion method) or M_{SV} (the molecular weight from sedimentation-viscosity method), and that the values of the exponent a_s range from 0.33 to 0.39, in fair agreement with Ishida *et al.*'s value. M_{SD} and M_{SV} are hydrodynamic averages,

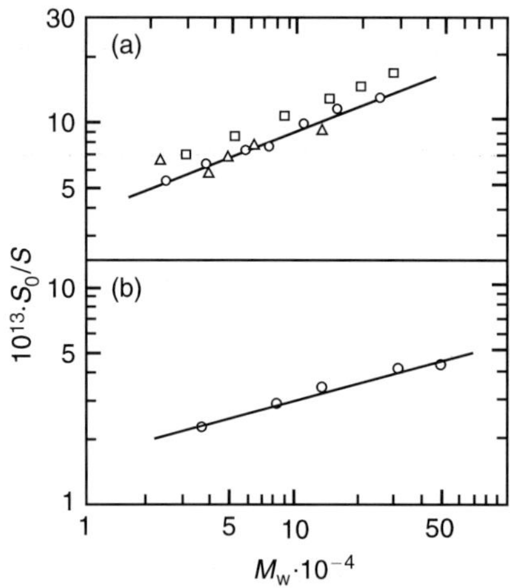

Figure 3.14.1   Molecular weight dependence of the sedimentation constant s_0: (a) cellulose acetate (DS 2.46) in acetone at 25 °C; Ishida *et al.*,[1,2] (O); Singer,[3] (●); Holmes and Smith,[4] (Δ); Golbev and Frenkel,[5] (□); (b) cellulose acetate (DS 2.92) in *N,N*-dimethylacetamide at 25 °C.

Table 3.14.1

Parameters K_s and a_s in eq. (3.14.1) for cellulose acetate in acetone

Degree of substitution	Temperature (°C)	$K_s \times 10^4$	a_s	N	M range $\times 10^{-4}$	Average	M_w/M_n	Reference
2.33	30	1.22	0.38	4	1.04–19.4	M_{SD}	1.27	3
2.31	20	2.21	0.33	9	2.12–12.1	M_{SD}	1.12–2.44	4
2.34	25	1.20	0.39	8	4.9–26.9	M_{SV}	1.15–1.99	5
2.46	25	1.02	0.39	7	2.42–24.9	M_{SV}	1.2–1.3	2

N: Number of sample fraction; M: (Mol wt range) $\times 10^{-4}$.

dependent on the solvent nature.[6] Until 1980s, there was no CA (DS 2.46) or CA (DS 2.92) sample in which M_w and M_n have been determined simultaneously. M_{SD} and M_{SV} are correlated with M_w for Schulz–Zimm distributions of the molecular weight by the relationships:[6]

$$M_{SD} = M_w\{1 - (1 + a)/[3(h + 1)]\} \tag{3.14.4}$$

$$M_{SV} = M_w[\Gamma(h + 1 - (h - a)/3]^{3/2}[\Gamma(h + 1 + a)]^{1/2}/\{(h + 1)[\Gamma(h + 1)]^2\} \tag{3.14.5}$$

Here, a is the exponent in the MHS equation (eq. (3.11.1)), $\Gamma(x)$, the gamma function of x and h the polydispersity parameter (eq. (3.5.5)). $a = 0.616$ was evaluated for CDA/acetone (eq. (3.11.9)). Then literature data on a and either M_{SD}/M_w or M_{SV}/M_n (M_n, number-average molecular weight by membrane osmometry) enables us to calculate h by using eqs. (3.14.4) or (3.14.5). Accordingly, we can convert M_{SD} or M_{SV} into M_w.

The estimated range of M_w/M_n is collected in the column 8 of Table 3.14.1. Among four fractions in Singer's paper[3] two fractions (fractions 9 and 15) showed $M_w/M_n < 0.9$ indicating that M_{SD} and/or M_w values are unreasonable. One fraction (fraction 2C) lacked an M_n value. Three fractions (B_2, C_2, and D_2) in Holmes and Smith's paper[4] had $M_w/M_n < 1$ and a further six fractions (from a total of 9 fractions) had broader MWDs ($M_w/M_n > 1.4$). A significant change in M_w/M_n with M_w was observed for a series of fractions in Golubev and Frenkel's paper.[5] Note that all the fractions employed in the literature[3–5] were prepared by the precipitation method. In Figure 3.14.1, the values of $M_w/M_n > 1$ are plotted for comparison. Equation (3.14.1) appears to fit the corrected data of Singer[3] and of Holmes and Smith[4] equally well.

If the data obtained in this study and in the literature (except for those of Golubev and Frenkel[5]) are taken into account, then the following relationship between M_w and s_0 for CA (DS 2.46) in acetone is obtained (s_0 in seconds):

$$s_0 = 1.70 \times 10^{-14} M_w^{0.345} \tag{3.14.10}$$

REFERENCES

1. S Ishida, H Komatsu, T Terakawa, Y Miyazaki and K Kamide, *Mem. Facult. Eng., Kanazawa Univ.*, 1979, **12**, 103.

2. S Ishida, H Komatsu, H Kato, M Saito, Y Miyazaki and K Kamide, *Makromol. Chem.*, 1982, **183**, 3075.
3. SJ Singer, *J. Chem. Phys.*, 1947, **15**, 341.
4. FH Holmes and DI Smith, *Trans. Faraday Soc.*, 1957, **53**, 69.
5. VM Golubev and SY Frenkel, *Visokomol. Soedin., Ser. A.*, 1968, **10**, 750.
6. See, K Kamide, in *Polymer Engineering*, **Vol. 4**, (ed. Soc. Polym. Sci. Japan), MWD and Properties of Polymers, Chizin Shokan, Tokyo, 1968, Chapter 19, p. 192; H-G Elias, R Bareiss and JG Watterson, *Fortschr. Hochpolym. Forch.*, 1973, **11**, 11.

3.15 FLORY'S VISCOSITY PARAMETER AND PARTIAL DRAINING EFFECT

3.15.1 Flory's viscosity parameter

Flory viscosity parameter Φ is defined by[1]

$$\Phi = [\eta]M_{\mathrm{w}}/(q_{\mathrm{w},z}^{-1}6^{3/2}\langle S^2\rangle_z^{3/2})\tag{3.15.1}$$

$q_{\mathrm{w},z}$ is a correction factor for polymolecularity. In the case when the MWD of the polymer sample obeys the Schulz–Zimm distribution $q_{\mathrm{w},z}$ is given by

$$q_{\mathrm{w},z} = (h+2)^{3/2}(h+1)^{-1}\Gamma(h+2)\Gamma(h+3/2)^{-1}\tag{3.15.2}$$

where Γ is gamma function and $h = \{(M_{\mathrm{w}}/M_{\mathrm{n}})-1\}^{-1}$.

The molecular weight dependence of Φ is expressed by the relationship of semi-empirical nature:

$$\Phi = K_\Phi M^{a_\Phi}\tag{3.15.3}$$

where K_Φ and a_Φ are the parameters for a given polymer solvent system.

Cellulose, amylose, and their derivatives[2]

The Flory viscosity parameter, Φ was estimated from viscosity and LS data of various cellulose,[3,4] and its derivatives such as cellulose trinitrate (CTN),[5,6] CN,[5] CA,[7] cellulose tricaproate (CTCp),[8] cellulose tricarbanilate (CTC),[9] methyl cellulose (MC),[10] ethyl cellulose (EC), sodium carboxymethyl cellulose (NaCMC),[11] HEC,[12] EHEC,[13] sodium cellulose xanthate (NaCX),[14,15] and amylose,[16] and amylose derivatives, including amylose triacetate (ATA),[17] and amylose tricarbanilate (ATC).[18]

Figure 3.15.1 depicts the results of the log–log plot of Φ and M_{w}.[2] As can be seen from the figure, Φ for cellulose, amylose, and their derivatives is strongly molecular weight dependent and can be empirically expressed by eq. (3.15.3) to a fairly good approximation over the entire experimentally accessible molecular weight range. a_Φ was determined in this manner from the slope of the plot in Figure 3.15.1 and summarized in column 11 of Table 3.15.1.

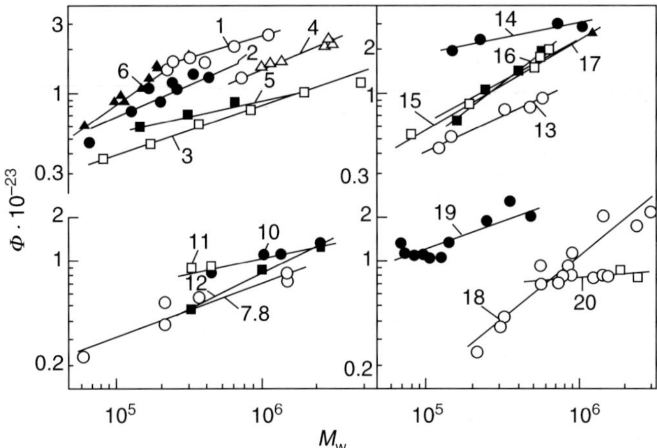

Figure 3.15.1 The molecular weight dependence of Flory's viscosity parameter Φ.[2] 1, cellulose in cadoxen;[3] 2, cellulose in FeTNa;[4] 3, CTN (N_c = 13.9%) in acetone;[5] 4, CTN (N_c = 13.6%) in acetone;[6] 5, CTN (N_c = 12.9%) in acetone;[5] 6, CA (acetyl content 55.3%) in acetone;[7] 7, CTCP in DMF;[8] 8, CTCP in 1-cl-N;[8] 9, CTCP in dioxane/water;[8] 10, CTC in acetone;[9] 11, CTC in cyclohexanone;[9] 12, CTC in dioxanone;[9] 13, MC in water;[10] 14, NaCMC in aq. NaCl;[11] 15, HEC in water;[12] 16, EHEC in water;[13] 17, NaCX in 1-MNaOH;[14,15] 18, amylose in DMSO;[16] 19, ATA in nitromethane;[17] 20, ATC in pyridine.[18]

CA (DS 0.49)/DMAc[24]

Flory's viscosity parameter Φ values for 10 CA (DS 0.49) fractions in DMAc are shown in Table 3.3.1. The Φ values are plotted against M_w as open circles in Figure 3.15.2,[24] where the data of CA (DS 2.46) and CA (2.92) fractions in the same solvent are included. The corresponding values for aqueous and FA solutions are shown in the seventh column of Table 3.5.1.[24] Here, $q_{w,z}$(= 1.41) was calculated assuming the Schulz–Zimm distribution. The Φ for CA (DS 0.49)/DMAc system is smaller than that for CA (DS 2.92) in the same solvent (as denoted by the open rectangles), which are 1.3–1.6 times smaller than the theoretical value $(2.87 \times 10^{23})^{25}$ for a nonpermeable polymer chain in the unperturbed state. The Φ values for CA (DS 0.49) in DMAc shows a significant molecular weight dependence. This may be expressed by eq. (3.15.3), and we obtain $K_\Phi = 0.61 \times 10^{23}$ and $a_\Phi = 0.103$, due to a relatively narrow range of M_w, are not accurate enough to estimate the unperturbed chain dimensions.

The open triangle in Figure 3.15.2 denotes CA (DS 2.46) in DMAc.

The Φ value for CA in DMAc decreases in the order: CA (DS 2.92) > CA (DS 0.49) > CA (DS 2.46). The Φ values for CA (DS 0.49) in water and FA are unexpectedly low, being approximately 1/16 and 1/30 the value in DMAc. This seems attributable to the large $\langle S^2 \rangle_z^{1/2}$ of CA in these solvents. Abnormally small values of Φ and its molecular dependence for CA (DS 0.49–2.92) undoubtedly indicate the draining effect.

The expansion factor α_s in these solutions was calculated, according to the procedure described in 3.13, using the penetration function ψ. The results[24] are listed in column 9 of Table 3.3.1 and in column 8 of Table 3.5.2.[24] The α_s values obtained remain almost

Table 3.15.1

Various parameters employed for evaluation of the unperturbed chain dimensions of cellulose, amylose, and their derivatives[2]

Polymer	Solvent	Sample number	M_w/M_n	α_s	a_2	K_m	a	$M_0 \times 10^{-5}$	$K_0 \times 10^{-5}$	a_Φ	a_1 Eq. (3.15.18)	a_1 Eq. (3.15.32)	Reference
Cellulose	Cadoxen	6	1.8–2.3	1.1–1.2	−0.216	0.0264	0.79_2	4.61	4.55	0.304	0.22	0.31	3
	FeTNa	9	2.0	1.0–1.2	−0.296	0.0494	0.77_9	2.01	5.01	0.429	0.17	0.29	4
CTN $N_c = 13.9\%$	Acetone	6	1.2	1.0–1.2	0	0.0076	0.90_3	5.58	6.36	0.379	0.05	<0.01	5, 19
13.6%	Acetone	9	2.0	1.0–1.1	0	0.0047	0.93_2	1.27	5.72	0.424	0	<0.01	6
CN 12.9%	Acetone	4	1.2	1.1–1.3	−0.024	0.0048	0.91_6	1.55	29	0.274	0.18	0.21	5, 19
CA AC = 55.3%	Acetone	9	1.29	1.0–1.1	−0.471	0.133	0.61_6	0.56	0.226	0.716	0.13	0.11	7
	THF	6	1.29	1.0–1.2	0	0.0513	0.68_8	0.56	573	0.105	0.150	0.08	7
CTCp	DMF	4	1.03–1.5	1.0	−0.248	0.268	0.49_5	2.86	4.17	0.377	0.01	−0.01	8
	1-Cl-N	4	1.03–1.5	1.0–1.1	−0.248	0.159	0.51_5	2.86	4.17	0.377	0.01	0.01	8
	Dioxane/water	4	1.03–1.5	1.0–1.1	−0.248	0.274	0.48_3	2.86	4.17	0.377	0.01	−0.02	8
CTC	Anisol	4	1.76–2.3	1.0	0	0.095	0.52_6	8.26	–	–	0	–	20
	Cyclohexanol	3	2.0–2.3	1.0	0	–	–	–	–	–	0	–	20
	Acetone	6	1.76–2.4	1.01–1.04	0	0.0012	0.91_2	8.26	60.6	0.21	0.09	0.20	9, 21
	Cyclohexanone	5	1.76–2.4	1.02–1.05	0	0.0026	0.83_7	8.26	0.75	0.508	−0.10	0.13	9, 21
	Dioxane	4	1.76–2.4	1.01–1.03	0	0.00082	0.97_1	8.26	1.52	0.462	−0.11	<0.01	9, 21
MC DS = 2	Water	5	2.4–2.9	1.01–1.04	−0.280	0.473	0.50_8	2.64	2.01	0.464	0.008	−0.09	10
NaCMC DS = 0.88	NaCl (I → ∞)	4	2.1–3.7	1.0	−0.072	0.173	0.60_4	3.95	1880	0.192	0.022	0.02	11
HEC DS = 1	Water	5	1.5–1.7	1.02–1.04	−0.128	0.0074	0.89_9	2.23	0.54	0.608	−0.004	−0.03	12
EHEC DS = 2	Water	4	3.07–3.23	1.01–1.10	−0.256	0.0406	0.75_4	3.02	0.207	0.688	0.08	−0.05	13
NaCX DS = 0.78	1-M NaOH	3	1.25	1.00–1.01	−0.240	0.0477	0.67_9	4.18	0.088	0.568	0.01	−0.10	14, 15
Amylose	DMSO	9	1.5	1.07–1.17	−0.336	0.0013	0.86_5	8.19	0.018	0.794	0.05	0.08	16
ATA	Nitromethane	12	1.5	1.04–1.33	0	0.0012	0.86_8	7.53	6.75	0.379	−0.10	−0.02	17
ATC	Pyridine/water	9	1.5	1.0	0	0.0039	0.59_9	11.8	218	0.144	−0.012	−0.015	22
	Pyridine	9	1.5	1.02–1.29	0	0.0012	0.86_4	8.55	35.7	0.056	0.13	0.31	18, 23

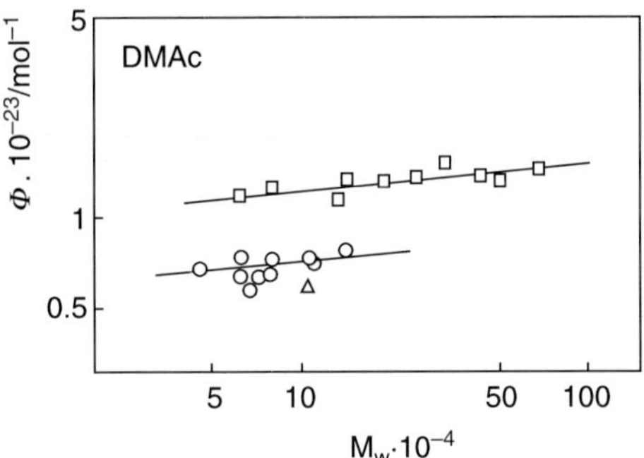

Figure 3.15.2 Molecular weight dependence of the Flory's viscosity parameter Φ for cellulose acetate in dimethylacetamide at 25 °C:[24] (O), cellulose acetate (DS 0.49);[24] (Δ), cellulose acetate (DS 2.46);[29] ($\square$), cellulose acetate (DS 2.92).[30]

constant within experimental error and are close to unity, suggesting that all the solvents employed are thermodynamically poor solvents for cellulose acetate. It should be remembered that any theory yields nearly the same relationships between ψ and the excluded volume parameter z for $\psi \leq 0.1$, which is the range in the CA (DS 0.49) solvent systems (column 8 of Table 3.3.1).

Using Φ and α_s the draining parameter X, defined by eq. (3.15.7) in the Kurata–Yamakawa theory,[26] was estimated (this method is referred to as the method 1A) and is given in the last columns of Table 3.3.1 and 3.5.1. For the CA (DS 0.49)/DMAc system, the X values obtained in this narrow molecular weight range are too small to permit examination of their molecular dependence. These values were found to be from 4.2 to 10.5 (average 7.3), and are smaller than $X \geq 10$ as is the case for many vinyl-type polymers.

CA (DS 1.75)/DMAc[27]

Column 10 of Table 3.3.2 shows Φ values of CA (DS 1.75) at 25 °C.[27]

CA (DS 2.46)/acetone, THF, DMAc[28,29]

Flory's viscosity parameters Φ for CDA solutions in acetone, THF, and DMAc at 25 °C is listed in Tables 3.15.2[28] and 3.15.3[29] and plotted against M_w in Figure 3.15.3.[29] Here, $q_{w,z}$ was calculated by assuming the Schulz–Zimm distribution for the CDA fractions used.

Obviously, Φ is smaller than the theoretical value for the nondraining limit in theta solvent ($\Phi(\infty) = 2.87 \times 10^{23}$), with a positive molecular weight dependence, expressed by eq. (3.15.3). For CDA, we obtain $a_\Phi = 0.716$ in acetone (see Table 3.15.1),[28] 0.105 in THF (see Table 3.15.1),[28] and 0.23 in DMAc (see Table 3.15.4).[29] The former is about three times larger than that theoretically calculated from $X = 1.48$ and the excluded volume effect parameter $a_1(= 3\epsilon$, where ϵ is defined by eq. (3.15.9)) as 0.10 with an assumption of the Gaussian chain ($a_2 = 0$; see eq. (3.15.10)) and the middle is in good agreement with an expected value from X and a_1. The experimental fact of $a_\Phi > 0$ has been investigated previously for various other cellulose derivatives

Table 3.15.2

Flory's viscosity parameter Φ, linear expansion factor α_s, and draining parameter X for cellulose diacetate in acetone and in tetrahydrofuran at 25 °C[28]

Polymer code	Acetone			THF		
	$\Phi \times 10^{-23}$	α_s	X from eq. (3.15.8)	$\Phi \times 10^{-23}$	α_s	X from eq. (3.15.8)
EF 2-10	0.63_5	1.01	0.52	–	–	–
EF 3-6	–	–	–	1.87	1.05	4.6
EF 2-11-1	0.83_2	1.02	0.82	–	–	–
EF 3-8	–	–	–	1.98	1.08	6.4
EF 3-10	0.96_0	1.02	0.90	2.01	1.05	6.0
EF 2-14	0.89_3	1.01	0.88	–	–	–
EF 3-12	1.04	1.02	1.10	1.93	1.07	5.8
EF 2-15	1.04	1.03	1.10	–	–	–
EF 3-13	1.27	1.04	1.50	2.10	1.09	8.5
EF 3-14	1.45	1.04	2.25	2.13	1.04	8.2
EF 3-15	1.78	1.09	4	–	–	–

(see column 11 of Table 3.15.1).[2] We note that the Φ value in DMAc (even for the highest molecular weight fraction studied) is about 70% smaller than the theoretical value 2.87×10^{23} for nondraining Gaussian coils. The Φ values in DMAc are much smaller than those in acetone and THF, as can be seen from Figure 3.15.3. Interestingly, Φ increases in the order of decreasing solvent polarity: DMAc < acetone < THF.

CA (DS2.92)/DMAc[30]

The values of Φ for CA (2.92) in DMAc are summarized in Table 3.3.5.[30] Figure 3.15.4(a) illustrates Flory's viscosity parameter Φ plotted as a function of the molecular weight for CTA solution in DMAc as open circle.[30] As noted previously for other cellulose derivatives (see Figure 3.15.1), the Φ value for CTA in DMAc reveals a significant molecular weight dependence, which may be semiempirically expressed by eq. (3.15.3).

Table 3.15.3

Flory's viscosity parameter Φ, linear expansion factor α_s, and draining parameter X for cellulose diacetate in dimethylacetamide at 25 °C[29]

Sample code	$M_w \times 10^{-4}$	$\Phi \times 10^{-23}$	α_s	X (by Method 1A)
EF 3-4	5.3	0.59	1.02	0.45
EF 3-6	7.3	0.57	1.02	0.43
EF 3-10	10.8	0.79	1.03	0.72
EF 3-13	15.6	0.79	1.03	0.71
EF 3-15	27.0	0.80	1.04	0.74

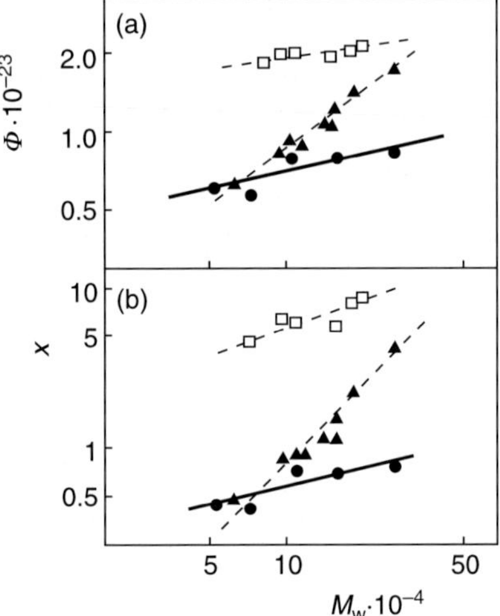

Figure 3.15.3 Molecular weight dependence of the Flory's viscosity parameter Φ (a), and the draining parameter X (b) for cellulose acetate (DS 2.46) in dimethylacetamide (circle),[29] acetone (triangle),[28] and tetrahydrofuran (rectangle).[28]

For CTA in DMAc, we obtain $K_\Phi = 0.35 \times 10^{23}$ and $a_\Phi = 0.106$. In addition, the magnitude of Φ is 1/1.8–1/2.6 of the theoretical value (2.87×10^{23}) at the unperturbed state. These experimental facts show that the draining effect may by no means be ignored, and that it plays an important role in the CTA/DMAc system at least over the entire molecular weight range in the experiment.

Table 3.15.4 lists the values of K_Φ and a_Φ, experimentally determined for CA with various $\langle\!\langle F \rangle\!\rangle$ solvent systems. The Φ values, calculated by using Nair *et al.*'s data,[31] are, surprisingly, 1/10 of those we obtained[2] and there are not examples of polymers, including cellulose derivatives/solvent system, showing such a small Φ values (e.g. see Figure 3.15.1).[2]

Table 3.15.4

Values of K_Φ, a_Φ and a_p for cellulose acetate and cellulose solutions

DS	Solvent	$K_\Phi \times 10^{-23}$	a_Φ	a_p
0.49	DMAc	0.61	0.103[24]	–
1.75	DMAc	0.26	0.12[27]	–
2.46	DMAc	515	0.23[29]	–
	Acetone	–	0.716[28]	0.31[35]
	THF	–	0.105[28]	–
2.92	DMAc	0.35	0.106[30]	0.25$_6$[35]
0	aq. LiOH	–	0.25[38]	–
	Cadoxen	–	0.43[2(3)]	–
	FeTNa	–	0.30[2(4)]	–

( ), literature in which experimental data are presented.

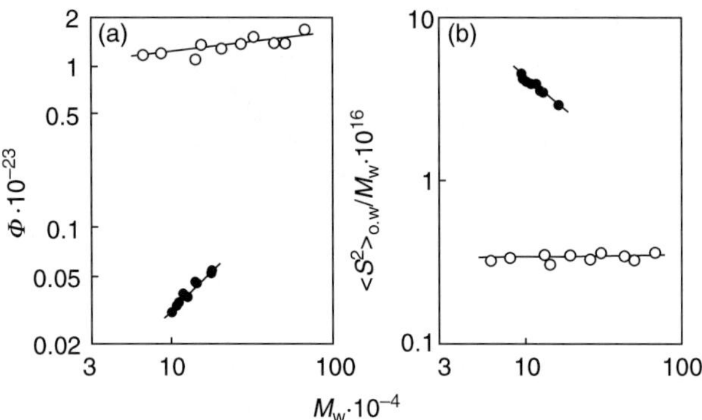

Figure 3.15.4   Molecular weight dependence of the Flory's viscosity parameter Φ (a) and of the ratio of the unperturbed radius of gyration $\langle S^2 \rangle_{0,w}$ to M_w for cellulose triacetate solutions (b):[30] (○), cellulose triacetate in dimethylacetamide;[30] (●), cellulose triacetate in dichloromethane/methanol (1:1 v/v; Nair *et al.*[31]).

Therefore, comparison of Figures 3.12.3 and 3.15.4(b) shows that Nair *et al.*'s data are not accurate enough to be analyzed further for establishing empirically the dependence of Φ on M_w. Nair *et al.* estimated the short-range interaction parameter A ($= (6\langle S^2 \rangle_0/M)^{1/2}$) according to the Stockmayer and Fixman plot (by our notation, method 2E),[25] which retains its validity only in the case where both a_Φ and a_2 are concurrently zero.

Cellulose, amylose, and their derivatives (Flory parameter P)

The parameter P, analogous to the Flory viscosity parameter Φ, is defined as[26]

$$P = q'_{w,z}(\xi/\eta_0)/(6^{1/2}\langle S^2 \rangle_z^{1/2}) \tag{3.15.4}$$

with

$$\xi = M(1 - v_p\rho_0)/s_0 N_A (\equiv \xi_{(S)}) \tag{3.15.5}$$

$$\xi = kT/D_0 (\equiv \xi_{(D)}) \tag{3.15.5'}$$

$q'_{w,z}$, is a polydispersity correction factor,[32] ξ, the friction coefficient, η_0 the viscosity of the solvent, v_p, the specific volume of the polymer, ρ_0, the density of the solvent and s_0, and D_0 the weight-average sedimentation coefficient and diffusion coefficient at infinite dilution. The sedimentation coefficient s_0 defined in eq. (3.15.5) coincides with that obtained by the momentum method within: $\pm 2\%$.[33] The latter is the weight-average s_0 (i.e. $s_{0,w}$) for a polydisperse polymer.[34] Then, the s_0 values obtained here can be approximated by $s_{0,w}$. The molecular weight dependence of Flory's parameter P is demonstrated in Figure 3.15.5. Clearly P in the above systems increases with increasing molecular weight.

The plots in the figure can be described empirically by the equation:

$$P = K_p M^{a_p} \tag{3.15.6}$$

Here, K_p and a_p are parameters characteristic for a given polymer solvent combination at constant temperature. The values of K_p and a_p thus determined are listed in Table 3.15.5.

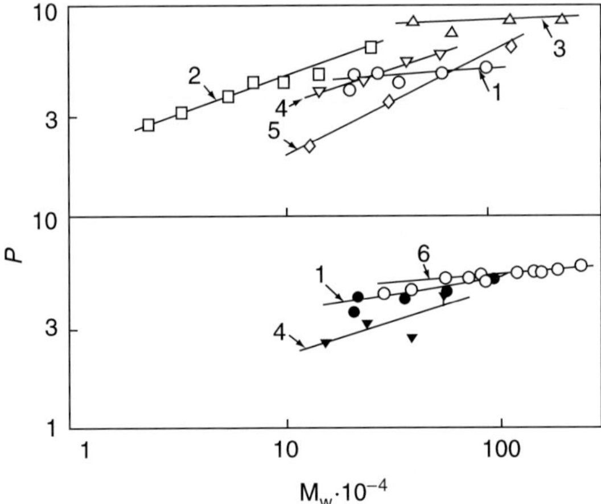

Figure 3.15.5 Molecular weight distribution of the Flory's viscosity parameter P: 1, cellulose in cadoxen; 2, cellulose diacetate in acetone; 3, cellulose tricarbanilate in cyclohexane; 4, ethyl hydroxyethyl cellulose in water; 5, sodium xanthate in 1-N NaOH; 6, amylose tricarbanilate in pyridine; (open mark), based on s_0 (using eq. (3.15.5)); (closed mark) based on D_0 (using eq. (3.15.5′)).

CA (DS 2.46; 2.92)/DMAc (Flory parameter P)[35]

In columns 10–12 of Table 3.7.2, $\xi_{w,w}(s)$, $q'_{w,z}$ and P values are summarized.[35] Here, we utilized eq. A. 14 of Ref. 32 in order to calculate $q'_{w,z}$.

Figure 3.15.6 shows the molecular weight dependence of P for CA (DS 2.46) in acetone and CA (DS 2.92) in DMAc.[35] We obtain $K_p = 0.122$ and $a_p = 0.311$ for CA (DS 2.46)/acetone and $K_p = 0.175$ and $a_p = 0.256$ for CA (DS 2.92)/DMAc. Experimental facts of $a_p \neq 0$ emphasize the role of the partially free-draining effect.

Table 3.15.5

Values of K_p and α_p for cellulose, amylose, and their derivatives[32]

Polymer	Solvent	S_0 or D_0	K_p in eq. (3.15.6)	a_p in eq. (3.15.6)	Reference
Cellulose	Cadoxen	S_0	2.10	0.067	3
		D_0	0.752	0.136	3
Cellulose diacetate	Acetone	S_0	0.122	0.311	35
Cellulose tricarbanilate	Cyclohexanone	S_0	5.98	0.024	36, 37
Ethylhydroxy-ethyl cellulose	Water	S_0	0.0755	0.328	13
		D_0	0.0695	0.300	13
Sodium cellulose xanthate	1-N NaOH	S_0	0.00558	0.504	14, 15
Amylose tri-carbanilate	Pyridine	S_0	2.10	0.064	18

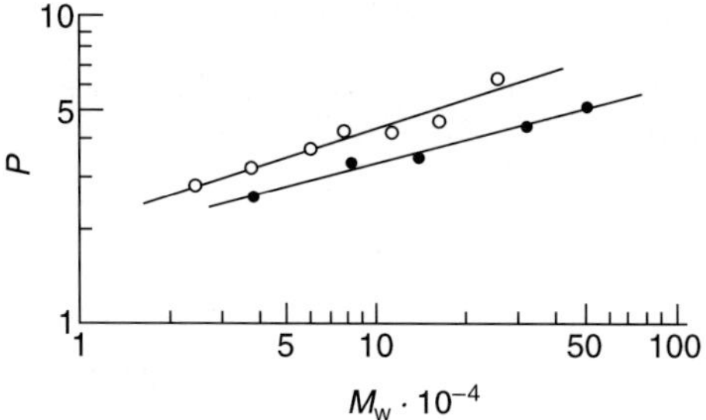

Figure 3.15.6 Molecular weight dependence of the Flory parameter P for cellulose acetate (DS 2.46) in acetate at 25 °C (○) and cellulose acectate (DS 2.92) in N, N-dimethylacetamide at 25 °C (●).

Cellulose/aq. LiOH[38]

Φ values of cellulose in aq. LiOH are collected in the eighth column of Table 3.5.4[38] and its molecular weight dependence is shown in Figure 3.15.7,[36] in which the data on cellulose in cadoxen and FeTNa are also included. Obviously, the Φ values are significantly smaller than the theoretical value at unperturbed, nondraining limit ($\Phi(\infty) = 2.87 \times 10^{23}$). a_Φ for aq. LiOH solution was 0.25, which is compared with 0.43[2] and 0.30,[2] previously evaluated for data on cadoxen[3] and FeTNa.[4]

In a previous paper,[39] we observed that specific interaction exists between CA polymer with $\langle\!\langle F \rangle\!\rangle$ of 0.49–2.92 and solvent, and the degree of the partial free drainage is larger for the systems with larger s_0. The draining effect was explained by the fact that boundary between solvated polymer molecules and nonsolvated solvent becomes obscure as s_0 becomes large. Analysis of the chemical shift of ^{1}H NMR spectrum and the adiabatic compressibility for cellulose/aq. LiOH system indicated that the solvation exists for the system, and s_0 was estimated to be four, assuming that only Li^+ solvated to cellulose.[40] For cellulose/aq. LiOH system, the draining effect may be closely correlated with the salvation.

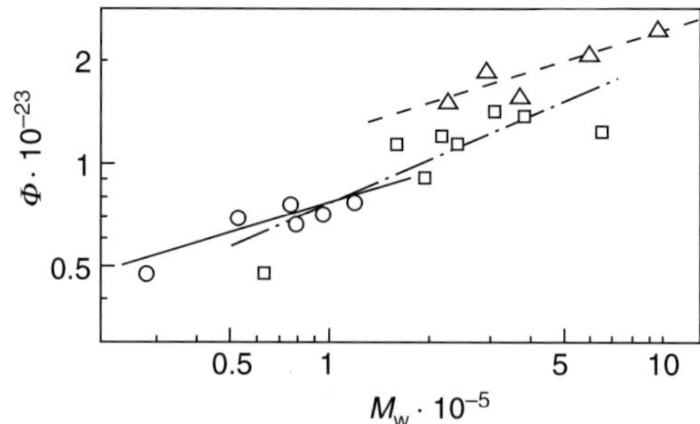

Figure 3.15.7 Molecular weight dependence of the Flory viscosity parameter Φ of cellulose in 6 wt% aq. LiOH (○), cadoxen: (△),[3] and iron sodium tartrate (□),[4] at 25 °C.[38]

Table 3.15.6

Flory's viscosity parameter Φ, the penetration function ψ, the expansion factor α_s and the draining parameter X for NaCS in 0.5 M aq. NaCl at 25 °C[41]

Sample code	$M_w \times 10^{-4}$	$\Phi \times 10^{-22}$	$\psi \times 10^{-2}$	α_s	X
CSK-1	150	16.4	3.71	1.02_0	2.9
CSK-2	116	15.4	8.06	1.07_1	2.7
CSK-3	74	13.1	6.35	1.05_1	1.7
CSK-4	53	10.3	6.56	1.05_1	1.1
CSK-5	40	10.2	6.95	1.05_7	1.1
CSO-2	24.4	9.2	7.39	1.06_2	0.9
CSK-6	8.0	5.3	4.90	1.03_6	0.4
CSK-7	7.2	4.7	5.38	1.04_1	0.34

Cellulose sulfate (DS 1.90)/aq. NaCl[41]

The Φ values estimated by eq. (3.15.1) are displayed in the third column of Table 3.15.6.[41] It should be noted that the Φ values depend on the MWD function employed. For example, the polymer with the normal Gaussian distribution ($M_w/M_n = 5$) gives Φ values about 5.1 times higher than those obtained by the Schulz–Zimm distribution. The former Φ values are for impermeable spheres in the unperturbed state. The assumption of MWD will lead to large errors in the calculated Φ values. The plot of Φ versus M_w (Figure 3.15.8)[41] leads to $K_\Phi = 5.74 \times 10^{20}$ and $a_\Phi = 0.40$.

3.15.2   Draining parameter

Theoretical background

Figure 3.15.9 illustrates of polymer molecules in flow of solvent. The draining parameter X is defined by the relationship:[26]

$$X = (1/2)(6\pi)^{1/2}(d/a')N^{1/2} \tag{3.15.7}$$

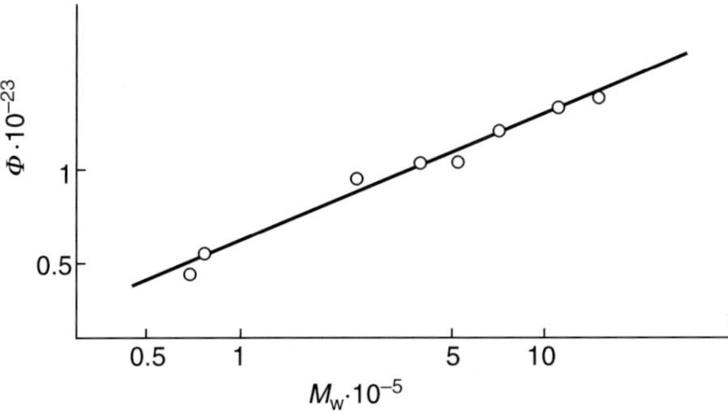

Figure 3.15.8   Molecular weight dependence of the Flory's viscosity parameter Φ for NaCS (DS 1.9) solution in 0.5 M aq. NaCl at 25 °C:[41] $\Phi = 5.74 \times 10^{20} M_w^{0.40}$.

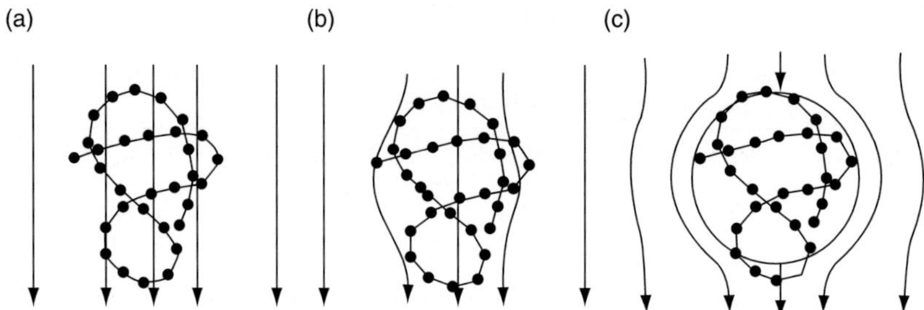

Figure 3.15.9 Polymer molecules in the Flow of solvent: (a), perfectly free draining; (b), partially free draining; (c), perfectly nondraining.

where a' is the length of a link (segment), d is the diameter of a hydrodynamic segment, and N the number of links connecting segments in one molecule. X can be evaluated by using the following methods:[42]

Method 1A. The Flory viscosity parameter Φ is related to the limiting viscosity number $[\eta]$, the weight-average molecular weight M_w and the z-average radius of gyration $\langle S^2 \rangle_z$ as

$$\Phi = [\eta]M_w/(\langle S^2 \rangle_z^{3/2} 6^{3/2})q_{w,z} = \Phi(X)a_s^{-(3-n(X))} \tag{3.15.1}$$

$$\alpha_s = \langle S^2 \rangle_z^{1/2}/\langle S^2 \rangle_0^{1/2} \tag{3.13.4}$$

where $q_{w,z}$ is a correcting factor for the polymolecularity of the sample. $\Phi(X)$ and $n(X)$ are functions of X defined by Kurata and Yamakawa (KY)[26] and $\Phi_0(X)$ denotes $\Phi(X)$ at $a_s = 1$. $\langle S^2 \rangle_0^{1/2}$ denotes $\langle S^2 \rangle^{1/2}$ in the unperturbed state (i.e. $a_s = 1$). X can be calculated from data of $[\eta]$, M_w, $\langle S^2 \rangle_z$ and a_s using eq. (3.15.1). In this section, the Schulz–Zimm-type distribution was assumed for the polymer samples and $\Phi_0(\infty) = 2.87 \times 10^{23}$ was utilized. Here, $[\eta]$ is expressed in terms of $cm^3\ g^{-1}$.

Method 1B. The molecular weight dependence of the limiting viscosity number can be expressed by the MHS equation,

$$[\eta] = K_m M^a \tag{3.11.1}$$

where K_m and a are parameters characteristic of a given polymer solvent system and temperature. The exponent a in eq. (3.11.1) can be divided into three parts[43]

$$a = 0.5 + v(X) + n(X)\epsilon + 1.5a_2 \tag{3.15.8}$$

where

$$\epsilon = d\ln a_s/d\ln M \tag{3.15.9}$$

$$a_2 = d\ln(\langle S^2 \rangle_0/M)/d\ln M \tag{3.15.10}$$

$v(X)$ appearing in eq. (3.15.8) has the significance defined in the KY treatment.[26] ϵ reflects the volume effect and a_2 represents the nonGaussian nature of the unperturbed polymeric chain.

When the sedimentation coefficient at infinite dilution s_0 is related to the molecular weight by

$$s_0 = K_s M^{a_s} \tag{3.14.3}$$

where K_s and a_s are parameters characteristic of a given polymer solvent system, it can be readily shown that the exponent a_s in eq. (3.14.3), is given by

$$a_s = 0.5 - \mu(X) + (1 - m(X))\epsilon - 0.5a_2 - \epsilon \tag{3.15.11}$$

$\mu(X)$ as well as $m(X)$ was defined in the KY theory.[26] A combination of eqs. (3.15.8) and (3.15.11) yields eq. (3.15.12):[42]

$$3a_s + a - 2.0 = v(X) - 3\mu(X) + (n(X) - 3m(X))\epsilon \tag{3.15.12}$$

Both X and ϵ should be determined from the measurement of a and a_s by using eqs. (3.15.8) and (3.15.12) when the value of a_2 is practically zero, or estimated by some other absolute method.

Method 1C. The molecular weight dependence of the diffusion coefficient at infinite dilution, D_0 is expressed as semiempirically by

$$D_0 = K_D M^{a_d} \tag{3.15.13}$$

where K_D and a_d are dependent on a given polymer solvent system and the exponent a_d is given by eq. (3.15.14).

$$a - 3a_d + 1 = v(X) - 3\mu(X) + (n(X) - 3m(X))\epsilon \tag{3.15.14}$$

On substituting numerical values for a, a_d, and a_2 in eqs. (3.15.8) and (3.15.14), values of X and ϵ are obtainable from experimental data.

Method 1D. The MWD of the radius of gyration of polymer chain is expressed empirically by

$$\langle S^2 \rangle^{1/2} = K_\lambda M^{(\lambda+1)/2} \tag{3.12.1}$$

where

$$\lambda = d \ln(\langle S^2 \rangle / M) / d \ln M \tag{3.15.15}$$

K_λ and λ depend on a given polymer solvent system, and exponent λ is given by

$$a - (3\lambda + 1)/2 = v(X) - (3 - n(X))((\lambda + 1)/2 - 0.5 - 0.5a_2) \tag{3.15.16}$$

Eq. (3.15.16) gives a method for evaluating X directly from the data of a, λ, and a_2. At the limit of $X = \infty$, and if $n(X) = 2.43$ is replaced by 3 to a first approximation, eq. (3.15.16) is reduced to the familiar expression,

$$a = (3\lambda + 1)/2 \tag{3.15.16'}$$

Method 1E. The concentration dependence of the sedimentation coefficient is expressed by the empirical equation of the form:

$$(1/s) = (1/s_0)(1 + k_s c)$$

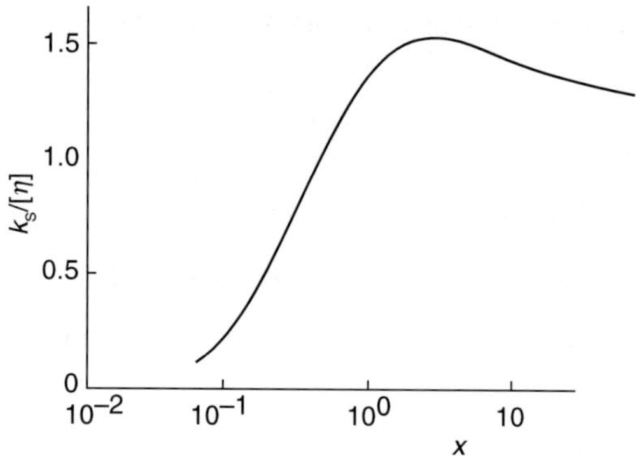

Figure 3.15.10   Plot of $k_s/[\eta]$ against X.

$$3a_s + a - 2.0 = v(X) - 3\mu(X) + (n(X) - 3m(X))\epsilon \tag{3.7.2}$$

where s is the sedimentation coefficient at concentration c, and k_s a parameter related to $[\eta]$ by the formula

$$k_s/[\eta] = (55/8)N_A(\Phi_0(X)^{1/3}P_0(X)^{-1})(16200\pi^2)^{-1}\alpha_s^{(m(X)-3n(X))} \tag{3.15.17}$$

where N_A is the Avogadro number and $P_0(X)$ is a function of X which was defined in the theory of KY.[26] The term $\alpha_s^{(m(X)-3n(X))}$ in the above equation can be approximated as unity under the conditions of $\alpha_s < 2$ within an error less than 10%. Hence, by eq. (3.15.17), we can determine X from data of limiting viscosity number $[\eta]$ and concentration dependence of sedimentation coefficient k_s, provided that $X < 1$ (Figure 3.15.10).

Method 1F. λ in eq. (3.12.1) can be written with the aid of eqs. (3.15.9) and (3.15.10) as

$$\lambda = a_2 + 2\epsilon \tag{3.15.18}$$

Eq. (3.5.18) provides a method for estimating ϵ from λ and a_2. Combining eqs. (3.15.18) with eq. (3.15.8), one obtains

$$a - 0.5 - 1.5\lambda = v(X) + (n(X) - 3)\epsilon \tag{3.15.19}$$

Then, substitution of eq. (3.15.19) into eq. (3.15.12) gives

$$3a_s + a - 2.0 = v(X) - 3\mu(X) + (n(X) - 3m(X)) \\ (0.5 + 1.5\lambda + v(X) - a)/(3 - n(X)) \tag{3.15.20}$$

The draining parameter X and ϵ can be determined from measurement of a, a_s, and λ by use of eqs. (3.15.19) and (3.15.20).

Method 1G. Parameter P, analogous to the Flory viscosity parameter Φ, is defined as[26]

$$P = q'_{w,z}(\xi/\eta_0)(6^{1/2}\langle S^2 \rangle_z^{1/2}) \tag{3.15.4}$$

where $q'_{w,z}$ = polymolecularity correction factor (in this case, D_0, M and s_0 are assumed to be the weight-average quantities), which is given by

$$q'_{w,z(s)} = [\Gamma(3 + h + a_2 + 2\epsilon)]^{1/2}\Gamma(1 + h + a_s)\Gamma(2 + h - a_s)$$
$$\times \{\Gamma(1.5 + h + 0.5a_2 + \epsilon)[\Gamma(2 + h)]^{3/2}\}^{-1} \tag{3.15.21}$$

$$q'_{w,z(s)} = [\Gamma(3 + h + a_2 + 2\epsilon)]^{1/2}\Gamma(1 + h + a_d)\Gamma(2 + h - a_d)$$
$$\times \{\Gamma(1.5 + h + 0.5a_2 + \epsilon)[\Gamma(2 + h)]^{1/2}\Gamma(1 + h)\}^{-1} \tag{3.15.22}$$

Eqs. (3.15.21) and (3.15.22) are derived by Kamide and Miyazaki[32] for polymers with the Schulz–Zimm-type MWD.

The frictional coefficient ξ of a flexible polymer in dilute solution is related to the draining parameter X and $(\langle S^2 \rangle_0/M)^{1/2}$ by

$$\xi/\eta = 6^{1/2}P(\langle S^2 \rangle_0/M)^{1/2}M^{1/2}\alpha_s \tag{3.15.23}$$

with

$$P = P_0(X)\alpha_s - (1 - m(X)) \tag{3.15.24}$$

$$P_0(X) = (3\pi^3/2)^{1/2}[XG_0(X)] \tag{3.15.25}$$

By using eq. (3.15.24), we can estimate the X value from P and α_s. The latter can be evaluated from a penetration function ψ. This method (method 1G) is analogous to method 1A.

In method 1A, the value of α_s should be determined in advance. Unfortunately, there have been only few LS measurements of $\langle S^2 \rangle_0^{1/2}$ for cellulose, amylose, and their derivatives in θ solutions. In addition, there is the possibility that the value of $\langle S^2 \rangle_0^{1/2}$ depends strongly on the solvent nature. Hence, we have to evaluate α_s indirectly for cellulose solutions. For this reason, α_s is determined from the penetration function ψ which is defined by[44]

$$\psi \equiv \bar{z}h_0(\bar{z}) = 0.746 \times 10^{-25}A_2M^2/\langle S^2 \rangle^{3/2} \tag{3.13.5}$$

where

$$\bar{z} = z/\alpha^3 \approx z/\alpha_s^3 \tag{3.13.8}$$

$$z = (1/8\pi^{3/2})BA^{-3}M^{1/2} \tag{3.15.26}$$

$$\alpha = \langle R^2 \rangle^{1/2}/\langle R^2 \rangle_0^{1/2} \tag{3.15.27}$$

$$A = (\langle R^2 \rangle_0/M)^{1/2} = (6\langle S^z \rangle_{w,0}/M_w)^{1/2} \tag{3.13.9}$$

$$B = \beta/m_0^2 \tag{3.13.10}$$

A_2 is the second virial coefficient, A and B, the short- and long-range interactions, $\langle R^2 \rangle$ and $\langle R^2 \rangle_0$, the mean square end-to-end distances of the chain in the perturbed and unperturbed state, β, the binary cluster integral, representing the interaction between the nonbonded segments of polymer chains, m_0, the molecular weight of a segment.

According to the KFSY theory,[45] $h_0(z)$ is related to $\bar{z}$ by

$$\bar{z}h_0(\bar{z}) = (1/5.047)\{1 - (1 + 0.683\bar{z})^{-7.89}\} \qquad (3.13.6)$$

The value of z can be determined from experimental data on A_2, M and $\langle S^2 \rangle_0$ by using eqs. (3.13.5) and (3.13.6). α_s is related to z through

$$\alpha_s^3 - 1 = 1.78z \qquad (3.13.11)$$

A combination of eqs. (3.13.6) and (3.13.11) offers a method for estimating α_s (accordingly, α) from $\bar{z}$. The coefficient (1.78) in eq. (3.13.11) differs slightly depending on the theory, but this variation does not afford a significant change in α_s in the range $\alpha_s^3 < 2$.

Methods 1A and 1D for estimating the draining effect were applied beforehand to some typical vinyl-type polymers in order to test their validity. The data from the literature[46−51] employed here are known to be accurate enough to judge the applicability of various theories of the excluded volume effect in detail. The results are summarized in Table 3.15.7. It is evident that for vinyl-type polymers, X exceed five. It should be noted here that the accuracy of X becomes rapidly low in the range $X > 10$ due to the incompleteness of the theory on Φ. In this section, we choose 2.87×10^{23}[,52,53] for $\Phi_0(\infty)$, but other values such as 2.84×10^{23}[,54] 2.82×10^{23}[,55] and 2.66×10^{23}[,56] have also been obtained theoretically.

The polymer chain of polychloroprene and polyisobutene can be, in general, treated as impermeable to solvent molecules. However, for poly(α-methylstyrene) and poly (p-methylstyrene), the contribution of the draining effect to $[\eta]$ is probably significant, although very small. This corresponds well to the fact of $a_\Phi \neq 0$ for these polymers. For poly (α-methylstyrene), Noda et al.[49] pointed out that $\alpha_\eta (\equiv ([\eta]/[\eta]_\theta)^{1/3}$; $[\eta]_\theta$ is $[\eta]$ in a θ solvent) cannot be expressed as a unique function of z and the nonunique dependence of α_η on z may be attributed to the partial draining of solvent through the polymer coils, but this may not be the only reason.

Application to experimental data

The literature on the viscosity, sedimentation, diffusion, and LS data of various cellulose and amylose derivatives were analyzed according to the methods described above. The polymers are cellulose,[3,4] CTN,[5,6,19,43,57−59] CN,[5,60] CA,[7,35,61,62] CTCp,[8] CTC,[9,63] MC,[10] EC,[64,65] sodium carboxymethyl cellulose (NaCMC),[11] HEC,[12,66] EHEC,[13] sodium cellulose xanthate (NaCX),[14] amylose,[16] ATA,[17] and ATC.[22,23]

Tables 3.15.8a and 3.15.8b summarize the X values thus estimated by methods 1A–1F. In method 1A, the ranges of the X value and its average value are shown. The figure in parentheses is the average. In method 1B–D, X is calculated by assuming $a_2 = 0$ at first, and then X is recalculated by using the a_2 value in Table 3.15.1 and shown in brackets. It is obvious that the effect of a_2 on an estimated value of X is very small. When 2.66×10^{23} is employed as $\Phi(\infty)$ in place of 2.87×10^{23}, the X value thus obtained is at most only 8% larger than that shown in the tables. Table 3.15.1 contains the sample number, the breadth in MWD of the samples (M_w/M_n), linear expansion factor α_s, estimated from Ψ, the parameter K_m, a and M_0 (see Section 3.16, method 2F).

The results of application of method 1G are summarized in Table 3.15.9, in which the data obtained by method 1A are also included. The average value of X determined by

Table 3.15.7

The draining parameter X and the unperturbed chain dimensions A for some typical vinyl-type polymers[2]

Polymer	Solvent	Range of M (10^{-4})	X		A_2	A_Φ	$A \times 10^8$ (cm)							σ	Reference
			1A	1D			2A	2B	2C	2E	2F	2G	Average		
Polychloroprene	MEK(θ)[a]	14.7–292	38–∞	>100	0	0	0.72_5	–	0.70_7	0.73_5	–	–	0.72_2	1.35	
	BA[b]	14.9–302	–[c]	>100	–	0	–	–[c]	0.78_1	0.75_2	0.79_4	–	0.77_6	1.45	46, 47
	CTC[d]	15.6–318	–[c]	>100	–	0	–	–[c]	0.86_5	0.79_5	0.84_9	–	0.83_6	1.56	
Poly(α-methyl-styrene)	*trans*-Decalin (θ)	76.8–747	3.2–5.4 (4.8)	∞	0	0.013	0.69_5	–	0.69_2	0.61_3	–	0.71_1	0.66_9[e]	2.36	
	Cyclohexane ($\theta = 34.5\,°C$)	34.2–747	4.3–7.1 (5.4)	∞	0	0.014	0.71_2	–	0.71_1	0.63_1	–	0.71_4	0.712_2[e]	2.51	48, 49
	Toluene (25 °C)	20.4–747	2.4–11 (6.4)	12	–	0.024	–	0.86_1	0.83_7	0.71_2	0.71_2	0.87_3	0.85_7[e]	3.02	
Poly(p-methyl-styrene)	DES[f] ($\theta = 16.4\,°C$)	16.3–197	7–11.7 (9.9)	10.2	0	0	0.67_5	–	0.68_6	0.61_3	–	–	0.65_8	2.32	
	Toluene	19.2–180	–[c]	6.2	–	0.044	–	–[c]	0.79_0	0.68_0	0.72_9	0.77_1	0.78_1	2.75	50
Polyisobutylene	IAIV[g°] ($\theta = 20\,°C$)	16.0–470	∞	∞	0	0	0.71_5	–	0.69_4	0.72_9	–	–	0.71_3	1.73	
	Cyclohexane	39.1–470	–[c]	17	–	0	–	–[c]	0.92_0	0.87_8	0.88_6	–	0.89_5	2.17	51
	Heptane	39.8–470	–[c]	100	–	0	–	–[c]	0.80_1	0.80_5	0.80_7	–	0.80_4	1.95	

[a]Methylethylketone.
[b]Butyl acetate.
[c]αB could not be evaluated by eq. (3.13.5) via Ψ.
[d]Carbontetrachloride.
[e]In the case of aΦ = 0, average value is caluculated from values by methods 2A, 2B, 2C, and 2F.
[f]Diethyl succinate.
[g]Isoamyl isovalerate.

Table 3.15.8a

The draining parameter X evaluated for cellulose, amylose, and their derivatives

Polymer	Solvent	X							Reference
		1A	1B	1C	1D	1E	1F	Other	
Cellulose	Cadoxen	2.2–9.2 (5.4)[a]	–	1.6 [0.8][b]	0.9 [0.3]	0.3	–	–	3
	FeTNa[c]	0.3–1.6 (1.0)	–	–	0.4	–	–	–	4
CTN $N_c = 13.9\%$	Acetone	0.33–2.4 (1.2)	–	–	0.7	–	–	–	5, 19
$N_c = 13.8\%$	–	–	–	–	–	–	–	0.36	57
$N_c = 13.8\%$	Acetone	–	0.66	1.13	–	–	–	–	58
$N_c = 13.6\%$	Acetone	2.6–30 (9.9)	–	–	0.2	–	–	–	6
$N_c = 13.5\%$	Ethyl acetate	–	0.73	–	–	–	–	0.2–0.4	43
$N_c = 13.3\%$	Ethyl acetate	–	–	–	–	0.17–0.62	–	–	59
$N_c = 13.3\%$	Acetone	–	–	–	–	0.13–0.29	–	–	59
CN $N_c = 12.9\%$	Acetone	0.65–1.5 (1.0)	–	–	1.8 [1.3]	–	–	–	5
$N_c = 12.1\%$	–	–	–	–	–	–	–	0.49	60
CA Acetyl content (AC) = 53.5%	Acetone	–	0.74	–	–	–	–	–	61
(AC) = 54%	Acetone	–	1.8	–	–	–	–	–	62
(AC) = 55.3%	Acetone	0.52–4.0 (1.4)	1.4 [0.25]	–	–	0.27–0.44	–	–	7, 35
CTCp	DMF	0.26–0.58 (0.41)	–	–	0.8 [0.32]	–	–	–	8
	1-CI-N[d]	0.15–0.7 (0.41)	–	–	1.1 [0.4]	–	–	–	8
	Dioxane/water	0.63	–	–	–	–	–	–	8

[a] Average value.
[b] a_2 in Table III in Ref. 2 is used.
[c] Iron sodium tartrate.
[d] I-Chloronaphthalene.

Table 3.15.8b

The draining parameter X evaluated for cellulose, amylose, and their derivatives[2]

Polymer	Solvent	X							Reference
		1A	1B	1C	1D	1E	1F	Others	
CTC	Acetone	0.4–1.8 (1.1)	–	–	0.9	–	–	–	9,63
	Cyclohexanone	0.54–1.6 (0.95)	–	–	4.0	–	–	–	9,63
	Dioxane	0.3–1.6 (0.77)	–	–	–	–	–	–	9,63
MC (DS = 2)	Water	0.3–0.8 (0.56)	–	–	0.8 [0.19]	0.6–2.7 (1.4)	0.21	–	10
EC	Acetone	–	2.5	2.0	–	0.09–1.3 (0.5)	–	–	64
	Acetone	–	1.1	1.1	–	–	–	–	65
NaCMC (DS = 0.88)	NaCl (I → ∞)	4.3–∞ (8.1)[a]	0.2 [0.3]	–	2.6	–	2.1	–	11
HEC (DS = 1)	Water	0.37–3.6 (1.9)	–	–	–	–	–	–	12, 66
	Cadoxen	0.49–4.2 (2.9)	–	–	0.13	–	–	–	12, 66
EHEC (DS = 2)	Water	0.48–4.3 (2.0)	0.12 [0.3]	0.1	–	–	–	–	13
NaCX (DS = 0.78)	1-M NaOH	0.1–0.18 (0.14)	0.03	0.39	0.01	0.31–∞	–	–	14
Amylose	DMSO	0.17–7.2 (1.8)	–	–	0.01	–	–	–	16
ATA	Nitromethane	0.19–6.4 (2.1)	–	–	0.1	–	–	–	17
ATC	Pyridine/water	0.9–1.5 (1.1)	0.36	–	6.0	–	–	–	22
	Pyridine	0.23–0.66 (0.5)	0.70	–	5.0	–	–	–	23

[a]The value of $X = \infty$ is excluded in averaging process.

Table 3.15.9

The draining parameter X evaluated from the Flory's parameters P and Φ[32]

Polymer	Solvent	S_0 or D_0	Method 1G (from P)	Method 1A (from Φ)[67]
			X	X
Cellulose	Cadoxen	S_0	1.4–12 (5.0)[a]	2.2–9.2 (5.4)[a]
Cellulose diacetate	Acetone	$\{D_0$ $\{S_0$	0.9–30 (6.7)[a] 0.5–[b*] (1.96)[c]	0.52–4.0 (1.45)[a]
Cellulose tricarbanilate	Cyclohexanone	S_0	*	0.54–1.6 (0.95)[a]
Ethylhydroxyethyl cellulose	Water	$\{S_0$ $\{D_0$	0.55–10.0 (3.7)[a] 0.38–1.5 (0.73)[a]	0.48–4.3 (2.0)[a]
Sodium cellulose xanthate	1-N NaOH	S_0	0.28–*(0.49)[c]	0.1–0.18 (0.14)[a]
Amylose tricarbanilate	Pyridine	S_0	2.1–*(13.2)[c]	0.23–0.66 (0.5)[a]

[a]Number in parenthesis means average.

[b]Asterisk denotes the case of $P > P_0(\infty)(\equiv 5.2)$.

[c]Number in parenthesis means average, which is obtained except for the case of $P_0 > P_0(\infty)(\equiv 5.2)$.

method 1G agrees satisfactorily with that of method 1A, except for CTC and ATC. The experimental data for the former polymer seem less reliable. Therefore, the conclusions drawn from Tables 3.15.8a and 3.15.8b do not contradict those from a_Φ and a_2.

CA (DA 0.49). The expansion factor α_s in DMAc, DMSO, water, and FA was calculated according to the procedure using the penetration function ψ. The results are listed in column 9 of Table 3.3.1[24] and in column 9 of Table 3.5.1.[24] The α_s values remain almost constant within experimental error and are close to unity, suggesting that all the solvents employed are thermodynamically poor solvents for CA.

It should be remembered that any theory yields nearly the same relationships between ψ and the excluded volume parameter z for $\psi \leq 0.1$, which is the range in the CA (DS 0.49)/solvent systems (column 8 of Table 3.5.1). Using Φ and α_s, the draining parameter X, defined in the KY theory[26] was estimated (method 1A), and is given in the final columns of Tables 3.3.1[24] and 3.5.1.[24]

For the CA (DS 0.49)/DMAc system, the X values obtained in this narrow molecular weight range are too small to permit examination of their molecular dependence. These values were found to be from 4.2 to 10.5 (average 7.3), and are smaller than $X \geq 10$, which is the case for many vinyl-type polymers.

CA (DS 1.75). The X values of CA (DS 1.75) in DMAc are collected in column 11 of Table 3.3.2.[27] The X values range from 0.8 to 1.3.

CA (DS 2.46). The X values for CA fractions in acetone and in THF, estimated by method 1A, are given in the fourth and seventh columns of Table 3.15.2.[28] In acetone, X increases gradually with M_w from 0.52 to 4.2. $X = 1.48$ and 6.6 are obtained as average for acetone and THF, respectively. Stein and Doty[67] determined M_w, A_2, and $\langle S^2 \rangle_z^{1/2}$ of the fractions in acetone, whose $[\eta]$ was also measured by Badgley and Mark.[68] By use of these data, we obtain $X = 0.04$–0.07. Using data from Tanner and Berry's work[69] for CDA in THF and trifluoroethanol (TFE) at 25 °C, the X value as evaluated by method 1A is found to be 0.70 in THF and 0.27–0.8 (average, 0.46) in TFE. It will be noted that contamination by gel-like materials leads to erroneously low X values.

Assuming $a_2 = 0$, method 1B was applied to the data of Holmes and Smith (HS),[61] Golbev and Frenkel (GF),[62] and Ishida *et al.* (I),[35] who employed the same fractions as those for sedimentation velocity measurement. The X value is evaluated as 0.75, 2.5, and 1.4 for HS, GF, and I works, respectively. If we put $a_2 = -0.471$ (see Table 3.15.1) for the data from Ishida *et al.*'s work on the CDA fractions in acetone ($a = 0.616$ and $a_s = 0.384$) into eq. (3.15.12), then we obtain $X = 0.24$ in place of 1.4. Holmes and Smith[61] carried out the diffusion measurement in acetone for the same CDA fractions as those cited in method 1B. $a_d = 0.70$ is obtained from their data. Using $a = 0.98$ and $a_d = 0.70$, and assuming $a_2 = 0$, we obtain $X = 0.74$ by method 1C. Ishida *et al.*[35] determined s for the same CDA fractions as those used here in acetone as a function of c. Using their data, we obtain $X = 0.27$–0.44 (average, 0.32) by method 1E. The X values thus determined are in the range of 0.3–7 as recorded in Table 3.15.10, suggesting that the partially free-draining effect on the viscosity, sedimentation, and diffusion coefficients can never be ignored, particularly s_0 in acetone.

We have already demonstrated for cellulose, amylose, and their derivatives that the X value is almost invariably less than 2.24 (see Tables 3.15.8a and 3.15.8b).[2]

For the CA (DS 2.46) DMAc system, using $a = 0,616$ and $a_s = 0,39$ for CA (DS 2.46)/acetone, we obtain $X = 0.26(a_2 = -0.471)$[28] and from $a = 0,75$ and $a_s = 0.24$ for CA (DS 2.92)/DMAc, we conclude that $X = 0.3(a_2 = 0)$.[30] For the latter system, the equality $a_2 = 0$ has been experimentally confirmed.[30]

We can determine X from data of $[\eta]$ and k_s, provided that $a_s < 2$ and $X < 1$ (method 1E). The ratio $k_s/[\eta]$ for CA (DS 2.46)/acetone and CA (DS 2.92)/DMAc systems is collected in the ninth column of Table 3.7.2.[35] This ratio ranges between 0.47 and 1.02, being significantly smaller than 1.5–1.7, which is usually evaluated for the vinyl-type polymers. The X values obtained by this method slightly increase with molecular weight. We obtain $X = 0$–0.44 (average 0.32) for CA (DS 2.46)/acetone and 0.32–0.98 (average 0.65) for CA (DS 2.92)/DMAc, respectively.[35]

The draining parameter X, determined for CDA/DMAc by method 1A, is collected in Table 3.15.3[29] and can be compared with those in acetone ($X = 1.48$) and THF (6.6; Table 3.15.2).[28] It should be noted that method 1A assumes that $a_2 = 0$ in the unperturbed state, although the CDA chain in acetone is nonGaussian (i.e. $a_2 < 0$). Thus, X for CDA in acetone evaluated by method 1A might be less accurate, but the values evaluated by methods 1B and 1E[2] lie between 0.2 and 0.3, and are much smaller than approximate values at the nondraining limit (20 to -50). The X value for CDA in DMAc is the smallest among the three solvents in which the draining effect cannot be ignored.

If we assume a molecular chain model, then a' and N can be separately evaluated and d from X data alone using eq. (3.15.7).

Table 3.15.10

Draining parameter X of cellulose diacetate[28]

Solvent	DS	Temper- ature °C	X				Remarks
			Method 1A	Method 1B	Method 1C	Method 1E	
Acetone	2.45	25	0.52–4.2 (1.48)[a]	–	–	–	Kamide *et al.*[28]
				1.40 [0.24][b]		0.27–0.44 (0.32)	Ishida *et al.*[35]
	2.43	25	0.04–0.07 (0.06)	–	–	–	Stein *et al.*[67]
	2.33	20	–	0.74	0.74	–	Holmes and Smith[63]
	2.33–2.37	25	–	1.8	–	–	Golub and Frenkel[62]
			⎰4.6–8.5 (6.6) ⎱0.7	–	–	–	Kamide *et al.*[28]
THF	2.45	25					Tanner and Berry[69]
TFE[c]	2.45	20	0.27–0.8 (0.46)	–	–	–	Tanner and Berry[69]

[a]Number in parenthesis means average.
[b]Number in bracket denotes the value when α_2 is taken into account.
[c]Trifluoroethanol.

Another approach for evaluating d when X is unknown is to use the first-order perturbation theory for volume effect[70]

$$(\alpha_s^2 - 1)/[1 - (\theta/T)] = (134/105)(6/\pi)^{1/2}(d/a')^3 N^{1/2} \qquad (3.15.28)$$

where θ is Flory's theta temperature and T is the temperature, both expressed in K. Eq. (3.15.28) is valid only in the vicinity of $\alpha_s \approx 1$, but should hold for cellulose derivative solutions (see Section 3.12). Substituting the experimental values for a_s, θ, and T into eq. (3.15.28) makes it possible to determine the d value, if a' and N are known in advance.

The hydrodynamical segment is assumed to be equivalent to the statistical chain segment (the statistical chain model) or to the monomer unit (the simple equivalent chain model). On the basis of these two molecular chain models, a' and N can be estimated.

For a statistical chain model,

$$a' = A^2 m_0/\ell \qquad (3.15.29)$$

and

$$N = M/(A m_0/\ell)^2 \qquad (3.15.30)$$

where A is the unperturbed chain dimensions, ℓ ($= 6.15$ Å for CDA) and m_0 ($= 264$ for CDA) are the length and the molecular weight of the monomer unit, respectively.

For a simple equivalent chain model,

$$a' = \ell \qquad (3.15.29')$$

and

$$N = M/m_0 \qquad (3.15.30')$$

It may be expected that if the X values obtained here (Table 3.15.2) are reasonable, then the d values evaluated from X data using eq. (3.15.7) should agree fairly well with those obtained by the thermodynamical method (eq. (3.15.28)). In order to confirm this expectation, the experimental data on CDA sample EF3-8 ($M_w = 9.4 \times 10^4$) in acetone were analyzed.[29] The X value for this sample was determined to be 0.95 by method 1A from Table II of Ref. 71, and α_s was calculated to be 1.02 at 25.4 °C. We also estimated θ (in this case, the lower critical solution temperature θ_{LCST}) to be about 163 K for sample EF3-8 in acetone by analyzing the temperature dependence of A_2 and also A to be 1.6 at same temperature by method 2B (see eq. (3.13.3)). The results are shown in Table 3.15.11.[29] Here, the d value corresponding to $X = 20$ is included for comparison.

Although rather large experimental errors are involved in the determination of θ_{LCST} for CDA in acetone, the d values determined on the basis of the statistical chain segment model by these two different methods are quite consistent. The use of $X = 20$ in eq. (3.15.1) yields an unreasonably large d value, nearly 30 times larger than the d value (24.6 Å) obtained by the thermodynamic method. This indicates that the draining effect cannot be neglected, and that the X values determined in this section are reliable. In addition, the simple equivalent chain model, along with eq. (3.15.7),

Table 3.15.11

Evaluation of hydrodynamical segment diameters of CDA and polystyrene[29]

Polymer	Solvent	Sample no.	$M_{\mathrm{w}} \times 10^{-4}$	$X_{\mathrm{E}}^{\mathrm{a}}$	$d \times 10^8$ (cm)					
					Statistical chain model			Simple equivalent chain model		
					$X = X_{\mathrm{E}}^{\mathrm{b}}$	$X = 20^{\mathrm{b}}$	c	$X = X_{\mathrm{E}}^{\mathrm{d}}$	$X = 20^{\mathrm{d}}$	e
CDA	Acetone[f]	EF3-9	9.6	0.95	33.5	712	24.6	0.5	9.5	0.3
	DMAc	EF3-13	15.6	0.71	37.0	1047	–	0.3	7.3	–
Polystyrene[g]	Cyclochexane	HB2-3	320	33.1	9.3	5.6	7.3[h]	0.3	0.2	1.6

[a]X values estimated by method 1A using experimental data.
[b]From eqs. (3.15.7), (3.15.29), and (3.15.30).
[c]From eqs. (3.15.26), (3.15.29), and (3.15.30).
[d]From eqs. (3.15.7), (3.15.29′), and (3.15.30′).
[e]From eqs. (3.15.26), (3.15.29′), and (3.15.30′).
[f]Ref. 28.
[g]Ref. 72.
[h]Ref 26.

provides an unexpectedly smaller d value (0.5 Å). Kurata and Yamakawa have shown for polystyrene in cyclohexane that the draining effect is negligible (i.e. $X \geq 20$) and that the statistical chain model gives the segment size a more reasonable value than the simple equivalent chain model.[26] We analyzed the same data[72] used by Kurata and Yamakawa,[26] and summarized the results in Table 3.15.10, which confirm the validity of their conclusion.

Using the X (Table 3.15.2) and A (Table 3.16.2) data for a CDA sample EF3-13 in DMAc,[29] we determined the d value on the statistical chain segment model by eq. (3.15.1) to be 37 Å, which is about the same as that in acetone, being five times larger than the molecular diameter of a glucopyranose unit, crystallographically determined.[73] The d value obtained on the statistical chain model by neglecting the draining effect for CDA in DMAc is more than 700 Å, as in the case of CDA in acetone. This is too large to be accepted as the diameter of a hydrodynamical segment. As long as the statistical chain model is employed, the draining effect should not be neglected.

As will be shown in a later section (3.17), cellulose and cellulose derivatives are dissolved in solvents with appreciable solvation,[74–77] and this is considered to be responsible for the serious discrepancy between d and the actual molecular diameter. The values of d for other CDA fractions in DMAc were estimated to be in the range of 30–45 Å, on the basis of the statistical chain model. These values are not too large, but exceed those anticipated from the experiments on solvation.[39,40]

When a_2 is not zero, d cannot be determined simply from eq. (3.15.7), since the coefficient appearing in this equation should differ from $(6/\pi)^{1/2}$, judging from eq. C. 10 in Ref. 26. Thus, the deduction of the correct d for cellulose derivatives is even now open for a further study.

CA (DS 2.92). The X values for CA (DS 2.92) in DMAc were estimated by method 1A and are shown in the last column of Table 3.3.5.[30] Here, the averaged X value was found to be 2.1. To check the reliability of the values of Φ and a_Φ, which are very closely connected both with the draining parameter, the parameter X value was also evaluated from the following equations:

$$a_\Phi = (1/2)[\mathrm{d}\ln \Phi(X)/\mathrm{d}\ln X] - \{3 - n(X)\}a_1/3 \qquad (3.15.31)$$

and

$$a_1 = a - 05 - a_\Phi - 1.5a_2 \qquad (3.15.32)$$

On substituting numerical values for the quantities a, a_Φ, and a_1, obtained for CA (DS 2.92)/DMAc system (i.e. $a = 0.750$, $a_\Phi = 0.106$, and $a_2 = 0$; see Figure 3.15.4b), into eqs. (3.15.31) and (3.15.32), X is found to be approximately 4. Therefore, the values of Φ and a_Φ are roughly concordant, although the values determined by method 1A are much more reliable.

The significant contribution of the draining effect had not attracted very much attention until similar results were reported very recently for CA (DS 2.46)[28] and other cellulose derivatives.[2] It is of interest that the a_Φ value of CA (DS 2.46) in acetone is known to be exceptionally large.[28]

Reliability of the methods

The considerable difficulty in determining X by method 1A lies in the fact that the samples used in the literature had wide MWDs and their accurate shape was not determined experimentally as shown in the fourth column of Table 3.15.1. This is a limitation of method 1A. Exceptional cases are CN, CTN,[5,19] and CA,[7,28] whose samples have relatively narrow MWD ($M_w/M_n \approx 1.2$). Even in these cases, the X values estimated by using method 1A are usually less than two. In fact, the magnitude of the X value obtained by using various methods for each polymer sample is susceptible to large error. An accumulation of knowledge of X values was necessary in order to obtain the definite conclusions on the draining effect. Such a compilation of X values is shown in Tables 3.15.8(a) and 3.15.8(b). Considering the experimental accuracy, we cannot evaluate the exact value of X, but different methods always give $X \leq 2$ with some exceptions, and this is considerably lower than that of usual vinyl-type polymers. In addition, the value of a calculated from λ by using eq. (3.15.16$'$) is, on average, only 40% of the experimental one. Therefore, the partially free-draining effect on $[\eta]$ cannot be ignored for cellulose, amylose, and their derivatives. This conclusion was drawn previously for CN by Kamide.[42]

REFERENCES

1. See, for example, K Kamide and T Dobashi, *Physical Chemistry of Polymer Solutions. Theoretical Background*, Problems 8–32, Problems 8–33, Problems 9–24, Problems 9–28, Elsevier, Amsterdam, 2000.
2. K Kamide and Y Miyazaki, *Polym. J.*, 1978, **10**, 409.
3. D Henley, *Ark. Kemi.*, 1961., **18**, 327.
4. L Valtasaari, *Markromol. Chem.*, 1971, **150**, 117.
5. GV Schulz and E Penzel, *Markromol. Chem.*, 1968, **112**, 260.
6. MM Huque, DA Goring and SG Mason, *Can. J. Chem.*, 1958, **36**, 952.
7. K Kamide, T Terakawa, Y Miyazaki, unpublished results.
8. WR Krigbaum and LH Sperling, *J. Phys. Chem.*, 1960, **64**, 99.
9. VP Shanbhag, *Ark. Kemi.*, 1968, **29**, 1.
10. WB Neely, *J. Polym. Sci., Part A*, 1963, **1**, 311.
11. W Brown and D Henley, *Markromol. Chem.*, 1964, **79**, 68.
12. W Brown, D Henley and J Öhman, *Markromol. Chem.*, 1963, **64**, 49.
13. RS Manley, *Ark. Kemi.*, 1956, **9**, 519.
14. B Das, AK Ray and PK Choudhury, *J. Phys. Chem.*, 1969, **73**, 3413.
15. B Das and PK Choudhury, *J. Polym. Sci., Part A-1*, 1967, **5**, 769.
16. JMG Cowie, *Markromol. Chem.*, 1961, **42**, 230.
17. JMG Cowie, *J. Polym. Sci.*, 1961, **49**, 455.
18. W Banks, CT Greenwood and J Sloss, *Eur. Polym. J.*, 1971, **7**, 263.
19. E Penzel and GV Schulz, *Markromol. Chem.*, 1968, **113**, 64.
20. VP Shanbhag and J Öhman, *Ark. Kemi.*, 1968, **29**, 163.
21. J Öhman, *Ark. Kemi.*, 1968, **31**, 125.
22. W Banks, CT Greenwood and J Sloss, *Makromol. Chem.*, 1970, **140**, 100.
23. W Banks, CT Greenwood and J Sloss, *Eur. Polym. J.*, 1971, **7**, 879.
24. K Kamide, M Saito and T Abe, *Polym. J.*, 1981, **13**, 421.
25. WH Stockmayer and M Fixman, *J. Polym. Sci., Part C*, 1963, **1**, 137.
26. M Kurata and H Yamakawa, *J. Chem. Phys.*, 1958, **29**, 311.
27. M Saito, *Polym. J.*, 1983, **15**, 249.

28. K Kamide, T Terakawa and Y Miyazaki, *Polym. J.*, 1979, **11**, 285.
29. K Kamide and M Saito, *Polym. J.*, 1982, **14**, 517.
30. K Kamide, Y Miyazaki and T Abe, *Polym. J.*, 1979, **11**, 523.
31. PRM Nair, RM Gohil, KC Patel and RD Patel, *Eur. Polym. J.*, 1977, **13**, 273.
32. K Kamide and Y Miyazaki, *Polym. J.*, 1978, **10**, 539.
33. S Tomita, Kobunshi Sokuteiho (ed. Society of Polymer Science, Japan), Ultracentrifugation method, Baifukan, **Vol. 1**, Tokyo, 1973, p. 19.
34. H Inagaki, Kobunshi Kagaku (II). In *Shin-Jikken Kagaku*, Society of Polymer Science, Japan, **Vol. 1**, Maruzen, Tokyo, 1978, p. 544.
35. S Ishida, H Komatsu, H Katoh, M Saito, Y Miyazaki and K Kamide, *Makromol. Chem.*, 1982, **183**, 3075.
36. VP Shanbhag, *Ark. Kemi.*, 1968, **29**, 33.
37. VP Shanbhag, *Ark. Kemi.*, 1968, **29**, 139.
38. K Kamide and M Saito, *Polym. J.*, 1986, **18**, 569.
39. K Kamide and M Saito, *Eur. Polym. J.*, 1984, **20**, 903.
40. K Kamide, M Saito, unpublished result.
41. K Kishino, T Kawai, T Nose, M Saito and K Kamide, *Eur. Polym. J.*, 1981, **17**, 623.
42. K Kamide, *Mokromol. Chem.*, 1969, **128**, 197.
43. ML Hunt, S Newman, HA Scheraga and PJ Flory, *J. Phys. Chem.*, 1956, **60**, 1278.
44. M Kurata, *Industrial Chemistry of High Polymers*. Modern Industrial Chemistry Ser. No.18 Vol. III, Chapter 4, Asakura, 1975.
45. M Kurata, M Fukatsu, H Sotobayashi and H Yamakawa, *J. Chem. Phys.*, 1964, **41**, 139.
46. T Norisuye, K Kawahara, A Teramoto and H Fujita, *J. Chem. Phys.*, 1968, **49**, 4330.
47. K Kawahara, T Norisuye and H Fujita, *J. Chem. Phys.*, 1968, **49**, 4339.
48. T Kato, K Miyaso, I Noda, T Fujimoto and M Nagasawa, *Macromolecules*, 1970, **3**, 777.
49. I Noda, K Mizutani, T Kato, T Fijimoto and M Nagasawa, *Macromolecules*, 1970, **3**, 787.
50. G Tanaka, S Imai and H Yamakawa, *J. Chem. Phys.*, 1970, **52**, 2639.
51. T Matsumoto, N Nishioka and H Fujita, *J. Polym. Sci., Part A-2*, 1972, 23.
52. JG Kirkwood and J Riseman, *J. Chem. Phys.*, 1948, **16**, 565.
53. PL Auer and CS Cardner, *J. Chem. Phys.*, 1955, **23**, 1546.
54. BH Zimm, *J. Chem. Phys.*, 1956, **24**, 269.
55. JE Hearst, *J. Chem. Phys.*, 1962, **37**, 2547.
56. CW Pyun and M Fixman, *J. Chem. Phys.*, 1966, **44**, 2107.
57. GP Pearson and WR Moore, *Polymer*, 1960, **1**, 144.
58. G Meyerhoff, *J. Polym. Sci.*, 1958, **29**, 399.
59. K Kamide, T Shiomi, H Ohkawa and K Kaneko, *Kobunshi Kagaku*, 1965, **22**, 785.
60. WR Moore and GD Edge, *J. Polym. Sci.*, 1960, **47**, 469.
61. FM Holmes and DI Smith, *Trans. Faraday Soc.*, 1956, **52**, 67.
62. VM Golub and SYa Frenkel, *Vysokomol. Sbedin., Ser A*, 1967, **9**, 1847.
63. VP Shanbhag, *Ark. Kemi.*, 1968, **29**, 139.
64. K Uda and G Meyerhoff, *Makromol. Chem.*, 1961, **47**, 168.
65. G Meyerhoff and N Sutterlin, *Macromol. Chem.*, 1965, **87**, 258.
66. W Brown and D Henley, *Macromol. Chem.*, 1967, **108**, 153.
67. RS Stein and P Doty, *J. Am. Chem. Soc.*, 1946, **68**, 159.
68. WJ Badgley and H Mark, *J. Phys. Chem.*, 1947, **51**, 58.
69. DW Tanner and GC Berry, *J. Polym. Sci., Polym. Phys.*, 1974, **12**, 941.
70. K Kurata, H Yamakawa and E Teramoto, *J. Chem. Phys.*, 1958, **28**, 785.
71. H Suzuki, Y Miyazaki and K Kamide, *Eur. Polym. J.*, 1980, **16**, 703.
72. WR Krigbaum and DK Carpenter, *J. Phys. Chem.*, 1955, **59**, 1166.
73. DW Jones, *J. Polym. Sci.*, 1950, **5**, 519.
74. K Kamide, K Okajima and M Saito, *Polym. J.*, 1981, **13**, 115.
75. WR Moore and BM Tidswell, *Markromol. Chem.*, 1965, **81**, 1.
76. WR Moore, *J. Polym. Sci., C*, 1967, **16**, 571.
77. WR Moore, Solution properties of natural polymers. *The Chem. Soc.*, 1968, 185.

3.16   UNPERTURBED CHAIN DIMENSIONS (UCD)

3.16.1   Methods for estimating unperturbed chain dimensions

The UCD A can be expressed in terms of a short-range interaction parameter given by
eq. (3.13.9′):

$$A = (\langle R^2\rangle_{w,0}/M_w)^{1/2} \tag{3.13.9′}$$

The conformation parameter σ and characteristic ratio C_∞ are defined by eqs. (3.16.1)
and (3.16.2), respectively.

$$\sigma = (\langle R^2\rangle_{w,0}/\langle R^2\rangle_{0f})^{1/2} = A/A_f \tag{3.16.1}$$

$$C_\infty = A_\infty^2 M_b/d \tag{3.16.2}$$

where $\langle R^2\rangle_{0f}^2 (= A_f M^{1/2})$ and A_f are the root mean square end-to-end distance and
the short-range interaction parameter of a hypothetical chain with free internal rotation,
A_∞, the asymptotic value of A at infinite molecular weight (in the case of $a_2 \neq 0$, A value
at $M_w = 1 \times 10^5$ is utilized as A_∞), M_b, the mean molecular weight per skeletal bond
and d', the mean bond length (5.47 Å for cellulose derivatives and 4.25 Å for amylose
derivatives). Here, the $C1$ ring conformation is assumed for the standard β-D-glucose
residue. The methods of estimating A are summarized below.

Thermodynamic approach

Method 2A.[1] When the unperturbed radius of gyration $\langle S^2\rangle_0^{1/2}$ can be measured in a
Flory's theta solvent, where $A_2 = 0$, $z = 0$, and $a_s = 1$, A is directly evaluated by use of
eq. (3.13.9).

Method 2B.[1] When the linear expansion factor α_s is determined from the function ψ, the
unperturbed dimension, $\langle S^2\rangle_0^{1/2}$, A can be evaluated from the experimental data of $\langle S^2\rangle^{1/2}$
in nonideal solvents using eq. (3.13.4). Only in the range $\alpha_s^3 > 2$, the value of α_s depends
significantly on the theory of A_2 adopted. $\langle S^2\rangle_0$ can be transformed to $\langle R^2\rangle_0$ by use of
eq. (3.16.3):[2]

$$\langle R^2\rangle_0 = 6\langle S^2\rangle_0 \tag{3.16.3}$$

In this section, eq. (3.16.3) is applied even in the case of $a_2 \neq 0$.

Method 2C.[1] If the factor 1.78 in eq. (3.13.11) is replaced by 2.0, which is widely used for
case, then we obtain[3]

$$\alpha_s^3 - 1 \approx 2.0z \tag{3.13.11′}$$

Eq. (3.15.26) can be recast with the aid of eq. (3.13.11′) as

$$\langle S^2\rangle_w^{3/2}/M_w^{3/2} = A^3/6^{3/2} + (1/4\pi^{3/2})BM_w^{1/2} \tag{3.16.4}$$

Plots of $\langle S^2\rangle^{3/2}/M_w^{3/2}$ against $M_w^{1/2}$ (the Baumann plot)[4] enables us to evaluate A from the
intercept at $M_w^{1/2} = 0$. In this case, the unperturbed chain is assumed to be a Gaussian

chain. The experimental data of the z-average radius of gyration $\langle S^2 \rangle_z^{1/2}$ were converted to the weight-average radius of gyration $\langle S^2 \rangle_w^{1/2}$.

Method 2D.[1] In the ease of $a_2 \neq 0$, the Baumann plot is not applicable and the following equation can be used in place of eq. (3.16.4):

$$\langle S^2 \rangle^{3/2}/M_w^{3(1+a_2)/2} = K_0^{3/2} + (1/4\pi^{3/2})BM_w^{(1-3a_2)/2} \tag{3.16.5}$$

where

$$K_0 = (\langle S^2 \rangle_0/M_w)/M_w^{a_2} \tag{3.16.6}$$

By use of eq. (3.16.5), the plots of $\langle S^2 \rangle^{3/2}/M_w^{3(1+a_2)/2}$ against $M_w^{(1-3a_2)/2}$ (Baumann–Kamide–Miyazaki (BKM) plot) for a given solvent should result in a straight line and its extrapolation to $M_w^{1/2} = 0$ should give the $K_0^{1/2}$ value and thus K_0.

Hydrodynamic approach

Method 2E. According to the KY theory,[5] the limiting viscosity number $[\eta]$ is given by eq. (3.16.7):

$$[\eta] = \pi^{3/2}N_AXF_0(X)(\langle S^2 \rangle_0^{3/2}/M)(1 + p(X)z\cdots) \tag{3.16.7}$$

where $F_0(X)$ and $p(X)$ are functions of the draining parameter X. When $a_2 = 0$, eq. (3.16.7) can be rewritten on the basis of the Kawai–Kamide treatment[6] as

$$[\eta]/f(X)M_w^{1/2} = K + \Phi_0(X)p(X)(3/2\pi)^{3/2}BM_w^{1/2} \tag{3.16.8}$$

where

$$K = \Phi_0(X_0)A^3 \tag{3.16.9}$$

$$f(X) = XF_0(X)/X_0F_0(X_0) \tag{3.16.10}$$

The value of X_0 in eq. (3.16.10) is determined by the equation:

$$v(X_0) = \epsilon(3 - n(X_0)) \tag{3.16.11}$$

The value of K (accordingly, A) can be determined as an intercept at $M_w^{1/2} \to 0$ of the plot of $[\eta]/f(X)M_w^{1/2}$ against $M_w^{1/2}$. The well-known Stockmayer–Fixman (SF; or Kurata–Stockmayer–Fixman–Burchard; KSFB) equation:[7]

$$[\eta]/M_w^{1/2} = K + 2(3/2\pi)^{3/2}\Phi_0(\infty)BM_w^{1/2} \tag{3.16.12}$$

is derived from eq. (3.16.7) by putting $X = \infty$ and neglecting terms higher than $M_w^{1/2}$. It should be noticed here that eq. (3.16.12) does not hold when $a_2 \neq 0$ or when X is finite or at least $X < X_0$. In the latter case, a_Φ in eq. (3.15.3) no longer becomes zero. The above mentioned limitations of eq. (3.16.12) were not taken into consideration when this equation was applied to cellulose and its derivative solutions. In the case $a_2 = 0$ and $a_\Phi = 0$, the plot of $[\eta]/M_w^{1/2}$ versus $M_w^{1/2}$ (SF plot) obtained from experimental data for a given polymer should give a straight line and its intercept at $M_w^{1/2} \to 0$ and its slope should give K and B, respectively. Here, after the correction of MWD of the sample was applied to K, A was evaluated from K value using eq. (3.16.9) ($X_0 = \infty$). Here, $\Phi_0(\infty) = 2.87 \times 10^{23}$ was adopted.

Method 2F.[1] The following relationship holds between parameters K_m, a in eq. (3.11.1), a_2 and X.

$$\log K_\mathrm{m} + \log[1 + 2(a - 0.5 - \Delta - 1.5a_2)^{-1} - 2^{-1}]$$

$$= \log Kf(X) + (a - 0.5 - \Delta - 1.5a_2)\log M_0 \tag{3.16.13}$$

Eq. (3.16.13) and similar equations were derived by Kamide, Kawai, and their coworkers.[8–12] Δ in eq. (3.16.13) is defined by eq. (3.16.14).

$$\Delta = v(X) - v(X_0) = v(X) - \epsilon(3 - n(X_0)) \tag{3.16.14}$$

M_0 is a parameter depending on the molecular weight range, M_1–M_2, to which eq. (3.11.1) applies. In a case where the draining effect is negligible, the geometric mean, $(M_1 M_2)^{1/2}$, can be regarded to a fairly good approximation as M_0. K can be determined from K_m, a, X (accordingly, Δ), and a_2. However, the accuracy of estimation of X by eq. (3.16.13) is inadequate. Putting $a_2 = 0$ and $\Delta = 0$ in eq. (3.16.13), we obtain:

$$-\log K_\mathrm{m} + \log[1 + 2\{(a - 0.5)^{-1} - 2\}^{-1}] = -\log K + (a - 0.5)\log M_0 \tag{3.16.15}$$

Eq. (3.16.15) is, in principle, equivalent to eq. (3.16.12). By use of eq. (3.16.15), K (accordingly, A) can be evaluated from experimental data of K_m, a, and M_0, provided that $a_2 = 0$ and $a_\Phi = 0$ (i.e. $X \approx \infty$). It has been ascertained that eq. (3.11.1) is quite useful for estimating the K value of vinyl-type polymers.[13]

Method 2G.[1] Eq. (3.15.1) can be rearranged as:

$$[\eta] = 6^{3/2}\Phi(\langle S^2\rangle_{\mathrm{w},0}/M_\mathrm{w})^{3/2}M_\mathrm{w}^{1/2}a_\mathrm{s}^3 \tag{3.15.1'}$$

As was demonstrated in Section 3.15, Φ of cellulose, amylose, and their derivatives in dilute solutions reveal a large molecular weight dependence. This can be interpreted as another indication of the draining effect. Thus, the molecular weight dependence of Φ is empirically approximated by eq. (3.15.3):

$$\Phi = K_\Phi M^{a_\Phi} \tag{3.15.3}$$

where K_Φ and a_Φ are parameters characteristic of a polymer-solvent system and the molecular weight range, in which eq. (3.15.3) holds. a_Φ is given by

$$a_\Phi = [v(X) - \epsilon(3 - n(X))]_{\mathrm{AV}}. \tag{3.16.16}$$

It is clear that for X_0, a_Φ became essentially zero and $a_\Phi \approx \Delta$ (see eq. (3.16.14)). The right-hand side term in eq. (3.16.16) is a kind of average value of $(v(X) - \epsilon(3 - n(X))$ over the molecular weight range concerned.

A combination of eqs. (3.15.1), (3.16.6), and (3.15.3) leads to the equation of $[\eta]$ for the polymer solution in which $a_2 \neq 0$ and $\alpha_\Phi \neq 0$ hold, as is the case of cellulose, amylose, and their derivatives, given by eq. (3.16.17):

$$[\eta]/M^{0.5 + a_\Phi + 1.5a_2} = 6^{3/2}K_\Phi K_0^{3/2} + 0.66K_\Phi BM^{(1 - 3a2)/2} \tag{3.16.17}$$

Eq. (3.16.17) is a straightforward generalization of eq. (3.16.12) in the most versatile form.

According to eq. (3.16.17), a graph of $[\eta]/M_\mathrm{w}^{0.5+a_\Phi+1.5a_2}$ as a function of $M^{(1-3a_2)/2}$ (Kamide–Miyazaki (KM) I plot) should be linear, with $6^{3/2}K_\Phi K_0$ (accordingly, K_0) as the intercept. Eq. (3.15.9) can be rewritten as follows

$$a = 0.5 + a_\Phi + a_1 + 1.5a_2 \qquad (3.15.32)$$

where

$$a_1 = 3\epsilon \qquad (3.16.18)$$

Thus, the value of a_1 can be roughly estimated from a, a_Φ, and a_2, through use of eq. (3.15.32).

Verification of methods 2A–2G as methods for estimation of the unperturbed chain dimensions

methods 2A–2G were applied to the literature data[14–19] of typical vinyl-type polymers and method 2G was applied to the data on poly(a-methylstyrene) and poly (p-methylstyrene), for which a_Φ is positive.[1] A much better agreement between A values evaluated by method 2G and methods 2A, 2B, and 2C is observed. $a_2 = 0$ and $a_\Phi = 0$ for polychloroprene and polyisobutylene were experimentally confirmed. In this case, method 2D is equal to method 2C, and method 2G reduces to method 2E. In good solvents where ψ, defined by eq. (3.13.5),[20] exceeds 0.2, we could not determine a_2 from eq. (3.13.6) because the theoretical limiting value of ψ is 1/5.047. Thus, method 2B is limited in its applicability to the small α_s region. This is the case for cellulose and its derivatives (see Table 3.15.1). The results indicate that no large difference exists between the five methods and all of these methods yield an A value identical within ± 0.025. This proves the validity of these methods. Methods 2A, 2C, and 2G give the same A value, which is slightly larger than that by method 2E (by about 0.05) and method 2F (by about 0.03). It is noteworthy that, if the comparison is limited to the case of $a_\Phi = 0$, then the difference in the A value determined by methods 2C and 2E decreases greatly to 0.01.

Method 2H.[21] Tanner and Berry[22] have derived the following equation for expressing the molecular weight M_w dependence of the limiting viscosity number $[\eta]$ for polymers with a nonnegligible free-draining effect in solutions:

$$M_\mathrm{w}^{1/2}/[\eta] = [K'(6\langle S^2\rangle_\mathrm{w}/M_\mathrm{w})^{3/2}]^{-1}(1 + A'M_\mathrm{w}^{-1/2}) \qquad (3.16.19)$$

where K' and A' are parameters depending on the models used and $\langle S^2\rangle$ is the mean square radius of gyration. Eq. (3.16.19), which corresponds to eq. (31 in Tanner and Berry's paper,[22] has been derived by postulating $a_2[\equiv \mathrm{d}\log(\langle S^2\rangle_0/M)/\mathrm{d}\log \times M] \neq 0$ and $a_\Phi(\equiv \mathrm{d}\log \Phi/\mathrm{d}\log M) \neq 0$. Using eq. (3.16.19) and appropriate values for the draining parameter X, and assuming that $a_1(\equiv \mathrm{d}\log a_\mathrm{s}/\mathrm{d}\log M) = 0$ (where a_s is the linear expansion factor), Tanner and Berry estimated $\langle S^2\rangle_\mathrm{w,0}/M_\mathrm{w}$ from the intercept of a plot of $M_\mathrm{w}^{1/2}/[\eta]$ versus $M_\mathrm{w}^{-1/2}$ (Tanner–Berry (TB) plot)[22] for some cellulose derivatives. Hereafter, this method is designated as method 2H.

Kamide and Miyazaki[1] have demonstrated that $a_2 > 0$ and $a_1 \geq 0$ for cellulose, amylose, and their derivatives in solvents. Figure 8 in Ref. 1 shows that the magnitude $a_\Phi + 1.5a_2$ influences the A value, as determined by the viscosity plot (method 2E), especially in the vicinity of $a_\Phi + 1.5a_2 \approx 0$. Therefore, it can also be expected that the intercept of the Tanner–Berry plot will be sensitively influenced by $a_\Phi + 1.5a_2$, which is completely neglected in eq. (3.16.19). When $a_1 \neq 0$ and $a_2 \neq 0$, eq. (3.16.19) should be rearranged into eq. (3.16.20):

$$M_{\mathrm{w}}^{a_1+1.5a_2+0.5}/[\eta] = (K'6^{3/2}K_0^{3/2}K_a)(1 + A'/M_{\mathrm{w}}^{-1/2}) \tag{3.16.20}$$

where

$$K_0 = (\langle S^2 \rangle_0/M_{\mathrm{w}})/M_{\mathrm{w}}^{a_2} \tag{3.16.6}$$

and

$$K_\alpha = \alpha_{\mathrm{s}}^3/M_{\mathrm{w}}^{\alpha_1} \tag{3.16.21}$$

The plot of $M_{\mathrm{w}}^{a_1+1.5a_2+0.5}/[\eta]$ versus $M_{\mathrm{w}}^{-1/2}$ should be linear and K_0 can be evaluated from its intercept at $M_{\mathrm{w}}^{-1/2} = 0$, provided that K_α is given in advance. Unfortunately, the experimental determination of a_1, a_2, and K_a (especially of a_1 and K_a) is not sufficiently accurate to employ eq. (3.16.20) for estimating K_0.

Method 2I and method 2J.[23] The frictional coefficient ξ is defined by eq. (3.15.5) or eq. (3.15.5′):

$$\xi = M_{\mathrm{w}}(1 - v_{\mathrm{p}}\rho_0)/s_0 N_{\mathrm{A}} \tag{3.15.5}$$

$$\xi = kT/D_0 \tag{3.15.5′}$$

Here, s_0 and D_0, sedimentation and diffusion coefficients at infinite dilution; k, Boltzmann constant; T, temperature (K); v_{p}, specific volume of polymer; ρ_0, density of the solvent; and N_{A}, Avogadro's number. We can estimate A values from the molecular weight dependence of the frictional coefficient ξ.

According to the KY theory,[5] ξ of a flexible polymer chain in dilute solution is related to the draining parameter X and $(\langle S^2 \rangle_0/M)^{1/2}$ by eq. (3.16.22):

$$\xi/\eta_0 = 3\pi^{3/2}[XG_0(X)](\langle S^2 \rangle_0/M)^{1/2}\alpha_{\mathrm{s}}^{m(X)} \tag{3.16.22}$$

where $\eta_0 =$ viscosity of the solvent, $G_0(X)$ and $m(X) =$ functions of X given in the KY theory, $\alpha_{\mathrm{s}} =$ linear expansion factor $(= (\langle S^2 \rangle/\langle S^2 \rangle_0)^{1/2}$, $\langle S^2 \rangle^{1/2} =$ radius of gyration at perturbed state). Now, parameter P, analogous to the Flory viscosity parameter Φ, is defined as

$$P = q'_{\mathrm{w},z}(\xi/\eta_0)/(6^{1/2}\langle S^2 \rangle_z^{1/2}) \tag{3.15.4}$$

where $q'_{\mathrm{w},z} =$ polymolecularity correction factor (in this case, D_0, M, and s_0 are assumed to be the weight-average quantities).

Eq. (3.16.22) can be rewritten with the aid of eq. (3.15.4) in the form:

$$\xi/\eta_0 = 6^{1/2}P(\langle S^2 \rangle_0/M)^{1/2}M^{1/2}\alpha_{\mathrm{s}} \tag{3.15.23}$$

with

$$P = P_0(X)\alpha_s^{-(1-m(X))} \tag{3.15.24}$$

$$P_0(X) = (3\pi^{3/2})^{1/2}[XG_0(X)] \tag{3.15.25}$$

In the case when the draining effect is nonnegligible (i.e. $X \leq 10$). P is expected, from eqs. (3.15.24) and (3.15.25), to be molecular weight dependent. This dependence can be semiempirically expressed, to a first approximation, by eq. (3.15.6):

$$P = K_p M^{a_p} \tag{3.15.6}$$

Furthermore, if the nonGaussian nature exists, then the term $(\langle S^2 \rangle_0/M)^{1/2}$ in eq. (3.15.23) is not constant, but depends on the molecular weight. That is, the following equation

$$(\langle S^2 \rangle_0/M)(\equiv A^2/6) = K_0 M_w^{a_2} \tag{3.16.23}$$

can be approximated.

α_s in eq. (3.16.22) is, according to the Fixman theory,[3] given by:

$$\alpha_s = 1 + 0.638z \tag{3.16.24}$$

where

$$z = (3/2\pi^{3/2})^{1/2}(\approx 0.330)BA^{-3}M^{1/2} \tag{3.16.25}$$

(see also eq. (3.13.7))

Substituting eqs. (3.15.6) and (3.16.23–3.16.25) in eq. (3.15.23) gives

$$\xi/\eta_0 = 6^{1/2}K_p K_0^{0.5} M^{0.5+0.5a_2+a_p} x(1 + 0.0144BK_0^{-1.5} M^{0.5-1.5a_2}) \tag{3.16.26}$$

Eq. (3.16.26) holds for the case $a_2 \neq 0$ and $a_p \neq 0$. A graph of $(\xi/\eta_0)/M_w^{0.5+0.5a_2+a_p}$ as a function of $M_w^{0.5-1.5a_2}$ (Kamide–Miyazaki (KM II plot) enables $6^{1/2}q_{w,z}'^{-1}K_p K_0^{1/2}$ (and accordingly K_0) to be evaluated from the intercept at $M_w^{0.5-1.5a_2} = 0$, where M_w is the weight-average molecular weight.

The K_0 thus estimated corresponds to $(\langle S^2 \rangle_{0,z}/M_w)$ $(\equiv A_{z,w})$ and was converted to the value corresponding to $(\langle S^2 \rangle_{0,w}/M_w)$ $(\equiv A_{w,w})$. We will call this method $2J^{23}$ hereafter.

In the case $a_2 = a_p = 0$, eq. (3.16.26) reduces to

$$\xi/\eta_0 = P_0(\infty)\alpha_s^{-0.348}AM^{0.5}(1 + 0.211\,BA^{-3}M^{0.5}...) \tag{3.16.27}$$

where $P_0(\infty) = P_0$ value at $X = \infty$ and 5.2.[5] If $\alpha_s^{-0.348} \approx 1$, eq. (3.16.30) can be written in the form

$$\xi/\eta_0 = P_0(\infty)AM^{0.5}(1 + 0.211BA^{-3}M^{0.5}...) \tag{3.16.26'}$$

which is almost the same equation as that derived by Cowie and Bywater.[24] A can be determined from the plot of $\xi/(\eta_0 M_w^{0.5})$ versus $M_w^{0.5}$ (Cowie–Bywater (CB) plot) as the ordinate at $M_w^{0.5} = 0$ after the conversion of $A_{z,w}$ to $A_{w,w}$. This method will be referred to as method 2I for convenience.

The following equation is analogous to the viscosity equation:

$$[\eta] = 6^{3/2}K_{\Phi}K_0^{1.5}M^{0.5+1.5a_2+a_{\Phi}}(1 + 2(3/2\pi)^{1.5}BA^{-3}M^{0.5})$$ (3.16.28)

which has been developed by Kamide.

Method 2K.[25] In general, $[\eta]$ can be expressed by the equation:[5]

$$[\eta] = 6^{3/2}\Phi(\langle S^2\rangle_{w,0}^{3/2}/M_w^{1/2})\alpha_s^3$$ (3.15.1)

where $\langle S^2\rangle_0^{1/2}$ is the radius of gyration in the unperturbed state, Φ is the Flory viscosity parameter, and a_s is the dimensionless expansion factor defined as the ratio of $\langle S^2\rangle^{1/2}$ to $\langle S^2\rangle_0^{1/2}$. Next, we consider the case when Φ and $\langle S^2\rangle_0/M$ show molecular weight dependence of the forms:

$$\Phi = K_{\Phi}M^{a_{\Phi}}$$ (3.15.3)

$$\langle S^2\rangle_0/M = K_0M^{a_2}$$ (3.16.6)

Here, K_{Φ}, a_{Φ}, K_0, and a_2 are empirical parameters for the polymer/solvent combination and the temperature. We assume that eqs. (3.11.1), (3.15.3), and (3.16.6) are applicable over the same molecular weight range M_1 to M_2.

Eq. (3.15.1) can be transformed by using eqs. (3.15.3) and (3.16.6) into

$$[\eta] = 6^{3/2}K_{\Phi}K_0M^{0.5+a_{\Phi}+1.5a_2}\alpha_s^3$$ (3.16.29)

a_s in eq. (3.16.32) can be given by a closed simple form of an excluded volume parameter z:

$$a_s^3 = 1 + 2z$$ (3.13.11′)

Combining eqs. (3.11.1), (3.13.11′), (3.16.26), and (3.16.29) yields

$$\{K_m/(6^{3/2}K_{\Phi}K_0^{3/2})\}M^{a-0.5-a_{\Phi}-1.5a_2} = 1 + CM^{(1-3a_2)/2}$$ (3.16.30)

where

$$C = (1/4\pi)^{3/2}K_0^{-3/2}B.$$ (3.16.31)

Writing M_0 for M, taking logarithms and differentiating with respect to $\log M_0$ gives

$$1 + CM_0^{(1-3a_2)/2} = (1 - 3a_2)/\{1 - 3a_2 - 2(a - a_{\Phi} - 1.5a_2 - 0.5)\}$$ (3.16.32)

Substitution of eq. (3.16.32) into eq. (3.16.30), and subsequent rearrangement gives

$$-\log K_m + \log\{(1 - 3a_2)/(2(1 - a + a_{\Phi}))\} + (3/2)\log 6 + \log K_{\Phi}$$

$$= -(3/2)\log K_0 + (a - 0.5 - a_{\Phi} - 1.5a_2)\log M_0$$ (3.16.33)

when $a_2 = 0$, eq. (3.16.33) reduces to eq. (3.16.34):

$$-\log K_m - \log\{2(1 - a + a_{\Phi})\} + (3/2)\log 6 + \log K_{\Phi}$$

$$= -(3/2)\log K_0 + (a - 0.5 - a_{\Phi})\log M_0.$$ (3.16.34)

Putting $a_\Phi = 0$ and $a_2 = 0$ in eq. (3.16.33) reduces to eq. (3.16.15), derived by Kamide and Moore.[10,26]

We can determine K_0 from K_m, a, K_Φ, a_Φ, a_2, and M_0 by using eq. (3.16.33) without the aid of θ solvent measurements (Kamide-Saito (KS) I plot). If K_0 does not vary with the solvent nature, then the experimental plots of the left-hand side of eq. (3.16.33) as a function of $a - 0.5 - a_\Phi - 1.5a_2$ for the polymer in various solvents should be linear with a slope of log M_0, yielding $-3/2$ log K_0 as an intercept. We use the description method 2K[25] for this method. From the K_0 value thus obtained, A can be calculated by eq. (3.16.35):

$$A = (6K_0 M^{a_2})^{1/2} \tag{3.16.35}$$

It should be noted that eqs. (3.16.34) and (3.16.35) are rigorously valid only at $a_2 = 0$ and should be regarded as being approximate if $a_2 \neq 0$.

Similar relationships with eq. (3.13.11'), in which a_s is a linear function of z, yield the same equation as eq. (3.16.33). Eq. (3.16.30) and thus eq. (3.16.33) restrict application to a system of comparatively small z region.[28] Fortunately, in solutions of cellulose and amylose derivatives, the excluded volume effect is not large.

method 2K requires a value of a_2, which can only be estimated from the molecular weight dependence of A. Nevertheless, the determination of A is the final purpose of method 2K. This seems self-contradictory. In order to overcome this difficulty, a rough estimate of a_2 can be made by a thermodynamic approach (method 2B) from the experimental $\langle S^2 \rangle_z^{1/2}$ and α_s values estimated with help of the penetration function ψ, which is defined by eq. (3.13.5).[27] As will be demonstrated further below, the A value estimated by method 2K is rather insensitive to the magnitude of a_2. In addition, for solutions of cellulose and amylose derivatives, a_2 is usually zero or close to it.

An alternative expression of eq. (3.16.33) is

$$-\log K_m + \log\{(1 - 3a_2)/2(1 - a + \Delta)\} + 3/2 \log 6 + \log \Phi_0(X_0) + \log f$$

$$= -3/2 \log K_0 + (a - 0.5 - 1.5a_2)\log M_0 \tag{3.16.36}$$

where

$$\Delta = v(X) - v(X_0) \approx a_\Phi \tag{3.16.14}$$

$$f = XF_0(X)/X_0F_0(X_0) \tag{3.16.10}$$

$\Phi_0(X)$ and $F_0(X)$ are functions of the draining parameter X defined by Kurata and Yamakawa. Then, eq. (41 in Ref. 1 should be read as eq. (3.16.36).

Methods 2K and 2G need not only viscosity data but also some thermodynamic data for estimation of K_Φ, a_Φ, and a_2. Then, these methods are not strictly hydrodynamic.

Method 2K is compared with other hydrodynamic approaches and the thermodynamic approaches.

Method 2L.[28] We generalize the concept underlying method 2K to the case of the MWD of the molecular frictional coefficient, deriving a new method for evaluating the unperturbed chain dimensions.

Eq. (3.16.22) can be rearranged in the form:

$$\xi/\eta_0 = 6^{1/2}P(\langle S^2\rangle_0/M)^{1/2}M^{1/2}\alpha_s \tag{3.16.37}$$

with

$$P = P_0(X)\alpha_s^{-(1-m(X))} \tag{3.15.24}$$

$$P_0(X) = (3\pi^{3/2})^{1/2}[XG_0(X)] \tag{3.15.25}$$

P is a parameter, analogous to the Flory viscosity parameter Φ and is experimentally determined by either eq. (3.16.38) or (3.16.39):

$$P = q'_{w,z(s)}(\xi_{w,w(s)}/\eta_0)/(6\langle S^2\rangle_z)^{1/2} \tag{3.16.38}$$

or

$$P = q'_{w,z(D)}(\xi_{w(D)}/\eta_0)/(6\langle S^2\rangle_z)^{1/2} \tag{3.16.39}$$

with

$$\xi_{w,w(s)} = (1 - v_p\rho_0)/s_{0,w}N_A \tag{3.15.5}$$

$$\xi_{w,(D)} = kT/D_{0,w} \tag{3.15.5'}$$

$q'_{w,z(s)}$ and $q'_{w,z(D)}$ are polymolecularity correction factors given in analytical forms by Kamide and Miyazaki.[23] $D_{0,w}$, M_w, and $s_{0,w}$ are weight-average quantities.

The molecular weight dependence of P, which is expected for a partially free-draining chain sphere, can be semiempirically expressed to a first approximation by

$$P = K_pM^{a_p}. \tag{3.15.6}$$

K_p and a_p are parameters characteristic of a given polymer/solvent combination and they were determined for solutions of some cellulose derivatives (see Table 3.15.4). For a nonGaussian chain, $(\langle S^2\rangle_0/M)^{1/2}$ in eq. (3.13.9) is not constant, but depends on the molecular weight. Kamide and Miyazaki approximated the following equation:[1]

$$(\langle S^2\rangle_0/M)(= A^2/6) = K_0M^{a_2} \tag{3.16.6'}$$

a_s in eq. (3.15.24) is given by:[2]

$$\alpha_s = 1 + 0.638z + \cdots \tag{3.16.24}$$

If the higher terms of z^2 in eq. (3.16.24) are neglected, then eq. (3.16.24) can be rewritten as

$$\alpha_s = 1 + CM^{(1-3a_2)/2}. \tag{3.16.40}$$

with

$$C = 0.638(1/4\pi)^{3/2}K_0^{-3/2}B. \tag{3.16.41}$$

The MWD of s_0 and D_0 can be semiempirically expressed by eqs. (3.14.3) and (3.15.13):

$$s_0 = K_sM^{a_s} \tag{3.14.3}$$

$$D_0 = K_DM^{a_d} \tag{3.15.13}$$

Then, ξ in eqs. (3.15.5) and (3.15.5$'$) can be expressed in a general form:

$$\xi/\eta_0 = K_\xi M^{a_\xi} \tag{3.16.42}$$

with

$$a_\xi = 1 - a_s = -a_d \tag{3.16.43}$$

Substitution of eqs. (3.15.6), (3.16.6$'$), (3.16.40), and (3.16.42) in eq. (3.16.37) gives

$$K_\xi M^{a_\xi} = 6^{1/2}K_p K_0^{1/2} M^{(a_p+a_2/2+1/2)} \times (1 + CM^{(1-3a_2/2)}) \tag{3.16.44}$$

Writing M_0 for M, taking logarithms and differentiating with respect to log M_0, we obtain

$$1 + CM_0^{(1-3a_2)/2} = [1 - \{2/(1 - 3a_2)\}(a_\xi - a_p - a_2/2 - 1/2)]^{-1} \tag{3.16.45}$$

Substitution of eq. (3.16.45) into eq. (3.16.44) and subsequent rearrangement gives

$$-\log K_\xi + \log(1 - 3a_2)/2(1 - a_\xi + a_p - a_2) + (1/2)\log 6 + \log K_p$$

$$= -(1/2)\log K_0 + (a_\xi - a_p - a_2/2 - 1/2)\log M_0 \tag{3.16.46}$$

M_0 is a parameter depending on the molecular weight in the range M_1–M_2, over which eq. (3.16.42) applies and the geometric mean $(M_1 M_2)^{1/2}$ is a good approximation (see Appendix of Ref. 28).

We can determine K_0 (and thus A) from K_ξ, K_p, a_ξ, a_p, a_2, and M_0 by using eq. (3.16.46). This method is referred to as method 2L[28] (KS II plot).

In the case of $a_2 = 0$ and $a_p = 0$, eq. (3.16.46) reduces to eq. (3.16.47):

$$-\log K_\xi - \log 2(1 - a_\xi) + (1/2)\log P$$

$$= -\log A + (1/2)\log 6 + (a_\xi - 0.5)\log M_0. \tag{3.16.47}$$

3.16.2   Application to experimental data

Methods 2A to 2H

Cellulose, amylose, and their derivatives. The viscosity, sedimentation, diffusion and LS data of cellulose in cadoxen,[29] FeTNa,[30] CTN in acetone,[31–33] CN in aceton,[31–33] CA (CDA) in acetone,[34] THF,[34] CTC$_p$ in DMF,[35] 1-Cl-N,[35] dioxane/water,[35] CTC in acetone,[36] cyclohexanone,[36] anisol,[37,38] cyclohexanol,[37,38] dioxane,[37,38] MC in water,[39] NaCMC in aq. NaCl,[40] HEC in water,[41] EHEC in water,[42] NaCX in 1-N NaOH,[43,44] amylose in DMSO,[45] ATA in nitrometane,[46] ATC in pyridine/wate,[47] pyridine[48,49] were analyzed to estimate the unperturbed dimensions.

The average values of A obtained by methods 2A and 2B are recorded in the third and fourth columns of Table 3.16.1a, respectively.[1] Figure 3.16.1 represents the Baumann plot constructed according to eq. (3.16.4). The values of A, determined from the intercepts of the plots, are tabulated in the fifth column of Table 3.16.1(a). The Baumann plot can be represented by a straight line for each polymer solvent system. However, the slope of the plot is very often negative, showing that the B value

Table 3.16.1a

The unperturbed chain dimensions A estimated by various methods for solutions
of cellulose derivatives

Polymer	Solvent	$A \times 10^8$ (cm)					
		Thermodynamic approach				Hydrodynamic approach	
		2A	2B	2C	2D	2E	2F
Cellulose	Cadoxen		1.53	1.83		1.21	1.20
	FeTNa		1.96	2.31	2.27	1.31	1.29
	Aq. NaOH		2.2				
CN ($N_c = 12.9\%$)	Acetone		1.84	1.84	1.85	0.52	0.74
CN ($N_c = 13.9\%$)	Acetone		2.41	2.43		0.79	1.02
CA (DS $= 2.46$)	Acetone		1.68	2.14	1.77	1.10	1.12
	THF		1.23	1.24		0.99	1.00
CA (DS $= 2.92$)	DMAc		1.43	1.49		0.89	1.00
CTC$_p$	DMF	1.80		2.14	1.94	0.96	0.92
	1-Cl-N		1.75	2.14	1.94	0.90	0.85
	Dioxane/water					0.929	0.956
CTC	Anisol	0.818				0.774	0.758
	Cyclohexane	0.841	1.36	1.43	(1.43)	0.66$_6$	0.671
	Acetone		1.43	1.43		0.56	0.56
	Dioxane		1.88	1.85		0.55	0.47
MC(DS $= 0.2$)	Water		2.05	2.56	2.34	1.23	1.25
NaCS (DS $= 1.9$)	0.5M NaCl		1.20	1.23	1.47	0.46	0.53
NaCMC (DS $= 0.88$)	NaCl ($1 \rightarrow \infty$)		1.42	1.50	1.47	1.26	1.25
HEC (DS $= 1$)	Water		2.10	2.40	2.25	0.93	0.96
EHEC (DS $= 1$)	Water		2.10	2.41	2.37	1.24	1.22
NaCX	1 N NaOH		3.30	4.04	3.86	1.03	1.03
Amylose	DMSO		1.48	1.33	1.62	0.60	0.56
ATA	Nitromethane		1.04	1.14		0.55	0.55
ATC	Pyridine/water	0.998		0.998	(0.998)	0.775	0.770
	Pyridine	1.14		1.08		0.57	

is negative (i.e. among 23 polymer solvent systems, 15 systems have negative B, four systems have $B = 0$). This contrasts sharply with the positive value of A_2 for these polymer solvent systems. The result that $B < 0$ arises sometimes from the nonGaussian nature of a polymer chain in unperturbed state.

In Figure 3.16.2, the molecular weight dependence of $\langle S^2 \rangle_{0,w} / M_w$, estimated by method 2A and or 2B, is demonstrated. Figure 3.16.2 shows that a_2 in eq. (3.15.10) can be empirically regarded as constant in the molecular weight range concerned. The a_2 values obtained thus are given in the sixth column of Table 3.15.1. It is of interest to note that a_2 values for cellulose, amylose, and their derivatives in solution are zero or negative. In the case of $a_2 \neq 0$, the Baumann plot is not strictly applicable. Thus, in this case, the experimental data are plotted according to eq. (3.16.5) (BKM plot) in Figures 3.16.3(a) and (b). It is particularly noteworthy in this figure that the BKM plot can be approximated with a very good straight line with a positive slope. The A values for $M_w = 1 \times 10^5$

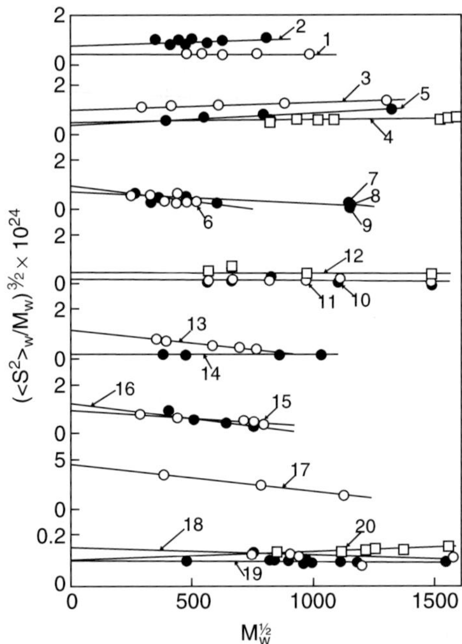

Figure 3.16.1   Baumann plots (eq. 3.16.4):[1] 1, cellulose in cadoxen;[29] 2, cellulose in iron sodium tartrate;[30] 3, cellulose triniterate ($N_c = 13.9\%$) in acetone;[31] 4, cellulose triniterate ($N_c = 13.6\%$) in acetone;[33] 5, cellulose nitrate ($N_c = 12.9\%$) in acetone;[31] 6, cellulose acetate (acetyl content 55.3%) in acetone;[34] 7, cellulose tricaproate in dimethyl formamide;[35] 8, cellulose tricaproate in 1-Cl-N;[35] 9, cellulose tricaproate in dioxane/water;[35] 10, cellulose tricarbanilate in acetone;[37] 11, cellulose tricarbanilate in cyclohexanone;[37] 12, cellulose tricarbanilate in dioxane;[37] 13, methylcellulose in water;[39] 14, sodium carboxy methylcellulose in aq. NaCl;[40] 15, hydroxyethyl cellulose in water;[41] 16, ethylhydroxyethyl cellulose in water;[42] 17, sodium cellulose xanthate in 1M NaOH;[43] 18, amylose in dimethyl sulphoxide;[45] 19, amylose triacetate in nitromethane;[46] 20, amylose tricarbanilate in pyridine.[49]

calculated from K_0 by eq. (3.16.6) (see Table 3.16.2) are included in the fifth column of Table 3.16.1(a).

The viscosity data are plotted according to eq. (3.16.12) (SF plot). As illustrated in Figure 3.16.4, the SF plot gives a good straight line, and A can be evaluated from its extrapolation to $M_w^{1/2} = 0$ (method 2E). The results are tabulated in the seventh column of Table 3.16.1(a). However, as was noted in the theoretical section, eq. (3.16.12) was based on the assumption that $a_2 = 0$, $a_\Phi = 0$ and $a < a^*$ are realized concurrently, where a^* is the upper applicable limit of a.[50]

As early as 1964, Kamide and Moore[11] applied the Stockmayer–Fixman method (eq. (3.16.12)) to cellulose trinitrate/ethyl acetate[51,52] and/acetone[52,53] systems, indicating that the plots for the trinitrate/ethyl acetate system are rectilinear and pass through the origin when extrapolated (Figure 3.16.5). The points for the trinitrate/ethyl acetate system suggest a curve that does not extrapolate to give an ordinate intercept greater than zero (Figure 3.16.6) in spite of the theoretical prediction of linearity and an ordinate intercept equal to K. This method, which assumes draining effects to be negligible, is clearly inapplicable to the systems considered.

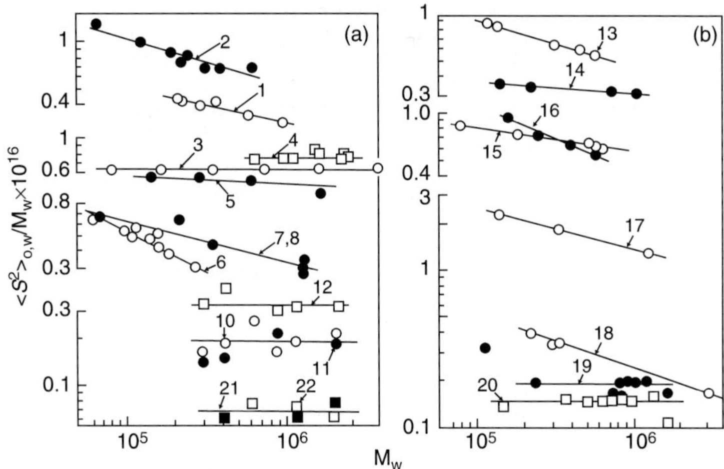

Figure 3.16.2   The molecular weight dependence of $\langle S^2 \rangle_{0,w}/M_w^1$. 1–20 on lines have the same meanings as those in Figure 3.16.1; 21, cellulose tricarbanilate in anisol;[37,38] 22, cellulose tricarbanilate in cyclohexanol.[37,38]

The fact of $a_2 \leq 0$ and $a_\Phi > 0$ for the almost cellulose, amylose, and their derivatives (Tables 3.15.1 and 3.15.4) indicate that eq. (3.16.12) is apparently inadequate for these polymers, as noted previously. Up until now, data for these polymers have been widely analyzed according to eq. (3.16.12) without reference to a_2 and a_Φ. The result of $a_\Phi > 0$ is compatible with X value ≤ 2, as shown in Tables 3.15.1 and 3.15.8.

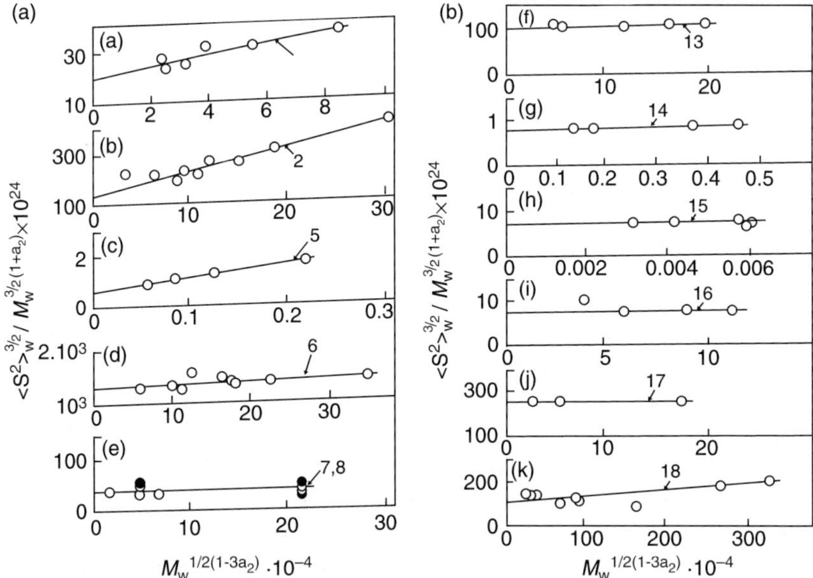

Figure 3.16.3   BKM plot (eq. (3.16.5)).[1] Numbers 1–20 on lines have the same meanings as those in Figure 3.16.1.

Table 3.16.1b

The unperturbed chain dimensions A estimated by various methods for solutions
of cellulose derivatives

Polymer	Solvent	$A \times 10^8$ (cm)					
				Hydrodynamic approach			
		2G	2H	2I	2J	2K	2L
Cellulose	Cadoxen	1.57	2.15	$1.43(s_0)$	1.92	1.56	1.70
				$1.18(D_0)$	2.04		
	FeTNa	2.17	2.34			2.07	
	aq. LiOH	2.13				2.13	
	aq. NaOH						
CN ($N_c = 12.9\%$)	Acetone	1.91	3.33			1.96	
CN ($N_c = 13.9\%$)	Acetone	2.45	3.33			2.50	
CA (DS = 2.46)	Acetone	1.75	1.45	1.72		1.76	1.80
	THF	1.23	1.37			1.26	
CA (DS = 2.92)	DMAc	1.47	1.59			1.46	1.68
CTC$_p$	DMF	1.99	1.03			2.00	
	1-Cl-N	1.82	0.96			1.80	
	Dioxane/water	1.90					
CTC	Anisol						
	Cyclohexane	1.25	1.50	1.96 (1.98)[a]	1.48		1.54
	Acetone	1.25	1.99			1.23	
	Dioxane	1.83	2.35			1.82	
MC (DS = 0.2)	Water	2.31	1.34			2.34	
NaCS (DS = 1.9)	0.5M NaCl	1.44	2.37			1.25	
NaCMC (DS = 0.88)	NaCl$(1 \rightarrow \infty)$	1.44	1.55			1.44	
HEC (DS = 1)	Water	2.26	2.63			2.26	
EHEC (DS = 1)	Water	2.32	2.06	$1.08(s_0)$	2.16	2.40	2.67
				$0.84(D_0)$	2.47		
NaCX	1N NaOH	3.96	1.47	0.73	3.63	3.97	4.31
Amylose	DMSO	1.76	1.52			1.80	
ATA	Nitromethane	1.07	1.29			1.14	
ATC	Pyridine/water	0.99					
	Pyridine	1.00		0.88	1.02	2.05	0.94

[a]The value obtained by Shanbhag.

By using the parameters K_m, a, and M_0 in Table 3.15.1 the Flory K value (and thus A)
was calculated from eq. (3.16.15). The results are shown in the last column of
Table 3.16.1(a) (method 2F). Figures 3.16.7(a) and (b) show the data graphed according
to eq. (3.16.17) (the KM I plot; method 2G) for the same viscosity data as those in
Figure 3.16.4. The plots are quite linear over the range examined and present no great
problem in extrapolation. The A values for $M_w = 1 \times 10^5$ are shown in Table 3.16.1(b).

Figure 3.16.8 illustrates the typical TB plots for cellulose, amylose, and their
derivatives.[21] The literature data used in Ref. 1 were also employed in Figure 3.16.8.

The plots can be reasonably approximated with straight lines, giving $[K'(6$
$\langle S^2 \rangle / M)^{3/2}]^{-1}$ as the intercepts. Since for these polymers, $X \leq 2$ has been confirmed

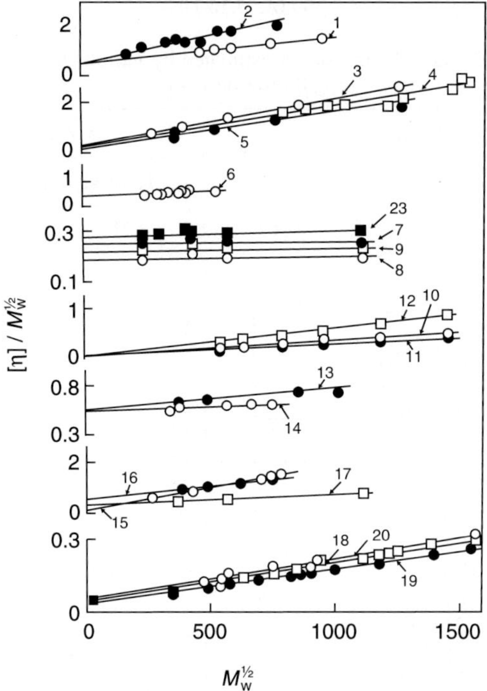

Figure 3.16.4   Stockmayer-Fixman plot (eq. (3.16.12)). Numbers 1–20 on lines have the same meanings as those in Figure 3.16.1; 23, cellulose tricaproate in dioxane.[35]

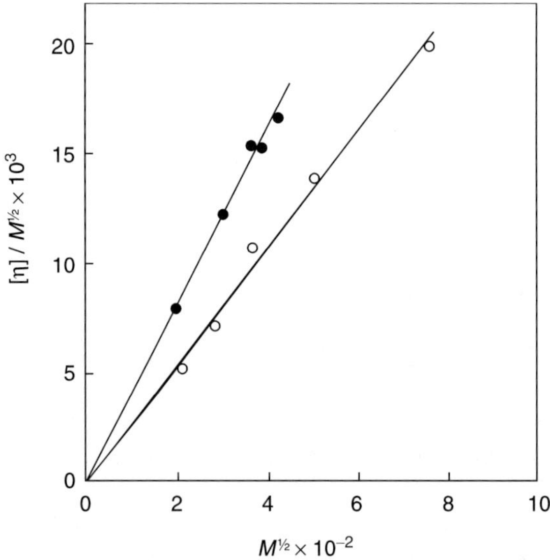

Figure 3.16.5 $[\eta]\,M^{-1/2}$ as a function of $M^{1/2}$ for cellulose trinitrate ethyl acetate systems.[11]

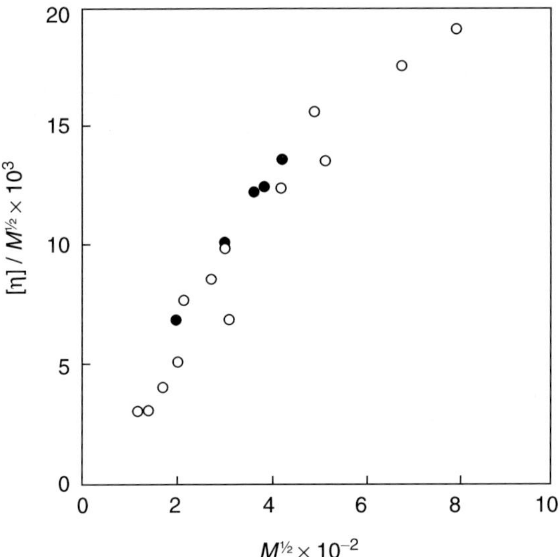

Figure 3.16.6 $[\eta]\,M^{-1/2}$ as a function of $M^{1/2}$ for cellulose trinitrate acetone systems.[11]

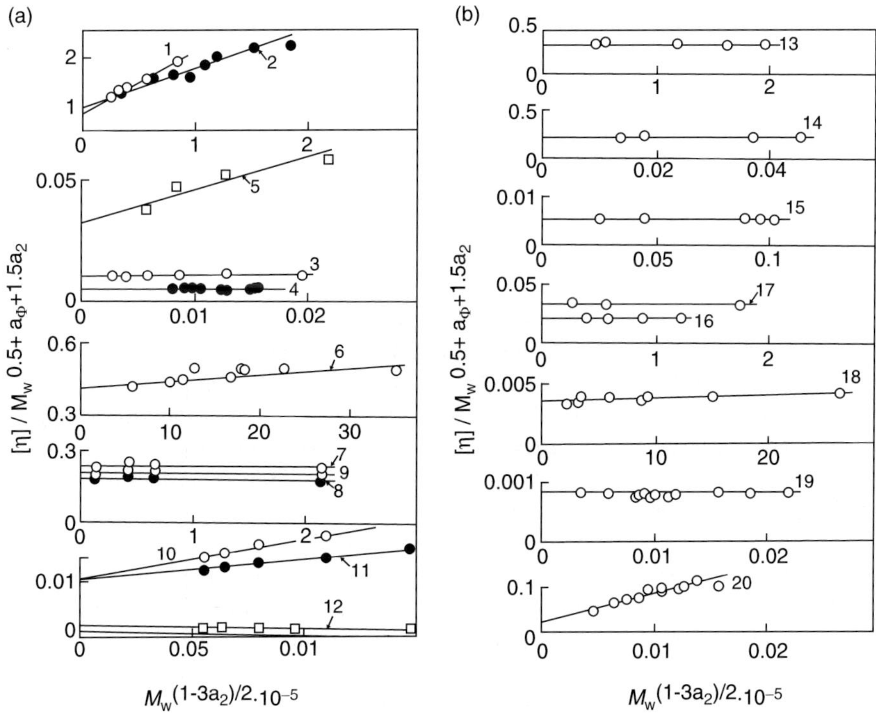

Figure 3.16.7   KM (I) plot (eq. (3.16.15)). Numbers 1–20 on lines have the same meanings as those in Figure 3.16.1.

Table 3.16.2

Unperturbed chain dimensions K_0 of cellulose, amylose, and their derivatives[1]

Polymer	Solvent	$K_0 \times 10^{18}$ (cm^3)		
		2D	2G	2J
Cellulose	Cadoxen	–	0.54_6	6.57 (s_0)
	FeTNa	28.2	23.7	57.1 (D_0)
CTN				
$N_c = 13.9\%$	Acetone	–	1.00	
13.6%	Acetone	–	4.38	
CN 12.9%	Acetone	0.75	0.80	
CA				
AC $= 55.3\%$	Acetone	118	116	112.1
CTC$_p$	DMF	65.4	75.4	
	1-Cl-N	65.4	62.8	
	dioxane/water	65.4	71.3	
CTC	Anisol	–	–	
	Cyclohexanol	–	–	
	Acetone	–	0.62	
	Cyclohexanone	–	0.55_9	0.318
	Dioxane			
MC DS $= 2$	Water	23.0	22.4	
NaCMC DS $= 0.88$	NaCL (I $\rightarrow \infty$)	0.83	0.79	
HEC DS $= 1$	Water	3.69	3.71	11.92 (s_0)
EHEC DS $= 2$	Water	17.8	17.1	11.80 (D_0)
NaCX DS $= 0.78$	1-M NaOH	39.4	41.4	33.51
Amylose DMSO		20.9	24.7	
ATA	Nitrometane	–	0.21_4	
ATC	Pyridine/water	–	0.16_8	
	Pyridine	–	0.16_8	0.154

DS = degree of substitution.

extensively by Kamide and Miyazaki,[1] K' can be taken as $2.87/1.259 \times 10^{23}$, using Table VIII of Ref. 22. The fourth column of Table 3.16.1 summarizes the A values thus determined ($A_{(2H)}$). The values $A_{(2H)}$ do not always coincide with the average of the A values obtained by methods 2A, 2B 2C, 2D, and 2G,[1,2] $A_{(m)}$. The ratio $A_{(2H)}/A_{(m)}$ increases significantly with increase in $a_1 + 1.5a_2$. The correlation coefficient γ is estimated to be 0.75 between $A_{(2H)}/A_{(m)}$ and $a_1 + 1.5a_2$. Considering the large experimental uncertainty in a_2, we can conclude that $A_{(2H)}$ is approximately equal to $A_{(m)}$ only at $a_1 + 1.5a_2 \approx 0$, as Kamide's theory predicts (eq. (3.16.20)). The samples in all the literature, except CA (CDA), cited in Table 3.15.1 had wide MWD. In contrast, relatively sharp fractions, which had been separated by the successive solutional method, were employed in a series of studies on cellulose acetates by Kamide *et al.* (see Section 3.2). Then, their data were examined in more detail. Note that CDA data in Table 3.15.1 was in press at the time of publication of Ref. 1.

Table 3.16.2 summarizes K_0 values, determined from the intercept of BKM (method 2D), KM (I) (method 2G), and KM (II; method 2J) plots.

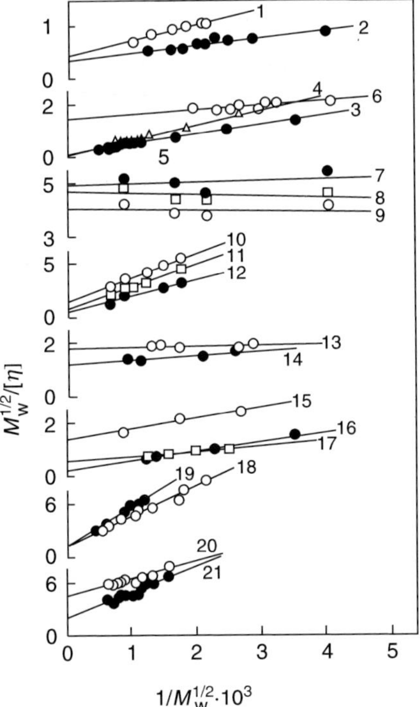

Figure 3.16.8 Plot of $M_w^{1/2}/[\eta]$ against $M_w^{-1/2}$ (Tanner–Berry plot). 1, cellulose in cadoxen; 2, cellulose in iron sodium tartrate;[30] 3, cellulose trinitrate ($N_c = 13.9\%$) in acetone; 4, CN ($N_c = 12.9\%$) in acetone;[31] 5, CA in acetone;[34] 6, CTC$_p$ in 1-Cl-N;[35] 7, cellulose tricaproate in dioxane/water;[35] 8, cellulose tricaproate in dimethyl fomamide;[35] 9, cellulose tricarbanilate in cyclohexanone;[37] 10, cellulose tricarbanilate in acetone;[37] 11, cellulose tricarbanilate in dioxane;[37] 12, methylcellulose in water;[37] 13, sodium carboxy methylcellulose in aq. NaCl solution;[40] 14, sodium cellulose xanthate in 1 M NaOH;[43,44] 15, hydroxyethyl cellulose in water;[41] 16, ethylhydroxyethyl cellulose in water;[42] 17, amylose in dimethyl fomamide;[45] 18, amylose triacetate in nitromethane;[46] 19, amylose tricarbanilate in pyridine/water;[47] 20, amylose tricarbanilate in pyridine.[49]

CA (DS 0.49). In Figure 3.16.9,[54] the ratio $\langle S^2 \rangle_{0,w}/M_w$ evaluated by method 2B is plotted against M_w for CA (DS 0.49)[55] in DMAc at 25 °C. The figure includes also data of CA (DS 1.75,[56] 2.46,[34,57] and 2.92[29]). The plotted points for CA (DS 0.49) scatter considerably around $\langle S^2 \rangle_{0,w}/M_w = 0.27 \times 10^{-6}$ cm^2 g^{-1} mol (accordingly, $A = 1.28 \times 10^{-8}$ cm) and the ratio $\langle S^2 \rangle_{0,w}/M_w$ can be regarded as approximately constant (i.e. $a_2 \equiv \mathrm{d}\ln(\langle S^2 \rangle_{0,w}/M_w)/\mathrm{d}\ln M_w \approx 0$). The CA (DS 0.49) chain in the unperturbed state can be considered Gaussian at a first approximation. The same method was applied to the data of CA (DS 0.49) aqueous and FA solutions in exactly the same manner, and A was found to be 2.51×10^{-8} cm and 3.03×10^{-8} cm in water and formamide, respectively, as indicated in the third and fourth columns of Table 3.16.3.

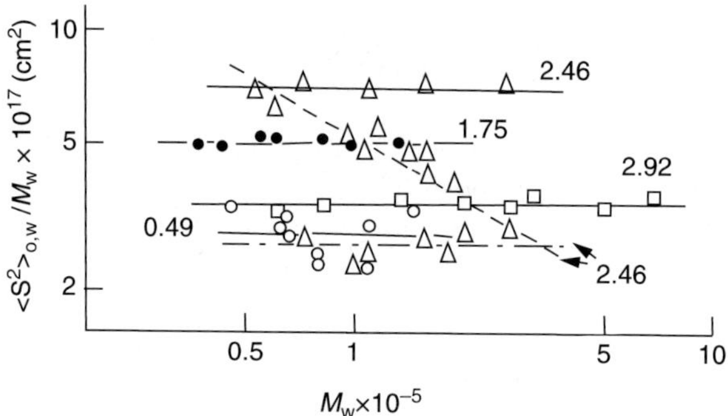

Figure 3.16.9   Molecular weight dependence of the mean square weight-average radius of gyration in the unperturbed state ($\langle S^2\rangle_{0,w}^{1/2}/M_w$) for cellulose acetate in various solvents.[54] The lines are determined by the least square method. (○), cellulose acetate (DS 0.49)-dimethylacetamide; (●), cellulose acetate (DS 1.75)-dimethylacetamide; (△), cellulose acetate (DS 2.46)-dimethylacetamide; (▲), cellulose acetate (DS 2.46)-acetone; (▲), cellulose acetate (DS 2.46)-tetrahydrofuran; (□), cellulose acetate (DS 2.92)-dimethylacetamide.

In general, the radius of gyration $\langle S^2\rangle_w^{1/2}$ can be expressed in terms of molecular weight by the relationship:

$$\langle S^2\rangle_w^{1/2} = K_\lambda M^{(\lambda+1)/2} \tag{3.12.1}$$

where,

$$\lambda = a_2 + 2\epsilon \tag{3.15.18}$$

$$\epsilon = \mathrm{d}\ln a_s/\mathrm{d}\ln M_w \tag{3.15.9}$$

$$a_2 = \mathrm{d}\ln(\langle S^2\rangle_{w,0}/M_w)/\mathrm{d}\ln M_w \tag{3.15.10}$$

For the CA (DS 0.49)/DMAc system λ was estimated to be 0.04 from the empirical relationship,

$$\langle S^2\rangle_z^{1/2} = 0.5 \times 10^{-8} M_w^{0.52} \ (\mathrm{cm}). \tag{3.16.48}$$

Putting $\lambda = 0.04$ and $a_2 = 0$ into eq. (3.15.18), we obtain $\epsilon = 0.02$, which is too small to detect experimentally from the molecular weight dependence of the expansion coefficient a_s. This means that the excluded volume effect on the MHS exponent a is negligibly small, being consistent with $a_s \leq 1.1-1.2$.

The exponent a in eq. (3.11.1) is given by the relationship:

$$a = 0.5 + a_\Phi + 3\epsilon + 1.5a_2. \tag{3.15.8'}$$

Putting $a = 0.60$, $a_\Phi = 0.103$, and $\epsilon = 0.02$ for the CA (DS 0.49)-DMAc system into eq. (3.15.8'), we obtain $a_2 = -0.04$, which is too small to detect by experiment. The main contributing factor to the a value is the partially free-draining effect and the volume effect is apparently minor.

Figure 3.16.10 shows the plots of Baumann, Stockmayer–Fixman, and Kamide *et al.* for CA (DS 0.49) in various solvents. The A and B values for (CADS 0.49) solutions

Table 3.16.3

Unperturbed chain dimensions A, conformation parameter σ, and characteristic ratio C_∞ of cellulose acetate in various solvents at 25 °C[54]

Methods	$A \times 10^8$ (cm)								
	$\langle\!\langle F \rangle\!\rangle^a = 0.49$			1.75	2.46			2.92	
	DMAc	Water	Formamide	DMAc	DMAc	Acetone	THF	DMAc	TCE
2B(α_s from ψ)	1.28	2.51	3.03	1.73	2.05	1.66^b	1.23	1.43	–
2C ($a_2 = 0$)	1.38	–	–	1.73	1.98	2.14	1.24	1.49	–
2D ($a_2 \neq 0$)	1.38	–	–	1.73	1.98	1.77^c	1.24	1.49	–
2E ($a_2 = a_\Phi = 0$)	1.12	1.17	1.19	1.12	0.79	1.10	0.99	0.89	0.82
2F($a_2 = a_\Phi = 0$)	1.14	1.22	1.22	1.16	0.92	1.12	1.00	1.00	0.85
2G ($a_2 \neq 0, a_\Phi \neq 0$)	1.40	–	–	1.79	1.90	1.75^c	1.23	1.47	–
2I ($a_2 \neq a_p = 0$)	–	–	–	–	–	1.11	–	0.57	–
2J ($a_2 \neq 0, a_p \neq 0$)	–	–	–	–	–	1.72	–	1.63	–
2K ($a_2 \neq 0, a_\Phi \neq 0$)	1.45	–	–	1.73	1.85	1.76	1.26	1.46	–
2L($a_2 \neq 0, a_p \neq 0$)	–	–	–	–	–	1.80	–	1.68	–
Most probable value	1.38	2.51	3.03	1.74	1.95	1.74	1.26	1.53	(1.27)
σ	2.58	4.69	5.66	3.69	4.39	3.92	2.84	3.57	2.97
C_∞	13.1	43.2	63.0	26.9	38.0	30.3	15.9	25.2	17.3

^{a}Total degree of substitution.
^{b}Averaged value over all fractions.
^{c}A value at $M_w \times 10^5$.

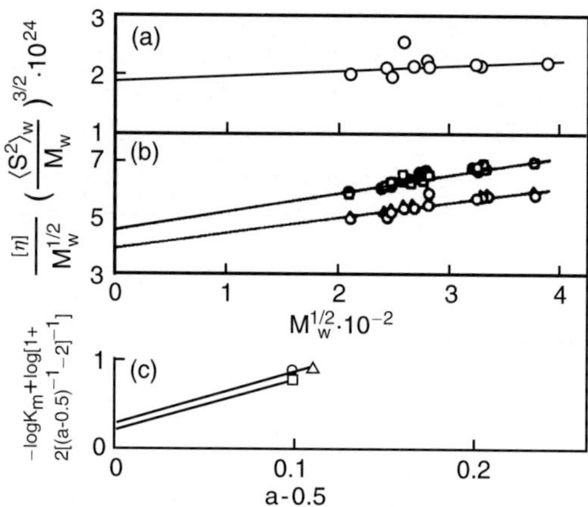

Figure 3.16.10   Baumann plot (a), Stockmayer–Fixman plot (b), and Kamide *et al.* plot (c) of cellulose acetate (DS 0.49) in dimethylacetamide (○); dimethyl sulfoxide (△); water (□); and formamide (●) at 25 °C.[55]

in DMAc, obtained from the intercepts and slopes of the above plots, are summarized in Tables 3.16.3 and 3.16.4, respectively.

The prerequisites for methods 2E and 2F are $a_2 = a_\Phi = 0$. The latter is evidently not acceptable from an experimental point of view (see Figure 3.15.2). Methods 2B and 2C are then preferable to methods 2E and 2F. The A values calculated from methods 2E and 2F are from some 20% (in DMAc) to 150% (in formamide) smaller than the most probable value, which is an average of the values obtained by the methods 2B and 2C. The large discrepancy in the A values derived by the thermodynamic (methods 2B and 2C) and hydrodynamic (methods 2E and 2F) approaches can be accounted for by taking the draining effect into consideration. Lines 2 to 4 of Table 3.16.3 show the A value in three solvents estimated by methods 2B to 2H.

Table 3.16.4

Long-range interaction parameter B of cellulose acetate in various solvents at 25 °C

Methods	$B \times 10^{27}$ (cm^3)				
	$\langle\langle F \rangle\rangle = 0.49$	$\langle\langle F \rangle\rangle = 2.46$			$\langle\langle F \rangle\rangle = 2.92$
	DMAc	DMAc	Acetone	THF	DMAc
2B (α_s from ψ)	–	–	–	–	–
2C ($\alpha_2 = 0$)	0.79	6.97	–2.45	1.67	3.9
2D ($\alpha_2 \neq 0$)	–	–	0.90	–	–
2E ($\alpha_2 = \alpha_\Phi = 0$)	3.70	6.90	1.90	2.71	4.3
2F ($\alpha_2 = \alpha_\Phi = 0$)	–	–	–	–	–
2G ($\alpha_2 \neq 0, \alpha_\Phi \neq 0$)	–	5.09	1.56	2.11	3.7

The conformation parameter σ and the characteristic ratio C_∞ of CA (DS 0.49) were calculated from the A values in the conventional manner, and are given in Tables 3.16.3. For CA (DS 0.49) in DMAc, $A = 1.38 \times 10^{-8}$ cm, $\sigma = 2.58$, and $C_\infty = 13.1$ were obtained as the most probable values. The σ value obtained here for this system is the minimum of these values in 25 systems of cellulose, amylose, and their derivatives in solvents,[1,27,34,55,56] and is not so different from those of the vinyl-type polymers.

CA (DS 1.75).[56] The MWD of $\langle S^2 \rangle_{0,w}/M_w$ was estimated by methods 2B, of CA (DS 1.75) in DMAc at 25 °C. The ratio $\langle S^2 \rangle_{0,w}/M_w$ can be regarded, within experimental error, as independent of M_w (Figure 3.16.9). The fifth column of Table 3.16.3 presents the A values of CA (DS 1.75) in DMAc at 25 °C, estimated by methods 2B to 2H.

CA (DS 2.46).[34,57] $\langle S^2 \rangle_0^{1/2}$ is calculated from the experimental $\langle S^2 \rangle_w^{1/2}$ data and a_s, which is estimated from ψ through use of eqs. (3.13.4–3.13.11). The value of A (i.e. $\langle S^2 \rangle_{0,w}/M_w$) thus obtained in acetone exhibits a rather striking molecular weight dependence, as shown in Figure 3.16.9 and a_2 amounts to -0.471 for CA (DS 2.46) in acetone (see Table 3.16.5). $a_2 < 0$ is not unusual for cellulose derivatives.[1] The average value of A in acetone is found to be 1.66×10^{-8} cm. In contrast, the A values in THF (1.23×10^{-8} cm) and in DMAc (2.05×10^{-8} cm) are almost independent of M_w (i.e. $a_2 = 0$). The negative correlations between a_Φ and a_2 have been demonstrated for numerous cellulose and amylose derivatives.[1]

The plot of $\langle S^2 \rangle_w^{3/2}/M_w^{3/2}$ versus $M_w^{1/2}$ (Baumann plot) yields $\langle S^2 \rangle_{0,w}^{3/2}/M_w^{3/2}$ as an intercept at $M_w^{1/2} = 0$ as shown in Figure 3.16.11(a). A line can be drawn through all the data points and A value is found to be 2.14×10^{-8} cm in acetone and 1.24×10^{-8} cm in THF, respectively. The slope of the plot is theoretically expected to reflect the sign of the long-range interaction parameter B (see eq. (3.16.5)). Accordingly, A_2 is negative when $B < 0$. However, a large negative slope of the plot for CDA in acetone sharply contradicts with the experimental fact of $A_2 > 0$ by LS and MO. This inconsistency can be resolved only by taking into account $a_2 \neq 0$ in the original Baumann's equation (eq. (3.16.4)). The A value thus estimated by this method for CDA in acetone is less accurate.

Using $a_2 = -0.471$, BKM plot for CDA in acetone is demonstrated in Figure 3.16.11(b), from which $K_0 = 1.18 \times 10^{-14}$ cm^2 is estimated.

Table 3.16.5

Evaluation of a_1 for cellulose acetate (DS 2.46 and 2.92) solutions at 25 °C[57]

DS	Solvent	a	a_Φ	λ	a_2	From a (eq. (3.15.30))	From λ (eq. (3.15.18))
							a_1
2.46	Acetone	0.616	0.716	-0.384	-0.471	0.107	0.13
	THF	0.688	0.105	0.116	0	0.083	0.17
	DMAc	0.738	0.23	0.06	0	0.008	0.09
2.92	DMAc	0.750	0.106	0.10	0	0.144	0.15

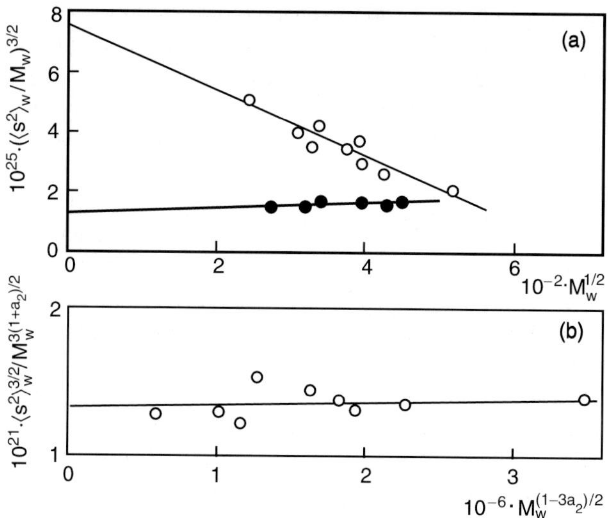

Figure 3.16.11   Baumann plot (a) and Baumann–Kamide–Miyazaki plot (b) of cellulose diacetate (CDA) in acetone (O) and tetrahydrofuran (●) at 25 °C.[34] In the BKM plot, $a_2 = -0.471$ is used for in acetone.

SF plots for CDA in acetone and THF are shown in Figure 3.16.12. Reasonably straight lines for each solution are obtainable. $A = 1.10 \times 10^{-8}$ cm and 0.99×10^{-8} cm are evaluated in acetone and THF, respectively, where $\Phi(\infty)$ was assigned the value 2.87×10^{23}. These values are less reliable, because for the CDA solution $a_\Phi > 0$ and $a_2 \leq 0$ are observed (see Table 3.15.4 and Fig. 3.16.9).

A plot of the left-hand side of eq. (3.16.15) as a function of $a - 0.5$ (Kamide plot) should be linear with a slope equal to log M_0. The Kamide plot for CDA in acetone and THF is illustrated in Figure 3.16.13. $K = 0.399$ (i.e. $A = 1.2 \times 10^{-8}$ cm) in acetone and 0.289 ($A = 1.00 \times 10^{-8}$ cm) in THF are determined by method 2F.

By using the values of a_2 and a_Φ together with K_Φ ($a_\Phi = 0.716$, $a_2 = -0.471$ and $K_\Phi \times 10^{-24} = 2.26 \times 10^{-5}$ for acetone, and $a_\Phi = 0.105$, $a_2 = 0$ and $K_\Phi \times 10^{-24} = 0.0573$

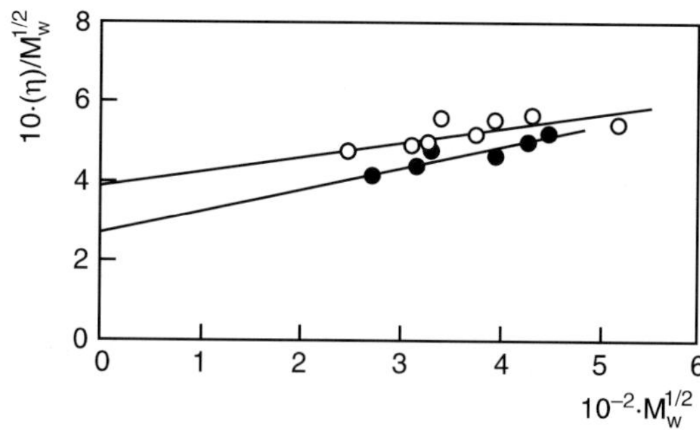

Figure 3.16.12   Stockmayer–Fixman plot according to eq. (3.16.12) for cellulose diacetate in acetone (O) and in tetrahydrofuran (●) at 25 °C.[34]

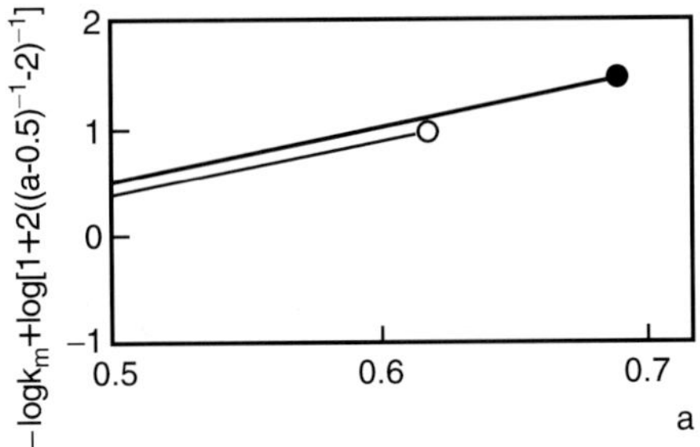

Figure 3.16.13   Kamide plot according to eq. (3.16.15) for cellulose diacetate in acetone (○) and in tetrahydrofuran (●) at 25 °C.[34] The lines, which have slopes of 1/2 log $(M_1 M_2)$, are constructed to pass through each data point.

for THF, the KM plot is constructed as shown in Figure 3.16.14, from which $K_0 = 1.16 \times 10^{-14}$ cm^2 in acetone and 2.50×10^{-17} cm^2 in THF are estimated.

Figure 3.16.15 shows the TB plot (method 2H; see eq. (3.16.19)) for CA (DS 2.46) in acetone and THF. We obtained $K' = (2.87/1.295) \times 10^{23}$ for the former and $K' = 2.87 \times 10^{23}$ for the latter from Table III of Ref. 34 and Table VIII of Ref. 22. A is found to be 1.45×10^{-8} cm in acetone and 1.37×10^{-8} cm in THF. For CDA in THF, Tanner and Berry have obtained $A = 1.91 \times 10^{-8}$ cm (this value is calculated from Table V of Ref. 19), which is about 40% larger than the value we obtained. The plot of their data scatters very much as shown in Figure 3.16.15 as closed rectangular, and its extrapolation contains much uncertainty.

Similar plots (Baumann, SF, Kamide, KM(I)) of CA (DS 2.46) in DMAc are shown in Figure 3.16.16. Table 3.16.4 presents the values of B for CA (DS 2.46) estimated from

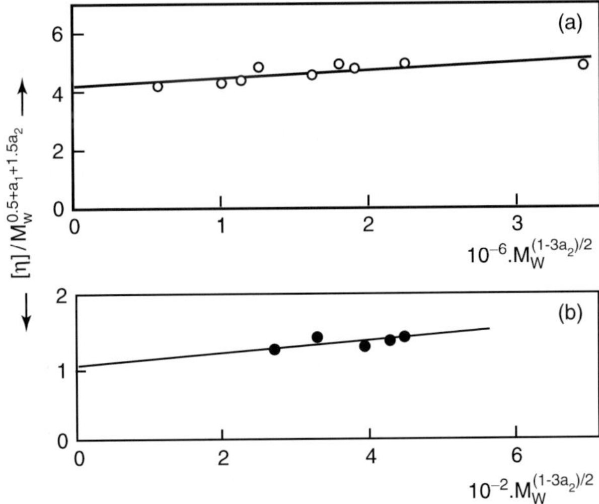

Figure 3.16.14   Kamide–Miyazaki plot according to eq. (3.16.17) for cellulose diacetate in acetone (○) and in tetrahydrofuran (THF) (●) at 25 °C.[34] The values of a_Φ and a_2 in Table IV of Ref. 34 were used.

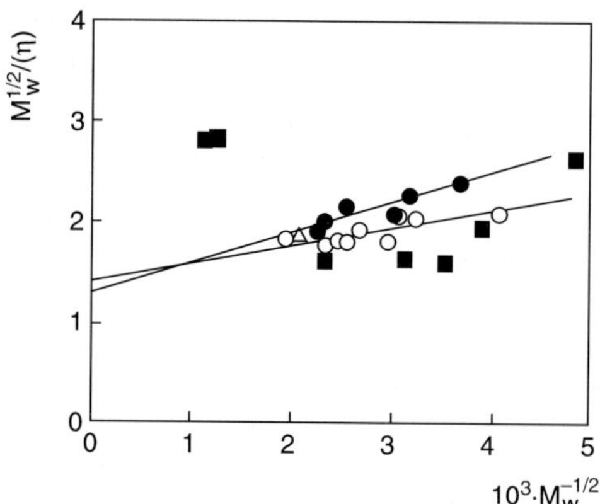

Figure 3.16.15   Tanner–Berry plot according to eq. (3.16.19) for cellulose diacetate in acetone (open mark) and in tetrahydrofuran (closed mark) at 25 °C;[34] (O) and (●);[34] (■), data from Tanner and Berry's work.[22]

Figure 3.16.11, 3.16.14, and 3.16.16. Table 3.16.3 also includes the values of the conformation parameter, σ and the characteristic ratio C_∞ for this system.

If the A values estimated by methods 2B to 2H are designated as $A_{(2B)}, \ldots, A_{\{2H\}}$ for convenience, the following relationships hold experimentally; for CA (DS 2.46)

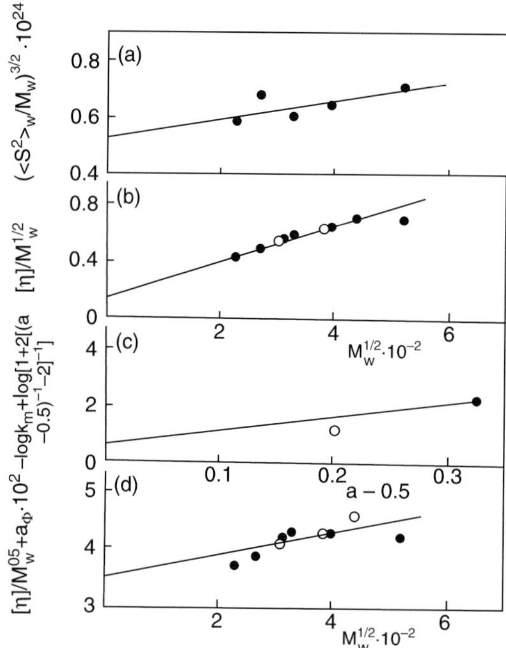

Figure 3.16.16   Baumann plot (a), Stockmayer–Fixman plot (b), Kamide plot (c), Kamide–Miyazaki (I) plot (d) for celluloses acetate (DS 2.46) in acetone at 25 °C. (●), M_w was determined in dimethylacetamide at 25 °C; (O), M_w was determined in acetone and in tetrahydrofuran.[57]

solutions:

$$A_{(2C)} > A_{(2B)} \approx A_{(2D)} \approx A_{(2G)} \gg A_{(2H)} > A_{(2E)} \approx A_{(2F)} \qquad \text{in acetone} \qquad (3.16.49)$$

$$A_{(2H)} > A_{(2B)} \approx A_{(2C)} \approx A_{(2G)} \gg A_{(2E)} \approx A_{(2F)} \qquad \text{in THF.} \qquad (3.16.50)$$

From a theoretical perspective, in the case of $a_2 \neq 0$ and $a_\phi \neq 0$, methods 2B, 2D, and 2G are expected to be the most reliable methods presently available for estimating the A value and in the case of $a_2 = 0$ and $a_\phi \neq 0$, methods 2B, 2C, and 2G are recommended.[1] The former corresponds to solution of CA (DS 2.46) in acetone and the latter to CA (DS 2.46) solution in THF. Agreement between methods 2E and 2F is excellent. However, the values estimated by these methods are 35% smaller than those by methods 2B, 2C (or 2D), and 2G, owing to the neglect of a_2 and a_ϕ. Method 2H overestimates or underestimates the A value depending on the magnitude of $a_1 + 1.5a_2$.[21] In fact, in acetone $a_1 + 1.5a_2$ is negative (-0.6), and method 2H is thus expected to underestimate A. In contrast, in THF $a_1 + 1.5a_2$ is positive (0.08) and so, method 2H may overestimate the A value. It has been observed that when $a_2 < 0$ method 2C has a tendency to overestimate A.[1] The relative order of the A value for CDA, as determined by the various methods, agrees well with the results obtained for cellulose, amylose, and their derivatives.[1]

Therefore, the most reliable A value ($A_{(m)}$) can be defined as $(A_{(2B)} + A_{(2D)} + A_{(2G)})/3$ for CA (DS 2.46) in acetone and $(A_{(2B)} + A_{(2C)} + A_{(2G)})/3$ for CA (DS 2.46) in THF, respectively. $A_{(m)} \times 10^8$ cm values thus calculated are 1.74 in acetone and 1.26 in THF and are also listed in Table 3.16.3. The value of σ corresponding to $A_{(m)}$ is not unusually high as compared with other cellulose derivatives.[1]

CA (DS 2.46) is considerably less flexible in acetone than in THF, suggesting that there is a specific solvent effect, such as a temperature effect or solvation, on a short-range interaction.

The contribution of the excluded volume effect to the exponent a, a_1 ($\equiv 3\epsilon$) can be roughly estimated from a, a_ϕ, and a_2 by eq. (3.15.30).

$$a = 0.5 + a_\phi + a_1 + 1.5a_2. \qquad (3.15.30)$$

The value of a_1 thus determined by eq. (3.15.30) is 0.107 in acetone and 0.083 in THF. The a_1 value can be also estimated by using an alternative equation

$$\lambda = a_2 + 2\epsilon = a_2 + (2/3)a_1. \qquad (3.15.18)$$

The experimental data of a, a_ϕ, λ, a_2, for CA (DS 2.46) in acetone, THF, and DMAc are summarized in Table 3.16.5.[57] The estimated values of a_1 using the above equations are collected in the seventh and eighth columns of the table. Both equations give almost the same magnitude of a_1, although a_1 values from λ are slightly larger than those calculated from a. a_1 values are not as large as 0.2. Acetone, THF, and DMAc are not good solvents, against CA (DS 2.46).

When the values of A and B are obtainable by methods 2C, 2D, 2E, and 2G, we can calculate A_2 from A, B, together with the experimental $\langle S^2 \rangle_{\text{w}}$ value, by using the relationship.[58–60]

$$A_2 = (N_A/2)Bh_0(z) \qquad (3.13.1)$$

In Figure 3.13.1, full (method 2C), dotted (method 2D), broken (method 2E), and chain (method 2G) lines are thus calculated by using eqs. (3.13.1) and (3.13.6) from A and B (by method 2C, 2D, 2E, and 2G; see Tables 3.16.3 and 3.16.4), and $\langle S^2 \rangle_z^{1/2}$ in Table 3.3.3. The A_2 value calculated by method 2C for acetone is a large negative value and is not shown in the figure. It is clear that method 2G gives the best fit for the experimental data. Consequently, the methods used here prove highly satisfactory for interpreting the dilute solution properties of CDA.

CA (DS2.92).[27] Values of $\langle S^2 \rangle_z^{1/2}$ in Table 3.3.5 were converted to values of $\langle S^2 \rangle_{0,w}$ $(= \langle S^2 \rangle_w / a_s^2)$ using these a_s values (method 2B). The ratio $\langle S^2 \rangle_{0,w} / M_w$, plotted in Figure 3.16.9 as a function of molecular weight, for CA (DS 2.92) solution in DMAc, is almost independent of M_w and accordingly, the CA (DS 2.92) chain in DMAc can be reasonably treated as a Gaussian chain (i.e. $a_2 = 0$), and the same applies for the CDA solution in THF[34] and in DMAc.[57] For the CTA/DMAc system we obtain $a_1 = 0.144$ putting $a_\Phi = 0.106$, and $a_2 = 0$ into eq. (3.15.32), which agrees well with 0.15 as calculated from eq. (3.15.18) (see Table 3.16.5). For the CTA/DMAc system, the partially free-draining effect is comparable to the excluded volume effect. It is therefore quite obvious that the dilute solution properties of CTA cannot be described adequately in terms of a simple two parameters theory.

The A value was evaluated for the CTA/DMAc system. Figure 3.16.17 represents the Baumann plot (method 2C), SF plot (method 2E), Kamide plot (method 2F), KM (I) plot (method 2G), and TB plot (method 2H). All of these methods (except 2E) gave a series of reasonably straight lines. All plots based on the Stockmayer–Fixman relationship showed downward curvature at high molecular weights. The A values estimated from method 2B to 2H for CA (DS 29.2) in DMAc and TCE are collected in columns 9 and 10 of Table 3.16.3 and the B values for the system are given in column 6 of Table 3.16.4.

The results in Table 3.16.3 are in particularly good agreement with the theoretical predictions. Methods 2E and 2F considerably underestimate the A value due to $a_\Phi \neq 0$[1] whereas method 2H overestimates the A value because $a_1 + 1.5a_2 > 0$.[21] Three methods (2E, 2F, and 2H), which rely upon viscosity data, are unlikely to provide values that are accurate in an absolute sense. As pointed out previously,[1,34] the SF plot, observed to hold almost universally for vinyl-type polymers, does not hold for cellulose derivatives.

The methods 2B, 2C, and 2G are the most reliable and promising in this respect. In fact, the A values determined by these three methods, within the limits of error, are almost the same. Thus as the most probable value, we estimated A of 1.46×10^{-8} cm.

The conformation parameter σ and the characteristic ratio C_∞ were calculated for CA (2.92) subsequently from the A value and tabulated in Table 3.16.3. The most probable σ and C_∞ values are found to be 3.57 and 25.2, respectively for the CA (DS 2.92)/DMAc system at 25 °C. These values should be compared with $\sigma = 4.39$ ($C_\infty = 28.0$) for CA (DS 2.46) in acetone and $\sigma = 2.84$ ($C_\infty = 15.9$) for the CA (DS2.46)/THF system.[34]

Table 3.16.6 shows the A values, estimated by methods 2E and 2F, of CA (DS2.46) in various solvents. The table demonstrates that the A values, estimated by using only viscosity data ($[\eta]$ and M_w, or K_m and a), have a large error of about $\pm 0.5 \times 10^{-8}$ cm. Thus, if only the viscosity data are available, using methods 2E and 2F we can estimate

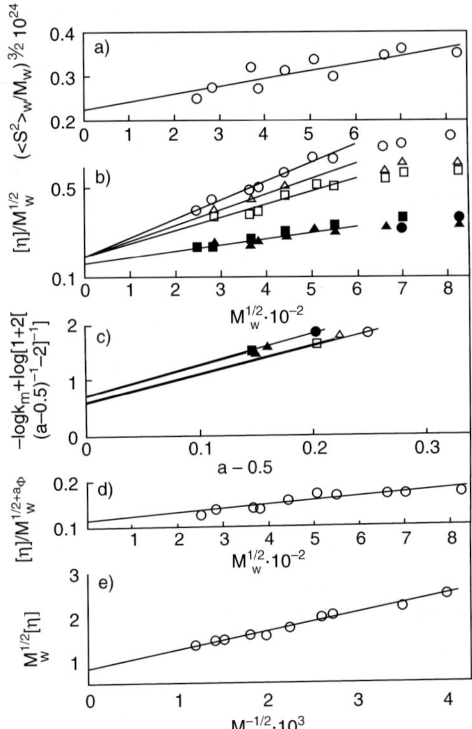

Figure 3.16.17 Baumann plot (a), Stockmayer–Fixman plot (b), Kamide plot (c), Kamide–Miyazaki (I) plot (d), and Tanner–Berry plot (e), for celluloses triacetate (DS 2.92) in dimethyl-acetamide (○), trifluoroacetic acid (△), acetone (□), dichloromethane (●), tetrachloroetane (■), and trichloromethane (▲).[27]

Table 3.16.6

Unperturbed chain dimensions A of cellulose triacetate, evaluated by methods 2E and 2F[27]

Method	$A \times 10^8$ (cm)					
	DMAc[a]	TFA[b]	Acetone	DCM[c]	TCE[d]	TCM[e]
2E	0.995	0.995	0.995	0.823	0.823	0.823
2F	0.995	1.00	0.995	0.853	0.876	0.884
Most probable value	1.46	–	–	–	–	–

[a]Dimethylacetamide.
[b]Trifluoroacetic acid.
[c]Dichloromethane.
[d]Tetrachloroethane.
[e]Trichloromethane.

only the comparable variation of the A value with the solvent nature. In Figures 3.16.17(b) and (c), the data points other than those for DMAc are plotted.

From Table 3.16.6, it is clear that DMAc, TFA, and acetone give almost the same A values for CA (DS 2.92). On the other hand, chlorinated hydrocarbons, like DCM, TCE, and TCM, yield slightly smaller A values. Comparison of Tables 3.16.3 and 3.16.6 thus allows a rough estimation of the true A value which is found to be 1.46×10^{-8} cm for the former solvents (group I) and $A \approx 0.87 \times 10^{-8} \times 1.46/0.977 = 1.27 \times 10^{-8}$ (Here, 0.87 is an average value of A estimated by method 2F for solutions in solvents of group II (DCM, TCE, and TCM) and 0.977 is a corresponding value for group I (DMAc, TFA, and acetone), respectively) for the latter group of solvents (group II), respectively. Here, these values were obtained on the usual assumption that the importance of the draining effect (i.e. a_Φ value) is independent of the solvent nature, but this is not necessarily so. The former is *ca.* 15% larger than the latter. The UCD of CTA, like a variety of other cellulose derivatives differing in chemical composition (for example, CTCp and cellulose tricarbanilate), varies with the nature of the solvent.[1]

Some vinyl-type polymers. The applicability of eq. (3.16.33) is not limited to cellulose, amylose, and their derivatives in solution. Kamide and Miyazaki have found $a_\Phi \neq 0$ for some vinyl-type polymer solutions, including poly(a-methyl styrene) in toluene[1] and poly(p-methyl styrene) in toluene.[1] In this section, methods 2A to 2G were applied to the above systems. The results are given in Table 3.15.7. Obviously, method 2G is preferable, in the case of $a_\Phi \neq 0$, to method 2F and the A values obtained by the former agree well with the averaged values of methods 2A, 2B, and 2C.

Methods 2I and 2J[23]

Literature data on cellulose,[29] CDA,[58,59] CTC,[60,61] EHEC,[42] and ATA[51] were analyzed extensively according to eqs. (3.16.26') and (3.16.26) in order to determine A or K_0.

The molecular weight dependence of Flory's parameter P is demonstrated in Figure 3.15.5, in which the data can be well represented by eq. (3.15.6) with a positive a_p. The values of K_p and a_p thus determined are also listed in the fourth and fifth columns in Table 3.15.5. Evidently, there is a close correlation, as expected, between a_p and a_Φ ($\equiv$ d $\times$ ln Φ/d ln M) (correlation coefficient $\gamma = 0.71$), which acts as a rough measure of the draining effect. That is to say, the facts $a_p > 0$ and $a_\Phi > 0$ mean that the draining effect for these polymers should not be ignored.

Figure 3.16.18 shows the CB plots.[23] Least square straight lines in the figure were drawn through data points. The data points based on s_0 are denoted by open marks and those on D_0 by closed marks.

The value of A, as estimated from the CB plot, is denoted by $A_{(2I)}$ and shown in the fifth column of Table 3.26.1b and Table 3.16.7, which also includes results obtained by Shanbhag for CTC in cyclohexanone.[63] There is excellent agreement between Shanbhag's result and the present result. Except in the CTC/cyclohexanone system, $A_{(2I)}$ (A evaluated by method 2I) is much smaller than the most reliable A value ($A_{(m)}$), as shown in Table 3.16.10. $A_{(2I)}$ is, on average, of the order of 50% of $A_{(m)}$. In other words, method 2I considerably underestimates A and is by no means applicable for cellulose, amylose, and their derivatives.[23]

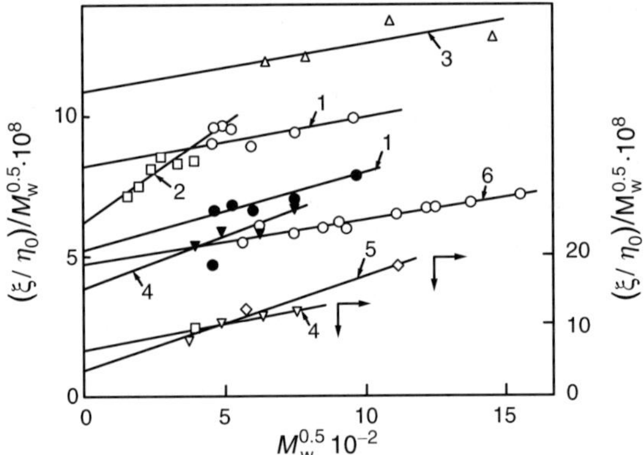

Figure 3.16.18   Plot of $(\xi/\eta_0)/M_w^{0.5}$ against $M_w^{0.5}$ (Cowie–Bywater plot)[23]: Numbers on curve have the same meanings as in Figure 3.15.5 and the ratio (ξ/η_0) is expressed in cm.

The KM (II) plot is displayed in Figure 3.16.19. All points can be represented by straight lines over the entire range of molecular weight experimentally accessible.

The values of K_0 together with the A values at $M_w = 1 \times 10^5$ ($A_{(2J)}$) are collected in the last two columns in Table 3.16.7 and in the sixth column of Table 3.16.1(b). The results indicate that no significant difference exists between $A_{(2J)}$ and $A_{(m)}$ (correlation coefficient $\gamma = 0.97$) if experimental error is considered. It is immediately evident from Table 3.16.7 that the inapplicability of method 2I to cellulose, amylose, and their derivatives can be regarded as mainly due to the neglect of a_2 and a_p. A comparison of eq. (3.16.26) and (3.16.26′) reveals that $A_{(2I)}$ may coincide with $A_{(2J)}$, and accordingly

Table 3.16.7

Unperturbed chain dimensions A and K_0 of cellulose and amylose derivatives[23]

Polymer	Solvent	The most reliable value $A_{(m)}$, 10^3 (cm)[a]	s_0 or D_0	$A_{(2I)}$, 10^8 (cm)	K_0, 10^{16} (cm^2) by method 2J	$A_{(2J)}$,[b] 10^5 (cm)
Cellulose	Cadoxen	1.71	s_0	1.43	6.57	1.92
			D_0	1.18	5.71	2.04
Cellulose diacetate	Acetone	1.84	s_0	1.11	112.1	1.72
Cellulose tricarbanilate	Cyclo-Hexanone	1.35	s_0	1.96 (1.98)[c]	0.318	1.48
Ethylhydroxyethyl cellulose	Water	2.30	s_0	1.08	11.92	2.16
			D_0	0.84	11.80	2.47
Sodium cellulose xanthate	1-N NaOH	3.79	s_0	0.73	33.51	3.63
Amylose carbanilate	Pyridine	1.07	s_0	0.88	0.154	1.02

[a]Value estimated in the previous paper.[1]
[b]Value obtained by Shanbhag.[64]
[c]Value at $M = 1 \times 10^5$.

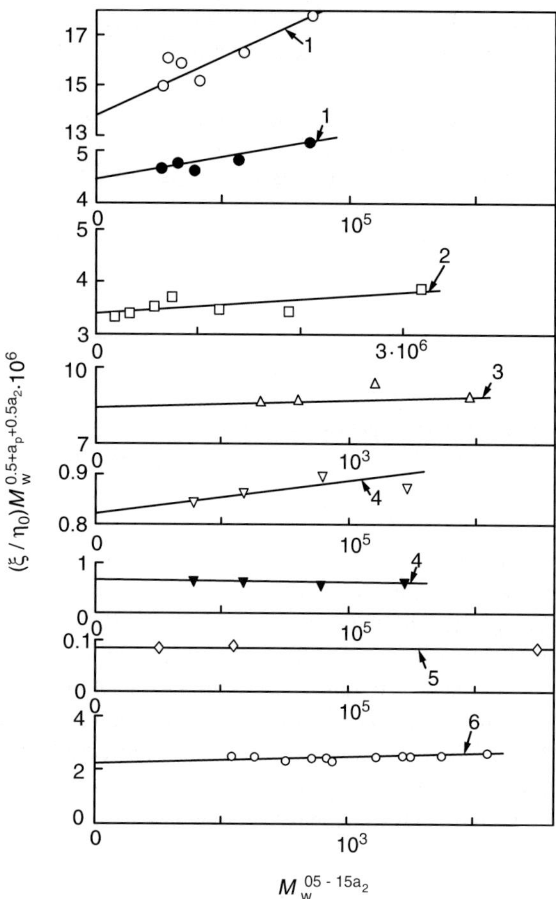

Figure 3.16.19 Plots of $(\xi/\eta_0)/M_w^{0.5+0.5a_2+a_p}$ against $M_w^{0.5-1.5a_2}$ (Kamide–Miyazaki (II) plot). Numbers on curve have the same meanings as in Figure 3.15.5 and the ratio (ξ/η_0) is expressed in cm.

with $A_{(m)}$, at $a_p + 0.5a_2 = 0$. This prediction is experimentally confirmed for cellulose and amylose derivatives, as shown in Figure 4 of Ref. 23, in which the ratio $A_{(21)}/A_{(2J)}$ decreases gradually with an increase in $a_p + 0.5a_2$. For the system of cellulose tricarbanilate in cyclohexanone, the large value of $A_{(21)}/A_{(2J)}$ cannot be interpreted in terms of $a_p + 0.5a_2(= 0.024)$.

The P values calculated for this system (7.36–8.49) are much larger than the theoretical maximum value $P_0(\infty)(= 5.2)$ (see Figure 3.15.5), suggesting that the experimental results are seriously questionable and that the application of method 2I to the above system leads unavoidably to an overestimation of the A value.

Therefore, we come to the conclusion that the UCD can be adequately estimated from the molecular weight dependence of s_0 and D_0 according to eq. (3.16.26), where both a_2 and a_p are satisfactorily taken into account. The slope of the KM (II) and CB plots affords us a long-range interaction parameter B (Table 3.16.4). Then, the second virial coefficient, A_2, can be readily calculated from A and B thus estimated by the plots, and from the experimental weight-average radius of gyration $\langle S^2 \rangle_w^{1/2}$ by using eq. (3.13.1), as demonstrated in Figure 3.13.2.

To summarize, the concept of a_2 and a_p (or a_Φ) explain the dilute solution properties of cellulose, amylose, and their derivatives in a consistent manner and the best and most widely used methods for estimating A from hydrodynamic data are methods 2G and 2J.

Method 2K[25]

In order to test the reliability of method 2K, eq. (3.16.33) was applied to experimental data for cellulose/cadoxen,[29] cellulose/FeTNa,[30] CN/acetone,[1,32] CTN/acetone,[1,31] CDA/acetone,[34] CDA/THF,[34] CTA/DMAc,[27] CTCp/DMF,[35] CTCp/1-chloronaphthalene (1-Cl-N),[35] CTC/acetone,[36,65] CTC/dioxan,[37,65] MC/water,[39] sodium carboxymethyl cellulose (NaCMC)/aq. NaCl,[40] sodium cellulose sulfate (NaCS)/0.5 aq. NaCl,[66] HEC/water,[41] EHEC/water,[42] sodium cellulose xanthate (NaCX)/1M NaOH,[43,44,63] amylose/dimethyl sulphoxide (DMSO),[45] and ATA/nitromethane.[46] For these systems (except for CDA, CTA, and NaCS) all parameters necessary for estimating A by eq. (3.16.33) have been determined by Kamide and Miyazaki,[1] and for CDA/acetone, CDA/THF, CTA/DMAc, and NaCS/aq. NaCl systems, the parameters were estimated in the original literature.[25]

For all the systems employed here, the partially free-draining effect cannot be neglected ($a_\Phi \neq 0$) and for some systems, the unperturbed chain is obviously nonGaussian ($a_2 < 0$). The A values estimated by method 2K are tabulated in the seventh column of Table 3.16.1(b).

Besides the A value, the B value can also be indirectly determined by use of methods 2C, 2D, 2E 2G, 2I, and 2J[1,23] (see Table 3.16.4). Methods 2G and 2J give the same B value as estimated by method 2C (when $a_2 = 0$) or 2D (when $a_2 \neq 0$). Moreover, the A_2 values calculated from A and B values (estimated by methods 2G and 2J) agree with the experimental values[1,23,24] (see Figures 3.13.1 and 3.13.2). In contrast, the B values estimated by methods 2E and 2I are much larger than those by method 2C, 2D, 2G, and 2J, giving A_2 values that are too large and not accommodated by the experimental A_2 value, as demonstrated above.[1,23]

Method 2L[28]

We apply method 2L (eq. (3.16.46)) to experimental data for cellulose/cadoxen,[29] CDA (DS 2.46)/acetone,[34,61,62] CTA (DS 2.92)/DMAc,[27] CTC/cyclohexanone,[63,64] EHEC/water,[42] NaCX/IN NaOH,[43] and amylose tricarbanilate (ATC)/pyridine.[48,49]

For these systems (except for CDA/acetone and CTA/DMAc), all parameters necessary for estimating A by eq. (3.16.46) were determined by Kamide and Miyazaki.[23] For CDA/acetone and CTA/DMAc systems, the parameters were estimated in the original literature (as summarized in Table I of Ref. 23). Note that for all the systems used here, the draining effect cannot be ignored ($a_\Phi \neq 0$) and for some systems, the unperturbed chain is obviously nonGaussian ($a_2 < 0$). We have demonstrated also for cellulose, amylose, and their derivatives $a_\Phi \neq 0$.[1,21,34,55,57,65]

The last column of Table 3.16.1(b) tabulates the A values, estimated by method 2L, for these systems.

Except for cellulose/cadoxen (s_0) and CTC/cyclohexanone systems, the following relationships are obtained:

$$A_{(2I)} \ll A_{(2J)} \cong A_{(2L)} \qquad (3.16.51)$$

where the suffix refers to the method employed for estimation of A. Method 2I, which assumes $a_2 = a_p = 0$, significantly underestimates the A value. The same tendency was observed between methods 2E, 2F, 2G, and 2K:

$$A_{(2E)} \cong A_{(2F)} \ll A_{(2G)} \cong A_{(2K)}. \qquad (3.16.52)$$

3.16.3 Consistency of the methods for estimating unperturbed chain dimensions

We attempted to check the consistency among the methods utilized for estimating the unperturbed chain dimensions. For this purpose, the correlation coefficient γ was calculated for any two methods arbitrarily chosen. Table 3.16.8 summarizes the correlation coefficient γ between the A values for cellulose and amylose derivatives.[25] The results are also schematically presented in Figure 3.16.20. Although methods 2B, 2C, and 2D are based on the thermodynamic approach and methods 2G, 2J, 2K, and 2L on the hydrodynamic approach, the A values obtained by methods 2B, 2C, 2D, 2G, 2J, and 2K are highly correlated ($\gamma > 0.95$) with each other (with the exception of a combination of methods 2C and 2D) and are accurate within 6%. The correlation coefficient γ between method 2C and 2D is estimated as 0.912. This value is significantly smaller than those corresponding to any combination among the methods 2B, 2C (or 2D), 2G, and 2K because, for solutions of cellulose and amylose derivatives, a_2 often deviates from zero.

On the other hand, the A value estimated by method 2E exhibits a high correlation with that by method 2F, as was the case for various vinyl-type polymers.[21] γ between 2D and 2G is larger than that between 2C and 2G, as is expected. Of course, the A values obtained by method 2D agree well with those by methods 2B, 2G, and 2K; γ value (0.960) between methods 2C and 2K is smaller than that (0.986) between methods 2D and 2K. The A values determined by methods 2E and 2F coincide. In the third to last line of Table 3.16.8, the A values estimated by a given method averaged over these systems are collected. The A values estimated by the former group (2B, 2C or 2D, 2G, 2J, and 2K) are twice as large as those by methods 2E and 2F, in which $a_2 \neq 0$ and/or $a_\phi \neq 0$ are not taken into consideration.

The A values obtained by methods 2B, 2C, 2D, 2G, 2J, and 2K are highly correlated ($\gamma > 0.90$) with those by method 2L. Method 2L is not significantly correlated with methods 2E, 2F, 2H, and 2I (i.e. $-0.5 < \gamma < 0.5$). Methods 2E, 2F, and 2I assume $a_2 = a_\phi = a_p = 0$ (Gaussian and nonpermeable chain sphere) and as pointed out previously, the A values estimated by methods 2E, 2F, and 2I are considerably smaller than those by the thermodynamic approach (methods 2B, 2C, and 2D).[1,21,22,34,54,62] This has been an unresolved contradiction in the past. Method 2L is apparently consistent with

Table 3.16.8

Correlation coefficient γ between the A values estimated for cellulose, amylose, and their derivatives by the two methods arbitrarily chosen[25]

Method	Thermodynamic approach			Hydrodynamic approach							
	2B	2C	2D	2E	2F	2G	2H	2I	2J	2K	
	$-$	0.965	0.993	0.302	0.349	0.984	0.223	-0.830	0.963	0.981	2B
		$-$	0.912	0.444	0.461	0.967	0.043	-0.813	0.950	0.960	2C
			$-$	0.207	0.212	0.993	-0.025	-0.839	0.987	0.986	2D
Correlationship				$-$	0.958	0.336	-0.274	0.447	-0.569	0.316	2E
coefficient					$-$	0.366	-0.086	0.512	-0.668	0.350	2F
γ						0.366	0.112	-0.393	0.963	0.998	2G
						$-$	$-$	0.501	-0.403	0.116	2H
								$-$	-0.793	-0.841	2I
									$-$	0.958	2J
										$-$	2K
Averaged A value $A \times 10^3$ cm	1.77	1.97	1.91	0.90	0.92	1.88	1.90	$-$	$-$	1.90	

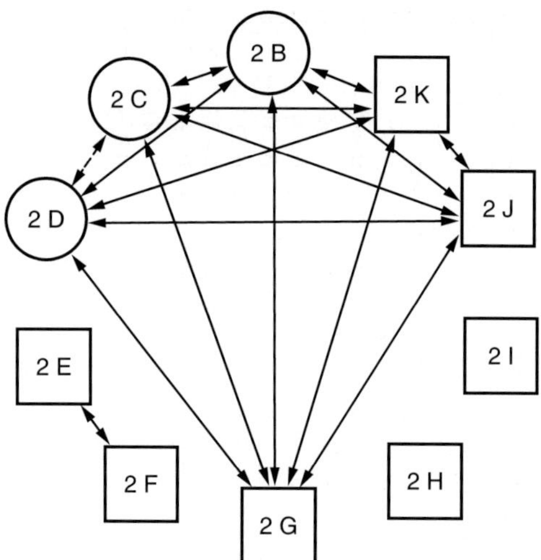

Figure 3.16.20   Correlation coefficient γ of A between the two methods arbitrarily chosen.[25] Circle: thermodynamic method; rectangle: hydrodynamic method; solid line: $\gamma > 0.95$; dotted line: $0.95 > \gamma > 0.90$.

thermodynamic and other hydrodynamic approaches if nonGaussian nature and the draining effect are properly taken into account.

A different conclusion should be drawn from the different group of the analytical procedures adopted. Examination of the literature shows that by using methods 2B and 2C, the polymer chain was inflexible, and with methods 2E and 2F, the chain was always deemed flexible. When methods 2B (or 2C) and 2E were applied using the same data, the results obtained by the former method were considered more reliable. Hence, the analytical method utilized is very important and should be carefully examined in advance. The confusion of the concept, in the past, to the flexibility and draining effect of cellulose and its derivatives (see Section 1.5) seems not to result from the inconsistence of the experimental data, but mainly from unreasonable usage of the analytical methods. In the past, conclusions with respect to chain rigidity of cellulose derivatives and the polymer–solvent interaction in these solutions differed greatly according to the method (thermodynamic approach or hydrodynamic approach (2E and 2F)) employed for the analysis.[1] This disagreement has been resolved completely by using methods 2G and 2J.[1,23] method 2K is consistent with the thermodynamic approach.

From the theoretical point of view, for $a_\Phi \neq 0$, methods 2B, 2C ($a_2 = 0$), and 2D ($a_2 \neq 0$), methods 2G, 2J, and 2K are reasonable and are recommended for use. Method 2K is a great improvement on method 2F. As predicted, method 2H overestimates A in the region $a_1 + 1.5a_2 < 0$ and underestimates in the region $a_1 + 1.5a_2 > 0$. This fact indicates that for solutions of cellulose and its derivatives, the excluded volume effect and/or the nonGaussian properties can never be ignored, when estimating unperturbed chain dimensions.

For cellulose, amylose, and their derivatives in solution, a_2 is zero or negative. This conclusion is not drawn by using erroneously an indirect method for evaluating

α_s, because $a_2 < 0$ was found even in theta solutions (e.g. CTCp in DMF[35] and MC in water[37]). This is then understood as a consequence of an increase in the chain stiffness with a decrease in its length. $a_2 < 0$ cannot be explained in terms of the statistics of the pearl necklace chain model.[66] Yamakawa and coworkers theoretically demonstrated, on the basis of the helical worm-like chain model, that a_2 can becomes negative.[67,68] The value of λ defined by eq. (3.12.1), is zero or negative for the majority of cellulose. The A values by method 2D, in which $a_2 < 0$ is taken into account, is in better agreement with A made by method 2G than that by method 2C. These are additional supports for $a_2 < 0$.

Kurata and Stockmayer described in their excellent review that Φ should remain essentially constant at its asymptotic value $\Phi_0(\infty)$.[69] Their prediction was shown to be incorrect (see Figures 3.15.1–3.15.8) by the analysis of numerous experimental data, which have been published after review.[69] In fact, in cellulose and its derivatives, an absolute value of Φ was not very accurate as compared with that of vinyl-type polymers. Instead, Φ can never be treated as constant over a wide range of molecular weight assessed. Evidently, a_Φ, which is a more rigorous criterion of the draining effect, is always distinctly positive (Tables 3.15.1 and 3.15.4). This agrees well with the results on the X values in Tables 3.15.6–3.15.8. Consequently, the experimental findings, $a_2 \leq 0$ and $a_\Phi > 0$, undoubtedly prove wrong the theoretical basis (i.e. the two parameter theory) of methods 2E and 2F and the conclusions drawn by using these methods are seriously called into questioned.

Of great experimental significance is the fact that methods 2A, 2B, 2C, 2D, 2G, 2K, and 2L give almost an identical value of A with an estimated uncertainty of $\pm 7\%$ (with an exception of the combination of methods 2B and 2C), which is smaller in magnitude by a factor of about two than that obtained by methods 2E and 2F. As is illustrated in Figure 3.16.20 and Table 3.16.8, the difference of the A value between group I (2B, 2C, 2D, 2G, 2K, 2L) and group II (2E, 2F) is highly significant at the 0.1% level. The group obviously underestimates A. An exception is NaCMC,[40] in which the draining effect contributes only slightly to $[\eta]$ ($X \approx 8.6$ by method 1A and $a_\Phi + 1.5a_2 \approx 0.08$) and the ratio of the values of A by method 2E to that of $A_{(m)}$, is found to be 0.88, where $A_{(m)}$, is the most reliable value, as will be discussed further below. The A value, directly measured in theta solvents, occasionally agrees with that evaluated by methods 2B, 2D, and 2G from the data in nontheta solvents (e.g. CTCp and ATC). This also provides evidence of the validity of methods 2B, 2D, and 2G.

All of the graphical procedures (methods 2C, 2D, 2E, 2 G, 2I, 2J, and 2K) are roughly linear and even in the large M_w range they do not show downward curvature (with the exception of CTN ($N_c = 13.9\%$) and CN ($N_c = 12.9\%$) in acetone). Hence, the linearity of the plots cannot be taken as a reasonable manifestation of the validity of the procedures. It appears to be true that the neglect of the higher terms of z in eq. (3.13.9), is reasonably acceptable. This agrees with $\alpha_s \approx 1$ and $z \approx 0$.

If $a_2 \neq 0$, method 2D yields an A value which is higher than that obtained by method 2C by about 0.1×10^{-8} cm and is only about 3% different from that by method 2G. In method 2D, the A value is less sensitive than B (when B is negative) to the a_2 value employed. The A value, determined by method 2G, is highly dependent on $a_\Phi + 1.5a_2$ value, especially in the range of $a_\Phi < 0.1$ and the B value, obtained by method 2G, is also highly sensitive to $a_\Phi + 1.5a_2$. For example, the A and B values were determined

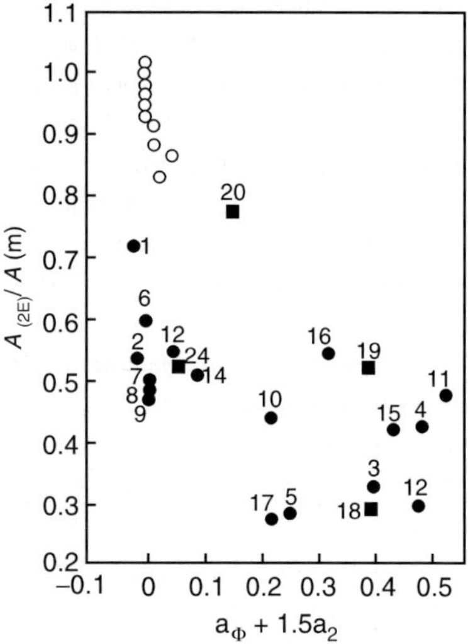

Figure 3.16.21   Plot of the ratio $A_{(2E)}/A_{(m)}$ against $a_\Phi + 1.5a_2$: ($\bigcirc$), vinyl-type polymers[12–17]; ($\bullet$), cellulose and its derivatives; ($\blacksquare$), amylose and its derivatives. Numbers 1–20 have the same meanings as those in Figure 3.16.1; 24 ATC in pyridine.[47,60]

for CN ($N_c = 12.9\%$) in acetone and for CA (acetyl content, 55.3%) in acetone by using methods 2D and 2G, assuming a_2 and a_Φ values. It should be noted that the slope of the BKM, SF, and KM plots for cellulose and amylose derivatives are often too small to allow for an accurate evaluation of B.

The A values estimated by methods 2E and 2F, $A_{(2E)}$ and $A_{(2F)}$, become smaller as the polymer chain has larger positive a_2 together with a larger positive a_Φ. This is illustrated in Figure 3.16.21, where the ratio of A estimated by method 2E to the most reliable A value is plotted against $a_\Phi + 1.5a_2$. It can be expected from the figure that the magnitude of $a_\Phi + 1.5a_2$ serves as a measure for reliability of methods 2E and 2F; that is, the disagreement with the A value by method 2E or 2F and that by other methods is greatest for large value of $a_\Phi + 1.5a_2$. As already pointed out, the values of A and B obtained by applying method 2E, 2F (or similar methods) lack reliability, especially at higher $a_\Phi + 1.5a_2$ values.

There is a crude negative correlation between a_2 and a_Φ ($\gamma = -0.63$). In other words, a polymer chain becomes less impermeable as the chain conformation deviates from Gaussian statistics and a_Φ has a tendency to compensate for a_2 in eq. (3.15.11).

In the case of $a_2 \neq 0$ and $a_\Phi \neq 0$, the exponent a in the MHS eq. (3.11.1) in Flory's theta solution, designated as a_θ, is not always 0.5. a_θ is generally given by

$$a_\theta = 0.5 + a_\Phi + 1.5a_2 \tag{3.16.53}$$

The value of a_θ was calculated from the values of a_2 and a_Φ given in Table 3.15.1 using eq. (3.16.53) and was compared with the experimental value. The results obtained were 0.505 (0.495) for CTPp[35] in DMF, ~(0.524) for CTC in anisol,[38] 0.544 (0.508) for MC in water,[39] and 0.644 (0.599) for ATC in pyridine/water.[47] The values in parentheses indicate the experimental values. The calculated a_θ value is in accordance with the experimental value if the error inherent in a_2 and a_ϕ is considered. Thus, there are actually several experimental cases of $a_\theta \neq 0.5$, which provide direct evidence of the existence of either a_2 or a_ϕ or both and even in the case $a_\theta = 0.5$ there is a possibility that $1.5 \, a_2$ compensates for a_ϕ.

In principle, both eqs. (3.16.8) and (3.16.13) give the correct value of K (accordingly, A), provided that a_2 equals zero (eq. (3.16.8)) or that a_ϕ is determined in advance (eq. (3.16.13)) and the draining parameter X is accurately evaluated. In these cases, the uncertainty of the X value plays an important role in the determination of A. However, the X value estimated for a given polymer solvent system scatters to a large extent (see Tables 3.15.6 and 3.15.7) although X is inversely proportional to a_ϕ. Assuming X to be constant, we determined the A value from the X value in Table 3.15.6 and 3.15.7 using eq. (3.16.8). The A value thus obtained decreases linearly with an increase in log X. The most reliable value of A falls often in the scatter of the A value estimated by eq. (3.16.8). This strongly suggests that the A value underestimated by method 2E (and probably also by method 2F) increases to the true value if the draining effect is taken into account. However, from a practical point of view, the use of eqs. (3.16.8) and (3.16.13) is not recommended for estimating A, because of the low accuracy in the estimation of X.

Hitherto, the A value obtained by methods 2B and 2C has been often regarded as less reliable, as compared with that by method 2E, because it was considered that the second virial coefficient A_2 has a large experimental uncertainty for cellulose, amylose, and their derivatives in solutions, for which A_2 is very small (in the order of 10^{-4} cm^3 mol $\times$ g^{-2}) and does not exhibit any significant molecular weight dependence except for CTN,[31,53] CN,[31,53] and CA.[34]

It has been widely recognized that the application of eq. (3.16.12) (or similar equations) to cellulose derivatives leads to a large expansion factor a_s and to relatively small unperturbed chain dimensions.[69] The inapplicability of eq. (3.16.12) (or similar equations) is clear from the fact that $a_2 \leq 0$ and $a_\phi > 0$, as, for example, shown in Figure 3.15.1 and 3.16.2. In conclusion, the expansion factor α_s is usually less than 1.3–1.4, even in good solvents. This is supported by estimating the excluded volume effect a_1 from a and a_2 using eq. (3.15.18), and from a, a_ϕ, and a_2 using eq. (3.15.30). The values of a_1 thus determined are summarized in Table 3.15.1 (see also Tables 3.16.5, 3.13.2, and 3.13.3). Among 23 polymer-solvent systems, a_1 is found to be practically zero for 19 systems, if eq. (3.15.18) is applied. This corresponds to a low a_1 value, a low z value, and $\alpha_s \approx 1.0$. For the other remaining three systems, $a_1 > 0.15$ is observed in cellulose in cadoxen ($a_1 = 0.22$), in FeTNa ($a_1 = 0.17$), and CN ($N_c = 12.9\%$) in acetone ($a_1 = 0.18$). When a large experimental error involved in a_1 (ca. 0.1) is considered, $a_1 > 0.15$ for these three systems is not exceedingly larger than that expected from α_s value. For example, we obtain $\alpha_s = 1.2$ for $a_1 \approx 0.16$ by using Voeks relationship.[70,71]

$$a_1 + 0.5 = (4\alpha_s^2 - 3)/(5\alpha_s^2 - 3) \tag{3.16.54}$$

Moreover, for these systems, it was confirmed that the $\alpha_{s(2G)}$ or $\alpha_{s(m)}$ value corresponds well with A_2 experimentally determined. If eq. (3.15.8) (which is less reliable than eq. (3.15.18)) is employed, then 16 systems exhibit $a_1 \approx 0$. Table 3.15.1 indicates a_ϕ to be a major contribution to a.

In the dilute solutions of cellulose, amylose, and their derivatives, the penetration function ψ is small (accordingly, $\alpha_s \sim 1.0$), but the excluded volume parameter z, evaluated from the slope of the SF plot, is unexpectedly large. This contradiction was long unresolved. It was shown by Kamide and Miyazaki[1] that all points for method 2E deviate largely from all the theoretical ψ–z curves, but the points of method 2G fall on any theoretical curve, providing a support for the superiority of method 2G. As such, if method 2G is utilized, then there is no inconsistency between ψ and z. The apparent contradiction mentioned above can be attributed to an erroneous usage of method 2E in the case of $a_2 \neq\sim 0$ and $a_\Phi \neq 0$. Since method 2G gives ψ and z which are near to zero, we can accurately determine a_s independently of the theory of excluded volume effect chosen (e.g. eq. (3.13.5)). In this connection, a large $[\eta]$ cannot be explained by a large expansion factor, but by large unperturbed chain dimensions.

As a further check on the validity of method 2G, we calculated $[\eta]$ from the molecular parameters such as A and B by method 2E, and K_Φ, a_Φ, a_2, K_0, and B by method 2G. The values of $[\eta]$ thus obtained which are noted as $[\eta]_{(2E)}$ and $[\eta]_{(2G)}$, respectively, are compared with the experimental value $[\eta]_{(exp)}$. The $[\eta]_{(2G)}$ agrees satisfactorily with $[\eta]_{(exp)}$ over the whole molecular weight ranges studied, supporting method 2G. $[\eta]_{(2E)}$ deviates to a large extent from $[\eta]_{(exp)}$ at a higher molecular weight.

A significant improvement in the agreement between the hydrodynamic approach (method 2G, 2K, and 2L) and the thermodynamic approach (methods 2C and 2D) is generally achieved for any actual system by correcting the nonGaussian nature of chain and the draining effect on $[\eta]$. In this sense, the most reliable value of A must be an average value of A estimated by methods 2A, 2B, 2C (in the case of $a_2 = 0$) or 2D (in the case of $a_2 \neq 0$) and 2G, 2K, and 2L. We denote this by $A_{(m)}$. The $A_{(m)}$ value thus determined is given in Table 3.16.3 for CA.

Table 3.16.9 summarizes A of UCD of a hypothetical freely rotating chain (A_f), in which the virtual mean bond length of C1 chain form of cellulose and its derivatives is 5.47 Å and that of amylose and its derivatives is 4.25 Å and glycosidic bridge angle is 110°.

Table 3.16.10 lists the most probable UCD $A_{(m)}$ and conformation parameter σ of cellulose, amylose, and their derivatives in solutions.

The values of σ for cellulose, amylose, and their derivatives in solutions are substantially larger than those ($\sigma < 3.0$) of the vinyl-type polymers. Exceptional cases are

Table 3.16.9

Unperturbed chain dimensions of a hypothetical free rotation chain A_f

Polymer	$A_f \times 10^8$ (cm)	Polymer		$A_f \times 10^8$ (cm)
Cellulose	0.614	NaCS	DS = 1.9	0.415
CT N_c = 13.9%	0.456	MC	DS = 2	0.567
13.6%	0.460	NaCMC	DS = 0.88	0.502
12.9%	0.470	HEC	DS = 1	0.494
CA DS = 2.92	0.465	EHEC	DS = 2	0.510
2.46	0.481	NaCX	DS = 0.78	0.505
CTC$_p$	0.366	Amylose		0.476
CTC	0.344	ATA		0.357
		ATC		0.267

Table 3.16.10

The most probable unperturbed chain dimensions $A_{(m)}$ and conformation parameter σ

Polymer	Solvent	Methods used for analysis	UCD $A_{(m)}$	Conformation parameter $\sigma(\equiv A_{(m)}/A_f)$
Cellulose	Cadoxen	B, C, G, J, K, L	1.69	2.79
	FeTNa	B, D, G, K	2.12	3.65
	Aq. LiOH	B, C, G, K	2.16	3.52
	Aq. NaOH	B	2.2	3.6
CN N_c = 13.9%	Acetone	B, C, G, K	2.45	5.37
13.6%	Acetone	B, C, G, K	2.08	4.5.2
12.9%	Acetone	B, D, G, K	1.89	4.02
CA DS = 0.49	DMAc	B, C, G, K	1.38	2.46
	Water	B	2.51	4.69
	Formamide	B	3.03	5.66
1.75	DMAc	B, C, G, K	1.75	3.47
2.46	Acetone	B, D, G, K	1.73	3.59
	THF	B, D, G, K, L	1.23	2.56
	DMAc	B, D, G, K	1.95	4.10
2.92	DMAc	B, C, G, J, K,L	1.46	3.14
CTC$_p$	DMF	A, D, G, K	1.93	5.27
	1-ClN	B, D, G, K	1.83	5.00
	Dioxane/water	G	1.90	5.19
CTC	Anisol	A	0.818	2.38
	Cyclohexanol	A	0.841	2.44
	Acetone	B, C, G, K	1.34	3.90
	Cyclohexanone	B, C, G, J, L	1.41	4.10
	Dioxane	B, C, G, K	1.85	5.38
NaCS DS = 1.9	Aq. Nacl	B, D, G, K	1.34	3.23
MC DS = 0.2	Water	B, D, G, K	2.26	3.99
NaCMC DS = 0.88	Aq. NaCl (I → ∞)	B, D, G, K	1.44	2.87
HEC DS = 1	Water	B, D, G, K	2.22	4.46
EHEC DS = 2	Water	B, D, G, K, L	2.37	4.64
NaCX	1 N NaOH	B, D, G, J, K, L	3.83	7.58
Amylose	DMSO	B, D, G, K	1.67	3.51
ATA	Nitromethane	B, C, G, K	1.10	3.08
ATC	Pyridine/water	A, C, G	0.992	3.72
	Pyridine	B, C, G, J, L	1.04	3.90

celluloses in cadoxen,[29] CTC in anisol[37,60] and *m*-cyclohexanol,[37,60] and NaCMC in aq. NaCl solution,[40] for which σ is in the range 2.0–3.0. From Table 3.16.10 it can be definitely concluded that cellulose, amylose, and their derivatives are semiflexible polymers, which have greatly extended unperturbed chain dimensions, owing to the markedly interrupted internal free rotation of the chain. Rather large differences in the UCD are observed in different solvents for CTC, where the solvent effect on A obtained here is completely the same as that derived by using method 2E or 2F.

The average value of C_∞ calculated from $A_{(m)}$, for cellulose and its derivatives, is 40.5 and that from A value by method 2E is 9. If the rigid C1 chair conformations are assumed for all glycosidic residue and approximate conformational energy calculations, with the statistical mechanical theory of chain configuration are employed, then the mean glycosidic bridge angle β corresponding to $C_\infty = 40.5$ is estimated from Figure 2 of Ref. 72 to be $117°$ and the β value for $C_\infty = 9°$ is $126°$. The former β value is in close agreement with the experimental data ($117.5°^{73}$ and $116.5°^{74}$), obtained by the crystallography, suggesting the high reliability of the $A_{(m)}$, values evaluated here. It should be remembered that the theoretical calculations of C_∞ were based on the unsubstituted cellulose. More detailed comparison of the theory and experiment on C_∞ is needed.

REFERENCES

1. K Kamide and Y Miyazaki, *Polym. J.*, 1978, **10**, 409.
2. See for example, K Kamide and T Dobashi, *Physical Chemistry of Polymer Solutions; Theoretical Background*, Problem 6-9, eq. (6.9.1, Elsevier, Amsterdam, 2000.
3. M Fixman, *J. Chem. Phys.,*, 1954, **23**, 1656.
4. H Baumann, *J. Polym. Sci., Polym., Lett.*, 1965, **3**, 1069.
5. M Kurata and H Yamakawa, *J. Chem. Phys.*, 1958, **29**, 311.
6. T Kawai and K Kamide, *J. Polym. Sci.*, 1961, **54**, 343.
7. WH Stockmayer and M Fixman, *J. Polym. Sci.*, 1963, **1**, 137, Part C.
8. K Kamide and T Kawai, *Kobunshi Kagaku*, 1962, **19**, 441.
9. K Kamide and T Kawai, *Kobunshi Kagaku*, 1963, **20**, 512.
10. K Kamide and WR Moore, *Kobunshi Kagaku*, 1964, **21**, 682.
11. K Kamide and WR Moore, *J. Polym. Sci.*, 1964, **2**, 1029, Part B.
12. K Kamide, Y Inamoto and G Livingstone, *Kobunshi Kagaku*, 1966, **23**, 1.
13. K Kamide, A Kataoka and T Kawai, *Makromol. Chem.*, 1970, **139**, 221.
14. T Norisuye, K Kawahara, A Teramoto and H Fujita, *J. Chem. Phys.*, 1968, **49**, 4330.
15. K Kawahara, T Norisuye and H Fujita, *Kobunshi Kagaku*, 1968, **49**, 4339.
16. T Kato, K Miyaso, I Noda, T Fujimoto and M Nagasawa, *Macromolecules*, 1970, **3**, 777.
17. I Noda, K Mizutani, T Kato, T Fujimoto and M Nagasawa, *Kobunshi Kagaku*, 1970, **3**, 787.
18. G Tanaka, S Imai and H Yamakawa, *J. Chem. Phys.*, 1970, **52**, 2639.
19. T Matsumoto, N Nishioka and H Fujita, *J. Polym. Sci.*, 1972, **2**, 23, Part A.
20. M Kurata, M Fukatsu, H Sotobayashi and H Yamakawa, *J. Chem. Phys.*, 1964, **41**, 139.
21. K Kamide and T Terakawa, *Polym. J.*, 1978, **10**, 559.
22. DW Tanner and GC Berry, *J. Polym. Sci., Polym. Phys.*, 1974, **12**, 941.
23. K Kamide and Y Miyazaki, *Polym. J.*, 1978, **10**, 539.
24. JMG Cowie and S Bywater, *Polymer*, 1965, **6**, 197.
25. K Kamide and M Saito, *Eur. Polym. J.*, 1981, **17**, 1049.
26. K Kamide and WR Moore, *J. Polym. Sci., Polym. Lett. Ed.*, 1964, **2**, 809.
27. K Kamide, Y Miyazaki and T Abe, *Polym. J.*, 1979, **11**, 523.
28. K Kamide and M Saito, *Eur. Polym. Sci.*, 1982, **18**, 661.
29. D Henley, *Ark. Kemi.*, 1961, **18**, 327.
30. L Valtasaari, *Markromol. Chem.*, 1971, **150**, 117.
31. GV Schulz and E Penzel, *Markromol. Chem.*, 1968, **112**, 260.
32. E Plenzel and GV Schulz, *Makromol. Chem.*, 1968, **113**, 64.
33. MM Huque, DA Goring and SG Mason, *Can. J. Chem.*, 1958, **36**, 952.
34. K Kamide, T Terakawa and Y Miyazaki, *Polym. J.*, 1979, **11**, 285.
35. WR Krigbaum and LH Sperling, *J. Phys. Chem.*, 1960, **64**, 99.
36. VP Shanbhag and J Ohman, *Ark. Kemi.*, 1968, **29**, 163.
37. VP Shanbhag, *Ark. Kemi.*, 1968, **29**, 1.
38. J Ohman, *Ark. Kemi.*, 1968, **31**, 125.

39. WB Neely, *J. Polym. Sci.*, 1963, **1**, 311, Part A.
40. W Brown and D Henley, *Macromol. Chem.*, 1964, **79**, 68.
41. W Brown, D Henley and J Ohman, *Macromol. Chem.*, 1963, **64**, 49.
42. RS Manley, *Ark. Kemi.*, 1956, **9**, 519.
43. B Das, AK Ray and PK Choudhury, *J. Phys. Chem.*, 1969, **73**, 3413.
44. B Das and PK Choudhury, *J. Polym. Sci.*, 1967, **5**, 769, Part A-1.
45. JMG Cowie, *Markromol. Chem.*, 1961, **42**, 230.
46. JMG Cowie, *J. Polym. Sci.*, 1961, **49**, 455.
47. W Banks, CT Greenwood and J Sloss, *Makromol. Chem.*, 1970, **140**, 109.
48. W Banks, CT Greenwood and J Sloss, *Eur. Polym. J.*, 1971, **7**, 879.
49. W Banks, CT Greenwood and J Sloss, *Eur. Polym. J.*, 1971, **7**, 263.
50. K Kamide and A Kataoka, *Makromol. Chem.*, 1969, **128**, 217.
51. ML Hunt, D Newman, H Sheraga and PJ Flory, *J. Phys. Chem.*, 1956, **60**, 127.
52. GP Pearson and WR Moore, *Polymer*, 1960, **1**, 144.
53. G Meyerhoff, *J. Polym. Sci.*, 1958, **29**, 399.
54. K Kamide and M Saito, *Eur. Polym. J.*, 1984, **20**, 903.
55. K Kamide, M Saito and T Abe, *Polym. J.*, 1981, **13**, 421.
56. M Saito, *Polym. J.*, 1983, **15**, 249.
57. K Kamide and M Saito, *Polym. J.*, 1982, **14**, 517.
58. PJ Flory, *Principles of Polymer Chemistry*, Chapter XIV, Cornel University Press, 1953.
59. WH Stockmayer, *Makromol. Chem.*, 1960, **35**, 54.
60. See for example, K Kamide and T Dobashi, *Physical Chemistry of Polymer Solutions; Theoretical Background*, see for example, eq. (5.20.2 and p. 266, Elsevier, Amsterdam, 2000.
61. S Ishida, H Komatsu, T Terakawa, Y Miyazaki and K Kamide, *Mem. Facul. Engnr., Kanazawa Univ.*, 1979, **12**, 103.
62. S Ishida, H Komatsu, H Katoh, M Saito, Y Miyazaki and K Kamide, *Makromol. Chem.*, 1982, **183**, 3085.
63. VP Shanbhag, *Ark. Kemi.*, 1968, **29**, 33.
64. VP Shanbhag, *Ark. Kemi.*, 1968, **29**, 139.
65. K Kishino, T Kawai, T Nose, M Saito and K Kamide, *Eur. Polym. J.*, 1981, **17**, 623.
66. H Benoit and P Doty, *J. Phys. Chem.*, 1953, **57**, 958.
67. H Yamakawa and M Fujii, *J. Chem. Phys.*, 1976, **64**, 522.
68. H Yamakawa, M Fujii and J Shimada, *J. Chem. Phys.*, 1976, **65**, 2371.
69. M Kurata and WH Stockmayer, *Fortschr. Hochpolym. Forsch.*, 1963, **3**, 196.
70. J Voeks, *J. Polym. Sci.*, 1959, **36**, 333.
71. K Kamide and T Kawai, *Kobunshi Kagaku*, 1963, **20**, 506.
72. KD Goebel, CE Halvie and DA Brant, *Appl. Polym. Symp.*, 1976, **28**, 671.
73. RA Jacobso, JW Wunderlich and WN Lipacomb, *Acta. Cryst.*, 1961, **14**, 606.
74. S Arnott and WE Scott, *J. Chem. Soc., Perkin Trans.*, 1972, **2**, 324.

3.17 SOLVATION

3.17.1 Determination of number of the solvated molecules per repeated unit[1]

The number of the solvated solvent molecules per gram of polymer, n in eq. (3.8.2) can be converted to the number of the solvated solvent molecules per repeating unit (s) through equation

$$s = (m_0/m_s)n \tag{3.17.1}$$

Here, m_0 and m_s are the molecular weights of the repeating unit of the polymer and the solvent, respectively. Except for the case of CA (DS 2.46)/DMAc and /DMSO systems, s does not markedly depend on concentration (s should be distinguished from the sedimentation coefficient s). By extrapolation of the plots of s versus concentration c to $c = 0$, we obtained the number of solvated solvent molecules per repeating unit s_0 at infinite dilution:

$$s_0 \equiv \lim_{c \to 0} s \qquad\qquad (3.17.2)$$

Table 3.17.1 shows the s_0 values of cellulose acetates (DS 0.49, 1.75, 2.46, and 2.92) in various solvents. Moore[2] estimated s_0 of CA (Section 2.25) whole polymer (viscosity average molecular weight $M_v = 1.1 \times 10^5$) acetone system as 7. Polystyrene (PS) methylethyl ketone (MEK) system obtained here is nearly equal to that of PS-toluene system ($s_0 = 0.38$) as reported by Bell and Pethrik.[3] s_0 of CA (Section 2.46) DMAc system is independent of M_w within experimental error. s_0 for CA (Section 2.46) DMAc system is almost five times larger than that of PS–MEK system.

The ratio of the increment of the density to that of the concentration at constant temperature $(\partial\delta/\partial c)_T$ and the similar ratio $(\partial V/\partial T)_c$ are independent of temperature and concentration, respectively, within experimental error. The sound velocity V increases very slightly in proportion to concentration and $(\partial V/\partial T)_c$ of acetone is -3.7×10^2 cm s^{-1} K. The temperature dependence of s and s_0 for CA (Section 2.46) acetone is shown in Figure 3.17.1.[1] s of the dilute solutions ($< 2\%$) and, of course, s_0 decreases with increase of temperature.

Figure 3.17.2 shows the relationship between s_0 of the CA whole polymer solvent systems at 25 °C and the dielectric constant ϵ of the solvent used.[1] While the dependence of s_0 on ϵ differs depending on $\langle\!\langle F \rangle\!\rangle$, s_0 of CA solvent systems, except for CA (DS 2.46) acetone, increases with increase of the polarity of the solvent in a similar manner as the chemical shifts of O-Ac and OH groups.[4]

Table 3.17.1

Number of solvated solvent molecules (s_0) at infinite dilution in cellulose acetate and polystyrene solution at 25 °C[1]

Solvent	Dielectric constants	s_0				PS
		CA				
		0.49[a]	1.75	2.46	2.92	
DMSO	40	1.1	1.4	2.2	1.6	–
DMAc	37.8	1.1	1.3	2.0	1.4	–
Acetone	20.7	–	–	2.8	–	–
THF	8.2	–	–	0.9	–	–
TCE	7.29	–	–	–	0.2	–
MEK[b]	15	–	–	–	–	0.4

[a]Total degree of substitution.
[b]Methylethylketone.

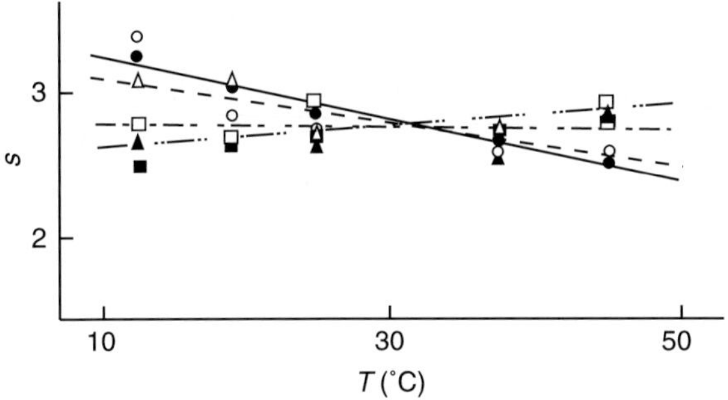

Figure 3.17.1 Temperature dependence of number of the solvated solvent molecules per glucose ring for cellulose acetate ($\langle\!\langle F \rangle\!\rangle$) DS 2.46/acetone solution with various concentration (g cm^{-3}). The lines are determined by least square method. (O), $c = 0$; (●), $c = 0.005$; ($\triangle$), $c = 0.01$; ($\square$), $c = 0.02$; (▲), $c = 0.03$; (■), $c = 0.04$.[1]

3.17.2 Effect of solvation on draining effect and unperturbed chain dimensions

Figure 3.17.3 shows the dependence of the unperturbed chain dimensions A and the characteristic ratio σ on number of the solvated solvent molecules (s_0) at 25°.[1]

All data points, except for CA (DA 2.46) acetone system, fall on a single curve, showing a close correlation between A or σ and s_0. This is indicative of the fact that the solvating solvent molecules prevent significantly the internal rotation of CA molecules, resulting in an increase in the chain rigidity. s_0 was evaluated by using Passynsky's theory in which the model is oversimplified and the magnitude of s_0 should be considered not to be an absolute number, but a parameter related to the number of the solvated solvent molecules.

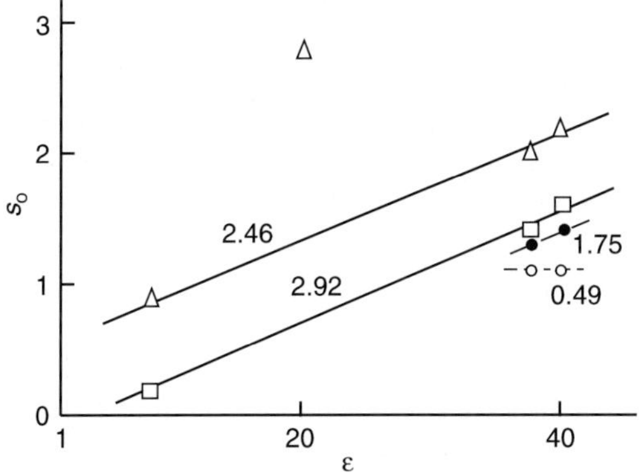

Figure 3.17.2 Relationship between the number of solvated solvent molecules per glucose ring at infinite dilution and the dielectric constant for cellulose acetate-solvent system at 25 °C: The numbers on the lines denote the total degree of substitution $\langle\!\langle F \rangle\!\rangle$ of CA. (O), CA (DS 0.49); (●), CA (DS 1.75); ($\triangle$), CA (DS 2.46); ($\square$), CA (DS 2.92).[1]

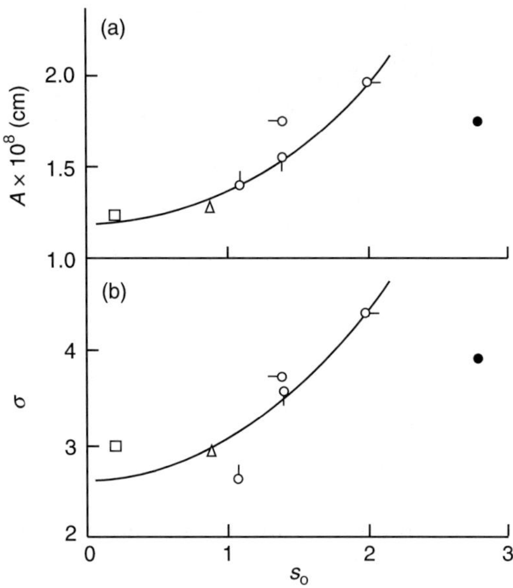

Figure 3.17.3   Plot of the unperturbed chain dimensions (a) and the conformation parameter (b) against number of the solvated solvent molecules at infinite dilution for cellulose acetate/solvent systems.[1] (Circle with upward tail), cellulose acetate (DS 0.49)-dimethylacetamide; (circle with left hand tail), cellulose acetate (DS 1.75)-dimethylacetamide; (circle with right hand tail), cellulose acetate (DS 2.46)-dimethylacetamide; ($\bullet$), cellulose acetate (DS 2.46)-acetone; ($\triangle$), cellulose acetate (DS 2.46)-tetrahydrofuran; (circle with downward tail), cellulose acetate (DS 2.92)-dimethylacetamide; ($\square$), cellulose acetate (DS 2.92)-tetrachloroethane.

Figure 3.17.4 shows the temperature dependence of A of CA (DS 2.46) fraction with M_w of 9.4×10^4 in acetone[5] and s_0 of CA (DS 2.46) whole polymer in acetone. Here, A was estimated by 2B method. A as well as s_0 decreases monotonically with rising temperature. At higher temperatures, the molecular force between O-Ac or OH group in CA molecule and the solvated solvent molecule will decay rapidly, leading to a decrease in s_0 and thus a decrease in A with rising temperature. Suzuki *et al.*[5] demonstrated the negative temperature dependence of A (see Section 3.18). It is now clear that $d[\eta]/dT$ can be finally interpreted in terms of ds_0/dT.

Figure 3.17.5 depicts the relationship between the draining parameter X and s_0 of cellulose acetate solvent systems. Here, X is taken for CA with M_w of 1×10^5. The draining effect becomes paramount with increasing number of the solvated solvent molecules for CA solutions.

When the diameter of a pyranose ring is assumed equal to the dimension of a unit cell (*a*-axis) of cellulose II crystal (8 Å) and the length of O-Ac group is taken as 5 Å, the hydrodynamic diameter d_h of one CA (DS 2.46) molecule solvated with two DMAc molecules (i.e. $s_0 \approx 2$) is estimated to be above 33 Å. This value is nearly the same as d_h (37 Å), estimated from X on the basis of the statistical chain model (see Section 3.15). This is additional evidence supporting the existence of the draining effect. Thus, it is clear that a single solvent molecular layer is bound to the naked

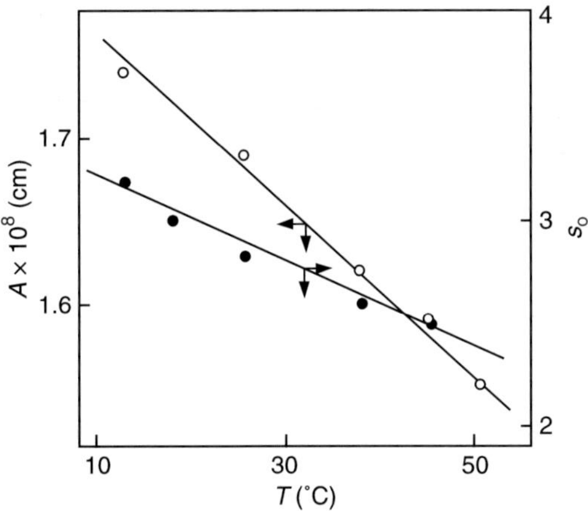

Figure 3.17.4 Temperature dependence of the unperturbed chain dimensions estimated by method 2B for cellulose acetate fraction with DS = 2.46 and $M_w = 9.4 \times 10^4$ in acetone (open mark) and number of the solvated solvent molecules at infinite dilution of a cellulose acetate (DS 2.46) whole polymer in acetone (closed mark). (O), A; (●), s_0.[1]

cellulose chain and this solvated cellulose chain behaves just like an isolated solute molecule. If this is true, then there is no distinct boundary between the solvents (including solvated and nonsolvated solvents) and the CA solute molecule in CA solution and, in this case, the solvent draining into the chain sphere can be theoretically assumed (see Figure 3.22.2).

The solvation is closely correlated with the thermodynamic properties together with the hydrodynamic properties of CA solutions and all the solution data are consistent when the solvation is taken into account.

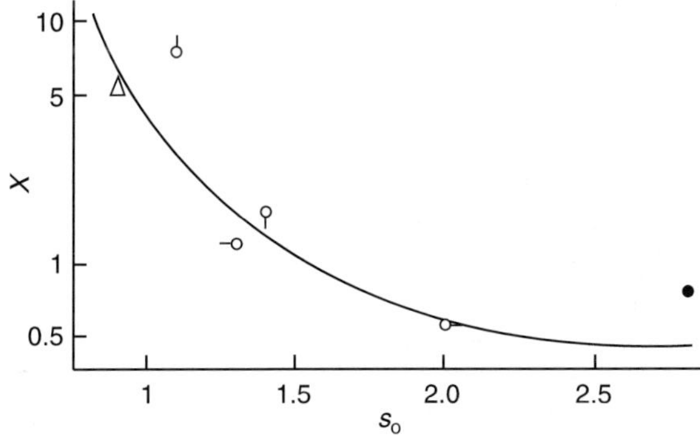

Figure 3.17.5 Correlation between the draining parameter at $M_w = 1 \times 10^5$ and number of the solvated solvent molecules at infinite dilution for cellulose acetatesolvent system at 25 °C.[1] (circle with upward tail), cellulose acetate (0.49)-dimethylacetamide; (circle with left hand tail) cellulose acetate (DS 1.75)-DM; (circle with right hand tail) cellulose acetate (DS 2.46)-dimethylacetamide; (●), cellulose acetate (DS 2.46)-acetone; (△), cellulose acetate (DS 2.46)-tetrahydrofuran; (circle with downward tail) cellulose acetate (DS 2.92)-dimethylacetamide.

REFERENCES

1. K Kamide and M Saito, *Eur. Polym. J.*, 1984, **20**, 903.
2. WR Moore, *J. Polym. Sci., C*, 1967, **16**, 571.
3. W Bell and RA Pethrik, *Eur. Polym. J.*, 1972, **8**, 927.
4. K Kamide, K Okajima and M Saito, *Polym. J.*, 1981, **13**, 115.
5. H Suzuki, Y Miyazaki and K Kamide, *Eur. Polym. J.*, 1980, **16**, 703.

3.18   EFFECT OF SOLVENT NATURE AND TOTAL DS

In this section, we intend to establish for CA the correlation between the molecular properties, $\langle\langle F \rangle\rangle$, and the degree of the solvation, by measuring the chemical shifts in NMR spectra and the adiabatic compressibility using an ultrasonic interferometer and by analyzing the solution data on CA with different $\langle\langle F \rangle\rangle$, given previously.[1-5]

3.18.1   Dielectric constant of solvents

The limiting viscosity number $[\eta]$ of CA (DS 0.49) is a unique function of the dielectric constant ε of the solvent in which the polymer is dissolved. Figure 3.18.1 is an example in which three kinds of CA fractions with the same molecular weight ($M_w = 1 \times 10^5$) and various DS (0.49, 2.46, 2.92) are shown.[3] $[\eta]$ increases with an increases in ε, approaching an asymptotic value that depends on DS ($\equiv \langle\langle F \rangle\rangle$) value of the polymer. The phenomenon cannot be interpreted in terms of the excluded volume effect (see Table 3.5.1).

Figures 3.18.2 and 3.18.3 show the variation in $[\eta]$, $\langle S_2 \rangle_z^{1/2}$, Φ, α_s, and X for the CA (DS 0.49) fraction ($M_w = 6.32 \times 10^4$) and CA (DS 2.46) ($M_w = 1 \times 10^5$) with the dielectric constant ε of the solvent.[3] Evidently, an increase in ε leads to larger values of $[\eta]$ and $\langle S^2 \rangle_z^{1/2}$ and smaller values of Φ and X. α_s is nearly independent of ε. This strongly suggests that at least for the observed marked variation in $[\eta]$, $\langle S^2 \rangle_z^{1/2}$, and Φ

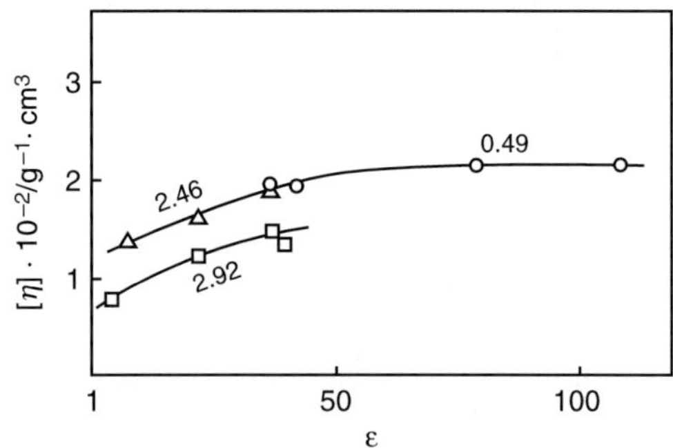

Figure 3.18.1   Limiting viscosity number $[\eta]$ of CA solutions, plotted against the dielectric constant ε of the solvent in which the polymers were dissolved: (○), cellulose acetate (DS 0.49);[3] (△), CA (DS 2.46);[4] (□), CA (DS 2.92).[3]

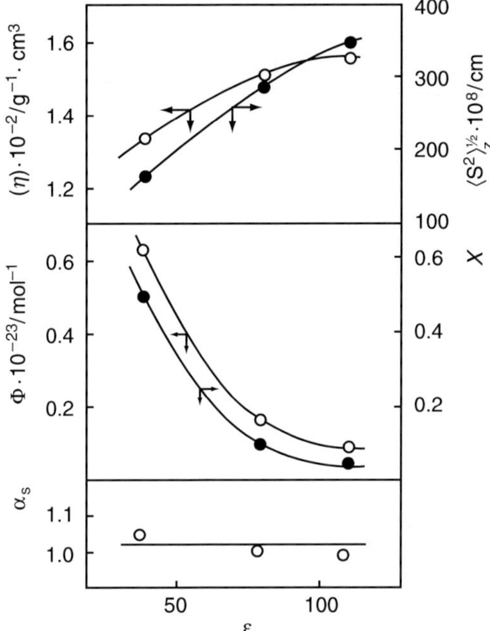

Figure 3.18.2   Dependence of the limiting viscosity number $[\eta]$, the radius of gyration $\langle S^2 \rangle_z^{1/2}$, Flory's viscosity parameter Φ, the expansion factor α_s and the draining parameter X of cellulose acetate (DS 0.49) fraction MW-5 on the dielectric constant of the solvent.[3]

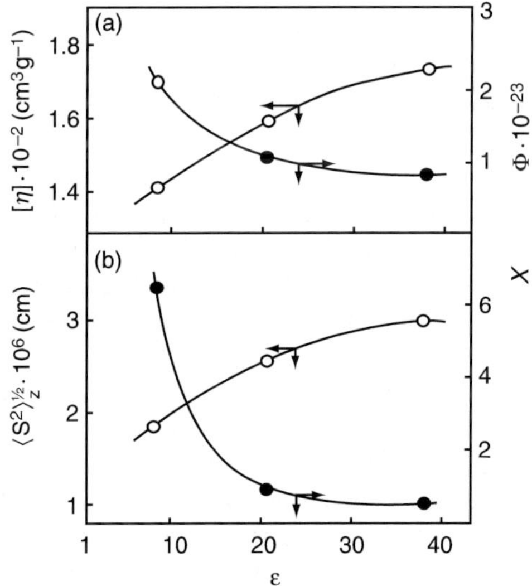

Figure 3.18.3   Effect of the dielectric constant of solvent ε on the limiting viscosity number $[\eta]$, Flory's viscosity parameter Φ (a) and the z-average radius of gyration $\langle S^2 \rangle_z^{1/2}$, the draining parameter X (b) for cellulose acetate (DS 2.46): $[\eta]$ and $\langle S^2 \rangle_z^{1/2}$ are for the polymer with $M_w = 1 \times 10^5$.[4]

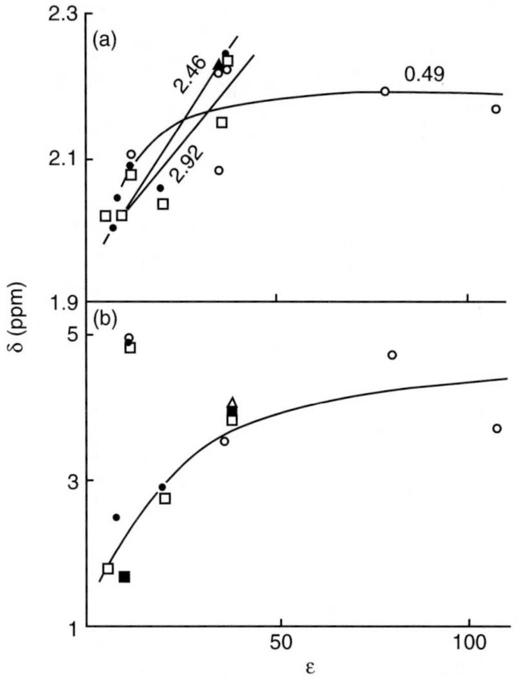

Figure 3.18.4   Plot of the chemical shift of O-acetyl proton (a) and hydroxyl proton (b) for cellulose acetate molecule in solvents against the dielectric constant. Numbers denote the total degree of substitution $\langle\!\langle F \rangle\!\rangle$. (O), cellulose acetate (0.49); (△), cellulose acetate (1.75); (●), cellulose acetate (2.46); (□), cellulose acetate (2.92).[6]

does not arise from variation in the long-range interaction parameter. Figure 3.18.4(a) and (b) shows the plots of the chemical shifts of O-Ac and OH group against ε of the solvent, respectively.[6]

Generally, both chemical shifts increase with increase in ε over the entire range of ε. This result indicates that solvents with high ε, such as TFA and DMAc, strongly interact with both O-Ac and OH groups of the CA molecule. The dependence of the O-Ac chemical shift on ε varies with $\langle\!\langle F \rangle\!\rangle$, being highest for CA (DS 2.46). In contrast, the chemical shift of the OH proton is a unique function of ε, irrespective of $\langle\!\langle F \rangle\!\rangle$.

Figure 3.18.5 shows the dependence of $A_{(m)}$ (hereafter simply referred to as A) on ε of the solvent.[7] The filled mark denotes the A value extrapolated to $\varepsilon = 1$. The UCD of CA molecules have a tendency to become larger in more polar solvent. Of course, the remarkable dependence of A on ε cannot be explained by the 'two parameter theory'. From Figure 3.18.5, it is expected that A of CA (DS 0.49) reduces to A of a hypothetical free rotating chain (A_f) at $\varepsilon \leq 20$. That is, if CA (DS 0.49) is dissolved in a solvent with $\varepsilon \approx 20$, then the CA (DS 0.49) chain should behave as freely rotating in the solvent. The extrapolated A value of CA with $\langle\!\langle F \rangle\!\rangle \geq 2.46$ to $\varepsilon = 1$ is larger than A_f.

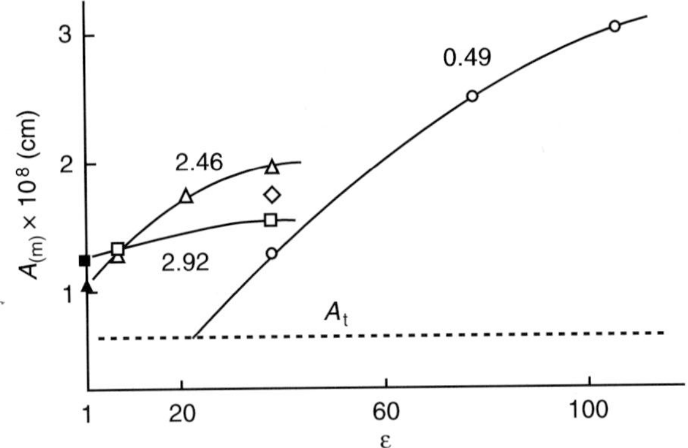

Figure 3.18.5   Dependence of the most probable unperturbed chain dimension on the dielectric constant of the solvent for cellulose acetate solvent systems. Broken line: the unperturbed chain dimension of hypothetical cellulose with free internal rotation. (open mark): the $A_{(m)}$ value estimated from experimental data; (closed mark): the asymptotic $A_{(m)}$ value at the limit of $\varepsilon = 1$. (O), cellulose acetate (DS 0.49); ($\Diamond$), cellulose acetate (DS 1.75); ($\triangle$), cellulose acetate (DS 2.46); ($\square$), cellulose acetate (DS 2.92).[7]

3.18.2   Total degree of substitution

Figure 3.18.6 shows the effect of the total DS of CA and CN with $M_w = 1 \times 10^5$ on their $[\eta]$ and conformation parameter σ (eq. (3.16.1)) in acetone and DMAc.[2] Here, the data points of CN are the results of Kamide and Miyazaki[8] in their paper, analyzed from works by Schulz and Penzel,[9,10] and by Huque et al.[11] $[\eta]$ and σ increase with DS in CN solutions, but decrease in CA solutions. A similar tendency for $[\eta]$ was observed by Howlette et al.,[12] Moore and Russel,[13,14] and Uda[15] for CA and by Warrow[16] and Linsley and Frank[17] for CN. These remarkable differences between CA and CN with respect to the effect of DS may be attributable to the fact that the CA molecule acts as proton acceptor to the solvents in question and the CN molecules act as electron acceptors. Note that in Figure 3.18.6, only two samples with different $\langle\!\langle F \rangle\!\rangle$ were utilized for analysis.

The effect of M_w on $[\eta]$-$\langle\!\langle F \rangle\!\rangle$ relationships for CA in DMAc is illustrated in more detail in Figure 3.18.7, in which the value CA (DS 0.49) with $M_w = 2 \times 10^5$ is calculated from eq. (3.11.2).[3] Other points for CA (DS 2.46) and CA (DS 2.92) are calculated from the MHS equations (eq. (3.11.7) and eq. (3.11.6)). The DS dependence of $[\eta]$ varies with M_w of the polymer; for relatively small M_w, $[\eta]$ decreases monotonously with an increase in DS, but for large M_w ($M_w > 2 \times 10^5$), $[\eta]$ reveals a maximum at a DS value. In the DS range of 0.49–2.46, $[\eta]$ is not very sensitive to DS, slightly increasing or decreasing as the DS increases. In the comparatively narrow DS range from 2.46 to 2.92, $[\eta]$ decreases remarkably with an increase in DS. This seems to indicate that the interaction between the OH groups remaining on the CA molecule and the solvent plays an important role in the solution viscosity.

Table 3.18.1[7] lists the values of the physical parameters, experimentally determined for CA with various $\langle\!\langle F \rangle\!\rangle$/solvent systems.[1–5] Here, $\gamma(\equiv d \log \langle S^2 \rangle_w^{1/2} / d \log M_w)$ is

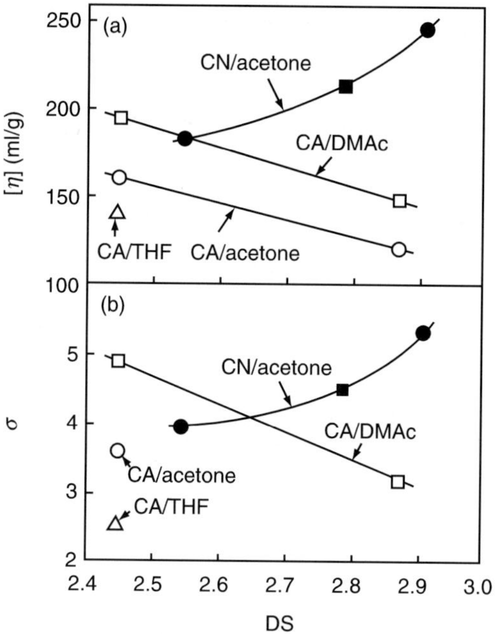

Figure 3.18.6 Effect of the degree of substitution on the limiting viscosity number $[\eta]$ of cellulose acetate and cellulose nitrate with $M_w \times 10^5$ in acetone, dimethylacetamide, and tetrahydrofuran and their conformation parameter σ^2: ($\square$), cellulose acetate in dimethylacetamide (Kamide *et al.*[2]); ($\bigcirc$), cellulose acetate in acetone (Kamide *et al.*[1,2]); ($\triangle$), cellulose acetate in tetrahydrofuran (Kamide *et al.*[1]); ($\bullet$), cellulose nitrate in acetone (data from Schulz and Penzel[9,10]); ($\blacksquare$), cellulose nitrate in acetone (data from Huque *et al.*[11]).

related to λ through the relationship,

$$\gamma = (\lambda + 1)/2 \tag{3.18.1}$$

The exponent a ranges from 0.60 to 0.75 for CA solvent systems, except for CA (2.46)/DMAc ($a = 0.82$). CA (DS 0.49) dissolves only in polar solvents with dielectric constant $\varepsilon > ca.$ 40, in which the exponent a is almost the same (0.60–0.61), independent of the solvent nature. CA (DS 2.92) dissolves in comparatively less polar solvents with $\varepsilon = 5$–40. Solubility behavior of CA (DS 2.46) is similar to that of CA (DS 2.92), except for the solvent with $\varepsilon < 10$, in which CA (DS 2.46) is insoluble at room temperature.

The γ value of any CA/solvent system is unexpectedly smaller than that of typical vinyl-type polymers dissolved in good solvents, such as PS/benzene system at 25 °C ($\gamma = 0.6$).[18] The Φ of CA/solvent systems depends significantly on the molecular weight as for many other systems of cellulose and its derivative with solvent.[8] The Φ value calculated using eq. (3.15.1) from $[\eta]$, M_w and $\langle S^2 \rangle_z^{1/2}$ for CA/DMAc systems is about 48% smaller than the theoretical Φ value of the unperturbed chain at non-draining limit ($\Phi(\infty)$) (2.87×10^{23}) (see Tables 3.3.1, 3.3.2, 3.3.5, 3.15.2, and 3.15.3). These experimental facts of the molecular weight dependence of Φ and the small magnitude of Φ mean that the molecule is partially free draining.

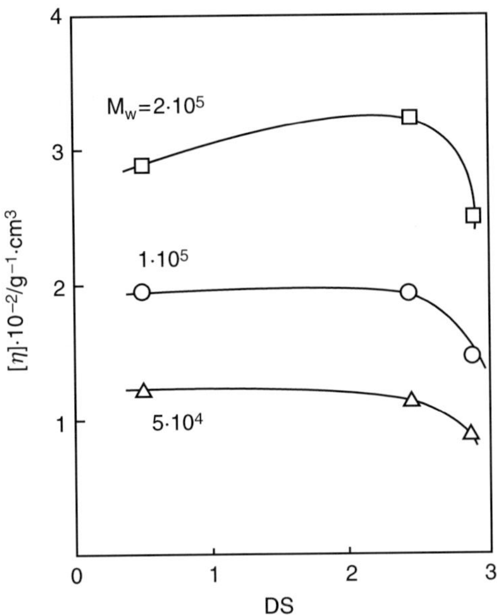

Figure 3.18.7   Effect of the degree of substitution on the limiting viscosity number $[\eta]$ of cellulose acetate in dimethylacetamide at 25 °C.[3]

Figure 3.18.8 demonstrates $\langle\langle F \rangle\rangle$ dependence of a, a_Φ, and a_2 of CA/DMAc systems. For this system, both exponent a in MHS equation and a_Φ reach a maximum at $\langle\langle F \rangle\rangle \approx$ 2.5, but a_2 is almost independent of $\langle\langle F \rangle\rangle$. Table 3.18.1 and Figure 3.18.8 indicate that the main factor contributing to a is the free-draining effect a_Φ (see Section 3.16). Using α_s, determined by eqs. (3.13.5–3.13.8) and Φ calculated through use of equation (3.15.1), X can be estimated (method 1A).[8]

Figure 3.18.9 depicts the log–log plots between X, determined by method 1A and M_w.[7] X for CA DMAc system is always <3 over the whole range of M_w studied, except for CA (DS 0.49). The extent of the draining effect in DMAc increases in the following order: CA (DS 0.49) $\ll$ CA (1.75) < CA (DS 2.92) $\ll$ CA (DS 2.46). In other words, the draining effect attains a maximum at $\langle\langle F \rangle\rangle = 2.46$.

Figure 3.18.10 shows the ^{1}H NMR spectra of CA (DS 0.49), CA (DS 1.75), CA (DS 2.46), and CA (DS 2.92) in DMAc.[6,7] Two or three separate signals due to methyl protons were detected between 1.9 and 2.3 ppm for CA (DS 1.75) and CA (DS 0.49) DMAc. For the CA (DS 0.49)/solvent system, it is expected from chemical analysis of CA that there is approximately one O-Ac group per two glucopyranose units. Therefore, the possibility of three types of acetylated glucopyranose units yields at least three nonequivalent methyl protons with different magnetic environments, assuming that the glucopyranose unit with no O-Ac group does not produce any effect on the methyl proton of O-Ac group magnetically. It should be noted that the OH proton signals of CA molecule occasionally overlap and are inseparable from water and HOD in solvents.[6] Table 3.9.1 summarizes the chemical shifts of O-Ac and –OH group of CA whole polymers with $\langle\langle F \rangle\rangle = 0.49$, 1.75, 2.46, and 2.92 in various solvents.[6]

Table 3.18.1

Various parameters of cellulose acetate-solvent systems at 25 °C[7]

Polymer ($\langle\langle F\rangle\rangle$)	Solvent	$K_m \times 10^2$ ($cm^3\,g^{-1}$)	a	γ	$K_\phi \times 10^{-23}$	a_ϕ	a_1	a_2	Molecular weight range[a] $\times 10^{-4}$	Reference
CA (0.49)	Formamide	20.9	0.60	–	–	–	–	–	4.55–14.5	3
	Water	20.9	0.60	–	–	–	–	–	4.55–14.5	3
	DMSO[b]	17.1	0.61	–	–	–	–	–	4.55–1.45	3
	DMAc[c]	19.1	0.60	0.52	0.61	0.10_2	0.06	0	4.55–14.5	3
CA (1.75)	DMAc	9.58	0.65	0.52	0.26	0.12_0	0.03	0	3.75–13.1	5
CA (2.46)	DMAc	1.34	0.82	0.53	0.51_5	0.23	0.09	0	5.3–27.0	4
	Acetone	0.13	0.61_6	0.30_8	2.26×10^{-4}	0.71_6	0.13	−0.42	6.1–26.5	1
	THF[d]	0.051_3	0.68_8	0.55_8	0.57	0.10_5	0.15	0	7.4–20.0	1
CA (2.92)	TFA[e]	3.96	0.70_6	–	–	–	–	–	6.36–69.0	2
	DMAc	2.64	0.75	0.55	0.35	0.10_6	0.15	0	6.36–69.0	2
	Acetone	2.89	0.75_5	–	–	–	–	–	8.22–69.0	2
	TCM[f]	4.54	0.64_9	–	–	–	–	–	8.22–69.0	2
	DCM[g]	2.47^h	0.70_4	–	–	–	–	–	6.36–69.0	2
	TCE[i]	3.93	0.66_2	–	–	–	–	–	6.36–69.0	2

[a]Molecular weight range.
[b]N,N-dimethylsulphoxide.
[c]N,N-dimethylacetamide.
[d]Tetrahydrofuran.
[e]Trifluoroacetic acid.
[f]Trichloromethane.
[g]Dichloromethane.
[h]Data at 20 °C.
[i]Tetrachloroethane.

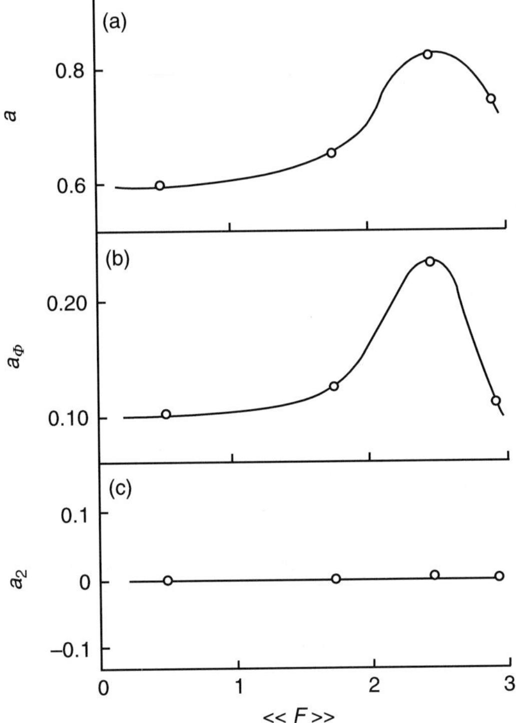

Figure 3.18.8   Dependence of the exponents a (eq. (3.11.1)), a_ϕ (eq. (3.15.3)), and a_2 (eq. (3.15.10)) on the total degree of substitution $\langle\!\langle F\rangle\!\rangle$ for cellulose acetate–DMAc system. (a) a; (b) a_ϕ; (c) a_2.[7]

In the CA (DS 2.92)-TFA, pyridine, CA (DS 2.46) DMAc, and THF systems, the chemical shift of the methyl proton shows no variation with the molecular weight. In contrast, the methyl proton signal of CA (DS 2.46) in TFA shifts toward lower magnetic field as M_w decreases. The possibility of interaction between solvent and polymer might be higher when the molecular weight of polymer is lower.

In Figure 3.18.11(a) and (b), the data from Figure 3.18.4 were used to show the effect of $\langle\!\langle F\rangle\!\rangle$ on the O-Ac and OH proton chemical shifts of CA in DMAc.[7] The figure contains the data of TFA and pyridine.[6] The chemical shift of the OH proton of CA in TFA overlapped that of carboxyl proton in TFA molecule, which appears in the range 9–11 ppm.[6] The chemical shift of O-Ac proton has a maximum for $\langle\!\langle F\rangle\!\rangle \approx 2.46$ in TFA and DMAc and the chemical shift of the OH proton attains maximum at $\langle\!\langle F\rangle\!\rangle \approx 1.7$ in DMAc, while the chemical shifts of O-Ac and OH proton in pyridine are almost independent of $\langle\!\langle F\rangle\!\rangle$. $\langle\!\langle F\rangle\!\rangle$ dependence of the O-Ac proton chemical shift is much smaller in TFA than in DMAc. The absolute magnitude of the O-Ac proton chemical shift in these solvents is in the following order: TFA > DMAc > pyridine, and that of the OH proton is in the following order: TFA > pyridine > DMAc. The large difference in $\langle\!\langle F\rangle\!\rangle$ dependence and in the absolute magnitude of the chemical shifts may be attributable to the difference in the solvent nature (see Table 3.17.1).

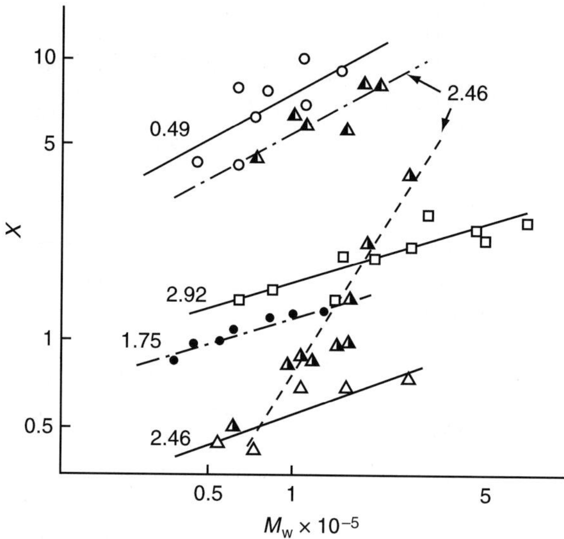

Figure 3.18.9 Log–log plot of the draining parameter X (defined in eq. (3.11.1)) against M_w for cellulose acetate/solvent systems. The lines were determined by the least square method. The numbers on the lines denote the total degree of substitution $\langle\!\langle F \rangle\!\rangle$ of cellulose acetate. (O), (CA 0.49)–dimethylacetamide; (●), cellulose acetate (DS 1.75)–dimethylacetamide; (△), cellulose acetate (DS 2.46)–dimethylacetamide; (half closed triangle (right)) cellulose acetate (DS 2.46)–dimethylacetamide; (half closed triangle (left)) cellulose acetate (DS 2.46)–tetrahydrofuran; (□), cellulose acetate (2.92)–dimethylacetamide.[7]

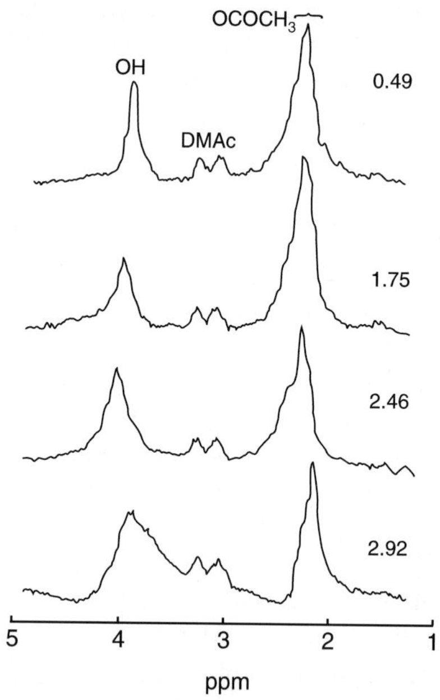

Figure 3.18.10 ^{1}H NMR spectra of cellulose acetate in dimethylacetamide.[7] The numbers attached to the spectra denote the total degree of substitution $\langle\!\langle F \rangle\!\rangle$.

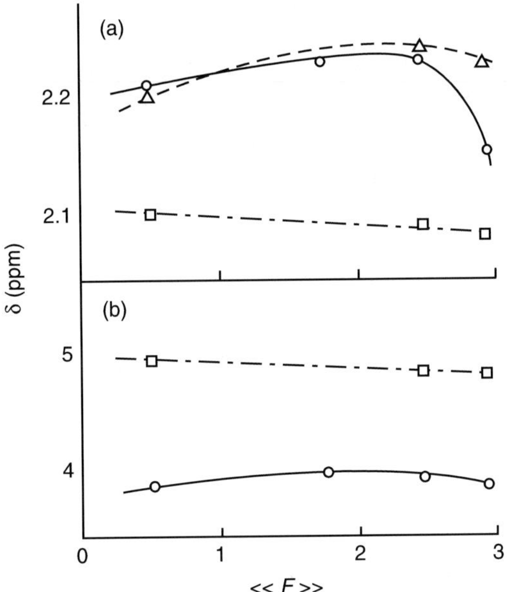

Figure 3.18.11   Effect of the total degree of substitution $\langle\!\langle F \rangle\!\rangle$ on the chemical shift of O-acetyl proton (a) and hydroxyl proton (b) for cellulose acetate-solvent systems. (O), dimethylacetamide; (Δ), trifluoroacetic acid; (□), pyridine.[7]

In Figure 3.18.12, the effects of $\langle\!\langle F \rangle\!\rangle$ on s_0 for CA/DMAc and CA/DMSO systems at 25 °C are shown.[7] In both systems, s_0 has a maximum at $\langle\!\langle F \rangle\!\rangle \approx 2.5$. As shown in Figure 3.18.12, the interactions between the O-Ac or the OH groups in CA molecules and DMAc are strongest at $\langle\!\langle F \rangle\!\rangle = 2.46$ among four CA polymers with different $\langle\!\langle F \rangle\!\rangle$, and this seems to cause the maximum in the number of solvated solvent molecules.

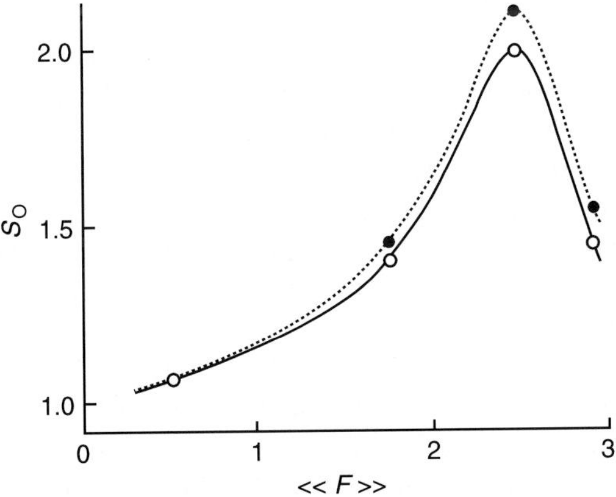

Figure 3.18.12   Effect of the total degree of substitution on number of solvated solvent molecule per glucose ring at infinite dilution for cellulose acetate–dimethylacetamide and cellulose acetate–dimethyl sulfoxide systems at 25 °C. (O), dimethylacetamide; (●), dimethyl sulfoxide.[7]

REFERENCES

1. K Kamide, T Terakawa and Y Miyazaki, *Polym. J.*, 1979, **11**, 285.
2. K Kamide, Y Miyazaki and T Abe, *Polym. J.*, 1979, **11**, 523.
3. K Kamide, M Saito and T Abe, *Polym. J.*, 1981, **13**, 421.
4. K Kamide and M Saito, *Polym. J.*, 1982, **14**, 517.
5. M Saito, *Polym. J.*, 1983, **15**, 249.
6. K Kamide, K Okajima and M Saito, *Polym. J.*, 1981, **13**, 115.
7. K Kamide and M Saito, *Eur. Polym. J.*, 1984, **20**, 903.
8. K Kamide and Y Miyazaki, *Polym. J.*, 1978, **10**, 409.
9. GV Schulz and E Penzel, *Makromol. Chem.*, 1968, **112**, 200.
10. E Penzel and GV Schulz, *Makromol. Chem.*, 1968, **113**, 64.
11. MM Huque, DA Goring and SG Mason, *Can. J. Chem.*, 1958, **36**, 952.
12. F Howlette, E Minshall and AR Urquhart, *Shirley Inst. Mem.*, 1941, **18**, 251.
13. WR Moore and J Russell, *J. Colloid Sci.*, 1953, **8**, 243.
14. WR Moore and J Russell, *J. Colloid Sci.*, 1954, **9**, 4.
15. K Uda, *Sen'i Gakkaishi*, 1962, **18**, 105.
16. HA Warrow, *Kolloid Z.*, 1943, **102**, 29.
17. CH Linsley and MB Frank, *Ind. Eng. Chem.*, 1953, **45**, 2491.
18. Y Miyaki, Y Einaga and H Fujita, *Macromolecules*, 1978, **11**, 1180 and references cited therein.

3.19　PERSISTENCE LENGTH

3.19.1　LS method (worm-like chain model)[1]

In Section 3.16, the unperturbed chain dimensions A for cellulose and its derivatives in solvents were evaluated by using various methods (2A–2L), all based on the pearl necklace model.[2–10] The A values obtained for these polymers are significantly larger than those for typical vinyl-type polymers[2,11] and indicate the semiflexibility of the cellulose chain. The equilibrium rigidity of a polymeric chain can also be represented by the persistence length (q).[12,13] Tsvetkov and coworkers[14,15] calculated the q values from sedimentation data of some cellulose esters by using the Hearst–Stockmayer (HS) theory.[16] Since the HS theory contains some mathematical errors as has been pointed out by Yamakawa and Fujii,[17] the q values obtained by Lyubina *et al.*[13] and Tsvetkov and coworkers[14,15] are not accurate, especially for the polymers with large q values. Burchard *et al.*[18–20] estimated the q values of CTC samples with narrow MWD in non- and (in part) theta solvents by LS, small angle X-ray, and neutron scattering.

In this section we attempt to determine, for cellulose and its derivatives, the following four kinds of persistence lengths q: (1) the q value at the perturbed state (q_{BD}) by using the Benoit–Doty (BD) theory,[21] modified by considering the MWD of the polymer samples; (2) the unperturbed q value (q_{BD}^0) by using the BD theory and the third power law-type excluded volume theory, which is derived by extending the fifth power law-type theory of Yamakawa and Stockmayer[22] for the worm-like chain; (3) the q value (q_{CL}^0) from A at coil limit; and (4) the q value (q_{YF}) from the limiting viscosity number data by the Yamakawa–Fujii (YF) theory.[23]

Theoretical background

Benoit–Doty's equation[21] for the unperturbed radius of gyration $\langle S_2 \rangle_w^{1/2}$ of monodisperse worm-like polymer chains can be generalized to the case of polydisperse worm-like chains with the Schulz–Zimm MWD in the form:

$$\langle S^2 \rangle_{0,z} = q^2 \left\{ \frac{(h+2)M_w}{(h+1)3qM_L} - 1 + \frac{2qM_L}{M_w} - \left(\frac{2(h+1)q^2M_L^2}{hM_w^2} \right) \right. $$
$$\left. \times \left[1 - \left(\frac{(h+1)qM_L}{(h+1)qM_L + M_w} \right) \right]^h \right\} \tag{3.19.1}$$

with

$$M_L = M_w/L \tag{3.19.2}$$

$$L = (M_w/m_0)\ell \tag{3.19.3}$$

Here, $h \equiv (M_w/M_n - 1)^{-1}$: M_L, shift factor; L, contour length; m_0, molecular weight of repeating unit (pyranose ring); ℓ, projection of pyranose ring onto the chain axis $(= 5.15 \times 10^{-8}$ cm). The suffix z means z-average. Assuming $\langle S^2 \rangle_{0,z} \approx \langle S_2 \rangle_z$ and putting the LS data and h, estimated experimentally, into eq. (3.19.1), we can calculate q (hereafter, referred to as q_{BD}).

The interpenetration function ψ for the worm-like touched beads model is related to the excluded volume effect parameter z by:[22]

$$\psi = A_2 M_w^2/(4\pi^{3/2} N_A \langle S^2 \rangle_w^{3/2}) = (\langle S^2 \rangle_{w,0}/\langle S^2 \rangle_{0,\infty,w})^{-3/2} \bar{z}h(\bar{z}) \tag{3.19.4}$$

where N_A is Avogadro's number; A_2, the second virial coefficient; $\langle S^2 \rangle_{0,\infty,w}$, $\langle S^2 \rangle_0$ at the limit of $M_w \to \infty$, a_s, the expansion factor. Yamakawa and Stockmayer (YS) derived the fifth power law-type expression of a_s and $\bar{z}h(\bar{z})$.[22] Kamide *et al.*[2–10] evaluated a_s using a penetration function of the third power law-type for a pearl necklace model. In order to compare their results, we derive, by modifying the YS theory, similar third power law-type equations. The results are:

$$a_s^3 - 1 = (3/2)(67/70K')z \tag{3.19.5}$$

$$\bar{z}h(\bar{z}) = [1 - \{1 + (3903/2865)Q\bar{z}\}^{-0.458}](1828/2865)Q \tag{3.19.6}$$

where Q is a function of L, the diameter of a bead d, and q, and K' is a function of L and q, and both Q and K' are given in the YS theory. Eqs. (3.19.5) and (3.19.6) correspond to the Fixman[24] and Kurata *et al.*[25] type equation, respectively, which are the third power law-type equation in the pearl necklace model. On the basis of eqs. (3.19.1–3.19.6) we can determine q (this is denoted by (q_{BD}^0)) and a_s concurrently, from the experimental data and the d value, evaluated by Ullman's theory.[26]

At the limit of infinite M_w any molecular chain can be regarded as a Gaussian chain and q is directly related to A as defined by the A value at the limit of $M_w = \infty$ in the form

$$q = M_L A_\infty^2/2. \tag{3.19.7}$$

We approximate the A value of the sample with the maximum molecular weight to be A_∞ and calculate q (q_{CL}^0) from A_∞ and M_L using eq. (3.19.7).

Yamakawa and Fujii[23] derived eq. (3.19.8) as an expression of the limiting viscosity number $[\eta]$, for the unperturbed worm-like chain, neglecting the draining effect term which appears in the Kirkwood–Riseman theory for the beads model[27]

$$[\eta] = \Phi(2q/M_L)^{3/2}M^{1/2} \tag{3.19.8}$$

where Φ is a function of q, L, and d and is tabulated in Table I of Ref. 23. In the strict sense, eq. (3.19.8) is only applicable in the case where the draining effect is negligible. Putting the experimental $[\eta]$ and M_w values and M_L, which is calculated using eq. (3.19.2) and d values into eqs. 37–39 of Ref. 23, we can determine q (referred to as q_{YF}) for each polymer sample.

Application to experimental data

Four kinds of the persistence length, q_{BD}, q_{BD}^0, q_{CL}^0, and q_{YF} were determined by applying eqs. (3.19.1–3.19.8) to literature data for cellulose and its derivatives.[6,7,9,10,28–33]

The values of q_{BD}, q_{BD}^0, and q_{YF} are shown in Figure 3.19.1(a–c) as a function of M_w. The MWD of q_{BD} can be classified into four categories:

(1) q_{BD} decreases with M_w. The slope as determined by d ln q_{BD}/d ln M_w, decreases in the order: CA (DS 2.46)/acetone[6] > EHEC/water[28] > HEC/water[29] > cellulose/ FeTNa.[30]

(2) q_{BD} is constant except for some of the lowest molecular weight samples, whose q_{BD} decreases with an increase in M_w: CA (DS 2,46)/N,N-DMAc[10] and CN(DS 2.91)/ acetone.[31]

(3) q_{BD} is practically constant over the whole M_w range: cellulose/cadoxen,[32] CA (DS 2,92)/DMAc,[7] CA (DS 2.46)/THF,[6] CA (DS 0.49)/DMAc,[9] and NaCMC (DS 0.88)/ NaCl aq. solution (at the limit of ion strength $I \rightarrow \infty$).[33]

(4) q_{BD} increases slightly with M_w: CN(DS 2.55)/acetone.[31]

The decrease in q_{BD} with M_w, observed in (1), strongly implies a negative value of a_2, defined as d $\ln(\langle S^2 \rangle/M)$/d ln M. In fact, Kamide and Miyazaki estimated a_2 to be -0.471 for CA (DS 2.46)/acetone, -0.256 for EHEC/water, -0.216 for cellulose/FeTNa, and 0.128 for HEC/water.[2] For these systems the molecular weight dependence of solvation should be considered.[9] Hypothetical employment of a constant q for the systems in group (2) over the whole range of M_w leads unavoidably to a significant molecular weight dependence of the ratio $\langle S^2 \rangle_w : M_w$ in the relatively low M_w range (in this case, $M_w < 10^5$). In contrast to this, the experimental data for all the systems belonging to group (3) lie well on the theoretical $\langle S^2 \rangle M_w$ versus M_w curve, calculated by putting the q_{BD} value for the sample with maximum M_w into eq. (3.19.1). CN(DS 2.55)/acetone is an exceptional case, where the contribution of the excluded volume effect to q_{BD} is not negligible, bringing about in a positive molecular weight dependence of q_{BD}.

For the polymer/solvent systems belonging to group (3), a_2 is found to be zero within the experimental error, suggesting that the unperturbed chain for these systems is a Gaussian chain. Figure 3.19.1(b) shows the plot of q_{BD}^0 against M_w. The molecular weight

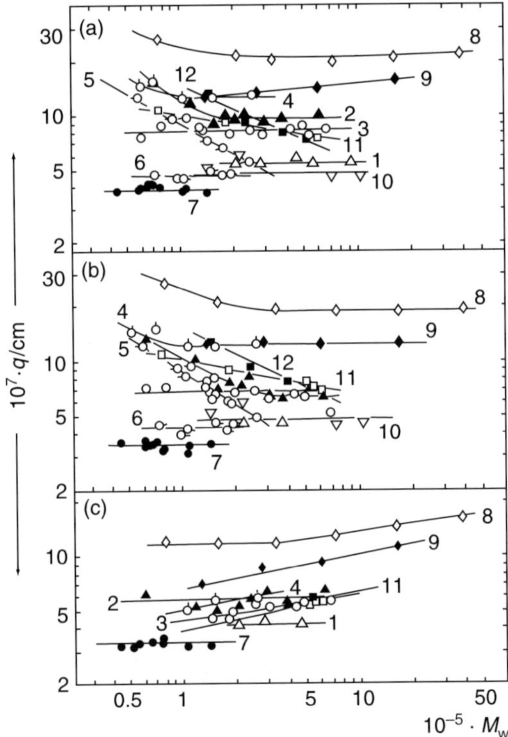

Figure 3.19.1   Plots of the perturbed Benoit–Doty persistence length, q_{BD} (a), the unperturbed Benoi–Doty persistence length, q_{BD}^0 (b), and the perturbed Yamakawa–Fujii persistence length, q_{YF} (c), as a function of the weight-average molecular weight M_w^1: curve 1: cellulose/cadoxen[32] ($\triangle$), curve 2: cellulose/iron sodium tartrate[30]; ($\blacktriangle$), curve 3: cellulose acetate (DS 2.92)/dimethylacetamide;[7] ($\bigcirc$), curve 4: cellulose acetate (DS 2.46)/dimethylacetamide;[10] (circle with upward tail), curve 5: cellulose acetate (DS 2.46)/acetone;[6] (circle with left-hand tail), curve 6: cellulose acetate (DS 2.46)/tetrahydrofuran[6] (circle with right-hand tail); curve 7: cellulose acetate (DS 0.49)/dimethylacetamide[9] ($\bullet$), curve 8: cellulose nitrate (DS 2.91)/acetone;[31] ($\diamond$), curve 9: cellulose nitrate (DS 2.55)/acetone;[31] ($\blacklozenge$), curve 10: sodium carboxymethyl cellulose/sodium chloride aq. solution (($\rightarrow \infty$)[33] (∇), curve 11: hydroxyethyl cellulose/water[29] ($\square$), curve 12: ethylhydroxyethyl cellulose/water[28] ($\blacksquare$).

dependence of q_{BD}^0 is very similar to that of q_{BD}, except for CN(2.55)/acetone whose q_{BD}^0 is essentially independent of M_w.

For all samples of cellulose derivatives employed here the following regression line is obtained, by the least squares method, between q_{BD}^0 and q_{BD}; $q_{BD}^0 = 0.954 \times q_{BD}$ (correlation coefficient $\gamma = 0.984$). In other words, the excluded volume effect is ascertained to play only a minor role in determining q, but nevertheless cannot be ignored.

Figure 3.19.2 shows the correlation between q_{BD}^0 of the samples with highest molecular weight and q_{CL}^0. A good agreement is obtained between them for all cellulose derivatives investigated. This means that the unperturbed chain dimensions A in the pearl necklace model is parallel to the unperturbed persistence length q in the worm-like chain model.

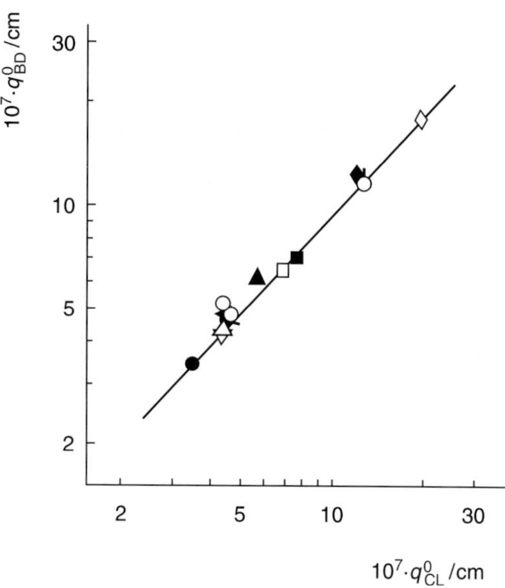

Figure 3.19.2   Correlation between the unperturbed Benoit–Doty persistence length, q_{BD}^0 and the unperturbed persistence length at coil limit, q_{CL}^0 for cellulose and its derivatives.[1] Here, both q_{BD}^0 and q_{CL}^0 are the average values over all the samples employed. Symbols have the same meanings as those in Figure 3.19.1.

Figure 3.19.1(c) shows q_{YF} determined by eq. (3.19.8) as a function of M_w. As we demonstrated before,[10] the d value for cellulose and its derivatives ranges widely between 2×10^{-7} cm and 4×10^{-7} cm. In the YF theory, an approximate equation of $[\eta]$ (eq. (39 of Ref. 23) contains typographical errors in the large $d/(2q)$ region and therefore we could not evaluate q_{YF} for the polymers with $d/(2q) > 0.1$. q_{YF} is always equivalent to or smaller than q_{BD} for all polymers examined. Yamakawa and Fujii considered that the draining effect does not exist for the worm-like cylinder model, if the flow rate is taken as zero at the surface of the worm-like cylinder. However, as shown in Figure 3.19.3, where the difference $q_{BD} - q_{YF}$ is plotted against the draining parameter X, evaluated by the KY theory,[34] $q_{BD} - q_{YF}$ exhibits a profound negative correlation versus X. We note that cellulose and its derivatives dissolve in solvents, with solvation[35] and for these cases the hydrodynamic segment should be the solvated (and not naked) polymer segment, for which the flow rate on the surface cannot be taken as zero.

Table 3.19.1 shows the q_{BD}^0 and A values, both averaged over all samples for the polymer/solvent systems. Here, the A values are determined by method 2B using LS data. The q_{BD}^0 values of cellulose and its derivatives are larger than those for typical vinyl-type polymers ($\approx 1.0 \times 10^{-7}$ cm), but markedly smaller than for typical stiff chain polymers, such as DNA.[26] Therefore, cellulose derivatives can be considered as semiflexible. It can be concluded that both the pearl necklace chain and the worm-like chain model are adequately applicable to these polymers.

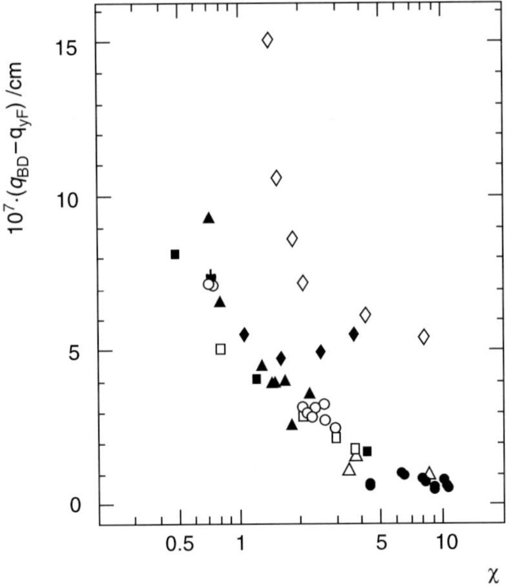

Figure 3.19.3   Plot of the difference between the Benoit–Doty persistence length and the Yamakawa–Fujii persistence length, $(q_{BD} - q_{YF})$ against the draining parameter X.[1] Symbols have the same meanings as those in Figure 3.19.1.

Table 3.19.1

The unperturbed persistence length q_{BD}^0 and unperturbed chain dimension A of cellulose and derivatives[1]

Polymer (DS)[a]	Solvent[b]	$q_{BD}^0 \times 10^8$ (cm)	$A \times 10^8$ (cm)[c]	Reference
Cellulose (0)	Cardoxen	42	1.63	32
	FeTNa	84	1.87	30
CA (2.92)	DMAc	67	1.40	7
CA (2.46)	DMAc	130	2.15	10
	Acetone	80	1.66	6
	THF	43	1.23	6
CA (0.49)	DMAc	34	1.38	9
CN (2.91)	Acetone	206	2.57	31
CN (2.55)	Acetone	123	2.07	31
NaCMC (0.88)	NaCl aq. solution ($I \to \infty$)	50	1.41	33
HEC (1)	Water	82	1.89	29
EHEC (2)	Water	92	1.88	28

[a]DS = degree of substitution.
[b]Cadoxen: cadmium ethylenediamine hydroxide; FeTNA: iron sodium tartrate; DMAc: N, N-dimethylacetamide; THF: tetrahydrofuran; I: iron strength.
[c]Estimated by Method 1B in Ref. 2.

3.19.2 Small angle X-ray scattering method

Cellulose acetate[36]

Figure 3.19.4 shows the effect of $\langle\!\langle F \rangle\!\rangle$ on q, determined from eq. (3.10.1) by small angle X-ray scattering method (SAXS) of CA in DMAc at 25 °C. In the figure the data of LS[3–5] analyzed by Saito[37] are also included. The persistence length (i.e. the chain rigidity) attains maximum at $\langle\!\langle F \rangle\!\rangle \approx 2.5$. This trend is in good agreement with that by LS analysis, although q by SAXS is some 10% larger than that by LS, compared at the same $\langle\!\langle F \rangle\!\rangle$. Note that CA samples used for LS measurements were carefully fractionated, sharp fractions, and those for SAXS were whole polymers. If we utilize 2.87 evaluated by Burchard and Kajiwara[18] for μ^*, then the difference in q between LS and SAXS becomes insignificant. Then, it is directly confirmed that the conclusion[39] on the rigidity of cellulose acetate, deduced from LS and hydrodynamic data, is unquestionably acceptable. As pointed out by Kamide and Saito, the $\langle\!\langle F \rangle\!\rangle$ dependence of q is quantitatively explained by the solvation.[6]

Cellulose nitrate[36]

A commercially available CN, with $\langle\!\langle F \rangle\!\rangle = 2.3$ by ^{13}C NMR analysis, manufactured by Asahi Chemical Ind. Co., Nobeoka, Japan, was employed for LS and SAXS measurements. Figure 3.5.3 shows Zimm plot of CN ((DS 2.3) in acetone at 25 °C. From the plot, we obtain $M_{\mathrm{w}} = 2.76 \times 10^5$, $\langle S^2 \rangle_z^{1/2} = 44.7$ nm, and $A_2 = 8.2 \times 10^{-4}$ cm^3 mol g^{-2} for the sample. By putting M_{w} and $\langle S^2 \rangle_z^{1/2}$, values into the equations eqs. (3.19.9)–(3.19.11), which were derived by Saito[37] with modification of the original Benoit–Doty equation for polydisperse polymers, $q = 12.0$ nm was obtained.[36,38]

$$\langle S^2 \rangle_z = q^2[M_{\mathrm{w}}(h+2)/\{3qM_L(h+1)\} - 1 + 2qM_L/M_{\mathrm{w}} - 2q^2M_{L^2}(h+1)$$
$$\times \{1 - (qM_L(h+1)/qM_L(h+1) + M_{\mathrm{w}})^h\}] \tag{3.19.9}$$

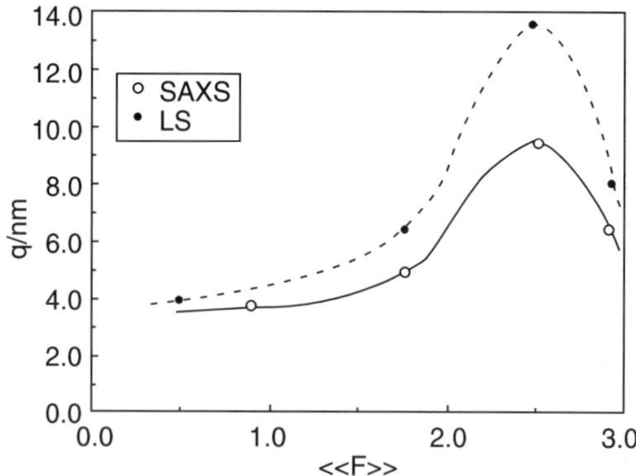

Figure 3.19.4 Relationship between the persistence length q of cellulose acetate in dimethylacetamide at 25 °C and its total degree of substitution $\langle\!\langle F \rangle\!\rangle$.[36]

Table 3.19.2

Persistence length q of cellulose nitrate in acetone at 25 °C[36]

$\langle\!\langle F \rangle\!\rangle$	Whole polymer or fraction	Experimental method	Analytical procedure	Persistence length q (nm)	Reference
2.3	W	SAXS	Kratky	9.6	39
	W	LS	Benoit–Doty	12.6	39
	F	V	Kurata–Stockmayer	1.9	40
2.5	W	SAXS	Kratky	10.0	41
	F	LS	Benoit–Doty	14.0	42
2.9	F	LS	Benoit–Doty	23.0	42

where

$$h^{-1} = (M_\mathrm{w}/M_\mathrm{n}) - 1 \tag{3.5.5}$$

$$M_L = M_\mathrm{w}/L \tag{3.19.2}$$

and L is the contour length.

From Kratky plot of the same CN sample in acetone at 25 °C (as shown in Figure 3.10.2), where 1 and 3 wt% solutions are employed, $k^* = 0.24$ nm^{-1} for 1 wt% and 0.27 nm^{-1} for 3 wt% solutions are estimated.[39] A slight concentration dependence of k^* is observed. On average, we obtain $q = 9.6$ nm for CN with $\langle\!\langle F \rangle\!\rangle = 2.3$ by the SAXS method.

Table 3.19.2 shows the persistence length q of CN samples with various $\langle\!\langle F \rangle\!\rangle$ in acetone. Except for Kurata–Stockmayer (KS)'s method[40] on viscosity data, q for CN lies between 9.6 and 23.0, indicating that CN molecules are semiflexible polymers. As Kamide and Moore pointed out for CN solutions as early as 1964,[43] a two parameter theory, on which KS procedure is based, cannot be applied to CD solutions (see Figures 3.16.5 and 3.16.6).

REFERENCES

1. K Kamide, M Saito and H Suzuki, *Makromol. Chem., Rapid Commun.*, 1983, **4**, 33.
2. K Kamide and Y Miyazaki, *Polym. J.*, 1978, **10**, 409.
3. K Kamide and Y Miyazaki, *Polym. J.*, 1978, **10**, 539.
4. K Kamide and M Saito, *Eur. Polym. J.*, 1981, **17**, 1049.
5. K Kamide and M Saito, *Eur. Polym. J.*, 1982, **18**, 661.
6. K Kamide, T Terakawa and Y Miyazaki, *Polym. J.*, 1979, **11**, 285.
7. K Kamide, Y Miyazaki and T Abe, *Polym. J.*, 1979, **11**, 523.
8. K Kishino, T Kawai, T Nose, M Saito and K Kamide, *Eur. Polym. J.*, 1981, **17**, 623.
9. K Kamide, M Saito and T Abe, *Polym. J.*, 1981, **13**, 421.
10. K Kamide and M Saito, *Polym. J.*, 1982, **14**, 517.
11. M Kurata, Y Tsunashima, M Iwama and K Kamada, in *Polymer Handbook.*, 2nd edn., (eds J Brandrup and EH Immergut), Wiley, New York, 1975, IV-34.
12. O Kratky and G Porod, *Recl. Trav. Chim.*, 1949, **68**, 1106.
13. SY Lyubina, SI Klenin, IA Sterlina and AV Troitskaya, *Vysokomol. Soedin., Ser. A*, 1973, **15**, 691.

14. LN Andreyeva, PN Lavrenko, EU Urinov, LI Kutsenko and VN Tsvetkov, *Vysokomol. Soedin., Ser. B*, 1975, **17**, 326.
15. Y Korneyeva, PN Lavrenko, E Urinov, AK Khripunov, LI Kutsenko and VN Tsvetkov, *Vysokomol. Soedin., Ser. A*, 1979, **21**, 1547.
16. J Hearst and W Stockmayer, *J. Chem. Phys.*, 1962, **37**, 1425.
17. H Yamakawa and M Fujii, *Macromolecules*, 1973, **6**, 407.
18. W Burchard and K Kajiwara, *Proc. R. Soc. London, Ser. A*, 1970, **316**, 185.
19. W Burchard, *Br. Polym. J.*, 1971, **3**, 214.
20. AK Gupta, JP Cotton, E Marchal, W Burchard and H Benoit, *Polymer*, 1976, **117**, 363.
21. H Benoit and PM Doty, *J. Phys. Chem.*, 1953, **57**, 958.
22. H Yamakawa and WH Stockmayer, *J. Chem. Phys.*, 1972, **57**, 2843.
23. H Yamakawa and M Fujii, *Macromolecules*, 1974, **7**, 128.
24. M Fixman, *J. Chem. Phys.*, 1962, **36**, 3123.
25. K Kurata, M Fukatsu, V Sotobayash and H Yamakawa, *J. Chem. Phys.*, 1964, **41**, 139.
26. R Ullman, *J. Chem. Phys.*, 1968, **49**, 5486.
27. JG Kirkwood and J Riseman, *J. Chem. Phys.*, 1948, **16**, 565.
28. RS Manley, *Ark. Kemi.*, 1956, **9**, 519.
29. W Brown, D Henley and J Ohman, *Makromol. Chem.*, 1963, **64**, 49.
30. L Valtasaari, *Makromol. Chem.*, 1971, **150**, 117.
31. GV Schulz and E Penzel, *Makromol. Chem.*, 1968, **112**, 260.
32. D Henley, *Ark. Kemi.*, 1961, **18**, 327.
33. W Brown and D Henley, *Makromol. Chem.*, 1964, **79**, 68.
34. M Kurata and H Yamakawa, *J. Chem. Phys.*, 1958, **29**, 311.
35. WR Moore and BM Tidswell, *Makromol. Chem.*, 1965, **81**, 1; WR Moore, *J. Polym. Sci., Part C*, 1967, **16**, 571; WR Moore, *Solution Properties of Natural Polymers: An International Symposium in Edinburgh*, The Chemical Society Edition, p. 185. The Chemical Society Burlington House, London, 1968; K Kamide, K Okajima and M Saito, *Polym. J.*, 1981, **13**, p. 115.
36. K Kamide and M Saito, *Makromol. Symp.*, 1993, **83**, 233.
37. M Saito, *Polym. J.*, 1983, **15**, 213.
38. K Kamide and M Saito, *Adv. Polym. Sci.*, 1987, **83**, 1.
39. K Kamide, Y Ito and M Saito, *Cellucon' 93*, 1993.
40. M Kurata and WH Stockmayor, *Fortschr. Hochpolym. Forsch.*, 1968, **3**, 196.
41. O Kratky, *Angew. Chem.*, 1960, **72**, 467.
42. GV Schulz and E Penzel, *Makromol. Chem.*, 1968, **112**, 260.
43. K Kamide and WR Moore, *J. Polym. Sci.*, 1964, **B2**, 809.

3.20 TEMPERATURE DEPENDENCE OF LIMITING VISCOSITY NUMBER AND RADIUS OF GYRATION

A recent systematic analysis[1–16] of the experimental data of solution viscosity, LS, sedimentation, and diffusion of cellulose and its derivative (CD) solutions indicated for the molecular properties of these polymers that the excluded volume effect is extremely small (Section 3.13.2), and that the unperturbed chain dimensions A varies depending on the polarity of the solvent (Section 3.18). However, in common solvents, the cellulose chain behaves semiflexibly (Section 3.18). As far as CD solutions are concerned, the above conclusions were derived at a constant temperature (near room temperature) and only fragmental knowledge of the temperature dependence of the molecular characteristics has been obtained up to the present, except CA in acetone. For example, only a few theta solvents were determined for CD as the solvent, in which A_2 becomes zero: tricaproate CTCp/DMF ($\theta = 314$ K),[17] CTC/anisole ($\theta = 94$ °C),[18] CTC/hexanol (73 °C),[18] and CDA-2-butanone (50 °C)[19] (see Table 3.21.3). Another example is two conflicting molecular explanations for a significantly negative temperature coefficient of the limiting viscosity number of CD solutions around room temperature.[17,20–32]

3.20.1   Limiting viscosity number

It is well known that the limiting viscosity number $[\eta]$ of most cellulose derivatives significantly decreases with increasing temperature.[17,20–30] Even in theta solvents, this effect has been observed for CTA[21] and CTCp.[17] This finding is in sharp contrast to those for ordinary vinyl polymers, for example, $[\eta]$ of polystyrene increases in poor solvents with increasing temperature, while it changes only slightly in good solvents.

CA (DS 2.46)/acetone[33]

The viscosity results on a CA (DS 2.46) fraction/acetone system are summarized in the second and third columns of Table 3.20.1.

They show that, when the temperature is raised from 12.6 to 50.3 °C, the value of $[\eta]$ decreases by 23%. This confirms the behavior frequently observed in most cellulose derivative/solvent systems. Figure 3.20.1 shows a plot of $\ln[\eta]$ versus T, from which the temperature coefficient of $[\eta]$ is to be determined. The plot is satisfactorily linear and $\mathrm{d}\ln[\eta]/\mathrm{d}T$ was estimated to be -6.9×10^{-3}. Surprisingly, this value is some 10 times those observed for ordinary polymers. On the other hand, the value for the Huggins constant increases slightly with increasing temperature. This phenomenon, although contrary to that of $[\eta]$, has often been found for many other cellulose derivatives.[23,29,30,34–37] The k' values of 0.5–0.6 are much larger than those of around 0.3 for vinyl polymers in good solvents. Such values are admittedly obtained in theta solvents, whereas acetone is not a theta solvent for this CDA fraction.

Cellulose/aq. LiOH[15]

Figure 3.20.2 shows the temperature dependence of $[\eta]$ of cellulose sample SA-3 in aq. LiOH (the concentration of LiOH $\phi_L = 6$ wt%) and the figure also includes the

Table 3.20.1

Viscosity and light scattering results on a cellulose diacetate fraction EF3-8 in acetone[33]

T (°C)	$[\eta]$ (cm^3 g^{-1})	k'	$A_2 \times 10^4$ (cm^3 mol g^{-2})	$\langle S^2 \rangle_z^{1/2}$ (Å)	
				436 nm	356 nm
12.6	168	0.51	4.1	243	241
25.4	153	0.54	3.8	234	235
37.7	141	0.53	3.6	225	225
45.0	134	0.54	3.5	220	221
50.3	130	0.60	3.4	215	216

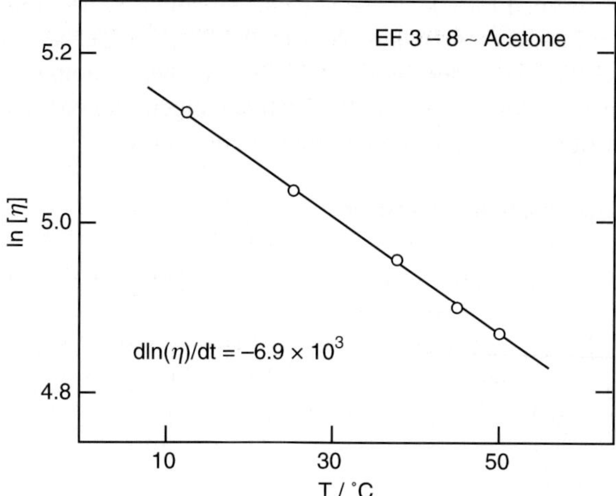

Figure 3.20.1   Temperature dependence of ln[η] for cellulose acetate (DS 2.46) fraction EF3-8 in acetone.[33]

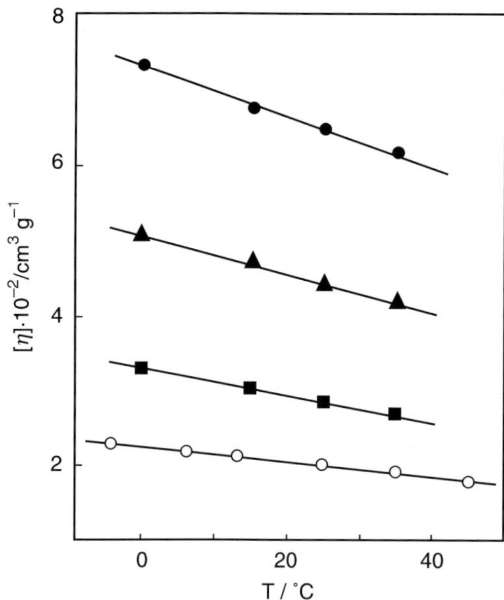

Figure 3.20.2   Temperature (T) dependence of the limiting viscosity number [η] of cellulose-solvent systems.[15] Open symbol: cellulose (sample code SA-3) -aq. LiOH;[15] closed symbol: cellulose-cadoxen system;[38] (●), sample code P1; (▲), P2; (□), P3.

literature data (sample code P1, P2, and P3) by Henley[38] on cellulose in cadoxen. $[\eta]$ for aq. cellulose/aq. LiOH system decreases monotonically with increasing temperature in a similar way as $[\eta]$ for cellulose in cadoxen.

Cellulose/aq. NaOH[16]

$[\eta]$, evaluated by extrapolating η_{sp}/c to $c = 0$, and Huggins coefficient k', determined from the slope of the plot (Figure 3.6.3), are plotted as a function of temperature in Figure 3.20.3. Here, in the figure, the literature data on cellulose (sample code SA3) 6 wt% aq. LiOH solution[15] (triangle), and CDA ($M_w = 9.4 \times 10^4$) acetone[33] (chain line) systems are included for comparison. Inspection of Figure 3.20.3 shows that $[\eta]$ of cellulose in 8 wt% aq. NaOH is smaller than that of cellulose in 6 wt% aq. LiOH at least in the range between 0.6 and 46.0 °C. In the range below 35 °C, $[\eta]$ in aq. NaOH decreases linearly with an increase in temperature as in the case of aq. cellulose/aq. LiOH and CDA-acetone systems. $d[\eta]/dT$ was found to be -1.2 cm^3 g^{-1} K^{-1} for aq. NaOH, of which the absolute magnitude was slightly larger than that for aq. LiOH (-1.02 cm^3 g^{-1} K^{-1}) and CDA-acetone (-1.01 cm^3 g^{-1} K^{-1}). $[\eta]$ in aq. NaOH decreased drastically in the range of 35–41 °C, suggesting that a certain conformational change of cellulose chain might occur in this narrow temperature range. Similar change in $[\eta]$ was observed in the region of helix–coil transition of the poly (γ benzyl L-glutamate)- dichloroacetic acid- ethylene dichloride system.[27] Huggins coefficients in aq. NaOH and aq. LiOH are about 0.8 and 0.6, respectively, which are not as sensitive to the temperature, slightly decreasing with increasing temperature, whereas the CDA-acetone system exhibited a positive temperature dependence of k'.

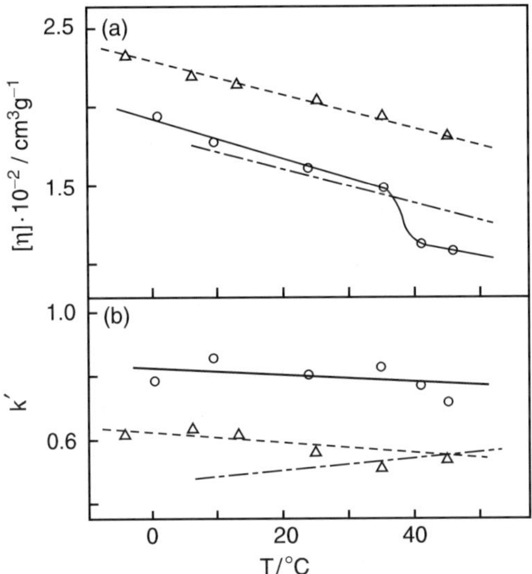

Figure 3.20.3 Plots of the limiting viscosity number $[\eta]$ (a) and Huggins' constant k' (b) against temperature T.[16] (O), cellulose/8% NaOH aq. solution (solid line);[16] (Δ), cellulose/6% LiOH aq. solution[15] (broken line); (chain line), cellulose diacetate/acetone system.[33]

3.20.2   Molecular weight

CA (DS 2.46)/acetone[33]

From the LS data at 25.4 °C, M_w of the fraction EF3-8 in acetone was found to be 9.4×10^4. This result is compared with that previously obtained in THF at 25 °C ($M_w = 10.0 \times 10^4$; see Table 3.5.3). Noting that both the solvent and the apparatus are different in the two measurements, one may consider a 5.3% difference in M_w as satisfactorily small. There is much closer agreements between the M_w values from GPC and viscosity measurements ($M_w = 9.5 \times 10^5$ and 9.5×10^5, respectively; see Table 3.5.2). The appearance of a neat reticulate diagram of a Zimm plot was kept almost unchanged even at 50.3 °C, which is only 3 °C below the boiling point of the solvent (acetone; see Figure 3.20.5). The intercept of this plot gives a value of 9.5×10^4 for M_w in reasonable agreement with that obtained at 25.4 °C.

CA (DS 2.46)/butanone[18]

Table 3.20.2 lists the value of M_w, A_2 and $\langle S^2 \rangle_z^{1/2}$ of CA (DS 2.46) in butanone at 30, 40, 50, and 60 °C. M_w of CA (DS 2.46) molecules in 2-butanone is almost constant over a range of 30–60 °C. The M_w value for sample EF3-12 in 2-butanone at 50 °C is in excellent agreement with that found in acetone in 25 °C (14.1×10^4).

Cellulose/aq. NaOH[16]

Figure 3.20.4 shows the temperature dependence of apparent weight average molecular weight M_w^*. Evidently, M_w^* of cellulose dissolved in aq. NaOH remains almost constant ($\sim 7.9 \times 10^4$) above 10 °C. M_w^* of the same sample in aq. LiOH (undialyzed) at 25 °C was $1.2 \times 10^{5,15}$ and is shown as the filled circle in the figure. This value is 1.5 times greater than that in aq. NaOH. M_w for cellulose (SA3) in cadoxen (8×10^4), indirectly evaluated by linear interpolation of two M_w data directly determined by the LS method for two other samples in dialyzed cadoxen,[15] is in fairly good agreement with M_w^* in aq. NaOH. Then, we can consider $M_w^* \approx M_w$ in aq. NaOH. In other words, the selective adsorption

Table 3.20.2

Light scattering results on the cellulose acetate (DS 2.46) sample in 2-butanone[19]

Sample code	T (°C)	$M_w \times 10^{-4}$	$A_2 \times 10^4$ (cm^3 mol g^{-2})	$\langle S^2 \rangle_z^{1/2}$ (Å)	$\langle S^2 \rangle_w^{1/2}$ (Å)
EF3-5	30	7.7	−0.50	375	346
	40	7.1	−0.25	353	326
	50	7.1	0	335	309
	60	7.1	0.25	316	291
EF3-9	40	9.2	−0.25	346	320
	50	9.2	0	308	285
EF3-12	50	14.1	−2.1	428	396

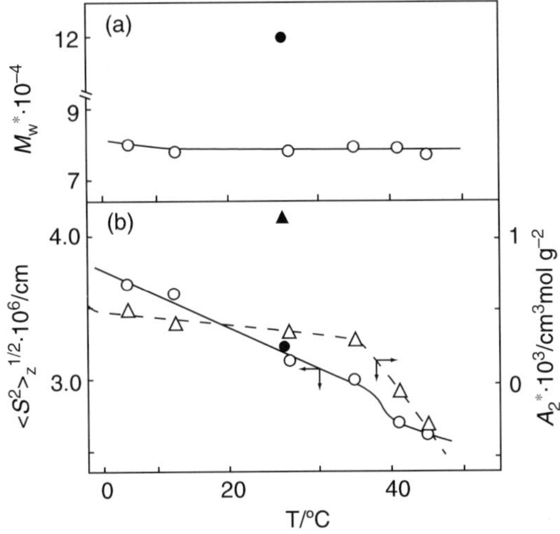

Figure 3.20.4 Plots of the apparent weight-average molecular weight M_w^* (a) and the apparent second virial coefficient A_2^* and the z-average radius of gyration $\langle S^2 \rangle_z^{1/2}$ (b) against temperature T. Open marks: cellulose/8% NaOH aq. solution; closed marks: cellulose/6% LiOH aq. solution system.[15]

does not occur above 10 °C for aq. cellulose/aq. NaOH system, but this cannot be ignored for aq. cellulose/aq. LiOH system. We can also conclude that cellulose dissolves molecularly into aq. NaOH, at least over a temperature range of 10–45 °C and the discussion on 'solubility' of cellulose in aq. NaOH made by Kamide et al.[39] is indeed acceptable.

3.20.3 Second virial coefficient and radius of gyration

CA (DS 2.46)/acetone[33]

By using the dn/dc values calculated from eq. (3.20.1),

$$dn/dc = 0.112 + 3.310^{-4}(T - 25) \qquad \text{(where T is the temperature in Celsius)}$$

$$(3.20.1)$$

the data at the other four temperatures were analyzed to determine the molecular parameters such as A_2, $\langle S^2 \rangle_z$, and M_w. As an example, the Zimm plot at 50.3 °C is depicted in Figure 3.20.5. The values of A_2 were determined from the concentration extrapolation of the Zimm plot and listed in the fourth column of Table 3.20.1. They are positive and sufficiently high. It should be noted that A_2 of the present system decreases with increasing temperature, which is opposite to the behavior observed by Tanner and Berry on the CDA-THF system.[40]

Angular intensity data obtained with 365 nm light were analyzed only for the radius of gyration of the sample. Instead of Kc/R_θ, values of $c/\Delta I(0, c)$ are plotted against $\sin^2(\theta/2) - 250c$, where c is the CDA concentration in g cm^{-3} units and ΔI is the scattered intensity difference between solvent and solution. The plot is adequately

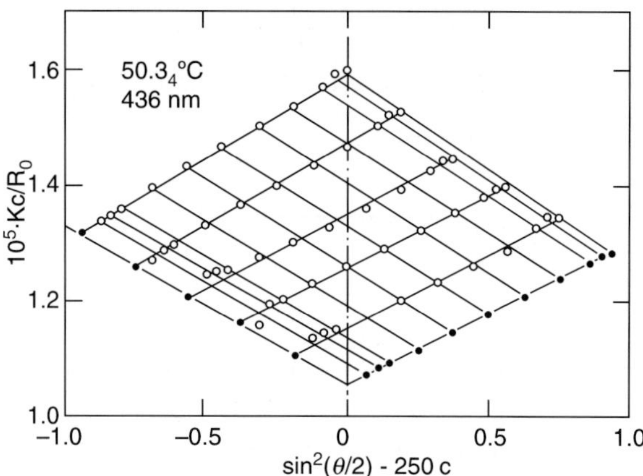

Figure 3.20.5   Zimm plots of light scattering data obtained with 436 nm light for cellulose acetate (DS 2.46) fraction EF3-8 in acetone at 50.3 °C.[33]

linear over the whole range of scattering angle, indicating that the molecular dimension of the sample is small when compared with the wavelength of the light. The shorter wavelength of this light considerably favored accurate determinations of the angular dependence of scattered light, because the scattered intensity was higher and more dependent on scattering angle. From the slope and the intercept of the plot in Figure 3.20.5, the $\langle S^2 \rangle_z^{1/2}$ value at infinite dilution was found to be 235 Å at 25.4 °C. This concurs with the blue light result of 234 Å. The same procedure was applied to the data at the other temperatures. All the results for $\langle S^2 \rangle_z^{1/2}$ are shown in Table 3.20.1. The two sets of those results are very similar, but the results obtained with 365 nm light are more accurate.

The following relationship below was established for the CDA fraction acetone system at 25 °C:

$$\langle S^2 \rangle_z^{1/2} = 7.39 \times 10^{-8} M_w^{0.308} \qquad \text{(cm)} \qquad (3.12.4)$$

Substitution of the M_w value of 9.4×10^4 (for EF3-8) into the above equation yields 252 Å for $\langle S^2 \rangle_z^{1/2}$. This exceeds the value of the present determination by 7%. The difference may be considered to correspond to the experimental errors in M_w determinations. Recalling the two-fold variation reported by Stein and Doty[41] on a CDA fraction with the almost same molecular weight, a 7% difference is hardly significant.

CA (DS 2.46)/2-butanone[19]

All values of A_2, shown in Table 3.20.2, are close to zero. Admittedly, they are subject to large error since the points extrapolated to $\theta = 0$ in the Zimm plot show some scatter. However, the Flory theta (θ) temperature, as defined by $A_2 = 0$, may be taken as 50 °C. This value is 13 °C higher than that determined by the Shultz–Flory plots (see Figure 3.21.4).

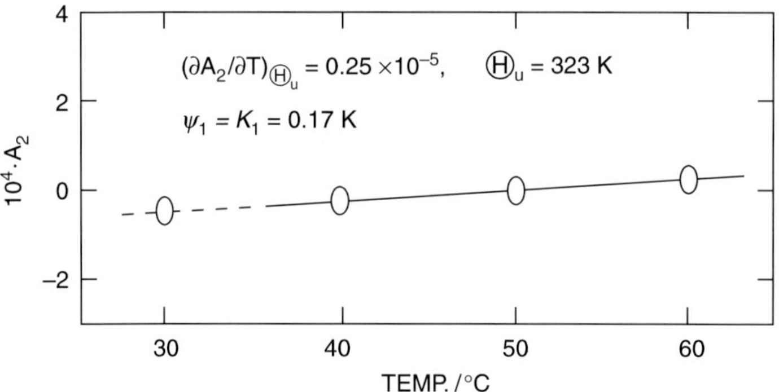

Figure 3.20.6   Plots of A_2 versus temperature from which θ_u and ϕ_1 are determined.[19]

Flory's entropy parameter ϕ at the θ temperature can be evaluated from the relationship

$$\phi = (V_0/v_2)\theta(\partial A_2/\partial T)_{T=\theta} \tag{3.20.2}$$

(V_0: molar volume of solvent; v_2: partial specific volume of polymer).

From plots of A_2 versus T (Figure 3.20.6), the temperature coefficient of A_2 is determined as 0.25×10^{-5} cm^3 mol g^{-2} deg. Substituting for the factors in the right-hand side of eq. (3.20.2) yields $\phi_1 = 0.17$.

Cellulose/aq. NaOH[16]

Figure 3.20.4(b) shows the temperature dependence of A_2^* (unfilled triangle) and $\langle S^2 \rangle_z^{1/2}$ (unfilled circle) of cellulose in aq. NaOH. The true second virial coefficient A_2 can be calculated from A_2^* by eq. (3.13.2):[42]

$$A_2 = (M_w^*/M_w)A_2^* \tag{3.13.2}$$

Using eq. (3.13.2), $A_2 = 4.91 \times 10^{-4}$ cm^3 mol g^{-2} was obtained at 3.5 °C and is shown in parenthesis in the fourth column of Table 3.5.7. Of course, at temperatures equal to or above 10 °C, A_2^* can be regarded as A_2 because $M_w^* \approx M_w$ holds. For comparison, the A_2 data on cellulose (sample code SA3) aq. LiOH at 25 °C are shown. The cellulose/aq. NaOH system has a remarkably smaller A_2 than the cellulose/aq. LiOH system, suggesting that aq. NaOH is poor compared with aq. LiOH. The temperature dependence of A_2 for cellulose/aq. NaOH system can be approximately represented by two straight lines. With an increase in temperature below 35 °C, A_2 decreases slowly, but above 35 °C. A_2 decreases remarkably, becoming zero at $T \geq 4$ °C. In other words, the 8 wt% aq. NaOH system is a Flory theta solvent found for cellulose. $\langle S^2 \rangle_z^{1/2}$ decreases also with an increase in temperature. A decrease in $\langle S^2 \rangle_z^{1/2}$ with temperature is significantly larger in the range between 35 and 41 °C than those in other temperature ranges (i.e. $5 \leq T \leq 35$ °C and $T \geq 41$ °C). The value of $\langle S^2 \rangle_z^{1/2}$ of the cellulose (SA3)/ aq. NaOH system at 26 °C is about 4% smaller than that of the same cellulose sample (SA3)/aq. LiOH system at 25 °C. $\langle S^2 \rangle_z^{1/2}$ of cellulose/aq. NaOH at 40 °C (i.e. the unperturbed z-average radius of gyration $\langle S^2 \rangle_{0,z}^{1/2}$) was estimated from Figure 3.5.6(b) to be 2.77×10^{-6} cm.

3.20.4   d ln a_s/dT, d ln$\langle S^2 \rangle_{0,w}$/dT, and d ln Φ/dT

The negative temperature coefficient of $[\eta]$ has long been considered as characteristic of cellulose derivative solution and has been discussed by several researchers. Mandelkern and Flory,[22] Flory $et\ al.$,[24] Krigbaum and Spuring[17] and Moore and Edge[27] ascribed it to the rapid decrease in the unperturbed molecular dimensions with increasing temperature. Kamide $et\ al.$[28] and Shanbhag[30] disagreed with this interpretation and preferred to consider volume effects rather than skeletal effects of the molecular chains. All of these deductions were made from indirect measurements of their molecular dimensions. Most researchers resorted to viscosity measurements and the universality of the Flory viscosity parameter Φ, which appears in eq. (3.15.1) below. Because it is evident that the Φ parameter for semiflexible polymers is smaller than that employed and varies by various factors, previous arguments may not be sound. In order to resolve the negative temperature coefficient of $[\eta]$, we must first determine the molecular dimensions by LS at the same temperature and in the same solvent as used for viscosity measurements. To the best of our knowledge, results from such a pair of measurements have not yet been published. Brown $et\ al.$[29] made both measurements on the hydroxyethyl cellulose/water system, but used different fractions in each experiment and thus the results are inappropriate for the present purpose.

This section presents the results of LS and viscosity measurements on the same fraction of CDA in acetone over a temperature range of some 48 °C.

Eq. (3.15.1) can be rewritten to give the limiting viscosity number $[\eta]$:

$$[\eta] = 6^{3/2} \Phi(\langle S^2 \rangle_{\mathrm{w}}^{3/2}/M_{\mathrm{w}}) \qquad (3.20.3)$$

or

$$[\eta] = 6^{3/2} \Phi(\langle S^2 \rangle_{0,\mathrm{w}}^{3/2}/M_{\mathrm{w}})a_{\mathrm{s}}^3 \qquad (3.20.3')$$

In order to examine the temperature coefficient d $\ln[\eta]/dT$, it is necessary to determine the value of each term in the right-hand side of the equation below:

$$\mathrm{d} \ln[\eta]/dT = \mathrm{d} \ln \Phi/dT + 1.5\mathrm{d} \ln\langle S^2 \rangle_{0,\mathrm{w}}/dT + 3\mathrm{d} \ln a_{\mathrm{s}}/dT \qquad (3.20.4)$$

CA (DS 2.46)/acetone[33]

$\langle S^2 \rangle_z$ values for CA in acetone was converted to $\langle S^2 \rangle_{\mathrm{w}}$, as listed in Table 3.20.3. Using data from the table, the penetration function ψ was evaluated and finally we obtained d ln α_{s}/dT from eqs. (3.13.4), (3.13.5), and (3.13.10). α_{s} values are listed Table 3.20.3. α_{s} values of CDA in acetone are very close to unity and it is concluded that d ln $\alpha_{\mathrm{s}}/dT \approx 0$.

The value of $\langle S^2 \rangle_{0,\mathrm{w}}^{1/2}$ is determined simply by dividing $\langle S^2 \rangle_{\mathrm{w}}^{1/2}$ by the factor of α_{s}. In the previous section, this procedure was referred to as method 2B[1] for estimating the unperturbed dimensions and has been satisfactorily applied.[1,5,6]

Figure 3.20.7 shows a plot of ln $\langle S^2 \rangle_{0,\mathrm{w}}^{1/2}$ versus T, demonstrating the decrease in the unperturbed dimensions of CDA molecules with increasing temperature. Indeed, such

Table 3.20.3

Conformation and hydrodynamic parameters of a cellulose diacetate fraction EF3-8 in acetone[33]

T (°C)	$\langle S^2\rangle_z^{1/2}$, (Å)	$\langle S^2\rangle_{0,w}^{1/2}$, (Å)	α_s	A (Å)	$\Phi \times 10^{-23}$
12.6	218	214	1.02	1.71	1.0
25.4	211	207	1.02	1.65	1.0
37.7	203	199	1.02	1.59	1.0
45.0	199	195	1.02	1.56	1.1
50.3	194	190	1.02	1.52	1.1

a phenomenon was indirectly deduced from viscosity measurements some 20 years ago. However, the decrease in $\langle S^2\rangle_{0,w}^{1/2}$, with increasing temperature has been unequivocally shown for the first time by LS. The slope of this plot gives -6.4×10^{-3} deg^{-1} for d ln $\langle S^2\rangle_{0,w}^{1/2}/dT$. When this value is compared with that observed for polystyrene,[43] -0.17×10^{-3} deg^{-1}, the rapid decrease in the unperturbed dimensions of CDA molecules is seen.

When connected with the observed high values of A_2, α_s is close to unity. This implies that CDA molecules are rather stiff in acetone solutions. Since the chain stiffness can be estimated from analysis of the unperturbed dimensions, we will briefly present some results of such an analysis. The unperturbed dimensions are usually expressed in the short-range interaction parameter A, which is given by

$$A = (\langle R^2\rangle_{0,w}/M_w)^{1/2} \qquad (\langle R^2\rangle_{0,w}^{1/2} \text{ is the weight} - \text{average end-to-end}$$
$$\text{distance at unperturbed state}) \tag{3.13.8$'$}$$

The A values thus calculated are listed in the fifth column of Table 3.20.3. It is understood that, when compared with the accepted value for polystyrene,[44] $A = (6 \times 7.6 \times 10^{-18})^{1/2}$ cm $= 0.68$ Å, a CDA molecule has a larger short-range interference.

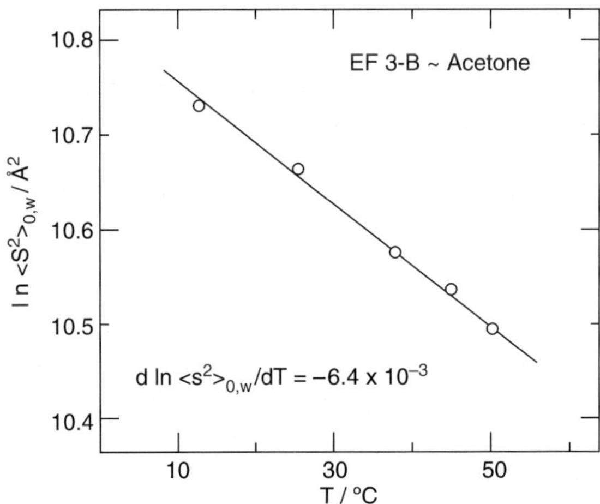

Figure 3.20.7   Temperature dependence of $\ln\langle S^2\rangle_{0,w}$ for cellulose acetate (DS 2.46) fraction EF3-8 in acetone.[33]

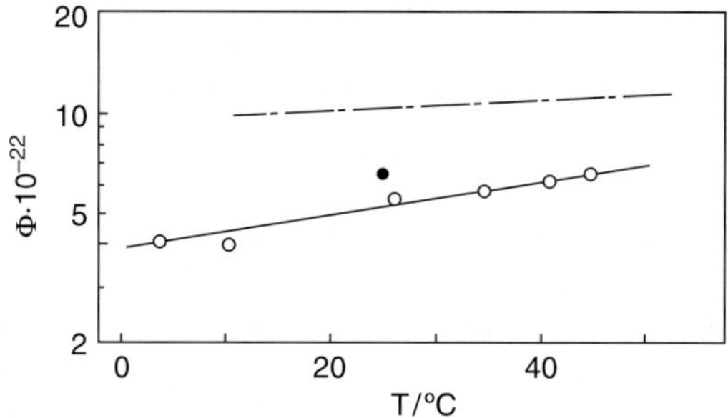

Figure 3.20.8   Plots of the Flory's viscosity parameter Φ against temperature T.[16] ($\bigcirc$) and solid line: cellulose/8% NaOH aq. solution; ($\bullet$) cellulose/6% LiOH aq. solution[15] and broken line: cellulose/diacetate acetone system.[33]

Next, we consider the Flory viscosity parameter Φ and its temperature coefficient. The Φ values are computed by eq. (3.15.1) and the necessary quantities given in the tables. They are listed in the last column of Table 3.20.3. It is seen that the Φ values of 1.05–1.15×10^{23} are, as for other cellulose derivatives, significantly smaller than the asymptotic value at nondraining limit, 2.87×10^{23}. In addition, Φ appears to increase slightly with increasing temperature, which is opposite to the usual behavior of most vinyl polymers.[45] A plot of $\ln \Phi$ versus T (Figure 3.20.8) gives $d \ln \Phi/dT = (2.8 \pm 0.5) \times 10^{-3}$ deg^{-1}.

All of the quantities necessary to examine the temperature coefficient of $[\eta]$ for CA (DS 2.46) acetone were obtained and are summarized below.

$$d \ln[\eta]/dT = -6.9 \times 10^{-3} \qquad (3.20.5)$$

$$d \ln \Phi/dT = (2.8 \pm 0.5) \times 10^{-3} \qquad (3.20.6)$$

$$d \ln\langle S^2 \rangle_{0,w}/dT = -6.4 \times 10^{-3} \qquad (3.20.7)$$

$$d \ln \alpha_s/dT \approx 0 \qquad (3.20.8)$$

Substituting these values into the corresponding terms in eq. (3.20.2), we find that this equality holds within experimental error. It can be deduced that the rapid decrease in unperturbed dimension with increasing temperature is the dominant cause of the large negative value for $d \ln[\eta]/ldT$. The effect of excluded volume is negligible. Indeed, a term $d \ln \Phi/dT$ acts positively, but such an opposite effect is suppressed by the large negative contribution of $(3/2) d \ln \langle S^2 \rangle_{0,w}/dT$. This view is in accord with that deduced by Flory $et\ al.$[24] in 1958, using their experimental results of $[\eta]$ and A_2 obtained by osmometry. They used the approximate theory of Orofino and Flory[46] which interrelates the quantities M_n, A_2, and $[\eta]$. Here, again, Φ was assumed to be practically equal to that at the nondraining limit. Although they employed an approximate theory and had no direct information on the molecular dimension, Flory $et\ al.$[24] obtained a correct insight into hydrodynamic behavior of cellulose derivatives. Historically, this view has been criticized by some researchers.[28,30,47] We recall that their arguments were based on the results of viscosity plots.[47,48] Accordingly, it is obvious from the low values of Φ and the more detailed analysis[5] that their incorrect explanations resulted from neglect of the draining effect.

Cellulose/aq. NaOH[16]

The Flory viscosity parameter Φ is defined by eq. (3.15.1).

$$\Phi = q_{w,z}[\eta]M_w/(6\langle S^2\rangle_z)^{3/2} \tag{3.15.1}$$

Here, $q_{w,z}$ is the polymolecularity correction factor, a prerequisite of the MWD function and breadth of MWD of the sample (see eq. (3.15.2)). The literature data[34,38] suggest that when the acid hydrolysis of cellulose proceeds to a greater extent, the products have $M_w/M_n = 2$. Then, in this study, we estimated $q_{w,z}(= 1.955)$ assuming the Schulz–Zimm distribution for MWD of sample code SA3 and 2 for its M_w/M_n. In the eighth column of Table 3.5.7, the values of Φ calculated by eq. (3.15.1) are shown.

Figure 3.20.8 shows the plot of Φ versus temperature.[16] In the figure, the data point for cellulose/aq. LiOH is included. The Φ value for cellulose/aq. NaOH, including the value at 40 °C (theta solvent), is slightly smaller than that for cellulose/aq. LiOH and is only 1/5 the theoretical value for the unperturbed nondraining chain (2.87×10^{23}). This means that the cellulose chain is partially free draining in aq. NaOH, as in the case of cellulose in aq. LiOH[15] and various cellulose derivatives in solvents.[1,10,49] Φ for cellulose/aq. NaOH increases monotonically with temperature as found for CDA-acetone system. $d \ln \Phi/dT$ was found to be 1.0×10^{-2} deg^{-1} for cellulose/aq. LiOH,[15] which is 3.5 times larger than that (2.8×10^{-3} deg^{-1}) for CDA-acetone.[33]

Combining eqs. (3.13.4–3.13.10) enables us to estimate α_s when α_s is not far from unity, using the LS data. Column 10 of Table 3.5.7 lists the α_s values determined thus. Here, $\langle S^2\rangle_z^{1/2}$ was converted to $\langle S^2\rangle_w^{1/2}$, assuming the Schulz–Zimm distribution and $M_w/M_n = 2$. α_s of cellulose in aq. NaOH solution below 35 °C lies between 1.00 and 1.01 almost independent of temperature. We can conclude that the excluded volume effect is extremely small in the cellulose/aq. NaOH system even below 35 °C. Then, $d \ln \alpha_s/dT \approx 0$.

Eq. (3.13.7) is valid only when the unperturbed chain is Gaussian and is only approximately valid when the unperturbed chain is not Gaussian, but the extent of departure from the Gaussian nature is not remarkable. Although, in the strict sense, the Gaussian nature of cellulose dissolved in aq. NaOH solution has not yet been experimentally confirmed, it is well known that the cellulose chain behaves as a Gaussian chain in cellulose/aq. LiOH systems and we can apply eq. (3.13.7) for cellulose in aq. NaOH solution. Here, $\langle S^2\rangle_{0,w}$ was evaluated from $\langle S^2\rangle_z$ and α_s (method 2B).

Figure 3.20.9 shows the temperature dependence of A. The temperature dependence of A is almost the same as that of $\langle S^2\rangle_w^{1/2}$ below 35 °C because α_s is practically independent of the temperature. In the figure, the value of A for cellulose in aq. LiOH at 25 °C is shown by the filled circle. A in aq. NaOH is nearly the same as that in aq. LiOH.

Each term in the right-hand side of eq. (3.20.4) can be estimated for cellulose/aq. NaOH below 35 °C. The results are:

$$d \ln[\eta]/dT = -7.42 \times 10^{-3} \tag{3.20.9}$$

$$d \ln \Phi/dT = 1.0 \times 10^{-2} \tag{3.20.10}$$

$$d \ln(\langle S^2\rangle_{0,w}^{1/2}/M_w)/dT = -1.26 \times 10^{-2} \tag{3.20.11}$$

$$d \ln \alpha_s/dT \approx 0 \tag{3.20.12}$$

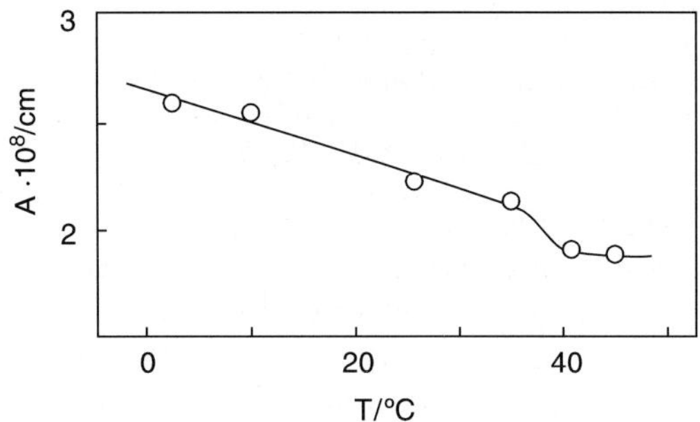

Figure 3.20.9 Plots of the unperturbed chain dimensions A of cellulose in aq. alkaline solvents against temperature T.[16] (O) and solid line: 8% NaOH aq. solution; (●): 6% LiOH aq. solution.[24]

Putting the values in eqs. (3.20.10–3.20.12) into the right-hand side of eq. (3.20.4), we obtain -8.90×10^{-1} which agrees fairly well with the experimental $d\ln[\eta]/dT$ value (-7.42×10^{-3}). Inspection of eqs. (3.20.10–3.20.12) shows that the temperature dependence of the unperturbed chain dimensions is a main factor governing the negative temperature coefficient of $[\eta]$ for cellulose/aq. NaOH, as was found for the CDA-acetone system.

The remarkable decrease in $[\eta]$ at the temperature range of 35–41 °C can also be explained by the large negative temperature dependence of the unperturbed chain dimensions. Kamide and Saito[49] indicated that there exists a strong correlation between the number of solvent molecules solvated per a pyranose ring at infinite dilution s_0 and A for the CDA-acetone system (see Figure 3.17.3) and the solvation makes the cellulose chain rigid. The negative temperature dependence of A, observed also for cellulose/aq. NaOH system, can be speculated to be due to the negative temperature dependence of the solvation.

REFERENCES

1. K Kamide and Y Miyazaki, *Polym. J.*, 1978, **10**, 409.
2. K Kamide and Y Miyazaki, *Polym. J.*, 1978, **10**, 539.
3. K Kamide and T Terakawa, *Polym. J.*, 1978, **10**, 559.
4. S Ishida, H Komatsu, T Terakawa, Y Miyazaki and K Kamide, *Mem. Facult. Tech., Kanazawa Univ.*, 1979, **12**, 103.
5. K Kamide, T Terakawa and Y Miyazaki, *Polym. J.*, 1979, **11**, 285.
6. K Kamide, Y Miyazaki and T Abe, *Polym. J.*, 1979, **11**, 523.
7. K Kamide and M Saito, *Eur. Polym. J.*, 1981, **17**, 1049.
8. K Kamide, M Saito and T Abe, *Polym. J.*, 1981, **13**, 421.
9. K Kishino, T Kawai, T Nose, M Saito and K Kamide, *Eur. Polym. J.*, 1981, **17**, 623.
10. K Kamide and M Saito, *Polym. J.*, 1982, **14**, 517.
11. S Ishida, H Komatsu, H Katoh, M Saito, Y Yamazaki and K Kamide, *Makromol. Chem.*, 1982, **183**, 3075.
12. K Kamide and M Saito, *Eur. Polym. J.*, 1982, **18**, 661.
13. K Kamide and M Saito, *Eur. Polym. J.*, 1983, **19**, 507.

14. M Saito, *Polym. J.*, 1983, **15**, 249.
15. K Kamide and M Saito, *Polym. J.*, 1986, **18**, 569.
16. K Kamide, M Saito and K Kowsaka, *Polym. J.*, 1987, **19**, 1173.
17. WR Krigbaum and LH Spurling, *J. Chem.*, 1960, **64**, 99.
18. VP Shanbhag and J Öhman, *Ark. Kemi.*, 1968, **29**, 163.
19. H Suzuki, Y Muraoka, M Saito and K Kamide, *Eur. Polym. J.*, 1982, **18**, 831.
20. DA Clibbins and A Creake, *J. Text. Inst.*, 1928, **19**, T77.
21. F Howlett, E Minshall and AR Urquhart, *J. Text. Inst.*, 1944, **35**, T133.
22. L Mandelkern and PJ Flory, *J. Am. Chem. Soc.*, 1952, **74**, 2517.
23. RSJ Manley, *Ark. Kemi.*, 1956, **9**, 519.
24. PJ Flory, OK Spurr Jr and DK Carpenter, *J. Polym. Sci.*, 1958, **27**, 231.
25. WR Moore and AM Brown, *J. Colloid Sci.*, 1959, **14**, 1.
26. WR Moore and AM Brown, *J. Colloid Sci.*, 1959, **14**, 343.
27. WR Moore and GD Edge, *J. Polym. Sci.*, 1960, **47**, 469.
28. K Kamide, K Ohno and T Kawai, *Kobunshi Kagaku*, 1963, **20**, 151.
29. W Brown, D Henley and J Öhman, *Makromol. Chem.*, 1963, **64**, 49.
30. VP Shanbhag, *Ark. Kemi.*, 1968, **29**, 1.
31. K Kamide, *Kobunshi Kagaku*, 1964, **21**, 152.
32. H Utiyama and M Kurata, *Bull. Inst. Chem. Res., Kyoto Univ.*, 1962, **42**, 128.
33. H Suzuki, Y Miyazaki and K Kamide, *Eur. Polym. J.*, 1980, **16**, 703.
34. W Brown and R Wikström, *Eur. Polym. J.*, 1965, **1**, 1.
35. WR Moore and J Russell, *J. Colloid Sci.*, 1953, **8**, 243.
36. WR Moore, JA Epstein, AM Brown and BM Tidswell, *J. Polym. Sci.*, 1957, **23**, 23.
37. P Howard and RS Parikh, *J. Polym. Sci.*, 1968, A-1 **6**, 537.
38. D Henley, *Ark. Kemi.*, 1961, **18**, 327.
39. K Kamide, K Okajima, T Matsui and K Kowsaka, *Polym. J.*, 1984, **16**, 858.
40. DW Tanner and GC Berry, *J. Polym. Sci., Polym. Phys. Ed.*, 1974, **12**, 941.
41. RS Stein and P Doty, *J. Am. Chem. Soc.*, 1946, **68**, 159.
42. MB Huglin, in *Light Scattering from Polymer Solutions* (ed. MB Huglin), Academic Press, London, 1972, Chapter 6, p. 192.
43. PJ Flory, *Statistical Mechanics of Chain Polymers*, Wiley, New York, 1969, p. 45.
44. GC Berry, *J. Chem. Phys.*, 1966, **44**, 4550.
45. H Yamakawa, *Modern Theory of Polymer Solutions*, Harper & Row, New York, 1971, p. 382.
46. TA Orofino and PJ Flory, *J. Chem. Phys.*, 1957, **26**, 1067.
47. M Kurata and WH Stockmayer, *Fortschr. Hochpolym. Forsch.*, 1963, **3**, 196.
48. WH Stockmayer and M Fixman, *J. Polym. Sci.*, 1963, C1 **1**, 37.
49. K Kamide and M Saito, *Eur. Polym. J.*, 1984, **20**, 903.

3.21   PHASE SEPARATION: CLOUD POINT CURVE (CPC) AND FLORY THETA SOLVENT

3.21.1   Theoretical background[1]

The Flory temperature θ (the critical miscibility temperature) and entropy parameter ϕ are experimentally determined from the critical solution point (CSP; temperature T_c and concentration v_p^c), using the Schultz-Flory (SF) method[2] for a polymer of infinite molecular weight through use of eqs. (3.21.1) and (3.21.2).

$$v_p^c = \frac{1}{1 + X_w^{1/2}}$$

(3.21.1)

$$1/T_{c} = 1/\{\,\theta\phi(1/X_{w}^{1/2} + 1/2X_{w})\} + 1/\theta \tag{3.21.2}$$

where X is the relative molecular volume ratio of the polymer to the solvent.

Note that eq. (3.21.2) is valid only for a monodisperse polymer/solvent system, in which the polymer–solvent thermodynamic interaction parameter χ (eq. (3.3.2)) is assumed to be independent of the molecular weight and concentration. Although being only approximate, the SF method has been used widely. For all polymer-solvent systems investigated, the SF method gave θ temperature, which agrees fairly well with that by the second virial coefficient method (that is, the temperature at which A_2 vanishes), but ϕ values estimated by the SF method are significantly larger than those by the A_2 method. Stockmayer pointed out the inadequacy of the basic (equation (3.21.2)) of the SF theory, proposing an alternative equation,[3]

$$v_{p}^{c} = 1/[1 + (X_{w}/X_{z}^{1/2})] \tag{3.21.3}$$

(X_{w} and X_{z} are the weight-average, and z-average X). Note that eq. (3.21.3) was rigorously derived for polydisperse polymer solutions.

Another factor that plays a much more important role is the concentration dependence of χ parameter, which was completely neglected in the SF and Stockmayer methods. To solve this problem, Koningsveld et $al.$ (KKR)[4] proposed a method for estimating θ and ϕ from data on critical concentration v_{p}^{c} and critical temperature T_{c} for a series of solutions differing in average molecular weight.[4] They employed a pair interaction parameter g, expanding as a series function of the polymer volume fraction v_{p}. They (KKR) assumed that the concentration independent parameter g_0 could be divided into the temperature-independent and -dependent components given by

$$g_{0} = g_{00} + g_{01}/T \tag{3.21.4}$$

Note that all parameters were evaluated such that the deviation of the experimental v_{p}^{c} and T_{c} data for all samples from the theory would be minimal. In their method, the curve fitting method was repeatedly used to evaluate g_{i}.

Kamide and Matsuda (KM)[5] proposed a method for estimating θ and ϕ from T_{c} alone, by taking into account the above two factors, independently determined by other methods or with v_{p}^{c} data, without assuming a specific temperature dependence of the χ-parameter. They (KM) showed that the concentration dependent parameters p_1 and p_2 in eq. (3.21.5):

$$\chi = \chi_{0}(1 + p_{1}v_{p} + p_{2}v_{p}^{2}) \tag{3.21.5}$$

(χ_0, concentration-independent parameter and v_{p}, polymer volume fraction) can be decided by comparing experimental v_{p}^{c} (v_{p}^{c}(exp)) and theoretical v_{p}^{c} (v_{p}^{c}(theo)), calculated on the basis of the neutral and spinodal conditions for the system, and that the critical χ_0, χ_0^{c}, which can be concurrently obtained, is related to $1/T_{c}$ through

$$1/T_{c} = (\chi_{0}^{c}/\theta\phi) + \theta^{-1}\{1 - (2\phi)^{-1}\} \tag{3.21.6}$$

Then, we can determine θ and ϕ from the plots of T_{c} (experimental) and χ_0^{c} (calculated from v_{p}^{c}(exp) using the KM method).

From a theoretical viewpoint, the concentration dependence of χ parameter (p_1, p_2) can also be determined by the osmotic pressure, vapor pressure, isothermal distillation, ultracentrifuge, phase equilibrium (the two phase volume ratio R, the partition coefficient σ, the polydispersity of the polymers in two phases, etc.), and the cloud point.[6] Except for the phase equilibrium and cloud point, all methods are experimentally limited to a relatively lower concentration range and do not enable us to evaluate p_2 accurately. p_1 and p_2 were determined by the KKS and KM methods using literature data of CPS for PS in 10 solvents and polyethylene in 16 solvents. Both the KKS and KM methods gave almost the same value[6] of p_1 for a given polymer/solvent system. Except for few solvents, the p_1 value for PS solutions can be regarded as constant, which is close to 2/3, which is theoretically predicted when $A_2 = A_3 = 0$ at θ (A_3 is the third-virial coefficient). ϕ estimated by KM method coincides with that by KKS method for a given polymer solvent system.

The Flory enthalpy parameter at infinite dilution κ_0 can be evaluated by:[7]

(1)   temperature dependence of vapor pressure, membrane osmotic pressure,
(2)   critical solution temperature T_c and critical solution concentration v_p^c for a series of solution of polymer (through use of the SF, Stockmayer, KKS, and KM treatments),

$$\kappa_0 = \phi_0 \theta / T \qquad (\phi_0 \text{ is } \phi \text{ at infinite dilution}) \qquad (3.21.7)$$

(3)   temperature dependence of the second virial coefficient in the vicinity of the Flory theta temperature θ (eq. (3.20.2)), and
(4)   calorimetry.

For literature data on polystyrene/cyclohexane systems, the above methods were applied.[7] Excellent agreement (± 0.02) was confirmed between κ_0 values at θ deduced by various methods if the KKS and KM methods are utilized and when M_w (M_n) of the polymer is the same. The SF and Stockmayer treatments overestimate ϕ values.

A theoretical method for calculating CPC and CPS of multicomponent polymer single solvent system was established by Kamide et al.,[8,9] considering the polydispersity of the polymer and the concentration and molecular weight dependences of the χ parameter.

Figure 3.21.1 illustrates the effect of the weight-average X of the polymer (X_w^0) on CPC for four polymers in a single solvent ($p_1 = 0.6$, $p_2 = 0$, and the molecular weight dependence parameter $\kappa_0 = 0$).[8] The unfilled circle in the figure is the critical point and the filled circle, the threshold cloud point. The critical point, independently calculated by a different procedure from CPC, lies just on the CPC. This verifies the reliability of the CPC calculation procedure used here. A monodisperse polymer solution with low X_w^0 shows CPC with approximate symmetry, which disappears with increasing X_w^0 and X_w^0/X_n^0 (here, the suffix 0 means the original sample). As X_w^0 increases, CPC shifts to lower v_p and higher T regions, showing skewness in the former. The critical points (the critical temperature T_c and the critical concentration v_p^c) coincides very well as expected from the threshold points (T_{tcp} and $v_{p,tcp}$) for solutions of strictly monodisperse polymers. As the polydispersity of the sample increases, the critical point moves to higher v_p region, but the threshold points shifts to lower v_p region causing them to differ significantly.

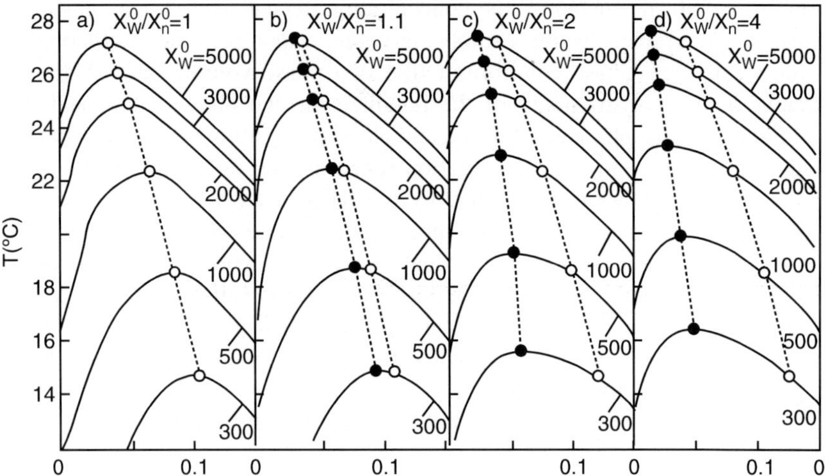

Figure 3.21.1   Effect of the weight average molar volume ratio X_w^0 of the polymer in a single solvent on cloud point curve ($p_1 = 0.5, p_2 = 0, k_0 = 0$)[8]: X_w^0/X_n^0. 1 (a), 1.1 (b), 2 (c), and 4 (d); ($\bigcirc$), critical point; ($\bullet$), precipitation threshold point; number on curve denotes X_w^0.

The deviation in the critical point from the threshold cloud point increases remarkably with an increase in X_w^0/X_n^0 and slightly with a decrease in X_w^0. Note that this difference is significant even for a polymer with $X_w^0/X_n^0 = 1.1$, which is often regarded as 'monodisperse polymer'. Very careful attention should be paid to this point in the case of low X_w^0 polymer. We examined two methods proposed by Kamide *et al.*[6] and Solč[10] for determining CPC of a quasibinary system consisting of a multicomponent polymer with the same chemical composition dissolved in a single solvent.[11] Kamide's procedures of direct integration employed were considered more general than the polynomial expansion procedure in the Solč method. Thus, the former method has the advantage of accurate calculation of the CPC in the low polymer concentration region.

3.21.2   Cellulose acetate/acetone and 2-butanone

Experiments[12,13]

For lower critical solution point (LCSP) measurements, the solution tubes were slowly warmed (*ca.* 1 °C/10 min) while stirring in an electric thermostat of the specially designed air bath type. The main body of the apparatus is made of copper block (diameter × height = 15 cm × 40 cm) with a hole at the center, in which the tube and its case can be situated. The copper black was heated by means of four thermocoax heaters imbedded therein. The upper critical solution point (UCSP) measurements were carried out by using a carbon dioxide-methanol bath over the temperature greater than −78 °C and liquid nitrogen at its boiling point. Cloud point was recorded as the temperature at which the solution became heavily turbid. However, in a CA (DS 2.46)-2-butanone system, any change in the solution transparency was very subtle, so that the determination of the cloud point for the system was liable to be less accurate.

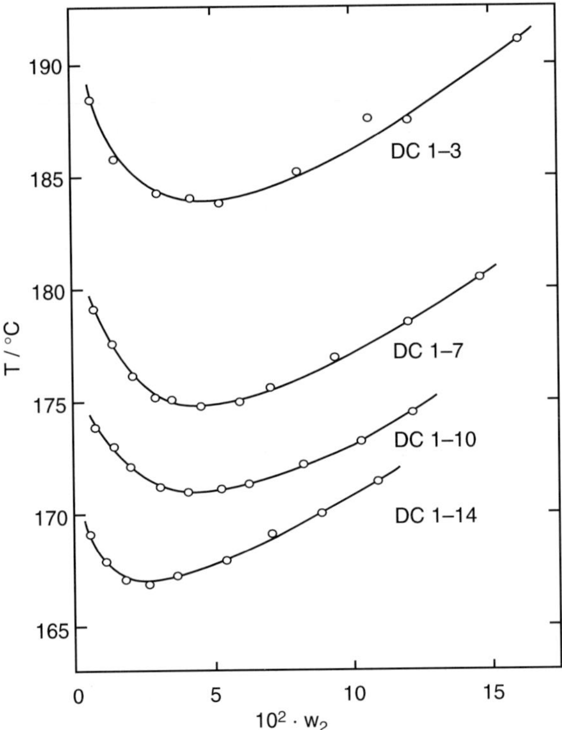

Figure 3.21.2   The phase diagram (temperature, weight fraction) for the cellulose diacetate (DS 2.46)/acetone system showing lower critical solution temperature for the fractions of indicated sample code.[14]

CPC

Figure 3.21.2[14] and Figure 3.21.3[13] depict the CPC (i.e. the temperature of cloud point versus the weight fraction w_2 of polymer) for each solution. For an example, in CA (DS 2.46)/acetone system, cloud points ranged from 166.9 to 191.0 °C, depending upon the molecular weight of a sample and the concentration of its solution. Naturally the cloud points for a lower molecular weight fraction were higher than for a higher molecular weight fraction.

The lower critical solution temperature (LCST) and upper critical temperature (UCST) were determined as the minimum and maximum temperatures (i.e. threshold temperatures) of each cloud point curve. Strictly speaking, these CST values are only apparent (see Figure 3.21.1), and the corrections for both polydispersity and pressure are necessary. CA (DS 2.46)-2-butanone system shows the existence of UCST, as well as LCST (Figure 3.22.3).

Table 3.21.1[12–14] shows the LCST and the UCST as determined from Figures 3.21.2 and 3.21.3 for CA (DS 2.46) solutions. In the table, the result of a preliminary experiment for whole polymer[12] is included.

The SF plots, based on eq. (3.21.1) for CA (DS 2.46) in acetone and 2-butanone, are shown in Figures 3.21.4 and 3.21.5, respectively. SF plots can be represented by linear

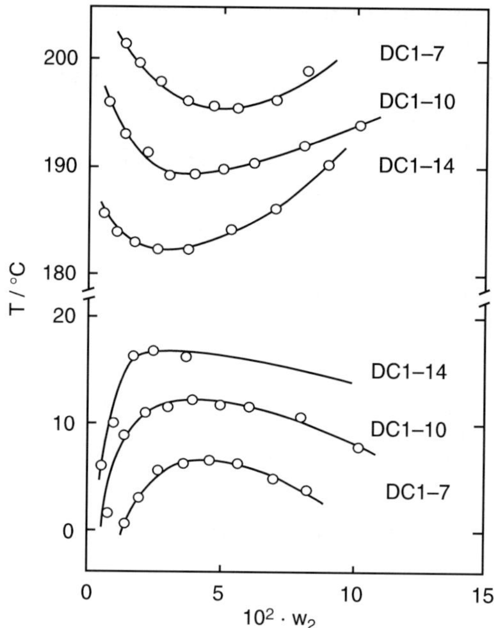

Figure 3.21.3   Plots of the cloud point against the weight fraction of cellulose diacetate for the fractionated samples in 2-butanone.[13]

line. We obtain $\theta_L = 428$ K and $\phi_L = -0.94$ for the former system (where the subscript L denotes LCST), and for the latter system $\theta_L = 433$ K, $\phi_L = -0.54$, $\theta_u = 310$ K, and $\phi_{1,u} = 0.34$ (where the subscript u denotes UCST).

Figures 3.21.2 and 3.21.3 were reanalyzed by Kamide and Matsuda according to the SF, KKR, and KM methods.[15]

Table 3.21.1

Critical solution temperatures for cellulose acetate (DS 2.46) in acetone and 2-butanone

Sample code	M_w ($\times 10^{-4}$)	Acetone[a]	2-Butanone[b]	
		LCST(K)	LCST(K)	UCST(K)
DC1-3	3.76[c]	457.1	–	–
DC1-7	7.55[c]	448.0	471.5	279.7
DC1-10	11.1[c] (10.6)[d]	444.2	465.0	284.5
DC1-14	17.5[c] (18.5)[d]	440.1	457.9	290.0
(whole polymer)	(12.0)[d]	438.3[e]	–	–

[a]Ref. 14.
[b]Ref. 13.
[c]By CPC.
[d]By LS.
[e]Ref. 12.

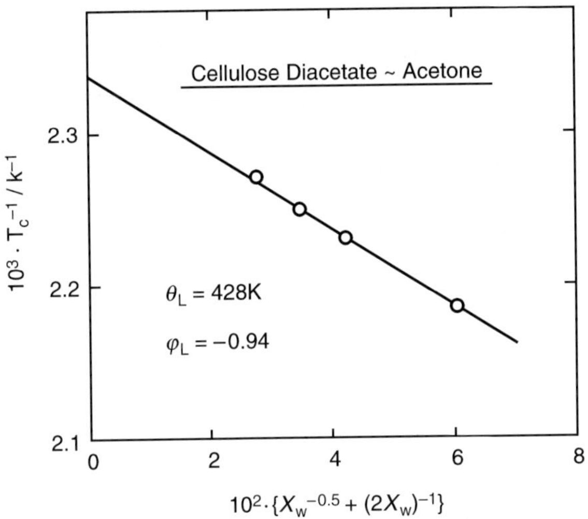

Figure 3.21.4   The Shultz–Flory plots of CA (DS 2.46)/acetone.[12]

The results are summarized in Table 3.21.2.[15] In the table, methyl ethyl ketone is an alternative notation for 2-butanone. All of these methods give almost the same θ temperature for each CA solution: For UCSP of CA (DS 2.46)-2-butanone system, $\theta = 312–313$ K is estimated by KKR and KM methods, which is about 10 °C lower than the θ value evaluated by the second virial coefficient method (*ca.* 323 K).

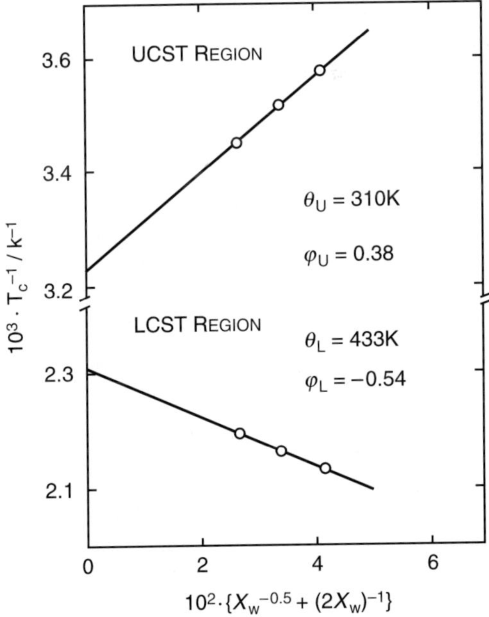

Figure 3.21.5   The Schultz–Flory plots of CA (DS 2.46)-2-butanone.[13]

Table 3.21.2

Concentration-dependent parameters p_1 and p_2 in the χ parameter, Flory's theta temperature θ, and the entropy parameter ϕ of cellulose diacetate, determined by different methods[15]

Solvent	UCSP or LCSP	Kamide and Matsuda[5]				Koningsveld et al.[4]				Shultz and Flory[2]	
		p_1	p_2	θ	ϕ	p_1	p_2	θ	ϕ	θ	ϕ
Acetone	LCSP	0.96	-21.4	433.1	-1.03	0.203	-11.78	429.7	-1.01	427.1	-0.78
Methyl-ethyl-ketone	UCSP	-0.67	2.52	312.2	0.47	-0.850	4.40	313.5	0.46	311.4	0.35
	LCSP	-0.96	7.38	432.0	-0.69	-1.170	9.56	430.7	-0.68	434.6	-0.51

Table 3.21.3

θ solvents reported on cellulose derivatives

Polymer $\langle F \rangle$	Molecular weight ($\times 10^{-4}$)	Solvent	Temperature (°C) UCST or LCST	Method	Reference
CDA[a]	–	Tetrachloroethane	56.6 (UCST)	A_2 of MO	17
CA (2.46)	6.9	2-Butanone	50 (UCST)	A_2 of LS	18
CTTCp[b]	13.2	Dimetylformamide	41 (UCST)	A_2 of MO	19
	20.6	Dioxane/water	43 (UCST)	A_2 of MO	
CTC[c]	43–217	Anisole	94 (UCST)	A_2 of LS	20
	43–217	Cyclohexanol	73 (UCST)	A_2 of LS	

[a]CDA, cellulose diacetate; $\langle F \rangle$ and molecular weight of the sample used were not indicated.

[b]Cellulose tricaproate.

[c]Cellulose tricarbanilate.

Low experimental accuracy of the latter method should be considered. ϕ values by KM and KKR methods coincides with each other, but differs significantly from the ϕ values by SF method, although the difference is not so remarkable as compared with in the case of vinyl-type polymer solutions: $\phi < 0$ for LCSP and $\phi > 0$ for UCSP are observed.

3.21.3 Flory theta solvents of cellulose derivatives

Table 3.21.3 shows some Flory theta solvents on cellulose derivatives.[16] According to the literature, no theta solvents are known for CDA except DCM[17] and 2-butanone.[18] Later, Kamide *et al.* disclosed that DCM was not a successful theta solvent because the CDA molecules did not dissolve molecularly at the reported theta temperature.[21]

REFERENCES

1. K Kamide and T Dobashi, *Physical Chemistry of Polymer Solution: Theoretical Background*, Problems 4-23a, Problems 4-23c, Elsevier, Amsterdam, 2000, pp. 157–172.
2. AR Shultz and PJ Flory, *J. Am. Chem. Soc.*, 1952, **74**, 4760.
3. WH Stockmayer, *J. Chem. Phys.*, 1949, **17**, 588.
4. R Konningsveld, LA Kleintjens and AR Shultz, *J. Polym. Sci.*, 1970, **8A-1**, 1261.
5. K Kamide and S Matsuda, *Polym. J.*, 1984, **16**, 825.
6. K Kamide, S Matsuda and M Saito, *Polym. J.*, 1985, **17**, 1013.
7. K Kamide, S Matsuda and M Saito, *Polym. J.*, 1988, **20**, 31.
8. K Kamide, S Matsuda, T Dobashi and M Kaneko, *Polym. J.*, 1984, **16**, 839.
9. K Kamide and T Dobashi, *Physical Chemistry of Polymer Solution: Theoretical Background*, Problem 4-21a, Problem 4-21d, Elsevier, Amsterdam, 2000, pp. 146–155.
10. K Solč, *Macromolecules*, 1970, **3**, 665.
11. K Kamide, S Matsuda and H Shirataki, *Eur. Polym. J.*, 1990, **26**, 379.
12. H Suzuki, K Ohno, K Kamide and Y Miyazaki, *Netsusokutei*, 1981, **8**, 67.
13. H Suzuki, M Muraoka, M Saito and K Kamide, *Br. Polym. J.*, 1982, **14**, 23.
14. H Suzuki, K Kamide and M Saitoh, *Eur. Polym. J.*, 1982, **18**, 123.

15. K Kamide and M Saito, *Adv. Polym. Sci.*, 1987, **82**, 1.
16. K Kamide, in *Wood and Cellulose*, **Vol. 17**, (eds DN-S Hon and N Shiraishi), 1990, Characterization of Chemically Modified Cellulose, Marcel Decker, New York, p. 801.
17. T Ikeda and H Kawaguchi, *Rep. Prog. Polym. Sci. Jpn.*, 1966, **9**, 23.
18. H Suzuki, M Muraoka, M Saito and K Kamide, *Br. Polym. J.*, 1982, **18**, 837.
19. WR Krigbaum and LH Sperling, *J. Phys. Chem.*, 1960, **64**, 99.
20. VP Shanbhang and J Öhman, *Ark. Kemi.*, 1968, **29**, 163.
21. K Kamide, T Terakawa and Y Miyazaki, *Polym. J.*, 1979, **11**, 285.

3.22 CONCLUDING REMARKS ON MOLECULAR PROPERTIES OF CELLULOSE CHAINS

Through a comprehensive experimental and analytical survey (3.1–3.21), the following findings have been confirmed as typical characteristics of cellulose and cellulose derivatives, especially cellulose acetate, in solutions.

1. Successive solutional fractionation (SSF; Figure 3.2.1) can be applied successfully to CA (CA) with DS $= 0.49$–2.92 to give a series of fractions with the breadth of molecular weight fractionation $M_w/M_n = 1.1 - 1.5$ (Table 3.2.2), which is practically independent of M_w CA (DS 0.49) in Table 3.3.1; CA (DS 1.75) in Table 3.3.2; CA (DS 2.46) in Tables 3.2.5, 3.3.3, Figure 3.2.4; CA (DS 2.92) in Table 3.3.5.

2. SSF is expected to be applicable to other cellulose derivatives than CA although its applicability to cellulose itself has not yet been verified (see 3.1.5a).

3. The number average molecular weight M_n of CA can be determined by membrane osmometry, VPO, and gel permeation chromatography. The values of M_n by the above three methods coincide with each other; CA (DS 2.46) in THF (Table 3.3.4).

4. M_n values of the CA samples in various solvents are the same; CA (DS 2.92) in DMAc, acetone, TCE, and TCM (Table 3.3.6).

5. The method for preparing optically clean solutions, which enable us to carry out LS measurement, was established. Then, LS can be applied to CA with DS $= 0.49$– 2.92 in water, acetone, THF, and DMAc; CA (DS 0.49) in Figure 3.5.1 and Table 3.3.1; CA (DS 1.79) in Table 3.3.2; CA (DS 2.46) in Figure 3.5.2 and Table 3.3.3 CN (DS 2.3) in acetone in Figure 3.5.3; NaCS in aq. NaCl (Table 3.3.7); and cellulose in aq. LiOH and aq. NaOH (Figure 3.5.4), respectively.

6. The weight-average molecular weight M_w values by LS of CA are the same in various solvents; CA (DS 0.49) in formamide, water, and DMAc (Table 3.5.1); CA (DS 2.46) in DMAc, acetone, and THF (Tables 3.3.3 and 3.5.2).

7. The reliability of LS was ascertained by measuring M_w of CA in solutions by different individuals, using different apparatus at different institutions under different operating conditions (CA (DS 2.46) in acetone and THF (Table 3.5.3)).

8. M_w value of CA (DS 2.46) remained constant over the temperature range 25.4– 50.3 °C in acetone (Tables 3.5.3 and 3.20.1) and 30–60 °C in 2-butanone (Table 3.20.2). M_w of cellulose in 8% NaOH is constant in the range of 3.5–45 °C (Table 3.5.8 and Figure 3.20.4).

Conclusions 3, 6, 7, and 8 indicate that CA and cellulose molecules dissolve in various solvents molecularly; in other words, CA and cellulose solutions studied here are molecular solutions.

9. The O-Ac proton chemical shift of CA (DS 0.49, 2.46, and 2.92) exhibits a linear correlation with dielectric constant ε of the solvent (Figure 3.9.4 and Table 3.9.1).

10. The exponent a in the MHS equation (eq. (3.11.1)), in which M_w by LS is used, is in the range of 0.6–0.82 for CA solutions (Table 3.11.2). All the literature data (other than those by Kamide *et al.*) for the CTA solutions fall in $a = 0.80-1.02$ (Table 3.11.1). The discrepancy can be explained by the effect of MWD of the breadth of fractions' MWD, which is the fatal disadvantage of SPF used for sample preparation in all literature other than those by Kamide *et al.* on the parameters K_m and a of the MHS equations, in which M_n was employed, and not by the intrinsic properties of cellulose chain (Figures 3.11.5 and 3.11e).

11. MHS equation of cellulose/aq. LiOH almost superposes with that of the cellulose/cadoxen system (Figure 3.11.6).

12. The contribution of the excluded volume effect a_1 to the exponent a is small (*ca.* 0.1 for CA (DS 2.46) and 0.15 for CA (DS 2.92)) as compared with the contribution of the draining effect a_Φ (Tables 3.15.1 and 3.16.5).

13. The radius of gyration $\langle S^2 \rangle_z^{1/2}$ of CA and cellulose in solvents is approximately proportional to $M_w^{1/2}$ with exception of CA (DS 2.46)/acetone (Table 3.12.1). The CA chains in solution do not deviate to a great extent from Gaussian chains.

14. The second virial coefficient A_2 of CA and CN solutions is generally in the order of 10^{-4} (mol cm^3 g^{-2}; see Section 3.13). The experimental A_2 data of CA and CN solutions can be reproduced in terms of the short-range and long-range interactions by methods 2J and 2G (Figure 3.13.4 for CN, Figure 3.13.1 and 3.13.2 for CA (DS 2.46)).

15. The expansion coefficient, α_s, evaluated through use of the penetration function Ψ from A_2, M_w and $\langle S^2 \rangle_z$ data, is roughly less than 1.2 (Tables 3.13.2, and 3.15.1–3.15.3), irrespective of all the theories connecting α_s with the excluded volume parameter z presented hitherto such as Flory, modified Flory, modified Flory–Krigbaum–Orofino, Zimm–Stockmayer–Fixman, Yamakawa–Tanaka, Kurata–Fukatsu–Sotobayashi–Yamakawa (KFSY-I and KFSY-II), Cassassa–Morravitz.

16. Conclusion 15 above on α_s does not change even if the worm-like chain model is employed in place of the pearl necklace model. That is, the expansion coefficient of the worm-like chain model $\alpha_s^w < 1.2$.

17. The molecular weight dependence of sedimentation coefficient at infinite dilution s_0 of CA and CN solutions can be expressed by eq. (3.14.1) (see Figure 3.14.1).

18. The Flory viscosity parameter Φ of all the cellulose derivatives solutions is dependent on M_w and expressed by eq. (3.15.3) (Figures 3.15.1–3.15.4 and 3.15.7). Φ is by far below an asymptotic value theoretically predicted.

19. The Flory parameter P, defined as analogous to Φ (eq. (3.15.4)), of cellulose derivative solutions also depends on M_w (Figure 3.15.5, Table 3.15.5).

Conclusions 18 and 19 indicate that in cellulose and cellulose derivative solutions Φ should not be taken as constant.

20. The validity of methods 1A and 1D for estimating the draining effect was ascertained with some typical vinyl-type polymers (Table 3.15.5).

21. The draining parameter X, (eq. (3.15.7)), evaluated for cellulose derivative solutions, is less than five with some exceptions, including cellulose/cadoxen (method 1A), CN ($N_c = 13.6\%$)/acetone (method 1A), CA (DS 2.46)/THF (method 1A), and NaCMC/ aq. NaCl (method 1A; Tables 3.3.5 and 3.15.6–3.15.9). This means that in cellulose derivative solutions, the draining effect can never be ignored on the hydrodynamic properties. That is, cellulose derivative molecules are semipermeable chains.

22. Thermodynamic (methods 2A–2D) and hydrodynamic (methods 2E–2L) approaches for estimating the UCD A were examined in detail (Tables 3.16.1a,b and 3.16.3, Figures 3.16.1–3.16.18). Method 2E and 2F, derived by neglecting the draining effect, yield the A value, which is significantly smaller than that estimated by the thermodynamic methods (2A–2D). The A values evaluated by methods 2G, 2J, 2K, and 2L, where the draining effect is taken into account, agree with those by methods 2A, 2B, 2C, and 2D (Table 3.16.8 and Figure 3.16.20).

23. The Stockmayer–Fixman plot (method 2E), the most extensively employed hitherto, and its analogous plots,[1] including Flory–Fox–Schafgen, Kurata–Stockmayer– Burchard, Bodanecky, and Inagaki–Suzuki–Kurata[1] (see Ref. 2), underestimate the A value of cellulose and cellulose derivative solutions. Erroneous application of the above methods to the cellulose derivatives solutions leads to the conclusion that cellulose (derivative) molecules are flexible and the solvents for these polymers are good. If the molecular weight dependence of Φ is properly taken into account in the theory, then the hydrodynamic approach (2G, 2J, 2L) agrees fairly well with the thermodynamic approaches (2A, 2B, 2C, or 2D). Then, the contradiction pointed out hitherto is completely dissolved.

24. The most probable values of A (A_m) and conformation parameter σ for cellulose and cellulose derivatives are tabulated (Table 3.16.9).

25. The number of solvated solvent molecules at infinite dilution per repeating pyranose ring unit, estimated from adiabatic compressibility measurements, s_0 is larger in the solvent with larger dielectric constant ε (Table 3.17.1). s_0 of CA (DS 2.46) in acetone is almost twice of the predicted one from an empirical $s_0 - \varepsilon$ relationships for other solvents (Figure 3.17.2).

26. With an increase in s_0, both A and σ increase (Figures 3.17.3 and 3.18.5) and X decreases (Figure 3.17.5). A and s_0 decrease with rising temperature (Figure 3.17.4 and Table 3.20.10). This means that the unperturbed chain expands with solvation and concurrently the solvated cellulose chain becomes semipermeable (draining). The important role of the solvation in A and X should be considered.

27. s attains maximum at $\langle\!\langle F \rangle\!\rangle \approx 2.5$ for CA/DMAc and CA/DMSO at $25\,°C$ (Figure 3.18.1).

28. The unperturbed persistence length q_{CL}^0 value at coil limit, evaluated from A for cellulose derivatives on the pearl necklace model, is parallel to the unperturbed persistence length q on the worm-like chain model (Figure 3.19.2). The persistence length evaluated using BD theory, q_{BD} and that by Yamakawa–Fujii theory, q_{YF} exhibit a profound negative correlation versus X (Figure 3.19.3).

29. The q value, determined by the LS and the SAXS of CA in DMAc revealed maximum at $\langle\!\langle F \rangle\!\rangle \approx 2.5$ (Figure 3.19.4). The value of q by LS for CN in acetone agrees with that by SAXS. However, for the above solution q, estimated by the KS viscosity plot is less than 25% of q estimated by LS and SAXS (Table 3.19.2).

This also suggests the inapplicability of the KS plot to CN solutions. q of CN in acetone increases with $\langle\langle F \rangle\rangle$ (Table 3.19.2).

Conclusions 24, 25, 29, and 30 indicate that the cellulose derivative chain is not so much flexible as Kurata and Stockmayer assumed on the basis of viscosity plot alone, but is rather semiflexible.

30. $[\eta]$ of high M_w CA-DMAc solution attains maximum at $\langle\langle F \rangle\rangle = 2.5$ (Figure 3.18.7). Generally, in the CA-DMAc system, a and a_Φ reached maximum at $\langle\langle F \rangle\rangle = 2.5$ (Figure 3.18.8), but a_2 (eq. (3.15.10)) is almost zero over an entire range of $\langle\langle F \rangle\rangle$ (Figure 3.18.8).

31. The temperature dependence of $[\eta]$ of cellulose and cellulose derivative solutions is, without exception, negative (Figure 3.20.1 and 3.20.2). The temperature dependence of unperturbed chain dimensions d $\ln\langle S^2 \rangle_0^{1/2}/$d $\ln T$ (Figure 3.20.8) and that of the Flory parameter d $\ln/$d $\ln T$ (Figure 3.20.9) are the main factors controlling d $\ln[\eta]/$dT. The temperature dependence of the excluded volume effect is minor (cellulose/aq. NaOH and CA (DS 2.46)/acetone).

32. The CPC for solutions of CA (DS 2.46) fractions in acetone and in 2-butanone can be determined experimentally (Figures 3.21.2 and 3.21.3). For the former system, the LCST was observed, and for the latter system, LCST and the UCST were detected (Table 3.21.2). The data were analyzed to evaluate Flory theta temperature θ using the FS (Figsures 3.21.4 and 3.21.5), KKR, and KM theories (Table 3.21.2).

3.22.1   Cellulose chain hypothetically dissolved in nonpolar solvent

The A value of CA dissolved in a common solvent depends significantly on $\langle\langle F \rangle\rangle$ as shown in Figure 3.22.1.[1] In the figure, the numbers denote the dielectric constant ε of the solvent and the solid line represents the relationship between A in DMAc and $\langle\langle F \rangle\rangle$. A of CA molecules in DMAc is maximum at $\langle\langle F \rangle\rangle \approx 2.5$ as for s_0 (Figure 3.18.2). The A value in DMAc extrapolated to $\langle\langle F \rangle\rangle = 0$ (i.e. cellulose; filled circle) is about 13 Å and the corresponding σ is 2.25. The open rectangles are extrapolated A value to $\varepsilon = 1$ given in Figure 3.18.5. In a hypothetical nonpolar solvent ($\varepsilon = 1$), A is expected to decrease monotonically with $\langle\langle F \rangle\rangle$ (chain line), approaching the value of the freely rotating cellulose chain (broken line) in the region of $\langle\langle F \rangle\rangle \leq 0.8$. Since any known solvent for cellulose, including cadoxen and iron sodium tartrate, are highly polar or electrolytic, the solution properties of cellulose in less polar solvents have never been investigated. An intercept of A at $\varepsilon = 1$ versus $\langle\langle F \rangle\rangle$ line (the chain line) at $\langle\langle F \rangle\rangle = 0$

$$\left(\text{i.e. } \lim_{\substack{\varepsilon \to 1 \\ \langle\langle f \rangle\rangle \to 0}} A\right)$$

should be the A value of cellulose molecules dissolved in a hypothetical nonpolar solvent.

$$\lim_{\substack{\varepsilon \to 1 \\ \langle\langle f \rangle\rangle \to 0}} A$$

represents the flexibility of the pure cellulose chain without any solvation. Figure 3.22.1 shows that cellulose dissolved in a hypothetical nonpolar solvent behaves as almost a

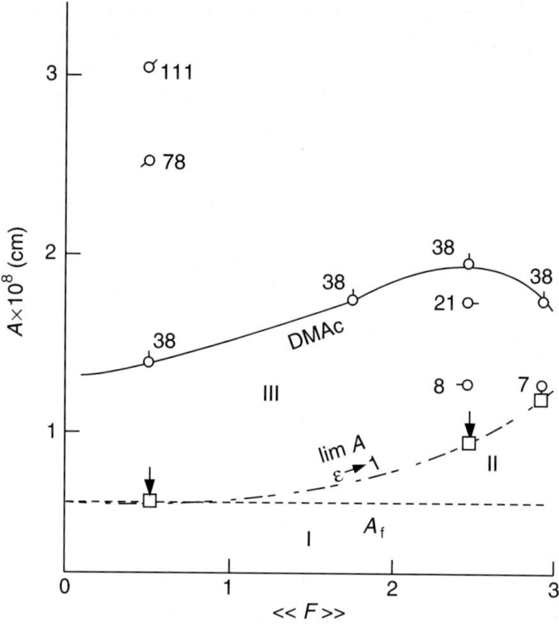

Figure 3.22.1   Plot of the unperturbed chain dimensions against the total degree of substitution for cellulose acetate-solvent systems. Solid line: CA-DMAc system; Chain line: asymptotic A at the limit of the dielectric constant $\varepsilon = 1$; broken line: the unperturbed chain dimensions of cellulose at the free rotational state A_f; (○), the asymptotic A value at the limit of $\varepsilon = 1$; (□), the asymptotic A value at the limit of $\varepsilon = 1$ and $\langle\!\langle F \rangle\!\rangle = 0$.[1]

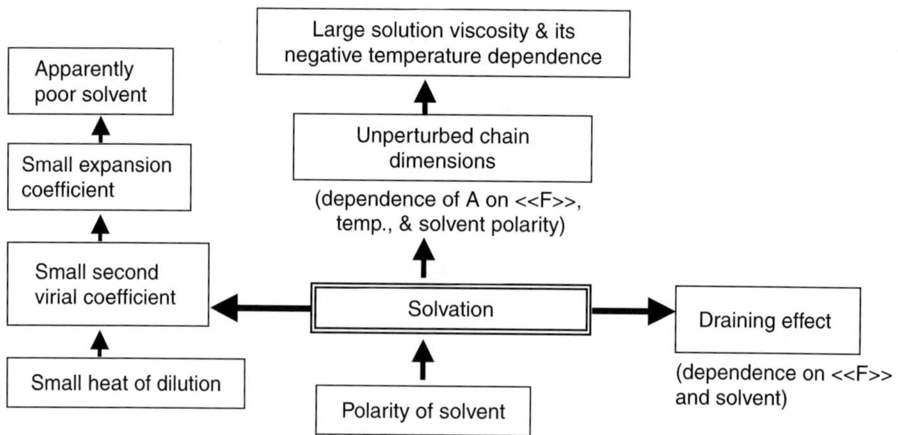

Figure 3.22.2   Correlation between solvation and some characteristic feature of cellulose derivative solutions.[4,5]

freely rotating chain; in other words, cellulose is intrinsically very flexible and the low degree of flexibility of the cellulose chain, deduced from the physical properties of cellulose solution and solid, is caused by the solvation or intra- or intermolecular hydrogen bond. The solvent dependence of A for cellulose can be estimated from Figure 3.18.5. The results are shown as closed marks in Figure 3.22.1.

CA can be considered to have the following three structural factors controlling the rigidity of the molecule. An interdependent rotational potential which controls the unperturbed dimensions in solution, by (1) the steric interactions between neighboring pyranose rings with and without substituent groups, (2) the intramolecular hydrogen bonds between C_3–OH and $O_5{}'$ (ring oxygen) and between C_6–OH and bridge oxygen, and (3) the steric hindrance of the solvated solvent molecules. In Figure 3.22.1, region I is that due to only the steric hindrance of the pyranose ring without substituent group. The existence of the intramolecular hydrogen bonds has been confirmed by Kamide and Okajima by NMR and IR methods in CA solids, but not by the NMR method in CA solutions,[3] indicating that hydrogen bonds of this kind are destroyed when CA is dissolved.

In addition, Tables 3.3.4, 3.3.6, and 3.5.1–3.5.3 reveal that CA dissolves molecularly in solvents with various ε values, giving the same M_w or M_n value, and Figure 3.18.5 shows that the flexibility of CA chain becomes larger in less polar solvents. These experimental results cannot be explained if the intermolecular hydrogen bonds are preserved more or less in CA solutions and if they contribute significantly to the rigidity of the chain. Hence, factor (2) can be ignored at least when CA solutions are concerned. Region II in Figure 3.22.1 represents qualitatively the contribution of steric hindrance due to the substituent group (*O*-Ac group), although the above three factors are not simply additive to A value. Region III can be roughly regarded as the contribution of the solvation to A because the steric effect of the solvents shown in the figure is of the same order. It can be seen that when CA is dissolved in a highly polar solvent, the solvation effect plays an important role in determining the rigidity of the CA molecule. Moore and Russell[4] assumed (without direct experimental evidence) that the degree of the destruction of the intramolecular hydrogen bonds depends on the acidity or basicity of the solvent and a basic solvent breaks the intramolecular hydrogen bonds, resulting in increase in the flexibility of CA molecules. This does not explain the dielectric dependence of the unperturbed chain dimension of the CA molecule, because the acidity or basicity of the solvent does not always correlate to the magnitude of its ε.

In summary, the characteristic feature of dilute solutions of cellulose and its derivatives, as mentioned above, can be reasonably and consistently explained by solvation (Figure 3.22.2).[5,6]

REFERENCES

1. K Kamide and M Saito, *Eur. Polym. J.*, 1984, **20**, 903.
2. K Kamide and T Dobashi, *Physical Chemistry of Polymer Solutions*, Elsevier, 2000, p541, Table 8–39.
3. K Kamide and K Okajima, unpublished results.
4. WR Moore and J Russell, *J. Colloid Sci.*, 1954, **9**, 338.
5. K Kamide and M Saito, *Adv. Polym. Sci.*, 1987, **83**, 5.
6. K Kamide and M Saito, *Adv. Polym. Mater.*, 1994, **83**, 233.

3.23   CELLULOSE LIQUID CRYSTAL

3.23.1   Formation of lyotropic liquid crystal from cellulose derivative solution in inorganic acid[1]

Recently, the formation of liquid crystalline systems from solutions in water and organic solvents of cellulose and its derivatives, including hydroxypropyl cellulose (HPC), ethyl cellulose (EC), cellulose di- and triacetates (CA) has been extensively reported by many investigators.[2–20] In Section 3.22, it was shown that the conformation parameter of CA in solution is larger in polar solvent (Figure 3.22.1), suggesting that the formation of cellulose liquid crystal of cellulose derivative in inorganic acid solvents as a high polar solvent is possible. We now turn to a consideration of this possibility.

Cellulose derivatives

Five commercial grade methyl cellulose MC samples with a total DS $\langle\!\langle F \rangle\!\rangle = 1.80$ and the viscosity- average degree of polymerization (DP) P_v ranging from 95 to 750 and two commercially available ethylcellulose (EC) samples ($\langle\!\langle F \rangle\!\rangle = 2.63$) with the solution viscosity of 43 and 87 cP in chloroform (1 wt% at 25 °C) were supplied by Kishida Chemicals Co., Japan, and were used as received. Four CA whole polymers with $\langle\!\langle F \rangle\!\rangle$ of 2.92, 2.46, 1.78, and 0.49 were synthesized in our laboratory by the methods described in previous papers.[21–25] Each CA whole polymer was fractionated using successive solution fractionation methods, already established by Kamide *et al.*[21–25] into seven fractions and the fractions with the viscosity-average DP P_v of 250 were employed for further study. A cyanoethyl cellulose (CyEC) with $\langle\!\langle F \rangle\!\rangle = 2.87$ was synthesized by reaction of alkalicellulose with acrylonitrile and four carboxyethyl/carbamoylethyl cellulose (CE/CmEC) samples with $\langle\!\langle F \rangle\!\rangle$ of 0.5–1.4 were prepared by hydrolysis of CyEC with 40 wt% aq. sodium hydroxide at 60 °C for given periods. $\langle\!\langle F \rangle\!\rangle$ of the carboxyl and carbamoylethyl group of CE/CmEC were determined by applying back titration method with 0.1 N hydrochloric acid and 0.1 N sodium hydroxide to the solution, and by elemental analysis on nitrogen. A hydroxypropyl cellulose (HPC) sample with the weight-average molecular weight M_w of 1.0×10^5 and the molecular substitution (MS) of 3, and a cellulose tripropionate (CP) sample were supplied by Scientific Product Inc., USA. Here, MS was defined as the average molar number of propylene oxide reacting with a glucose unit. Four sodium cellulose sulfate (NaCS) samples with $\langle\!\langle F \rangle\!\rangle = 1.67–2.46$, prepared by the method of Schweiger[26] as employed by Kamide *et al.*,[27] were used.

Preparation of solutions

To 0.5 g of cellulose derivative in a glass vessel, a given amount of acid was added, cooled to 0 °C, and the mixture was quickly mixed with a microspatula, and was stocked at 5 °C for 24 h.

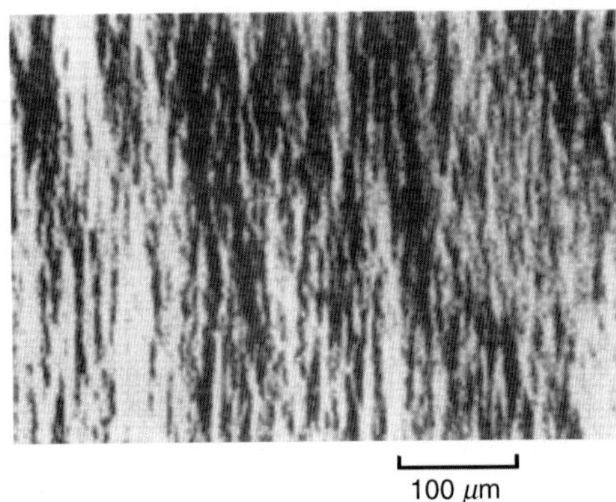

Figure 3.23.1 A photograph of a typical lyotropic liquid crystal for cellulose acetate ($\langle\!\langle\!\langle F \rangle\!\rangle\!\rangle =$ 2.46, $P_v = 250)/65$ wt% nitric acid system at 25 °C.[1] Polymer concentration, 45 wt% (×230).

Lower critical concentration

Thus prepared, 0.03 ml of the solution was inserted between a slide glass of 7.6×2.6 cm^2 (length × width) and a cover glass of 1.8×1.8 cm^2 (length × width). Shearing stresses were applied to the solution at 25 °C, and the lower critical concentration for the appearance of lyotropic liquid crystals C_L was determined 10 min after application of the shearing stresses as the minimum concentration at which the optical birefringence was observed under crossed polarizers. A typical liquid crystal system (cellulose acetate/nitric acid) was photographed with a polarizing microscope.

Figure 3.23.1 shows a typical photograph of the lyotropic liquid crystal of CA ($\langle\!\langle\!\langle F \rangle\!\rangle\!\rangle =$ 2.46, $P_v = 250)/65$ wt% nitric acid (polymer concentration, 45 wt%) taken at 25 °C.

Figure 3.23.2 illustrates an example of the viscosity of the solution η plotted as a function of polymer concentration C_p for the case of CA ($\langle\!\langle\!\langle F \rangle\!\rangle\!\rangle = 2.46$, $P_v = 250)/$ 65 wt% nitric acid at 25 °C. In this system, the lower critical concentration C_L (31 wt%),

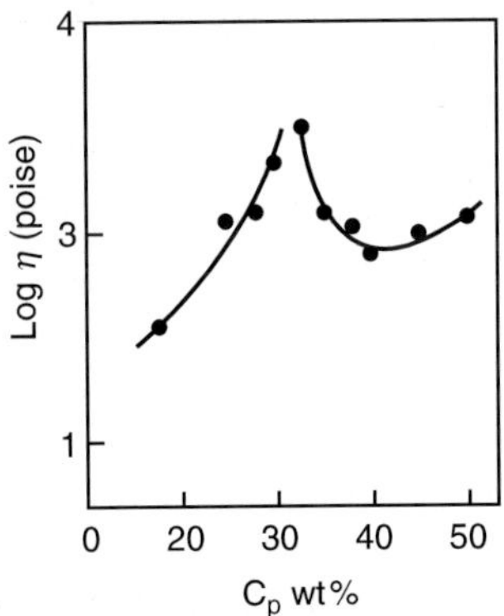

Figure 3.23.2 Plots of log η versus polymer concentration C_p (wt%) for cellulose acetate ($\langle\!\langle\!\langle F \rangle\!\rangle\!\rangle =$ 2.46, $P_v = 250)/65$ wt% nitric acid system at the shear rate of 20 s^{-1} and at 25 °C.[1]

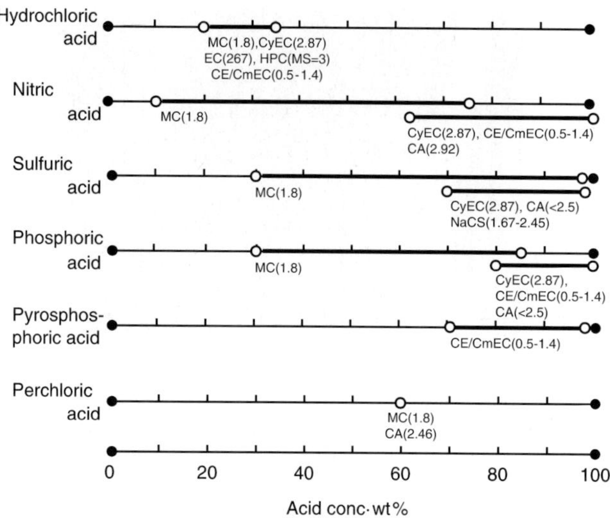

Figure 3.23.3   Schematic expression of the liquid crystal formation of cellulose derivative/inorganic acid systems. The numbers on the axis denote the concentrations of inorganic acid in water (wt%) as solvent.[1]

as determined microscopically, was found to coincide with the concentration at which viscosity of the solution began to decrease (31 wt%).

Figure 3.23.3 summarizes the suitable combination of some cellulose derivatives and inorganic acids, giving lyotropic liquid crystals. The concentration range, in which lyotropic liquid crystal is formed, is shown as the bold line ending with open circles or as a closed circle. Obviously, $\langle\langle F \rangle\rangle > 1$, as claimed by Panar et al.[6] in their open (but not registered) patent, is not the necessary condition for cellulose derivatives to form liquid crystals. All HPC (MS = 3.0), MC($\langle\langle F \rangle\rangle$ = 1.8) and CyEC ($\langle\langle F \rangle\rangle$ = 2.87) samples were found to form liquid crystals in almost all inorganic acids utilized here. However, EC ($\langle\langle F \rangle\rangle$ = 2.67) was found to become liquid crystals only in hydrochloric acid. A possible thermodynamic interaction between O-methyl group in MC and the acids is considered not to be so different from that between O-ethyl group in EC and acids. Therefore, a significant difference in the ability to form liquid crystals between MC and EC samples may be accounted for by the differences in the number of residual OH groups per given glucopyranose ring (i.e. $3 - \langle\langle F \rangle\rangle$). The difference between EC ($\langle\langle F \rangle\rangle$ = 2.67) and CyEC ($\langle\langle F \rangle\rangle$ = 2.87) cannot be interpreted by the degree of unsubstitution ($3 - \langle\langle F \rangle\rangle$), but by the difference in the polarity between cyanoethyl group in CyEC and ethyl group in EC. The former is much more polar than the latter and an interaction of cyanoethyl group with the inorganic acid is strong enough to lead finally to liquid crystals. CA with $\langle\langle F \rangle\rangle$ = 2.92 was dissolved in nitric acid and phosphoric acid to yield liquid crystals and showed flow birefringence, respectively. All CA samples with $\langle\langle F \rangle\rangle$ of less than 2.5 form liquid crystals in nitric, sulfuric, perchloric, and phosphoric acids.

Figure 3.23.4 shows the effects of the average molecular weight as denoted by P_v on C_L for MC with $\langle\langle F \rangle\rangle$ = 1.8 in 35 wt% hydrochloric, 72 wt% sulfuric, and 83 wt% phosphoric acids. C_L decreases monotonically with an increase in P_v, approaching an

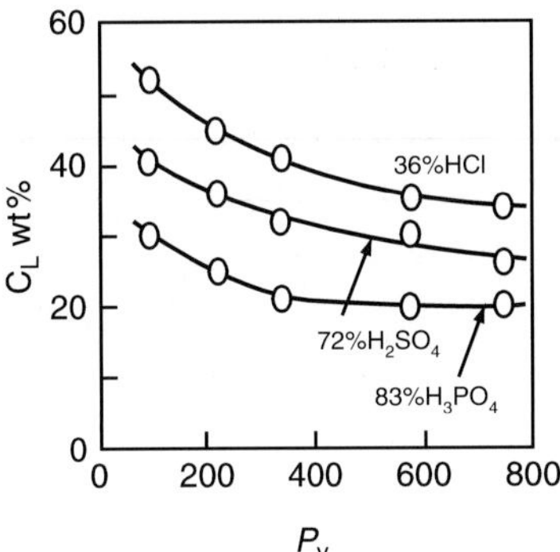

Figure 3.23.4   Dependence of the lower critical concentration C_L on degree of polymerization P_v for methylcellulose/inorganic acid systems.[1]

asymptotic value. C_L of MC with the same P_v increases with an increase in electroconductivity (EC) or with a decrease in pH of the solvent. In other words, a less acidic and less electroconductive solvent is preferable to form the liquid crystalline solution. These features are clearly illustrated in Figure 3.23.5.

Figure 3.23.6 shows the effects of $\langle\!\langle F\rangle\!\rangle$ on C_L for CA with $P_v = 250$ in 65 wt% nitric acid. The figure also contains the results (as filled marks) for CA in DMAc solution.[28] C_L decreases with increasing $\langle\!\langle F\rangle\!\rangle$ in both nitric acid and DMAc where $\langle\!\langle F\rangle\!\rangle < 2.46$. C_L for both nitric acid and DMAc reveals minimum at $\langle\!\langle F\rangle\!\rangle \approx 2.46$. Kamide and Saito demonstrated that the time averaged flexibility of CA chain attains maximum at $\langle\!\langle F\rangle\!\rangle = 2.46$ in DMAc.[29] Then, it can be said that the rigidity of the CA chains is closely correlated with C_L as was theoretically predicted by Flory.[30]

A more detailed account of the fine structure of the liquid crystals formed from acidic solution of cellulose derivatives will be given in Section 3.23.2.

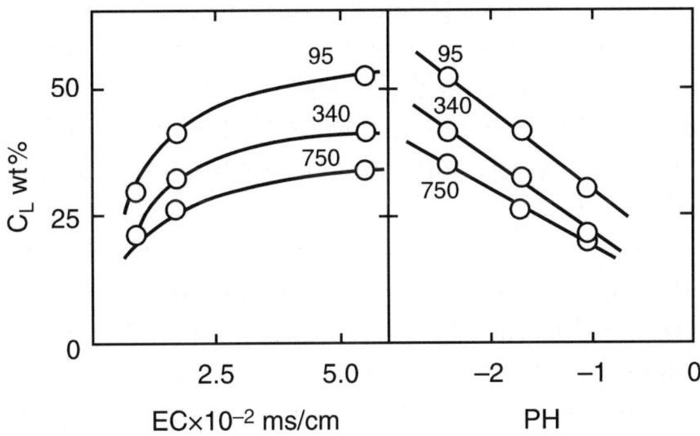

Figure 3.23.5   Dependence of the lower critical concentration C_L on electroconductivity and pH of the solvent for methylcellulose/inorganic acid systems.[1] The numbers on the curves denote P_v of methylcellulose.

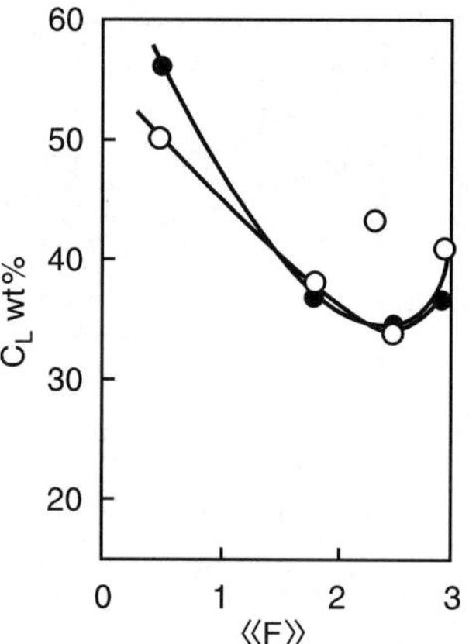

Figure 3.23.6   Dependence of the lower critical concentration C_L on total degree of substitution $\langle\!\langle F\rangle\!\rangle$ for cellulose acetate/65 wt% nitric acid (O) and cellulose acetate/dimethylacetamide (●) systems.[1]

3.23.2   Formation and properties of the lyotropic mesophase of the cellulose/mixed inorganic acid system[31]

Solvent systems in which cellulose can form liquid crystal have been reported as follows: N-methylmorpholine N-oxide (MMNO)/water,[18] dimethylacetamide/lithium chloride,[32] trifluoroacetic acid (TFA)/halogenated hydrocarbons (HH),[19] liquid ammonia (liq. NH_3)/ammonium thiocyanate (NH_4SCN)/water.[33] Patel and Cuculo *et al.* pointed out that the cellulose liquid crystal in liquid NH_3/NH_4SCN/water exhibited both nematic and cholesteric structures depending on the composition of the solvent component.[34] They also noted that cellulose molecules in the film coagulated and regenerated from anisotropic solution are oriented in the direction of shear imposed on the solution.[33] Patel and Gilbert[19] showed through an optical rotatory study that cellulose dissolved in TFA/halogenated hydrocarbons was the cholesteric liquid crystal. As an extension of Section 3.23.1, the lyotropic mesophase of cellulose in mixed inorganic acid is discussed in this section.

Figure 3.23.7 shows phase diagrams for the solubility of cellulose (the polymer concentration $C_p = 5$ wt%) in the sulfuric acid (SA)/polyphosphoric acid (PPA)/water (W) system at various storage time t_s. The insoluble region ($W \geq 0.4$) did not significantly alter with storage time (t_s) of the mixture. The region in which cellulose degraded expanded with an increase in t_s to the region with lower SA (higher PPA) composition, and the region in which cellulose dissolved narrowed. At $t_s = 400$ h, the composition of the SA/PPA/W system exhibiting solubility of cellulose was limited in the range of higher PPA (*ca.* 0.6) and lower SA (*ca.* 0.2). These findings suggested that in the SA/PPA/W system with higher PPA content, the decomposition of cellulose is suppressed during storage. We examined the ability to form mesophase solution for the SA/PPA/W system with the compositions of SA ≤ 0.5, PPA ≥ 0.3, $W \leq 0.4$.

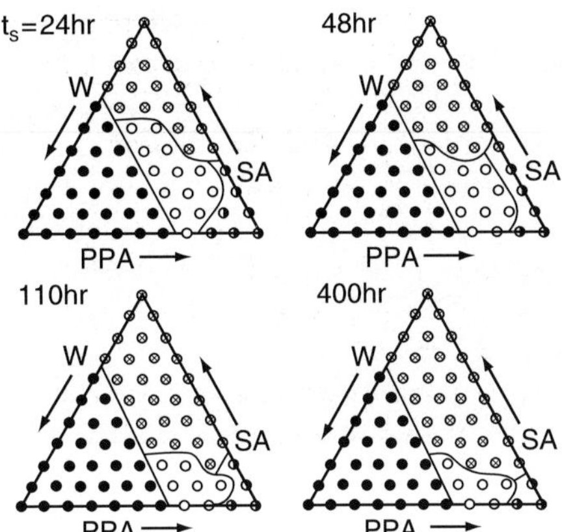

Figure 3.23.7   Phase diagrams for solubility of cellulose in sulfuric acid/polyphosphoric acid/ water system (C_p = 5 wt%) at various storage times t_s[31]: (O) soluble, partial soluble; (●) insoluble; (⊗) degraded.

Figure 3.23.8 shows the phase diagram of cellulose/SA/PPA/W system (C_p = 18 wt%) at t_s = 48 h. A bright view was observed under the shear in relatively wide area of SA/PPA/W composition (as shown by the shadowed area in the figure). With increasing PPA content in the solvent mixture, the relaxation time increased and at a very specific composition (SA/PPA/W = *ca.* 1:8:1 w/w/w (closed circle)), birefringence did not relax within at least 60 s. Over the whole area shadowed in Figure 3.23.8, except for the closed circle corresponding to the above specific composition, birefringence quickly relaxed within 0.1 s. Only at a specific composition, cellulose could give anisotropic solution. The photomicrograph of this anisotropic solution is shown in Figure 3.23.9. Meeten and Navard[35] reported for a CTA/TFA system that the relaxation time (τ) of bright view induced by shear is remarkably different between anisotropic and isotropic solutions (i.e. $\tau \approx$ 65 s for the former and $\tau <$ 0.2 s for the latter).

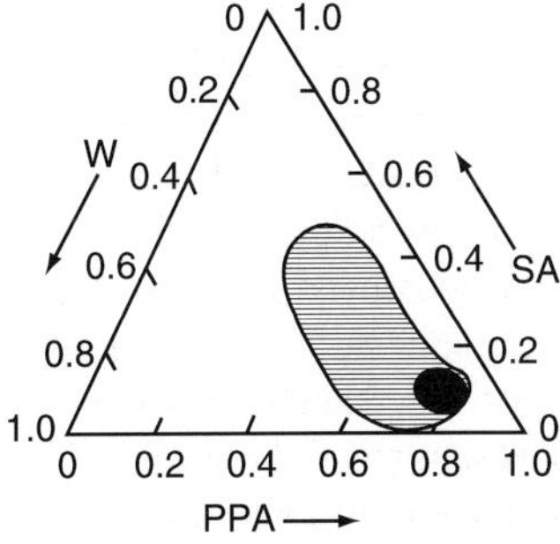

Figure 3.23.8   Phase diagram of cellulose in sulfuric acid/polyphosphoric acid/water system (C_p = 18 wt%):[31] shadowed area: isotropic phase; closed area: anisotropic phase (sulfuric acid/ polyphosphoric acid/water = 1:8:1 w/w/w, t_s = 48 h).

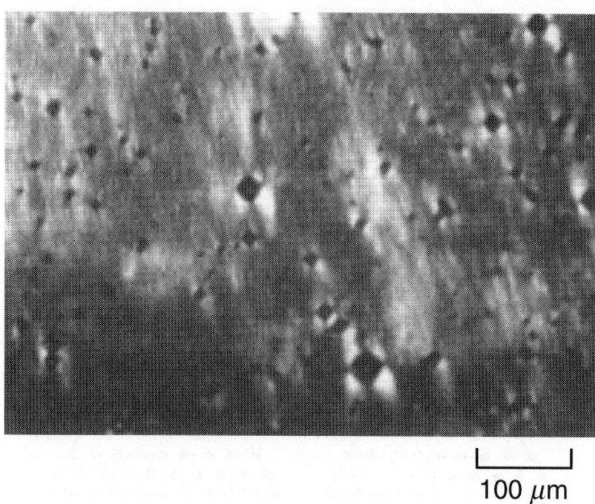

Figure 3.23.9   Optical microgram of anisotropic solution of cellulose (C_p = 18 wt% in sulfuric acid/polyphosphoric acid/water 1:8:1 w/w/w) system.[31]

The minimum cellulose concentration (C_{pmin}) needed for the formation of anisotropic phase at the composition of SA/PPA/W = 1:8:1 was 16 wt%. Cellulose in the solvent mixtures of other compositions never gave the anisotropic phase even if C_p was increased to 20 wt%. Therefore, it was experimentally confirmed that the composition of the solvent mixture is closely related to the formation of the isotropic mesophase of cellulose.

The mixed solvent with the specific composition (SA/PPA/W = 1:8:1 w/w/w) proved to form easily a white viscous precipitate on standing at $-10\,°C$, while different compositions did not bring about such a precipitate. This strongly suggests that the SA/PPA/W mixed solvent with the composition of SA/PPA/W = 1:8:1 w/w/w may form a characteristic structure.

Figure 3.23.10(a) shows Raman spectra of PPA/W mixtures with three different compositions (a-ii–a-iv) and phosphoric acid (a-i). Figure 3.23.10(b) shows Raman spectra of SA/PPA/W mixtures with three compositions (SA/PPA/W = 1:8:1, 2:7:1, and 1:7:2 w/w/w) (b-i, b-iii). A Raman band at 730 cm^{-1} may be characteristic of polymeric P–O stretching because no band at 730 cm^{-1} was observed for phosphoric acid (see Figure 3.23.10(a-i)). With decrease in PPA concentration, the intensity of the 730 cm^{-1} peak decreased, but the peak did not disappear in the range of PPA concentration larger than 40 wt%. On the other hand, the SA/PPA/W (SA/PPA/W = 2:7:1 and 1:7:2 w/w/w) showed no band around 730 cm^{-1} (see Figure 3.23.10(b-ii, b-iii)), suggesting that sulfuric acid in the system facilitates the hydrolysis of polymeric P–O bond. However, a band at 730 cm^{-1} was observed only in the SA/PPA/W system (SA/PPA/W = 1:8:1 w/w/w) (see Figure 3.23.10(b-i)), which is capable of forming lyotropic mesophase of cellulose. These findings indicate that sulfuric acid in SA/PPA/W mixture might be fixed in PPA matrix at specific composition and does not play as a catalyst for the hydrolysis of PPA. Such a supermolecular structure of the SA/PPA/W system may be the principal criterion for the formation of lyotropic cellulose mesophase and direct experimental analysis of such structure requires further study.

Kamide *et al.*[1] first demonstrated that cellulose derivatives with total DS $\langle\!\langle F \rangle\!\rangle < 1$ could form lyotropic liquid crystals in inorganic acids (Section 3.23.1) and there remains

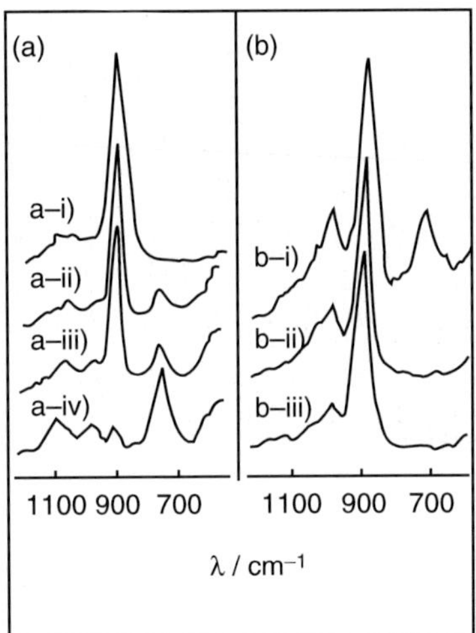

Figure 3.23.10 Raman spectra of aqueous polyphosphoric acid and phosphoric acid (a): and sulfuric acid/polyphosphoric acid/water mixture (b): (a-i) phosphoric acid; (a-ii) 40 wt% aqueous polyphosphoric acid; (a-iii) 51 wt% aqueous polyphosphoric acid; (a-iv) 100 wt% polyphosphoric acid; (b-i) sulfuric acid/polyphosphoric acid/water = 1:8:1; (b-ii) sulfuric acid/polyphosphoric acid/water = 2:7:1; (b-iii) sulfuric acid/polyphosphoric acid/water = 1:7:2 (w/w/w).[31]

the possibility that cellulose is partially derivatized in the SA/PPA/W liquid crystal system of a specific composition (SA/PPA/W = 1:8:1 w/w/w) and the cellulose derivatives formed in SA/PPA/W system build up the mesophase.

Figure 3.23.11 shows Cross-Polarization Magic Angle Sample Spinning (CP/MAS) solid state ^{13}C NMR spectra of the cellulose films (anisotropic (a)-film and isotropic (i)-film) prepared from anisotropic solutions and isotropic solutions by coagulating with methanol. CP/MAS spectra for a- and i-films are principally the same and these spectra indicate that both films are pure cellulose (C_1: 105 ppm; C_4: 83 ppm; C_2, C_3, and C_5: 74 ppm; C_6: 63 ppm). The possibility of derivatization is thus denied. For these cellulose films, the degree of breakdown in $O_3H\cdots O_5'$ intramolecular hydrogen bond $\chi_{am}(C_3)$ was estimated to be about 90%. In other words, the coagulation method employed here gives a cellulose solid containing only a few intramolecular hydrogen bonds. Kamide *et al.*[36,37] verified experimentally that the solubility of cellulose II is governed by $\chi_{am}(C_3)$ independent of DP ranging from 100 to 300. Then, the above two cellulose samples recovered from SA/PPA/W system are expected to dissolve in a 9.1 wt% NaOH solution at low temperature, although DP of these samples was determined by the viscometric method to be 270 and 230 for a- and i-films, respectively. The method adopted for preparation of a- and i-films may provide a powerful means to give alkali soluble cellulose. In fact, the above two recovered films (a- and i-film) easily dissolved in aq. 9.1 wt% NaOH at 4 °C.

In order to further examine the possibility of derivatization of cellulose in SA/PPA/W system, the solutions were subjected to ^{13}C NMR measurements. The results are shown in Figure 3.23.12, in which the spectrum of the steam exploded pulp in aq. 9.1 wt% NaOH is included for reference. ^{13}C NMR spectra of a- and i-film in a 9.1 wt% NaOH

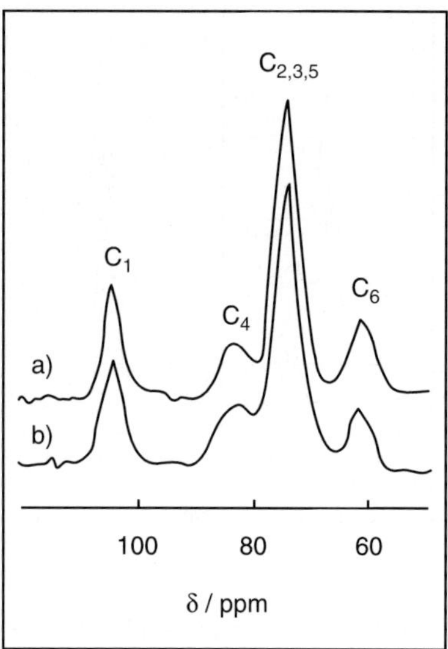

Figure 3.23.11   MAS ^{13}C NMR spectra of the cellulose recovered from optically anisotropic solution (a) and isotropic solution (b): $C_p = 18$ wt%, sulfuric acid/polyphosphoric acid/water = 1:8:1 w/w/w; (b) $C_p = 18$ wt%, sulfuric acid/polyphosphoric acid/water = 3:4:3 w/w/w.[31]

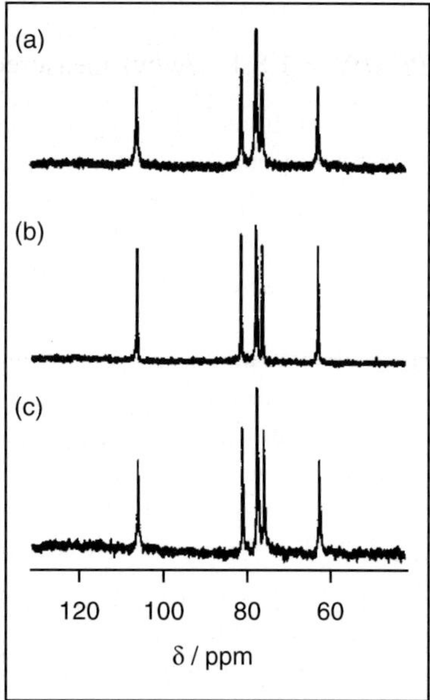

Figure 3.23.12 ^{13}C NMR spectra of recovered cellulose and starting cellulose dissolved in aqueous 9.1 wt% sodium hydroxide: (a) cellulose recovered from anisotropic solution ($C_p = 18$ wt%, sulfuric acid/polyphosphoric acid/water = 1:8:1 w/w/w, $t_s = 48$ h); (b) cellulose recovered from isotropic solution ($C_p = 18$ wt%, sulfuric acid/polyphosphoric acid/water = 3:4:3 w/w/w, $t_s = 48$ h); (c) starting cellulose.[31]

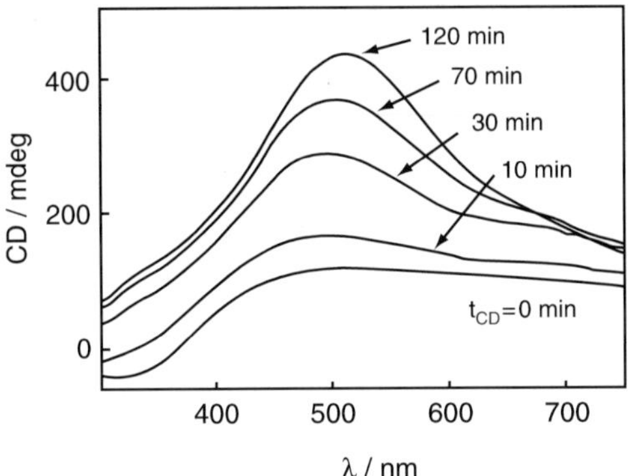

Figure 3.23.13 Circular dichroism spectra of the lyotropic mesophase of cellulose sulfuric acid/polyphosphoric acid/water mixture (C_p = 18 wt%, 1:8:1 w/w/w) at 0 °C; standing time t_{CD} is given in the figure.[31]

solution show C_1 peak at 106.5 ppm, C_4 peak at 82.0 ppm, C_3, C_5 peak at 78.5 or 79.0 ppm, C_2 peak at 77.0 ppm, and C_6 peak at 63.5 ppm, exhibiting almost the same spectra of the starting material cellulose. If cellulose could be derivatized in the SA/PPA/ W system, then cellulose sulfate or cellulose phosphate would be a possible candidate. If this is true, then a new peak should appear at around 101–103 or 68–70 ppm. However, no such peak was detected. This is the second experimental evidence for no derivatization of cellulose in SA/PPA/W system.

Figure 3.23.13 shows circular dichroism (CD) spectra of cellulose lyotropic meso-phase (C_p = 18 wt%, SA/PPA/W = 1:8:1 w/w/w) measured at various standing times t_{CD}. It is known that the appearance of CD peak provides direct evidence for the cholesteric mesophase.[3] With increase in t_{CD}, the spectrum for the present system became sharp and the intensity of the CD peak around at 510 nm (= λ_0) increased, approaching an asymptotic value at t_{CD} = 120 min. This indicates that cellulose mesophase in the A/PPA/W system has a cholesteric structure. Yamagishi[38] observed that it takes a very long time for the thermotropic mesophase of cellulose derivatives to complete the formation of the cholesteric structure mainly due to its extremely high viscosity (e.g. about 1 week at 60 °C for cellulose-$O(CH_2CH_2O)_3$–CH_3). The lyotropic mesophase of cellulose/SA/PPA/W system also has very high viscosity and thus a few hours are necessary to recover the cholesteric structure after distortion by shear stress. The value of cholesteric pitch (P') was calculated from eq. (3.23.1):

$$P' = m\lambda_0/n \sin\theta \qquad (3.23.1)$$

where m is an integer, λ_0 is the maximum wavelength of the observed CD peak, n, average refractive index of the mesophase solution, θ, an angle between the direction of the incident beam and the plane of the cholesteric layer using λ_0 = 510 nm and n = 1.459 to be 350 nm. Obviously, λ_0 lies in the visible light region, corresponding to the appearance of the rainbow color of the mesophase solution. The CD peak in the positive side means the left-handed cholesteric structure.

Table 3.23.1 shows assignments and molecular orientation factor f^{11} of the characteristic IR bands for a- and i- films. A slight difference in peak positions of IR

Table 3.23.1

Infrared dichroism of cellulose film recovered from sulfuric acid/polyphosphoric acid/water system[31]

Wave number (cm^{-1})	Assignment	f^{11}[a]	
		a-Film[b]	i-Film[c]
3475	OH (intramolecular hydrogen bond)[d]	0.336	
3444	OH (intramolecular hydrogen bond)[d]	0.344	0.342
3440	OH[e]	0.353	
3357	OH (intermolecular hydrogen bond)	0.354	0.350
2927	CH_2 antisynmetric stretching[f]	0.342	
2899	CH stretching[f]	0.346	0.350
1430	CH_2 bending[e]	0.350	
1375	CH bending[e]	0.322	0.334
1335	OH in plane bending[e]	0.300	
1161	CO (bridge)[e]	0.304	0.334
1108	CO stretching[e]	0.273	0.347
1069	CO stretching[e]	0.304	0.352
1029	CO stretching[e]	0.314	0.348
899	C1 group stretching[g]	0.304	
895	C1 group stretching[g]		0.344

[a]The molecular orientation factor f^{11} in parallel to the shear direction for the characteristic IR band (see eq. 5 of Ref. 31).
[b]Film cast from anisotropic solution.
[c]Film cast from isotropic solution.
[d]Ref. 39.
[e]Ref. 40.
[f]Ref. 41.
[g]Ref. 42.

band (e.g. OH stretching region and C_1 group stretching) between a-film and i-film was detected. Differences in dichroism of IR band between a-film and i-film are considerable. The IR band at 3475 cm^{-1} is responsible for the $O_3H\cdots O_5'$ intramolecular hydrogen bond,[39] (this peak could be detected although its intensity was very weak) showing weak parallel dichroism for a-films. The IR band at 3444 cm^{-1} shows parallel dichroism for both i-film and a-film. On the other hand, antisymmetrical bridge C–O vibration (*ca.* 1161 cm^{-1}), antisymmetrical in-phase ring stretching vibration (*ca.* 1108 cm^{-1}) and C–O stretching vibration (1069 cm^{-1}) show perpendicular dichroism for a-film, while these IR bands reveal parallel dichroism for i-film. The latter is commonly observed for stretched cellulose film made from isotropic solution.

Thus, the cellulose film recovered from the anisotropic solution system has unique molecular chain orientation different from the film prepared from isotropic solution. It is especially interesting that the bridge C–O vibration, which gives typical IR band characterizing the main axis of cellulose molecules, shows perpendicular dichroism, while the $O_3H\cdots O_5'$ intramolecular hydrogen bond remains parallel to the shear direction (see Table 3.23.1). Similar results have been already reported by the Okajima *et al.*[49] on lyotropic cellulose acetate 70 wt% nitric acid system (Section 3.23.3). The distorted helical structure of a unit molecular chain of cellulose due to the intermolecular interaction is one possibility to explain the above.

In summary, an acid mixture with a specific composition (sulfuric acid/polyphos phoricacid/water $=$ *ca.* 1:8:1 w/w/w) forms cellulose liquid crystal with cholesteric structure as detected by circular dichroism and the cellulose is not derivatized in the system. A Raman spectroscopic study of the solvent mixture revealed that only at a specific composition, the hydrolysis of PPA proved to be restricted. A polarized FT-IR study showed that both bridge C–O and in-phase ring C–O of the cellulose molecules in the film prepared from the anisotropic solution are perpendicularly orientated to the shear direction imposed on the mesophase.

3.23.3   Anisotropic mesophase of CA in inorganic solvent mixture[43]

In Section 3.23.2, it was shown that cellulose in the mixture of SA, PPA, and W (SA/PPA/W $=$ 1:8:1 w/w/w) generates lyotropic liquid crystal above some cellulose concentration, depending on the DP. This type of solvent mixture is advantageous with respect to low volatility, ease of handling, and low degree of decomposition of polymer in relatively low temperature range. Successively, CAs with $\langle\!\langle F \rangle\!\rangle$ ranging from 0.5 to 2.9 were found to be dissolved into the solvent mixture up to 30 wt% around which lyotropic mesophase with cholesteric texture forms. In this section, an attempt was made to disclose effects of $\langle\!\langle F \rangle\!\rangle$ and temperature on the cholesteric structure of the CA and cellulose mesophases. For this purpose, we prepared the mesophases of cellulose and CA with various $\langle\!\langle F \rangle\!\rangle$ (0.5–2.9) in the mixture of SA/PPA/W $=$ 1:8:1 w/w/w, and determined the cholesteric pitch of the mesophases using circular dichroism method.

CA samples

The M_v and $\langle\!\langle F \rangle\!\rangle$ of the CA samples including cellulose are listed in Table 3.23.2. According to Kamide and Saito,[44] the DS at C_2, C_3, and C_6 positions ($\langle\!\langle f_2 \rangle\!\rangle$, $\langle\!\langle f_3 \rangle\!\rangle$, $\langle\!\langle f_6 \rangle\!\rangle$, respectively) is almost equal for each CA prepared by the above-mentioned method.

Table 3.23.2

Total degree of substitution $\langle\!\langle F \rangle\!\rangle$ and the viscosity-average of molecular weight M_v of the samples[43]

Sample number	$\langle\!\langle F \rangle\!\rangle$	$M_v \times 10^{-4}$ (DP[a])
CA-0	0.00	5.4(330)
CA-1	0.66	5.1(300)
CA-2	0.90	4.8(240)
CA-3	1.70	6.5(350)
CA-4	1.75	4.5(190)
CA-5	2.46	9.6(360)
CA-6	2.89	23.0(810)

[a]The viscosity-average degree of polymerization.

Solution of CA in SA/PPA/W

The cellulose and CA samples were mixed with the solvent mixture of SA/PPA/W at the polymer concentration (C_p) up to 30 wt%. In order to avoid the decomposition of the polymer, the mixture of polymer and solvent was immediately transferred to a refrigerator-controlled temperature at $-5\,°C$ and then stirred occasionally by hand. The maximum time required for making clear solution was *ca.* 40 h. The solution thus prepared served for polarizing the microscope observation and CD measurement at 48 and 100 h after mixing, respectively.

An image of the polarizing microscope of the CA with $\langle\langle F\rangle\rangle = 2.89$ in a mixture of SA/PPA/W = 1:8:1 w/w/w) is shown in Figure 3.23.14. This solution gives typical colored image peculiar to mesophase.

Figure 3.23.15 shows phase diagrams determined by polarizing microscopic observation for the CA SA/PPA/W system at constant polymer concentration (25 wt%). This figure also includes the result of the cellulose SA/PPA/W system (polymer concentration is 18 wt%) for comparison. Here, 16 mixtures with different composition of SA/PPA/W, which did not give rise to any decomposition of the polymer for at least 10 h, were chosen as solvent of the samples. With the decrease of $\langle\langle F\rangle\rangle$, the region of optically anisotropic mesophase becomes narrower, indicating that CA with higher $\langle\langle F\rangle\rangle$ readily tends to form liquid crystal in the SA/PPA/W system. The solvent mixture with SA/PPA/W = 1:8:1 is a common solvents to form liquid crystal for cellulose and CA with $\langle\langle F\rangle\rangle$ ranging from 0.9 to 2.9.

The lower critical concentration of the CA in the common solvent to form liquid crystal, C_p^*, is plotted against $\langle\langle F\rangle\rangle$ in Figure 3.23.16. In the figure, the results of CA aq. nitric acid solution system[45] are also depicted for comparison. C_p^* of the CA-SA/PPA/W system, which shows slight increase with $\langle\langle F\rangle\rangle$, is far lower than that of the CA aq. nitric acid system. The effect of the molecular weights of the CA samples must not be so serious because Patel and Gilbert[46] have already reported that the molecular weight of CTA does not affect the critical volume fraction points in case of CTA/TFA liquid crystal system.

Based on the lattice model, Flory proposed[47] a theoretical relationship between the lower critical volume (v_2) of the polymer to generate anisotropic phase and an aspect ratio

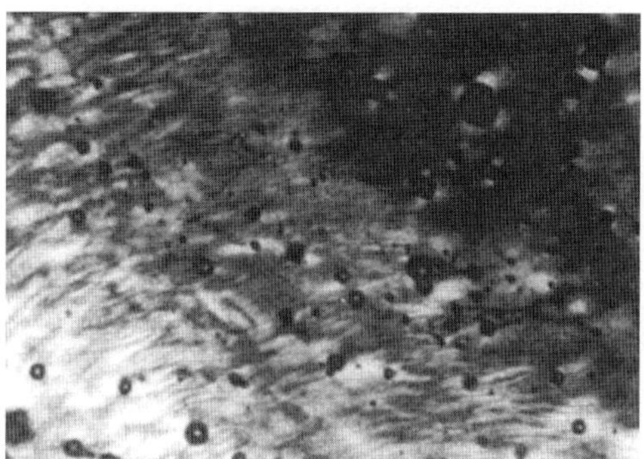

Figure 3.23.14 Polarized photomicrograph of mesophase of cellulose acetate/sulfuric acid/polyphosphoric acid/water system. $\langle\langle F\rangle\rangle = 2.89$, $C_p = 30\%$.[43]

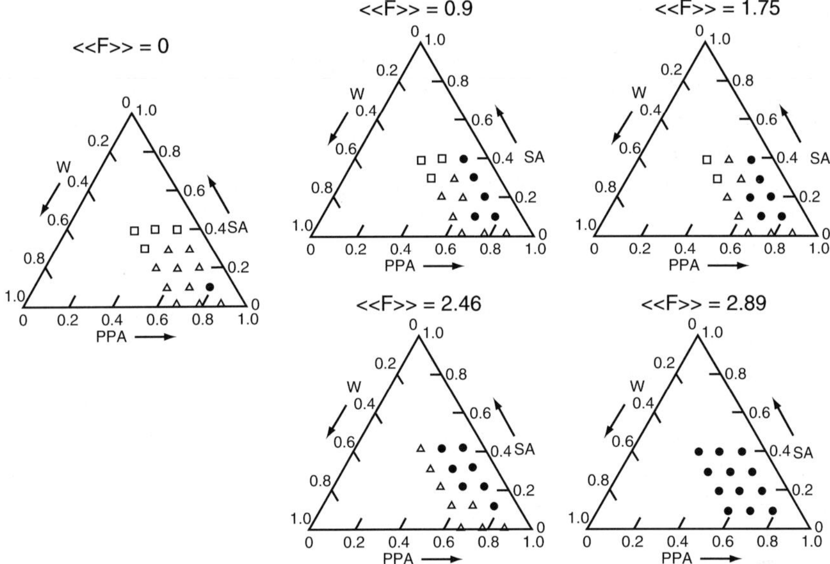

Figure 3.23.15 Phase diagrams of cellulose acetate/sulfuric acid/polyphosphoric acid/water system.[43] (●) Optically anisotropic solution; (△) flow orientation solution; (□) optically isotropic solution.

(χ') of Kuhn statistical segment of the polymer;

$$v_2 = 8/\chi'(1 - 2/\chi') \qquad (3.23.2)$$

Here, χ' can be calculated from a persistence length q and diameter d of the polymer chain. Kamide and Saito[29] reported that the q of CA in DMAc, determined by LS method,

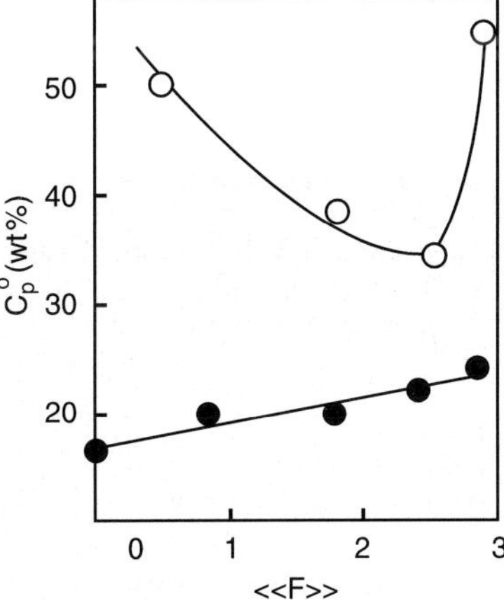

Figure 3.23.16 Lower critical concentration C_p^* of mesophase solution of cellulose acetate/sulfuric acid/polyphosphoric acid/water system.[43] (●) Cellulose acetate/sulfuric acid/polyphosphoric acid/water system; (○) cellulose acetate/aqueous nitric acid system.

becomes maximum at $\langle\!\langle F\rangle\!\rangle = 2.5$ due to specific polymer–solvent interaction (solvation). From this fact, total DS dependence of C_p^* of CA-aq. nitric system was reasonably explained in terms of chain rigidity of CA. The quite different dependence of C_p^* in SA/PPA/W system against $\langle\!\langle F\rangle\!\rangle$ and lower values of C_p of this system suggest that CA molecular chains are far more rigid in the mixture than aq. nitric acid solution due to strong interaction between CA chains and SA/PPA/W molecules. The chain rigidity in the mixture may monotonically decrease with an increase of $\langle\!\langle F\rangle\!\rangle$.

All of the mesophase with 30 wt% at 0 °C show positive dichroism in the wavelength range observed, indicating that the CA with $\langle\!\langle F\rangle\!\rangle = 0.66$–$2.46$ forms left-handed cholesteric liquid crystal. λ_0 of the CA with $\langle\!\langle F\rangle\!\rangle = 0.9$–$1.75$ increases monotonically with $\langle\!\langle F\rangle\!\rangle$. The CD peak of the CA with $\langle\!\langle F\rangle\!\rangle = 0$ ($\equiv$ cellulose), 0.66, and 2.46, 2.89 was out of detection in the range of 250 nm $< \lambda_0 < 800$ nm.

Table 3.23.3 summarizes refractive index n, λ_0 and the cholesteric pitch (P') calculated using eq. (3.23.1) for each sample. n shows slight increase with $\langle\!\langle F\rangle\!\rangle P'$ of the CA with $\langle\!\langle F\rangle\!\rangle$ more than 2.46 is expected larger than 570 nm because λ_0 of these polymer is higher than 800 nm. Evidently, P' of cellulose and CA mesophases concomitantly increases with $\langle\!\langle F\rangle\!\rangle$, coinciding with the results of acetylethylcellulose (AEC)/chloroform systems reported by Guo and Gray.[48]

Reciprocal P' of CA-SA/PPA/W system is plotted against $\langle\!\langle F\rangle\!\rangle$ (see Figure 6 of Ref. 43), where the pitch of left-handed cholesteric texture is taken as negative for convenience. $\langle\!\langle F\rangle\!\rangle$ dependence of reciprocal P' of the two systems (CA-SA/PPA/W and AEC/chloroform (by Guo and Gray[48])) is in sharp contrast each other.

The inversion of the helicoidal sense from left-handed to right-handed is observed in AEC/chloroform system at around $\langle\!\langle F\rangle\!\rangle = 2.7$. On the other hand, from the plot of $1/P'$ versus $\langle\!\langle F\rangle\!\rangle$, the data of CA-SA/PPA/W system does not seem to cross the transversal axis even if the line is extended to $\langle\!\langle F\rangle\!\rangle \approx 2.9$. Guo and Gray[48] suggested that a significant dependence of cholesteric pitch of AEC/chloroform system on minor changes in $\langle\!\langle F\rangle\!\rangle$ originates from the properties of helical chain conformation of cellulose molecule in the solution, which controls the torsional power between neighboring cholesteric layers. Miyamoto et al.[43] suggested that no observation of helicoidal sense inversion in CA-SA/PPA/W system may relate to the slight dependence of polymer–solvent interaction on

Table 3.23.3

Refractive index n, peak wave λ_0 in circular dichroism spectra and calculated pitch P' for different cellulose acetate samples in sulfuric acid/polyphosphoric acid/water at 0 °C

Sample number	$\langle\!\langle F\rangle\!\rangle$	n	λ_0 (nm)	P' (nm)
CA-0	0.00	–	–	67[a]
CA-1	0.66	–	–	158[a]
CA-2	0.90	1.438	455	316
CA-3	1.70	1.440	547	380
CA-4	1.75	1.444	650	450
CA-5	2.46	–	>800	>570
CA-6	2.89	–	>800	>570

[a]Estimated by the extrapolation method using the polymer concentration dependence of λ_0.

$\langle\!\langle F \rangle\!\rangle$ as mentioned before, which affects the properties of helical conformation of cellulose molecule.

The peak wavelength in the CD spectra of the CA liquid crystal system shifts to the longer side with increasing temperature and this dependence of λ_0 is thermally reversible. Because n of the system was almost constant, this λ_0 dependence means that P' increases with temperature. Temperature dependence of the pitch becomes smaller with $\langle\!\langle F \rangle\!\rangle$. Guo and Gray have reported that temperature dependence of the cholesteric pitch is originated to the change of the 'twist angle' between the cholesteric layers.[48] We suggest that the high steric hindrance of O-Ac groups with high $\langle\!\langle F \rangle\!\rangle$ sample cause the insensitiveness of the twisting power between layers to temperature.

In summary, CAs with various $\langle\!\langle F \rangle\!\rangle$ (0.52–2.89) can form lyotropic liquid crystal in the mixture of SA, PPA, and W of a specific composition (SA/PPA/W = 1:8:1 w/w/w). CD spectra of the CA mesophase indicated that the texture is left-handed cholesteric irrespective of $\langle\!\langle F \rangle\!\rangle$. The cholesteric pitch determined by the CD method remarkably depends on polymer concentration and temperature, and monotonically increases with an increase of $\langle\!\langle F \rangle\!\rangle$ at constant temperature and polymer concentration. $1/P'$ almost depends linearly on temperature and the cube of the polymer concentration.

3.23.4 Liquid crystal spinning of CA lyotropic solutions[49]

The possibility of the production of fiber with high breaking strength and high Young's modulus from lyotropic liquid crystals by applying the liquid crystal spinning method attracts keen interest in industry. Extensive studies have been conducted on cellulose liquid crystal spinning. For example, Quenin et $al.$[50] prepared a cellulose fiber with the tensile strength (TS) of 6 g d^{-1} of fiber with 9000 m in length) from cellulose/N-methylmorpholine N-oxide (MMNO)/water system by a dry jet, wet spinning method. Yang et $al.$[51] obtained a cellulose fiber with TS of 3 g d^{-1} from cellulose/liq. NH$_3$/NH$_4$SCN system by the same spinning method as that used by Quenin. A regenerated cellulose fiber with high TS (18 g d^{-1}) was successfully produced by O'Brien who saponified the CTA fiber (TS = 15 g d^{-1}) spun from CTA/trifluoroacetic acid (TFA)/halogenated hydrocarbon (HH) liquid crystal system.[52] However, the organic solvents used for the formation of cellulosic liquid crystals have serious drawbacks such as high cost, high toxicity, and difficulty of solvent recovery.[1]

In Section 3.23.1, it was indicated that almost cellulose derivatives form lyotropic crystal in inorganic acid systems,[1] and C_L for cellulose acetate/65% nitric acid and CA/DMAc reveals maximum at $\langle\!\langle F \rangle\!\rangle \approx 2.46$ (see Figure 3.23.6). Note that, in this case, the CA samples employed had $\langle\!\langle f_6 \rangle\!\rangle/\langle\!\langle F \rangle\!\rangle \approx 0.34$ and $M_v = (10 \pm 2) \times 10^4$. Monochloracetic acid is also effective in giving lyotropic crystal.[53] In this section, an attempt is made to carry out liquid crystal spinning of CA DMAc and 70% nitric acid.[49]

Lyotropic liquid crystal of CA in DMAc and 70% nitric acid

Table 3.23.4 lists the molecular characteristics of CA samples and their lower critical concentration C_L, defined in Section 3.23.1 as the minimum concentration at which the optical birefringence was observed under cross polarizers in DMAc and 70% nitric acid. Note that C_L is not LCST in Section 3.21. Here, CA samples are prepared by two-step

Table 3.23.4

Lower critical concentration C_L of cellulose acetate-dimethylacetamide and-*ca.* 70% nitric acid system at 25 °C[49]

Sample code	$M_v \times 10^{-4}$	$\langle\!\langle F \rangle\!\rangle$	$\langle\!\langle f_6 \rangle\!\rangle / \langle\!\langle F \rangle\!\rangle$	C_L at 25 °C (%)	
				DMAc	70% nitric acid
CDA	12	2.46	0.34	36.5	38.0
CMA II-1	9	1.20	0.34	48.9	–
CMA II-2	9	1.05	0.35	49.1	–
CMA II-3	8	0.93	0.32	47.4	–
CMA II-4	8	0.79	0.35	44.0	–
CMA II-5	8	0.69	0.37	46.4	–
CMA II-6	7	0.54	0.49	24.8	–
CMA I-1	28	1.20	0.56	18.9	–
CMA I-2	26	0.68	0.60	23.8	–

method (CA with $\langle\!\langle F \rangle\!\rangle = 2.46$ in acetic acid solution was homogeneously hydrolyzed with hydrochloric acid; CMA II) or by a one-step method (wood pulp ($M_v = 16 \times 10^4$) in DMAc/LiCl was directly acetylated by acetic anhyride/pyridine; CMA I).[54] For CA with relatively higher $\langle\!\langle f_6 \rangle\!\rangle / \langle\!\langle F \rangle\!\rangle$ (0.56–0.60) and higher M_v (26–28 × 10⁴), C_L is lower than CA samples with $\langle\!\langle f_6 \rangle\!\rangle / \langle\!\langle F \rangle\!\rangle \approx 1/3$ and lower M_v. C_L for CA ($\langle\!\langle F \rangle\!\rangle = 2.46$)/70 wt% nitric acid is approximately the same as that for CA ($\langle\!\langle F \rangle\!\rangle = 2.46$)/DMAc.

Okajima *et al.* found by IR analysis that (1) molecular orientation phenomena of CA ($\langle\!\langle F \rangle\!\rangle = 2.46$)/DMAc lyotropic system significantly differ when shear is imposed on the system, depending on both $\langle\!\langle F \rangle\!\rangle$ and $\langle\!\langle f_6 \rangle\!\rangle$. (2) Orientation characteristics of the lyotropic crystals of CA with the same $\langle\!\langle F \rangle\!\rangle$ and $\langle\!\langle f_6 \rangle\!\rangle$ differ depending on the solvent employed (e.g. DMAc and nitric acid):[49] CA/nitric acid lyotropic system showed a kind of regularity when sheared, perpendicular to the shear direction (Figure 3.23.3), but CA/DMAc lyotropic system did not. This feature may be reserved in the fiber structure spun from the above lyotropic system. (3) CDA/70 wt% nitric acid lyotropic system exhibits a characteristic IR perpendicular dichroism band corresponding to ring CO stretching band which ordinary shows parallel dichroism for cellulose. (4) Kevlar-like structure is expected for the fiber spun from CDA/70% nitric acid lyotropic system.

Liquid crystal spinning.[49] Two kinds of lyotropic crystals of CA are prepared: $\langle\!\langle F \rangle\!\rangle$ of the polymer, and polymer concentration are summarized in Table 3.23.5. Unsolved particles in the liquid crystals were excluded by centrifuging the dope at 6000 rpm for 20 min, immediately before spinning. The dope thus prepared was extruded through a screw-type extruder (single die; diameter 0.25 mm) at output 2 cm³ min⁻¹ into air (air gap, i.e. the distance of die surface and coagulation bath; 1.5–10 cm) at 10 °C and then wound up at speed of 13.7–29.6 m min⁻¹. The spinning conditions are listed in the fifth, sixth, and seventh columns of Table 3.23.5.

Table 3.23.5

Spinning conditions of cellulose acetate liquid crystals[49]

Fiber sample	$\langle\!\langle F \rangle\!\rangle$	Solvent at 10 °C	Polymer concentration (%)	Air gap[a] (mm)	Spinning speed (m min^{-1})	Coagulant
DMAc-1	2.46	DMAc	50	10.0	13.7	Water
DMAc-2	2.46	DMAc	50	2.0	13.7	Water
DMAc-3	2.46	DMAc	40	10.0	29.6	Water
DMAc-5	2.46	DMAc	40	1.5	29.6	Water
DMAc-6	2.46	DMAc	40	1.5	13.6	Water
HNO$_3$-1	2.20	HNO$_3$[b]	50	10.0	20.2	Water
HNO$_3$-2	1.78	HNO$_3$	53	5.0	20.2	Water
HNO$_3$-3	1.72	HNO$_3$	53	1.5	20.2	Water

[a]Distance between die surface and coagulation bath.
[b]70 wt% HNO$_3$.

Mechanical properties of fibers from liquid crystals.[49] Some mechanical properties of the cellulose fibers spun from lyotropic crystals are summarized in Table 3.23.6. The corresponding data of commercial CA ($\langle\!\langle F \rangle\!\rangle = 2.46$) fibers, spun by dry spinning of isotropic acetone solutions are shown in the bottom row of the table. The fibers spun from CA ($\langle\!\langle F \rangle\!\rangle = 2.46$)/70% nitric acid lyotropic system exhibit always superior mechanical properties such as tensile strength (3.2–4.0 g d^{-1}) and Young's modulus (140–174 g d^{-1}) to those of the fibers from CA ($\langle\!\langle F \rangle\!\rangle = 2.46$)/DMAc lyotropic system.

The significant difference in properties of fibers, which are produced from anisotropic solutions, and those from isotropic solution are evident. Fibers spun from CA/DMAc anisotropic system have larger TS, shorter tensile elongation, and larger Young's modulus

Table 3.23.6

Mechanical properties of cellulose acetate fibers spun from lyotropic crystals[49]

Fiber sample	Denier (d)	Tensile strength[a] (g d^{-1})	Tensile elongation[a] %	Young modulus[b] (g d^{-1})	T_{max}^{c} (°C)	Melting point[d] (°C)	Heat of fusion[d] (J g^{-1})
DMAc-1	125.3	1.66	4.1	86.9	180	238.8	12.5
DMAc-2	151.1	1.67	4.2	91.9	183	238.8	13.4
DMAc-3	73.8	2.96	6.1	116.1	185	238.1	13.3
DMAc-4	154.3	2.80	6.5	112.5	196	238.1	11.1
DMAc-5	79.4	2.25	4.4	106.3	188	238.8	13.4
DMAc-6	151.9	2.80	5.5	114.2	190	238.7	13.9
HNO$_3$-1	61.9	4.01	5.2	173.5	210	233.7	6.7
HNO$_3$-2	83.0	3.22	4.3	139.5	208	235.7	6.5
HNO$_3$-3	106.9	3.49	4.6	142.8	208	235.7	6.5
Blank[e]	75.3	1.32	28.7	36.8	176	231.3	4.8

[a]Japan Industry Standard (JIS) L–1013–1981–7.5.
[b]JIS L–1013–1981–7.10.
[c]Peak temperature of mechanical loss tangent–temperature curve (110 Hz, heating rate (HR) 10 °C min^{-1}).
[d]Differential scanning calorimetry (HR 10 °C min^{-1}).
[e]Commercial fiber (manufactured by Asahi) for textiles.

than those from CA/acetone isotropic system. The mobility of amorphous chains, deduced by peak temperature of mechanical loss tangent–temperature curve T_{max} (see Table 3.23.6), is restricted more tightly in the following order: CA/acetone (isotropic) < CA/DMAc (anisotropic) < CA/nitric acid (anisotropic).

REFERENCES

1. K Kamide, K Okajima, T Matsui and S Kajita, *Polym. J.*, 1986, **18**, 273.
2. GS Gaspmok, NN Zegalova, BV Vasilev and OG Trankanov, *Polym. Sci., USSR*, 1969, **11**, 2468.
3. RS Werbowyi and DG Gray, *Mol. Cryst. Liq. Cryst.*, 1976, **34**, 97, (Letters).
4. J Maeno, Japan Laid Open Patent 52-94882, 1977.
5. NG Belnikevich, LS Bolotnikova, ES Edilyan, YV Brestkin and SY Frenkel, *Vysokomol. Soedin., Ser. B*, 1976, **18**, 485.
6. M Panar, OB Willcox, Japan Laid Open Patent, 53-96229, 1978.
7. T Asada, *Polym. Prepr. (Am. Chem. Soc., Div. Polym. Chem.)*, 1979, **20**, 70.
8. JS Aspter and DG Gray, *Macromolecules*, 1979, **12**, 562.
9. SM Abaroni, *Mol. Cryst. Liq. Cryst.*, 1980, **56**, 237.
10. RS Werbowyi and DG Gray, *Macromolecules*, 1980, **13**, 69.
11. J Behda, JF Fellers and JL White, *Colloid Polym. Sci.*, 1980, **258**, 1333.
12. Y Onogi, JW White and JF Fellers, *Non-Newtonian Fluid Mech.*, 1980, **7**, 121.
13. T Tsutsui and R Tanaka, *Polym. J.*, 1980, **12**, 473.
14. K Shimamura, J White and JF Fellers, *J. Appl. Polym. Sci.*, 1981, **26**, 2165.
15. Y Onogi, JW White and JF Fellers, *J. Polym. Sci., Polym. Phys. Ed.*, 1980, **18**, 663.
16. S Tseng, A Valente and DG Gray, Post-Graduate Research Laboratory Report PGRL 210, Pulp Paper Research Institute, Canada.
17. H Chanzy, M Dube and RH Marchessault, *J. Polym. Sci., Polym. Lett. Ed.*, 1979, **17**, 219.
18. H Chanzy, A Peguy, S Chaunis and P Monzie, *J. Polym. Sci., Polym. Phys. Ed.*, 1980, **18**, 1137.
19. L Patel and RD Gilbert, *J. Polym. Sci., Polym. Phys. Ed.*, 1981, **19**, 1231.
20. L Patel and RD Gilbert, *J. Polym. Sci., Polym. Phys. Ed.*, 1981, **19**, 1949.
21. K Kamide, T Terakawa and Y Miyazaki, *Polym. J.*, 1979, **11**, 285.
22. K Kamide, Y Miyazaki and T Abe, *Makromol. Chem.*, 1979, **180**, 2801.
23. K Kamide, M Saito and T Abe, *Polym. J.*, 1981, **13**, 421.
24. K Kamide and M Saito, *Polym. J.*, 1982, **14**, 517.
25. M Saito, *Polym. J.*, 1983, **15**, 249.
26. R Schweiger, *Carbohydr. Res.*, 1972, **21**, 219.
27. K Kamide, K Okajima, T Matsui, M Ohnishi and H Kobayashi, *Polym. J.*, 1983, **15**, 309.
28. K Kamide, K Okajima and K Kowsaka, unpublished results.
29. K Kamide and M Saito, *Eur. Polym. J.*, 1984, **20**, 903.
30. JP Flory, *Proc. R. Soc. London, Ser. A*, 1956, **234**, 60.
31. K Kamide, I Miyamoto and K Okajima, *Polym. J.*, 1993, **25**, 453.
32. E Bianchi, A Ciferri, G Conio, A Cosani and M Terbojevich, *Macromolecules*, 1985, **18**, 646.
33. YS Chen and JA Cuculo, *J. Polym. Sci., Polym. Chem. Ed.*, 1986, **24**, 2075.
34. K.-S Yang, MH Theil and JA Cuculo, In *Polymer Association Structure* (ed. MA El-Nokaly), ACS Symposium Series 384, The American Chemical Society, Washington, DC, 1989, p. 156.
35. GH Meeten and P Navard, *Polymer*, 1982, **23**, 1727.
36. K Kamide and K Okajima, *Polym. J.*, 1984, **16**, 857.
37. T Yamashiki, T Matui, M Saito, Y Matuda, K Okajima, K Kamide and T Sawada, *Br. Polym. J.*, 1990, **22**, 201.
38. T Yamagishi, D. Eng. Thesis Kyoto University, 1989, p. 37.
39. J Hayashi, A Sueoka and T Watanabe, *J. Chem. Soc. Jpn.*, 1974, 1320.
40. HG Higgins and CM Stewart, *J. Polym. Sci.*, 1961, **51**, 59.

41. R Schneider and J Vodnanski, *Collect. Czech. Chem. Commun.*, 1963, **28**, 2060.
42. RT O'Corner and EF Dupre, *Text. Res. J.*, 1958, **28**, 382.
43. I Miyamoto, T Matsui, M Saito and K Kamide, *Polym. J.*, 1996, **28**, 937.
44. K Kamide and M Saito, *Advanced Polymer Science*, Springer, Berlin, 1987, p. 30.
45. K Kamide and K Okajima, unpublished data.
46. DL Patel and RD Gilbert, *J. Polym. Sci., Polym. Phys. Ed.*, 1983, **21**, 1079.
47. PJ Flory, *Proc. R. Soc. London, Ser. A*, 1956, **234**, 73.
48. JX Guo and DG Gray, *Macromolecules*, 1989, **22**, 2086.
49. K Okajima, T Kuriki, S Kajita and K Kamide, *J. Text. Mach. Soc. Jpn.*, 1987, **42**, T47.
50. I Quenin, H Chanzy, M Paillet and A Peguy, In *Integration of Fundamental Polymer Science and Technology* (eds LA Kleinties and PJ Lemstra), Elsevier, Amsterdam, 1986, p. 57.
51. KS Yang, MH Theil and JA Cuculo, In *Polymer Association Structure* (ed. MA E1-Nokaly), ACS Symposium Series 384, The American Chemical Society, Washington, DC, 1989, p. 156.
52. OJP O'Brien, US Patent No. 4,464,323; No. 4,501,886, 1984.
53. S Kajita and K Okajima, Japan Open Patent, No. 58-125701, 1983.
54. K Kamide, K Okajima and K Kowsaka, *Polym. J.*, 1987, **19**, 1405.

3.24   PREPARATION OF POROUS CELLULOSE MEMBRANE BY THE PHASE SEPARATION METHOD: PHENOMENOLOGICAL EFFECTS OF SOLVENT CASTING CONDITIONS ON PORE CHARACTERISTICS[1]

Among numerous methods proposed hitherto for preparing porous polymeric membranes, such as solvent casting, radiation track etching, stretching, and sintering, the solvent casting method is of immense technological importance, as it enables us to produce membranes with a wide range of mean pore size. In the solvent casting method, polymer solution cast on a flat glass plate is phase separated into polymer-rich and lean phases by evaporating good volatile solvent from the solution or by immersing the cast solution into coagulation solution containing nonsolvent. Kamide and Manabe[2] noticed an important role of 'particle growth concept' in membrane formation mechanism in the solvent casting method (i.e. the phase separation method).

Recently, Kamide and collaborators (KI)[3–8] reported, based on their particle growth concept, theoretical studies, as well as computer experiments, on both the mechanism of formation of porous polymeric membranes prepared by the phase separation method and the characteristics of their membrane structure. The detailed steps of the formation of polymeric membranes by the phase separation method are illustrated in Figure 3.24.1. If the initial polymer volume fraction v_p^0 is less than the polymer volume fraction at a CPS v_p^c, then the polymer-rich phase separates first as nuclei (steps a and b of Figure 3.24.1), which grow to the primary particles (steps i–d in Figure 3.24.1),[4] and the primary particles amalgamate into the secondary particles (steps d–f in Figure 3.24.1).[5]

In the subsequent steps, the secondary particles contact with each other to form gel membranes, which become dried membrane through desolvation and drying (steps g–i in Figure 3.24.1). For the latter steps (steps g–i in Figure 3.24.1), KI-derived equations giving pore size distribution $N(r)$ (r, radius of pore) and porosity P_r as functions of phase volume ratio $R(\equiv V_{(1)}/V_{(2)}$; $V_{(1)}$ and $V_{(2)}$ are volumes of polymer lean and rich phases, respectively) and radius of secondary particles S_2.[6] Furthermore, they clarified by

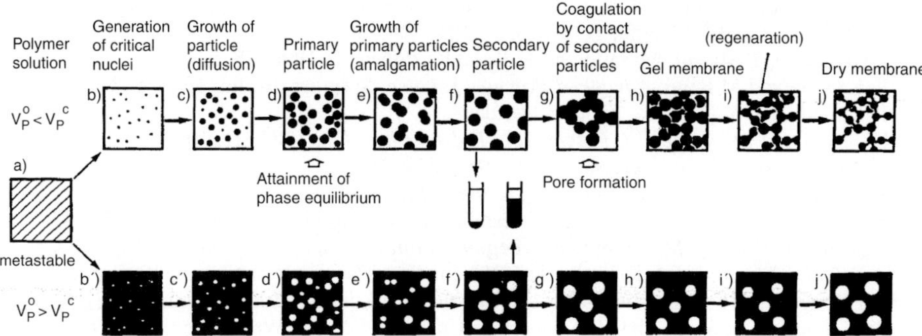

Figure 3.24.1 Elementary steps in porous polymeric membrane formation by the phase separation method. v_p^0: initial polymer volume fraction of the solution when the phase separation occurs; v_p^c: polymer volume fraction at the critical solution point.

computer simulation experiments[7] the applicability limits of the above theoretical equation of $N(r)$ to practice and the effects of the structural contraction occurred, more or less, during the coagulation process (steps f–h in Figure 3.24.1) on the resultant dry membrane structure.

Until recently, no one knew that truly liquid–liquid two-phase separation occurs in the system of cellulose solution and nonsolvent. Accordingly, a rather complicated method has been proposed in patent literatures to produce porous cellulose membranes with mean pore diameter of 0.2–3.0 μm. In 1981, Iwata *et al.*[9] discovered empirically that acetone is suitable as nonsolvent which causes liquid–liquid two-phase separation in cuprammonium cellulose solution. Based on their discovery, porous regenerated cellulose membranes with the mean pore diameter, as measured by the water flow rate method $2r_f$, ranging from 15 nm to 1 μm have been commercially produced. KI confirmed the growth of the primary particles to the secondary particles in the above system of cuprammonium cellulose solution/acetone/water.[5]

In this section, an attempt was made (1) to prepare porous regenerated cellulose membranes by casting cellulose cuprammonium solutions and then immersing them into aqueous acetone solutions as coagulant, and (2) to investigate membrane characteristics such as radius of secondary particles S_2 on the surfaces of the membranes, mean pore diameter measured by the water flow rate method $2r_f$, membrane porosity by apparent density method $P_r(d_3)$, and membrane thickness of dry membrane L_d, and (3) to clarify phenomenological effects of solvent-casting conditions on pore characteristics of the membrane formed and to explain the effect in terms of the particle growth theory proposed previously by KI.

Figure 3.24.2 represents a semiquantitative phase diagram for cellulose/cuprammonium solution/acetone system. Note that this system is not a simple system of polymer/solvent/nonsolvent, and that it is impossible to draw a two-dimensional phase diagram for this system. Accordingly, this figure is not a quantitative phase diagram obtained by phase separation experiments, but a general, qualitative one modeled after a phase diagram for a system of polymer/solvent/nonsolvent. In the figure, a thick curve and a thin curve represent the CPC and gelation lines, respectively. An open circle and a closed

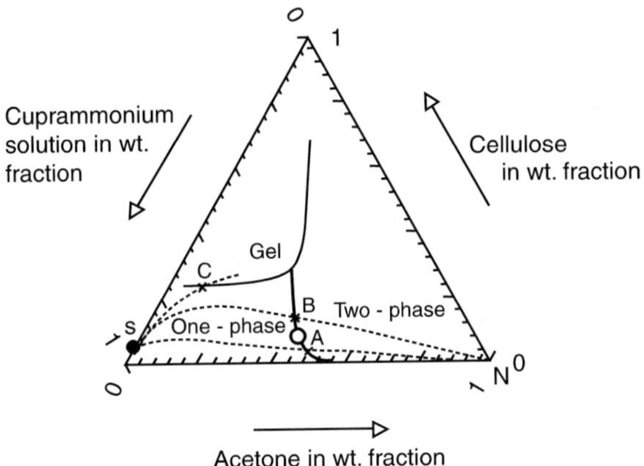

Figure 3.24.2 Qualitative phase diagram for a system of cellulose/cuprammonium solution/acetone.[1] Thick full curve, cloud point curve; thin full curve, gelation line, open circle, critical dilution point; closed circle, starting solution (i.e. cast cellulose solution).

circle represent composition of CPS (volume fraction of polymer, v_p^c) and that of starting solution (i.e. cast cellulose solution; volume fraction of polymer, v_p^s), respectively.

In practice, water fraction w_{H_2O} values of the cast solutions are nearly equal to those of coagulation solutions with the acetone fraction $w_{acetone}$ ranging from 0.20 to 0.25. Therefore, water molecules scarcely move across an interface between the cast cellulose solution and the coagulation solutions. Composition of the cast solution changes along curve SAN; it crosses the CPC at point A, where initial polymer volume fraction v_p^0 is less than v_p^c.

The results obtained are summarized briefly as follows:

(1) Surfaces of membranes prepared by immersing cast solutions in coagulation solution with weight fraction of acetone $w_{acetone}$ below 0.30 consisted of the secondary particles of the polymer-rich phase (referred to as the 'polymer particle').

(2) As the ammonia concentration in the system was higher, polymer particles of polymer rich phase grow faster. Ammonia was one of the most important factors dominating the size of cellulose particles, composing the membranes.

(3) Membrane thickness of dry membrane L_d was in proportion to the cellulose weight fraction w_{cell} of cast solutions. On the contrary, $P_r(d_3)$ and $2r_f$ were inversely proportional to w_{cell}. These experimental findings suggest strongly that density of dried polymer particles increases in proportion to w_{cell} in the solutions.

(4) Pore shape in a whole body of a membrane changed drastically from noncircular pores to circular pores when $w_{acetone}$ in coagulation solutions exceeded 0.30, indicating that $w_{acetone}$ dominates phase separation conditions such as phase volume ratio R ($\equiv V_{(1)}/V_{(2)}$; $V_{(1)}$ and $V_{(2)}$ are volumes of polymer lean and rich phases, respectively), and compositions of phase separation points.

(5) Membranes consisting of large secondary particles have larger mean pore size, and this fact agrees well with the prediction by KI's lattice theory on pore size distribution.

(6) Change in pore shape of membrane surface from noncircular to circular with the casting conditions corresponds well to drastic changes in $2r_f$ and tensile strength TS of the membrane.

REFERENCES

1. H Iijima, M Iwata, M Inamoto and K Kamide, *Polym. J.*, 1997, **29**, 147.
2. K Kamide and S Manabe, In *Material Science of Synthetic Membranes* (ed. DR Lloyd), ACS Symposium Series, No. 269, American Chemical Society, Washington, DC, 1985, pp. 197–228.
3. K Kamide, *Thermodynamics of Polymer Solutions: Phase Equilibria and Critical Phenomena*, Elsevier, Amsterdam, 1990, Chapter 6.
4. K Kamide, H Iijima and S Matsuda, *Polym. J.*, 1993, **25**, 1113.
5. K Kamide, H Iijima and H Shirataki, *Polym. J.*, 1994, **26**, 21.
6. H Iijima, S Matsuda and K Kamide, *Polym. J.*, 1994, **26**, 39.
7. K Kamide, H Iijima and A Kataoka, *Polym. J.*, 1994, **26**, 623.
8. H Iijima, A Kataoka and K Kamide, *Polym. J.*, 1995, **27**, 1033.
9. M Iwaya, S Manabe and M Inoue, Japanese Patent 1,439,049 (1988); Japanese Patent 1,434,154 (1988); Japanese Patent 1,473,266 (1988); US Patent 4,581,140 (1986); US Patent 4,604,326 (1986); US Patent 4,822,540 (1989).

<h1 style="text-align:center">– 4 –</h1>

Cellulose in Aqueous Sodium Hydroxide

4.1 SOLUBILITY OF CELLULOSE IN AQUEOUS ALKALI SOLUTION

4.1.1 Some experimental findings[1]

Since Mercel discovered the process of the mercerization of cellulose—a process pertaining to cotton fabrics by treating cellulose with aqueous (aq.) alkali solution under tension—the swelling phenomena of cellulose with aq. alkali solution has long been studied in some detail by many investigators. For example, for natural cellulose, about 8–10 wt% aq. sodium hydroxide (NaOH) at low temperature was found experimentally to be the most powerful swelling agent[2,3] in which only a small part (probably, a low molecular weight compound) of cellulose can be dissolved.[4–7] Without giving the details, Staudinger *et al.*[6] stated that cotton and mercerized cotton can dissolve in a 10% (w/v) NaOH solution when their viscosity—average degrees of polymerization (P_v)—are below 400, and that regenerated cotton with a P_v of less than 1200 can also dissolve in 10% (w/v) NaOH. The solubility behavior of the regenerated cellulose observed by Staudinger *et al.* was unfortunately not reproducible. Viscose and cuprammonium rayons never dissolve completely in 10% (w/v) aq. NaOH. The words 'löslichkeit in 10-iger NaOH,' which Staudinger used in his study, should be understood to mean partial dissolution. The experimental fact that cellulose contains the alkali soluble part was used to evaluate the lateral order distribution of cellulose fibers without seriously considering the molecular weight fractionation effect by the alkali. In this case, it has been accepted without direct evidence that only the difference in the aggregate state of cellulose molecular chains influences predominantly the solubility of cellulose towards alkali. From a theoretical viewpoint, the cellulose molecular chains influence predominantly the solubility of cellulose in water if there are no intra- and intermolecular hydrogen bonds. Discussion of the solubility of cellulose from the viewpoints of both the aggregate state of the molecular chain and hydrogen bonding seems useful, but needs further detailed experimental evidence. Recent advances in high resolution nuclear magnetic resonance (NMR) techniques, especially in the cross polarization magic angle sample spinning (CP-MAS) ^{13}C NMR technique for solid cellulose,[8–16] will facilitate the solution of the problem of the solubility of cellulose in aq. NaOH.

In this section, we show that regenerated cellulose prepared under special conditions is completely soluble in a 10 wt% aq. NaOH at 4 °C, and discuss the solubility of cellulose in terms of intramolecular hydrogen bonding using CP-MAS ^{13}C NMR and deuteration infrared (IR).

Cellulose samples

Four regenerated cellulose sample films were prepared as follows. A cellulose solution (cellulose concentration = 8 wt%) was made by dissolving purified cotton (viscosity average molecular weight $M_v = 19.4 \times 10^4$) in cuprammonium solution. The resulting solution composition (molar ratio) was cellulose/NH_3/Cu/H_2O = 1.0/6.4/1.0/75.6. The solution was cast on a glass plate and then immediately dipped into a 5 wt% aq. sulfuric acid (H_2SO_4) bath at 28 °C, washed, and dried in air. During this procedure, coagulation and regeneration seem to occur almost simultaneously. The film thus prepared was coded as BRC-1. Sample BRC-2 and BRC-4 films were obtained by casting the cellulose solution on glass plates and partially evaporating ammonia from the cast solutions in air at 25 °C for 3 and 120 min, respectively, dipping them in a 2 wt% aq. H_2SO_4 at 25 °C for 20 min, followed by washing with water and drying in air. We obtained the BRC-3 film by dipping BRC-2 film into water at 25 °C for 24 h, followed by drying in air. A regenerated cellulose solid mass sample coded as BRC-5 was obtained as follows. The cellulose solution was poured into acetone with strong agitation, and the coagulated precipitate was separated by filtration. The precipitate was regenerated with a 2 wt% aq. H_2SO_4 at 25 °C for 20 min, followed by washing with water and drying in air. BRC-1 to BRC-4 were subjected to X-ray diffraction, deuteration IR spectroscopy, CP-MAS ^{13}C NMR spectroscopy, and a solubility test. BRC-5 was subjected only to the X-ray diffraction and the solubility test.

Three acid hydrolyzed cotton samples (CL-1, CL-2, CL-3) treated with 6 wt% aq. H_2SO_4 at 60 °C for 0–120 min and the powdered pulp (PC-1) obtained by ball milling were also subjected to X-ray diffraction and the solubility test. The CP-MAS ^{13}C NMR spectrum was recorded for PC-1.

For polarized IR spectroscopy, the oriented film samples for BRC-1 and BRC-4 were prepared as follows. A wet BRC film containing 300% water was stretched in one direction by 200%, fixed in that state, and then dried in air. BRC-4 oriented film was obtained by the procedure described above except that it was stretched 300%.

Viscosity average molecular weight

M_v of the samples and the starting cellulose materials were calculated from following equation:[17]

$$[\eta] = 3.85 \times 10^{-2} M_w^{0.76} \qquad \text{in cadoxen at 25 °C} \qquad (4.1.1)$$

Here, M_w is the weight average molecular weight.

X-ray diffraction

X-ray diffraction patterns of the samples were recorded by the transmission method with a RU-200PL type X-ray diffractometer (Rigaku Denki Co., Japan). The crystallinity

$\chi_c(X)$ of the samples was a calculated by the equation proposed by Segal:[18]

$$\chi_c(X) = (I_{002} - I_{am})/I_{002} \qquad (4.1.2)$$

where I_{002} is the diffraction intensity for the peak ($2\theta = 21.7°$) of the plane (002), and I_{am}, the intensity at $2\theta = 16°$ for cellulose II. For cellulose I, I_{002} is the intensity for the peak at $\sim 2\theta = 22.6°$, and I_{am}, that at $2\theta = 19°$. We defined the degree of amorphous content of the samples $\chi_{am}(X)$ by eq. 4.1.3:

$$\chi_{am}(X) = 1 - \chi_c(X) \qquad (4.1.3)$$

Deuteration IR

IR spectra of the sample and partially deuterated sample films were recorded with an IR spectrometer model 430 special (Shimadzu Co., Japan), using a deuteration cell designed, and constructed at our laboratory for this purpose. Deuteration of the sample was carried out by passing D_2O vapor into the cell for 180 min to attain deuteration equilibrium. During deuteration the sample holder was kept at 100 ± 0.5 °C. The fraction of the accessible part χ_{ac} (IR) at equilibrium was calculated by the equations proposed by Mann and Marrinan.[19]

$$\log(I_0/I)_{OD}/\log(I_0/I)_{OH} = 1.11 \times C_{OD}/C_{OH} \qquad (4.1.4)$$

$$C_{OD} + C_{OH} = 1 \qquad (4.1.5)$$

Here, $(I_0/I)_{OH}$ and $(I_0/I)_{OD}$ are the maximum optical density at the OH stretching region (~ 3440–3370 cm^{-1}) and OD stretching region (~ 2490–2520 cm^{-1}), respectively. C_{OD} and C_{OH} are the fractions of the accessible and inaccessible parts at equilibrium, respectively. Hence, C_{OD} directly gives $\chi_{ac}(IR)$.

NMR spectroscopy

CP-MAS ^{13}C NMR spectra of the cellulose samples were recorded with an FT NMR spectrometer FX-200 (50.1 MHz; JEOL, Japan) under the following conditions: 90° pulse, pulse width; 4.5 μs, pulse repetition; 5 s, pulse interval (cross polarization contact time); 1–2 μs, data point; 8 k.

For the BRC-4 solution in NaOH–D_2O (10:90, w/w; polymer concentration C_p; 5 wt%), the ^{13}C NMR spectrum was obtained using the same spectrometer and operating conditions as above. The relative amount of the higher field peaks χ_h (NMR) of the C_4 carbon peaks was estimated from eq. (4.1.6):

$$\chi_h(NMR) = I_h/(I_1 + I_h) \qquad (4.1.6)$$

Here, I_1 and I_h denote peak areas under the peaks centered at 87.9 and 84 ppm, respectively, and the line shapes were approximated as the Loreazian type.

Solubility test

The cellulose film samples were cut into small pieces (5×5 mm^2) and vacuum dried at 40 °C for 18 h. 1–5 parts of the cellulose were dispersed into 99–95 parts of 1–5 wt% aq. NaOH precooled at 4 °C for 1 h and quickly distributed using a home mixer intermittently in 1 min to minimize the rise in local temperature. The dispersed solution was then ultracentrifuged at 20,000 rpm at 4 °C for 45 min, followed by measuring the amount of cellulose (m_G) in the remaining gelatinous layer (lower liquid phase) by regeneration. This procedure was also applied to the BRC-5 and natural cellulose samples. The solubility of cellulose S_a is defined by eq. (4.1.7):

$$S_a = 100 \times (m_0 \times m_G)/m_0 \qquad (4.1.7)$$

where m_0 denotes the amount of cellulose originally used in this test.

Figure 4.1.1 shows the X-ray diffraction patterns of the cellulose samples. As expected, all of the regenerated cellulose samples had the crystalline form of cellulose II, showing typical diffraction peaks for the plane (002) at $2\theta = 21.7°$ (Figure 4.1.1a) and the acid-hydrolyzed cotton (CL series) and the powdered pulp (PC-1) (Figure 4.1.1b) was comprised of the cellulose I type crystals with typical diffraction peaks for the plane (002) at $2\theta = 22.6°$.

The second and fifth columns of Table 4.1.1 show the results for the crystal forms and the amorphous content $\chi_{am}(X)$, respectively. $\chi_{am}(X)$ for the BRC series samples ranges from 0.46 to 0.88 and that of the CL series is about the same (0.23 ± 0.03). PC-1 was

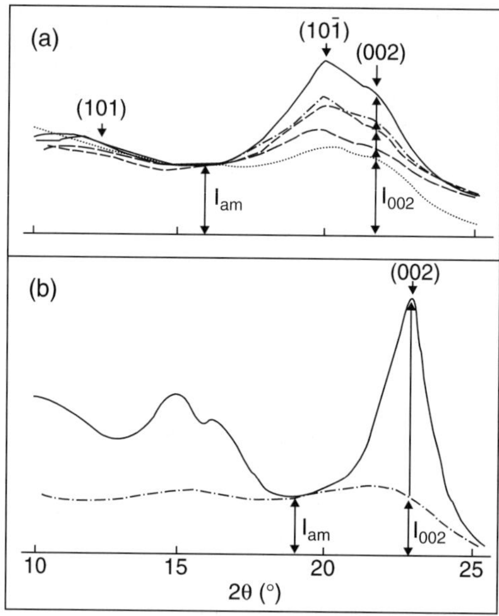

Figure 4.1.1   X-ray diffraction patterns for regenerated (a) and natural (b) cellulose.[1] (a) full line, BRC-1; chain line, BRC-2; double chain line, BRC-3; broken line, RC-4; dotted line, BRC-5; (b) full line, CL-1; chain line, PC-1.

Table 4.1.1

Molecular characteristics of cellulose samples[1]

Sample code	Crystal form	$M_v \times 10^{-4}$	S_a (%)	$\chi_{am}(X)$	$\chi_{ac}(IR)$	$\chi_h(NMR)$
BRC-1	II	8.1	52	0.46	0.59	0.56
BRC-2	II	7.9	62	0.69	0.68	0.62
BRC-3	II	7.9	81	0.88	0.64	0.79
BRC-4	II	7.9	90	0.73	0.68	0.87
BRC-5	II	7.9	100	0.94	–	–
CL-1	I	19.4	9	0.21	–	–
CL-2	I	8.3	31	0.25	–	–
CL-3	I	5.8	30	0.26	–	–
CL-4	I	1.9	58	0.92	–	–

nearly amorphous ($\chi_{am}(X) = 0.92$). The M_v data appear in the third column of Table 4.1.1. M_v of the regenerated samples are nearly all about the same ($8.0 \pm 0.1 \times 10^4$). Those of the CL series range from 5.8 to 19.4×10^4, with that of PC-1 as low as 1.9×10^4.

Figure 4.1.2 shows the effect of NaOH concentration C_a (wt%) of the aq. NaOH solution on S_a at the initial polymer concentration $C_p = 5$ wt% at 4 °C of samples BRC-4 (unfilled) and PC-1 (filled). Obviously, the 10 wt% aq. NaOH is the most potent cellulose solvent. The NaOH concentration dependence of S_a seems independent of the crystalline form.

Figure 4.1.3 shows the dependence of S_a on M_v for the CL (cotton linter) series having nearly the same $\chi_{am}(X)$ value for the 10 wt% aq. NaOH at 4 °C. Numbers denote the polymer concentration C_p (wt%) initially charged in the alkali. S_a drops below 15% irrespective of C_p when M_v of the polymer increases. S_a increases with a decrease in M_v and C_p: S_a for cotton having $M_v = 5.8 \times 10^4$ was 72.5% at $C_p = 1$ wt%. In the figure,

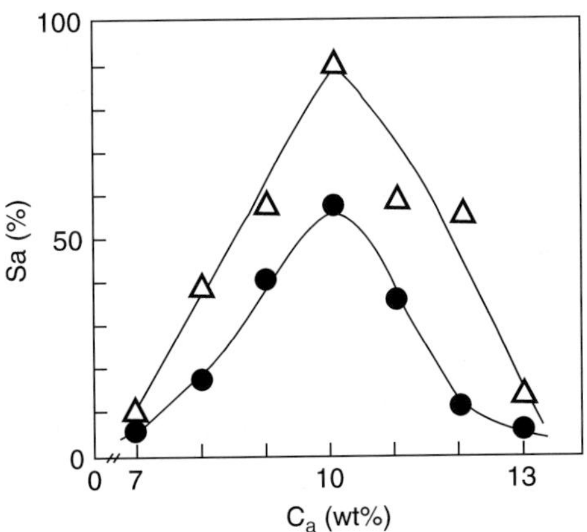

Figure 4.1.2   Relationship between solubility S_a of cellulose at 4 °C and NaOH concentration C_a (wt%) in water as the solvent:[1] (Δ), BRC-4; (●), PC-1.

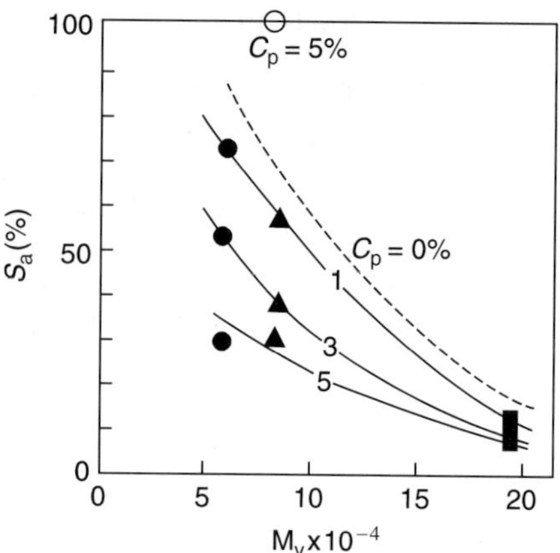

Figure 4.1.3   Dependence of solubility S_a of natural cellulose in a 10 wt% aq. NaOH at 4 °C on the viscosity average molecular weight M_v^1 : (■),CL-1; (▲), CL-2; (●), CL-3; (○), BRS-5. C_p on the lines denotes the initial concentration (wt%) of cellulose in 10 wt% aq. NaOH and the broken line was drawn by extrapolating C_p to 0, from Figure 4.1.4.

the broken line was drawn from a line of S_a extrapolated to $C_p = 0\%$. Thus, the solubility of natural cellulose in alkali strongly depends on both M_v and C_p.

Figure 4.1.4 shows the dependence of S_a on the C_p of cellulose samples for a 10 wt% aq. NaOH at 4 °C. In the figure, the open marks denote BRC series samples and the closed marks, the cotton (CL series). The dependence of BRC series is not as large as that of the CL series samples. BRC series samples prepared in this study had higher alkali solubility than cotton when compared at the same M_v (about 8×10^4). In particular, sample BRC-5 was completely soluble when the C_p below 5 wt%. In the fourth column of Table 4.1.1, S_a for a 10 wt% aq. NaOH at $C_p = 5$ wt% at 4 °C is shown. The regenerated cellulose, prepared by changing the vaporization time of ammonia from the cuprammonium

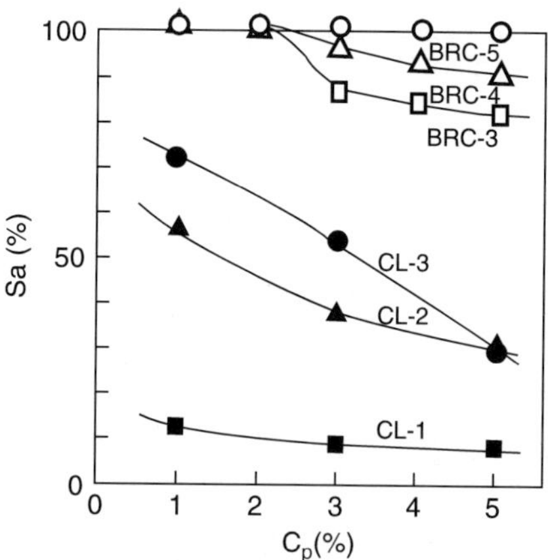

Figure 4.1.4   Initial concentration C_p dependence of solubility S_a of natural cellulose in a 10 wt% aq. NaOH at 4 °C[1]: (□), BRC-3; (Δ), BRC-4; (○), BRC-5; (■), CL-1; (▲), CL-2; (●), CL-3.

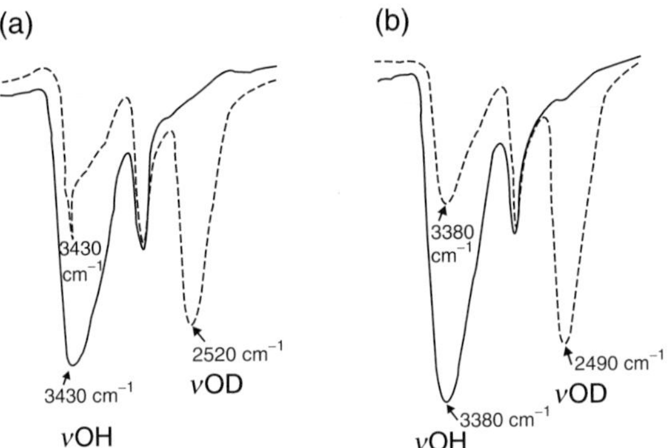

Figure 4.1.5   IR spectra of OH and OD stretching region for BRC-1 (a) and BRC-4 (b):[1] solid line, before deuteration; broken line, after deuteration.

cellulose solution and the precipitation method of cellulose reveals a large variety of S_a. That is, S_a decreases from 100 to 52% in the following order: BRC-5 > BRC-4 > BRC-3 > BRC-2 > BRC-1. S_a of the natural cellulose was at most 58% (for PC-1). The difference in S_a for these regenerated samples might be correlated to the difference in the supermolecular structure of the samples.

Figure 4.1.5 exemplifies typical infrared spectra of the deuterated and nondeuterated samples BRC-1 and -4. $\chi_{ac}(IR)$, evaluated using eqs. (4.1.4) and (4.1.5) from spectra is summarized in the sixth column of Table 4.1.1. $\chi_{ac}(IR)$ ranges from 0.59 to 0.68 and is essentially the same for any samples.

Figure 4.1.6 shows the CP-MAS ^{13}C NMR spectra for the samples. Assignment of the spectra was made with the aid of previous works by Earl and Van der Hart[9,10] and is given in Table 4.1.2. Depending on the preparative method, C_1, C_4, and C_6 carbon peaks

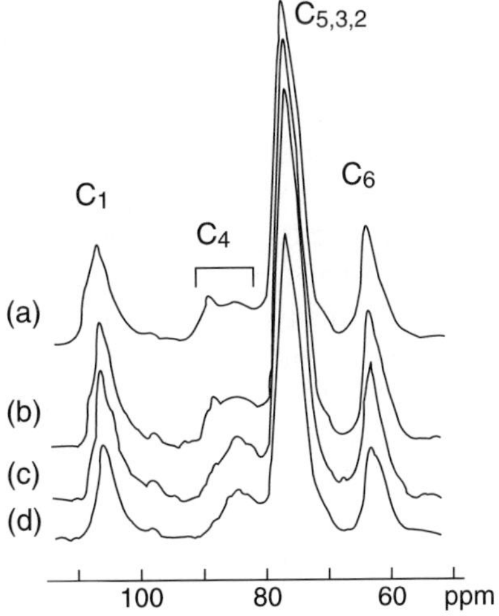

Figure 4.1.6   CP-MAS ^{13}C NMR spectra of BRC series samples[1]: (a), BRC-1; (b), BRC-2; (c), BRC-3; (d), BRC-4.

Table 4.1.2

NMR peak assignments (in ppm)[1]

Sample	Peak position[a]			
	C_1	C_4	C_5, C_3, C_2	C_6
BRC-1	107.3, 105.4, 97.5	87.9*, 83.8, 81.7	75.1(broad)	63.1, 61.0, 60.0
BRC-2	107.3, 105.3, 104.3	88.8*, 87.9*, 84.3	75.2(broad)	63.0
		84.3, 83.6, 82.8		
BRC-3	107.3, 105.3, 98.0	89.8*, 87.7*, 86.4	75.1(broad)	62.9, 60.5
		85.8, 85.0, 84.0		
		82.3, 81.7, 80.6		
BRC-4	107.3, 105.2, 100.5	87.8*, 85.3, 84.2	75.3(broad)	62.8, 61.8, 60.7
	97.3	82.6, 81.8		60.0
PC1	105.4, 104.4	90.0*, 88.8*, 87.3		
		86.8, 85.8, 84.7		
		83.2, 80.9		
Cellulose solution from BRC-4	104.7	79.9	76.4, 75.0	61.9

[a]Peaks marked by * arise from the highly ordered region.

change systematically and, in particular, variation in the C_4 carbon peak is remarkable. Therefore, we analyzed the C_4 carbon peak region using the method described in the experimental section.

Figure 4.1.7 shows the magnified ^{13}C NMR spectra in the range of 80–90 ppm for the C_4 carbon peak region, which can be approximately divided into two envelopes, as shown by the broken lines in the figure. χ_h(NMR) was calculated by eq. (4.1.6) and is shown in the seventh column of Table 4.1.1. For the BRC-1 to BRC-4 samples χ_h(NMR) was in the range of between 0.5 and 0.87 and that for PC-1 was 0.70.

Naturally occurring cellulose has very low solubility in aq. alkali. Cotton dissolves by only a few percent in a 16 wt% aq. NaOH and pulp has about 10% solubility in such a solution at 4 °C, depending on M_v and C_p. Following the dissolution, the solutions become turbid within several minutes at 20 °C. For the first time, we were able to make celluloses, having fairly high molecular weights, dissolve to a large extent in a dilute alkali solution at 4 °C. In fact, sample BRC-5 have a completely homogeneous solution which remained transparent and extremely stable at 20 °C.

A cellulose sample, prepared physically from pulp, dissolved by as much as 58% (at $C_p = 5\%$) in alkali and a significant increase in amorphous content was observed, although its M_v was as low as 1.9×10^4 (Table 4.1.1). Accordingly, an increase in the solubility of cellulose having crystal form I seems to reflect an increase in amorphous content or a decrease in M_v. However, according to our experience over the years, no physical treatment can produce cellulose with S_a (at $C_p = 5$ wt%) $> 85\%$. Sample PC-1 was almost amorphous cellulose, as indicated by X-ray diffraction, but its S_a was only 58%. Thus, χ_{am}(X) is not the only factor governing S_a.

Next, a correlation among S_a, χ_c(X), χ_{ac}(IR), and χ_h(NMR) was examined among samples with cellulose crystal form II to clarify the factors contributing to S_a. In this case,

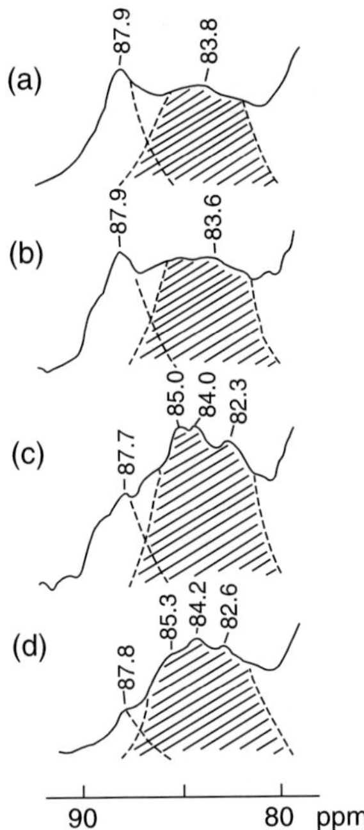

Figure 4.1.7   CP-MAS ^{13}C NMR spectra of C$_4$ carbon peak region for BRC series samples:[1] symbols are the same as in Figure 4.1.6. Numbers on the peaks denote peak values in ppm and the broken lines denote separation of peak areas under the peaks. Hatched line shows peak areas for higher magnetic field components.

sample BRC-5 was excluded due to lack of χ_h(NMR) data. The results are shown in Figure 4.1.8. The numbers are the correlation coefficients γ between two arbitrarily chosen parameters, γ for $S_a \sim \chi_h$(NMR) is as high as 0.998, and those for $S_a \sim \chi_c$(X) and $S_a \sim \chi_{ac}$(IR) are 0.777 and 0.603, respectively. Thus, χ_h(NMR) is most closely related to S_a. In other words, the solubility behavior of cellulose cannot be explained by only the

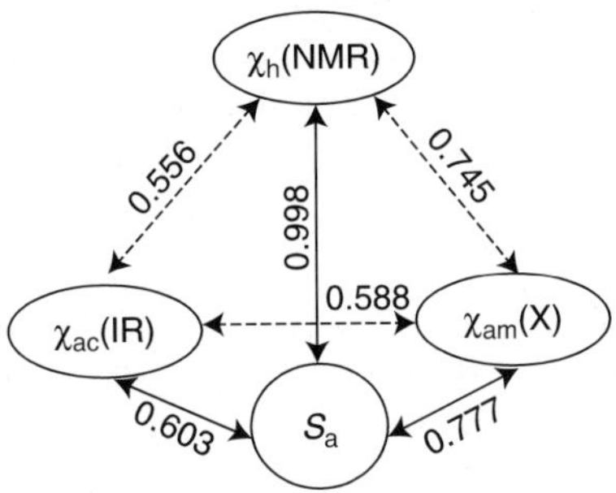

Figure 4.1.8   Correlations among solubility S_a and the so-called amorphous content evaluated by X-ray diffraction, IR and NMR.[1] Numbers are the correlation coefficients γ for two arbitrarily chosen parameters.

concepts of 'crystal amorphous' or 'accessible/inaccessible.' The correlation coefficients between the amorphous content parameters are as follows: $\chi_{am}(X) \sim \chi_h(NMR)$, 0.745; $\chi_h(NMR) \sim \chi_{ac}(IR)$, 0.556; $\chi_{am}(X) \sim \chi_{ac}(IR)$, 0.588.

$\chi_{am}(X)$ and $\chi_h(NMR)$ shown in Table 4.1.1 are not well correlated and are clearly inconsistent with those of Hirai *et al.*[13] who reported the fraction of sharp peak component in the C_4 carbon region (i.e. $1 - \chi_h(NMR)$) for natural and hydrolyzed cotton and rayon, and regenerated cellulose from a dimethylsulfoxide–paraformaldehyde solution, evaluated assuming the Lorenzian type line shape, closely correlates with the degree of crystallinity determined by X-ray diffraction. It should be noted here that the NMR spectra do not directly reflect the supermolecular structure.

We shall now discuss the driving force responsible for splitting the C_4 carbon peaks into two envelopes, and its physical meaning. The peak height at 87.9 ppm becomes lower for samples in the following order: BRC-1 > BRC-2 > BRC-3 > BRC-4. The peak position (87.9 ppm) for these regenerated samples is in good agreement with literature data.[8–16]

Figure 4.1.9 shows the chemical structure of a cellobiose unit. All C_1 and C_4 carbons are linked by oxygen atoms, which are the center of rotation accompanied by conformational change in the pyranose ring. Thus, one major factor influencing the magnetic properties of C_1 and C_4 carbons is conformational change about the C_1–O–C_4 sequence. This change is predominantly controlled by intramolecular hydrogen bonds between the neighboring glucopyranose units in a cellobiose molecule.[20,21] An intramolecular hydrogen bond between the hydrogen attached to the C_3 carbon ring and oxygen in the neighboring glucopyranose ring causes the formation of a seven membered cyclic ether, whose C_3 and C_4 carbons are located at the terminals to constitute a kind of conjugate system of π electrons on the oxygen atoms and σ electrons on the C–O orbitals. If the electrons in this system are mobile, then C_4 carbon is cationized and C_3 carbon is anionized. If the intermolecular interaction, which can be disregarded as negligible even if it exists, then is not taken into account, the electron density in the C_4 carbon becomes lower than that on the C_4 carbon not participating in the seven membered ether on formation of the intramolecular hydrogen bond, resulting in a shift of the C_4 carbon NMR peak to a lower magnetic field. Hence, we can consider a partial breakdown of the intramolecular hydrogen bonds that existed in the cellulose sample to create a magnetic field about the C_4 carbon heterogeneously, so that its peak (and possibly also the C_1 carbon) produce new components in a higher magnetic field, compensating for the sharp component of the C_4 carbon peak. This change results in the formation of two envelopes in the C_4 carbon peak region. The systematic change observed in Figure 4.1.7 can be reasonably interpreted by the above mechanism.

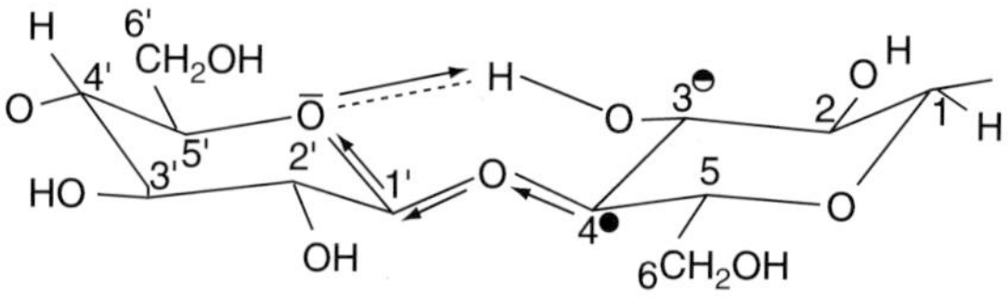

Figure 4.1.9 Schematic representation of the $\pi - \sigma$ electron conjugate system in a cellobiose unit.[1]

It may be concluded that a low magnetic field envelope in the C_4 carbon peak can be assigned to cellobiose units having fewer intramolecular hydrogen bonds (i.e. the region about the aforementioned cellobiose units in cellulose molecules has a very disordered conformation). We cannot simply conclude that because there are absolutely no intramolecular hydrogen bonds, the peak narrows significantly, since cellobiose units in cellulose molecules in the solid state can take on a large number of conformations.

On the basis of the above mechanism, variation in the lowest peak intensity among the C_1 carbon peaks between cellulose samples can be explained in the same manner as that in the case of the C_4 carbon peak (i.e. the breakdown of intramolecular hydrogen bonds). The lowest peak (107.3 ppm) for the C_1 carbon becomes weak in the following order: BRC-1 > BRC-2 > BRC-3 > BRC-4 (Figure 4.1.6 and Table 4.1.2).

Figure 4.1.10 shows the ^{13}C NMR spectrum for the solution of BRC-4 in NaOH ~ D_2O (1:9 w/w). Except for the C_3 and C_5 carbon peaks, all others obviously shift to a magnetic field higher than that for solid BRC-4. One C_4 carbon peak for the solution is located at 79.9 ppm, which is a much higher magnetic field than that of the original BRC-4 solid (87.9 and *ca.* 84 ppm). BRC-4 dissolves molecularly in a 10 wt% aq. NaOH without forming an alcoholate and solvation with solvent[22] and, hence, no intramolecular hydrogen bonds should exist in the solution. The appearance of the C_4 carbon peak at 79.9 ppm as a single sharp peak in the solution is quite reasonable considering that the C_4 carbon peak must shift to a higher magnetic field when the intramolecular hydrogen bonds are destroyed.

In view of the chemical structure of the two carbon–oxygen sequences about the C_4 and C_5 carbons, as illustrated in Figure 4.1.11, the deshielding effect on the C_4 and C_5 carbons is estimated to be of the same order when the intramolecular hydrogen bonds are completely destroyed. Thus, the C_4 carbon peak should appear near the C_5 carbon peak. As is evident from Table 4.1.2, the C_4 carbon peak for the BRC-4 solution was observed at 79.95 ppm, and this is very close to the C_5 carbon peak at 76.5 ppm. Destruction of the intramolecular hydrogen bonds is also expected to cause the C_3 carbon peak for solid

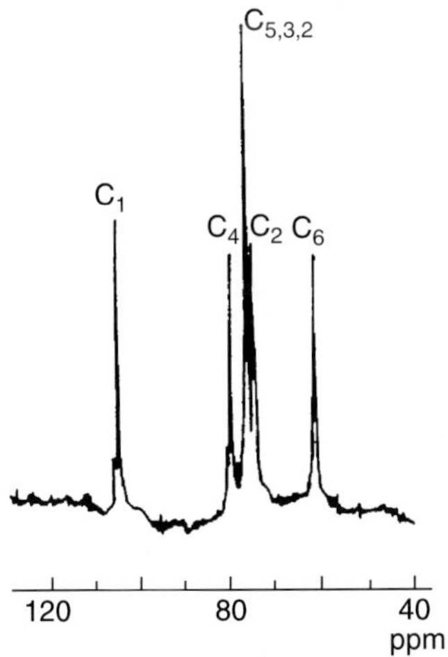

Figure 4.1.10 ^{13}C NMR spectrum for BRC-4 solution in NaOH–D_2O (1:9 w/w).[1]

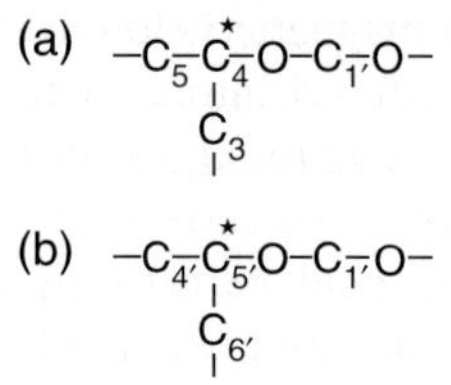

(a) $-C_5\bar{C_4}^*O-C_{1'}-O-$
            $|$
            C_3

(b) $-C_{4'}\bar{C_{5'}}^*O-C_{1'}-O-$
            $|$
            $C_{6'}$

Figure 4.1.11   Chemical structures of two carbon–oxygen sequences about C_4 (a) and C_5 (b) following destruction of intramolecular hydrogen bonds.[1]

cellulose to shift to a lower magnetic field, which is the reverse of that for the C_4 carbon peak. The main C_3 carbon peak for the original BRC-1 is seen at 75.1 ppm, while the C_3 carbon peak for the BRC-4 solution in alkali is observed at 76.4 ppm. The latter is slightly lower than the former.

Additional experimental evidence supporting the formation of a higher field component of the C_4 carbon NMR peak by the destruction of intramolecular hydrogen bonds was obtained by IR. Figure 4.1.12 shows the polarized IR spectra of BRC-1 (a) and BRC-4 (b) (solid line; parallel, broken line; perpendicular) films after deuteration. The BRC-1 whose $S_a = 52\%$ (at $C_p = 5$ wt%) in a 10 wt% aq. NaOH at 4 °C shows two parallel dichroism bands at 3480 and 3430 cm^{-1}. These two bands are responsible for the two types of intramolecular hydrogen bonds between $O_3-H\cdots O_5'$ with different bond lengths (2.78 and 2.82 Å), as observed for cellulose crystal II. At a lower wave number, some perpendicular dichroism bands responsible for intermolecular hydrogen bonds appeared in sample BRC-1. In contrast, sample BRC-4 with $S_a = 90\%$ does not have parallel dichroism bands at 3480 and 3430 cm^{-1}, but rather a broad parallel dichroism

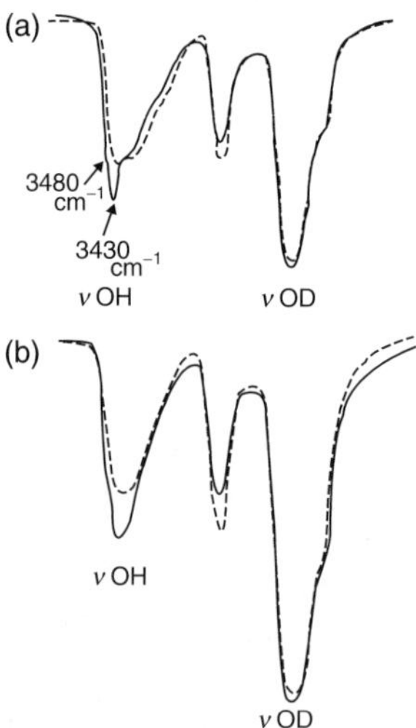

Figure 4.1.12   Polarized IR spectra for BRC-1 (a) and BRC-4 (b):[1] solid line, parallel; dashed line, perpendicular.

band at $3380\,cm^{-1}$, attributable to destruction of intramolecular hydrogen bonds. Because the IR spectra were taken after deuteration, the remaining OH stretching regions of the spectra shows the inaccessible part of the cellulose. Thus, Figure 4.1.12 shows that intramolecular hydrogen bonds are destroyed even in the inaccessible part of sample BRC-4. It may thus be concluded that almost all intramolecular hydrogen bonds are destroyed in solid sample BRC-4. Note that χ_h(NMR) for BRC-4 is nearly unity (0.87). The findings also strongly suggest that the higher field component of the C_4 carbon peak appeared as a result of intramolecular hydrogen bond destruction.

In conclusion, it was demonstrated that cellulose whose intramolecular hydrogen bonds are completely broken or weakened, dissolve in aq. alkali and the solution thus prepared is stable over a long period of time. For a more detailed understanding of the mechanism underlying the dissolution of the cellulose reported here in alkali, additional research will be presented in later sections.

4.1.2 An explanation of aq. alkali solubility by intramolecular hydrogen bond concept[23]

Solid state high resolution ^{13}C NMR spectra were observed, using the CP-MAS technique, for various cellulose samples of different crystalline forms, crystallinity, and supermolecular structures, revealing the usefulness of CP-MAS NMR method for analyzing the crystalline and noncrystalline solid structure of celluloses.[1,8–10,15,21,24–26] Each peak, assigned to C_1, C_4, and C_6 carbons in anhydroglucopyranose ring, was observed to consist of one or two sharp resonance(s) and one broad one, respectively. The relative content of the sharp and broad components, estimated from the integral, depended significantly on crystal structure, crystallinity, and morphology.[10] In particular, the remarkable variation in the shape of the C_4 carbon peak at 80–90 ppm (tetramethylsilane was used as a reference standard) has attracted the attention of many investigators. On the basis of a linear relationship between the integrated fraction of the downfield sharp component of C_4 carbon peaks, obtained in CP-MAS ^{13}C NMR spectra, and the crystallinity determined by X-ray diffraction for regenerated cellulose sample with different crystallinity and cotton, Horii *et al.*[25] concluded that the downfield sharp and upfield broad components are contributions from the crystalline and noncrystalline components, respectively. Earl and VanderHarts[10] concluded from an analysis of the CP-MAS ^{13}C NMR spectra of native cellulose from several origins that the broad peaks of the C_4 and C_6 carbons are attributed to anhydroglucose on the surface of cellulose elementary fibrils in addition to noncrystalline and narrow C_4 and C_6 peaks (at 66 and 90 ppm) are due to anhydroglucose underneath the elementary fibrils considered to be completely perfect crystalline components.

As is well known, the crystallinity of cellulose, χ_c, varies significantly according to the measuring method employed. For example, cotton has $\chi_c = 73\%$ by the X-ray diffraction method, 64% by the density method, and 58% by the method of water accessibility.[27] Thus, it is of no purpose to discuss the numerical coincidence of parameters (content of sharp components) evaluated from NMR with χ_c estimated by other popular methods. By X-ray diffraction, the regular part in the packing between molecular chains is measured and the length should be more than 30 Å. In contrast, the ^{13}C NMR method seems almost independent of molecular chain packing (that is, intermolecular interaction), but is very sensitive to the conformation of the cellobiose units of a molecular chain.

In Section 4.1.1, we found that the appearance of the C_4 carbon peak of a cellulose solid over a wide range from 90 to 80 ppm can be adequately explained simply by considering the intramolecular hydrogen bonds between the C_3 hydroxyl and O_5 ring oxygen, correlating the content of the broad component of C_4 carbon peak (χ_h(NMR)) with the solubility of cellulose in alkali solution.[1] A sharp component in the C_4 carbon region was found attributable to the C_4 carbon in a cellobiose unit, in which a $O_3-H\cdots O_5'$ intramolecular hydrogen bond is strongly formed, and a broad component reflects the distribution of the bond strength of an incompletely destroyed intramolecular hydrogen bond.

Obviously, all references[2,8–10,15,21,24–26] are concerned with the peaks of only the C_1 carbon and no further study has been carried out on the theoretical analysis of the ^{13}C NMR peaks of carbons other than the C_4 carbon. This section attempts to analyze the CP-MAS NMR peaks of each carbon of various samples having the crystalline form of cellulose I or II, in view of the presence of intra- and intermolecular hydrogen bonds.

Cellulose

Wood pulp cellulose (C-1) with a viscosity average molecular weight $M_v = 2.1 \times 10^5$, a crystallinity χ_c by X-ray diffraction of 78% and the crystal form of cellulose I was used. A 100 weight part of C-1 was immersed in a 2000 weight part of 18 wt% aq. sodium hydroxide solution at 30 °C for 30 min and then hand pressed to exclude excessive alkali from the sample. The alkalicellulose thus prepared was converted to cellulose in a 1 wt% hydrochloric acid solution and dried *in vacuo* and designated as sample code C-2, with $M_v = 1.65 \times 10^5$, $\chi_c = 55\%$ and the crystal form of cellulose II. C-1 and C-2 were ground in a ball mill at 25 °C for 8 h to give cellulose powders and were designated as C-1b ($M_v = 2.0 \times 10^4$, $\chi_c = 8\%$) and C-2b ($M_v = 1.6 \times 10^4$, $\chi_c = 0\%$), respectively. Here, M_v was determined from the solution viscosity of the cadoxen solution at 25 °C[17] and χ_c was evaluated by Segal's method.[18] χ_h(NMR), defined in Section 4.1.1 for C-1b and C-2b, were 62.4 and 91.8%, respectively. These values correlated to their solubility in a 10 wt% aq. alkali at 4 °C.

CP-MAS ^{13}C NMR

CP-MAS ^{13}C NMR spectra were recorded by a JEOL FX-200 type FT-NMR spectrometer under the following operating conditions: resonance frequency, 50.1 MHZ; cross polarization contact time, 2 ms; repetition time, 5 s; rotational velocity of magic angle sample, 3–3.5 kHz; measuring temperature, room temperature. The scan number is shown in Figure 4.1.13.

Figure 4.1.13 shows the CP-MAS NMR spectra of four cellulose solid samples. Table 4.1.3 shows the peak positions (tetramethylsilane as reference standard) and integrated peak intensity. In this table, the suffixes s and b are the shoulder and broad peaks, respectively. The NMR data on cellulose in aq. sodium hydroxide solution obtained in Section 4.1.1 were also included in the table for comparison. Peak assignment

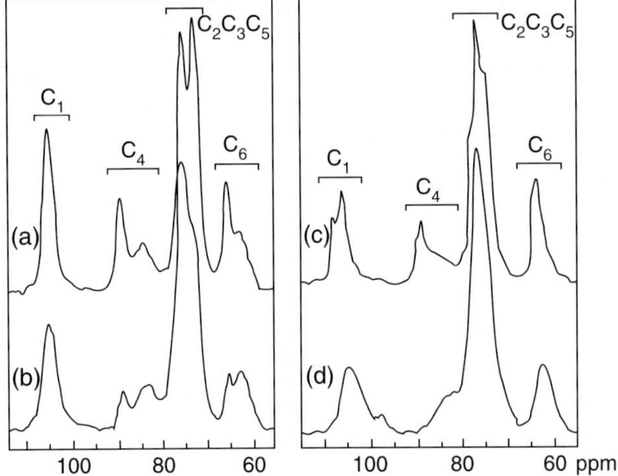

Figure 4.1.13 CP-MAS ^{13}C NMR spectra of cellullose samples:[23] (a), C-1 (150 × scan); (b), C-1b (200 × scan); (c), C-2 (495 × scan); (d), C-2b (400 × scan).

was carried out using the results of Voelter *et al.*[28] for β-glucose and cellobiose in D$_2$O for our reference. C$_1$, C$_4$, and C$_6$ carbon peaks were easily assignable since they were almost completely separated from each other and other carbon peaks as well. The C$_2$, C$_3$, and C$_6$ carbon peaks overlapped to some extent with each other in the range 77–73 ppm.

Table 4.1.3

Peak assignment of cellulose samples

Sample	Carbon position (ppm)								
	C$_1$			C$_4$			C$_2$, C$_3$, C$_5$		C$_6$
C-1	105.4			89.1	84.5^b		75.3 72.7		65.5 63.0^b
	(1.00)			(0.47)	(0.53)		(1.57) (1.51)		(0.55) (0.38)
C-1b	105.4 104.4^s			90.0	83.5^b		75.0^b 72.9^s		65.4 62.6
	(1.00)			(0.29)	(0.61)		(3.16)		(0.29) (0.55)
C-2	107.4 105.4 102.7^s			89.1	87.9	84.6^b	77.0^s 75.2 73.4^s		63.0 62.0^s
	(0.19) (0.81)			(0.13)	(0.27)	(0.60)	(3.13)		(0.88)
C-2b	104.3^b 97.4^b			82.2^b			75.5^b		62.2^b
	(0.87) (0.13)			(0.82)			(3.72)		(0.86)
Cellulose in NaOH aq.^{a}	104.7			80.0			76.4 75.0		61.9
	(1.00)			(0.94)			(2.44) (1.00)		(1.06)

Integrated intensity (in parenthesis) was normalized by C$_1$ relative intensity; b denotes a broad peak and s, the shoulder peak.
^{a}Data were reproduced from Ref. 1.

The data of Figure 4.1.13 and Table 4.1.3 lead to the following conclusions:

(1) For all cellulose samples, except amorphous cellulose (C-2b), C_4 and C_6 carbon peaks split into one or two sharp lower magnetic field components and a broad, higher magnetic field component, respectively.

(2) The sharp components in the C_4 and C_6 carbons decrease and broad components increase by ball milling.

(3) The C_1 peak broadens by ball milling and its center position shifts to a higher magnetic field by *ca.* 1 ppm.

(4) For C-2 (cellulose having crystal form of cellulose II), the C_1 and C_4 carbon peaks each have two sharp components, but the C_1 and C_4 carbon peaks each have only one sharp component for C-1 (cellulose having crystal form of cellulose I).

(5) For C-2b (amorphous cellulose), all carbon peaks including the C_4 carbon peak contain only broad components and their peak (central) positions are near those for cellulose II dissolved in alkali solution.

(6) The C_4 carbon peak for C-2b is observed at 80–84 ppm, which is slightly lower than that (80 ppm) of cellulose II dissolved in alkali, but appreciably higher than the sharp peaks (87.9, 89.1 ppm) for the C-2 sample.

(7) For C-2, a new small peak lacking in the spectra of both the C-2 solid and cellulose dissolved in alkali, is observed at 97.4 ppm.

Conclusions (1)–(3) are valid, regardless of crystal form and (1)–(4) have already been reported elsewhere.[8–10,15,21,24,25]

Sample C-2b has $\chi_c = 0\%$, $\chi_h(NMR) = 91.8\%$ and is an amorphous cellulose. In its CP-MAS NMR spectrum, the small peak at 97.4 ppm (experimental fact (7)) can be considered the C_1 anomeric carbon peak for cellulose oligomers formed by oxidative degradation during ball milling. Taking into consideration experimental data (5) and (6), along with the fact that cellulose dissolved in aq. sodium hydroxide has no inter- and intramolecular hydrogen bond[1] since the C_4 carbon peak becomes singlet and sharp at 80 ppm, we can conclude that the amorphous cellulose solid contains many cellobiose units, with very weak intramolecular hydrogen bonds, assuming that they exist at all. The intermolecular hydrogen bond randomly formed between two or more cellobiose units may exist, but may not conspicuously influence the chemical shift of the NMR spectrum, as already described. To clarify all the features of the NMR spectra for cellulose, it is adequate as a first approximation to consider the electron charge density on given carbon atoms in view of the possible presence of intramolecular hydrogen bonds. $O_3-H\cdots O_5{}'$ (a), $O_2-H\cdots O_6{}'$ (b), and $O_6-H\cdots O_2{}'$ (c) in cellulose solid.[29–32] If a cellobiose unit takes on the form (a) shown in Figure 4.1.14, then the C_3 and C_4 carbons will occupy ionic and cationic ends, respectively, owing to the formation of a seven-membered π–σ electron conjugate system. The C_1 carbon may be slightly cationized, being near the cationic end C_4 carbon. In consideration of the shielding effect induced by the electron density on the NMR spectrum, the C_1 and C_4 carbon peaks must shift towards a lower magnetic field and the C_3 carbon toward a higher magnetic field, compared to those for amorphous cellulose assumed to have no intramolecular hydrogen bonds. For the forms (b) and (c) shown in Figure 4.1.14, such a π–σ electron conjugate system as that in form (a), does not come about. Thus, to anticipate the shielding effect, only the electron density on the C_2 and C_6

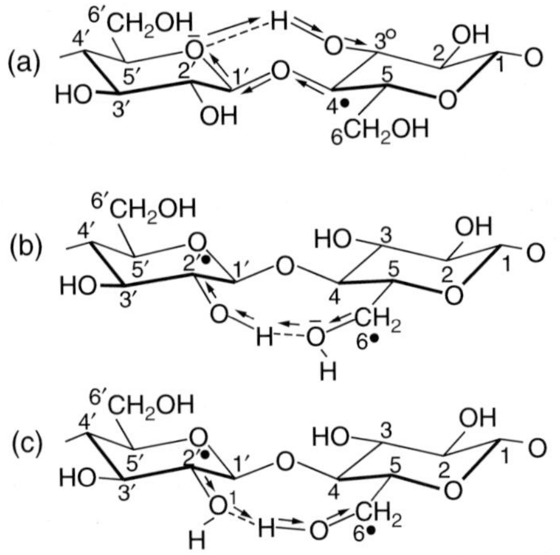

Figure 4.1.14 Schematic representation of intramolecular hydrogen bonds in cellobiose units:[23] (a) O_3–H$\cdots O_5'$; (b), O_2–H$\cdots O_6'$; (c) O_6–H$\cdots O_2'$; arrow denotes electron localization.

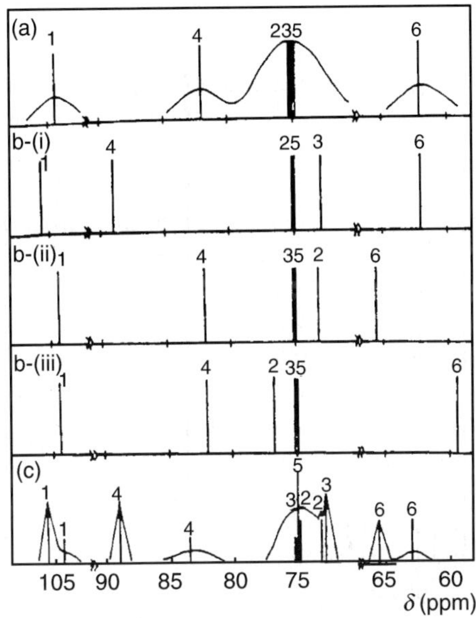

Figure 4.1.15 Schematic representations of experimental and hypothetical CP-MAS spectra of cellulose:[23] (a), amorphous (C-2b, experimental); (b) (i) cellulose I with the formation of a perfect O_3–H$\cdots O_5'$ intramolecular hydrogen bond (hypothetical), (ii) cellulose I with the formation of a perfect O_2–H$\cdots O_6'$ intramolecular hydrogen bond (hypothetical), (iii) cellulose I with the formation of a perfect O_6–H$\cdots O_2'$ intramolecular hydrogen bond (hypothetical); (c), cellulose I (C-1, experimental).

carbons directly involved in the forms (b) and (c) need be considered. In the form (b), the cationized C_6 carbon must shift towards a higher magnetic field. The reverse applies for form (c).

Figure 4.1.15(a) and (c) show a schematic representation of the NMR spectra for the C-2b sample as amorphous cellulose and the C-1 sample as cellulose with cellulose I crystals, respectively. Both were deducted from Figure 4.1.13. In the following, we shall calculate the NMR spectrum of each cellulose I sample. We assume that (1) all the cellulose molecules in a given sample solid, regardless of whether the molecules belong to the crystalline or amorphous region, can take only one of three possible types of intramolecular hydrogen bonds (see Figure 4.1.14), (2) in the crystalline region of cellulose I, only a specific value is allowed for the bond length of each type of intramolecular hydrogen bond (i.e. $O_3-H\cdots O_5$, $O_2-H\cdots O_6'$, and $O_6-H\cdots O_2'$), and (3) in the amorphous region, the distance of the intramolecular hydrogen bonds can be varied in a limited range from those in the crystal region. Figure 4.1.15b(i)–(iii) show the calculated line spectra of cellulose with the cellulose I crystalline form and three possible intramolecular hydrogen bonds. A comparison of Figure 4.1.15c with Figure 4.1.15b(i)–(iii) shows that cellulose of crystal form I consists of three types of cellobiose units with $O_3-H\cdots O_5'$, and $O_2-H\cdots O_6'$, intramolecular hydrogen bonds, and units without intramolecular hydrogen bonds.

Figure 4.1.16(a–c) show similar schematic NMR spectra for cellulose with cellulose II type crystals. Here, note that in cellulose II crystals, two different types of $O_3-H\cdots O_5'$

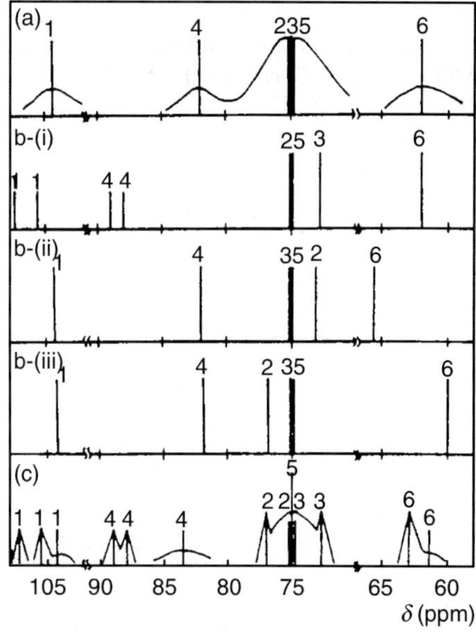

Figure 4.1.16   Schematic representations of experimental and hypothetical CP-MAS spectra of cellulose:[23] (a), amorphous (C-2b, experimental); (b) (i) cellulose II with the formation of a perfect $O_3-H\cdots O_5'$ intramolecular hydrogen bond (hypothetical), (ii) cellulose II with the formation of a perfect $O_2-H\cdots O_6'$ intramolecular hydrogen bond (hypothetical), (iii) cellulose II with the formation of a perfect $O_6-H\cdots O_2'$ intramolecular hydrogen bond (hypothetical); (c), cellulose II (C-2, experimental).

intramolecular hydrogen bonds exist, corresponding to two types of chain conformations of 'bent' and 'twist'[21] or two types of chain packings 'parallel' and 'antiparallel'. The other assumptions are the same as described for cellulose I. On the basis of a comparison between Figure 4.1.16(c) and Figure 4.1.16(a,b (iii)), we can conclude that the cellulose with cellulose II type crystals contains four types of cellobiose units: units with O_3–H$\cdots O_5'$, O_2–H$\cdots O_6'$, and O_6–H$\cdots O_2'$ intramolecular hydrogen bonds, and units without intramolecular hydrogen bonds. It should be noted that O_6–H$\cdots O_2'$ intramolecular hydrogen bond, lacking in cellulose I, exists in cellulose II. This is probably related to the two 'bent' and 'twist' conformations or chain packings of 'parallel' and 'antiparallel' for cellulose II. The former may support the observations in this study since, in the latter case, rotational conformational change about C_5–C_6 is quite large and requires the considerable NMR peak shift of C_6 carbon not observed experimentally. The C_6 carbon peak can be considered to shift from the hypothetical line NMR peak positions by some specific intermolecular hydrogen bond.

REFERENCES

1. K Kamide, K Okajima, T Matsui and K Kowsaka, *Polym. J.*, 1984, **16**, 857.
2. See for example, SM Neals, *J. Text. Inst.*, 1920, **20**, T373.
3. C Beadle and HP Stevens, The Eighth International Congress on Applied Chemistry, 1912, **13**, 25.
4. H Dillenius, *Kunstseide Zellwolle*, 1940, **22**, 314.
5. E Schwart and W Zimmerman, *Melliands Textilber. Int.*, 1941, **22**, 525.
6. H Staudinger and R Mohr, *J. Prakt. Chem.*, 1941, **158**, 233.
7. O Eisenluth, *Cellul. Chem.*, 1941, **19**, 45.
8. RH Atalla, *J. Am. Chem. Soc.*, 1980, **102**, 3249.
9. WL Earl and DL Vander Hart, *J. Am. Chem. Soc.*, 1980, **102**, 3251.
10. WL Earl and DL Vander Hart, *Macromolecules*, 1981, **14**, 570.
11. F Hirai, A Horii and R Kitamaru, Preprint, 45th Chemical Society of Japan, Annual Meeting, 1982, 1172.
12. F Hirai, A Horii and R Kitamaru, *Polym. Prepr. Jpn.*, 1982, **31**, 842.
13. F Hirai, A Horii, A Akita and R Kitamaru, *Polym. Prepr. Jpn.*, 1982, **31**, 2519.
14. F Hirai, A Horii and R Kitamaru, Preprint, 47th Chemical Society of Japan, Annual Meeting, 1983, 1392.
15. J Hayashi, M Takai, R Tanaka, M Hatano and S Nozawa, Preprint, 47th Chemical Society of Japan, Annual Meeting, 1983, 1393.
16. GE Maciel, *Macromolecules*, 1982, **15**, 686.
17. W Brown and R Wikström, *Eur. Polym. J.*, 1966, **1**, 1.
18. L Segal, *Text. Res. J.*, 1959, **29**, 786.
19. J Mann and HJ Marrinan, *Trans. Faraday Soc.*, 1956, **52**, 492.
20. J Hayashi, Preprint, Cellulose Micro Symposium, Sapporo, 1983, 21.
21. See for example, J Blackwell and RH Marehessault, In *Cellulose and Cellulose Derivatives, Part IV*. (eds NM Bikales and L Segal), Chapter XIII, Wiley, New York, 1971, p. 18.
22. K Kamide, K Okajima, T Matsui and T Abe, unpublished results.
23. K Kamide, K Okajima, K Kowsaka and T Matsui, *Polym. J.*, 1985, **17**, 701.
24. CA Fyfe, RL Dudley, PJ Stephenson, Y Deslandes, GK Hamer and RH Marchessauht, *J. Macromol. Sci., Rev. Macromol. Chem. Phys.*, 1983, **C232**, 187.
25. F Horii, A Hirai and R Kitamaru, *Polym. Bull.*, 1982, **8**, 163.
26. J Kunz, C Scheler, B Schröter and B Philipp, *Polym. Bull.*, 1983, **10**, 56.

27. VW Tripp, in *Cellulose and Cellulose Derivatives, Part IV.* (eds NM Bikales and L Segal), Chapter XIII-C, Wiley, New York, 1971, p. 305.
28. W Voelter, E Breitmaier and G Jung, *Angew. Chem.*, 1971, **83**, 1011.
29. DW Jones, In *Cellulose and Cellulose Derivatives, Part IV.* (eds NM Bikales and L Segal), Chapter XIII-C, Wiley, New York, 1971, p. 117.
30. C Woodcock and A Sarko, *Macromolecules*, 1980, **13**, 1183.
31. A Sarko and R Muggi, *Macromolecules*, 1974, **7**, 486.
32. PJ Kolpak and J Blackwell, *Macromolecules*, 1976, **9**, 273.

4.2   INTRAMOLECULAR HYDROGEN BONDS AND SELECTIVE COORDINATION OF ALKALICELLULOSE[1]

Alkalicellulose is chemically highly reactive so that cellulose ethers (such as carboxy methylcellulose, hydroxyethylcellulose, hydropropylcellulose, methylcellulose, ethyl cellulose, and their coethers) are at present commercially synthesized on a large scale via alkalicellulose as an intermediate compound from cellulose. Various crystal forms including alkalicellulose I, II, III, and IV have been found to exist by X-ray diffraction and for each form the chemical composition of the sodium ion Na^+, water and anhydrous glucose unit was determined by the chemical analysis.[2–5]

It is widely recognized that the reactivity of alkalicellulose against other chemical reagent and the distribution of substituent groups of cellulose derivatives formed thus depend remarkably on the solid state structure (crystal form, crystallinity, and morphology) of alkalicellulose. Alkalicellulose I (Na-cell I) readily reacts with carbon disulfide to give cellulose xanthate and the ratio of the probability of carboxymethylation at the hydroxyl group attached C_6 position $\langle\!\langle f_6 \rangle\!\rangle$, to that of total degree of substitution $\langle\!\langle F \rangle\!\rangle$ depends on the crystal form of the starting cellulose, if the operating conditions of carboxymethylation reaction are kept the same.[6]

However, there are unfortunately few methods available to directly obtain detailed information on the chemical structure of cellulose solid. Wide angle X-ray diffraction provides information only on the crystal structure. Recently, we demonstrated that the CP-MAS ^{13}C NMR technique is particularly useful to analyze the hydrogen bonds of a cellulose solid, and concluded that the intramolecular hydrogen bond plays an important role in the solubility of cellulose in dilute alkali.[7] Applying the same technique to alkalicellulose, Kunze *et al.*[8,9] demonstrated (although very qualitatively due to the low resolution of their NMR spectra) the change in the crystal form with alkali concentration.

In this section, an attempt is made to determine, using CP-MAS ^{13}C NMR, whether the intramolecular hydrogen bond and selective coordination of Na^+ exist in Na-cell I, prepared by dipping cellulose I in a 18 wt% aq. sodium hydroxide solution.

Alkalicellulose

A 100 weight part of cellulose was immersed in a 2000 weight part of an 18 wt% aq. sodium hydroxide solution at 30 °C for 30 min and then pressed between polyacrylo-nitrile cloth and cellulose filter paper by a hand press to give a *ca.* 280 weight part of alkalicellulose. Alkalicellulose thus prepared was referred to as AC-1. AC-1 was crushed

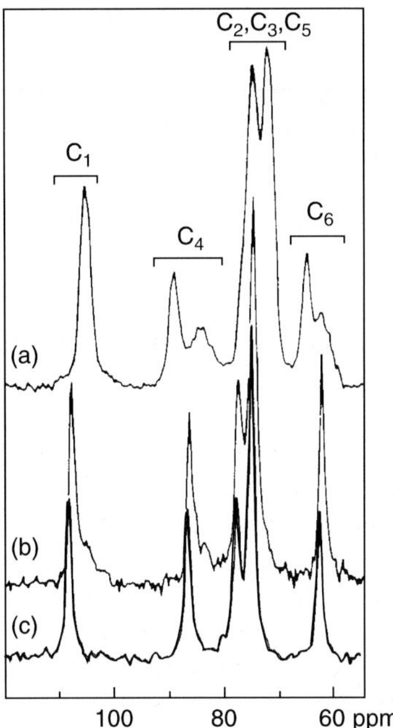

Figure 4.2.1 CP-MAS ^{13}C NMR spectra of cellulose and alkalicelluloses. (a) cellulose (C-1); (b) alkalicellulose without aging (AC-1); (c) aged alkalicellulose(AC-1a).[1]

in a mixer so that its apparent specific volume became 15 ml g^{-1} and stored in a closed vessel containing air at 49 °C for 30 h. The sample thus aged was referred to as AC-1a.

Figure 4.2.1 shows the CP-MAS ^{13}C NMR spectra of cellulose sample C-1 (Figure 4.2.1(a)), borrowed from,[10] and of its alkali celluloses AC-1 and AC-1a (Figure 4.2.1(b) and (c), respectively). Table 4.2.1 shows the peak positions (TMS as reference material) and integrated peak intensity (in blanket) of NMR spectra in

Table 4.2.1

Peak assignment of cellulose and alkalicellulose[1]

Sample	Carbon position (ppm)							
	C_1		C_4			C_2, C_3, C_5		C_6
C-1	105.4^b		89.1	84.5^b		75.3	72.2	65.5	63.0^b
	(1.00)		(0.47)	(0.53)		(1.57)	(1.51)	(0.55)	0.38
AC-1	107.8	105.3^s	86.4	85.4^s	83.8^b	77.7	74.9	62.8
	(1.00)		(0.71)		(0.29)	(0.98)	(2.10)	(0.98)
AC-1a	108.3		86.8			78.0	75.1	63.0
	(1.00)		(1.02)			(1.10)	(2.10)	(0.98)

b, denotes broad peaks; s, shoulder peak and the value in parenthesis are peak intensities.

Figure 4.2.1. Suffix b refers to a broad peak. The half value width of all NMR peaks of alkalicellulose was only 1/2–1/4 that of cellulose. The peak positions of cellulose were almost similar to those for alkalicellulose. Peaks at 108, 90–80, and 63 ppm were assigned for C_1, C_4, and C_6 carbons. In Figure 4.2.1, the C_1 and C_4 carbon peaks for AC-la are seen as singlet at 108.3 and 86.3 ppm, respectively. However, the AC-1 sample has a small higher field component in both the C_1 and C_4 carbon peak regions, showing AC-la has weaker parts in intramolecular hydrogen bonding compared with that of AC-1.

From Figure 4.2.1, it is obvious that the C_4 and C_6 carbon peaks for cellulose show broad distribution depending on the wide distribution of the bond strength of the intramolecular hydrogen bond (Figure 4.2.l(a)). In contrast, all carbon peaks for alkalicellulose are simple and sharp (Figure 4.2.1(b) and (c)). This is particularly apparent in AC-la. This strongly suggests that in alkalicellulose, the distance between molecular chains is widened due to swelling action of alkali against cellulose such that as many conformations of glucoside linkage as in the solution are possible, and that the rate of interconversion of these conformations is relatively rapid. This leads to a single and sharp time averaged NMR peak (i.e. AC-1a) due to each carbon's relaxation time. In this sense, alkalicellulose has the uniform solid structure. Each carbon peak position does not always coincide with that for cellulose dissolved in aq. sodium hydroxide. The C_4 carbon peak in solution, where O_3–H$\cdots O_5'$ intramolecular hydrogen bonds are considered to be completely broken, resonates at $ca.$ 80 ppm[7] and the corresponding peak, originating from the region in which the intramolecular hydrogen bonds are strongly formed, is observed at 89 ppm. The C_4 carbon peak of alkalicellulose is observed at 86–87 ppm. These experimental facts show that even in alkalicellulose, weak intramolecular hydrogen bonds exist. C_1 carbon peak of cellulose varies between 104 and 106 ppm, depending on the strength of O_3–H$\cdots O_5'$ intramolecular hydrogen bonds.[10] Unexpectedly, the C_1 carbon peak of alkalicellulose is sharp at 108 ppm and significantly outside the possible variation of the C_1 carbon peak of cellulose. This must be explained on a basis other than the intramolecular hydrogen bonding. This situation is the same for the peak at 78 ppm for alkalicellulose. To explain the appearance of this peak, the abstraction of an electron from the carbon adjacent to the hydroxyl group, induced by selective coordination of Na^+ to the hydroxyl group, should be considered. As is well known, alkalicellulose I has the chemical composition of $C_6H_{10}O_5\cdot NaOH\cdot 3H_2O$, meaning that Na^+ is coordinated to a glucose ring.

Figure 4.2.2 shows the possible chemical structure of alkalicellulose I when Na^+ is selectively coordinated to the hydroxyl groups at C_2, C_3, and C_6, respectively. Whether the states shown in Figure 4.2.2 are as stable as alcoholates is not clear, but the electron density on the carbons located at α and β position from the hydroxyl groups selectively coordinating with Na^+ must diminish. Thus, the α and β carbon NMR peaks must shift towards a lower magnetic field, compared with the corresponding carbon peaks when the aforementioned selective coordination of Na^+ is not considered. The degree of cationization of α carbon is larger than that of β carbon.

For the form (a) shown in Figure 4.2.2, C_2 carbon might be considerably cationized as may the C_3 carbons. In this connection, the degree of cationization of C_1 carbon is larger than that of the C_3 carbons because, in alkalicellulose, the remaining O_3–H$\cdots O_5'$ intramolecular hydrogen bonding may anionize the C_3 carbon. Thus, for the form (b), the C_3 carbon is considerably cationized and C_2 and C_4 slightly cationized. For the form (c),

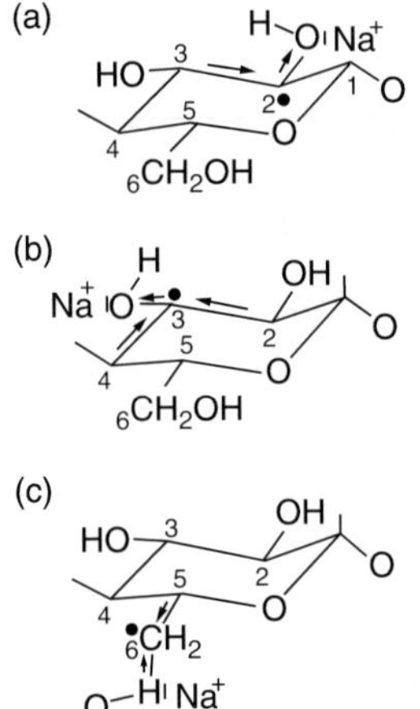

Figure 4.2.2 Possible chemical structure of alkalicellulose.[1] Alkalicellulose with selective coordination of a sodium ion to the hydroxyl oxygen at the C_2 (a), C_3 (b), and C_6 (c) positions; arrow denotes location of the electron.

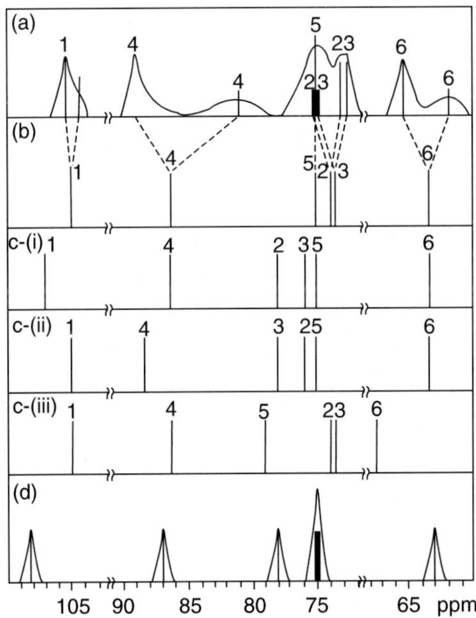

Figure 4.2.3 Schematic representation of CP-MAS ^{13}C NMR spectra of cellulose:[1] (a), cellulose I experimentally observed; (b), alkalicellulose when selective coordination of the sodium ion does not occur; (c) (i) alkalicellulose with selective coordination of the sodium ion at C_2, (ii) alkalicellulose with selective coordination of the sodium ion at C_3, (iii) alkalicellulose with selective coordination of the sodium ion at C_6; (d) aged alkalicellulose (AC-1a), experimentally observed.

the C_6 and C_5 carbons are cationized. Thus, corresponding to the forms, shown in Figure 4.2.2, the C_1, C_2, C_3, C_4, C_5, and C_6 carbon NMR peaks must shift towards a lower magnetic field, compared with the corresponding carbon peaks when the coordination of Na^+ is not considered.

Figure 4.2.3(a) shows a schematic representation of the CP-MAS ^{13}C NMR spectrum of cellulose I, experimentally determined. Figure 4.2.3(b) shows a hypothetical spectrum of alkalicellulose I having an O_3–H$\cdots O_5'$ intramolecular hydrogen bond when the effect of the coordination of Na^+ is not considered. Figure 4.2.3(c) (i)–(iii) show hypothetical line spectra of alkalicelluloses, having the chemical structure shown in Figure 4.2.2(a–c), respectively. A comparison of these spectra with the spectrum in Figure 4.2.3(b), shows (l) the C_1, C_2, and C_3 carbon peaks to be located in lower magnetic fields when Na^+ coordinates to the hydroxyl group at C_2 position (Figure 4.2.3(c) (i)), (2) all C_2, C_3, and C_4 carbon peaks for alkalicellulose when Na^+ is assumed to coordinate to the hydroxyl group at C_3 position to shift towards a lower magnetic field (Figure 4.2.3(c) (ii)), and (3) the C_5 and carbon peaks also to shift towards a low magnetic field when Na^+ coordinates to the hydroxyl group at the C_6 position (Figure 4.2.3(c) (iii)). The degree of this shift is expected to be greater for α carbons than for β carbons. Figure 4.2.3(d) shows a schematic spectrum, experimentally observed for sample AC-la. A comparison of Figure 4.2.3(d) with Figure 4.2.3(c) (i)–(iii) indicates that selective coordination Na^+ to the hydroxyl group at C_2 (Figure 4.2.2(a)) is the most probable.

To summarize, in an aged alkalicellulose solid, the molecular motion is relatively rapid as compared with a cellulose solid and has intramolecular hydrogen bondings and a sodium ion selectively coordinates to the hydroxyl oxygen at the C_2 position of cellulose.

REFERENCES

1. K Kamide, K Kowsaka and K Okajima, *Polym. J.*, 1985, **17**, 707.
2. K Kiessig, H Hess and H Sobue, *Z. Phys. Chem.*, 1939, **43**, 309.
3. JO Warwicker, In *Cellulose and Cellulose Derivatives, Part IV.* (eds NM Bikales and L Segal), Chapter XIII-H, Wiley, New York, 1971, p. 325.
4. J Hayashi, T Yamada and S Watanabe, *Sen-i Gakkaishi*, 1974, **30**, T-190.
5. J Hayashi and T Yamada, *Nippon Kagaku Kaishi*, 1975, **96**, 544.
6. K Kamide, K Okajima, T Matsui, K Kowsaka and S Nomura, *Polym. Prepr. Jpn.*, 1984, **33**, 589.
7. K Kamide, K Okajima, T Matsui and K Kowsaka, *Polym. J.*, 1984, **16**, 857.
8. J Schröter, J Kunze and B Philipp, *Acta Polym.*, 1981, **32**, 732.
9. J Kunze, A Ebert, B Schröter, K Frigge and B Philipp, *Polym. Bull.*, 1981, **5**, 399.
10. K Kamide, K Okajima, K Kowsaka and T Matsui, *Polym. J.*, 1985, **17**, 701.

4.3   SOME CHARACTERISTIC FEATURES OF DILUTE AQUEOUS ALKALI SOLUTION[1]

For many years, it has been widely known that natural or regenerated cellulose swell, at least partially in aq. alkali solutions. In this case, cellulose shows maximum swelling at a specific concentration of aq. alkali solution: 8–10 wt% for aq. sodium hydroxide (NaOH) and 5–10 wt% for aq. lithium hydroxide (LiOH) at comparatively lower temperature.[2–7] Recently, Kamide *et al.* demonstrated that cellulose samples, regenerated under certain conditions, can completely dissolve in 8–10 wt% aq. NaOH solution at 4 °C and in 6 wt% aq. LiOH.[8] They successfully correlated the solubility of cellulose in aq. alkali solution with the degree of breakdown of an intramolecular hydrogen bond ($O_3 \cdots O_5'$) of cellulose as determined by CP-MAS ^{13}C NMR method.[8] When exploded or extruded after treatment with water under high pressure for a relatively short time,[9–11] natural cellulose can also dissolve perfectly into aq. NaOH solution below 7 °C. The alkaline solubility of the treated natural cellulose is also closely correlated with the degree of breakdown of an intramolecular hydrogen bond[12] and it seems that almost complete dissolution of cellulose into aq. alkali with specific concentration at low temperature requires the breakdown of intramolecular hydrogen bonds at C_3 and C_6, as determined by CP-MAS ^{13}C NMR, by at least 45%. Therefore, one of the most important factors controlling the maximum swelling or solubility of cellulose in aq. alkali solution at specific alkaline concentration is the structure of the aq. alkali solution.

In this section, some structural characteristics of aq. alkali solutions with specific concentration in which the maximum swelling or maximum solubility of cellulose is attained will be demonstrated, and a tentative structure of cellulose/aq. alkali solution systems will be proposed.

Measurements

Electrical conductivity (σ) was measured for aq. alkali as functions of alkali concentration C_a and temperature. We recorded ^1H and ^{23}Na NMR spectra for various aq. alkali solutions on a FX-200 type FT-NMR spectrometer (JEOL, Japan) with multinuclei measurement system. The optical rotatory angle α was measured as a function of C_a and temperature for various cellobiose aq. alkali systems (20 g of cellobiose in 100 g of aq. alkali) at the wave number of the incident light $\lambda = 436$ nm (Hg lamp) was also measured by a DIP 181 type instrument (Nihon Bunkou Co., Japan). For these systems, α has proven to decrease with time, almost obeying first-order kinetics and, therefore, the measurements, which required 2 s, were done 2.5 min after the solution preparation. A specific rotatory angle $[\theta]$ was calculated from α, using the relationship $[\theta] = \alpha(l \cdot c)$, where l and c represent the length of the pass (cm) and cellobiose concentration (mol dl^{-1}). Raman spectra were recorded for aq. alkali solutions with various C_a on a spectrophotometer (JRS-400D Laser Raman Spectrophotometer, JOEL, Japan) at room temperature. Half value width $\Delta_{1/2}$ of the Raman absorption band in question was determined. Sound velocity v was measured with a Pierce type ultrasonic interferometer, constructed by Nomura and Miyahara,[13] operating at 4.999985 MHz. The temperature of the solution was controlled to ± 0.005 °C by circulating water around

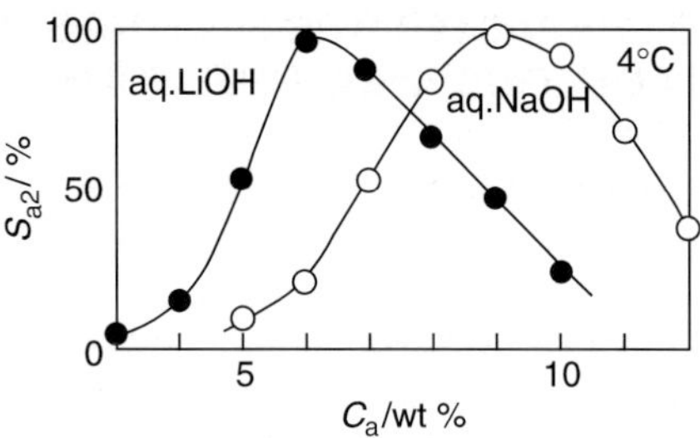

Figure 4.3.1   Solubility S_a of cellulose at 4 °C as a function of alkali concentration C_a^l: (O), aq. NaOH; (●), aq. LiOH.

the cell of the interferometer. The adiabatic compressibility β was calculated by the Laplace equation 4.3.1.[14]

Figure 4.3.1 shows the solubility S_a at 4 °C of a cellulose sample as a function of alkali concentration C_a. For this purpose, an alkali soluble pulp with cellulose I crystal (the viscosity average degree of polymerization $P_v = 380$, determined from the Mark–Houwink–Sakurada equation established for cellulose/cadoxen system,[15] and the degree of breakdown of intramolecular hydrogen bond at C_3 and C_6 $\chi_{NMR}(C_3 + C_6) = 48\%$, as determined by CP-MAS ^{13}C NMR), was prepared by physical treatment[9–11] to break down the intramolecular hydrogen bond to some extent. S_a was measured as follows: 5.43 g of cellulose (water content 8 wt%) were dispersed in 95 g of a given concentration of aq. alkali, precooled at 4 °C, stood for 8 h with occasionally mixing, and the insoluble part was separated by centrifuging at 3000 rpm for 20 min. The insoluble part was neutralized with hydrochloric acid to precipitate the cellulose completely. The cellulose was washed with water, dried first in air and then *in vacuo*, and was then weighed. S_a was defined as $100 \times (5 - w)/5(\%)$. S_a was found to attain almost 100% at specific concentrations of aq. LiOH (*ca.* 5.8 wt%) and NaOH (about 9 wt%)

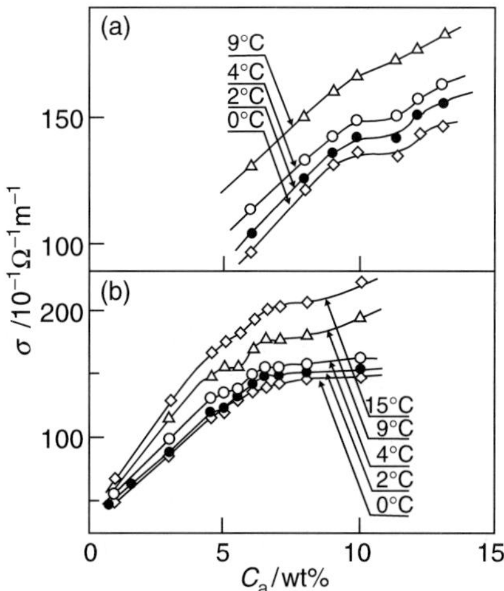

Figure 4.3.2   Dependence of electrical conductivity σ of aq. solution of NaOH (a) and LiOH (b) on alkali concentration C_a.[1]

solutions. Interestingly, these two concentrations fall on about 2.5 mol dl^{-1}. We believe that these are the first experimental results demonstrating that cellulose having crystal form I dissolves completely in aq. alkali solution if the intramolecular hydrogen bond of cellulose sample is appropriately broken down in advance.

Figure 4.3.2 shows the electrical conductivity σ (expressed in SI unit $10^{-1}\ \Omega^{-1}\ m^{-1}$) of aq. solution of NaOH (a) and LiOH (b) plotted against the concentration of alkali C_a at a given temperature. The positive C_a and temperature dependences of σ ($\delta\sigma/\delta C_a \geq 0$ and $\delta\sigma/\delta T > 0$) were observed over the C_a range of 0–15 wt%. However, the gradient of σ versus C_a plot became nearly zero in *ca.* 9–12 wt% of aq. NaOH in a lower temperature range, whereas cellulose exhibited the maximum swelling or solubility. Aq. LiOH exhibited the same phenomena.

Figures 4.3.3(a) and (b) show the proton chemical shift δ_H (a) and ^{1}H spin lattice relaxation rate $1/T_{1H}$ (b) of aq. NaOH solutions as a function of C_a at 4 °C (circle) and 10 °C (triangle), respectively. The figure also contains the data (closed symbols) for a cellobiose solution in the alkali. C_a dependence of δ_H is almost the same, but its temperature dependence is reserved as observed for the σ–C_a relationship. In contrast, $1/T_{1H}$ of aq. NaOH exhibits a remarkable sharp increase in almost the same C_a region observed for δ_H–C_a relation. When cellobiose was added, a gradient of δ_H versus C_a plot increased abruptly, approaching nearly zero in the above plateau region. The region

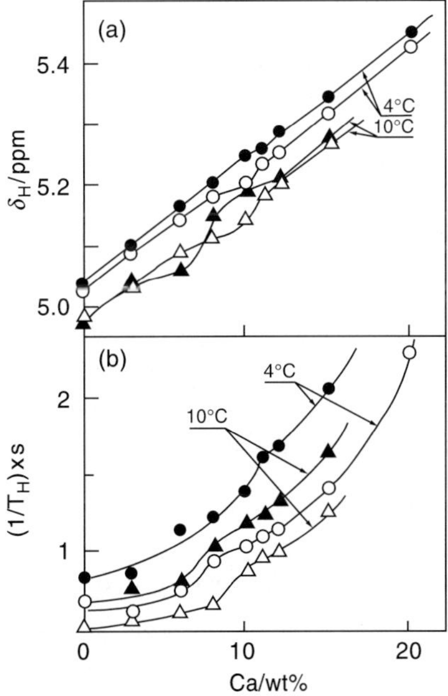

Figure 4.3.3   Dependence of proton chemical shift δ_H (a) and ^{1}H spin lattice relaxation rate $1/T_{1H}$ (b) of aq. NaOH solutions and cellobiose/aq. NaOH system on alkali concentration C_a^1 : (○), aq. NaOH at 4 °C; (●), cellobiose/aq. NaOH at 4 °C; (△), aq. NaOH at 10 °C; (▲), cellobiose/aq. NaOH at 10 °C.

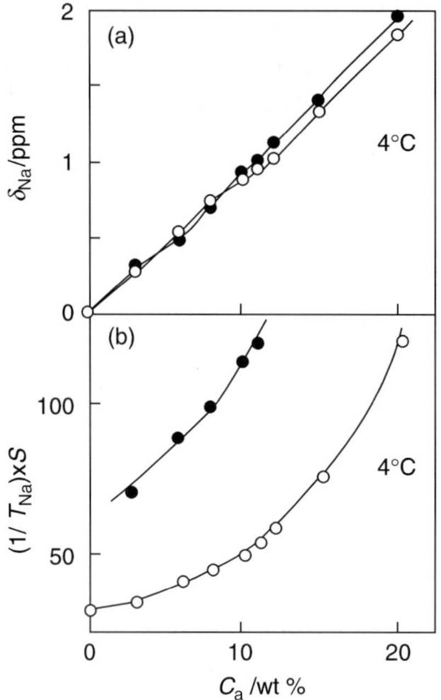

Figure 4.3.4 Dependence of ^{23}Na chemical shift δ_{Na} (a) and ^{23}Na spin lattice relaxation rate $1/T_{1Na}$ (b) of aq. NaOH at 4 °C on alkali concentration C_a^1: (○), aq. NaOH; (●), cellobiose/aq. NaOH.

of sharp $1/T_{1H}$ increase shifted to higher C_a at 4 °C and to lower C_a at 10 °C. Both δ_H and $1/T_{1H}$ values of the cellobiose/aq. NaOH system at 4 °C were always higher than those of aq. NaOH at 10 °C over all C_a region.

Figures 4.3.4(a) and (b) show ^{23}Na chemical shift δ_{Na} (a) and ^{23}Na spin-lattice relaxation rate $1/T_{1Na}$ of aq. NaOH (b) as a function of C_a at 4 °C. δ_{Na} shows a similar C_a dependence to that observed for δ_H. A reflection point appeared in the C_a range of 8–14 wt%. When cellobiose was added, δ_{Na} almost linearly increased with C_a and δ_N values below $C_a = 8$ wt% were almost equal to those for the aq. NaOH system. $1/T_{1Na}$ values for both aq. NaOH and cellobiose/aq. NaOH systems increased monotonically with an increase in C_a over the entire C_a range examined, but the absolute value for the lattice system is considerably larger than that for the former system. Absolute $1/T_{1Na}$ was found to be far larger than $1/T_{1H}$.

Figure 4.3.5 shows the number of water molecules solvated to an NaOH molecule S as a function of C_a at 4 °C (open circle) and 20 °C (open triangle). Here, S was calculated from adiabatic compressibility β using Passynsky's equation:[14]

$$S = \{1 - (\beta/\beta_s)\}(100\rho - C_a')/C_a' \qquad (4.3.1)$$

Here, β_s denotes the adiabatic compressibility of water and C_a' is the concentration of the alkali expressed in mol l^{-1}. S generally decreased linearly with increasing in C_a at 20 °C. However, C_a dependence of S at 4 °C was roughly approximated with two linear lines above 11 wt% and below 6 wt% with different slopes. A significant deviation from these two lines attained a maximum at $C_a = 8$–9 wt%. It can be concluded that one mole of NaOH solvates

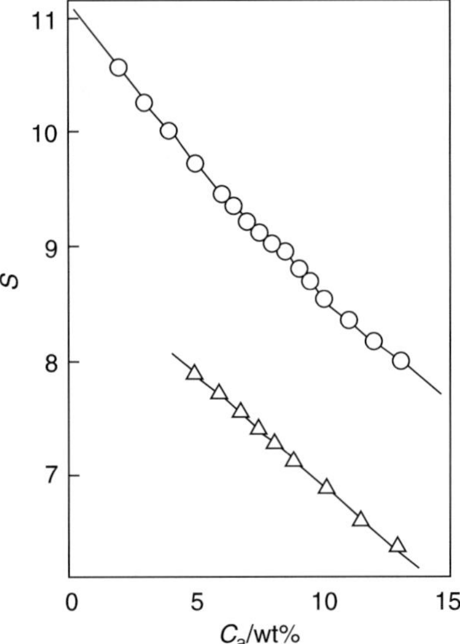

Figure 4.3.5   Number of water molecules solvated to an NaOH molecules S as a function of alkali concentration C_a^1; (○), 4 °C; (△), 20 °C.

with about 8–9 mol of water, which is larger by about 0.2 mol than the value expected from the two linear C_a dependence lines, over the C_a range of 8–10 wt% at 4 °C.

   Figure 4.3.6 shows the specific rotatory angle $[\theta]$ of the cellobiose/aq. alkali systems as a function of C_a at 4, 9, and 15 °C. $[\theta]$ seems to decrease with increasing C_a but exhibits a relatively sharp peak at $C_a = 9$ wt% for aq. NaOH and 5 wt% for aq. LiOH. Absolute $[\theta]$ values were always higher at lower temperatures when compared at the same C_a level.

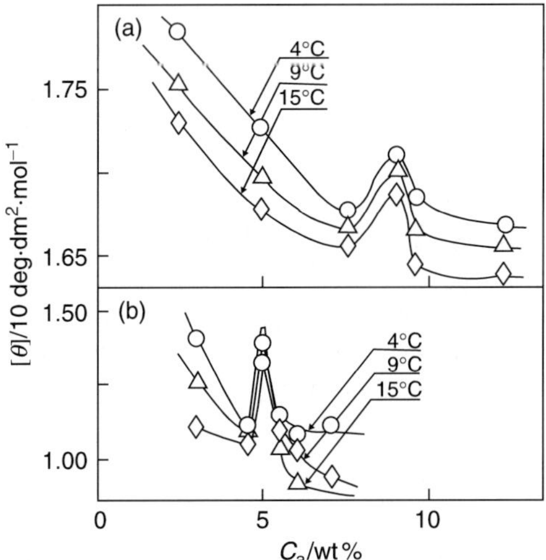

Figure 4.3.6   Dependence of the specific rotatory angle $[\theta]$ of the cellobiose/aq. alkali systems on alkali concentration C_a^1: numbers denote temperature.

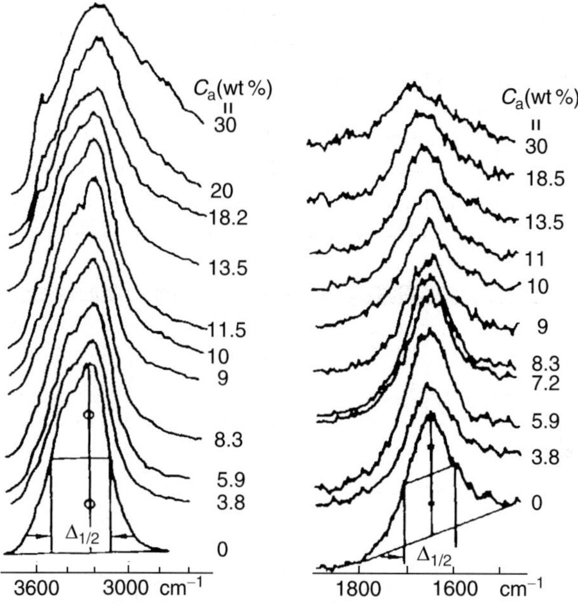

Figure 4.3.7   Raman spectra in the OH stretching (a) and deformation (b) regions for aq. NaOH:[1] numbers denote C_a (wt%).

Figures 4.3.7(a) and (b) show the Raman spectra in the OH stretching and deformation regions for aq. NaOH with various C_a, respectively. The maximum scattering peak positions and the spectra shapes change remarkably depending on C_a. In the OH stretching region, the shoulder peak located at a higher wavenumber region in the

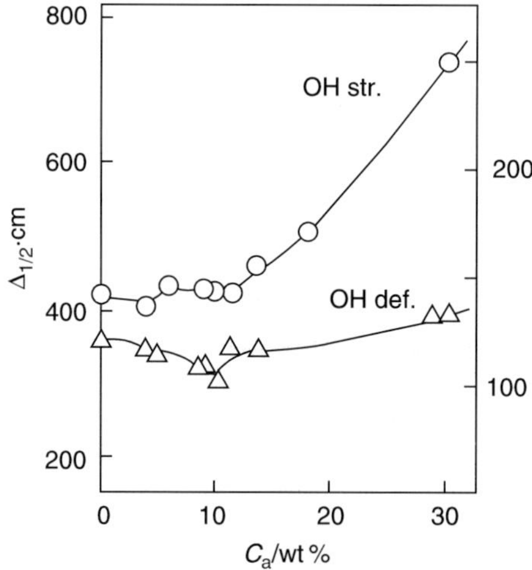

Figure 4.3.8   Dependence of the half value width $\Delta_{1/2}$ of scattering peaks for OH stretching region (O) and OH deformation region ($\triangle$) on alkali concentration C_a complicated when C_a is around 7–9 wt%.[1]

maximum peak became weak over the C_a range of 8–9 wt% and again became distinct in the C_a range 10–11.5 wt%. In the OH deformation region, the shape of peaks is very complicated when C_a is around 7–9 wt%.

In Figure 4.3.8, the half value widths $\Delta_{1/2}$ of scattering peaks for OH stretching region and OH deformation region are plotted against C_a. In the OH stretching region, $\Delta_{1/2}$ remained practically constant up to *ca.* 10 wt%, and particularly in the range of 5–11 wt%. $\Delta_{1/2}$ increased linearly with an increase in C_a in the range $C_a > 11$ wt%. In contrast to this, in OH deformation, a gradual decrease of $\Delta_{1/2}$ was observed for $C_a < 10$ wt% and $\Delta_{1/2}$ increased slightly after passing a minimum $\Delta_{1/2}$ at $C_a = 8$–10 wt%.

Figure 4.3.9 shows the dependence of wavenumbers at peak $1/\lambda$ corresponding to the OH stretching (denoted as circles) and the OH deformation (denoted as triangles) on C_a. The $1/\lambda$ of the OH stretching decreased with increase of C_a until $C_a = 20$ wt% and, after passing through the minimum, increased greatly. The peak position of OH deformation clearly shows a minimum at $C_a = 9$ wt%. The shift in width of the peak position gives a maximum intensity for OH deformation range of as much as 40 cm^{-1}. For example, the aforementioned peak appeared at 1627 cm^{-1} and 1670 cm^{-1} at $C_a = 9$ and 30 wt%, respectively. The former peak is also located at a considerably lower wave number compared with that for pure water (1640 cm^{-1}).

The characteristic features of aq. alkali solutions observed in this section are summarized in Table 4.3.1. The experimental facts observed here indicate that one of the most important factors controlling the dissolution of the cellulose, whose intramolecular hydrogen bond has been adequately broken, into aq. alkali is the structure of the alkali with specific concentration. Therefore, it is useful to clarify its structure. Hereafter, emphasis is placed on aq. NaOH.

In the C_a and temperature range where cellulose shows maximum swelling or dissolution, σ is around 130–150 Ω^{-1} cm^{-1}, irrespective of alkali species. However, an

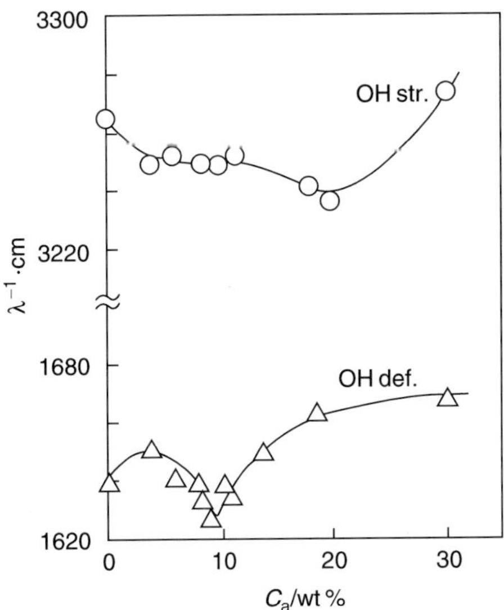

Figure 4.3.9 Dependence of the maximum peak position (wave number λ) in the OH stretching (O) and OH deformation ($\triangle$) on C_a.[1]

Table 4.3.1

Specific features observed for aq. NaOH solutions[1]

Physical properties	Temperature (°C)	Specific C_a range (wt%)		Figures	Remarks
		NaOH	LiOH		
Solubility of cellulose (S_a)	4	9.0	5.8	Figure 4.3.1	
Electrical conductivity (σ)	0–15	9.0–12.0	4.5–6.0	Figure 4.3.2	Low cationization of Na^+
^{1}H Chemical shift of H_2O (δ_H)	4	7.5–12.0		Figure 4.3.3(a)	Weak hydrogen bond of OH^-
	10	7.5–12.0			
^{1}H Chemical shift of H_2O with cellobiose (δ_H)	4	8.0–11.0		Figure 4.3.3(a)	Strong hydrogen bond of cellobiose
	10	7.5–12.0			
^{1}H Spin-lattice relaxation rate of H_2O ($1/T_{1H}$)	4	6.5–8.0		Figure 4.3.3(b)	Strong interaction (hydrogen bond) of OH^-
	10	8.0–10.0			
^{1}H Spin-lattice relaxation rate of H_2O with cellobiose	4	10.0–12.0		Figure 4.3.3(b)	Strong hydrogen bond of cellobiose
	10	7.0–9.0			
^{23}Na Chemical shift of Na^+ (δ_{Na})	4	8.0–14.0		Figure 4.3.4(a)	Weak hydrogen bond of Na^+
^{23}Na Spin-lattice relaxation rate of NaOH ($1/T_{1Na}$)	4	–		Figure 4.3.4(b)	
Solvated water molecules (S)	4	8.0–9.0		Figure 4.3.5	Extra solvated H_2O molecules
	20	–			
Specific rotatory angle of cellobiose (θ)	4–15	9.0	5.0	Figure 4.3.6	Specific configuration of cellobiose
HVW of Raman peaks ($\Delta_{1/2}$)				Figure 4.3.8	
(–OH stretching)	25	–			
(–OH deformation)	25	8.0–10.0			Long lifetime
Wavenumber of Raman peaks (λ)				Figure 4.3.9	
(–OH stretching)	25	–			
(–OH deformation)	25	9.0			Weak hydrogen bond

explicit appearance of plateau region (about 2.5 mol mol 1^{-1}) in σ–C_a relationships (Figure 4.3.2) implies that there is some metastable state that restricts ionic transportation to some extent. The restriction of ionic transportation means that the degree of cationization or anionization of the masses in the system is smaller than those expected from their C_a dependence in the outside of the plateau region. A higher absolute (σ) value at higher temperature is probably due to a larger dissociation constant of the system at higher temperature.

H^+ and OH in H_2O and NaOH exchange rapidly. This means that we cannot distinguish these hydrogen species separately; we can only determine the time average state. Two different physical meanings of proton chemical shift δ_H can be considered. δ_H is a measure of the average hydrogen bond strength or the degree of cationization of the hydrogen. In the first case, it is expected that the higher the δ_H, the stronger is the hydrogen bond. The plateau region of the δ_H versus C_a plot suggests the existence of a state where the strength of the hydrogen bonds of the system became less than the value extrapolated from the experimental relation in the region of C_a below 7.5 wt% or more than 12 wt%. In the second case, cationization of hydrogen, judging from the results for δ_H, should be higher at lower temperature, which is in sharp contrast to the results obtained by electrical conductivity measurements. Thus, a simple cationization concept cannot be accepted. The existence of plateau region for T_{1H} in the same C_a region observed for δ_H is an indication that a rotational motion of the proton as a whole is restricted and the lifetime of a given state is shortened more than anticipated from the extrapolation of experimental T_{1H}–C_a relationships below $C_a = 7.5$ wt%. In other words, in this C_a region, a proton excited by an NMR field more easily transfers its energy to the lattice (e.g. surrounding H_2O molecules). Therefore, it is probably true to say that if cellulose is added to the aq. NaOH solution with a specific concentration, then cellulose can easily accept energy from the system.

^{23}Na chemical shift δ_{Na} versus a C_a plot again reveals a small plateau as observed for δ_H. The overall absolute value of δ_{Na} at 10 °C, obtained in a separate experiment, was almost equal to that at 4 °C at the same C_a, showing that δ_{Na} is independent of temperature. Thus, it is conceivable that δ_{Na} is not a measure of the degree of cationization but rather a measure of the degree of hydrogen bond strength of the whole system made through interactions of the solvated water molecules. Since $1/T_{1Na}$ monotonically increases with C_a, the system has no specific state where ^{23}Na excited by an NMR field more easily transfers its energy to the lattice in a specified C_a range. The lifetime of Na^+ in a given ionic structure is far shorter than that of H^+, judging from $1/T_{1Na}$ and $1/T_{1H}$. Therefore, it is obvious that Na ions do not play a very important role in the dissolution of cellulose in aq. alkali.

The number of solvated water molecules S on Na^+ in the C_a range of 6–11 wt% is slightly larger (about 0.2 mol/mol of NaOH) than that extrapolated from the experimental relationships over the range of C_a below 6 wt% and more than 11 wt%. This explains the results obtained for σ, δ_H, T_{1H}, and δ_{Na}. Solvation of extra water molecules on the totally cationized or anionized mass (Na^+ or OH^- solvated with a definite number of water molecules) through hydrogen bonding may decrease its cationic density and weaken the hydrogen bond, resulting in a reduction of σ and δ_H. In this connection, H^+ and OH^- ions in solvated water molecules, are to some extent, mutually interchangeable and can also easily interchange with free water molecules. The extra solvated water then makes weak hydrogen bonds against original free water in the system and this leads to decrease in T_{1H} of the system through dipole–dipole interactions between OH^- and H^+. The extra solvated water molecules can thus act as an energy transfer media between the masses and free water.

The experimental results for the half value widths of the maximum Raman peak $\Delta_{1/2}$ of OH stretching and deformation indicate that these vibration states have a longer lifetime at $C_a < 10$ wt% than those at $C_a > 10$ wt%. However, similar results on the maximum Raman peak $1/\lambda$ of the OH deformation suggest that the strength of hydrogen bond formed in the direction of OH deformation vibration is weakest at $C_a = 9$–10 wt%. Raman spectra, due to the OH deformation, reveal that there are several OH vibration states, each having a different strength of hydrogen bond at $C_a = 7$–10 wt%. These results are comparable to the results for σ, δ_H, T_{1H}, and S_a.

When cellobiose, as the cellulose model, is added into aq. alkali solution, the total strength of the hydrogen bond of the system becomes greater, as judged from an increase in δ_H. This tendency is somewhat more pronounced in the specified C_a range where cellulose shows maximum dissolution. At 4 °C, where cellulose shows more preferred dissolution than at 10 °C, 16 T_{1H} versus C_a plot gives almost a linear relationship over the all C_a region, excluding the existence of the specific C_a region observed for aq. alkali system without cellobiose. This means that cellobiose takes part in the hydrogen bond formation with the cationized or anionized mass strongly solvated with water excluding a part of the extra solvated water molecules. In this connection, an acid–base reaction concept for the dissolution of cellulose can be denied because if such a reaction should take place, δ_H would decrease, thus reflecting a decrease in the strength of the base. The existence of cellobiose in the system also lowers T_{1H}, suggesting that cellobiose can easily form hydrogen bonds with surrounding water resulting in a shorter lifetime of the existing state of the proton in the total system. The use of cellobiose as cellulose model is reasonable since we discuss only a time average state of the system induced by the change in total hydrogen bonds.

It is not clear at present whether the abrupt increase in specific rotatory angle $[\theta]$ of cellobiose aq. alkali solution systems in the specific C_a range suggests that cellobiose takes a specific conformation in the system. However, if this is so, the aq. alkali with specific concentration may have some specific interaction with intramolecular hydrogen bond in celluloses. In the sixth column of Table 4.3.1, the above discussion is briefly summarized. A tentative structure of aq. NaOH solution with specific concentration (9 wt%) and a dissolved state of cellulose in the system are illustrated in Figure 4.3.10. A total of 8 mol of water, solvate with a Na^+ ion and a OH^- ion (i.e. 4 mol/each ion) form the cationic and anionic masses, respectively. Around these masses, 0.2 mol of water on the average forms the second solvation area (extra water solvation area) and the outer of the area is surrounded by about 22.8 mol of free water. Since any cellulose will swell considerably with the alkali and will be converted to alkali cellulose, intermolecular hydrogen bonds might be ruptured easily. For the complete dissolution of cellulose, the breakdown of intramolecular hydrogen bonds is the minimum necessary condition. In the first approximation, the interactions of aq. alkali with specific concentration with OH groups in cellulose, which do not participate in intramolecular hydrogen bonding, might determine the dissolution of cellulose into the aq. alkali. Thus, as shown in Figure 4.3.10, cellulose strongly interacts with both cationic and anionic masses, and also forms relatively strong hydrogen bonds

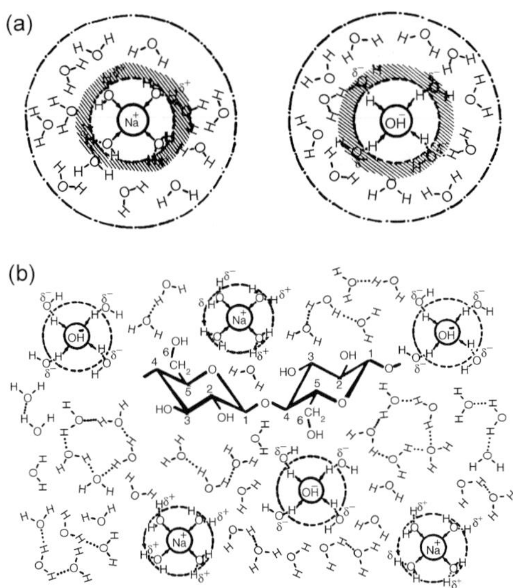

Figure 4.3.10   Schematic representation of tentative structure of aq. NaOH solution with specific concentration (9 wt%) and the dissolution state of cellulose in the system.[1]

with free water originally existing in aq. NaOH. The free water has a tendency to make hydrogen bonds with each other through interaction between cellulose and the ionic masses. In this process, it is plausible that the existence of a second solvation layer in the original aq. NaOH solution plays a very important role in maintaining the electrical and hydrogen bonding nature of the system and in replacing the layer with cellulose.

Na cations do not play a direct role in the dissolution of cellulose and this has been proven by Kamide *et al.*,[16] who experimentally verified that Na ions did not show affiliation to the specific OH of cellulose in its aq. alkali solution and, in contrast to this, the alkali cellulose has Na ions to the OH group at the C_2 position of a glucopyranose unit (Section 4.2).

To summarize, we confirmed the existence of the structure of aq. alkali solution with specific alkali concentration (2.5 mol l^{-1}), in which cellulose shows maximum swelling or maximum dissolution, by analyzing the alkali concentration dependence of electrical conductivity, ^{1}H and ^{23}Na NMR, solvation and Raman spectra of aq. alkali.

REFERENCES

1. T Yamashiki, K Kamide, K Okajima, K Kowsaka, T Matsui and H Fukase, *Polym. J.*, 1988, **20**, 447.
2. C Beadle and HP Stevens, The Eighth International Congress on Applied Chemistry, 1912, **13**, 25.
3. SM Neals, *J. Text. Inst.*, 1920, **20**, T373.
4. H Dillenius, *Kunstseide Zellwolle*, 1940, **22**, 314.
5. H Staudinger and R Mohr, *J. Prakt. Chem.*, 1941, **158**, 233.
6. E Schwart and W Zimmerman, *Melliands Textilber. Inst.*, 1941, **22**, 525.
7. O Eisenluth, *Cellul. Chem.*, 1941, **19**, 45.

8. K Kamide, K Okajima, T Matsui and K Kowsaka, *Polym. J.*, 1984, **16**, 857.
9. T Yamashiki, T Matsui, M Saito, K Okajima and K Kamide, *Br. Polym. J.*, 1990, **22**, 73.
10. T Yamashiki, T Matsui, M Saito, K Okajima and K Kamide, *Br. Polym. J.*, 1990, **22**, 121.
11. T Yamashiki, T Matsui, M Saito, T Matsuda, K Okajima and K Kamide, *Br. Polym. J.*, 1990, **22**, 201.
12. K Kamide and K Okajima, 1986, US Patent, 4,634,470.
13. Y Miyahara and Y Matsuda, *J. Chem. Soc. Jpn., Pure Chem., Sect.*, 1960, **81**, 692.
14. A Passynsky, *Acta Physicochem. USSR*, 1947, **22**, 137.
15. W Brown and R Wikström, *Eur. Polym. J.*, 1966, **1**, 1.
16. K Kamide, K Kowsaka and K Okajima, *Polym. J.*, 1985, **17**, 707.

4.4   STRUCTURAL CHANGE IN CELLULOSE SOLID DURING ITS DISSOLUTION INTO AQ. ALKALI

4.4.1   X-ray diffraction study[1]

Structural change in cellulose after aq. NaOH treatment has long been the subject of extensive studies, mainly using the X-ray diffraction method. The achievements attained by previous studies[2] are summarized very briefly as follows. Natural cellulose maintains its original crystal structure for aq. NaOH with the alkali concentration C_a up to 9 wt%. In the range of 9–12 wt%, a new crystalline form begins to appear and, above 12.5 wt%, the X-ray diffraction pattern relating to the original cellulose disappears, and only that resulting from the new crystalline (alkali-cellulose I (Na-cell I)) is observed. Note that the numerous past works were carried out only for original, alkali insoluble or only partially soluble natural cellulose solid, isolated after dipping in aq. NaOH, and no attention was paid to its dissolution in aq. alkaline solution.

This section intends to clarify the dependence of 'alkali soluble' cellulose structural change on the alkaline concentration (i.e. change in crystalline structure) using the X-ray diffraction method. The alkali soluble cellulose remains at an undissolved state when it is mixed with aq. NaOH solution with various alkaline concentrations. This section also aims to further clarify the dissolution mechanism of the alkali soluble cellulose in aq. alkaline solution.

Cellulose samples

An alkali soluble cellulose (crystal form: cellulose I) was prepared from spruce pulp by a steam explosion treatment.[3–5] An alkali soluble, regenerated cellulose sample prepared as in Section 4.1[7] was acid hydrolyzed to give the same P_v value (in this case, $P_v = 331$) as that of the alkali soluble cell I sample, and was employed as an alkali soluble cell sample for further study.

Some physical properties of the sample were determined by the following methods. The viscosity average degree of polymerization P_v was determined to be 338 from the limiting viscosity number $[\eta]$ in cadoxen at 25 °C using the equation[6] below:

$$[\eta] = 1.84 P_w^{0.78}$$

$$(4.4.1)$$

where P_w is the weight average degree of polymerization. The amorphous content determined by the X-ray method $\chi_{am}(X)$,[7] and the degree of breakdown of intramolecular hydrogen bonds at C_3, as determined by CP-MAS ^{13}C NMR $\chi_{am}(C_3)$,[7] were 21 and 47%, respectively, according to the procedure described elsewhere[7,8] (see Section 4.1.1).

Solubility of the cellulose sample S_a in 7–12 wt% aq. NaOH solution at different temperatures (5–30 °C) was determined with an initial polymer concentration of 5 wt% by the method described in the previous section (Section 4.1.1).[7]

One gram of the alkali soluble cellulose sample was mixed with a predetermined amount (4, 6.7, and 9 g) of aq. NaOH solution with C_a ranging from 3 to 15 wt% at treating temperature $T = -6, 4, 10, 15, 20,$ and 50 °C, controlled within ± 0.1 °C. In this case, the alkali soluble cellulose solid sample and the aq. NaOH solution were maintained at the temperature for 60 min. Then, the paste mixture was mounted on the sample holder of the X-ray diffraction apparatus. To avoid absorbance and diffraction of X-ray via a window in the sample holder, such as a glass plate, the paste mixture was not covered by any window. X-ray diffraction of this mixture was measured by reflection method on an X-ray diffractometer (Rotor Flex RU200 PL; Rigaku Denki. Japan; 37.5 kV × 40 mA) with a position sensitive proportional counter (PSPC). The time required for a measurement was 5 min. The dependence of the X-ray diffraction pattern of the mixture (cellulose/9.1 wt% aq. NaOH) on the keeping time t (expressed in minutes) was also measured. The intensities (ΔI_1 and ΔI_2, respectively; see Figure 4.4.2) of two peaks at the diffraction angles $2\theta = 22.4$–$23.1°$, corresponding to cellulose I crystal (hereafter denoted as cell I) and $2\theta = 20.5°$, for 'soda cellulose' (hereafter denoted as Na-cell I), were conventionally represented in terms of their photon count numbers. They were subtracted from the photon count numbers of the amorphous part at each angle, estimated from the linear baseline connecting the photon count number at $2\theta = 19°$ and that at 26°. ΔI_1, ΔI_2, and their corresponding peak angles for alkali soluble cellulose/aq. NaOH mixtures, were evaluated as functions of the alkali concentration C_a, the keeping time t, keeping temperature T, and cellulose concentration.

Figure 4.4.1 shows the solubility (S_a) of the alkali soluble cell I and cell II at 4, 15, and 30 °C as a function of the alkali concentration, C_a. Both alkali soluble celluloses can be completely dissolved in 8.5–9% aq. NaOH solution at 4 °C, while the alkali soluble cell II showed higher solubility than alkali soluble cell I in the C_a range, less than 10 wt% at

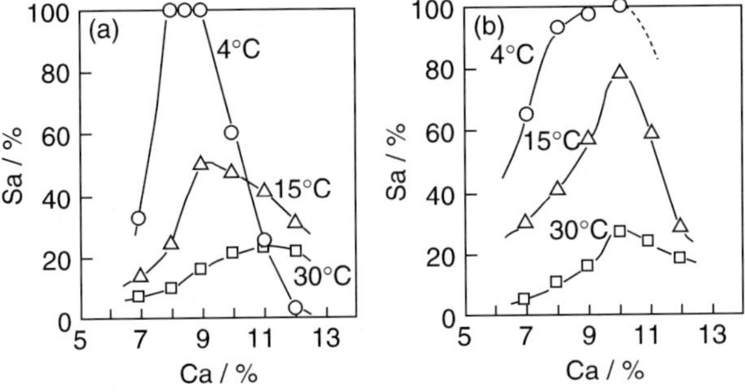

Figure 4.4.1   Solubility (S_a) of cell I and cell II samples against aq. NaOH solutions with various concentrations C_a at 4, 15, and 30 °C: (a) cell I, (b) cell II.[1]

4 and 15 °C. For example, at 4 °C, S_a of cell I is as low as 30% against 7 wt% aq. NaOH, while that of cell I is 65% in the same solvent. C_a gives a maximum dissolution for cell I, has a tendency to increase with increasing the dissolving temperature, but at maximum S_a, C_a for cell II remains roughly constant ($C_a = 10$ wt%). At a temperature higher than 15 °C, S_a of the cell I sample has a C_a more than 11 wt% higher than S_a at 4 °C, but the absolute S_a is below 40%.

Figure 4.4.2 shows typical X-ray diffraction patterns of the mixtures of cellulose and aq. NaOH system (cellulose/aq. NaOH = 1/4, w/w) prepared at 15 °C after the keeping time $t = 6$ min. Obviously, the peak responsible for Na-cell I is not detected in the range of C_a less than 9 wt%, and becomes distinct in the C_a range higher than 11 wt%. It has been experimentally found that aq. NaOH solution with $C_a = 15$ wt% completely converts alkali soluble cellulose to Na-cell I. The peak responsible for cell I ((002) plane) becomes broad and less intense as C_a increases from 8 to 11%. Close inspection of the figure reveals a systematic shift of the peak responsible for cell I towards the lower diffraction angle in the range $C_a \leq 9$ wt%. Similar phenomena were also observed at 4 °C.

Figure 4.4.3 shows the relationship between X-ray diffraction peak intensity (ΔI_1, ΔI_2) for the mixtures after keeping time $t = 60$ min and C_a. Circles denote cell I and rectangles denote Na-cell I. Unfilled symbols show the results at 15 °C and filled marks show those at 4 °C. At 4 °C, ΔI_1 (at $2\theta = 22.9$–$23.1°$), cell I continuously decreases with an increase in C_a (up to *ca.* 10 wt%), and then rapidly decreases to zero at higher C_a. At 15 °C, ΔI_1 ($2\theta = 22.9$–$23.2°$) remains almost constant in the range $C_a \leq 9$ wt%, but decreases dramatically with C_a from 10 to 12 wt%, rapidly approaching zero. The difference in the peak intensity for cell I in the C_a range of between 8 and 10 wt% at 4 and 15 °C corresponds to the result shown in Figure 4.4.1.

The cellulose/aq. NaOH system with higher S_a gives the weaker diffraction peak intensity for cell I. On the one hand, the peak for Na-cell I was very weak and lay within the realm of experimental uncertainty until $C_a = 9$ wt%, independent of the temperature

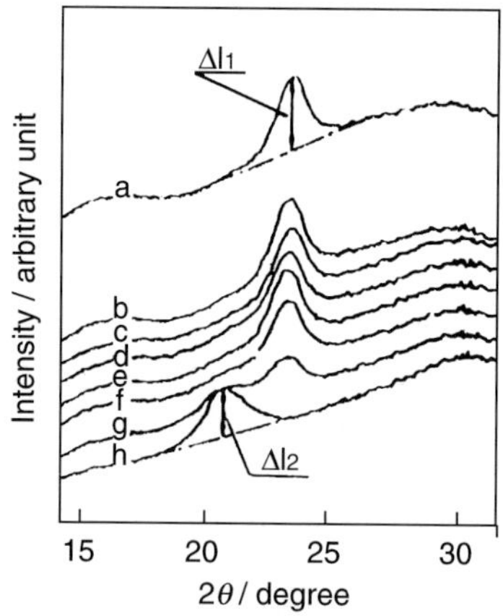

Figure 4.4.2   Typical X-ray diffraction patterns of the mixtures of alkali soluble cellulose aq. NaOH system at 15 °C:[1] (a) $C_a = 0$ wt%, (b) 3 wt%, (c) 7 wt%, (d) 8 wt%, (e) 9 wt%, (f) 10 wt%, (g) 11 wt%, (h) 15 wt%: the ratio of cellulose/aq. NaOH solution was kept to be 1/4 (w/w).

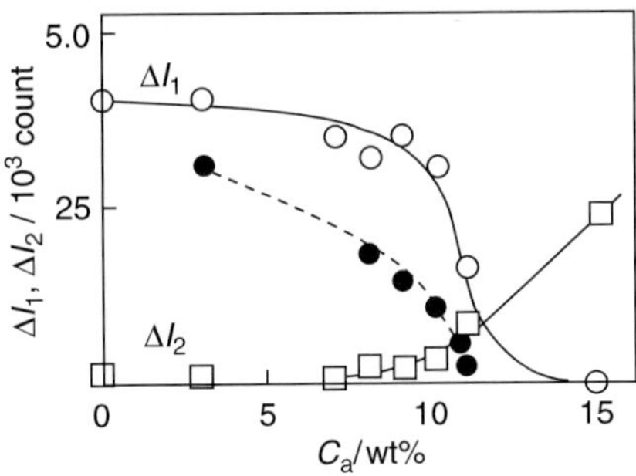

Figure 4.4.3   The relation between the corrected photon count number for cell I (ΔI_1), Na-cell I (ΔI_2), and aq. NaOH concentration C_a; circle,[1] ΔI_1; rectangle, ΔI_2; unfilled mark shows the results at 15 °C and filled mark those at 4 °C.

studied. These results explicitly indicate that the dissolution of cellulose into aq. NaOH solution with $C_a = 8-9$ wt% occurs without any conversion of cellulose to Na-cell I. ΔI_2 for Na-cell I increases at $C_a \geq 11$ wt%, finally approaching the value for complete Na-cell I. Thus, the very low S_a value of the cellulose against aq. NaOH solution with $C_a > 12$ wt% might be closely related to production of Na-cell I.

We then carried out solubility test of the completely converted Na-cell I against aq. NaOH solution where $C_a = 8-10$ wt% at the temperature range between -5 and 30 °C, and we found that the Na-cell I was not soluble in these alkaline solutions. This seems to be related to the fact that Na-cell I has a very uniform structure in view of CP-MAS ^{13}C NMR spectrum (see Figure 1 of Ref. 8). In addition, the strength of intramolecular hydrogen bond at C_3, estimated from the position of its chemical shift of C_4 carbon (*ca.* 87 ppm), is stronger than that of the region of weaker intramolecular hydrogen bond (*ca.* 84 ppm) in cellulose solid.[8] Here, it should be noted that the strong intramolecular hydrogen bond of $O_3 \cdots O_5$ for cellulose with cellulose I crystal gives C_4 carbon NMR peak at *ca.* 89 ppm and, of course, the cellulose with a higher fraction of this NMR peak is difficult to dissolve in aq. 9.1 wt% NaOH.

Figure 4.4.4 shows the relationship between X-ray diffraction angles at the peaks for cell I and Na-cell I for the mixtures at $t = 60$ min and C_a at the treating temperature $T = 15$ °C. Obviously, peak positions for cell I of the mixtures of alkali soluble cellulose/ aq. NaOH lies in the range of $2\theta = 22.9-23.1°$, which is higher than that for Na-cell I ($2\theta = 20.5°$). The peak responsible for cell I for the mixture of alkali soluble cellulose/ 9.1 wt% aq. NaOH is significantly lower than other mixtures of cellulose/aq. NaOH ($C_a \neq 9$ wt%), showing that the distance between the crystalline lattice of cell I in this specified mixture is slightly wider than those for other mixtures. This phenomenon was found to be reproducible and we can therefore conclude that a specific structure of the 9.1 wt% aq. NaOH solution proposed in Section 4.3[9] is one of the factors bringing about this phenomenon.

Figure 4.4.5(a) and (b) shows the changes in the peak position corresponding to (002) plane (a) and ΔI_1 (b) of cell I for the mixtures of alkali soluble celllose/9.1 wt% aq. NaOH (1/4,w/w: ○ and 1/5.7, w/w: △) as a function of the keeping time t at 4 °C. Immediately after the cellulose sample comes into contact with aq. NaOH, the peak at $2\theta = 23.10°$ of

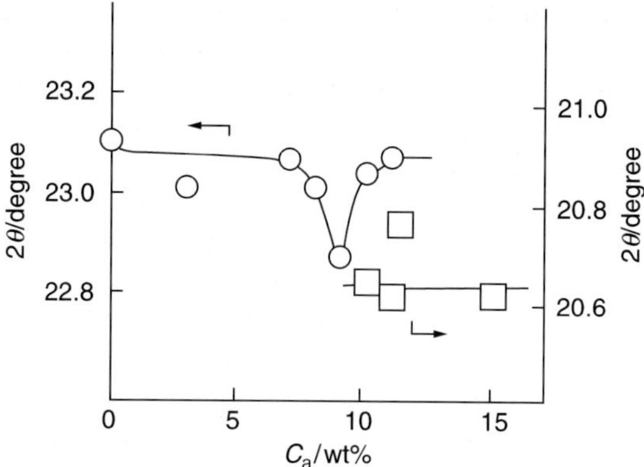

Figure 4.4.4 The relation between the crystalline peak position 2θ of cell I and Na-cell I for the mixture of alkali soluble cellulose/aq. NaOH and C_a;[1] (O), Cell I; (□), Na-cell I.

cell I shifts to a lower angle of $2\theta = 22.80°$ (in the case of cellulose concentration of the resulting mixture, 20 wt%) and $22.45°$ (in the case of cellulose concentration, 15 wt%). As the keeping time t becomes longer, the peak again shifts to a larger angle and then to a smaller angle after passing through a maximum value. This was observed irrespective of the cellulose concentration of the system. For the system of cellulose concentration 20 wt%, the maximum peak angle ($2\theta = 22.95°$) is attained at $t = 60$ min, and for the system of cellulose concentration 15 wt%, the maximum peak angle ($2\theta = 22.65°$) is lower and is attained at shorter t (30 min) than that of the system of cellulose whose concentration is 20 wt%. This means that the contact of a 9.1 wt% aq. NaOH to alkali soluble cellulose instantly widens the average distance between crystalline lattice of (002) plane for cell I and, as a result, the crystalline part with wider crystalline lattice dissolves, resulting in an increase in the peak angle (see Figure 4.4.5a). Then the distance between the crystalline lattice in the remaining crystalline part may very gradually

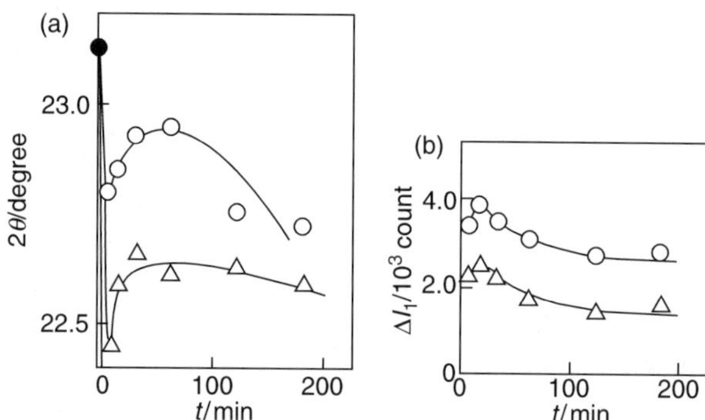

Figure 4.4.5 Dependence of peak position (a) and ΔI_1 (b) of cell I in the system of cellulose/9.1 wt% aq. NaOH on the treating time, t:[1] (O), cellulose/aq. NaOH (1/4, w/w); (Δ), cellulose/aq. NaOH (1/5.7, w/w). Crystalline peak of cell I for original alkali soluble cellulose is shown as corresponding to the closed mark.

increase with the invasion of aq. NaOH. It is thus reasonable to consider that a larger amount of NaOH in the system makes the distance between crystalline lattice much wider during the contact period examined here. ΔI_1 also decreases at the instance of contacting aq. NaOH due to the strong swelling of cellulose. When 5 min $\leq t \leq$ 15 min, ΔI_1 clearly increases until $t = 15$ min and then decreases monotonically. Note that the peak position representing ΔI_1 varies depending on the treating time t. The increase in ΔI_1 at 5 min $\leq t \leq$ 15 min may reflect the real increase in the fraction of the crystalline part of which crystalline lattice distance is slightly wider than that for the original cellulose crystal I. This is because the solid state ^{13}C NMR analysis shows that sufficient water molecules contained in cellulose with cellulose I crystal release the strain that existed in the original dry cellulose.[10] In the present system, the aq. NaOH solution in cellulose solid might behave in the same manner as water contained in cellulose solid described above. In this case, it can be concluded that in the process of dissolution of the alkali soluble cellulose, the aq. NaOH with specific concentration penetrates into amorphous part as well as into crystalline (ordered) part. This dissolves the amorphous part with the incorporation of some remaining amorphous parts into the 'ordered region,' and then dissolution proceeds into the resulting ordered part.

Figure 4.4.6 shows a schema of dissolving process of alkali soluble cellulose into aq. NaOH solution with specific C_a, which is deduced from the above experimental results and discussion.

Figure 4.4.7 shows the treating temperature dependence of diffraction peak intensities ΔI_1 (circle; cell I) and ΔI_2 (rectangle; Na-cell I) for cellulose/9 wt% aq. NaOH (1/9, w/w)

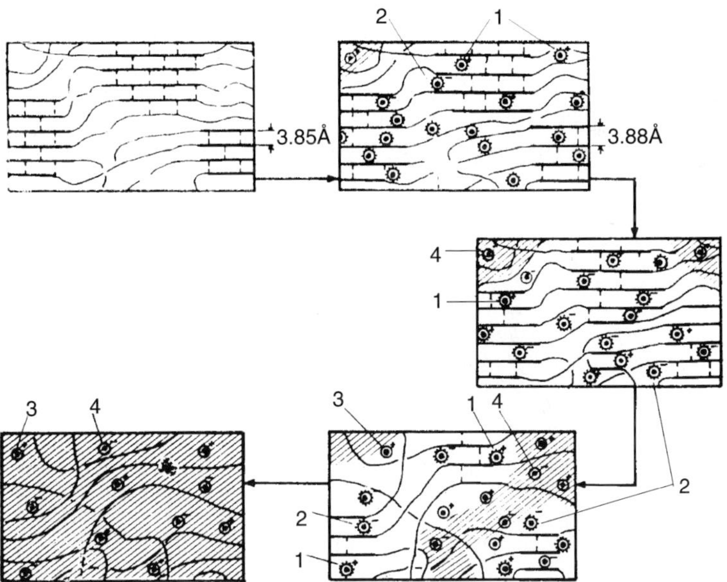

Figure 4.4.6   A schema of the dissolving process of alkali soluble cellulose into aq. NaOH with specific concentration $C_a = 9$ wt%[1]; ladder-like part, ordered region; fine line, cellulose chains in disordered region; (1) Na$^+$ with specific solvated structure; (2) OH$^-$ with specific solvated structure; (3) Na$^+$ with normal structure; (4) OH$^-$ with normal structure; hatched area, dissolved part.

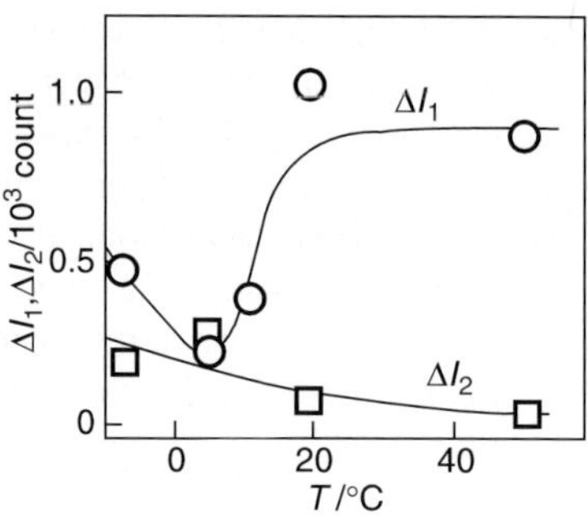

Figure 4.4.7   The treating temperature dependence of ΔI_1 and ΔI_2 for cellulose/9 wt% aq. NaOH (1/9, w/w) system:[1] circle, ΔI_1; rectangle, ΔI_2.

system. Obviously, the peak intensity responsible for Na-cell I, ΔI_2, is very weak in all of the temperature ranges examined here, and the peak intensity for cell I, ΔI_1, showed a minimum at 4 °C. The existence of minimum peak intensity for cell I explains the experimental fact that solubility of the alkali soluble cellulose attains a maximum at 4 °C.

Figure 4.4.8 shows the polymer concentration (in this case, cellulose fraction) dependence of diffraction peak intensities ΔI_1 (circle; cell I) and ΔI_2 (rectangle; Na-cell I) for cellulose/9 wt% aq. NaOH system at 4 °C. ΔI_1 for Na-cell I is very low and practically constant, irrespective of the cellulose concentration in the system. On the other hand, ΔI_1 for cell I linearly increases with an increase in cellulose concentration. The value of cellulose concentration at an intercept of the above linear line, where baseline ($\Delta I_1 = 0$) denotes an upper limit of cellulose concentration below which the alkali soluble cellulose completely dissolves in 9 wt% aq. NaOH solution at 4 °C.

In summary, as illustrated in Figure 4.4.9, we found that (1) alkali soluble cellulose shows no structural change when contacted with aq. NaOH solution with C_a less than 7 wt%, resulting in no dissolution into these aq. alkaline solution, (2) mixing with

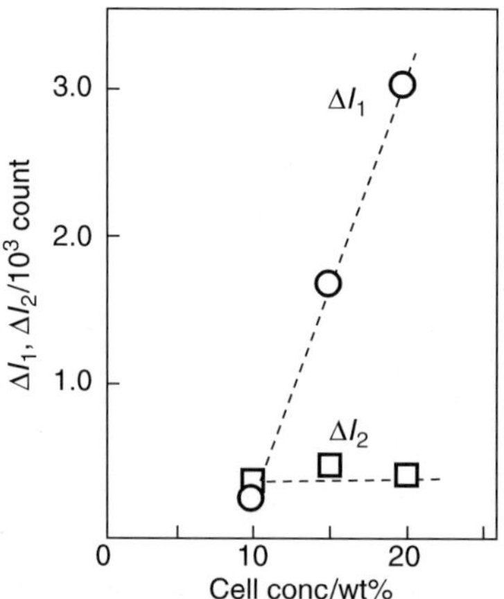

Figure 4.4.8   The cellulose concentration dependence of ΔI_1 and ΔI_2 for cellulose/9 wt% aq. NaOH system at 4 °C:[1] circle, ΔI_1; rectangle, ΔI_2.

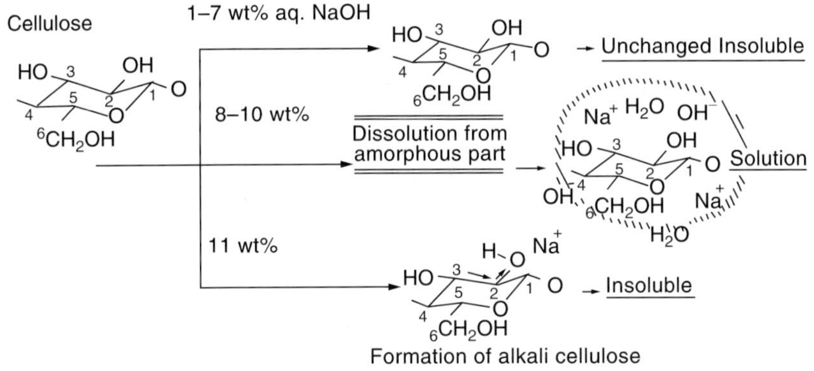

Figure 4.4.9   The illustration of structural change of alkali soluble cellulose concentrated with aq. NaOH solutions with different alkaline concentrations C_a.

aq. NaOH solution where C_a is more than 11 wt% brings about the conversion of cellulose to the structurally stable Na-cell I, which results in a low degree of dissolution of cellulose into these aq. alkaline solutions, (3) only an aq. NaOH solution with specific $C_a = 8$–9 wt%, which can dissolve alkali soluble cellulose, serves to widen the crystalline lattice (002) plane of cellulose I crystal slightly, and (4) the dissolution of alkali soluble cellulose into above aq. NaOH solution might occur first in the amorphous part, degrading the supermolecular structural regularity in the neighboring ordered part and, finally, giving a solution without conversion of the cellulose to Na-cell I.

4.4.2   Nuclear magnetic resonance study[11]

Recent advances in solid state NMR techniques will enable us to obtain more detailed and physically different information, as compared with X-ray diffraction, for elucidating the dissolving mechanism of alkali soluble cell II, as well as alkali soluble cell I. This subsection describes the results of CP-MAS ^{13}C NMR analysis of alkali soluble cell I and II mixed with aq. alkali solution.[11]

As an approximate measure expressing the degree of breakdown in intramolecular hydrogen bonds of cell I in cell I/aq. NaOH mixture, the structural parameters $\chi_{am}(C_k)$ ($k = 2, 3$ and 6) and $\chi_{am}(C_2 + C_3)$ were calculated from NMR spectra by use of eqs. (4.4.2)–(4.4.5).[2,7,12]

$$\chi_{am}(C_3) = 100 \times \{I_3/(I_2 + I_3)\} \ (\%) \tag{4.4.2}$$

$$\chi_{am}(C_6) = 100 \times \{I_7/(I_6 + I_7)\} \ (\%) \tag{4.4.3}$$

$$\chi_{am}(C_2 + C_3) = 100 \times [\{I_3/(I_2 + I_3)\} - 1]/2 \ (\%) \tag{4.4.4}$$

$$\chi_{am}(C_2) = 2\chi_{am}(C_2 + C_3) + \chi_{am}(C_3) \ (\%) \tag{4.4.5}$$

Here, I_k is the peak intensity for the peaks numbered from lower magnetic field side (see Figure 4.4.11). For the cell II/aq. NaOH system, only $\chi_{am}(C_3)$ was able to be calculated because of the overlapping of the peaks necessary for estimating $\chi_{am}(C_2)$ and $\chi_{am}(C_6)$.

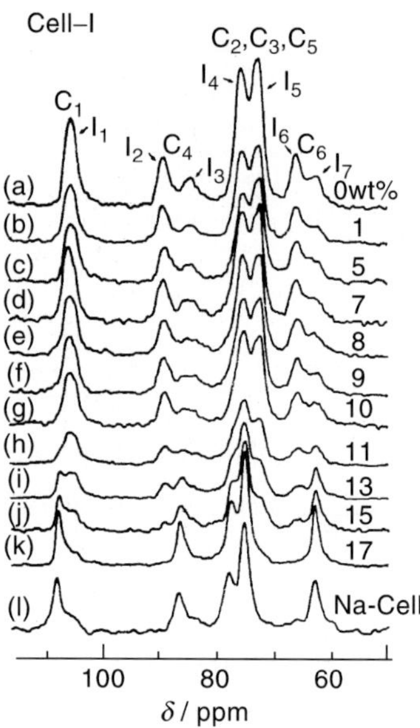

Figure 4.4.10 CP-MAS ^{13}C NMR spectra of cell I sample in aq. NaOH solution with various alkaline concentrations C_a:[1] (a) $C_a = 0$ wt%, (b) 1 wt%, (c) 5 wt%, (d) 7 wt%, (e) 8 wt%, (f) 9 wt%, (g) 10 wt%, (h) 11 wt%, (i) 15 wt%, (k) 17 wt%, (l) Na-cell I (from Ref. 8).

Figures 4.4.10 and 4.4.11 show CP-MAS ^{13}C NMR sepctra for the systems of cell I/aq. NaOH and cell II/aq. NaOH, respectively. In the figures, the NMR spectrum of the alkali cellulose from cell I reported by Kamide *et al.*[8] is also included. The C_1 carbon peak for cell I is observed as a broad envelope centered at 105.4 ppm, while the C_1 carbon peak for cell II centered at 105.4 ppm has two clearly distinguishable shoulder peaks at 107.0 and 104.0 ppm. The peaks in the C_4 carbon region of cell I are separated into two envelopes centered at 89.1 and 84.5 ppm, and the former peak has been assigned due to the strong intramolecular hydrogen bond ($C_3 \cdots O_5'$).[7,13] For cell II, the spectrum in the C_4 carbon region is composed of two envelopes, although not highly distinct. Note that cell II exhibited a relatively sharp peak in the same position (80 ppm) as that observed for alkali soluble cellulose/aq. NaOH solution. Previously, the peak was assigned to C_4 carbon peak[2] with no $O_3 \cdots O_5'$ intramolecular hydrogen bond. Therefore, the peak at 80 ppm for cell II might be responsible for the C_4 carbon in cellobiose unit with the completely destructed $O_3 \cdots O_5'$ intramolecular hydrogen bond. Two large envelopes centered at 75.3 and 72.4 ppm are observed in the region for cell I system, and the C_2, C_3, and C_5 peak region for cell II system is explicitly composed of three envelopes peaked at 77.0 (sharp shoulder), 75.2, and 73.4 ppm. The C_6 carbon peak region is also separated into two envelopes (65.5 and 62.3 ppm) for cell I, while three peaks occur at 63.5, 62.6, and 61.5 ppm for cell II. Both the positions of NMR peaks characteristic for the systems of cell I/aq NaOH with $C_a = 1-9$ wt% and for the systems of cell II/aq. NaOH with $C_a = 1-8$ wt% are almost equivalent to those for the system of cell I (or cell II)/water. That is, all peaks observed in the above cellulose/aq. NaOH systems are characteristic of pure cellulose.

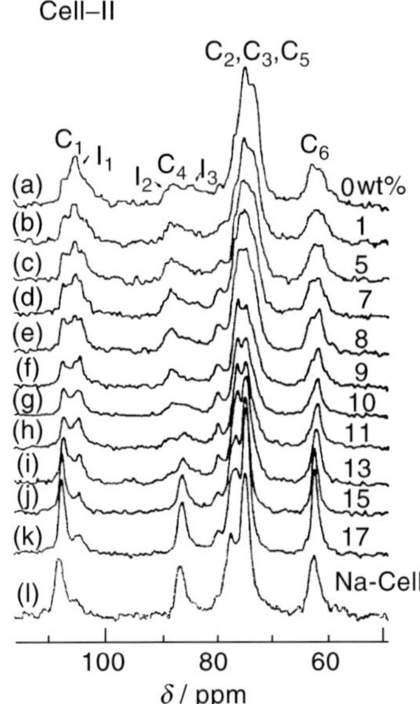

Figure 4.4.11   CP-MAS 13NMR spectra of cell II sample in aq. NaOH solution with various alkaline concentrations C_a:[1] (a) $C_a = 0$ wt%, (b) 1 wt%, (c) 5 wt%, (d) 7 wt%, (e) 8 wt%, (f) 9 wt%, (g) 10 wt%, (h) 11 wt%, (i) 15 wt%, (k) 17 wt%, (l) Na-cell II.

When C_a ($C_a \geq 10$ wt%) is increased, the NMR spectra of cell I and cell II/aq. NaOH solution systems were observed to change in the following manner:

For cell I: C_1: appearance of a new peak at 107.4 ppm together with decrease in the intensity of peak at 105.4 ppm. C_4: appearance of a new peak at 86.3 ppm, accompanied by a decrease in the intensities of both peaks at 89.1 and 84.5 ppm. C_2, C_3, C_5: appearance of new peaks at 77.5 and 74.8 ppm (near the original peak at 75.3 ppm) with a diminishing of the peak at 72.4 ppm. C_6: appearance of a new peak at 62.6 ppm (near the original peak at 62.3 ppm) with diminishing of the peak at 65.5 ppm.

For cell I: C_1: apparent increase in the peak intensity of 107.4 ppm peak and decrease in peak intensities of the peaks at 105.4 and 104.0 ppm. C_4: similar to cell I. C_2, C_3, C_5: appearance of new peaks at 77.5 ppm (near the original peak at 77.0 ppm) and 74.8 ppm (near the original peak at 75.2 ppm) with the disappearance of the peak at 73.4 ppm. C_6: apparent increase of the peak intensity of 62.6 ppm with a diminishing of the peaks at 63.5 and 61.5 ppm.

Note that the intensity of a peak detected at 80 ppm for cell II increased with C_a in its range less than 10 wt% and then decreased, with a further increase in C_a. NMR patterns for both cell I and cell II, mixed with aq. NaOH having $C_a = 17$ wt%, are quite similar to each other. These are also similar to that of alkali cellulose, showing a C_1 carbon peak at 107.4 ppm, a C_4 carbon peak at 86.3 ppm, and a C_6 carbon peak at 62.6 ppm. Therefore, all these spectral changes observed in the range $C_a \geq 10$ wt% are concluded due to the formation of alkali cellulose.

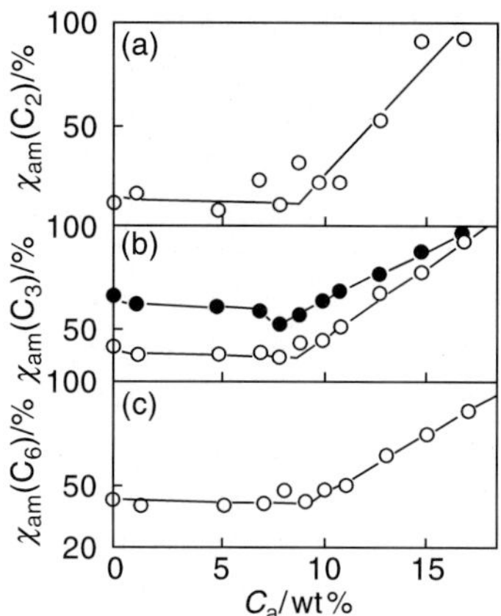

Figure 4.4.12   Change in $\chi_{am}(C_k)$s of cell I/aq. NaOH system against NaOH concentration C_a: (a), $\chi_{am}(C_2)$; (b), $\chi_{am}(C_3)$; (c), $\chi_{am}(C_6)$: In (b) data points for cell II are included as closed circles.

Figure 4.4.12 shows the plots of $\chi_{am}(C_2)$ (a), χ_{am} (C$_3$) (b), and χ_{am} (C$_5$) (c), as determined by eqs. (4.4.2)–(4.4.5), against alkali concentration (C_a) for cell I/aq. NaOH system. In Figure 4.4.12(b), the results for cell I/aq. NaOH system are also included. Here, it should be noted that eqs. (4.4.2)–(4.4.5) are applicable only to cellulose solid and, therefore, $\chi_{am}(C_k)$ is estimated for the system where cellulose and Na-cell I coexist. However, this does not simply express the degree of breakdown in intramolecular hydrogen bond. $\chi_{am}(C_2)$, $\chi_{am}(C_3)$ and $\chi_{am}(C_6)$ for cell I/aq. NaOH systems have almost the same dependence on C_a: $\chi_{am}(C_k)$, being roughly kept constant in the C_a range of 1–8 (or 9) wt% and then increased linearly with an increase in C_a. For cell II/aq. NaOH system, the C_a dependence of $\chi_{am}(C_3)$ is similar with that for cell I/aq. NaOH system. Detailed examination revealed a small minimum in χ_{am} (C$_3$) at $C_a = 7$–9 wt%.

Close inspection of Figure 4.4.12 and Figure 4 of Ref. 11 highlights that four different C_a regions can be classified, corresponding to the changes in the supermolecular structures of cellulose:

For cell I: region 1: $0 < C_a \leq ca.$ 9 wt%: region 2: *ca.* $9 < C_a \leq 11$ wt%: region 3: $11 < C_a \leq 15$ wt%: region 4: $15 < C_a \leq 17$ wt%:
For cell II: region 1; $0 < C_a \leq ca.$ 8 wt%; region 2; *ca.* $8 < C_a \leq ca.$ 10 wt%; region 3; *ca.* $10 < C_a \leq 15$ wt%; region 4; $15 < C_a \leq 17$ wt%.

In *region 1* (i.e. $0 < C_a \leq ca.$ 9 wt% for cell I, $0 < C_a \leq ca.$ 8 wt% for cell II) the components with a relatively weak intramolecular hydrogen bond originally existing in the amorphous region may be plasticized by penetration of hydrated alkali ions, resulting in lowering of CP-MAS ^{13}C NMR signal. Note that the amorphous region is plasticized but is not gelatinized or dissolved out. The slight decrease in $\chi_{am}(C_3)$ and $\chi_{am}(C_6)$, when mixed with aq. NaOH with $C_a = 1$ wt%, can be interpreted by the plasticizing effect. In addition, the constancy of these supermolecular parameters in this C_a region strongly suggests that structural changes are limited within the amorphous region.

In *region 2* (i.e. *ca.* $9 < C_a \leq 11$ wt% for cell I and *ca.* $8 < C_a \leq ca.$ 10 wt% for cell II) most of the original amorphous part is converted either to a gel state or to a solution state and, at the same time, some of the crystalline parts of the cellulose is converted to an amorphous state, which progresses further into either a gel or solution state. All of these changes result in a considerable lowering of NMR signal intensity.

In *region 3* (i.e. $11 < C_a \leq 15$ wt% for cell I and *ca.* $10 < C_a \leq 15$ wt% for cell II) solid-to-solid phase transition of cellulose to Na-cell I and solid-to-gel or solid-to-solution phase transition occur without concurrent chemical change. Thus, the unchanged cellulose crystalline region is susceptible to CP-MAS ^{13}C NMR spectra. This explanation is verified by the experimental facts that the ratio $I_{C_1}:I_{C_1}0$ (I_{C_1} is the absolute intensity of total C_1 envelope in the case $C_a > 1$ wt% and $I_{C_1}0$ is for $C_a = 0$ wt%) is constant in this C_a region, whereas with $I_{C_1}(1)/I_{C_1}0$ ($I_{C_1}(1)$, the fraction of the envelope centered at 107.4 ppm in C_1 carbon peak region), starts to increase slightly.

In *region 4* (i e. $15 < C_a \leq 17$ wt% for both systems) the Na-cell I formation occurs much faster than the solid-to-gel (or solution) phase transition of cellulose, mainly due to low solubility power of aq. alkali with this range of C_a (see Figure 4.4.1) leading to large $I_{C_1/C_1}0$. As early as 1939, Sobue *et al.* observed that the lower limit of C_a of aq. NaOH, in which the formation of Na-cell I was detected at 40 °C, was 12 wt%.[14] Their findings fortify the above reasoning.

From the experimental results obtained here it can be concluded that, when the alkali soluble cellulose is dispersed at a polymer concentration around 5 wt% into aq. NaOH with $C_a = 8-11$ wt% at low temperature, the alkali (hydrated hydroxyl and hydrated sodium ions) firstly completely destroys the region consisting of cellulose molecules with weak intramolecular hydrogen bond (i.e. amorphous region), which results in the activation of the molecular mobility. Second, owing to this molecular activation, the molecular gap in the neighboring crystalline region widens (the unzipping effect), allowing the alkali to further penetrate into the subsequent crystalline region neighboring the destroyed crystalline region. This results in the total destruction of the crystalline part. Finally, the cellulose sample dissolves completely into aq. NaOH, without forming Na-cell I. This was previously proposed by Kamide *et al.*[1] to explain the dissolution of cell I in aq. NaOH. In the C_a region, where formation of Na-cell I starts, the solubility of cellulose is low, even if the temperature is kept low. The Na-cell I formed in aq. NaOH with $C_a = 17$ wt% proved to swell slightly but it did not dissolve, even when C_a of the alkali coexisting with the Na-cell I is adjusted to 9 wt% by diluting with aq. NaOH with low C_a at a low temperature, and the polymer concentration is reduced as low as 3 wt%.

The reason why Na-cell I does not show complete dissolution in aq. NaOH is not evident at present. However, it may be related to the well-known facts that Na-cell I has a stoichiometrical composition (1 mol of NaOH/1 mol of glucoside residue) between cellulose and aq. NaOH. It also has highly uniform $O_3\cdots O_5'$ intramolecular hydrogen bond, as judged from the existence of a single and relatively sharp C_4 carbon peak at 87 ppm, as compared with that of ordinary cellulose solid (roughly showing two peaks at 89 and 83 ppm) in its CP-MAS ^{13}C/MMR spectrum,[8] rejecting the unzipping by excess NaOH.

REFERENCES

1. K Kamide, K Yasuda, T Matsui, K Okajima and T Yamashiki, *Cellul. Chem. Technol.*, 1990, **24**, 23.
2. See for example, M Fujii, In *Cellulose Handbook*. (eds H Sobue and N Migita), Chapter 9, Asakura, 1958, p. 268; K Hess and C Trogus, *Z. Physik. Chem.*, 1931, **11**, 381 Tabelle 3 for ramie fiber.
3. T Yamashiki, T Matsui, M Saitoh, K Okajima, K Kamide and T Sawada, *Br. Polym. J.*, 1990, **22**, 73.
4. T Yamashiki, T Matsui, M Saitoh, K Okajima, K Kamide and T Sawada, *Br. Polym. J.*, 1990, **22**, 121.
5. T Yamashiki, T Matsui, M Saitoh, Y Matsuda, K Okajima, K Kamide and T Sawada, *Br. Polym. J.*, 1990, **22**, 201.
6. W Brown and R Wikström, *Eur. Polym. J.*, 1965, **1**, 1.
7. K Kamide, K Okajima, T Matsui and K Kowsaka, *Polym. J.*, 1984, **16**, 857.
8. K Kamide, K Okajima and K Kowasaka, *Polym. J.*, 1985, **17**, 707.
9. T Yamashiki, K Kamide, K Okajima, K Kowsaka, T Matsui and H Fukase, *Polym. J.*, 1988, **20**, 447.
10. F Horii, A Hirai and R Kitamaru, *Macromolecules*, 1987, **21**, 17.
11. H Yamada, K Kowsaka, T Matsui, K Okajima and K Kamide, *Cellul. Chem. Techn.*, 1992, **26**, 141.
12. K Kamide, K Okajima, K Kowsaka and T Matsui, *Polym. J.*, 1985, **17**, 701.
13. K Kamide, K Okajima and K Kowsaka, *Polym. J.*, 1992, **24**, 71.
14. H Sobue, H Kissing and K Hess, *Z. Phys. Chem. B*, 1939, **34**, 309.

4.5 FLOW BIREFRINGENCE AND VISCOSITY[1]

In this section, we study the hydrodynamic properties of cellulose dissolved in aq. sodium hydroxide solution and in cadoxen at the polymer concentration of *ca.* 1–5 wt%, mainly employing flow birefringence and viscosity methods, to determine the dissolved state of cellulose in these solvents from dynamic light scattering (DLS) and the spin lattice relaxation time measurements by ^{13}C NMR.

Table 4.5.1 shows the preparative method and M_v of cellulose samples employed. Hereafter, we simply designate the cellulose samples obtained from the conifer pulp as cellulose I and the regenerated samples as cellulose II.

Preparation of cellulose solutions for flow birefringence
and viscosity measurements

First, the specially prepared cellulose sample flakes were dissolved in a 9 wt% aq. NaOH solution at 4 °C by agitating for 1 min with a homogenizer. For the purpose of eliminating undissolved residue, the solution was centrifuged in an ultracentrifuge with acceleration of $1.5 \times 10^4 g$ (*g*, standard gravity) for 1 h in a Hitachi model 55p-7 automatic preparative ultracentrifuge. Subsequently, the cellulose concentrations C_p of the solutions were adjusted to 1.75–5.1 wt% by dilution with a 9 wt% aq. NaOH solution and stocked at 5 °C in a dark place.

Second, the cellulose samples were dispersed in the cadoxen at $C_p = 5$ wt%. The system of cellulose and cadoxen was stored at 5 °C for 1 week with occasional agitation.

Table 4.5.1

Preparative method and M_v and χ_{am} cellulose employed[1]

Sample code	Crystalline form	Cellulose source	Preparative method	$M_v \times 10^4$ [a]	χ_{am} [b] (%)
P1	Cellulose I	Conifer pulp	Steam explosion[c]	5.4	52
P2				6.9	57
P3				9.7	48
B1	Cellulose II	Acid hydrolyzed cotton	Cuprammonium	13.0	78
B2				5.2	72
B3				7.3	73
B4				9.7	75

[a] Viscosity average molecular weight in cadoxen.[2,3]
[b] Ref. 4.
[c] Ref. 5.

Centrifugation of the solution under $G = 2 \times 10^3 g$ for 1 h gave a clear and transparent solution as the supernatant phase.

Flow birefringence

For the measurement of extinction angle χ, an improved version of Micro FBR Mark II manufactured by Waken-Yaku Co. (Kyoto, Japan) was used. A schematic diagram of the apparatus is illustrated in Figure 4.5.1. Here, clearance between the stationary inner cylinder and the outer rotor was chosen as 0.2 mm in order to suppress possible turbulent flow and to obtain sufficient brightness of transmitted light through the flowing solution layer.

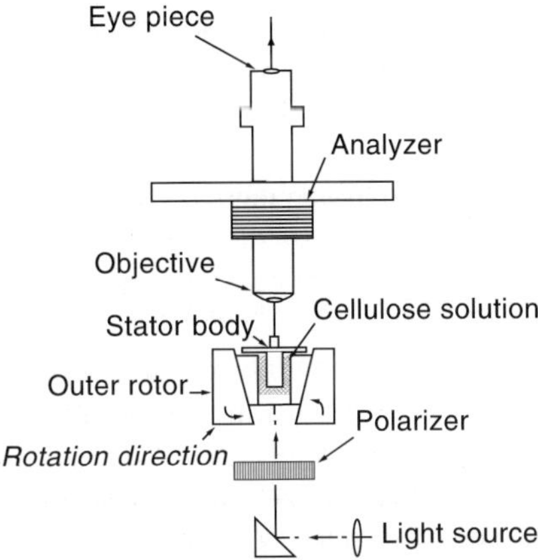

Figure 4.5.1 Schematic diagram of Micro FBR Mark II manufactured by Waken-Yoku Co.[1] A xenon lamp was utilized as the light source. The cellulose solution (*ca.* 100 μl), stored at 5 °C for *ca.* 1 h after preparation, was placed in an annular space between the two concentric cylinders. The cell of the apparatus was maintained under thermostat at 20 ± 0.01 °C during the measurements. χ was observed using a microscope and cross polarizers.

Solution viscosity

The solution viscosity η was measured for the samples B2 and P1 in a 9 wt% NaOH solution at 5 °C as a function of shear rate $d\gamma/dt$ ranging from $C_p = 1 \times 10^2 - 2 \times 10^5 \text{ s}^{-1}$. An automated high-shear capillary viscometer (HVA 6; Anton Paar K.G., Graz, Austria)[6] was utilized with a capillary of inner diameter × length = 0.8 × 100 mm.

Preparation of cellulose solutions for dynamic light scattering measurements

The samples P1 and B2 were dissolved in cadoxen at the polymer concentrations of 5 and 0.5 wt%. The cellulose solution ($C_p = 5$ wt%) (S5) was centrifuged under an acceleration ranging from 1 g (without centrifugation) to $6.36 \times 10^4 g$ at 4 °C in a Hitachi model 55p-7 automatic preparative ultracentrifuge ($S_{5\rightarrow c}$). The upper two-thirds of the supernatant phase was mixed with cadoxen to prepare a 0.5 wt% solution, shaken often by hand ($S_{5\rightarrow c\rightarrow 0.5}$), and stocked for 1–5 days at 25 °C ($S_{5\rightarrow c\rightarrow 0.5(1)}$ and $S_{5\rightarrow c\rightarrow 0.5(5)}$).

The solutions $S_{5\rightarrow c\rightarrow 0.5(1)}$ and $S_{5\rightarrow c\rightarrow 0.5(5)}$ were employed for DLS measurements. The 0.5 wt% solution ($S_{0.5}$) was employed directly for DLS measurements without further purification.

Dynamic light scattering (see Section 3.5.5)

^{13}C-Nuclear magnetic resonance

From the spectra, the spin lattice relaxation time T_1 of the cellulose in cadoxen-deuterated water was estimated by the inversion recovery method.[7]

Figure 4.5.2(a) shows the effect of the cellulose concentration on the shear rate $d\gamma/dt$ dependence of χ for cellulose II (sample B2, $M_v = 5.2 \times 10^4$) in a 9 wt% aq. NaOH solution at 20 °C. Flow birefringence was not detected for the solution with C_p below 3.4 wt% in the lower shear rate region. While χ of the solution with C_p below 3.4 wt%, observed in the higher shear rate region, monotonically decreases with an increase in $d\gamma/dt$, in $C_p > 4.3$ wt% solutions, with an increase in $d\gamma/dt$, χ decreases at lower $d\gamma/dt$ and increases in the region of higher $d\gamma/dt$. This anomalous behavior of χ (i.e. the positive dependence of χ on $d\gamma/dt$) is not caused by turbulent flow, because even at the maximum $d\gamma/dt$ value used (*ca.* $3 \times 10^3 \text{ s}^{-1}$), the Reynolds number of the solution with $C_p > 4$ wt%, calculated from the data on the average velocity of the fluid (30 cm s^{-1}), density (1.02 g cm^{-3}), clearance (0.02 cm), and viscosity (4.0 dyne cm^{-2} s), is 0.154, which is always far less than the critical value (2100) at which the laminar flow changes to turbulent flow. Similar anomalies of χ have also been reported on solutions of some flexible polymer solvent systems, such as polystyrene in dioxan and poly (para-tert-butyl phenyl methacrylate) in tetrachloromethane. All of the systems showed a negative sign of segmental anisotropy.[8] In the low $d\gamma/dt$ region, χ of the solution with a higher C_p is smaller, indicating that cellulose chains tend to align in the direction of flow. Here, note that the extinction angle observed in laminar flow is equal to the average orientation angle of polymer chains.

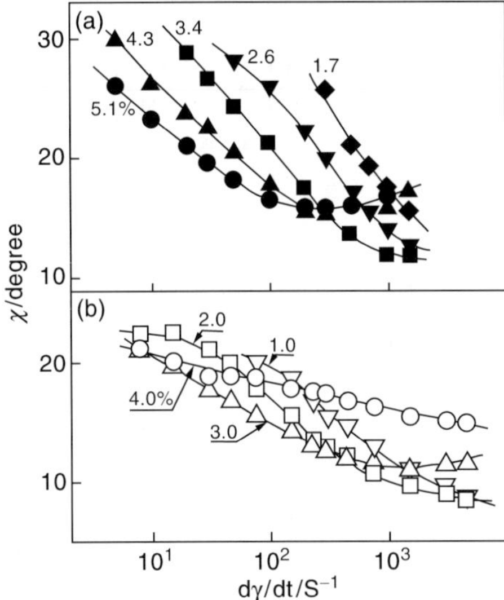

Figure 4.5.2   Shear rate dγ/dt dependence of extinction angle χ for: (a) cellulose I (sample code B2); (b) cellulose I (sample code P1)—9 wt% NaOH system at 20 °C.[1] Numbers attached to the line denote cellulose concentration (wt/wt%).

Figure 4.5.2(b) shows the effect of the cellulose concentration on the shear rate dγ/dt dependence of χ for cellulose I (sample P1, $M_v = 5.4 \times 10^4$) in a 9 wt% aq. NaOH solution at 20 °C. Contrary to the case of the cellulose solutions, χ of the cellulose I solutions does not depend on the polymer concentration, even in the lower shear rate range.

Figure 4.5.3(a) depicts the effect of the molecular weight M_v on the dγ/dt dependence of χ of the cellulose solutions in 9 wt% aq. NaOH ($C_p = 2.2$–2.3 wt%). In the region of dγ/d$t < 500$ s^{-1}, the system of the polymer with higher molecular weight has a smaller χ value.

Figure 4.5.3(b) shows the effect of the molecular weight M_v on the dγ/dt dependence of χ of the cellulose I in a 9 wt% aq. NaOH solution ($C_p = 2.3$ wt%). With an increase in M_v, the χ versus dγ/dt relationship slightly shifts to the lower χ axis. In contrast to the case of the cellulose solutions, χ of the cellulose I solutions show no anomalies in the whole range of dγ/dt investigated.

As illustrated in Figure 4.5.4, χ is larger in cellulose II solutions than in cellulose I, if the solutions of samples with the same M_v are compared at the same dγ/dt. At $M_v \approx 1.3 \times 10^5$, the same χ value is obtained for both solutions. The molecular weight dependence of $\chi \sim$ dγ/dt relation for cellulose I solutions is significantly smaller than that for cellulose solutions. Such a small dependence cannot be expected for molecularly dispersed polymer solutions. These facts suggest strongly that in the semidilute solution regime the dissolved state of cellulose solute in a 9 wt% aq. NaOH solution differs depending on whether the sample is natural cellulose or regenerated cellulose.

To examine whether the significant difference of the molecular weight dependence of the χ versus dγ/dt relation between natural and regenerated cellulose solutions observed in aq. NaOH is a specific characteristic of aq. NaOH solution, cadoxen was employed

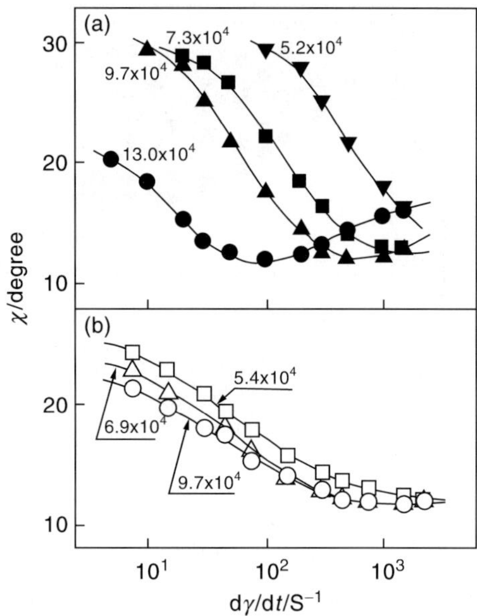

Figure 4.5.3 Shear rate $d\gamma/dt$ dependence of extinction angle χ for: (a) cellulose II ($C_p = 2.2$–2.3 wt%); (b) cellulose I (2.3 wt%) with various viscosity average molecular weight (M_v) ~ 9 wt% NaOH system at 20 °C.[1] Numbers attached to the line denote M_v.

as solvent for this purpose. The $d\gamma/dt$ dependence on χ of the cellulose II and I with various molecular weights in cadoxen is shown in Figure 4.5.5(a) and (b), respectively. χ of the cellulose solutions decreases with an increase in $d\gamma/dt$, and at constant $d\gamma/dt$ (e.g. 1×10^3 s^{-1}), the sample with higher M_v reveals lower χ. On the other hand, χ of the cellulose I solutions is smaller than 30°, showing, except for the solution of cellulose I

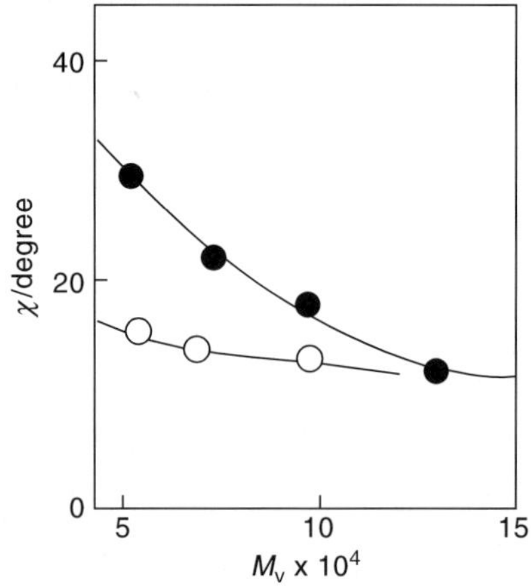

Figure 4.5.4 Molecular weight M_v dependence of extinction angle χ for solutions of cellulose II ($d\gamma/dt = 1.0 \times 10^2$ s^{-1}) (●), and cellulose I ($d\gamma/dt = 1.2 \times 10^2$ s^{-1}) (○), in 9 wt% NaOH at 20 °C ($C_p = 2.2$–2.3 wt%).[1]

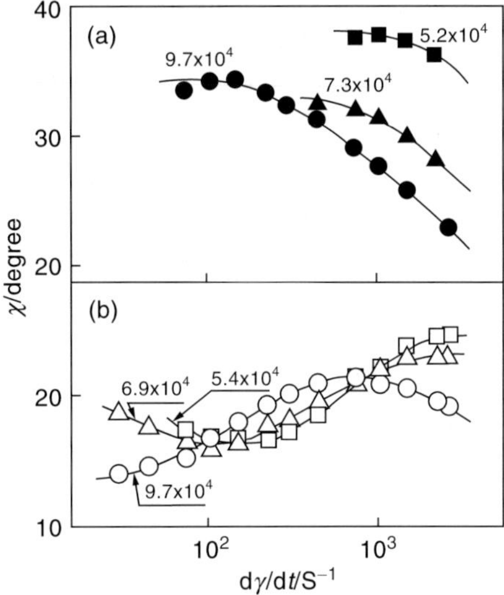

Figure 4.5.5   Shear rate $d\gamma/dt$ dependence of extinction angle χ for cellulose II (a) and I (b) (5 wt%) with various viscosity average molecular weight (M_v)—cadoxen system at 20 °C.[1] Numbers attached to the line denote M_v.

with $M_v = 9.7 \times 10^4$, the minimum and maximum at $d\gamma/dt \sim 3000$ and 110 s^{-1}, respectively. Evidently, the shear rate dependence of χ of the cellulose I and II solutions are quite different even if cadoxen is employed as solvent, which is a better solvent for cellulose than aq. alkali solution.

Figure 4.5.6 shows the plots of steady-state viscosity η of cellulose II (sample code B2, $M_v = 5.2 \times 104$) and cellulose I (P1, $M_v = 5.4 \times 10^4$) in a 9 wt% aq. NaOH solution with $C_p = 5.0$ wt% at 20 °C as a function of log $d\gamma/dt$. η of the cellulose solution at a given $d\gamma/dt$ ($2 \times 10^2 \sim 1 \times 10^4 \text{ s}^{-1}$) is nearly twice that of cellulose I solution at 20 °C. This indicates that, with a given combination of M_v, C_p, $d\gamma/dt$, and temperature, the cellulose solution is more viscous than the cellulose I solution. In other words, the dissolved state of semidilute cellulose solutions is different depending on the preparation

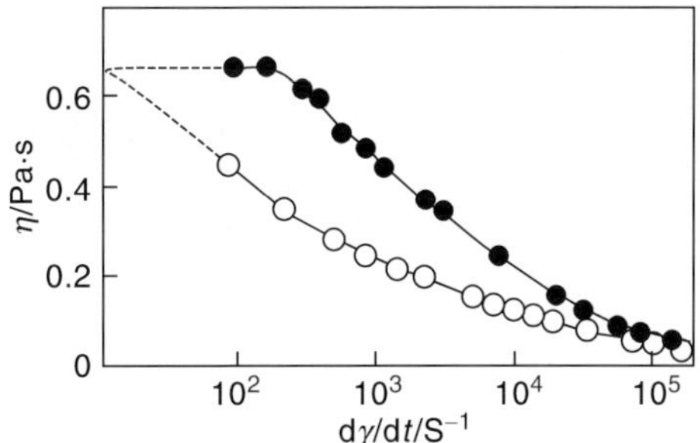

Figure 4.5.6   Shear rate $d\gamma/dt$ dependence of viscosity η of cellulose II (sample code B2) (●) and I (sample code P1) (○) ($C_p = 5.0$ wt%) in aq. 9 wt% NaOH at 20 °C.[1]

method of the cellulose samples used. The shear rate at which transition from Newtonian to non-Newtonian flow occurs, is estimated to be about 2×10^2 s^{-1} for cellulose solution and less than *ca.* 1×10^2 s^{-1}–1×10 s^{-1} for cellulose I solution. The cellulose solution behaves roughly as a Newtonian fluid, at least in the range $d\gamma/dt < 200$ s^{-1}, while cellulose I solutions are explicitly non-Newtonian in the whole range of $d\gamma/dt$ investigated ($90 < d\gamma/dt < 1.6 \times 10^5$ s^{-1}).

Under the conditions that polymeric chains in the solutions are Gaussian, and the contribution of the viscosity of the solvent η_0 to that of the solution η is negligible (i.e. $\eta \gg \eta_0$), Kriegel[9] showed that cot 2χ of the system in laminar shear flow is proportional to the product, $\eta(d\gamma/dt)d\gamma/dt$:

$$\cot 2\chi \propto \eta(d\gamma/dt)d\gamma/dt \tag{4.5.1}$$

Accordingly, if the solution can be regarded as a Newtonian liquid, eq. (4.5.1) can be simplified to

$$\cot 2\chi \propto d\gamma/dt \tag{4.5.2}$$

The validity of eq. (4.5.2) has been experimentally proven over wide $d\gamma/dt$ range at various temperatures for the solution of typical flexible polymers, such as polystyrene.[10]

Figure 4.5.7 shows the plot of cot 2χ of the cellulose II in a 9 wt% aq. NaOH solution against $d\gamma/dt$. In the lower $d\gamma/dt$ region, cot 2χ is not proportional to $d\gamma/dt$ but to $(d\gamma/dt)^{0.4}$, irrespective of M_v. In the higher $d\gamma/dt$ region, data points of all solutions deviate downward from the straight line, which represents the relationship ($\cot 2\chi \propto (d\gamma/dt)^{-0.4}$). In the case of the cellulose II with $M_v = 5.2 \times 10^4$, the $d\gamma/dt$ value (at which the deviation starts to occur) is in the vicinity of that at which the solution starts to behave as a non-Newtonian fluid, as indicated in Figure 4.5.6.

Table 4.5.2 summarizes T_1 of the samples P1 and B2 solutions in cadoxen ($C_p = 4$ wt%). T_1 of the C_6 carbon of cellulose I and II is the smallest among those of all the carbons. T_1 of the C_6 carbon of cellulose II is 1.26 times larger than that of cellulose I, suggesting that regardless of crystal polymorph, the hydroxyl group at the C_6 position of cellulose I molecules tends to form inter- or intramolecular hydrogen bonds.

Figure 4.5.8 shows the plot of logarithmic ratio of hydrodynamic diameter d/d_0 of cellulose I (circle) and II (triangle) solution against centrifugal acceleration G for the

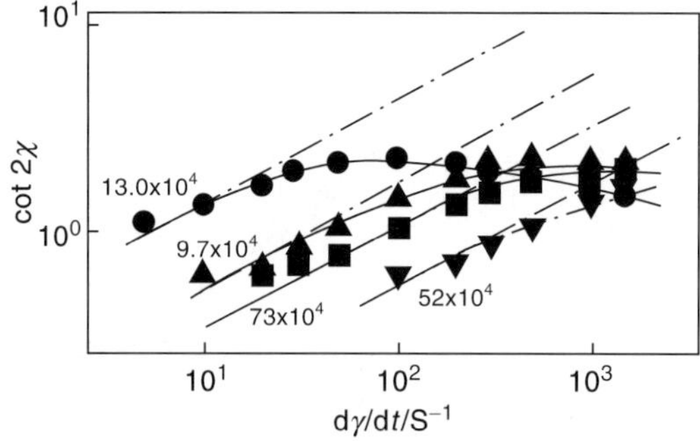

Figure 4.5.7 Shear rate $d\gamma/dt$ dependence of cot 2χ of cellulose II ($C_p = 2.2$–2.3 wt%) with various viscosity average molecular weight M_v—aq. 9 wt% NaOH solution systems.[1] Numbers attached to the line denote M_v.

Table 4.5.2

Spin lattice relaxation time T_1, as determined by ^{13}C NMR inversion recovery method, of C_1–C_6 carbons on a pyranose unit of cellulose (P1, B2)/cadoxen (D$_2$O) (C_p = 4 wt%) at 30 °C[1]

Chemical shift (ppm)	Carbon	T_1{cell I (sample P1)}(s)	T_1{cell II (sample B2)}(s)	T_1(cell II)/ T_1(cell I)
105.9	C_1	154	166	1.07
80.8	C_4	150	155	1.03
78.6	C_3	148	130	0.88
78.1	C_5	157	148	0.94
77.0	C_2	136	135	0.99
63.3	C_6	53	67	1.26

cellulose cadoxen solution. Here, d_0 is taken for the hydrodynamic diameter of the solute in the solution ($S_{0.5}$) of cellulose I and II. The d of cellulose solutions is obviously constant, regardless of the storage period after dilution and the acceleration in ultracentrifugation, which is almost the same as that (see column 8 of Table 3 of Ref. 1) for a 0.5% solution, directly prepared. Kamide and Saito[4] prepared 0.6% cellulose II ($M_v = 4.12 \times 10^4$) solutions in cadoxen, centrifuged at $6.36 \times 10^4 g$, ultrafiltered through fluoride membranes (Fluoropore R; Sumitomo Electric Co., Japan; mean pore size $0.1 \sim 0.45$ nm) to give the optically clear solutions, for which the z average radius of gyration, $\langle S^2 \rangle_z^{1/2}$, was measured by the static light scattering method eq. (3.5.4). $\langle S^2 \rangle_z^{1/2}$ obtained was converted to d (50–60 nm), which is approximately 10 times larger than d estimated by the DLS method. Unfortunately, we have no sound reason for the above significant difference in d. The reproducibility of the DLS experiments was ascertained. It is then better to consider d as an experimental parameter, representing the particle size, even if the absolute magnitude does not mean the dimension itself.

The constancy of d for cellulose II cadoxen solutions over a wide range of preparative conditions indicates that the solutions do not contain gel-like particles and in the 0.5% solution cellulose is dissolved molecularly. In contrast to this, for cellulose I solution at

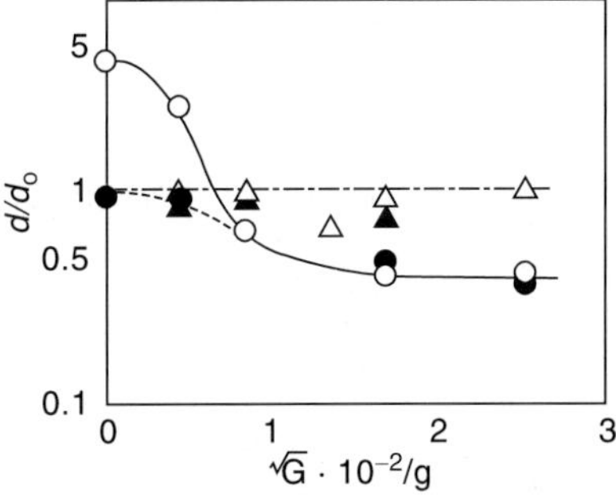

Figure 4.5.8   Centrifugal acceleration G dependence of ratio of hydrodynamic radius of diluted solution (d) to that of undiluted solution (d_0).[1] (Circle), cellulose I solution; (triangle) Δ cellulose II solution. (Unfilled symbol), 1 day after dilution; (filled symbol), 5 days after dilution.

5 days after dilution ($S_{5 \to c \to 0.5(5)}$), the ratio d/d_0 at $G = 0$ is much larger than unity (*ca.* 4.5), suggesting that the dissolved state of cellulose in the 0.5 wt% solution, diluted from the concentrated solution, is different from that of undiluted 0.5 wt% solution ($S_{0.5}$). In other words, the supramolecular structure of cellulose I solid is not completely destroyed in cadoxen to give a molecularly dispersed 0.5 wt% solution, but the destruction is suspended at an intermediate state. There is a significant dependence of d on the acceleration G in an ultracentrifuge for cellulose I solution, meaning that in cellulose I solution larger supramolecular aggregates exist, which can be excluded by centrifugation, resulting in a smaller d for the solution treated at higher G.

The associated polymer chains in the solution readily move to the direction of solvent flow, even under lower shear field. The observed low extinction angle of cellulose I solution may be plausibly explained in terms of the associated chains.

In summary, the extinction angle χ of aq. alkali solutions of the cellulose I sample is significantly less shear rate dependent as compared to that of the cellulose II sample. In the latter system, the χ versus $d\gamma/dt$ relations for a given cellulose sample shifted to the higher $d\gamma/dt$ side with decrease in the average molecular weight. The viscosity of the cellulose II sample in aq. sodium hydroxide solutions is approximately twice that of the cellulose I sample in the same solvent if compared at the same molecular weight, same concentration, and same temperature. The latter solution showed a non-Newtonian property at relatively smaller $d\gamma/dt$ than the former solution did. Spin lattice relaxation time T_1 (by ^{13}C NMR) of cellulose in cadoxen solution was smaller in cellulose I, suggesting the existence of intra- and intermolecular hydrogen bonds at the C_6 position of cellulose molecules in cellulose I solution.

A DLS study on cellulose in cadoxen showed that in a 5 wt% solution of cellulose I, cellulose particles are dispersed over time into smaller particles, and that the larger particles could be excluded by ultracentrifugation. In cellulose I solutions, the cellulose particles had almost the same size during storage. The above findings indicate that in 5 wt% cellulose I solutions in aq. alkali or in cadoxen, cellulose I is not dissolved molecularly, but a supramolecular structure of the solid is at least partly reserved in the above solutions.

REFERENCES

1. K Yasuda, M Saito and K Kamide, *Polym. Intern.*, 1993, **30**, 393.
2. D Henley, *Ark. Kemi*, 1961, **18**, 327.
3. W Brown and R Wikström, *Eur. Polym. J.*, 1966, **1**, 1.
4. K Kamide, K Okajima, T Matsui and K Kowsaka, *Polym. J.*, 1984, **16**, 857.
5. T Yamashiki, T Matsui, M Saitoh, K Okajima, K Kowsaka, K Kamide and T Sawada, *Br. Polym. J.*, 1990, **22**, 73.
6. K Lederer and J Schurz, *J. Rheol. Acta*, 1975, **14**, 252.
7. K Kamide, M Saito, K Kowsaka and K Okajima, *Polym. J.*, 1987, **19**, 1337.
8. VN Tsvetkov, *Rigid-Chain Polymers*, Consultants Bureau, New York, 1989, Chapter 5, p. 270.
9. HJ Kriegel, *Adv. Polym. Sci.*, 1969, **6**, 170.
10. W Philipp, *J. Appl. Phys.*, 1965, **36**, 3033.

4.6   GELS OF CELLULOSE AQUEOUS ALKALI SYSTEM[1]

Gelation phenomena of cellulose derivative solutions, for example, the methyl cellulose water system on cooling[2–12] and the cellulose nitrate organic solvent system on warming,[13] have been studied so far. Especially for the former system, detailed conditions of gelation and some thermal properties of these gels have emerged. With respect to cellulose solutions, it has long been well known that the gel arises by vaporization of ammonia from a concentrated cellulose cuprammonium solution. However, in marked contrast with cellulose derivatives, a systematic investigation on the gelation behavior of cellulose solutions has never before been performed. The fact that a stable and simple solvent, in which cellulose can be dissolved molecularly without formation of any complex, has been rarely found, seems to prevent us from studying this subject at advanced levels.

It was also observed by Kamide *et al.* that when the temperature of the cellulose solution elevates or falls from room temperature (in other words, with a concomitant decrease of the solvent power), gelation occurs at specific temperatures. The cellulose gels thus obtained are shown in Figure 4.6.1. While various systems that form gels by cooling or heating are already known, no polymer solution which has two gelation points near room temperature has yet been found, except for the cellulose aq. NaOH solution system.

This section is principally devoted to disclosing, for regenerated cellulose in aq. alkaline solutions, the detailed gelling conditions and the viscoelastic and thermal properties of the cellulose gels formed.

Preparation of cellulose samples and their solutions

Hydrolysis of purified cotton linter proceeded in a cuprammonium solution (Cu, 11; NH_3, 200; water, 1000 g) by storing in a dark place at 20 °C for *ca.* 1 week.[14,15] Addition of a 20 wt% aq. H_2SO_4 solution to the above solution brought about a regenerated cellulose sample (sample code C1) as a precipitate. Nine regenerated cellulose samples (code C2–C10) with different molecular weights were prepared from the sample C1 slurry, including 25 wt% aq. H_2SO_4 solution at 30 °C, by altering storage time. The viscosity average molecular weight (M_v) of these samples, as listed in Table 4.6.1, was estimated

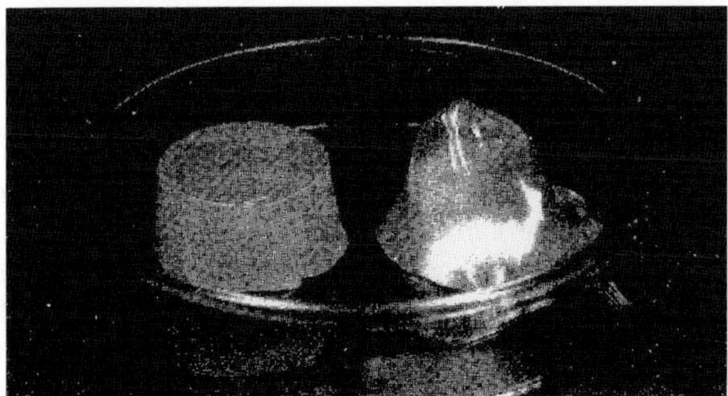

Figure 4.6.1   Gels of cellulose 9 wt% aq. NaOH solution system at 20 °C. Right side gel, which is partially melting, generated at −9 °C for 70 h and left side gel generated at 40 °C for the same period as the right side gel.[1]

Table 4.6.1

Molecular characteristics of cellulose samples used in this section[1]

Sample code	Molecular weight $\times 10^{-4}$	Experimental method utilized
C1	12.0 (M_w)[a]	Starting sample, LS[b]
C2	4.0 (M_v)[c]	LS, KV[d], BD[e], VE[f], SY[g]
C3	5.2 (M_v)	KV, VE, SW[h], SY
C4	7.3 (M_v)	KV, VE, SW
C5	7.8 (M_v)	LS, SW, SY
C6	1.9 (M_v)	LS, SW
C7	5.9 (M_v)	BD, VE
C8	2.4 (M_v)	SY
C9	6.4 (M_v)	SY
C10	9.7 (M_v)	

[a]Weight average molecular weight.
[b]Light scattering.
[c]Viscosity average molecular weight.
[d]Kinematic viscosity.
[e]Ball drop method.
[f]Viscoelasticity.
[g]Syneresis.
[h]Swelling.

through the use of the Mark–Houwink–Sakurada equation (eq. (4.6.1)) determined by Brown–Wikström[16] using cadoxen as a solvent.

$$[\eta] = 3.85 \times 10^{-2} M_w^{0.76} \ (cm^3 \ g^{-1})(25\,^\circ C) \tag{4.6.1}$$

where $[\eta]$ is the limiting viscosity number and M_w is the weight average molecular weight.

We dissolved the cellulose sample flake in 9 wt% aq. NaOH solution at 5 °C (dissolving temperature T_p) by agitating for 1 min with a home mixer, followed by additional agitation for the same period (15 h) later. In order to eliminate undissolved residue, the solution was centrifuged under 1.5×10^4 gravity for 1 h, until transparent and clean cellulose solutions were finally obtained. The solutions were stocked in a dark place maintained at 5 °C, where the solutions were ascertained to be stable for several months.

Figures 4.6.2(a) and (b) show time (t) dependences of the kinematic viscosity ν ($\nu = \eta/\rho$: ρ, density) of sample C3 ($M_v = 5.2 \times 10^4$) solution with cellulose concentration $C_p = 5$ wt%. Kinematic viscosity ν of the solution treated (annealed before quenching) above 25 or below -2 °C monotonically increased after the solution temperature was changed. In the figure, the treatment temperature is shown. In the case of the solution maintained at 2 and 0 °C, we observed a slight increase of ν at $t > 22$ and $t > 21$ h, respectively. Moreover, at $t = 24$ h, all of the solutions examined (except for those held at 5–20 °C) exhibited a significant increase in turbidity. Judging from these results, it can be said that cellulose chains in the solution treated below 2 or above 25 °C

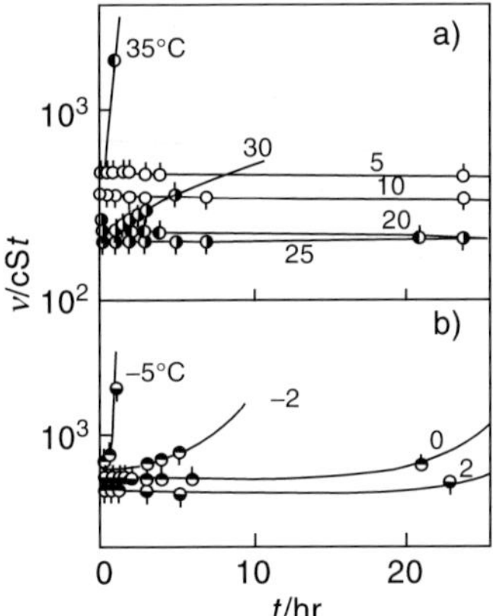

Figure 4.6.2   Time t dependence of the kinematic viscosity ν of cellulose (sample code C3, $M_v = 5.2 \times 10^4$) solution (the polymer concentration $C_p = 5$ wt%; dissolving temperature $T_p = 5\,°C$).[1] The numbers attached to the line denote the temperature.

tend to aggregate together (at least within 24 h), and as more time elapses, all these solutions are eventually supposed to approach gel, which is defined by Ferry[17] as the solution that exhibits no steady state flow (i.e. $\nu \rightarrow \infty$). Following this definition, the cellulose ~ 9 wt% aq. NaOH solution system has two gelation points, both very close to room temperature. We define the gel formed on warming, and that formed on cooling, as higher temperature gel (HTG) and lower temperature gel (LTG), respectively.

Figure 4.6.3 depicts the time dependence of ν of the sample C3 solutions with C_p ranging from 3 to 6 wt% at 30 °C (a) and 0 °C (b). Below $C_p = 4$ wt%, ν does not show

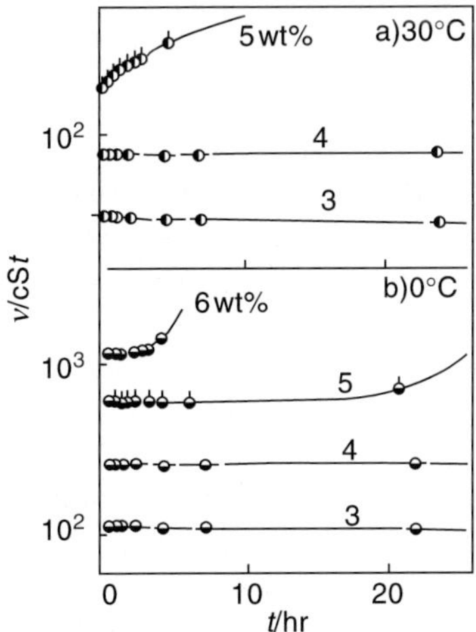

Figure 4.6.3   Time t dependence of the kinematic viscosity ν of cellulose (sample code C3) solution with various concentration (the dissolving temperature $T_p = 5\,°C$) at 30 °C (a) and at 0 °C (b).[1] The numbers attached to the line denote the wt%.

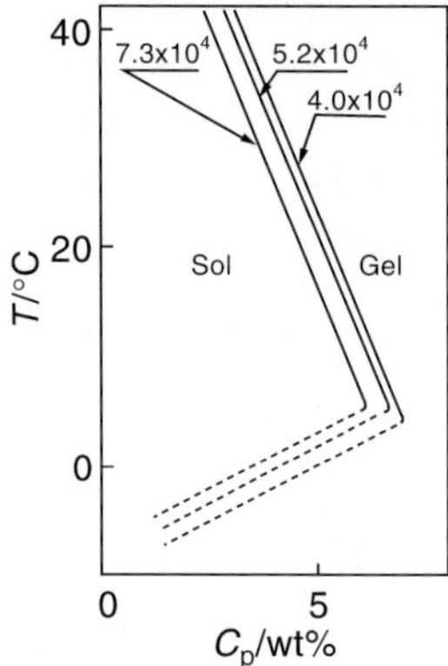

Figure 4.6.4   Upper gelation temperature (solid line) and lower gelation temperature (broken line) of cellulose ~ 9 wt% aq. NaOH solution systems plotted as a function of the cellulose concentration.[1] The numbers attached to the line denote the viscosity average molecular weight of the cellulose samples.

any remarkable change with storage time at either temperature, indicating that there exists a critical concentration above which the solution transforms to the gel.

Figure 4.6.4 shows a liquid-gel phase diagram determined by the kinematic viscosity method in the temperature jumping method for the solutions of three cellulose samples ($M_v = 4.0$, 5.2, and 7.3×10^4). As expected, the gel region becomes narrower as the molecular weight and the polymer concentration increases. The relationship between the glass transition temperature by the jumping method (T_j^G) and polymer concentration for a given cellulose sample can be represented by two straight lines (HTG, solid line; LTG, broken line), of which the slopes are practically independent of the molecular weight of cellulose. The experimental finding that the upper and lower gelation lines cross at 5 °C suggests that the cellulose aq. NaOH solution system with relatively high polymer concentration ($C_p = 6$–7 wt%) cannot exist as the liquid state.

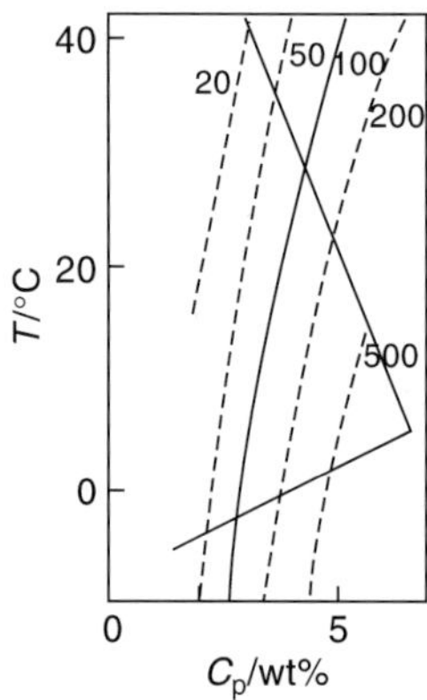

Figure 4.6.5   Isokinematic viscosity line of cellulose sample C3 9 wt% aq. NaOH solution system.[1] Numbers attached to the lines denote the kinematic viscosity (cSt). Heavy solid lines denote upper and lower gelation temperature lines.

In Figure 4.6.5, isokinematic viscosity curves of the sample C3 solution are plotted against C_p. The value of ν in the figure corresponds to that measured at 10 min after the viscometer was immersed in a water bath. Every isokinematic viscosity curve traverses two gelation lines corresponding to HTG and LTG of the same sample solution, implying that the absolute ν value of the solution does not serve effectively as a measure of generating the gel from the cellulose solution.

Figure 4.6.6(a) shows the change in ν of the cellulose C3 solution, which was prepared at $T_p = 30\,°C$ for t_a (min) and then quickly cooled down to and kept at the temperature $T_j = 5\,°C$, with C_p of 4.6 wt% during the heat treatment (temperature T_a; time t_a). For the sake of comparison, data of untreated solutions were also plotted. Irrespective of t_a, ν of the solution ($T_p = 5\,°C$, $T_a = 30\,°C$) at quenching and keeping temperature $T_j = 5\,°C$ declined rapidly for about the initial 2 h, attaining a constant value which is apparently higher for longer annealing time. It became clear that the aggregation structure formed at $30\,°C$ remains partially undestroyed even at $5\,°C$, which is sufficiently below the upper transition temperature. In other words, the gel formed at the higher temperature region is thermally irreversible. Previously, we suggested that in the dilute solution region of the cellulose in aq. NaOH solution there possibly occurs a specific transition of cellulose chain conformation, such as coil-globule transition[14] (Figure 3.20.4). The experimental fact that when the cellulose aq. NaOH solution system is heated, the radius of gyration of the chain decreases abruptly at $40\,°C$ suggests the possibility of the occurrence of this kind of transition. For the aggregated structure in the gel of naturally occurring polysaccharides, several models have been proposed, for example, the 'egg-box' type for polyelectrolytes such as poly-L-guluronate (alginate) and its optical isomer in water containing alkali metal ions,[18–20] and the double helical form for carrageenan and agarose

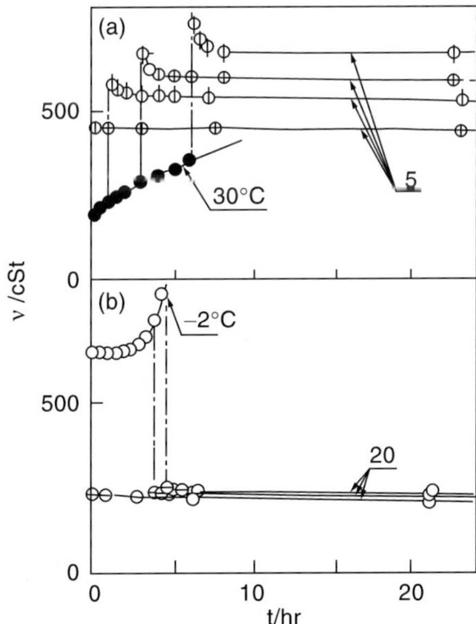

Figure 4.6.6   Change in ν for cellulose C3 solution with time.[1] Number in the figure means T_a (°C). The detailed caption is given in Figure 6 of Ref. 1.

in water.[21] However, with respect to cellulose gel, cellulose chains must not aggregate in any regular form, such as egg-box or helix, because cellulose chains in the solution or in the solid state have never been reported to be conformed in helical strands. However, unfortunately, detailed information on the microstructure of cellulose gels is still lacking. Considering the fact that in the relatively limited range of around room temperature, solvent power gets larger as the temperature falls,[15] an initial declination of ν (as shown in Figure 4.6.6(a)) is most likely caused by the partial melting of the gel.

Figure 4.6.6(b) is another example of the change in ν for the cellulose C3 solution ($C_p = 4.6$ wt%, $T_p = 5\,°C$, $T_a = T_j = 20\,°C$). In this case, too, just after the temperature reached T_j, ν of the system immediately falls on the line of the untreated solution. It is interesting to note that the LTG unquestionably has thermal reversibility, unlike HTG.

Figure 4.6.7 shows the relationship between the scattering light intensity difference at scattering angle 90° ΔI_{90} of the cellulose sample C1 solution with the concentration C_p ranging from 0.6 to 3 wt%, and the temperature, which was raised at the rate of $4\,°C\,min^{-1}$. Every line in the figure has an upward bending point. Over the temperature range of 4–35 °C, the second virial coefficient, A_2, and the scattering function at scattering angle 90°, P(90) of the dilute solution of the C1 sample, had no inflection point and the drastic change in the volume of the cellulose solution was also not observed. Therefore, the upward inclination of ΔI_{90} curves in the figure may be attributed to an increase in apparent molecular weight with an onset of association of the polymer chains. The temperature at which ΔI_{90} curves bend upward (indicated with arrows) is defined as the gelation point at slow heating process, T^G. In Figure 4.6.7, the results on C1 solution with $C_p = 1.9$ wt% held at $30 \pm 0.1\,°C$ were demonstrated and, in general, ΔI_{90} of the solution held below T^G remained almost constant for at least 1 day.

Gelation points of the sample with the same M_w experimentally observed (except for the range of $C_p < 1$ wt%) are far below the cloud point curves, calculated on the basis of

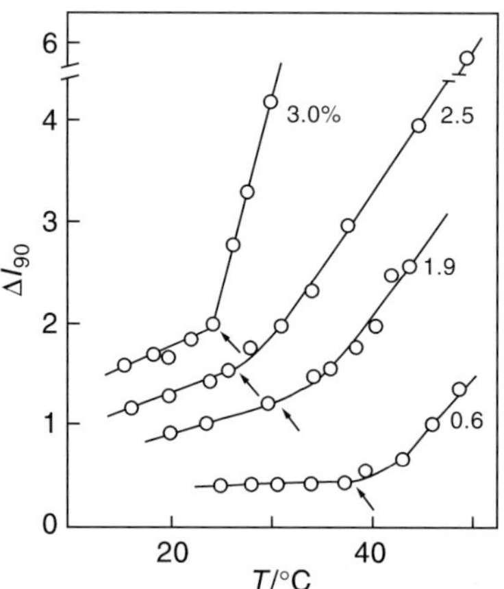

Figure 4.6.7 Scattered light intensity difference at scattering angle 90 °C, ΔI_{90}, of cellulose sample C1 solutions plotted against the temperature T, when the solutions were heated at constant rate of $4\,°C\,h^{-1}$.[1] Numbers in the figure denote the cellulose concentration (wt%) and arrows indicate the gelation temperature (T^G (°C)).

Kamide *et al.*'s theory[22–24] using the experimental data of $\phi = -0.77$, $\theta = 313$ K (from analysis of Figure 6 of Ref. 15: Figure 3.20.4) and assuming the first-order concentration dependence parameter $p_1 = 0$ or $p_1 = 0.67$ (theoretical value predicted for nonpoplar polymer nonpolar solvent system[22]) and $M_w/M_n = 2$, so that gelation of this system seems not to be correlated with the phase separation of the solution. This is a reasonable explanation why cloud point is not experimentally observed for the cellulose aq. NaOH solution system. However, it is not yet clear to what kind of thermodynamic transition the onset of gelation of cellulose solution at the higher temperature region corresponds. The melting of LTG occurs at a higher temperature as the concentration and the molecular weight increase. The heat of fusion ΔH^G mole^{-1} of cross-linkage in these LTGs can be roughly estimated from the slope of the plots of C_p versus the reciprocal melting temperature T_m for each sample. For this purpose, the Eldridge–Ferry (EF) equation,[25] which was derived by assuming that only a binary association of polymer chains is responsible for the gelation of the solution, was applied. The slopes of the lines, obtained for two cellulose samples, are practically equal, making $\Delta H^G = 31$ kJ mol^{-1}. ΔH^G of cellulose LTG is very close to those of atactic vinyl type polymer gels, including polyvinylalcohol,[26] polyvinylchloride,[27] and polystyrene.[28] When we take into account the possible contribution to the hydrogen bonds existing in the cellulose gel, the intermolecular association energy should not be larger than 31 kJ mol^{-1}. This value is small enough to allow the phase transition of LTG to sol.

Figure 4.6.8 shows the time course of the storage modulus G$'$ for the C7 solution ($C_p = 5$ wt%) at five different annealing temperatures (T_a). At a given T_a, G$'$ shows a remarkable increase over the initial state. According to Kuhn's theory[29] on the viscoelasticity of the polymer network, G$'$ is directly related to the number of cross-linkages in unit volume of the polymer solution. Inspection of Figure 4.6.8 indicates that the number of clusters consisting of the cross-linkings of the cellulose chains is larger at higher T_a. All of the curves in the figure can be superimposed on a single master curve.

Effects of the cellulose concentration and the cellulose molecular weight on the aging time dependence of G$'$ are very similar with the temperature effect. All curves in

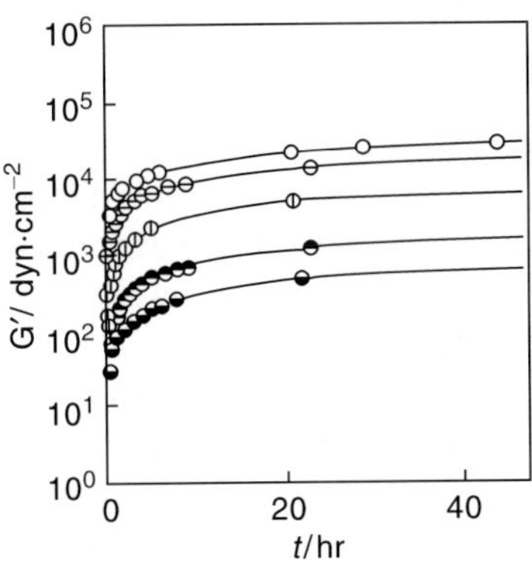

Figure 4.6.8   Time course of the storage modulus G$'$ of cellulose ($M_v = 5.9 \times 10^4$) solutions with $C_p = 5$ wt%.[1] (Lower half filled circle), heat treatment temperature $T_a = 33\,°$C; (upper half filled circle), 35 $°$C; (vertically halved circle), 40 $°$C; ($\ominus$), 45 $°$C; ($\bigcirc$), 50 $°$C.

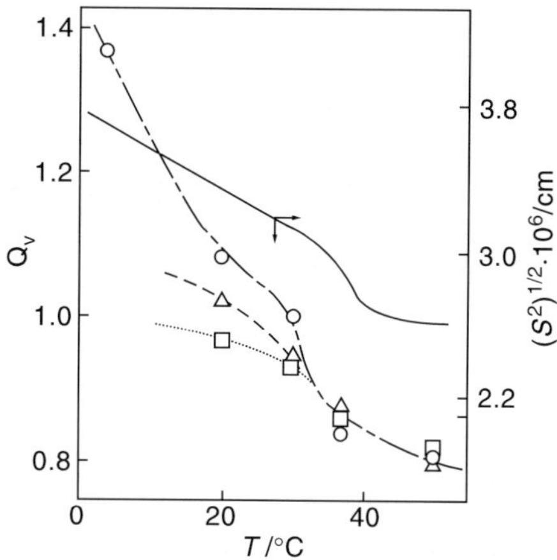

Figure 4.6.9 Aging temperature dependence of the volume expansion ratio Q_v of HTG ($M_v = 5.3 \times 10^4$, $C_p = 6.0$ wt%) in 9 wt% aq. NaOH solution medium.[4] (O), aging time 1 day; (Δ), 3 days; ($\square$), 7 days. Solid line, the z average radius of gyration $\langle S^2 \rangle_z^{1/2}$ cited from Figure 5 of Ref. 14.

log G'–log t plots can be readily superimposed with respect to the polymer concentration or the molecular weight. Evidently, these curves almost coincide, indicating that the number of clusters in the solution is larger in more concentrated polymer solutions with higher molecular weight, and the aggregation rate, evaluated from the slope of the master curves, is quite independent of these three factors.

Figure 4.6.9 shows the annealing temperature dependence of the volume expansion factor (as defined by the ratio of the volume of the annealed gels to that of the unannealed gels) Q_v of the sample C3 gel suspended in 9 wt% aq. NaOH solution medium at annealing time for 1 (chain line), 3 (broken line), and 7 days (dotted line). After annealing for 1 day, the cellulose gel maintained at a lower temperature region tends to swell because of the stronger solvent power at a lower temperature. At about 32 °C, the gel starts to shrink drastically. Above 35 °C, the degree of shrinkage does not markedly depend on the annealing time. The decrease of Q_v seem to be accompanied by chain conformation changes, such as coil-globule transition, because the radius of gyration, determined by the light scattering method in dilute regime, decreased remarkably in almost the same temperature range.[14] At annealing temperatures below 30 °C, in which 9 wt% aq. NaOH solution behaves as a good solvent toward cellulose, gel swelling occurs initially, followed by volume contraction.

The phase diagram of the cellulose aq. NaOH solution system was constructed as a function of polymer concentration, molecular weight, annealing temperature, and aging time. Evidently, longer annealing at a higher temperature brings about syneresis in the area of lower polymer concentration and lower molecular weight. In addition, cellulose gel, formed from its solution in aq. NaOH solution, never attains an equilibrium state even if kept longer than 1 month. In this sense, HTG is observed here as only in a thermally pseudoequilibrium state.

4.6.2 Conclusions

1. The cellulose aq. NaOH solution system has two gelation temperatures near room temperature.
2. While HTG is thermally irreversible, LTG has thermal reversibility.
3. Cellulose gel suspended in the solvent shrinks drastically at specific temperatures.
4. Cellulose gel is in a thermally pseudoequilibrium state.

REFERENCES

1. K Kamide, M Saito and K Yasuda, Formation and viscoelastic properties of cellulose gels in aq. alkali. In *Viscosity of Biomaterials*, (eds WG Glasser and H Hatakeyama), American Chemical Society, Washington, DC, 1992, Chapter 12, p. 184.
2. E Heymann, *Trans. Faraday Soc.*, 1935, **31**, 846.
3. AB Savage, *Ind. Eng. Chem.*, 1957, **49**, 99.
4. W Kuhn, P Moser and H Mijer, *Helv. Chim. Acta*, 1961, **44**, 770.
5. WB Neerly, *J. Polym. Sci., C*, 1963, **1**, 311.
6. DA Rees, *Adv. Carbohydr. Chem. Biochem.*, 1969, **26**, 267.
7. ED Klug, *J. Polym. Sci., C*, 1971, **36**, 491.
8. A Kagemoto, Y Baba and R Fujishiro, *Makromol. Chem.*, 1972, **154**, 105.
9. Y Baba and A Kagemoto, *Kobunshi Runbunshu*, 1974, **31**, 446.
10. Y Baba and A Kagemoto, *Kobunshi Runbunshu*, 1974, **31**, 528.
11. T Kato, M Yokoyama and A Takahashi, *Colloid Polym. Sci.*, 1978, **256**, 15.
12. S Nakura, S Nakamura and Y Onda, *Kobunshi Runbunshu*, 1981, **38**, 133.
13. S Newman, WR Krigbaum and DK Carpenter, *J. Phys. Chem.*, 1956, **60**, 648.
14. K Kamide, M Saito and K Kowsaka, *Polym. J.*, 1987, **19**, 1173.
15. K Kamide and M Saito, *Polym. J.*, 1986, **18**, 569.
16. W Brown and R Wikström, *Eur. Polym. J.*, 1965, **1**, 1.
17. JD Ferry, *Viscoelastic Properties of Polymers*, 3rd edn., Wiley, New York, 1980, p. 537.
18. DA Rees and EJ Welsh, *Angew. Chem. Int. Ed.*, 1977, **16**, 214.
19. R Kohn, *Pure Appl. Chem.*, 1975, **30**, 371.
20. ER Morris, DA Rees and D Thom, *Chem. Comm.*, 1969, 701.
21. DA Rees, IW Steele and FB Williamson, *J. Polym. Sci., C*, 1969, **29**, 261.
22. K Kamide, *Thermodynamics of Polymer Solutions*, Elsevier, Amsterdam, 1990, Chapter 2.
23. K Kamide, K Matsuda, T Dobashi and M Kaneko, *Polym. J.*, 1984, **16**, 839.
24. K Kamide, S Matsuda and M Saito, *Polym. J.*, 1985, **17**, 1013.
25. JE Eldridge and JD Ferry, *J. Phys. Chem.*, 1954, **58**, 992.
26. H Maeda, T Kawai and R Kashiwagi, *Kobunshi Kagaku*, 1956, **58**, 992.
27. MA Hanson, PH Morgan and GS Park, *Eur. Polym. J.*, 1972, **8**, 1361.
28. HM Tan, A Moet, A Hiltner and E Baer, *Macromolecules*, 1983, **16**, 28.
29. W Kuhn, *Kolloid Zeits*, 1936, **76**, 258.

4.7 CELLULOSE FIBERS AND FILMS WET SPUN FROM CELLULOSE–AQUEOUS ALKALI SOLUTION

4.7.1 Overview

Since Kamide and collaborators first demonstrated that cellulose samples, regenerated from their cuprammonium solution under specialized conditions, can completely dissolve in 8–10 wt% aq. sodium hydroxide (NaOH) solution at 4 °C, and that the solubility of the cellulose against aq. NaOH solution is mainly governed by the degree of the breakdown of intramolecular hydrogen bonds[1] (see Section 4.1), much effort has been made to prepare such an alkali soluble cellulose directly and economically from natural cellulose. Although the steam explosion technique has been widely applied to wood chips and bagasse in order to separate ligno components from cellulosic components, this technique has not been applied to pure cellulose.[2] Kamide *et al.* succeeded in preparing alkali soluble cellulose by applying the steam explosion treatment as a method for the breakdown of intramolecular hydrogen bonds on almost pure celluloses, such as soft wood pulp and hard wood pulp[3,4] (see Section 5.1). They also clarified the specific solvation structure of 8–10 wt% aq. NaOH solution as a cellulose solvent[5] (see Section 4.3), and the structural change of the alkali soluble cellulose during its dissolution into the aq. NaOH solution[6] (see Section 4.4). In addition, Kamide *et al.*[1,7] proved that in the alkali soluble cellulose ~ aq. NaOH and the alkali soluble cellulose–aq. lithium hydroxide systems cellulose was dissolved molecularly without forming a derivative or complex (see Section 3.5).

In the regenerated cellulose fiber industry, viscose rayon and cuprammonium rayon, which were industrialized during the early 1900s, still occupy an exclusive position even though there are problems with the possible discharge of toxic gases and substances from their spinning systems. To meet these environmental problems, many organic solvent systems such as dimethylformamide (DMF)–nitrogen oxide,[8] N-methyl-morpholine N-oxide–water,[9] dimethylsulfoxide–para-formaldehyde,[10] liquid ammonia–ammonium thyocyanate–water,[11] chloral–DMF–pyridine,[12,13] and dimethyl-acetamide–lithium chloride[14] systems aiming at a closed process for the regenerated fiber production have been investigated. Although an industrial spinning system has been developed using N-methyl-morpholine N-oxide–water as cellulose solvent, problems including the production of the explosive by-products, and the difficulty in solvent recovery are still not completely solved. In addition (with some exceptions in most of these organic solvents in viscose and cuprammonium cellulose solutions), cellulose dissolves as a derivative or complex, requiring chemical regenerating processes besides neutralization and refining processes when these systems are employed to produce the regenerated cellulose fiber. This situation creates an interesting and important research field; that is, the utilization of the novel cellulose–aq. alkali solution, which does not seem to bring about any serious hazards and requires no chemical regeneration process in producing fibers, films, and so on. If a new class of fibers, films, membrane, and other products could be obtained by using the novel cellulose solution system, then it would have a paramount impact on the cellulose chemical industry.

Kamide and colleagues attempted to establish a new process using the novel cellulose solution by a wet spinning method with acid coagulation bath to elucidate some characteristic features of the new fiber in its morphology, mechanical properties, and supermolecular structure, in comparison with some commercially available regenerated cellulose filament-type fibers.

Figure 4.7.1 illustrates the flow chart of the wet spinning process of the cellulose–aq. NaOH system.

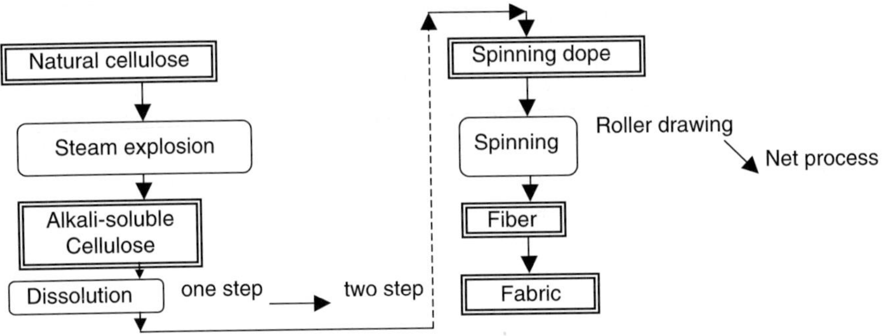

Figure 4.7.1 Flow chart of wet spinning of cellulose from cellulose—aq. NaOH system.

Research and development of the process were carried out through following four stages:

Stage 1: preliminary wet spinning test[15]
Stage 2: laboratory scale wet spinning[16,17]
Stage 3: small size bench scale wet spinning[18–20]
Stage 4: bench plant wet spinning (net process)[21,22]

The progress of new process is briefly summarized in Table 4.7.1 on alkali soluble cellulose, in Table 4.7.2 on cellulose dope for wet spinning, in Table 4.7.3 on wet spinning conditions, and in Table 4.7.4 on structure and mechanical properties of the fibers.

4.7.2 Stage 1: preliminary wet spinning test[15]

Properties of various cellulose dopes

We dispersed 50 g dry base of alkali soluble cellulose (water content *ca.* 8–12 wt%) into 950 g of a given concentration of 9.1 wt% aq. NaOH solution. The solution was

Table 4.7.1

Alkali soluble cellulose

Stage	Starting raw material	Steam explosion		P_v	Alkali soluble cellulose			
		Pressure P	Time t (s)		Crystal form	χ_{am} (%)	S_a (%)	Particle size (μ)
1		2.9 MPa	30	331	I	–	c.100	–
2	Sulfite pulp (white spruce)	2.9 MPa	30	331	I	–	c.100	–
3	$P_v = 1060$ α-cellulose	9–30 kg/cm^2	–	290–409 320	I I	>0.34	>99.97	12
4			–	320	I		99.9	12

Table 4.7.2

Cellulose dope for wet spinning

Stage	Alkali concentration C_a (wt%)	Cellulose concentration C_p (wt%)	Temperature (°C)	Purification and degasification method	Viscosity (P)	Preparative method
1	9.1	5	4	Centrifugation 10,000 rpm, 60 min	–	Single-step
2	9	5	4	C[a] 7,5000 rpm, 60 min	–	Single-step
3	7.6	5	4	Filtration with sintered metal filter (15 μ m cut)	10	Two-step
4	7.6	5	4	F[b]	10	Two-step

[a]Centrifugation.
[b]Filtration.

precooled at 4 °C and left for 8 h with intermittent mixing by a home mixer. The resultant solution was subjected to centrifugation at 10,000 rpm for 60 min in order to exclude the slightly remaining undissolved part and to carry out degasification at 4 °C. The original solution (hereafter referred to as dope 1) thus obtained was employed as a spinning dope and was used to prepare the gelatinized dopes, by treating dope 1, introduced in a glass syringe (inner volume: 10 ml) under the following conditions: dope 2 at -5 °C for 4 h; dope 3-1–3-4 at 32 °C for 3 h, 1, 2, and 3 days, respectively.

Figure 4.7.2 shows differential optical micrographs of cellulose dopes ((A) dope 1; (B) dope 2; (C) dope 3-1) when given shear rate ($\mathrm{d}\gamma/\mathrm{d}t = 0 - 4400\ \mathrm{s}^{-1}$) were imposed on the dopes. For each starting dope, the value of dynamic rigidity G' was shown under the micrograph. The degree of the unevenness of the starting cellulose dopes detected by this

Table 4.7.3

Wet spinning conditions

Stage	Method	Nozzle (hole) diameter (mm)	Number	Spinning velocity (m min^{-1})	Coagulant (%)	Denier
1	Pooly process	0.20–0.59	1	0.12–1.2	20% H_2SO_4 (5 °C)	monofilament
2	Roller process	0.06[a]	50 100	5.5–24.0	20% H_2SO_4 (5 °C)	57–84
3	Roller process	0.67	40	60	20–65% H_2SO_4 (5 °C)	75
4	Net process	0.05	33	100	20% H_2SO_4 (-7 °C)	75

[a]Au–Pt alloy.

Table 4.7.4

Structure and mechanical properties of fibers

Stage	Denier/filament number	$P_v{}^a$	$\chi_c(X)^b$ (%)	ACS^c (Å)	$f_b{}^d$ (%)	$f_c{}^e$ (%)	$\chi_{am}(C_3)^f$ (%)	TS^g (g d^{-1})	TE^h (%)
1	Monofilament	–	64	39	19.5	–	–	0.84	5.3
2	53/50–57/100	–	65–67	32–34	–	79–83	40–49	1.53–1.82	4.3–7.3
3	75/3.3	–	–	–	–	–	–	c.2.1	8
4	7.6/3.3	304	45.5	–	–	75	42.2	1.8	21
								1.9	15

[a] Viscosity average degree of polymerization of fibers.
[b] Crystallity by X-ray diffraction method.
[c] Apparent crystal size.
[d] Total molecular orientation by optical birefringence.
[e] Crystalline orientation parameter.
[f] Degree of break down in C_3–C_5 intramolecular hydrogen bond.
[g] Tensile strength.
[h] Tensile elongation.

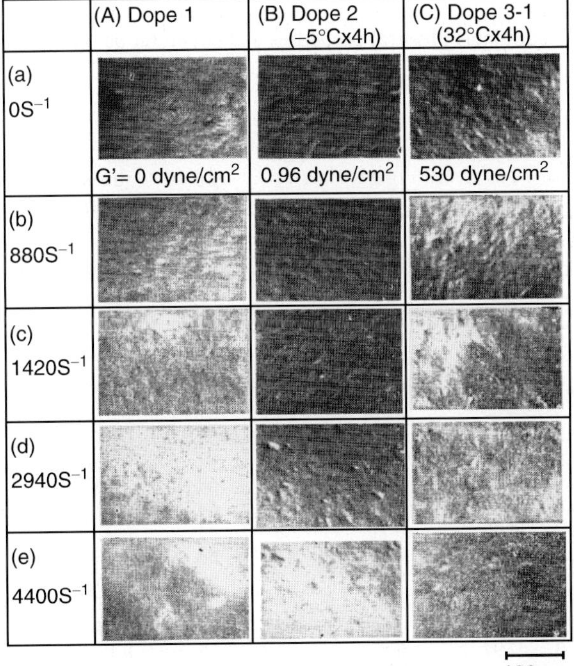

Figure 4.7.2   Differential interference optical micrographs of various cellulose dopes when given shear stresses imposed on the dopes:[15] (A) dope 1; (B) dope 2; (C) dope 3-1; (a) shear rate $d\gamma/dt = 0\,s^{-1}$; (b) $d\gamma/dt = 880\,s^{-1}$; (c) $d\gamma/dt = 1420\,s^{-1}$; (d) $d\gamma/dt = 2940\,s^{-1}$; (e) $d\gamma/dt = 4440\,s^{-1}$.

method decreases in the order: dope 3-1 (gelatinized at 32 °C for 3 h) > dope 2 (gelatinized at -5 °C for 4 h) > dope 1 (original dope). The order is comparable to the value of G' of the dopes. Imposing the shear stress on the gelatinized dopes tends to lower the degree of unevenness of the dopes, showing that the cellulose gels are mechanically reversible to a certain extent. This tendency is most significant for dope 3-1.

Wet spinning

The dopes in a glass syringe (inner volume: 10 ml) with a needle (length 10 mm, inner diameter: 0.20–0.59 mm) as a spinneret were extruded with a plunger/piston type extruder (microfeeder, Furue Science Co., Japan) into an aq. 20 wt% H_2SO_4 bath (length 43 cm) controlled at 5 °C at the extruding speed (v_1) of 51–660 cm min^{-1} (extruding volume: 0.14–0.23 ml min^{-1}) and a resultant monofilament was rolled up on a bobbin (made from polyvinylchloride) dipped in 20 wt% aq. H_2SO_4 bath (dipping time 3–10 min) at the take up speed (v_2) of 120–1200 cm min^{-1}. The resultant fiber, rolled up on a bobbin, was washed with tap water until the water became neutral. This was followed by air drying. The proper combination of the above spinning conditions was carefully chosen to adjust the shear rate $d\gamma/dt$ imposed on the dopes to 530–4400 s^{-1} and to adjust

Table 4.7.5

Spinning conditions, physical properties and structural parameters of fibers[15]

Spinning conditions			Physical properties				Structural parameters		
Dope	D_r	$d\gamma/dt$ (10^3 s^{-1})	TS (g d^{-1})	TE (%)	E (g d^{-1})	E_{nb} (g cm d^{-1})	χ_c (%)	ACS (Å)	f_b (%)
Dope 1	0.9	0.53	0.74	5.2	62	0.32	65	40	4.0
Dope 1	0.9	2.94	0.77	8.9	64	0.32	60	39	16.9
Dope 1	0.9	4.40	0.80	6.3	62	0.61	63	39	15.0
Dope 2	0.9	4.40	0.84	5.3	64	0.36	64	33	19.5
Dope 3-1	0.9	0.53[a]							
Dope 3-1	0.9	0.88[b]							
Dope 3-1	0.9	1.42	0.44	0.75	57	0.23	55	40	16.0
Dope 3-1	0.9	2.94	0.76	4.4	54	0.26	60	44	17.7
Dope 3-1	0.9	4.40	0.70	4.7	59	0.37	63	41	18.4
Dope 3-1	1.3	2.94	0.64	4.3	57	0.23	65	37	18.7
Dope 3-1	1.9	2.94	0.65	2.1	62	0.10	62	39	18.5
Dope 3-1	2.4	2.94	0.69	1.1	68	0.036	54	39	15.5
Dope 3-1	0.6	4.40	0.67	3.0	54	0.19	62	39	19.2
Dope 3-1	1.7	4.40	0.71	2.8	64	0.22	64	39	20.2
Dope 3-1	1.9	4.40	0.74	2.2	69	0.16	52	43	16.1
Dope 3-1	1.9	4.40	0.72	1.5	69	0.06	45	47	22.3
Dope 3-2	0.9	2.94	0.73	3.5	68	0.23	55	39	16.8
Dope 3-3	0.9	2.94	0.66	2.8	62	0.14	63	44	14.7
Dope 3-4	0.9	2.94	0.57	2.9	55	0.12	62	41	16.5
Viscose rayon			1.44	18.1	91	2.3	60	35	–
Cuprammonium rayon			2.05	9.1	117	1.6	61	54	48

[a]Pulverized during drying.
[b]Very weak.

the draft ratio D_r ($\equiv v_2/v_1$) to 0.9–1.9. For the wet spinning of dope 1 (original cellulose solution), 3 wt% aq. H_2SO_4 was employed as a proper coagulant for monofilament preparation.

The spinning conditions (starting cellulose dope, D_r and $d\gamma/dt$) employed for each experiment are summarized in the first three columns of Table 4.7.5.

Structural characteristics of monofilament obtained from gelatinized cellulose dopes

The monofilament from dope 1 has even and fine stripes parallel to the fiber axis and its cross-section seem uniform. Dope 2, which is not so seriously gelatinized, gives a monofilament with even (but not so fine) stripes parallel to the fiber axis. On the other hand, for the monofilament obtained from dope 3-1, which is considerably gelatinized, the stripes in the side face are coarse and disordered, the cross-section revealing defects. This means that the degree of gelation of the cellulose dope heavily influences the appearance of the fiber.

Tensile strength (TS), tensile elongation (TE), Young's modulus (E), the energy required for breaking (En_b), and molecular orientation parameter (f_b) (determined by birefringence method) of the fibers obtained are listed in the fourth, seventh, and tenth columns of Table 4.7.5, respectively.

Figure 4.7.3 shows X-ray diffraction patterns of the typical monofilament fibers obtained from three starting dopes under constant spinning conditions ($D_r = 0.9$, $d\gamma/dt = 2940$ s^{-1}), along with those of two commercially regenerated cellulose fibers (viscose rayon and cuprammonium rayon) as reference. All fibers obtained are crystallographically cellulose II, showing characteristic peaks at $2\theta = 20.5°$ (($10\bar{1}$) plane) and 22.0° (002) plane). These X-ray diffraction patterns are not very different from that of cuprammonium rayon, but the peak separation between $2\theta = 20.5°$ and 22.0° is somewhat better than those for the commercially regenerated cellulose fibers.

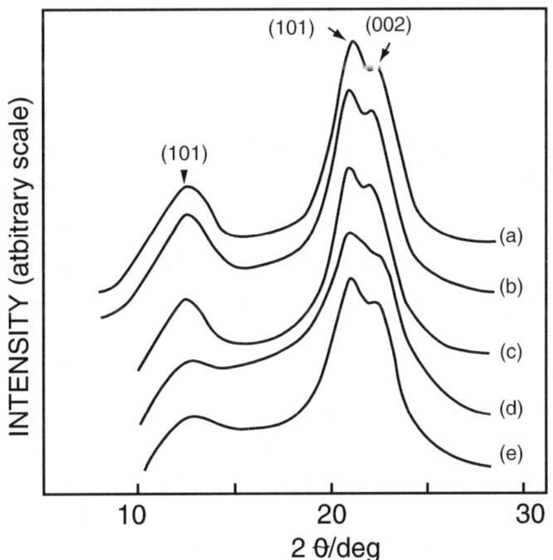

Figure 4.7.3 X-ray diffraction patterns of the typical cellulose monofilaments:[15] (a) from dope 1 (original cellulose dope); (b) from dope 2 (gelatinized at 5 °C for 4 h); (c) from dope 3-1 (gelatinized at 32 °C for 3 h); (d) viscose rayon; (e) cuprammonium rayon.

X-ray crystallinity $\chi_c(X)$ and apparent crystal size (ACS) of these fibers and other fibers obtained under different spinning conditions are given in the eighth and ninth columns of Table 4.7.5. Here, the crystallinity $\chi_c(X)$ was estimated by Segal's method,[23] using the following relations:

$$\chi_c(X) = 100[I(002) - I_{am}]/I(002) \tag{4.7.1}$$

Here, I (002) and I_{am} mean the peak intensities, corresponding to (002) plane ($2\theta = 21.7°$) and amorphous ($2\theta = 16.0°$). ACS was estimated through use of Scherrer's equation:[24]

$$ACS = 0.9\lambda/\cos\theta \times \beta \tag{4.7.2}$$

with

$$\beta = (B^2 - b^2)^{1/2} \tag{4.7.3}$$

Here, λ is the wavelength of the incident X-ray (1.5418 Å); θ, the diffraction angle corresponding to the (002) and (10$\bar{1}$) planes; b, the instrumental constant (0.2°), and B', the half value width in radians of the diffraction angle of the (002) and (10$\bar{1}$) planes.

IR analysis has some preferable features compared with CP-MAS ^{13}C NMR analysis, enabling us to estimate the degree of formation of both intra- and intermolecular hydrogen bonds, providing special information on intramolecular hydrogen bond, according to our concept.[25] Obviously the profiles of absorbing bands at 1050–1000 cm^{-1} and 3400–3200 cm^{-1} for the films obtained from gelatinized dopes are considerably different to that for the film obtained from the original dope. The relative absorbance of the CO stretching bands at 1092 (ring), 1048, and 1002 cm^{-1} for films from gelatinized dopes tends to lower as the extent of gelation of the dopes increases. The relative absorbance of the peaks for intramolecular hydrogen bonds in OH stretching (3480 and 3446 cm^{-1}) of these films is significantly high, while the intermolecular hydrogen bonding of all films seem to be in the same order. In the previous study, we pointed out that the degree of the $O_3 \cdots O_5{'}$ intramolecular hydrogen bond formation of the multifilament obtained from the cellulose dope (same as dope 1 in this study) with 20 wt% aq. H_2SO_4 (5 °C) as a coagulant is much higher than the commercially regenerated cellulose fibers. This previous experimental fact and the present result led us to the conclusion that the intramolecular hydrogen bonds already existed in the gelatinized cellulose dopes, and their existence may be one of the central forces making the cellulose solution gelatinous.

Figure 4.7.4 shows IR spectra of the films prepared from the three starting cellulose dopes under the same coagulation conditions used for the wet spinning of these dopes.

Influence of spinning conditions on the fiber properties

Effect of the shear rate, the draft ratio D_r, and the gelation time t on the physical and structural properties such as tensile strength (TS), tensile elongation (TE), Young's modulus (E), breaking energy (En_b), crystallinity $\chi_c(X)$), ACS, and molecular orientation parameter (f_b) was studied extensively on the monofilament fibers produced from dope 1 (original dope; A) and dope 3-1 to dope 3-4. The molecular orientation of the filament spun from a gelatinized cellulose dope is quite high (even at low shear rate)

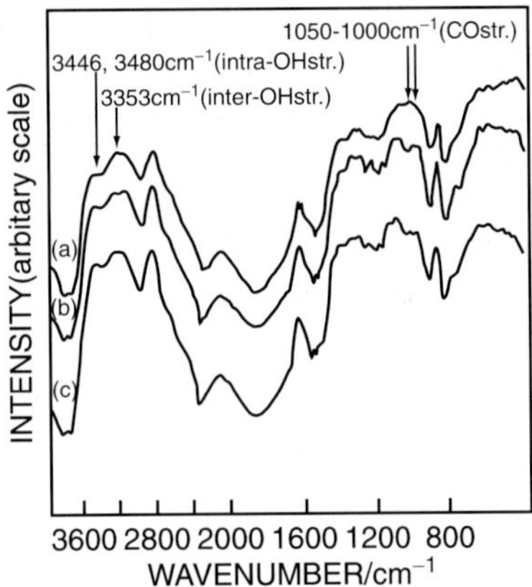

Figure 4.7.4 IR spectra of the films prepared from three starting cellulose dopes under the same coagulation conditions used for the wet spinning of these dopes:[15] (a) from dope 1 (original cellulose dope); (b) from dope 2 (gelatinized at $-5\,^{\circ}$C for 4 h); (c) from dope 3-1 (gelatinized at $32\,^{\circ}$C for 3 h).

compared with the filament from the nongelatinized cellulose dope. Thus, some molecular force may exist that makes the molecular chain stiff in the gelatinized cellulose dope. Except for ACS and f_b, other physical properties have a similar dependence on D_r in two spinning conditions differing shear rate: (1) TS is almost independent of D_r, (2) TE and En_b tend to decrease with an increase in D_r, and (3) $\chi_c(X)$ seems to be constant to some threshold value of D_r, after which decreases rapidly. ACS and f_b of the filament, produced at $d\gamma/dt = 2940$ s^{-1}, tend to decrease as D_r increases. The reverse is the case for the filament spun at $d\gamma/dt = 4400$ s^{-1}. Clearly, TS, TE, En_b, and f_b have the tendency to decrease with an increase in gelation time, while $\chi_c(X)$ and ACS seem independent of the gelation time. Young's modulus E increases with the gelation time and decreases in the longer time region. Gelation adversely influences the physical properties of the resultant filament when spun at a given spinning condition.

As seen from Table 4.7.5, all the data on the physical properties of the filaments prepared from the cellulose dope are at a very low level compared with those of commercially regenerated cellulose fibers. This implies that the results obtained here are only of a laboratory experimental level and therefore that more preferable spinning conditions for these dopes may exist.

In summary, the gelatinized cellulose dopes could be spun into monofilament if the proper shear stress is imposed on the dopes. However, the physical properties of these filaments are not very satisfactory.

4.7.3 Stage 2: laboratory scale wet spinning[16]

Selection of coagulants

In order to select roughly the appropriate coagulants for multifilament spinning, the cellulose solution prepared earlier was introduced into a glass syringe (inner diameter:

11 mm ϕ) with a needle (length: 10 mm; inner diameter: 0.255 mm ϕ) as a spinneret. Then, the solution was extruded with the aid of a plunger piston-type extruder (Microfeeder, Furue Science Co., Japan) into a coagulation bath (length: 43 cm) containing various coagulants at the extruding speed v_1 of 120 cm min^{-1}, and a resultant monofilament was rolled up in a water bath at the take up speed v_2 of 120 cm min^{-1} (draft ratio $D_r = v_2/v_1 = 1.0$). Aq. solutions of sulfuric acid (H_2SO_4), hydrochloric acid (HCl), acetic acid (CH_3COOH), and phosphoric acid (H_3PO_4), and their several salts were examined as coagulants. In most cases, the temperature of the coagulation bath was about 15 °C because higher temperatures tended to gelatinize the cellulose solution. For the strong acids such as aq. H_2SO_4 and HCl, the spinning test was also carried out at a lower temperature (*ca.* 5 °C) of the coagulation bath. During the experiments, coagulation state and spinnability were observed for each coagulant. The total assay was made as follows: (×) too weak coagulation with no spinnability; (△) weak coagulation but forming spinnable fiber (unstable taking up); (○) proper coagulation and good spinnability; (××) too strong coagulation with forming frail coagulated fiber (impossible taking up).

Table 4.7.6

Results of the coagulation and spinnability tests[16]

Anion	Concentration (wt%)	Cation							
		H^+		Na^+	NH_4^+	Mg^{2+}	Ca^{2+}	Zn^{2+}	Al^{3+}
		5 °C	15 °C	15 °C	15 °C	15 °C	15 °C	15 °C	15 °C
SO_4^{2-}	1	×	×	×	×	×	–	×	×
	3	×	○	×	△	×	–	△	△
	5	△	○	△	○	△	–	△	△
	10	○	○	○	○	△	–	××	××
	15	○	○	○	○	××	–	××	××
	20	○	××	○	○	××	–	××	××
	30	○	××	○	○	××	–	××	××
Cl^-	1	×	×	×	×	×	×	×	×
	3	○	○	×	△	×	×	△	○
	5	○	○	△	○	△	△	△	××
	10	○	○	△	○	××	△	△	××
	15	○	××	○	○	××	××	××	××
	20	××	××	○	××	××	××	××	××
	30	××	××	××	××	××	××	××	××
CH_3COO^-	5	○	○	△	△	××	–	△	××
	7	○	○	○	△	××	–	△	××
	10	○	○	○	○	–	××	××	××
	15	○	○	○	○	–	××	××	××
	20	○	××	○	○	–	××	××	××
P_4^{3-}	5	○	○	△	△	△	–	△	××
	10	○	○	○	○	△	–	△	××
	15	○	○	○	○	××	–	××	××
	20	○	××	○	○	××	–	××	××

Table 4.7.6 displays the results on coagulation and spinnability tests on cellulose monofilament spinning from the novel cellulose solution. Clearly, a wide variety of coagulants can be employed for this novel cellulose solution, although the coagulation and spinnability are strongly dependent on the spinning conditions, such as the extruding speed, hole number, and its diameter of spinneret, the length of coagulation bath, and the take up speed.

When we use the coagulation bath controlled at $15\,°C$ aq. H_2SO_4 with the acid concentration of coagulant $C_{sa} = 3-15$ wt%, aq. HCl with $C_{sa} = 3-10$ wt%, aq. CH_3COOH with $C_{sa} = 5-15$ wt%, and aq. H_3PO_4 with $C_{sa} = 5-15$ wt% proved to be applicable to the wet spinning of the novel cellulose solution. Aq. solutions of the sodium and ammonium salts of these acids with a wide range of concentration also enabled us to make new cellulose fiber. Magnesium, calcium, and zinc salts of these acids do not seem to be effective coagulants. At a coagulation temperature of $5\,°C$, the proper acid concentration range of the strong acids as coagulant seemed to shift to a higher concentration region. For example, the proper concentration range for aq. H_2SO_4 lies between 10 and 30 wt%. Because the novel cellulose solution must be neutralized in the process of the wet spinning, and the stock temperature of the cellulose solution is *ca.* $5\,°C$, a 20 wt% aq. H_2SO_4 (controlled at $5\,°C$) was adopted as the coagulant for the multifilament spinning of the novel cellulose solution.

Multifilament spinning

Multifilament spinning of the cellulose solution was carried out by using a wet spinning machine (see Figure 4.7.5), constructed by the authors. ① It is a plunger piston-type extruder with a stainless cylinder (inner diameter: 50 mm) and a cooling jacket (JP-H model microfeeder, Furue Science Co., Japan). ② The first coagulation bath length is 80 cm, ③ the second coagulation bath length is 50 cm, ④ the water bath length is 100 cm, ⑤ the boiling water bath length is 50 cm, ⑥ with an oiling roll, ⑦–⑩ Nelson-type rollers (diameter: 98 mm φ), ⑪ roll heaters (diameter: 150 mm φ), ⑫ and a take up device. Two spinnerets with different hole numbers (50 and 100) but with the same diameter (0.06 mm) for each hole were used. The cellulose solution prepared at $4\,°C$ was introduced into the cylinder cooled at $5\,°C$, and the air was excluded carefully from the system, then extruded at the speed v_1 (2.55–22.5 ml min^{-1}) into the first coagulation bath (20 wt% aq. H_2SO_4 at $5\,°C$). The coagulated fiber was rolled up at the speed of v_2 (5.5 – 24.0 ml min^{-1}) at the Nelson roller, ⑦ dipped into the second coagulation bath (water: $20\,°C$), passed through the washing baths again (water: 20 and $95\,°C$), and after oiling the fiber, was dried at four roll heaters (first roller: $180\,°C$; second roller: $130\,°C$; third roller: $120\,°C$; fourth roller: $30\,°C$), and finally was taken up.

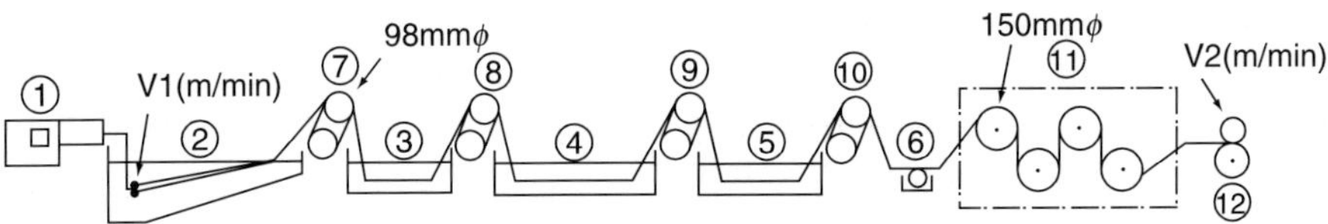

Figure 4.7.5   Schema of multifilament spinning machine.[16]

Table 4.7.7

Spinning conditions of cellulose fibers from cellulose/aq. NaOH dope

Sample code	Spinneret (n holes^{-1})	v_1 (m min^{-1})	v_2 (m min^{-1})	D_r	$d\gamma/dt$ (10^3 s^{-1})	P_v	Denier
F-1	100	2.55	6.7	2.6	5.8	312	57
F-2	50	4.5	5.5	1.2	10	306	53
F-3	100	4.5	10.8	2.4	10	313	69
F-4	50	22.5	24.0	1.1	50	326	84
VR[a]						381	74
CuR[b]						752	75

Sample code	TS (g d^{-1})	TE (%)	χ_c(X) (%)	ACS (Å) (002)	ACS (Å) (10 $\bar{1}$)	f_c (%)	χ_{am}(C$_3$) (%) Dry	χ_{am}(C$_3$) (%) Wet	$\Delta\chi$
F-1	1.56	6.3	65	34	35	83	40	37	0.93
F-2	1.82	7.3				81	49	41	0.84
F-3	1.71	4.3	67	32	34	83	45	36	0.80
F-4	1.53	7.3				79	42	39	0.93
VR[a]	1.44	18.1	60	29	33	84	54	52	0.96
CuR[b]	2.05	9.1	61	36	37	87	76	73	0.96

[a]Viscose rayon.
[b]Cuprammonium rayon.

Table 4.7.7 shows the spinning conditions (extruding speed v_1, take up speed v_2, draft ratio D_r, and shear rate $d\gamma/dt$). Also shown are the viscosity average degree of polymerization P_v, mechanical properties [denier (d), tensile strength (TS), and tensile elongation (TE) at dry state] and structural characteristics (χ_c(X), ACS, crystalline orientation factor f_c, χ_{am}(C$_3$) (at dry and wet), $\Delta\chi[= \chi_{am}(C_3)_{wet}/\chi_{am}(C_3)_{dry}]$) of the new fiber obtained. For comparison, the mechanical and structural properties of typical commercially regenerated fibers (viscose rayon and cuprammonium rayon) are also shown.

Here, the crystalline orientation for the fiber bundle sample placed on fiber specimen attachment was estimated by the X-ray transmission method. In this procedure, the scintillation counter was held constant at $2\theta = 20.0°$ [(10$\bar{1}$) plane] and the fiber bundle sample was revolved perpendicular to the direction of the incident X-ray, and the X-ray diffraction intensities at azimuth angles ψ were recorded.

Crystalline orientation parameter f_c was estimated by the following equation:[26]

$$f_c = \{1 - (\psi_{1/2}/180)\} \tag{4.7.4}$$

Here, $\psi_{1/2}$ denotes the half value width of the azimuth angle, expressed in degrees.

From the spectra the degree of breakdown of the intramolecular hydrogen bond at C$_3$ hydroxyl groups in glucopyranose unit, χ_{am}(C$_3$) was estimated by the following equations proposed in our previous work[27,28]

$$\chi_{am}(C_3) = 100 \times I_h(C_4)/\{I_h(C_4) + I_l(C_4)\} \tag{4.7.5}$$

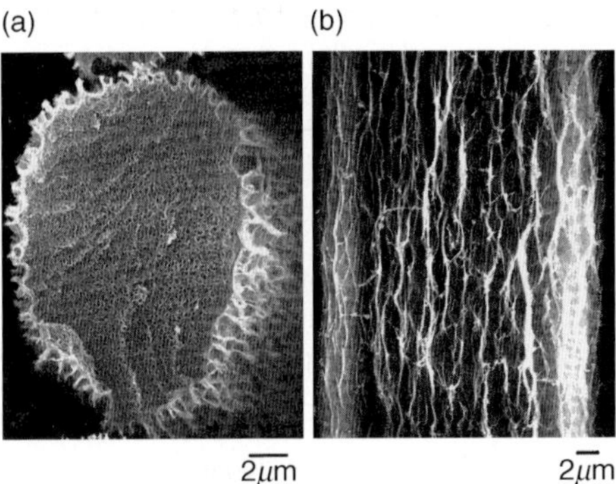

Figure 4.7.6 SEM micrographs of the lyophilized fiber of as spun fiber F-4: (a) cross-sectional view, (b) side face view.[16]

Here, I_1 and I_h are the fractions of lower and higher magnetic peaks in the C_4 carbon peak region (see Section 4.1). The ratio of the value at dry and wet states $\Delta\chi[= \chi_{am}(C_3)_{wet}/\chi_{am}(C_3)_{dry}]$ was also calculated as a measure of molecular mobility in the fiber when wet.

Physical and mechanical properties

As seen in Figure 4.7.6, the cross-section of the lyophilized new cellulose fiber (F-4) is almost a true circle.

Its very thin and coarse skin is not the structure constituted of densely coagulated molecules as seen in viscose rayon, and it is rather more porous than the inner part. The inner structure of the new fiber is also very porous with an average pore size of 110 nm. The whole appearance of the other fibers spun here was almost the same as those shown in the figure. This means that, as spun, new cellulose fiber might be a kind of coagulated gel containing a large amount of solvent.

P_v of the fibers is of the same order as that of the starting alkali soluble cellulose. TS of the new fibers ranges from 1.53 to 1.82 g d^{-1}, which is in the same range as that of regenerated fibers. However, the TE is smaller than those for the regenerated fibers. The shear rate $d\gamma/dt$ imposed on the solution, coming out through the spinneret, tends to result in the higher TS in the $d\gamma/dt \leq 10^4$ s^{-1}, but an excess shear rate seems not to be preferable for obtaining the fiber with the higher TS. The TS of the fiber may not be influenced by draft ratio $D_r(= v_2/v_1)$ within these experimental conditions.

Structural characteristics

A typical new fiber (F-1) is crystallographically cellulose II, revealing two characteristic X-ray diffraction peaks at 20.0° for $(10\bar{1})$ plane and at 21.7° for (002) plane, as observed for viscose rayon and cuprammonium rayon (Figure 4.7.7). Table 4.7.7 (columns 11–14) shows that (1) the crystallinity χ_c of the new fibers is relatively high ranging from 0.65 to 0.67, compared with those (*ca.* 0.6) of the commercially regenerated fibers; (2) the new fibers have orientation parameter f_e ranging from 0.79 to 0.83, which is considerably

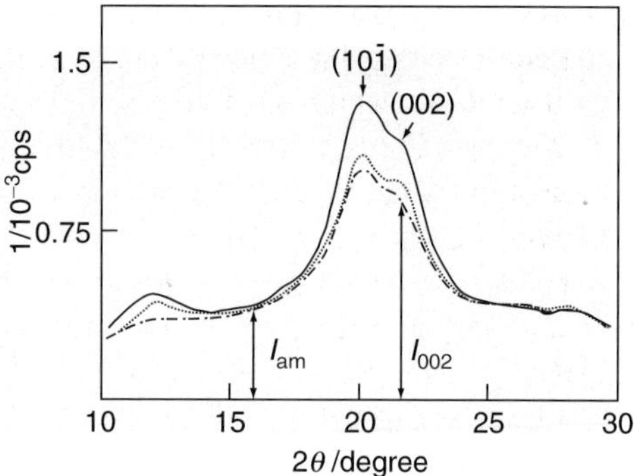

Figure 4.7.7   X-ray diffraction patterns of a typical new fiber (F-1) (full line); viscose rayon (chain line); and cuprammonium rayon (dotted line).[16]

lower than those of commercially regenerated fibers; and (3) ACS increases in the order: cuprammonium rayon > new fiber > viscose rayon.

Figure 4.7.8 (a) and (b) show CP-MAS ^{13}C NMR spectra of a typical new fiber, viscose rayon, and cuprammonium rayon at dry and wet states, respectively.

The spectral shape in the NMR spectra of these fibers at dry state is different from sample to sample. This tendency is most distinct for C_4 carbon peak region, suggesting that there are several molecular packing (or ordering) states in the molecules. C_4 carbon peak region might reflect the possibility of the formation of $O_3 \cdots O_5'$ intramolecular hydrogen bond, as pointed out previously.[1] A close analysis revealed that the degree of the breakdown of intramolecular hydrogen bond at C_3 position $\chi_{am}(C_3)$ for the new fiber is notably smaller, indicating the higher degree of formation of $O_3 \cdots O_5'$ intramolecular hydrogen bond, as shown in column 15 of Table 4.7.7. This is comparable to the results obtained by X-ray analysis, which revealed higher crystallinity for the new fiber. When measured at wet state, the peak separation of all samples becomes better than that at dry state. For example, two sharp peak components at 107.4 and 105.4 ppm for C_1 carbon peak region, and 89.2 and 88.1 ppm for C_4 carbon peak region, are clearly detectable.

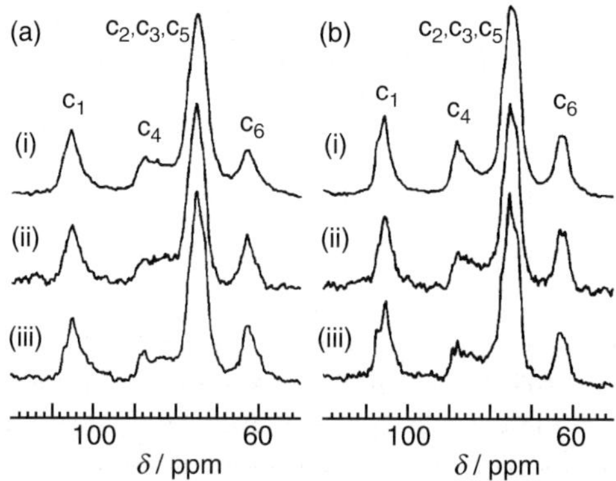

Figure 4.7.8   CP-MAS ^{13}C NMR spectra of cellulose fibers at (a) dry and (b) wet state:[16] (i) typical new fiber (F-1), (ii) viscose rayon, and (iii) cuprammonium rayon.

C_6 carbon peak region of viscose rayon splits into two peaks by wetting. The whole spectral shape for C_4 carbon peak region of the new fiber at wet state is relatively sharper than that of others. As seen in column 16 of Table 4.7.7, $\chi_{am}(C_3)$, for all samples, decreases more or less by wetting. This indicates that $O_3 \cdots O_5'$ intramolecular hydrogen bond in the new fiber, as well as the commercially regenerated cellulose fibers, apparently increases, although the degree of increase is very conspicuous for the new cellulose fiber obtained here. It is quite astonishing that the wet treatment facilitates the molecular mobility in the disordered region of new fibers, even if they have higher crystallinity and higher degree of intramolecular hydrogen bond. This result may correspond to the experimental facts that the new fiber is very porous. Horii *et al.*[29] reported the similar NMR spectral change in natural cellulose by wetting. They suggested that the molecular chains in amorphous region of natural cellulose are partially ordered in the water swollen state, but those chains are distorted through drying, and the strain can be released when cellulose becomes wet again. On the contrary, they found that the CP-MAS ^{13}C NMR spectrum of cellulose II did not change by wetting. Horii *et al.* explained this phenomenon by assuming that the structure of the amorphous region in cellulose II has no regularity in both the wet and dry state. However, NMR spectra of the new cellulose fibers spun in this study (cellulose II) show remarkable change in C_4 region by wet treatment. These spectral changes may be caused by the temporary formation of intermolecular hydrogen bond between O_5 oxygen and water molecules or by the intramolecular hydrogen bond formation accompanying rearrangement of cellulose molecules in a wet state. Note that the NMR spectral change caused by wet treatment also takes place for the commercially regenerated fibers. This seemingly opposes the results reported by Horii *et al.*,[29] but we believe that the higher order of coagulated structure may play an important role on the molecular mobility induced by the interaction with polar solvent.

In summary, a cellulose dope prepared from a novel alkali soluble cellulose 9 wt% aq. sodium hydroxide solution was subjected to wet spinning, using aq. H_2SO_4 as a coagulation bath. The novel cellulose fiber reveals cellulose II crystal form from X-ray analysis. While the crystallinity is higher, its crystalline orientation is slightly lower than the commercially regenerated fibers. The degree of breakdown of intramolecular hydrogen bonds at $C_3[\chi_{am}(C_3)]$ of the cellulose fiber, determined from CP-MAS ^{13}C NMR, is much lower than others, and the CP-MAS ^{13}C NMR spectra of its dry and wet state were significantly different from each other. This indicated that the cellulose molecules in the novel cellulose fiber are quite mobile. This phenomenon has not been reported for regenerated cellulose fibers.

Films coagulated from alkali soluble cellulose I and cellulose II[17]

In this section, an attempt is made to clarify the difference in structure and morphology of the cellulose films coagulated from the alkali soluble cellulose I and cellulose II solutions by using sulfuric acid in various concentrations as a fundamental basis from which to establish an appropriate wet spinning method for these novel cellulose solutions.

Alkali soluble cellulose samples. An alkali soluble cell I sample (the viscosity average degree of polymerization $P_v = 331$, $\chi_{am}(C_3) = 46\%$, solubility $S_a = ca.\ 100\%$) was prepared by applying the steam explosion treatment on a soft wood (mainly white spruce) pulp (Alaska pulp (USA), manufactured by Alaska Pulp Co.; α-cellulose content: 90.1 wt%, $P_v = 1060$) under the conditions of water vapor pressure $P = 2.9$ MPa and the treating time $t = 30$ s with water content of the original cellulose of 80%, as described in previous papers[1–3] (see Section 5.1). An alkali soluble cell II sample ($P_v = 453$, $\chi_{am}(C_3) = 92\%$, $S_a = 100\%$) was prepared as follows. A purified cotton linter (α-cellulose content: 95.7%, $P_v = 1279$) was dissolved in a cuprammonium solution ($NH_3/Cu/H_2O = 7.0/3.6/89.4$, w/w/w) at polymer concentration of 8 wt%. The solution was poured into acetone and the resultant precipitates were regenerated by 2 wt% aq. H_2SO_4 for 3 h, followed by neutralization, washing, and drying.

Preparation of cellulose/aq. alkali solutions. The given amount of the alkali soluble cellulose (water content: *ca.* 8–12 wt%) was dispersed into 8.65 wt% aq. NaOH solution, precooled at 4 °C, stood for 8 h with intermittent mixing with a home mixer (Type SM, Sanyo Electric Co., Ltd, Japan) to give polymer concentration $C_p = 3.4$–5.6 wt%, and the resultant solution was subjected to centrifugation (55P Type centrifugal apparatus, Hitachi Machinary Co., Ltd, Japan) at 10,000 rpm for 60 min in order to exclude the remaining undissolved part and to carry out the degasification at 4 °C. The solution thus obtained was immediately subjected to film preparation by wet coagulation method.

Preparation of coagulated films. Cellulose solutions ($C_p = 3.4$–5.6 wt%) were cast on a glass plate (10×10 cm^2) to a thickness of 1 mm and the glass plate was immersed gently into aq. H_2SO_4 solutions (concentration $C_{sa} = 20$–80 wt%) controlled at -6–40 °C for 30 min. The coagulated films were washed thoroughly with water for a sufficient period of time at 20 °C, and the resultant wet gel films were wrapped in aluminum foil, followed by freezing in liquid nitrogen. The frozen films were lyophilized with an apparatus (Type FD-1, Tokyo Rika Machinery Co., Ltd, Japan). The gel films coagulated with $C_{sa} = 80$ wt% could not be recovered owing to strong dissolving action of the coagulant against the coagulated films.

Influence of concentration of coagulant on the coagulation state of the films. Figure 4.7.9 shows SEM micrographs of surface and inner phase of the lyophilized film (A. cell I ($C_p = 4.1$ wt%) system; B. cell II ($C_p = 4.7$ wt%) system) coagulated by aq. H_2SO_4 with $C_{sa} = 20$–70 wt% at 5 °C for 5 min). For the inner phase, a magnified ($\times 30,000$) SEM micrographs are also shown. Average pore size ($2r$) of the films obtained from cell I system (hereafter denoted simply as cell I film(s)) seems smaller for the surface than the inner phase but does not show a clear skin structure. In contrast, the films obtained from cell II system (hereafter denoted as cell II film(s)) exhibited clear skin structure (thickness: 0.5–1.5 μ m, as shown by the mark↔; $2r$, *ca.* 0 nm) when aq. H_2SO_4 solutions with $C_{sa} = 40$–60 wt% were used as coagulant but aq. H_2SO_4 with $C_{sa} \approx 65$ wt% gave somewhat porous and denatured surface probably due to the strong dissolving action against the films.

The inner phase of cell I films coagulated by aq. H_2SO_4 with $C_{sa} = 20$–65 wt% seems to be basically composed of the backbone structure with very thin membrane, as shown by the arrow, which has defects (pore with diameter of 20–100 nm; this structure is hereafter denoted as UP + M structure: UP, uncircular pore). According to Kamide and coworkers,[30–39] the mechanism of membrane formation from polymer solution

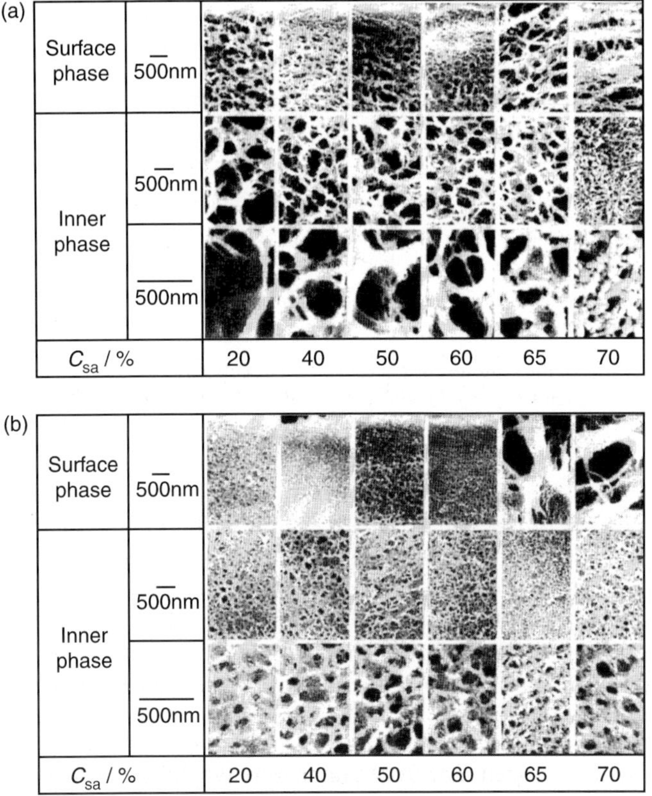

Figure 4.7.9   SEM micrographys of surface and inner of the lyophilized films coagulated by aq. H_2SO_4 with concentration $C_{sa} = 20$–70 wt% at 5 °C for 5 min: (a) cell I ($C_p = 4.1$ wt%) system; (b) cell II ($C_p = 4.7$ wt%) system. For the inner phase, magnified ($\times30{,}000$) SEM micrographs are also shown.

by the solvent casting method is that, at the early stage of phase separation and with the lapse of time growing, the primary particles with a diameter of 10–30 nm generate the secondary particles with diameter of 50–600 nm^{40} (see Figure 3.24.1). The amalgamation of the secondary particles brings about the porous polymeric membrane. Therefore, the network-like backbone structure indicated here is thought to be made by collision of the second particles (diameter $d_p = ca.$ 50 nm). Aq. H_2SO_4 with $C_{sa} = 70$ wt% gave a noncircular pore (UP)^{30} structure made by the collision of secondary particles with $d_p = 20$–100 nm.

The average pore size ($2r$) of the inner phase of cell I films is summarized as a function of C_{sa} as follows: $2r = ca.$ 1000 nm at $C_{sa} = 20$ wt%, $2r = ca.$ 500 nm at $C_{sa} = 40$–55 wt%, $2r = ca.$ 100 nm at $C_{sa} = 70$ wt%. Note that a clear secondary particle with a diameter $d_p = ca.$ 40 nm was first observed for cell I films when aq. H_2SO_4 with $C_{sa} = 65$ wt% is used as a coagulant, and that the thin membrane accompanying the backbone structure disappeared at $C_{sa} = 70$ wt%. For the cell II solution system, all aq. H_2SO_4 solutions employed here gave the UP structure to cell II

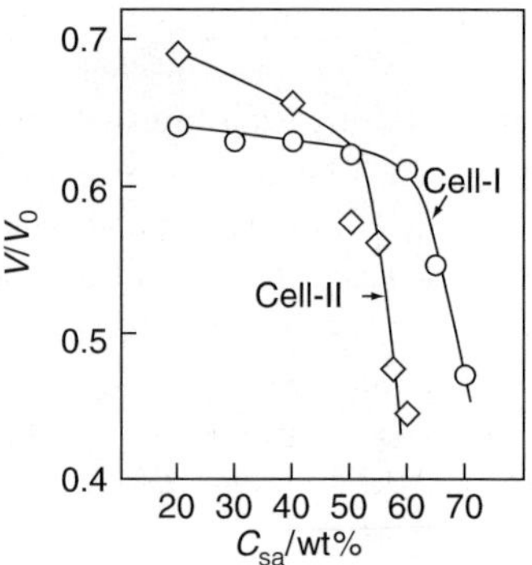

Figure 4.7.10   Plots of the volume contraction V/V_0 of cellulose solution systems as a function of concentration of coagulant C_{sa}:[17] (○), cell I ($C_p = 4.1$ wt%); (◇), cell II ($C_p = 4.7$ wt%).

films with a far smaller pore size than that for cell I, compared at the same C_{sa} level. With an increase in C_{sa}, $2r$ of inner phase of cell II films is almost independent of C_{sa}, except for that of $C_{sa} = 65\%$, which exhibits a smaller pore size (*ca.* 50 nm).

The secondary particle size (d_p) never grows larger than 30 nm. The considerable difference in coagulation state observed for cell I and cell II films might closely relate to their dissolved state in aq. NaOH solution. One plausible hypothesis may be that parts of the alkali soluble cell I intrinsically associates in a concentrated solution with each other, constituting some fixed structure in its dissolved state in the solution, leading to very porous and larger secondary particles in coagulation process.

Figure 4.7.10 shows the volume contraction V/V_0 of cell I ($C_p = 4.1$ wt%) and cell II ($C_p = 4.7$ wt%) solution systems as a function of concentration of coagulant C_{sa}. Here, V_0 is initial volume of solution ($V_0 = 8.85$ cm^3) at 4 °C; V, volume of coagulate film). The cell I system revealed no significant V/V_0 change ($0.64 - 0.61$) in the range of $C_{sa} = 20–60$ wt% but V/V_0 abruptly decreased to 0.47 at $C_{sa} = 70$ wt%. In contrast, cell II system V/V_0 decreased with an increase in C_{sa}, showing a relatively sharp decrease in the range of $C_{sa} = 50–60$ wt%, and the evaluation of V/V_0 at $C_{sa} = 70$ wt% became impossible because of the dissolution of the polymer phase by the coagulant. These facts closely correspond to the coagulation state observed in Figure 4.7.9. For both systems, aq. H_2SO_4 solutions with specific C_{sa} (70 wt% for cell I; 60 wt% for cell II) have an ability to give a dense coagulation state of cell I and cell II films.

Aq. H_2SO_4 solutions can be divided into three categories according to C_{sa} (or dissociation state of H_2SO_4): *region I*, $C_{sa} = 100–80$ wt% where undissociated H_2SO_4 abruptly diminished while HSO_4^- ion steeply increases but without any existence of SO_4^{2-} ion; *region II*, $C_{sa} = 80–60$ wt% where HSO_4^- ions and SO_4^{2-} ions rapidly approach their equilibrium states; *region III*, $C_{sa} = 40–0$ wt% where HSO_4^- and SO_4^{2-} ions are in the almost equilibrium state. Note that formation of HSO_4^- ion at $C_{sa} = 97–80$ wt% and that of SO_4^{2-} at $C_{sa} = 75–40$ wt% takes place independently owing to the large difference in their dissociation constant K_1 and K_2:[41]

$$H_2SO_4 \rightarrow H^+ + HSO_4^-, \quad K_1 = 2000; \qquad HSO_4^- \rightarrow H^+ + SO_42^-, \quad K_2 = 0.02^{41}$$

Thus, when aq. H_2SO_4 solutions are employed as coagulant, their coagulation ability should be considered in view of their dissociation state.

From Raman spectra analysis on material transportation between aq. H_2SO_4 solution with $C_{sa} = 70$ wt% and 8.65 wt% aq. NaOH solution separated by a Teflon membrane, it is concluded that concentrated H_2SO_4 solution penetrates to aq. NaOH solution and the amount of H_2SO_4 is far larger than that expected for neutralization, and that aq. H_2SO_4 with $C_{sa} = 70$ wt% allows only water from aq. NaOH solution to penetrate; that is, the H_2SO_4 solution exhibits a strong dehydration action. Aq. H_2SO_4 has a strong ability to give a densely coagulated state to the cellulose films exists in region II, and these H_2SO_4 solutions are in a transition state to easily produce SO_4^{2-} ion by absorbing water from other media.

Figure 4.7.11 shows CP-MAS ^{13}C NMR spectra of the cellulose films used in Figure 4.7.9 ((a) cell I films; (b) cell II films).

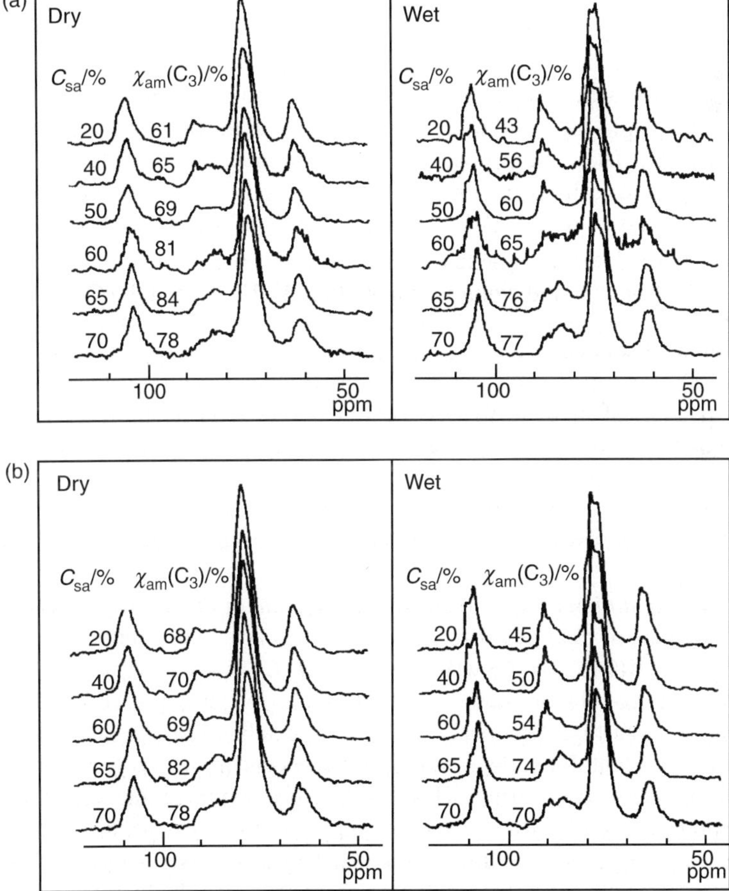

Figure 4.7.11   CP-MAS ^{13}C NMR spectra of the cellulose films used in Figure 4.7.9:[17] (a) cell I films; (b) cell II films. $\chi_{am}(C_3)$ values are also shown.

The spectra were recorded both in dry and wet state and $\chi_{am}(C_3)$ values, which means an extent of breakdown of intramolecular hydrogen bond, such as O_3–$OH\cdots O_5'$,[28] are also shown in the figure. Obviously, the patterns of NMR spectra for cell I films at dry state abruptly changed at $C_{sa} = 50$–60 wt%, revealing a sudden increase in the $\chi_{am}(C_3)$ value, and corresponding changes were observed at $C_{sa} = 60$–65 wt% for cell II films. This means that the formation of intramolecular hydrogen bond at C_3 position becomes hard beyond the threshold value of C_{sa} (60 wt% for cell I system; 65 wt% for cell II system) of aq. H_2SO_4 used as coagulant. The NMR spectra at wet state and the value of $\chi_{am}(C_3)_{wet}/\chi_{am}(C_3)_{dry}$ clearly point a critical C_{sa} value (65 wt%) for both systems, beyond which $\chi_{am}(C_3)_{wet}/\chi_{am}(C_3)_{dry}$ becomes almost unity and the appearance of sharp peaks in the C_4 carbon peak region by wetting is strongly depressed. These results indicate that aq. H_2SO_4 solution has the ability to coagulate cellulose film densely and is also able to produce the structure in which cellulose molecules are not mobile when wet,[29,42,43] although the development of O_3–$H\cdots O_5'$ intramolecular hydrogen bond formation is quite low.

Influence of cellulose concentration C_p on the coagulation state of lyophilized cellulose films. Figure 4.7.12 shows SEM micrographs of the lyophilized films prepared from cellulose solutions with various C_p using aq. H_2SO_4 solutions at 5 °C for 5 min.

Aq. H_2SO_4 with $C_{sa} = 20$ wt% did not give skin structure for both systems, irrespective of C_p. The average pore size ($2r$) of the inner phase of films became smaller (1000 → 200 nm for cell I film; 400 → 200 nm for cell II film) for both systems with an increase in C_p, and $2r$ of cell II film was smaller than that of cell I film if the C_p was constant. When aq. H_2SO_4 solution with higher C_{sa} is employed as coagulant, the average pore size ($2r$) of the inner phase of cell I films became smaller (300 nm for $C_{sa} = 20\%$; 200 nm for $C_{sa} = 70\%$ at $C_p = 4.8\%$). The surface of the cell I films, however, became more porous with an increase in C_p owing to the strong dissolving ability of aq. H_2SO_4 with higher C_{sa}. This might be explained by the fact that the cell I system with low C_p has larger content of alkali by which concentrated H_2SO_4 is neutralized, leading to the restriction of dissolving action of the acid on the cellulose film produced. Aq. H_2SO_4 with high C_{sa} gave a skin structure to cell II films, irrespective of C_p. The pore size of the inner phase of cell II films became smaller (300 → 200 nm) with an increase in C_p, but the size of secondary particles remained almost constant ($d_p = ca.$ 30 nm).

Influence of coagulation time on the coagulation state of the films. Figure 4.7.13 shows SEM micrographs of the lyophilized cellulose films coagulated under given conditions (cell I: $C_p = 4.1$ wt%, $C_{sa} = 65$ wt%, 5 °C; cell II: $C_p = 4.7$ wt%, $C_{sa} = 65$ wt%, 5 °C) as a function of coagulation time t_c. For both films, longer t_c resulted in the films with somewhat porous surface. Obviously, cell II films exhibited similar (and the most dense) coagulation states both for surface and inner phase concurrently at $t_c = 1$ min, with skin structure (thickness, *ca.* 5 μm). This result corresponds to the result on material transportation pictured by Raman spectroscopy. On the other hand, it was difficult to determine the coagulation conditions for cell I films to give similar and most dense coagulation states both for surface and inner phase concurrently. In a separate experiment using aq. H_2SO_4 with $C_{sa} = 20$ wt%, the following was clarified: (1) The pore size of the inner phase of cell I films was almost constant (*ca.* 400 nm) at $t_c \geq 7$ min and, similarly,

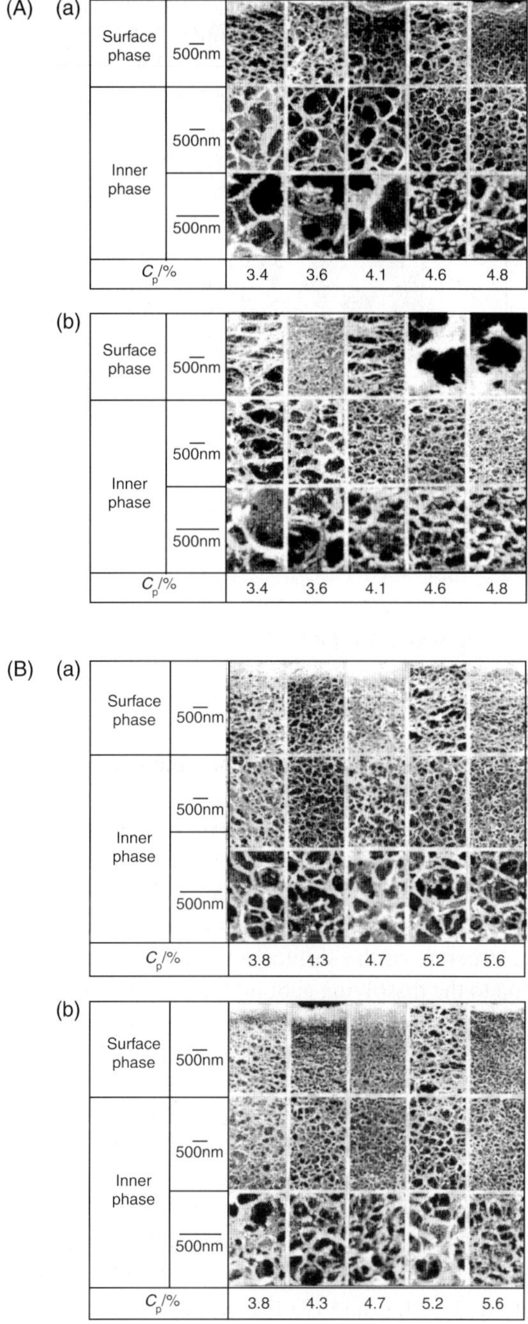

Figure 4.7.12 SEM micrographs of the lyophilized films from cellulose solutions with various cellulose concentration C_p using aq. H_2SO_4 solution at $5\,^{\circ}C$ for $5\,min$: (A) cell I films (a) $C_{sa} = 20\,wt\%$; (b) $C_{sa} = 70\,wt\%$); (B) cell II films (a) $C_{sa} = 20\,wt\%$; (b) $C_{sa} = 60\,wt\%$).[17]

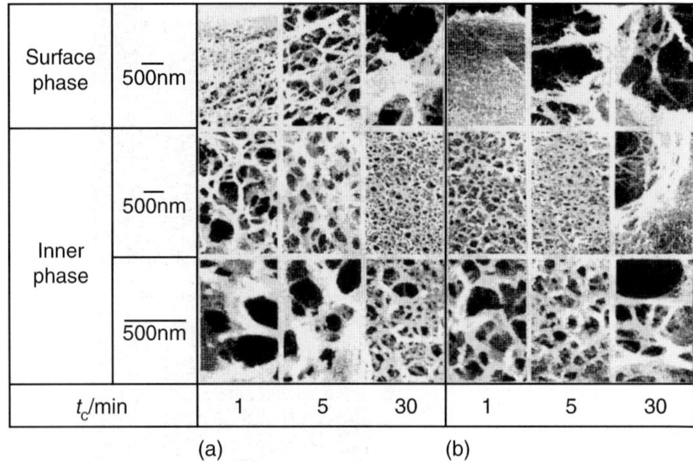

Figure 4.7.13   SEM micrographs of the lyophilized films coagulated under given conditions (cell I: $C_p = 4.1$ wt%, $C_{sa} = 65$ wt%; cell II: $C_p = 4.7$ wt%, $C_{sa} = 65\%$) as a function of coagulation time t_c: (A) cell I films; (B) cell II films.[17]

the pore size of the surface was constant (300 nm) at $t_c \geqq 5$ min (2) The surface of cell II films became most dense ($2r = ca.$ 100 nm) at $t_c \geqq 5$ min, and the pore size of inner phase kept constant ($2r = ca.$ 200 nm), irrespective of t_c.

Influence of coagulation temperature on the coagulation state of the films. Figure 4.7.14 shows SEM micrographs of the lyophilized cellulose films coagulated under given conditions (cell I: $C_p = 4.1$ wt%, $C_{sa} = 65$ wt%, $t_c = 5$ min; cell II: $C_p = 4.7$ wt%, $C_{sa} = 65$ wt%, $t_c = 5$ min) as a function of coagulation temperature T_c (-6–40 °C). For both films, $T_c = 5$ °C resulted in the most dense inner phase (cell I: $2r = ca.$ 500 nm; cell II: $2r = 50$ nm). Obviously, the surface of cell I film exhibited porous structure at higher T_c, while the porous structure of the surface was observed at lower T_c for cell II films. This might correspond to the dissolving action of the coagulant at lower T_c for cell II film and to the decomposing action of the coagulant at higher T_c for cell I film.

The phenomenological difference observed for cell I and cell II films by coagulation are summarized in Table 4.7.8.

In summary, several phenomenological differences were detected for cell I and cell II systems. SEM observation on the lyophilized coagulated cellulose films revealed the following differences: (1) for the alkali soluble cell II system, the existence of secondary particles was evident in the range of $C_{sa} \geq 20$ wt% and the most dense structure was given when $C_{sa} = 65$ wt%, (2) For the alkali soluble cell I system, the secondary particles became detectable at $C_{sa} \geq 65$ wt% and the coagulant with $C_{sa} \geq 65$–70 wt% gave the most dense structure of the film, and (3) the particle size constituting the most dense structure of the films is smaller for cell II system than cell I system.

A strong dehydrating action from cellulose solutions was confirmed for the coagulant with $C_{sa} \geq 60$ wt% by Raman spectroscopy and the neutralization rate of cell II system was much higher than cell I system. CP-MAS ^{13}C NMR analysis showed that both

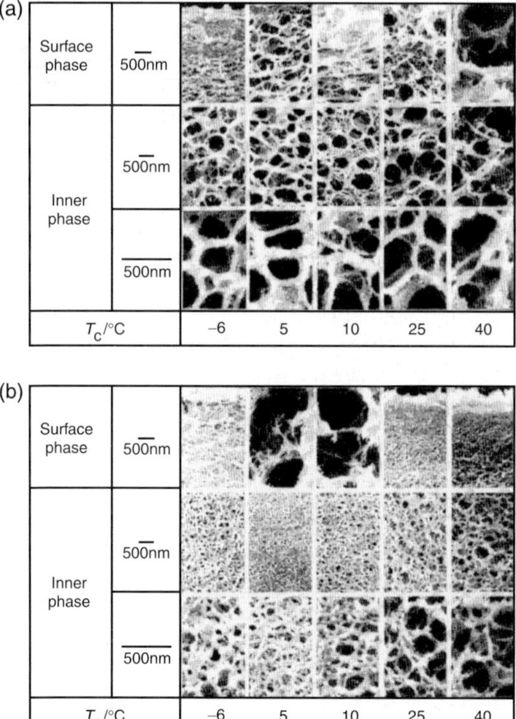

Figure 4.7.14 SEM micrographs of the lyophilized films coagulated under given conditions (cell I: $C_p = 4.1$ wt%, $C_{sa} = 65$ wt%; cell II: $C_p = 4.7$ wt%, $C_{sa} = 65\%$) as a function of coagulation temperature T_c (-6–40 °C): (a), cell I films; (b), cell II films.[17]

densely coagulated films developed practically no intramolecular hydrogen bonds at the C_3 position.

4.7.4 Stage 3: small size bench scale wet spinning[18–20]

Research findings, such as 'complete dissolution of cellulose into aq. NaOH' does not simply lead to the industrialization of fiber or film production using the cellulose–aq. alkali system without further effort. For example, the method of dissolution employed in the laboratory is too crude for plant operation, and the method for determining the solubility S_a is not adequate to qualify the spinning dope.

Two-step dissolution method[18]

Further improvement of solubility of cellulose into aq. NaOH is a necessary requirement for spinning stability because undissolved particles may clog dye, resulting in breaking of the running, not fully regenerated filaments. Even if these particles passed through dye, they will indisputably worsen the evenness of the filament.

A starting cellulose resource sulfite pulp from white spruce (α-cellulose content 90.1 wt%, $P_v = 1060$) in saturated water vapor of 15 kg cm^{-2} was exploded with water

Table 4.7.8

Phenomenological difference in coagulated state of celluloses obtained from cell I/NaOH, cell II/NaOH systems under acid coagulation conditions[17]

Conditions (phenomena)	Cell I	Cell II
$C_{sa} = 20{-}70\%$ (5 °C, 5 min)		
Existence of skin phase	Unclear	$C_{sa} = 40{-}60\%$ Thickness: 0.5–1 μ m $2r = 0$ nm
Outlook of inner phase	$C_{sa} = 20{-}65\%$: UP + M $C_{sa} = 70\%$: UP	UP
Most dense state of inner phase	$C_{sa} = 70\%$ $2r = 100$ nm $2d = 20{-}100$ nm	$C_{sa} = 65\%$ $2r = 50$ nm $2d = 30$ nm
Volume contraction	$C_{sa} = 60 \rightarrow 70\%$: 0.47	$C_{sa} = 50 \rightarrow 60\%$: 0.45
CP-MAS spectra	$C_{sa} = 50 \rightarrow 60\%$ Considerable destruction of O_3–O_5' intramolecular hydrogen bond	$C_{sa} = 60 \rightarrow 65\%$
$C_p = 3.4{-}5.6\%$ (5 °C, 5 min)		
Outlook of inner phase	$C_{sa} = 20\%$ UP + M $C_{sa} = 70\%$ $C_p \leq 3.6\%$: UP + M $C_p \geq 4.1\%$: UP	$C_{sa} = 20\%$; UP + M $C_{sa} = 60\%$ $C_p \leq 4.7\%$: UP + M $C_p \geq 5.6\%$: UP
Most densely coagulated state	$C_p = 4.8\%$ $2r = 100$ nm $2d = 20{-}100$ nm	$C_p = 5.6\%$ $2r = 100$ nm $2d = 20{-}50$ nm

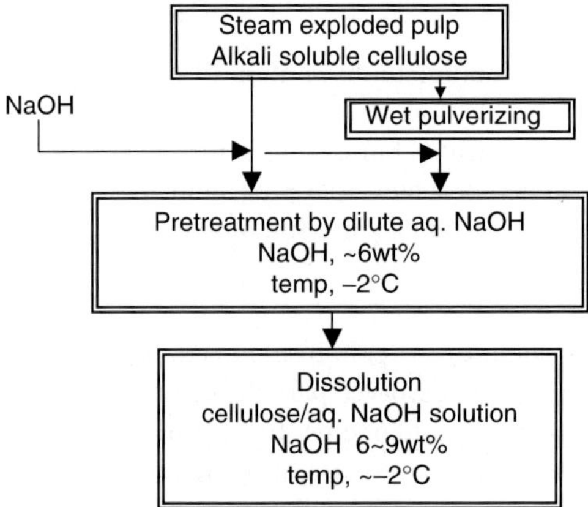

Figure 4.7.15   Flow diagram of dissolution of alkali-soluble cellulose into aq. NaOH.[18]

into atmosphere to give cellulose with $P_v = 290-409$. The exploded pulp was wet pulverized with media-type wet mill.

Figure 4.7.15 shows the industrial preparative procedure of the spinning solution (i.e. an improved two-step dissolution method).

To evaluate the solubility S_a of cellulose in aq. alkali solution, the centrifugal method was employed in laboratory test.[1,25]

This method is not accurate in the range $S_a \geq 99\%$, due to difficulty of separation of undissolved materials remaining in the solution. The Coulter counter method was introduced into the bench plant operation. It involves counting the number of undissolved particles as function of particle size (10 μm intervals) by counter (Counter Electric Inc., USA) and then calculating the volume of undissolved materials R_c (ml/20 l solution). R_c can be converted to S_a through the equation,

$$S_a = 100 - R_c \times S_1/(S_2 \times 20 \times 10^3 \times C_p) \tag{4.7.6}$$

Here, S_1 is cellulose density (see Section 1.4), S_2 the dope density (see Section 1.1), and C_p the cellulose concentration (wt%). Undissolved material of exploded pulp in 9 wt% NaOH is the material that is converted from cellulose I to cellulose II (Figure 4.7.16).

Figure 4.7.17 shows the effect of alkali concentration at pretreatment stage in Figure 4.7.15 on the undissolved volume R_c (a) and the effect of final alkali concentration of spinning dope on R_c and gelation time of the final dope (the cellulose concentration $C_p = 5$ wt% and $C_a = 7.6$ wt%) (b).

R_c is reduced to one-third of that obtained without the pretreatment step. R_c is independent of C_a in the range $C_p = 2-6$ wt% and gelation is suppressed most effectively at $C_a \approx 7-8$ wt%. The solubility of exploded pulp was conspicuously improved by using the two-step dissolution method.

From an analysis of change in dissolving velocity of cellulose in aq. alkali with time, Yamane et al.[18] demonstrated that the dissolution is governed by diffusion. They concluded that physical removal of highly viscous, dissolved phase (concentrated solution) from the surface of cellulose particles by applying intense agitation is essential to make cellulose particle solid in the dope as small as possible.

Figure 4.7.18 exemplifies the effect of dissolving conditions, including the rotating rate R of mixer (a) ($T = -2\,°C$, $t = 4$ min), dissolving time t (b) ($R = 12,000$ rpm, $T = -2\,°C$) and temperature T (c) ($R = 12,000$, $t = 4$ min). R_c decreases from 314.16 ml ($S_a = 60.02\%$) at 1000 rpm to 33.29 ml ($S_a = 95.76\%$) at 12,000 rpm, and

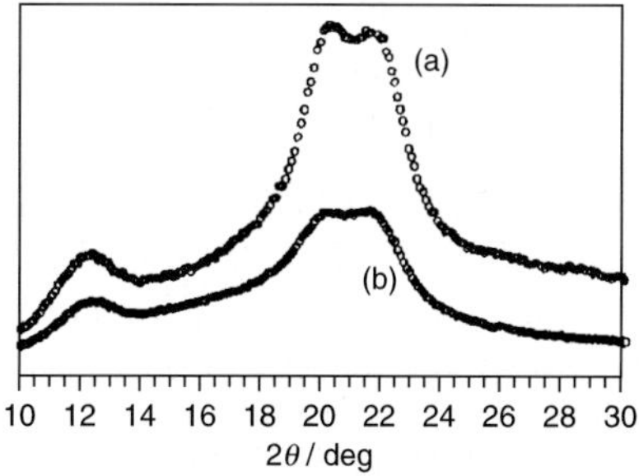

Figure 4.7.16   X-ray diffraction diagram of insoluble cellulose residue (a) and viscose rayon (b).[18]

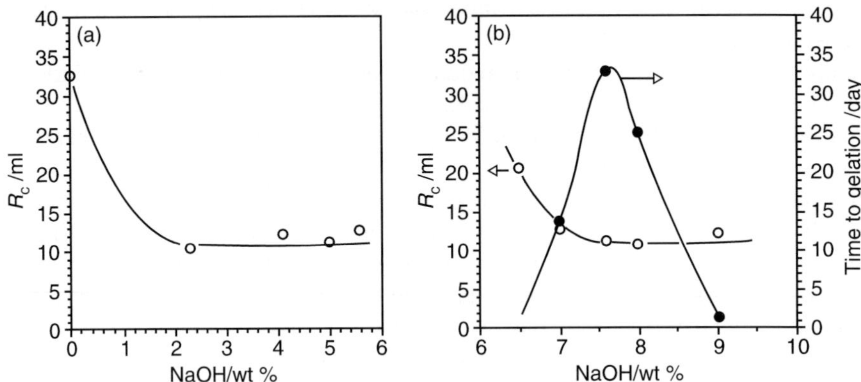

Figure 4.7.17   Effect of sodium hydroxide concentration at pretreatment stage on volume of insoluble residue R_c (a) and that of sodium hydroxide concentration of final dope on R_c and time to gelation (b).[1]

within 4 min, R_c attains asymptotic value. At 40 °C, cellulose is almost insoluble ($R_c = 962.98$ ml; $S_a = 9.05\%$). Temperature dependence of R_c is closely correlated with temperature dependence of solvent power, as expressed by the second virial coefficient A_2 of cellulose–aq. alkali solution (see Figure 3.20.4).

Figure 4.7.19 shows the effect of wet pulverization time of an exploded pulp on mean particle size (a), surface area (b), slurry viscosity (c), P_v (d), and R_c (e) of spinning dope ($C_p = 5$ wt%, two-step dissolution, C_a (step 1) $= 5.0$ wt% and C_a (step 2) $= 7.6$ wt%). The mean size decreased to 12 μm after 15 min pulverization and below 10 μm after 25 min (a). The surface area of particles increased approximately 4.6 times of the original area after 20 min (b). This roughly corresponds to reduction of the mean particle size to one-third of the original particles during the same period of pulverization. The viscosity of slurry increased markedly with time (c). The slurry had appearance of emulsion. P_v decreased slightly by pulverization. R_c approaches to the level below 1 ml ($S_a > 99.9\%$). Following Kamide *et al.*'s experiments, Laszkiewicz and Cuculo[44] observed that S_a of natural cellulose into 8.5 wt% NaOH is only 34.56%. This should be compared with the value higher than 99.9%, attained by Yamane et al.[18]

Table 4.7.9 summarizes the effect of the mean diameter and P_v of wet pulverized cellulose particles after explosion on solubility into 7.6 wt% NaOH. From the table, the following empirical relation was established.

$$R_c = 1.76 \times 10^{-12} \times D^{3.44} \times P_v^{3.26} \qquad (4.7.7)$$

where D is the mean particle size.

In summary, (1) alkali soluble, steam exploded pulp is wet pulverized, (2) pulverized cellulose is pretreated with 2–5 wt% solution, (3) the final 7–9 wt% solution is prepared to give spinning dope (7.6% solution is effective to suppress gelation).

Stricter criterion on the number and the size of undissolved gel particles in the unit volume of the spinning solutions, which are allowed to flow smoothly through dye at higher spinning velocity, are adopted. Now, we can prepare the fine dope of cellulose on a commercial scale with $P_v = 300$ and $S_c \geq 99.9\%$ in aq. alkali solution for wet spinning.

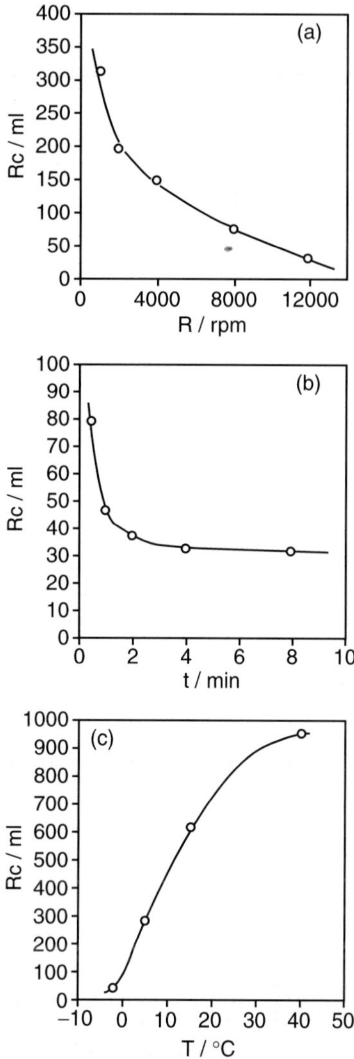

Figure 4.7.18   Effect of dissolving conditions on volume of insoluble cellulose residue R_c: (a) revolution number of agitator R; (b) dissolving time t; (c) dissolving temperature T (cellulose; steam exploded sulfite pulp from spruce).[18]

Cellulose resources suitable for wet spinning[19]

Six kinds of steam-exploded, sulfite pulps, and two kinds of kraft pulps with P_v lower than 600, are dissolved in aq. NaOH solution by the two-step method (Figure 4.7.15). Table 4.7.10 shows the structural parameters of original and steam exploded pulps. The sulfite pulp prepared from broadleaf and coniferous trees showed far lower R_c value than that of the kraft pulp.

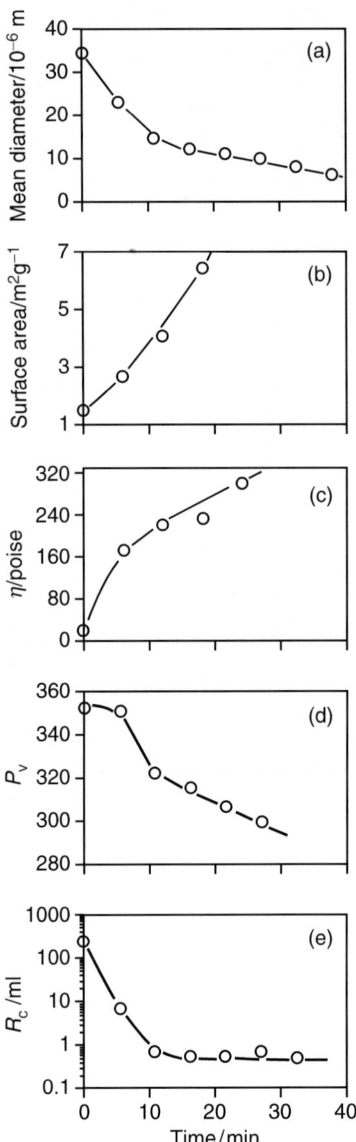

Figure 4.7.19 Effect of wet pulverizing of exploded pulp on mean diameter (a), surface area (b), slurry viscosity (c), P_v (d), and insoluble cellulose residue R_c in the aq. NaOH solution (e).[18]

High solubility of the sulfite pulp was explained in terms of the high degree of breakdown of intramolecular hydrogen bonds, estimated by CP-MAS ^{13}C NMR ($\chi_{am}(C_3) > 0.34$) in cellulose solid. R_c of the sulfite pulp to aq. 7.6 wt% NaOH solution is well described experimentally as

$$R_c = 6.82 \times 10^{-26} \times P_v^{8.9} \times C_p^{5.7} \qquad P_v = 215-320, \; C_p = 3-7\text{wt\%} \qquad (4.7.8)$$

Table 4.7.9

Effect of mean diameter of pulverized cellulose with various P_v on solubility[18]

Sample number	Pulverized condition		Pulverizing time (min)	Mean diameter (μm)	Isoluble residue R_c (ml/20 l dope)	Soluble fraction (%)
	P_v after Pulverized	P_v before pulverized				
1	283	409	40	7	0.14	99.98
2	281	314	16	10	0.47	99.94
3	292	349	26	11	0.71	99.91
4	294	314	8	16	2.50	99.68
5	357	409	20	12	1.90	99.76
6	356	382	10	20	25.43	96.76

The sulfite pulp with $P_v = 200-320$ is confirmed useful for commercial wet spinning to produce new cellulose fiber using aq. NaOH.

Spinning of alkali soluble cellulose–aq. NaOH system using sulfuric acid as coagulant[20]

This subsection intends to demonstrate an application of the results obtained in membrane formation study (see Films coagulated from alkali soluble cellulose I and cellulose II) to wet spinning of fibers. Steam exploded sulfite pulp was wet pulverized and dissolved by the two-step dissolution method into 7.6 wt% aq. NaOH to give a 5.0 wt% dope (viscosity: 10 P). The solubility of cellulose S_a was larger than 99.97%.

Figure 4.7.20 depicts schematic representation of spinning machine. Fibers of 75 d/45 filaments were spun by using horizontal drawing systems (65 wt% H_2SO_4 at $-7\,°C$) with stationary bath.

Table 4.7.10

Structural parameters of original and steam exploded pulp[19]

Pulping process	Wood species	Brand name of pulp	Original pulp				Exploded pulp	
			$\chi_{am}(X)$ (%)	ACS^c (Å)	$\chi_{am}(C_3)$	$\chi_{am}(C_6)$	$\chi_{am}(C_3)$	$\chi_{am}(C_6)$
Sulfite	N^a	ALAPUL-T	27	50	38	34	41–48	45–59
	L^b	SAICCOR-R	21	46	37	35	43–47	44–50
	N	TEMSOLV	18	49	39	37	42–50	46–51
	N	TEMSUPR	18	51	38	36	40–49	45–50
Kraft	L	SULFATATE-H-J	17	56	32	30	42	38
	N	BUKEYE-V5	18	52	35	31	35–40	35–38

[a]N: needle leaf tree.
[b]L: broad leaf tree.
[c]ACS: apparent crystal size.

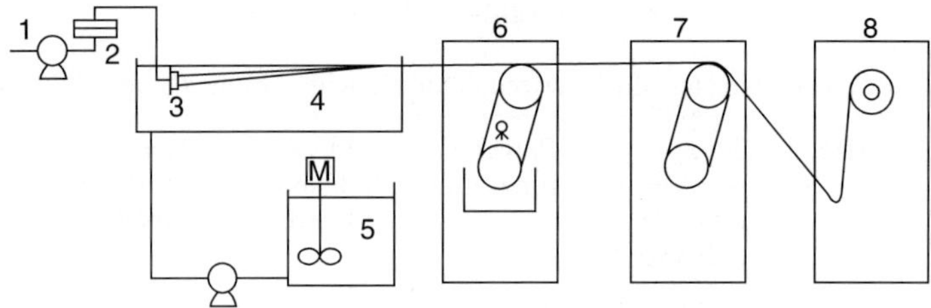

Figure 4.7.20   Schematic representation of spinning machine.[20] (1) Gear pump equipped with jacket for cooling; (2) sintered metal filter with mesh size of 15 μm; (3) spinneret nozzle with 40 holes of 0.07 mm diameter; (4) coagulant bath; (5) reservoir of coagulant; (6) Nelson type roller for washing; (7) Nelson type roller for drying; (8) winder.

Figure 4.7.21 shows the effect of H_2SO_4 concentration C_{sa} and length of coagulation bath on the structural parameters (crystalline orientation factor f_c and optical birefringence Δn) of the new fibers. Yamane *et al.* observed that the stress against volume contraction in fiber direction (i.e. force emerged on the running filament due to volume contraction by desolvation) F_{cont} increased remarkably with C_{sa} in the range

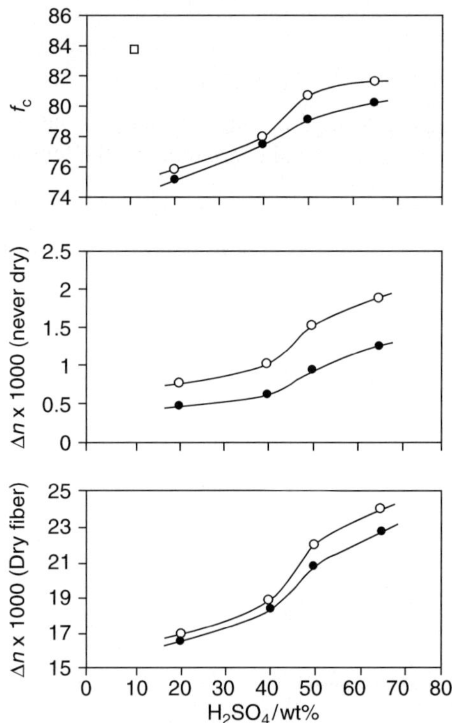

Figure 4.7.21   Degree of the crystal orientation determined by X-ray method f_c and birefringence Δn as a function of sulfuric acid concentration: (○), length of coagulation path $L = 0.1$ m; (●), L $= 0.05$ m; (□), data of commercially available viscose rayon.[19]

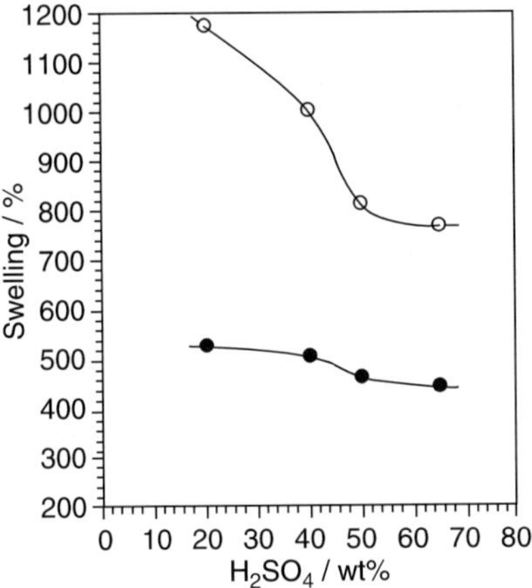

Figure 4.7.22   Effect of sulfuric acid concentration on the swelling of never-dry fibers:
(O) volumetric method; (●) gravimetric method; length of coagulation path $L = 0.1$ m.[19]

$C_{sa} > 40$–50 wt%, in which significant volume contraction occurred concurrently.[17]
The results in Figure 4.7.21 correspond well with the above findings.

Figures 4.7.22 and 4.7.23 show the effects of C_{sa} of sulfuric acid on the swelling
property of never-dry filament and on the shrinkage of filament running at drying stage,
respectively. The dehydration of sulfuric acid gives rise to high shrinking stress on
coagulated gel yarn, leading to high orientation and packing density.

Figure 4.7.24 demonstrates the tensile strength and the tensile elongation of the fibers
spun as a function of the length of coagulation bath L (i.e. coagulation time).

In summary, cellulose filament with tensile strength at dry state higher than that of
regular viscose rayon was successfully obtained using 50–60 wt% sulfuric acid.

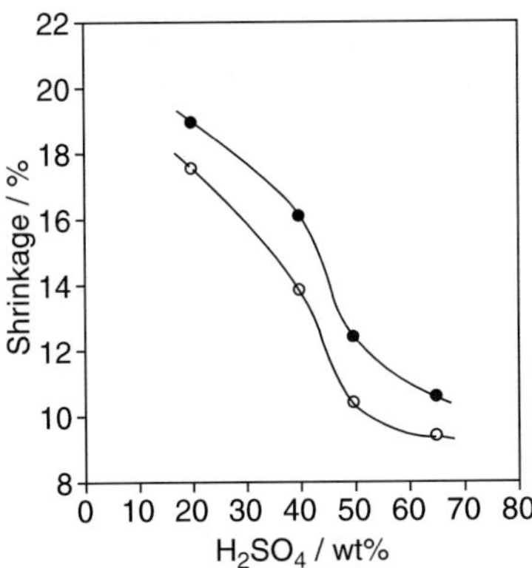

Figure 4.7.23   Effect of sulfuric acid concentration on the degree of shrinkage at drying process:
(O) length of coagulation path $L = 0.1$ m; (●) 0.05 m.[19]

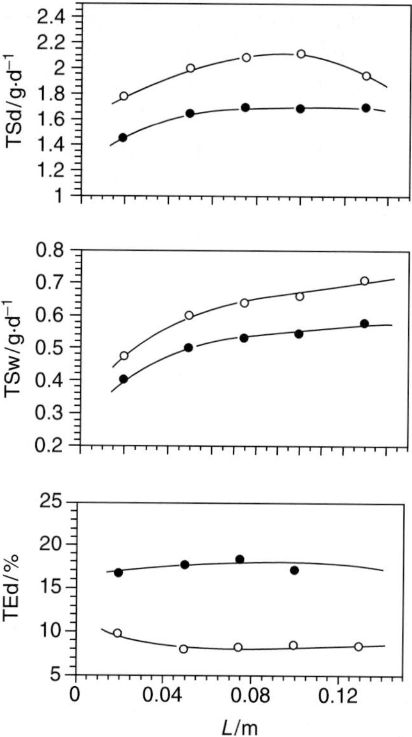

Figure 4.7.24   Effect of length of coagulation path L on the tensile strength in dry and wet state (TSd, TSw) and the tensile elongation in dry state TEd: (O) drying under constant yarn length condition; (●) drying under tension free condition.[19]

4.7.5   Stage 4: bench plant wet spinning: net process[21,22]

Wet spinning[21]

Figure 4.7.25 illustrates the bench plant of continuous multifilament spinning machine constructed by Asahi. Net process was originally developed by Asahi between the 1970s

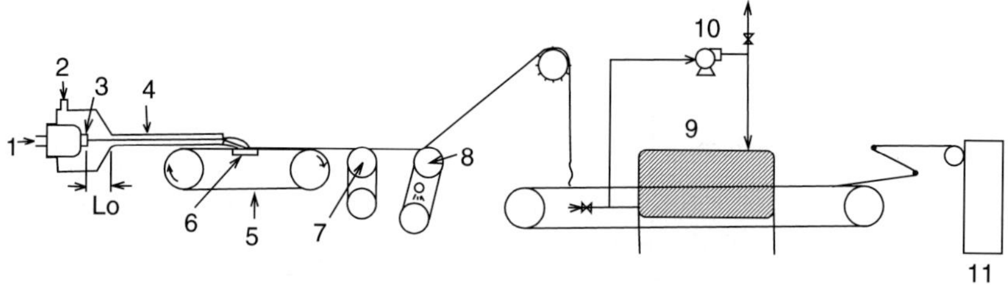

Figure 4.7.25   Schema of continuous multifilament spinning machine.[21] 1, dope; 2, coagulant, (20 wt% H_2SO_4, -7 °C); 3, spinning nozzle; 4, coagulation funnel; 5, wire net; 6, suction box; 7, pick up rollers (Nelson type); 8, washing and drawing rollers (Nelson type); 9, dryer; 10, blower; 11, take up device.

and the 1990s in cuprammonium rayon process.[45,46] Net process enables us to keep the tension on coagulated gel filament at an extremely low level. The process consists of a cylindrical spinning funnel invented by Asahi for viscose rayon and a net as supporter of coagulated gel (filament), which runs at the same speed of winding roller. In this process, less hydraulic resistance arises from coagulant and high speed spinning is thus attainable.

As prepared in Section 4.7.4, dope was subjected to the bench scale net process spinning.

From systematic and detailed analysis of the wet spinning process, it became clear that in order to spin better quality fibers at high speed spinning, the following factors contributed significantly:

(1) Lowering draft ratio by increasing extrusion velocity v_1 at spinneret (use of a narrower hole dye).
(2) Reducing hydraulic resistance at early stage of coagulation by shortening the distance from the spinnerets to the inlet of funnel.
(3) Reducing hydraulic resistance in the coagulation funnel by minimizing velocity difference between the coagulant gel and the filament (i.e. surface of take up role).
(4) Drawing never-dried filaments by 1.2–1.3 after completion of coagulation.

The spinning apparatus was designed and constructed to meet the above requirement (Figure 4.7.25).

Table 4.7.11 illustrates the effect of spinning velocity on mechanical properties of the fibers. Here, the length of coagulation funnel and the spinning tension are designed to keep constant.

The spinning conditions employed for long-run spinning are the same as those employed for preliminary experiments (as shown in Table 4.7.12), except for the following items: P_v (304), length of coagulation funnel, and spinning velocity. Mechanical properties are almost unchanged in the range of spinning velocity of 100–160 m min^{-1}, provided that both the coagulation time and the spinning tension are kept constant (Table 4.7.11). Here, the spinning velocity is defined as the winding velocity (surface velocity) of take up role. It is shown that using the net process high tensile strength/high elongation filaments can be produced at a commercial level. The spinning velocity is a good measure representing total level of the production technology, which cannot be attained by improving the spinning process alone, but by creating numerous elements other than a winding machine.

Table 4.7.11

Mechanical properties of fibers spun at various spinning velocity in preliminary experiments[21]

Spinning velocity (m min^{-1})	Length of coagulation funnel (mm)	Tensile strength (g d^{-1})	Tensile elongation (%)
100	300	1.81	21.0
120	360	1.80	19.8
140	420	1.89	20.4
160	480	1.91	18.9

Table 4.7.12

The long-run spinning conditions for net process[22]

Dope	
Cellulose	$P_v = 320$
Cellulose concentration	$C_p = 5$ wt%
Alkaki concentration	$C_a = 7.6\%$
Nozzle	0.05 mm diameter × 33 holes
Coagulation	5.8 mm diameter × 33 min
Velocity difference[a]	$\Delta v = 20$ m min^{-1}
Draft ratio	1.25
Spinning velocity	100 m min^{-1}
Fiber (denier/filament number)	75 d/33 f

[a]Difference between surface velocity (v_1) of role and mean velocity (v_L) of coagulant flowing inside coagulation funnel ($\Delta v = v_1 - v_L$).

Structural and mechanical properties of fibers and fabrics[22]

Table 4.7.12 shows the spinning conditions utilized for long run continuous multi-filaments wet spinning.

Crystalline and intramolecular hydrogen bond natures of new cellulose fibers

Table 4.7.13 summarizes the structural parameters of new cellulose fiber (SK-14) and some commercial fibers. $\chi_c(X)$ of the new cellulose fiber is similar in magnitude to that of organic spun rayon. However, it is larger than those of polynosic, cuprammonium rayon, and viscose rayons. A similar tendency is also seen for $1 - \chi_{am}(C_3)$. Thus, the new cellulose fiber (SK-14) is structurally akin to organic spun rayon, rather than cuprammonium rayon and viscose rayon. This feature might come from the dissolving state of cellulose in these solvents. In former two solvents (aq. NaOH and aq. N-methyl morpholine N-oxide), cellulose is molecularly dispersed without forming derivatives or complex,[28,47] but in the later two solvents, cellulose is dissolved in the state of complex or a derivative. Presumably, the dissolved state of cellulose strongly influences the coagulated state of cellulose. For example, Yamane *et al.* have found that the morphology of regenerated cellulose membranes recovered from its cuprammonium cellulose solution is predominantly determined by the difference in complex forms caused by various coagulants.[48] All fibers in Table 4.7.13 are obtained by aq. coagulation systems. As a crystallization mechanism of regenerated cellulose, Takahashi proposed that ($1\bar{1}0$) plane with hydrophobic nature is the plane formed first, and then this plane is piled up, forming the crystalline lattice.[49] Since hydroxyl groups in glucopyranose units of cellulose take equatorial positions, the axial direction is in hydrophobic nature. In the case of aq. coagulation system, the surface energy of this hydrophobic plane might be diminished so as to facilitate the piling up of glucopyranose rings, favoring the sheet structure of ($1\bar{1}0$) planes. The firm sheet structure tends to be formed when coagulants, which can have strong interactions with hydroxyl groups in cellulose such as water, are utilized and also when cellulose in solution does not exist as derivative or complex.

Table 4.7.13

Structural parameter[22]

Sample yarn	Starting material	P_v	Crystallinity $\chi_c(X)$ (%)	Orientation f_c (%)	$1 - \chi_{am}(C_3)$ (%)
New cellulosic filament yarn (75 d/33 f)	Pulp	304	45.5	75	42.2
Viscose conventional[a]	Pulp	268	29.2	86	35.0
Viscose HSS[b]	Pulp	268	22.1	84	30.9
Viscose LS[c]	Pulp	268	21.5	86	28.9
Cupro conventional[d]	Cotton linter	822	43.4	90	41.5
Cupro HSS[e]	Cotton linter	822	40.9	89	
Cupro LS[f]	Cotton linter	822	38.2	91	36.7
Organic spun rayon[g] staple	Pulp	594	45.9	91	41.0
Polynosic Fujibou	Pulp	506	43.8	90	40.8

[a]Viscose rayon filament, conventional spinning, 75 d/26 filaments.
[b]Viscose rayon filaments, high speed spinning, 75 d/33 filaments.
[c]Viscose rayon filament, low speed spinning, 75 d/33 filaments.
[d]Cuprammonium rayon filament, conventional spinning, 75 d/45 filaments.
[e]Cuprammonium rayon filament, high speed spinning, 75 d/54 filaments.
[f]Cuprammonium rayon filament, low speed spinning, 75 d/54 filaments.
[g]1.5 d.

This might be one of the reasons why new cellulose fiber (as well as organic spun rayon) has large crystallinity. In fact, X-ray crystallinity $\chi_c(X)$ of regenerated celluloses obtained by organic solvents such as acetone and toluene has proved to be very low.[50]

The most considerable structural difference between the new cellulose fiber and others, including organic spun rayon, is that the crystalline orientation f_c, is far lower than those for organic spun rayon, cuprammonium rayon, and polynosics. This feature is clearly reasoned by the difference in draft or stretching imposed during each fiber spinning.

The draft and stretching on new cellulose filament is less than 1 and 1.25, respectively, which are equivalent to or smaller than those for viscose rayon. The draft on organic spun rayon and cuprammonium rayon is larger than 10 and 100, respectively.[51,52] For organic spun rayon, liquid crystalline nature or flow birefringent nature is also influenced. Even for the present cellulose/aq. NaOH solution system, if the coagulant with stronger the dehydration power such as 65 wt% aq. sulfuric acid is used, f_c might increase by the dehydration orientation principle.[20]

New cellulose fiber is characterized as having higher intramolecular hydrogen bonding with lower intermolecular hydrogen bonding in its amorphous region. For such regenerated fiber, it is possible to assume the existence of plane lattice or sheet-like structures in amorphous region, proposed by Hayashi *et al.*[53] and Hermans.[54] For example, (1$\bar{1}$0) sheet itself is formed with hydrophobic interaction by piling up of glucopyranose rings, and then, if such a structure is assumed in amorphous structure, then the amorphous region might be composed of highly developed intramolecular hydrogen bonding and immature intermolecular hydrogen bonding. The existence of such sheet-like structure might be affirmed by the structure forming mechanism proposed by

Takahashi[49] and by the results of Yamane et al. that the structural change induced by water takes place in the perpendicular direction to ($1\bar{1}0$) plane.[50]

The cross-section and longitudinal section of TEM micrographs show that the new cellulose filament has no morphological distribution in the fiber radius direction and is composed of short fibrils. Three phase morphology (skin phase in outermost layer, network with no considerable fibril structure in the medium layer, and the well-developed fibrillar structure in the most inner layer) was observed for cuprammonium rayon. For cuprammonium rayon, the enzyme etching works most effectively on the inner part remaining in the skin part, indicating the rather low crystallinity of the inner part. The skin core structure of viscose rayon is characterized by numerous voids with different sizes, but without fibrillar structure. Organic spun rayon has a similar, even structure and morphology to that of new cellulose fiber, but with well-developed fibrils. The above experimental facts that new cellulose and organic spun rayon have one phase morphology and viscose and that cuprammonium rayons have multiphase morphology might be determined by whether chemical structural change takes place or not during the coagulation procedure.

Mechanical and topical properties of new cellulose fiber and its woven fabrics[22]

Fabrics were woven, dyed, and finished using the new fiber (SK-14) under the conditions summarized in Table 4.7.14. Table 4.7.15 illustrates some mechanical properties of new cellulose fiber together with other commercially regenerated cellulose fibers. The properties of the new cellulose fiber treated with steam are also included. The steam

Table 4.7.14

Preparation of fabric from new cellulose filament[22]

Fabric	Weave pattern	Plain weave
	Fabric count	
	Warp	132 yd in.$^{-1}$
	Weft	83 yd in.$^{-1}$
	Air jet loom	Z-A 103, Zudakoma Industry Co., Ltd.
	Revolving speed	600 rpm
Dyeing condition		Open width dyeing, 24 °C × 20 h
	Pressure of squeeze rolls	5 kg cm^{-2}
	Dyes	Reaction dyes
	Dyeing assistants	NaOH, Na$_2$CO$_3$, urea
Condition of resin finish		Open width padding followed by dry curing
	Pressure of squeeze rolls	5 kg cm^{-2}
	Dry cure	100 °C × 1 min to 160 °C × 3 min
	Resin	'Yunika resin' GS-15(di-methylol di-hydroxy ethylene urea)
	Catalyst	'Yunika catakyst' MC-109 (MgCl$_2$ + organic acid)
	Softening agents	Methylol amide

Table 4.7.15

Mechanical properties of cellulose fibers[22]

Sample yarn	Conditioned tenacity (g d^{-1})	Conditioned elongation (%)	Wet tenacity (g d^{-1})	Wet elongation (%)	Knot tenacity (g d^{-1})	Knot elongation (%)	Conditioned modulous (g d^{-1})	Wet modulous (g d^{-1})
New cellulosic filament yarn (75 d/33 f)	1.8	21	0.5	23	1.3	17	110	2.8
New cellulosic filament yarn steam treated (75 d/33 f)	1.9	15	0.7	16	1.3	12	146	7.1
Viscose conventional	1.73	19.1	0.83	30.3	1.49	14.4	76	3
Viscose HSS	1.83	17.3	0.9	23.1	1.75	15.0	85	
Viscose LS	1.92	17.0	0.82	34.4	1.62	13.5	96	4
Cupro conventional								
Cupro HSS	2.85	13.4	1.68	16.9	2.2	9.8	105	
Cupro LS	2.73	11.6	1.83	31.3	2.28	6.3	136	20
Organic spun rayon staple	4.60	13.0	4.00	17.5			410	31
Polynosic Fujibou	3.80	13.0	2.6					14

Sample yarn	Shrinkage in boiling water (%)	Whiteness	Yellowness	Luster	Swelling capacity (%)	Fibrillation	Dryestuff absorption (%)
New cellulosic filament yarn (75 d/33 f)	2.8	87.1	8.2	14.0	80	95	78.1
New cellulosic filament yarn steam treated (75 d/33 f)	1.6	86.0	11.3	14.6	50	95	55.8
Viscose conventional	2.5		6.7	44.5	102	100	53.0
Viscose HSS	1.6		13.5	12.4	106	100	
Viscose LS	6.3		14.5	50.0	120	100	
Cupro conventional							
Cupro HSS	0.4		14.1		88	85	
Cupro LS	3.6		16.8		80	80	
Organic spun rayon staple	0.2				65	50	
Polynosic Fujibou					65	60	

treatment on new cellulose fiber brought about the lowering in elongation, swelling capacity, dyestuff absorption, and the increasing in modulus, all of which are related to molecular ordering. New cellulose fiber shows similar tensile properties to viscose rayon, and although less lustrous, it is superior to that of viscose rayon. Dimensional stability of new cellulose is higher than that of viscose rayon, judged by swelling capacity and this tendency becomes definite for the steam-treated new cellulose, of which swelling capacity is the lowest among all regenerated cellulose fibers. The resistance to fibrillation, which is a key property for judging the applicable field of end usages, is also superior to other fibers, except for viscose rayon.

Table 4.7.16 shows typical properties of woven fabrics made from new cellulose fibers with or without steam treatment, together with those from viscose and cuprammonium rayons. The table contains the results with and without resin finishing. The fabrics woven from the steam treated new cellulose yarn without resin finishing gives some softness, compared with other commercial fabrics, and this tendency does not alter after resin finishing. Abrasion resistance of the fabrics of the new steam-treated cellulose is highest among fabrics tested here, and is the most hard to wrinkle. Colorfastness of the new cellulose fabrics is in the same order as the others.

Table 4.7.16

Properties of plain weave fabrics[22]

Sample	Resin finish	Color fastness to rubbing in wet state	Color fastness to light	Surface abrasion resistance	Bending resistance (warp) (cm)	Wrinkle recovery (%)
New cellulosic filament yarn (75 d/33 f)	O	2–3	5	2900	3.6	45
New-cellulosic filament yarn, steam treated (75 d/33 f)	O	2–3	4	3700	3.8	59
Viscose rayon (75 d/33 f)	O	2–3	5	3300	4.4	47
Cuprammonium rayon (75 d/54 f)	O	3	5	2100	4.4	56
New-cellulosic filament yarn (75 d/33 f)	–	1–2	5	2100	3.7	
New-cellulosic filament yarn, steam treated (75 d/33 f)	–	1–2	4	5400	3.7	
Viscose rayon (75 d/33 f)	–	1–2	5	6000	4.2	
Cuprammonium rayon (75 d/54 f)	–	1–2	5	2600	4.3	

O: resin finish was applied.

In summary, cellulose multifilaments with mechanical properties, equally matched to those of commercially available viscose filaments, can be produced by net process at the spinning velocity almost equal to that of viscose (100 m min^{-1}). The X-ray crystallinity index $\chi_c(X)$ of the new cellulose filament was far higher than those of other commercial rayons and was slightly lower than that of the organic spun rayon. The new cellulose filament showed the lowest degree of crystal orientation among these three kinds of cellulose fibers. In the new filament, the degree of the intramolecular hydrogen bonding $(1 - \chi_{am}(C_3))$ developed. The tensile strength and elongation of the new filament were comparable to those of regular viscose rayon.

The new filament showed a lower swelling ratio and lower fibrillation nature in water. Woven fabrics made from the filament gave some softness and high abrasion, compared with other commercial ones and were hard to wrinkle. Thus, new cellulose fiber (SK-14) is an environmentally friendly innovation, offering outstanding potential as an alternative to the present commercially regenerated cellulose fiber, especially viscose and cuprammonium rayons.

REFERENCES

1. K Kamide, K Okajima, T Matsui and K Kowsaka, *Polym. J.*, 1984, **16**, 797.
2. For example, International Conference on Proceedings of the International Workshop on Steam Explosion Techniques: Fundamentals and Industrial Applications, Milano, Italy, 1988, October 20–21.
3. T Yamashiki, T Matsui, M Saitoh, K Okajima, K Kamide and T Sawada, *Br. Polym. J.*, 1990, **22**, 73.
4. T Yamashiki, T Matsui, M Saitoh, K Okajima, K Kamide and T Sawada, *Br. Polym. J.*, 1990, **22**, 121.
5. T Yamashiki, K Kamide, K Okajima, K Kowsaka, T Matsui and H Fukase, *Polym. J.*, 1988, **20**, 447.
6. K Kamide, K Yasuda, T Matsui, K Okajima and T Yamashiki, *Cellul. Chem. Technol.*, 1990, **24**, 23.
7. K Kamide and M Saito, *Polym. J.*, 1986, **18**, 569.
8. H Williams, US Patent 3,236,669, 1966.
9. C Graenacher, R Sallman, USP 2,179,181, 1939; D Johnson, British Patent 1,144,048, 1969.
10. S Hudson and J Cuculo, *J. Polym. Sci. Polym. Chem. Ed.*, 1980, **18**, 3469.
11. D Johnson, M Nicolson and F Haigh, *Appl. Polym. Symp. Proc. Cellul. Conf.*, 1975, **8**, 28.
12. K Kamide, K Okajima, T Matsui and S Manabe, *Polym. J.*, 1980, **12**, 521.
13. K Kamide, T Matsui, K Okajima and S Manabe, *J. Soc. Text. Mach. Jpn*, 1980, **33**, T111.
14. A Turbak, A El-Kafrawy, F Snyder and A Auerbach, US Patent 4,302,252, 1981.
15. T Yamashiki, M Saito, K Yasuda, K Okajima and K Kamide, *Cellul. Chem. Technol.*, 1990, **24**, 237.
16. T Yamashiki, T Matsui, K Kowsaka, M Saitoh, K Okajima and K Kamide, *J. Appl. Polym. Sci.*, 1992, **44**, 691.
17. T Matsui, T Sano, C Yamane, K Kamide and K Okajima, *Polym. J.*, 1995, **27**, 797.
18. C Yamane, M Saito and K Okajima, *Sen-i Gakkaishi*, 1996, **52**, 310.
19. C Yamane, M Saito and K Okajima, *Sen-i Gakkaishi*, 1996, **52**, 318.
20. C Yamane, M Saito and K Okajima, *Sen-i Gakkaishi*, 1996, **52**, 369.
21. C Yamane, M Saito and K Okajima, *Sen-i Gakkaishi*, 1996, **52**, 378.
22. C Yamane, M Mori, M Saito and K Okajima, *Polym. J.*, 1996, **28**, 1039.
23. L Segal, J Greely, A Martin and C Conrad, *Text. Res. J.*, 1959, **29**, 786.
24. P Scherrer, *Göttinger Nachr.*, 1918, **2**, 98.

25. K Kamide, K Okajima and K Kowsaka, *Polym. J.*, 1985, **17**, 701.
26. E Otto and M Spurlin, in *Cellulose and Cellulose Derivatives*, **Vol. II**, 2nd Edn., (eds E Otto and M Spurlin), Interscience, New York, 1954.
27. K Kowsaka, K Kamide and K Okajima, Abstract International Symposium on New Functionalization, Development in Cellulosics & Wood, Cellucon 88, 1988, p. 18.
28. K Kamide, K Okajima, T Matsui and K Kowsaka, *Polym. J.*, 1984, **16**, 857.
29. F Horii, A Hirai and R Kitamaru, *Ann. Rep. Res. Inst. Chem. Fibers, Jpn*, 1985, **42**, 41.
30. K Kamide, *Thermodynamics of Polymer Solutions. Phase Equilibria and Critical Phenomena*, Elsevier, Amsterdam, 1990, Chapter 6.
31. K Kamide, K Yasuda, M Saito and K Okajima, *Polym. Prepr. Jpn*, 1989, **38**, 1126.
32. K Kamide, M Saito and K Yasuda, in *Viscoelasticity of Biomaterials* (eds WG Glasser and H Hatakeyama), ACS Symposium Series, No. 489, The American Chemical Society, Washington, DC, 1992, Chapter 12.
33. K Kamide and S Manabe, Role of micro-phase separation phenomena. In *Formation of Porous Polymeric Membrane*, ACS Symposium Series, No. 269, The American Chemical Society, Washington, DC, 1985, p. 197.
34. K Kamide, H Iijima and S Matsuda, *Polym. J.*, 1993, **25**, 1113.
35. K Kamide, H Iijima and H Shirataki, *Polym. J.*, 1994, **26**, 21.
36. H Iijima, S Matsuda and K Kamide, *Polym. J.*, 1994, **26**, 439.
37. K Kamide, H Iijima and A Kataoka, *Polym. J.*, 1994, **26**, 623.
38. H Iijima, A Kataoka and K Kamide, *Polym. J.*, 1995, **27**, 1033.
39. H Iijima, K Sogawa and K Kamide, *Polym. J.*, 1996, **28**, 808.
40. K Kamide, S Manabe, T Matsui, T Sakamoto and S Kajita, *Kobunshi Ronbunshu*, 1977, **34**, 205.
41. TF Young, LF Maranville and HM Smith, In *Structure of Electrolytic Solutions*. (ed. WJ Hammer), Wiley, New York, 1959, p. 48.
42. F Horii, A Hirai, R Kitamaru and I Sakurada, *Cellul. Chem. Technol.*, 1985, **19**, 513.
43. F Horii, A Hirai and R Kitamaru, in *The Structures of Cellulose Characterization of the Solid States*. (ed. RH Atalla), ACS Symposium Series, No. 340, The American Chemical Society, Washington, DC, 1987, p. 119, Chapter 6.
44. B Laszkiewicz and A Cuculo, *J. Appl. Polym. Sci.*, 1993, **50**, 27.
45. K Kamide and K Nishiyama, In *Regenerated Cellulose Fibers*. (ed. C Woodings), 5 Cuprammonium Processes, Woodhead Publishers, Cambridge, 2001, p. 88.
46. K Kamide, *J. Ind. Econ. Nara Sangyo Univ.*, 2001, **154**, 81.
47. D Gagnaire, *J. Polym. Sci. Polym. Chem. Ed.*, 1980, **18**, 13.
48. M Inamoto, I Miyamoto, M Iwata, K Okajima, unpublished results.
49. T Takahashi, *Sen-i Gakkaishi*, 1969, **25**, 80.
50. T Hongoh, C Yamane, M Saitoh and K Okajima, *Polym. J.*, 1996, **28**, 1077.
51. D Loubinoux and S Chaunis, *Text. Res. J.*, 1987, **57**, 61.
52. E Thiele, Deutshe Patent 154507, 157157, 1901.
53. J Hayashi, J Masuda and Y Watanabe, *Nippon Kagaku Kaishi*, 1974, **5**, 948.
54. PH Hermans, *J. Polym. Sci.*, 1949, **4**, 145.

– 5 –

Solubilization and Structural Factors Governing Solubility and the Dissolved State

5.1 SOLUBILIZATION OF CELLULOSE BY THE STEAM EXPLOSION METHOD

5.1.1 The influence of cellulose resources[1]

Because the density of hydroxyl groups in cellulose molecules is quite high, it would not be surprising if cellulose was completely water soluble. However, the existence of intra- and intermolecular hydrogen bonds in the cellulose solid interrupt the dissolution of the cellulose solid into a water solvent. For many years it has been well known to cellulose chemists that natural and regenerated celluloses swell, at least partially, in aqueous alkali solutions.[2-7] Recently, Kamide *et al.*[8] have demonstrated that cellulose samples regenerated from cuprammonium solution under specialized conditions can completely dissolve in 8–10 wt% aqueous sodium hydroxide (NaOH) solution at 4 °C (see, Section 4.1). They successfully correlated the solubility of cellulose in aqueous alkali solution with the degree of breakdown of an intramolecular hydrogen bond at $O_3 \cdots O_5'$ of cellulose, as determined by solid state cross-polarization/magic angle sample spinning (CP/MAS) ^{13}C NMR.[8] If such an alkali-soluble cellulose could be produced directly and economically from natural cellulose by a simple, physical treatment, it would be of paramount importance to the cellulose chemical industry.

The steam explosion method involves the abrupt discharge of the materials penetrated with high-temperature water vapor into open air and can potentially destroy the inner and outer structure of any material and inspire desired properties into materials. This method has been widely utilized in biomass conversion technology for cattle feed production,[9] a new pulping application,[10] and gasification of biomass.[11]

When cellulose is treated by the steam explosion method, even the intramolecular hydrogen bonds may break down. This has been shown to be the key to producing alkali soluble cellulose.[12] Since naturally occurring celluloses have differing intrinsic supermolecular structures, due to the differences in their cell and basic tissue systems and the associated structure of surface skin and fibrillar bundle, steam explosion treatment on different species of cellulose is expected to bring about different influences on the supermolecular structures.

In this section, we review the steam explosion method as applied to almost pure cellulose, such as wood pulp and purified cotton linter, and study the influence of cellulose resources on the changes in morphology, degree of polymerization DP, solubility S_a, and solid structure during steam explosion.

Cellulose sample

The starting samples used included a soft wood pulp (mainly white spruce, Alaska pulp (USA), manufactured by Alaska Pulp Co., USA: α-cellulose content, 90.1 wt%; viscosity-average degree of polymerisation $P_v = 1060$), a hard wood pulp (mainly eucalyptus) (supplied by Rio Grande Companhia de Celulose Do Sol-Riocell (Brasil): α-cellulose content, 91.7 wt%; $P_v = 994$), and a purified cotton linter (*Gossypium hirsutum*, imported from USA and purified by Asahi Chemical Ind. Co. Ltd: α-cellulose content 95.7 wt%, $P_v = 1279$). The process of purification of the raw cotton linter involved alkali cooking (3.5 wt% aq. NaOH, heated to 172 °C for 90 min), bleaching with chlorine (270 ppm, at 25 °C for 45 min), and removal of lignin with aqueous alkaline solution, followed by washing and drying. Soft wood and hard wood pulps were cut into small pieces ($10 \times 5 \times 1.5$ mm^3) before being subjected to steam explosion.

Steam explosion apparatus and explosion procedure

The steam explosion apparatus is illustrated in Figure 5.1.1. The apparatus was originally designed by Sawada[13] and constructed by Japan Chemical Engineering & Machinery Co. Ltd (Japan).

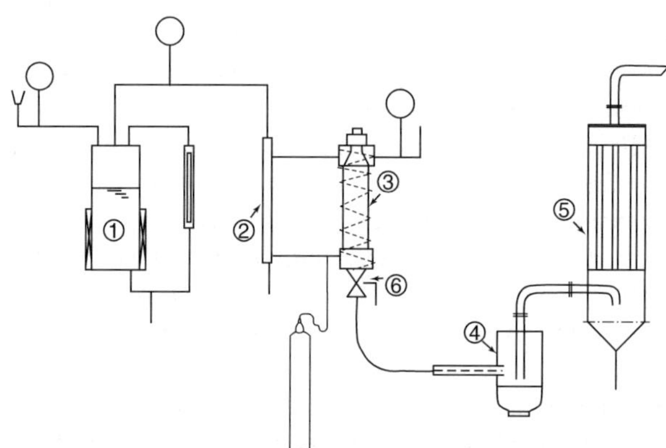

Figure 5.1.1   Steam explosion apparatus.[1] ①, Heater (volume 14 l; pressure resistance 5.9 MPa; temperature resistance 275 °C); ②, steam header (volume 0.5 l); ③, main digester vessel (volume 1.2 l; pressure resistence 5.4 MPa; temperature resistance 270 °C); ④, cyclon receiver (volume 6 l); ⑤, condenser with a silencer (area of heat conductance 0.7 m^2); ⑥, ball valve.

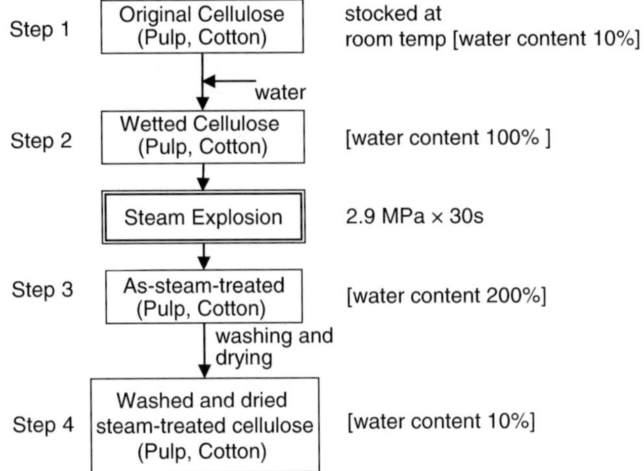

Figure 5.1.2   Schematic representation of the steam explosion process.

Prior to the steam explosion experiments, the weight ratio of water:dry cellulose (stocked at room temperature: Step 1; see Figure 5.1.2) was adjusted to 1:1 as follows: one part of cellulose was immersed in 20 parts of distilled water at 25 °C for 15 h, the excess water was then removed by centrifuging at 2000 rpm (170 g) for 5 min using a KM-18 centrifuge manufactured by Matsumoto Machinery Co. Ltd, Japan. The steam explosion experiments involved: 200 g (dry base) of cellulose sample (prepared as stated in Step 2; see Figure 5.1.2) put into a digester vessel, which was then closed and heated to a predetermined temperature. A saturated water vapor, controlled to a given steam pressure P (1.0–4.9 MPa, corresponding to a saturated steam temperature T in a vessel of 456–525 K), was then introduced into the vessel for a given treatment time t (15–300 s). After closing the vapor input valve, the ball valve connected to the discharging nozzle was quickly opened and the treated cellulose was discharged, with the aid of the pressure difference, into the cyclon receiver. The treated cellulose (Step 3; see Figure 5.1.2) was then gathered, packed into a sealed polyethylene bag, and allowed to stand at room temperature. Part of the treated sample was subjected to ^{13}C NMR measurement, while the remainder was thoroughly washed with water that had been displaced by acetone, air dried (Step 4; see Figure 5.1.2) and subjected to CP/MAS ^{13}C NMR, X-ray diffraction and scanning electron microscopic (SEM) analyses, P_v, and solubility S_a measurements. The procedure for the steam explosion is illustrated in Figure 5.1.2.

(a) *Viscosity-average DP* of cellulose P_v was determined according to the procedure in Section 4.1.

(b) *Solubility* (S_a) of cellulose was evaluated by the method described in Section 5.2.1, but, in this section the insoluble part was excluded by centrifuging the mixture at 7500 rpm for 60 min. The insoluble, swollen cellulose with aqueous alkali was neutralized with 1.3 wt% aqueous hydrochloric acid to completely precipitate the cellulose. The insoluble cellulose was then washed then with water, dried in air and *in vacuo*, and then weighed. The solubility S_a was calculated as a percentage.

(c) *SEM* observations were made according to Section 5.2.1.[14]
(d) *X-ray diffractometry* was carried out according to the method described in Section 5.2.1. The amorphous fraction $\chi_{am}(X)$ of the sample was calculated from the crystallinity $\chi_c(X)$, which was estimated by Segal's method.[15]
(e) *Apparent crystal size (ACS)* was estimated through use of Scherrer's equation.[16]
(f) *Solid-state CP/MAS ^{13}CNMR measurement.*

From ^{13}C NMR, the range of degrees of breakdown of the intramolecular hydrogen bonds at a given C_k ($k = 2$, 3 and 6) position among three positions of the hydroxyl groups in the glucopyranose unit, $\chi_{am}(C_k)$, was estimated by eqs. (5.1.1)–(5.1.3) (see eqs. (4.4.2)–(4.4.4)). Note that $\chi_{am}(C_3)$ and $\chi_{am}(C_6)$ can be independently estimated. $\chi_{am}(C_2)$ may have some uncertainty owing to the lack of precise peak assignment in C_2, C_3 and C_6 carbon regions.

$$\chi_{am}(C_3) = 100 I_h(C_4)/\{I_h(C_4) + I_l(C_4)\} \tag{5.1.1}$$

$$\chi_{am}(C_6) = 100 I_h(C_6)/\{I_h(C_6) + I_l(C_6)\} \tag{5.1.2}$$

$$\chi_{am}(C_2) = 100[3 I_l(C_{2,3,5})/\{I_h(C_{2,3,5}) + I_l(C_{2,3,5})\} - 1] - \chi_{am}(C_3) \tag{5.1.3}$$

where I_h and I_l are the fractions of higher and lower magnetic peaks in the carbon peak region given in parentheses.

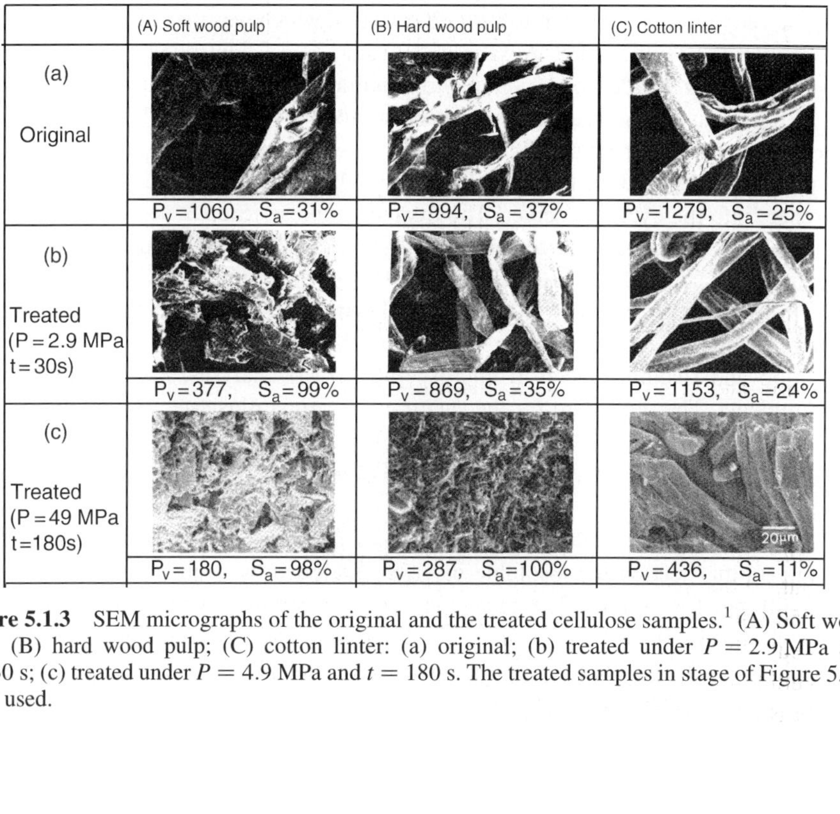

	(A) Soft wood pulp	(B) Hard wood pulp	(C) Cotton linter
(a) Original			
	$P_v = 1060$, $S_a = 31\%$	$P_v = 994$, $S_a = 37\%$	$P_v = 1279$, $S_a = 25\%$
(b) Treated ($P = 2.9$ MPa $t = 30$s)			
	$P_v = 377$, $S_a = 99\%$	$P_v = 869$, $S_a = 35\%$	$P_v = 1153$, $S_a = 24\%$
(c) Treated ($P = 49$ MPa $t = 180$s)			
	$P_v = 180$, $S_a = 98\%$	$P_v = 287$, $S_a = 100\%$	$P_v = 436$, $S_a = 11\%$

Figure 5.1.3   SEM micrographs of the original and the treated cellulose samples.[1] (A) Soft wood pulp; (B) hard wood pulp; (C) cotton linter: (a) original; (b) treated under $P = 2.9$ MPa and $t = 30$ s; (c) treated under $P = 4.9$ MPa and $t = 180$ s. The treated samples in stage of Figure 5.1.2 were used.

Figure 5.1.3 shows SEM photographs of the original and the treated cellulose samples. The apparent fiber thicknesses of the original samples are as follows: soft wood pulp $44-15$ μm; hard wood pulp $16-5$ μm; cotton linter $26-9$ μm. From the magnified SEM photographs (Figure 5.1.4), the surface of the original soft wood pulp seems uneven and there appear to be pores (bordered pit) on the cell wall and the fibrillar bundles are aligned irregularly. There are no obvious pores on the original hard wood pulp or the cotton linter. The fibrillar bundle of the hard wood pulp seems to be regularly aligned, almost parallel to the fiber axis, but that of cotton linter gives a network structure and its fibril density seems higher than that of the hard wood pulp. After the steam explosion at $P = 2.9$ MPa ($T = 508$ K) and $t = 30$ s, a significant morphological change is detected only for soft wood pulp. In this case, there is a shortening of the fibers and partial fibrillation can be observed. Similar morphological changes are found for hard wood pulp only after treatment at $P \geq 4.9$ MPa ($T = 525$ K) and $t \geq 180$ s. However, even under such drastic steam explosion conditions, cotton linter does not exhibit any significant morphological change. P_v and S_a of the samples are shown in Figure 5.1.3 and Table 5.1.1. Under the above explosion conditions ($P = 2.9-4.9$ MPa ($T = 508-525$ K), $t = 30-180$ s) P_v of soft wood pulp decreases from 1060 to $377-180$ and, probably at least in part corresponding to such a P_v decrease, S_a increases from 31% to about 99%.

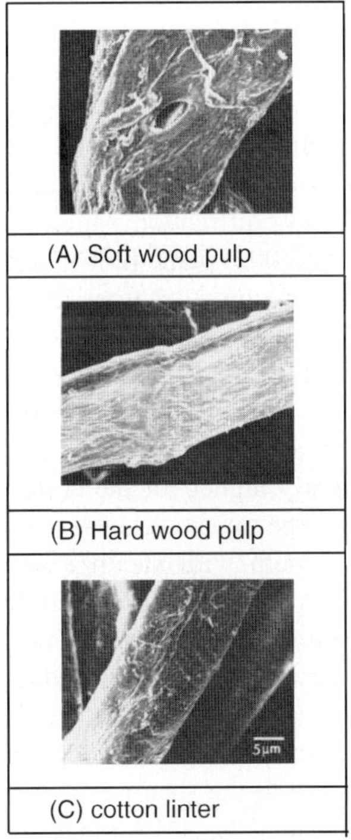

Figure 5.1.4   Magnified SEM micrographs of the original cellulose samples.[1] (A) Soft wood pulp; (B) hard wood pulp; (C) cotton linter.

Table 5.1.1

Changes in P_v, S_a, $\chi_{am}(X)$ and $\chi_{am}(C_k)$ ($k = 2, 3, 6$) of the cellulose samples by a typical steam explosion treatment

Original cellulose	Steam explosion conditions	P_v	S_a (%)	$\chi_{am}(X)$ (%)	ACS (Å)	$\chi_{am}(C_2)$ (%)	$\chi_{am}(C_3)$ (%)	$\chi_{am}(C_6)$ (%)
Soft wood pulp	Original	1060	31	27	50	20	37	35
	2.9 MPa × 30 s	377	99	20	54	1	48	38
	4.9 MPa × 30 s	180	98	22	54	5	45	39
Hard wood pulp	Original	994	37	30	50	5	48	38
	2.9 MPa × 30 s	869	35	26	56	8	46	33
	4.9 MPa × 30 s	287	100	21	54	9	44	33
Cotton linter	Original	1279	25	19	69	26	37	31
	2.9 MPa × 30 s	1153	24	20	69	26	32	34
	4.9 MPa × 30 s	436	11	15	72	15	32	26

The P_v values of hard wood pulp (994) and cotton linter (1279) show a slight decrease (approximately 120), but no significant change in S_a (hard wood pulp about 35%, cotton linter approximately 25%) is observed for both samples when treated at $P = 2.9$ MPa ($T = 508$ K) and $t = 30$ s. More drastic steam explosion conditions ($P = 4.9$ MPa ($T = 525$ K), $t = 180$ s) result in a considerable lowering in P_v for hard wood pulp and cotton linter, and S_a of the treated hard wood pulp increases, approaching approximately 100%. In contrast, S_a of the treated cotton linter unexpectedly decreases despite the considerable lowering of P_v. These results indicate that sensitivity of S_a or P_v to steam explosion decreases in the following order: soft wood pulp > hard wood pulp > cotton linter.

Figure 5.1.5 shows typical X-ray diffraction patterns of the original and the treated cellulose samples. The peak of the (002) plane ($2\theta = 22.5°$) in the diffraction patterns of all the treated cellulose samples becomes sharper than the original samples, and the shoulder peak at $2\theta = 20.5°$ has a tendency to increase in intensity with the steam explosion treatment. Cotton linter, in its original state, has a sharper peak for the (002) plane than the untreated soft wood and hard wood pulps. This indicates the higher perfection of the crystal lattice, at least regarding the (002) plane. The peak for the (101) plane ($2\theta = 14.7°$) is slightly higher for the cotton linter samples than the wood pulps. Note that the separation of the two peaks observed at $2\theta = 14.7°$ and $16.3°$ ((101) plane) increases for soft wood pulp with steam explosion, although the degree of separation is clearly lower than for the cotton linter samples, including the original cotton linter. $\chi_{am}(X)$ and ACS of the samples are given in Table 5.1.1.

Table 5.1.1 shows that for soft wood and hard wood pulps, the steam treatment at $P = 2.9$ MPa ($T = 508$ K) and $t = 30$ s decreases $\chi_{am}(X)$ and increases ACS, and for cotton linter, $\chi_{am}(X)$ and ACS do not change significantly with the treatment, indicating that cotton linter is highly resistant to the steam explosion.

Figure 5.1.6 shows some typical ^{13}C NMR ranges of the original and the treated cellulose samples. There, the treated samples are those treated with washing and air-drying (Step 4). There are several obvious changes in NMR range after the steam

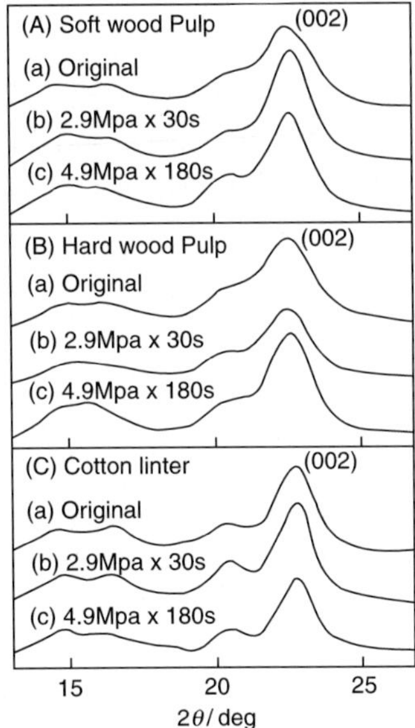

Figure 5.1.5   X-ray diffraction patterns of the original and the treated cellulose samples.[1] (A) Soft wood pulp; (B) hard wood pulp; (C) cotton linter:[1] (a) original; (b) treated under $P = 2.9$ MPa and $t = 30$ s; (c) treated under $P = 4.9$ MPa and $t = 180$ s. Treated samples are the same shown Figure 5.1.3.

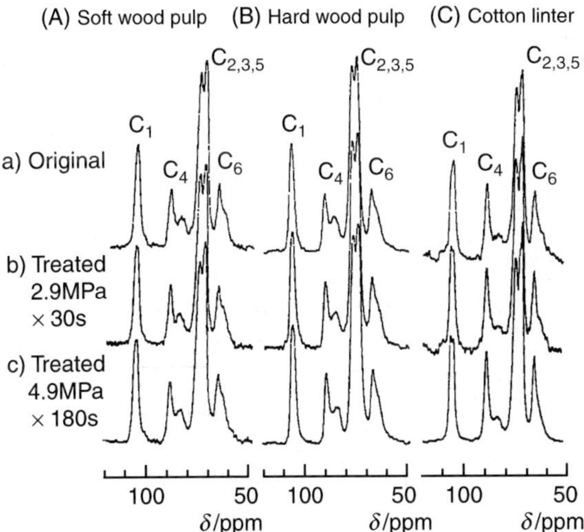

Figure 5.1.6   CP/MAS ^{13}C NMR range of the original and the treated cellulose samples.[1] (A) Soft wood pulp; (B) hard wood pulp; (C) cotton linter:[1] (a) original; (b) treated under $P = 2.9$ MPa and $t = 30$ s; (c) treated under $P = 4.9$ MPa and $t = 180$ s. Treated samples are the same as those shown Figure 5.1.3.

explosion treatment: (1) a lower magnetic field peak for the C_4 carbon of all the samples seems to slightly sharpen, and, contrary to this, a lower magnetic field peak for the C_6 carbon seems to merge into the higher magnetic field peak. (2) For the cotton linter sample the C_1 carbon peak splits into two peaks, and the higher magnetic field peak sharpens for C_2, C_3, and C_5 carbons. The values of $\chi_{am}(C_k)$ ($k = 2$, 3 and 6) for C_2, C_3, and C_6, estimated from the ^{13}C NMR range, are listed in Table 5.1.1. Application of the steam explosion method to soft wood pulp causes an increase in $\chi_{am}(C_3)$ of approximately 30%, a slight increase in $\chi_{am}(C_6)$ and a significant decrease in $\chi_{am}(C_2)$. For hard wood pulp and cotton linter, $\chi_{am}(C_3)$ and $\chi_{am}(C_6)$ tend to decrease slightly and $\chi_{am}(C_2)$ reveals no significant changes.

Table 5.1.1 indicates that the solubility S_a of the treated cellulose samples in a 9.1 wt% aq NaOH solution varies depending on both sample species and P_v, and that the almost complete dissolution of the treated soft wood and hard wood pulp samples with $P_v < 400$ in 9.1 wt% aq NaOH solution can be attained if the cellulose has $\chi_{am}(C_3) > 44\%$ and $\chi_{am}(C_6) > 33\%$. Note that Figures 5.1.3–5.1.6 do not necessarily show supermolecular structural changes of the sample resulting from the steam explosion treatment because the measurements were not taken for treated samples but for dried samples These were prepared by washing the treated samples with water and subjecting them to the air drying process. The drying process may cause significant structural changes.

Figure 5.1.7 shows the CP/MAS ^{13}C NMR range of the samples obtained at some typical stages in the steam explosion treatment. Clearly, wet treatment of any original sample causes an obvious peak separation in the C_4 carbon peak region. The degree of the peak separation in the C_6 carbon peak region by wetting of the original samples depends strongly on sample species and decreases in the following order: soft wood pulp > hard wood pulp > cotton linter. In the ^{13}C NMR range of the steam treated samples, hard

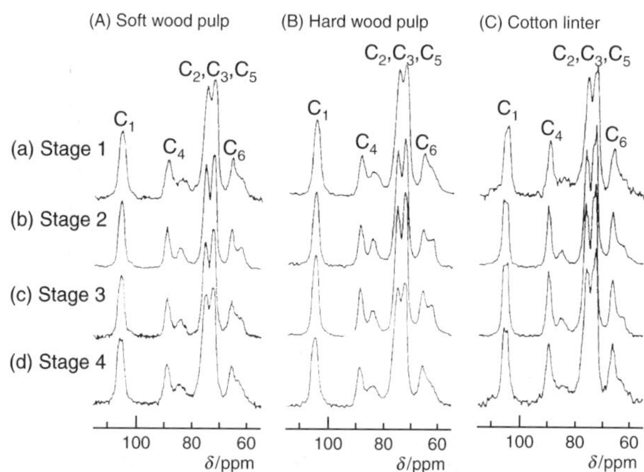

Figure 5.1.7   CP/MAS ^{13}C NMR range of the cellulose samples obtained in each step of the steam explosion process shown in Figure 5.1.2.[1] (A) Soft wood pulp; (B) hard wood pulp: (C) cotton linter; (a) Stage 1 (original; water cotton 10 wt%); (b) Step 2 (pretreated by water; water content 100 wt%); (c) Step 3 (steam exploded ($P = 2.9$ MPa and $t = 30$ s; water content $ca.$ 200 wt%)); (d) Step 4 (washed and dried; water content 10 wt%).[1]

wood pulp and cotton linter show clear peak separation in the C_6 carbon peak region, but for soft wood pulp the separation is somewhat poor. The ^{13}C NMR range of the treated samples (Step 4) shown in Figure 5.1.7(d) are apparently much the same as those for the original samples. The values of $\chi_{am}(C_k)$, estimated from Figure 5.1.7, for cellulose samples at the four typical steps in the steam explosion process are summarized in Table 5.1.2.

$\chi_{am}(C_3)$ and $\chi_{am}(C_6)$ of the wet (Step 2) and the steam treated (Step 3) samples of soft wood pulp increase substantially compared with the original sample, even after the washing and drying stage (Step 4). $\chi_{am}(C_3)$ of the treated sample is larger than that of the original. $\chi_{am}(C_3)$ of hard wood and cotton linter does not show any significant change during the steam explosion process. $\chi_{am}(C_6)$ of hard wood pulp shows the same tendency as that observed for soft wood pulp, but that of cotton linter shows no significant change. In other words, neither $\chi_{am}(C_3)$ nor $\chi_{am}(C_6)$ of cotton linter changes significantly. $\chi_{am}(C_2)$ is readily decreased for all samples by wetting, and remains at the lower level in the steam explosion process. It is interesting to note that $\chi_{am}(C_2)$ of steam exploded hard wood pulp and cotton linter samples with washing and drying recovers to the level of the original samples. The changes in these structural parameters representing the amorphous region in the three original samples correspond well to the changeability of the sample characteristics towards the steam explosion, as judged from changes in morphology, P_v and S_a. $\chi_{am}(C_k)$ for cellulose with crystal form I was defined by Kowsaka et al.[12] by assuming that only $O_3-H\cdots O_5'$ and $O_2-H\cdots O_6'$ type intramolecular hydrogen bonds are allowed to form,[17] as shown in Figure 5.1.8. If this is true, $\chi_{am}(C_2)$ and $\chi_{am}(C_6)$ should coincide with each other (i.e. $\chi_{am}(C_2) = \chi_{am}(C_6)$) and the negative $\chi_{am}(C_2)$ values observed in Table 5.1.2 are apparently unrealistic.

Table 5.1.2 shows that $\chi_{am}(C_2)$ is nearly equal to $\chi_{am}(C_6)$ for the original cotton linter and these two values are quite different for soft wood pulp and hard wood pulp, indicating

Table 5.1.2

Changes in $\chi_{am}(C_k)$ of the cellulose samples during a typical steam explosion treatment at 2.9 MPa, 30 s[1]

Original cellulose	Step	Water content (%)	$\chi_{am}(C_2)$ (%)	$\chi_{am}(C_3)$ (%)	$\chi_{am}(C_6)$ (%)
Soft wood	Original	10	20	37	35
	Original, wetted	100	− 36	43	53
	Steam exploded	200	0	41	40
	Washed and air-dried	10	1	48	38
Hard wood	Original	10	5	48	38
	Original, wetted		− 13	49	46
	Steam exploded	200	− 15	46	47
	Washed and air-dried	10	8	46	33
Cotton linter	Original	10	28	37	31
	Original, wetted	100	− 2	35	35
	Steam exploded	200	− 10	33	33
	Washed and air dried	10	26	32	34

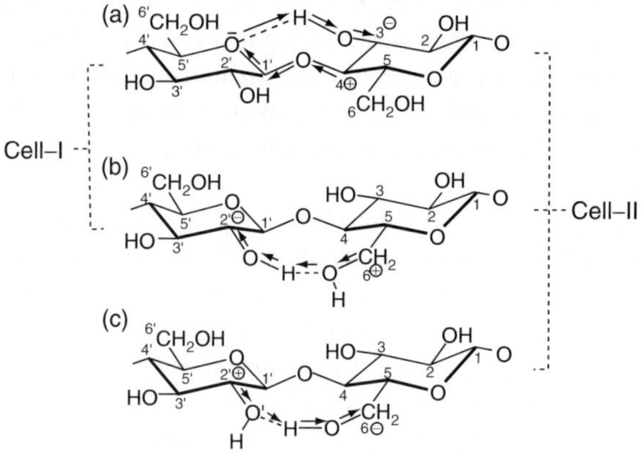

Figure 5.1.8   Assumed intramolecular hydrogen bonds for cellulose I and cellulose II crystals for ^{13}C NMR peak assignment of ring carbons in glucopyranose unit, which is a basis to estimate $\chi_{am}(C_k)$: (a) $O_3-H\cdots O_5'$; (b) $O_6\cdots H-O_2$; (c) $O_6\cdots H\cdots O_2'$. $\ominus$ and $\oplus$ mean anionized and cationized carbons, compared with the ring carbons in cellobiose unit which have no intramolecular hydrogen bonds, resulting in higher and lower magnetic field shift of the corresponding ^{13}C NMR peak positions, respectively.[1,17]

that the above assumption does not always hold. In addition, $\chi_{am}(C_2)$ was found to be often negative, especially for the cellulose with relatively high water content. These results require the amendment of the above assumption and one possibility is that in the amorphous region of the cellulose having crystal form I (i.e. natural cellulose) there also exists a molecular packing which tends to form $O_2-H\cdots O_6'$ (one of the characteristic intramolecular hydrogen bonds for cellulose II) besides $O_2-H\cdots O_6'$ (see Figure 5.1.8). In this respect, $\chi_{am}(C_2)$ may be underestimated and $\chi_{am}(C_6)$ may be overestimated. Other reasons for the discrepancy between $\chi_{am}(C_2)$ and $\chi_{am}(C_6)$ for almost dried cellulose samples include: (1) the low accuracy in resolving the peaks in the C_2, C_3, and C_5 regions, bringing about the relatively low reliability of integral peak intensities, (2) ignoring the effect of intermolecular hydrogen bonds on chemical shifts on C_2 and C_6 carbons, and (3) the difference in Overhauzer effect for carbon peaks because NMR operating conditions are chosen to always give the larger integral intensity per unit carbon in the C_2, C_3, and C_5 carbon peak region than in other carbon peak regions (C_1, C_4, and C_6). The reason for the negative value of $\chi_{am}(C_2)$ when cellulose is considerably wet is not clear at present, but the difference in spin-lattice relaxation time of each carbon when wet needs to be considered.

The experimental data on morphology and P_v of the treated cellulose indicate that (1) wood pulp (soft wood, hard wood) is readily changed by steam explosion treatment as compared with seed fiber (cotton linter), and (2) among wood pulps a hard wood pulp is more resistant than a soft wood pulp to the steam explosion treatment. These experimental facts are expected to be closely related to the supermolecular structure of the original cellulose. Cotton linter is almost pure cellulose (approximately 91%), wood plants, however, consist of approximately 50% cellulose and approximately 50% other foreign materials (lignin, pentosan, and resins). Wood pulp is produced principally by extracting these foreign materials from wood plants, which inevitably leads to a higher porosity of wood pulp than that of cotton linter. In fact, SEM observations confirm that fibrillar packing density of cotton linter is highest among the three natural cellulose resources employed here. Furthermore, wood plants have their own characteristics in supermolecular structures, mainly due to differences in their cell and basic tissue systems and in their associated structure between surface skin and fibrillar bundle. The cell constituting conifer (soft wood) is mostly built up by tracheids, which are mutually connected by rather large bordered pits. The existence of small pores in soft wood pulp was ascertained by SEM observations. Broad leaved trees (hard wood) consist mainly of libriform wood fiber, the cell wall of which is very thick with only a very narrow simple pit. Consequently, the porosity of the natural celluloses decreases in the following order: soft wood pulp > hard wood pulp > cotton linter. This may explain the experimental results of morphological changes during steam explosion. High-temperature water vapor penetrates into the voids of cellulose, destroying the supermolecular structure of cellulose by rapid expansion of water vapor when discharged abruptly into open air. Therefore, the degree of morphological change must depend on the porosity and this expectation was experimentally proven. Since high-pressure water vapor may behave as acid,[18] a kind of acid hydrolysis may occur during the steam explosion, resulting in a decrease in P_v. The porosity of the sample may also play an important role in this reaction during the steam explosion procedure. In fact, the degree of decrease in P_v was experimentally proven to obey the order of porosity.

A tentative mechanism for the changes in morphology and P_v of cellulose resources, resulting from the steam explosion treatment, is illustrated in Figure 5.1.9.

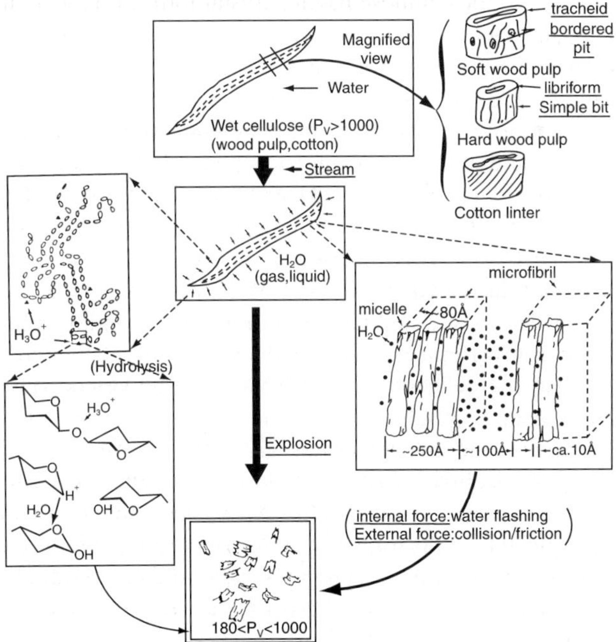

Figure 5.1.9 Schematic representation of the steam explosion mechanism for changes in morphology and degree of polymerization of cellulose samples.[1]

Penetration of acidic water into the internal space between microfibrils, which consist of micelles and acidic scission of polymer chains by the penetrated water, and physical scission of fibrillar bundles by exploded water is shown. The degree of these scissions is obviously determined by the ease of penetration, and this is demonstrated by the structural view of the magnified cell wall of a fiber for each resource.

The changes in the structural parameter $\chi_{am}(C_k)$, especially $\chi_{am}(C_2)$, for different cellulose resources during the steam explosion seem to correlate closely with the degree of porosity of the samples.

Marchessault and St-Pierre[19] reported that the crystallinity of cellulose in wood chips treated by the steam explosion method showed no appreciable change. Tanahashi and Higuchi,[20] who carried out the same kind of experiments, observed an increase in the crystallinity and ACS of cellulose in wood, probably during the digestion period with high-pressure water vapor. The steam explosion experiments on almost pure cellulose coincide with the results obtained by Tanahashi and Higuchi.[20]

Although the reason is not fully clear at present, it is plausible that water and high-temperature water vapor penetrates into the paracrystalline part of cellulose, as well as the amorphous part, and recrystallizes the paracrystalline part on drying. In other words, water molecules reorganize intermolecular regularity by, for example, releasing the strain existing in the original cellulose solid. This tentative mechanism for changing the supermolecular structure of soft wood cellulose during the steam explosion process is shown in Figure 5.1.10.

In this connection, it is interesting to note that, depending on the cellulose resources, $\chi_{am}(C_k)$ more or less changes when swollen with water and in particular water and by steam explosion (before drying), irrespective of cellulose resources. This indicates that the structural change (reorganization of intermolecular regularity) of natural cellulose with the aid of existing water molecules may most likely occur with the formation of a hydrogen bond in which C_2 hydroxyl groups participate when the cellulose molecules in the disordered region move to constitute some ordered structure, although it is not yet clear what kind of structural change really takes place. Kitamaru *et al.*[21,22] pointed out that fractions of higher magnetic field peaks and lower magnetic field peaks of each C_1 and C_6 carbon peak for natural celluloses (cotton and ramie) changed reversibly from dried to wet states of cellulose when the water content of the cellulose in the wet state was carefully kept constant during CP/MAS ^{13}C NMR measurements. They also found that the carbon peaks in the C_2, C_3, and C_5 carbon regions for cotton cellulose are clearly classified into two categories (*ca.* 200 s and *ca.* 160 s) relating to the spin lattice relaxation time when the cellulose is wet. This may support our prediction that the structural change of natural cellulose is most likely to occur where a C_2 hydroxyl group participates. The latter experimental results given by Kitamaru may account for the negative value of $\chi_{am}(C_2)$ obtained when cellulose is wet. Our results also coincide well with the X-ray diffraction results obtained by Sarko and Muggi[23] that only C_3 and C_6 hydroxyl groups participate in the formation of intermolecular hydrogen bonds of natural cellulose (ramie) and also with our previous experimental result,[24] suggesting that formation of alkali cellulose from natural cellulose (wood pulp) occurs exclusively at a C_2 hydroxyl group.

The cellulose II crystal is already known to be energetically more stable than the cellulose I crystal because cellulose II crystals cannot be converted to a cellulose I crystal

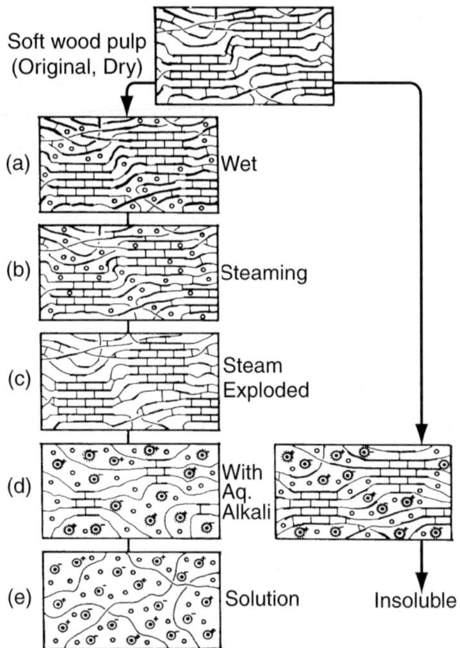

Figure 5.1.10   Schematic representation of supermolecular structural changes of soft wood pulp by steam explosion and the dissolution mechanism of the steam-exploded pulp into aq. NaOH with specific concentration.[1] Ladder-like part, ordered region (straight vertical line means intermolecular hydrogen bond); fine line, cellulose chain with no intramolecular hydrogen bond; thick line, cellulose chain with intermolecular hydrogen bond: (O), free water molecule; ($\diamond^{+}$), Na^+ with specific solvated structure; ($\diamond^{-}$), OH^- with specific solvated structure; ($\odot^{+}$), Na^+ with normal structure; ($\odot^{-}$) OH^- with normal structure.

by any treatment,[25] and if the change of a cellulose I crystal to a cellulose II crystal occurs a $O_2-H\cdots O_6'$ type intramolecular hydrogen bond must change into another type of $O_2-H\cdots O_6'$ and $O_2\cdots H-O_6'$. Therefore, at least hydroxyl groups at C_2 or C_6 should move during the course of crystal change of cellulose I to cellulose II.

It is currently unknown why $\chi_{am}(C_k)$ ($k = 3$ and 6) increases (that is, the disordered fraction by NMR analysis increases) in some cases in spite of an increase in X-ray crystallinity of steam-exploded cellulose, although the X-ray crystallinity estimated here is not an absolute value, but rather a relative one. A very tentative mechanism for the dissolution of the resultant cellulose in aq. NaOH with specific structure[26] is also illustrated in Figure 5.1.10, and shows, in summary, (1) wood pulps were more effectively treated than cotton linter in view of changes in the morphology, P_v and S_a, reflecting supermolecular structural characteristics governing the ease of water penetration, (2) $\chi_{am}(C_k)$ ($k = 3$ and 6) at a given C_k position of the glucopyranose ring is a highly reliable measure to predict the solubility of the treated cellulose if P_v is less than 400, (3) the water treatment of solid cellulose prior to steam explosion initiated the change in $\chi_{am}(C_k)$ ($k = 2$, 3 and 6), especially $\chi_{am}(C_2)$, and (4) the structural change of pulp cellulose during steam explosion may occur at C_2 hydroxyl groups.

5.1.2　Effect of treatment conditions[27]

In Section 5.1.1 we first demonstrated that completely alkali soluble cellulose could also be obtained directly from natural cellulose (soft and hard wood pulps) by physical treatment (steam explosion). In this subsection, we intend to clarify the influence of the steam explosion conditions on the changes in morphology, viscosity-average degree of polymerization P_v, solubility S_a, and structure of the treated soft wood pulp cellulose.

Cellulose sample

The initial samples used were of a soft wood pulp (white spruce, Alaska pulp (USA), manufactured by Alaska Pulp Co., Ltd (USA): α-cellulose content 90.1 wt%, $P_v = 1060$). The pulp was cut into small pieces ($10 \times 5 \times 1.5$ mm^3), and then subjected to the steam explosion experiments.

Steam explosion apparatus and explosion procedure

Prior to the steam explosion treatment, the water content of cellulose was adjusted in the following way: one part of cellulose was immersed in 20 parts of distilled water at 25 °C for 15 h, then the water content of cellulose was set to a value ranging from 60 to 300% by centrifuging the wet chips (KM-18 centrifuge, Matsumoto Machinery Co., Ltd, Japan). Cellulose with low water content (10%) was prepared by further drying the centrifuged chips in a hot wind dryer (Type HLK-2, Tabai ESPEC Co., Ltd, Japan).

A steam explosion apparatus as previously described[2] (Section 5.1.1) was utilized without serious modification. Steam explosion experiments were carried out as follows: after 50–200 g (dry base) of cellulose with known water content was put into the digester vessel which had been heated in advance to a desired temperature, the vessel was closed tightly. A saturated water vapor was then controlled at a given steam pressure ($P = 1.0$–4.9 MPa gauge pressure, corresponding to saturated steam temperature T in the vessel of 456–525 K) was carefully introduced into the vessel, which was then allowed to stand for a given time ($t = 15$–300 s). After closing the vapor input valve, the ball valve connected to the discharging nozzle was quickly opened and the treated cellulose was discharged into the cyclon receiver, with the aid of pressure difference. The treated cellulose was then gathered and washed with water. The water in the sample cellulose was carefully displaced by a large amount of acetone several times in order to avoid the drastic structural change that would occur with drying, and finally the sample was dried in air.

(a)　*DP of cellulose* was determined according to the procedure described in Section 5.2.1.[14]

(b)　*Solubility* (S_a) of cellulose in aq. NaOH was evaluated by the method described in Section 5.2.1.[14,28]

(c)　*SEM observation* was made according to the procedure in Section 5.2.1.[14]

(d) *X-ray diffractometry* was performed by the method described in Section 5.1.1.[1]
(e) *Solid-state CP/MAS ^{13}C NMR measurement* was made according to the method described in Section 5.2.1.
(f) *Yield of the exploded sample:* the yield (Y, expressed as %) was estimated. The steam explosion treatment has the potential ability to hydrolyze cellulose to give components with very low P_v. These components may be water soluble and, as a result, give a relatively low yield. This is a major drawback of the steam explosion method.

Figure 5.1.11 shows SEM photographs of the soft wood pulp treated under conditions of water content in cellulose = 100%, $P = 1.0–4.9$ MPa ($T = 456–525$ K) and $t = 30$ s. S_a and P_v of the samples are given below each photograph. When P is less than 1.9 MPa ($T = 487$ K), pulp fiber length is slightly shortened due to the treatment. At $P = 2.4$ MPa ($T = 498$ K), the shortening of fiber length is distinct and at $P = 2.9$ MPa ($T = 508$ K), fibrillation (the separation of a fibrillar bundle into finer bundles) takes place as well as shortening of the fibers. The length of fibers of the sample treated at $P = 2.9$ MPa

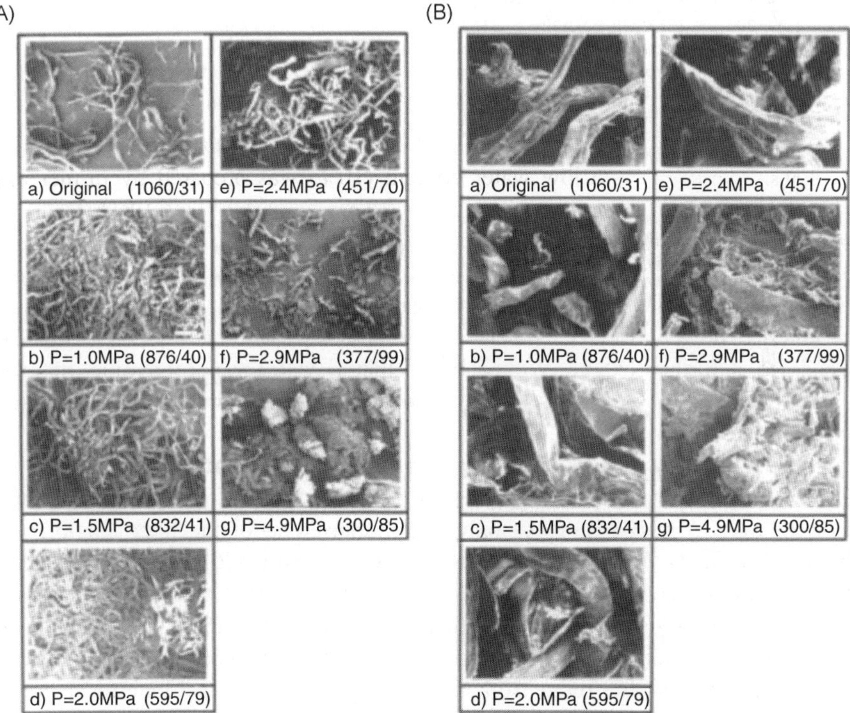

Figure 5.1.11   SEM photographs of the surface of the original and treated soft wood pulps under steam explosion conditions of constant treatment time $t = 30$ s and various pressure P:[27] (A) magnification 48; (B) magnification 480. (a) Original; (b) treated at $P = 1.0$ MPa ($T = 456$ K); (c) treated at $P = 1.5$ MPa ($T = 473$ K); (d) treated at $P = 2.0$ MPa ($T = 487$ K); (e) treated at $P = 2.4$ MPa ($T = 498$ K); (f) treated at $P = 2.9$ MPa ($T = 508$ K); (g) treated at $P = 4.9$ MPa ($T = 525$ K).

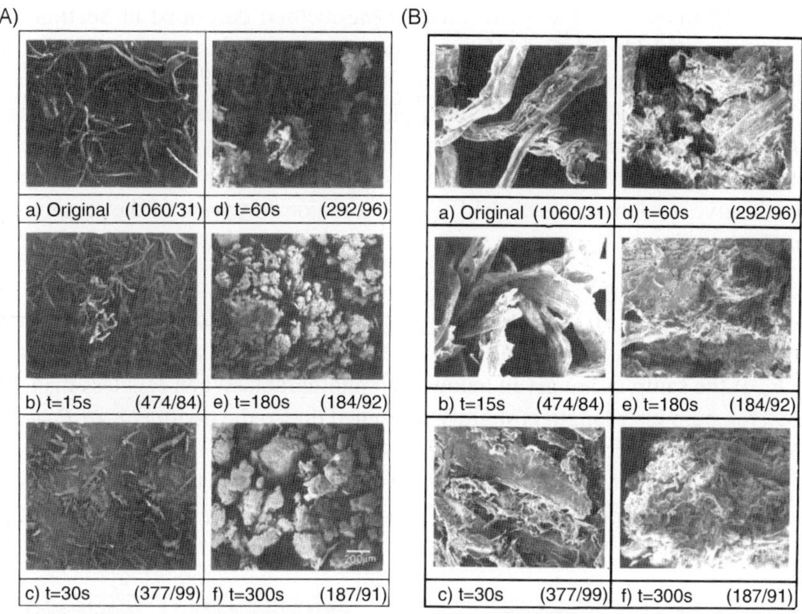

Figure 5.1.12 SEM photographs of the surface of the original and treated soft wood pulps under steam explosion conditions of constant pressure ($P = 2.9$ MPa, $T = 508$ K), and various treatment times, t:[27] (A) magnification 48; (B) magnification 480. (a) Original; (b) treated at $t = 15$ s; (c) treated at $t = 30$ s; (d) treated at $t = 60$ s; (e) treated at $t = 180$ s; (f) treated at $t = 300$ s.

($T = 508$ K), $t = 30$ s, range from 10 to 200 μm, being significantly shorter than the fibers in the original sample (1000–4000 μm). The apparent fiber thickness of the fibrillated fiber for the same sample is usually less than 0.4–10pm (original sample fiber thickness: 15–44 μm). When P reaches 4.9 MPa ($T = 525$ K), fibrillation proceeds to a further extent. We conclude that a water vapor pressure of at least 2.9 MPa ($T = 508$ K) seems necessary to bring about morphological change of the wood pulp fiber at constant t (30 s).

Figure 5.1.12 shows SEM photographs of the soft wood pulp treated for various times ($t = 15$–300 s) at constant P (2.9 MPa; $T = 508$ K). A treatment time t of 15 s at $P = 2.9$ MPa, $T = 508$ K, gives only slight shortening of the fiber length, but the shortening becomes more detectable at $t = 30$ s, at which time fibrillation of pulp fibers also takes place. At $t = 60$ s ($P = 2.9$ MPa, $T = 508$ K), shortened fibers with considerable fibrillation are obtained. When t increases beyond 180 s, the steam exploded cellulose gives much finer fibrils accompanying the shortening of fibers. The diameter of the fibrillated fiber bundles range from 0.1 to 10 μm, which is finer than for the original sample (15–4 μm). Brownish and highly swollen cellulose masses appear in samples treated under the conditions of $t \geq 60$ s and $P = 2.9$ MPa ($T = 508$ K), and these masses are solidified during the drying process after washing, as shown in Figure 5.1.12. When significant morphological change of wood pulp is observed ($P \geq 2.9$ MPa ($T \geq 508$ K) at $t = 30$ s, and $t \geq 30$ s at $P = 2.9$ MPa ($T = 508$ K)), P_v of the pulp decreases to less than 400.

Table 5.1.3

The yield, P_v, S_a, $\chi_{am}(C_k)$ of the original cellulose and steam exploded celluloses under various conditions[27]

Original cellulose	Steam explosion conditions ($P \times t$)	Yield (%)	P_v	S_a (%)	$\chi_{am}(X)$ (%)	$\chi_{am}(C_2)$ (%)	$\chi_{am}(C_3)$ (%)	$\chi_{am}(C_6)$ (%)
Soft wood	Original	–	1060	31	27	20	37	35
Soft wood	1.0 MPa × 30 s	97.4	876	40	29	2	46	46
	2.0 MPa × 30 s	97.3	595	79	25	0	43	50
	2.4 MPa × 30 s	96.8	451	70	24	8	46	46
	2.9 MPa × 30 s	96.6	377	99	20	1	48	47
	4.9 MPa × 30 s	96.5	300	85	24	20	41	37
Soft wood	2.9 MPa × 15 s	97.1	474	84	24	9	45	51
	2.9 MPa × 60 s	95.3	292	96	26	12	46	42
	2.9 MPa × 180 s	94.5	184	92	26	13	44	41
	2.9 MPa × 300 s	70.2	187	91	24	20	41	42

Table 5.1.3 shows P_v, S_a, $\chi_{am}(X)$, and $\chi_{am}(C_k)$ of the samples treated at $P = 1.0$–4.9 MPa ($T = 456$–525 K) with constant $t = 30$ s, and at $t = 15$–300 s with constant $P = 2.9$ MPa ($T = 508$ K). Water content of the original sample was kept at 100%.

Figure 5.1.13(a) and (b) show the relationship between S_a of the treated samples and the steam explosion conditions (P and t) (a), and between P_v and the steam explosion conditions (b), respectively.

Figure 5.1.14 shows the projections of Figure 5.1.13 on the P–t plane. The contour lines are constant S_a (Figure 5.1.14(a)) or constant P_v (Figure 5.1.14(b)). The water content of the original soft wood pulp was kept at 100% until immediately before application of the steam explosion treatment. The numbers in Figure 5.1.14(a) and (b) are

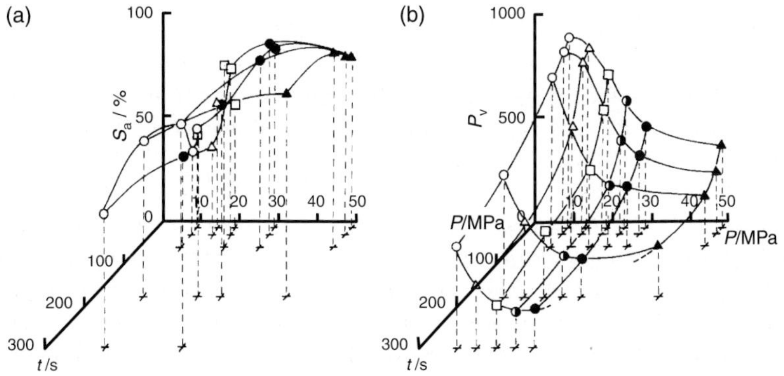

Figure 5.1.13 Dependence of (a) the solubility S_a and (b) the viscosity-average degree of polymerization P_v of the original and treated soft wood pulps on treatment pressure P and treatment time t in steam explosion experiments.[27]

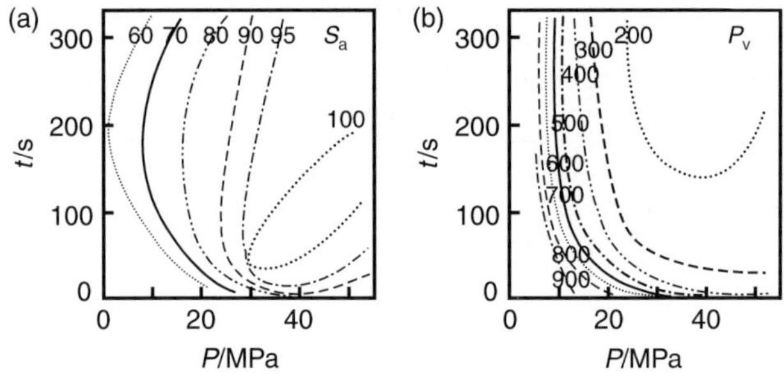

Figure 5.1.14   Contour lines of (a) the solubility S_a and (b) the viscosity-average degree of polymerization P_v as functions of treatment pressure P and treatment time t.[27]

the S_a and P_v values, respectively, of the treated cellulose. S_a increases rapidly with increase in P, approaching an asymptotic value at $P = 2.9$ MPa ($T = 508$ K), over a wide range of t, if $t > 90$ s, and the dependence of S_a on P seems irrespective of t. In other words, a further increase in P, beyond $P > 2.9$ MPa ($T > 508$ K), gives only a minor effect on S_a. At $P = 1.0$ MPa the maximum S_a (70%) is attained at $t = 180$ s, and when P exceeds 2.9 MPa ($T = 508$ K) the treatment with longer t has a small effect on S_a of the treated cellulose (Figure 5.1.14(a)). It should be noted that the maximum S_a attainable at P lower than 2.9 MPa ($T = 508$ K) is considerably smaller than that obtained at $P = 2.9$ MPa.

From Figure 5.1.14(b) it can be seen that t and P contribute equally to the decrease in P_v of the treated sample when $t \leq 180$ s and $P \leq 2.9$ MPa ($T \leq 508$ K). When P exceeds 2.9 MPa ($T = 508$ K) t is the single dominant factor controlling the decomposition of the cellulose chain length during treatment. The contour line for $P_v = 200$ apparently shows that at the same t (*ca.* 180 s) there are two values of P (*ca.* 2.9 and *ca.* 4.9 MPa) giving the same P_v of the treated samples (after washing with water and acetone). However, as will be described later, the yield of the treated pulp (after washing with water and acetone) obtained under conditions of $P = 4.9$ MPa (525 K) and $t = 180$ s is considerably lower (*ca.* 70%) than that of the treated sample obtained under milder conditions (*ca.* 95%). In the former case, a substantial part of the treated cellulose with much lower DP is washed out, resulting in overestimation of the apparent P_v.

Figure 5.1.15 demonstrates a relationship between S_a and P_v of the treated samples prepared under a wide range of operating conditions. The straight line drawn in the figure is calculated by the least squares mean method and is expressed by:

$$S_a(\%) = -0.08P_v + 116.6 \tag{5.1.4}$$

Therefore, S_a of the cellulose generally tends to increase with a decrease in P_v. While, the experimental data scatter around the line given by eq. (5.1.4), all data lie within $\pm 10\%$ limits, which are shown in Figure 5.1.15 by the broken lines. It is obvious from Figure 5.1.15 that P_v is an important factor governing S_a, but other factors besides P_v might exist. A possible factor is the degree of breakdown in intramolecular hydrogen bonds of the steam exploded samples ($\chi_{am}(C_k)$).

Figure 5.1.16 shows the effect of repetition of the steam treatment, N, on S_a and P_v of the soft wood pulp sample. The circles denote a series of the explosion runs under

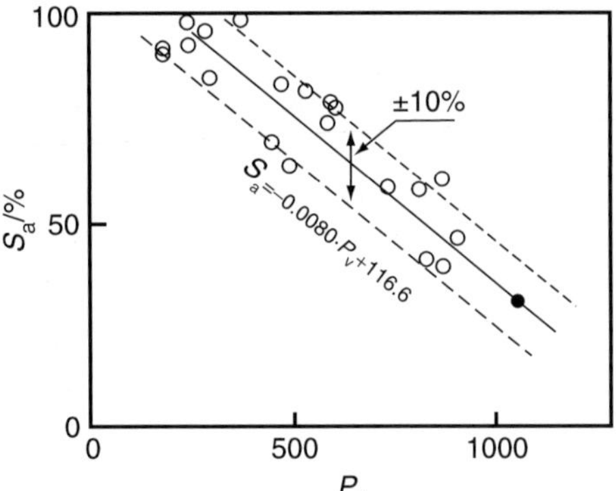

Figure 5.1.15   Relationship between the solubility S_a and the viscosity-average degree of polymerization P_v of the original and the treated soft wood pulps.[27]

the conditions of $P = 2.9$ MPa ($T = 508$ K) and $t = 30$ s, and the rectangles are a series of runs performed under the conditions of $P = 2.9$ MPa ($T = 508$ K) and $t = 15$ s. In the first run in both series, the water content of cellulose was adjusted to 100% but in the subsequent runs the water content of the samples used for repeated steam explosions was kept at approximately 200%. In both series the first run gives a profound increase in S_a and further increase in S_a is observable only when a short t is utilized. P_v of the samples is greatly lowered by the first run for both conditions and the repeated treatment brings about a slight decrease in P_v, approaching a plateau in P_v at $N = 2$, which is higher for the shorter treatment time. The morphology of the sample obtained at the second run under the conditions of $P = 2.9$ MPa ($T = 508$ K) and $t = 30$ s (total $t = 60$ s) is similar

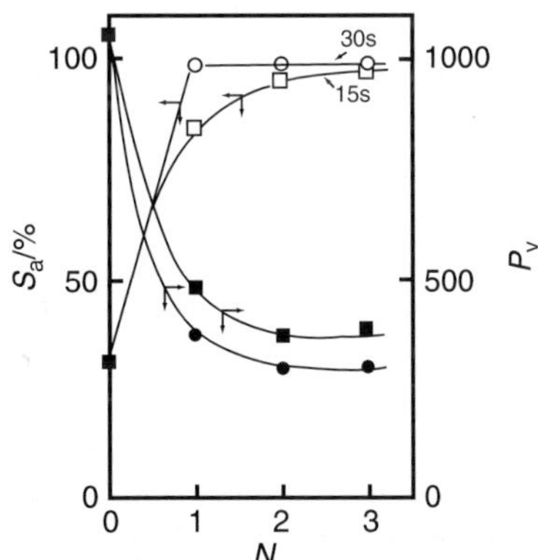

Figure 5.1.16   Effect of number of repeat runs of the steam explosion treatment on the changes in the solubility S_a and the viscosity-average degree of polymerization P_v of the treated soft wood pulps:[27] circles denote $P = 2.9$ MPa ($T = 508$ K) and $t = 30$ s;[27] rectangles denote $P = 2.9$ MPa ($T = 508$ K) and $t = 15$ s. Black symbols show P_v; open symbols show S_a. Data for the original soft wood pulps are included.

to that obtained by a batch explosion under the conditions of $P = 2.9$ MPa ($T = 508$ K) and $t = 300$ s, and the resulting P_v and S_a of the sample in the former case are 300 and 99%, respectively. The S_a value is found to be higher than that for the sample ($P_\mathrm{v} = 292$, $S_\mathrm{a} = 96\%$) obtained by a batch explosion under the conditions of $P = 2.9$ MPa ($T = 508$ K) and $t = 60$ s. This means that the repeated explosion method gives higher S_a and a more fibrillated sample than the corresponding batch explosion.

Under the steam explosion conditions $P = 1.0\text{–}2.9$ MPa ($T = 456\text{–}508$ K), $t = 15\text{–}300$ s, and $P = 4.9$ MPa, $t \le 60$ s, except for the conditions of $P = 4.9$ MPa and $t = 180$ s, the yield of cellulose (after washing with water and acetone) is always at 94.5–97.5% and there is no P_v dependence of the cellulose yield. The fact that the α-cellulose content of original soft wood pulp utilized is 90.1 wt% indicates that at least there is no actual loss of α-cellulose during the steam explosion treatment. The yield of the treated sample obtained under the conditions of $P = 4.9$ MPa and $t = 180$ s was as low as *ca.* 70%.

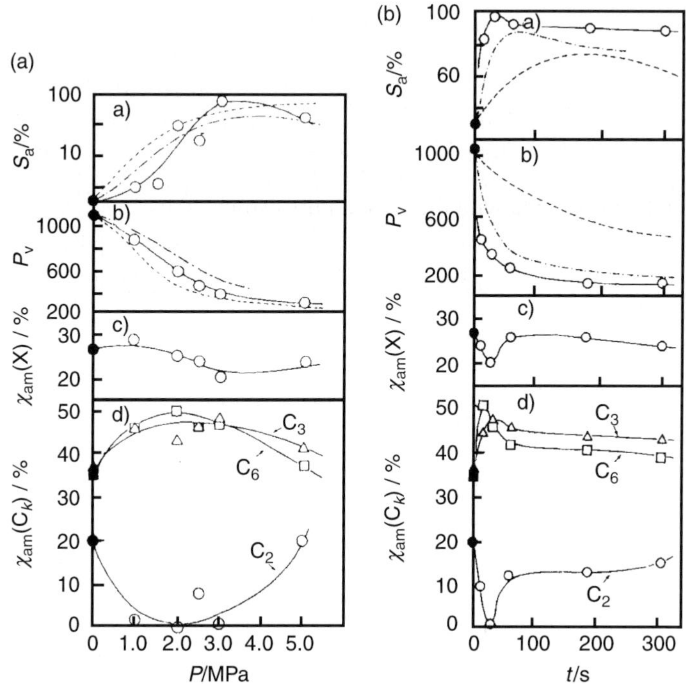

Figure 5.1.17   Changes in the solubility S_a and the viscosity-average degree of polymerization P_v, the amorphous content determined by X-ray diffraction $\chi_\mathrm{am}(X)$, and the degree of breakdown of intramolecular hydrogen bonds at C_k positions determined by solid state CP/MAS ^{13}C NMR $\chi_\mathrm{am}(C_k)$, of the treated soft wood pulps as functions of steam explosion conditions (treatment pressure P (a) and treatment time t (b)).[27] For S_a and P_v, the P dependence at other constant treatment times ($t = 15$ s, chain line; 60 s, broken line) and their t dependence at other constant treatment pressures ($P = 1.0$ MPa, broken line; 2.0 MPa, chain line) are also shown in the figures.

Figure 5.1.17 shows the dependence of S_a, P_v, $\chi_{am}(X)$, and $\chi_{am}(C_k)$ on P at constant t (30 s) and on t at constant P(2.9 MPa ($T = 508$ K)). For S_a and P_v their P dependence at constant treatment times $t = 15$ and 60 s and their t dependence at constant $P = 1.0$ and 2.0 MPa are also shown. S_a increases gradually as a function of P until $P = 2.9$ MPa ($T = 508$ K), where it levels off. The time dependence of S_a is somewhat different: S_a rapidly approaches a maximum at $t = 30$ s and then levels off at an S_a level slightly lower than the maximum. Appearance of the maximum in S_a as a function of t is also observed for the curves obtained at constant $P = 1.0$ and 2.0 MPa and the t giving the maximum tends to shift to longer t with decrease in P_v. Generally, P_v decreases almost linearly as a function of P until it levels off, but it decreases gradually as a function of t and before leveling. When P_v has leveled off ($= 200$) as a function of t at constant P (2.9 MPa), it is considerably lower than the leveled P_v ($= 320$) obtained as a function of P at constant t (30 s).

The former curve is very similar to that obtained for acid hydrolysis of cellulose at constant strength of acid.[29] The latter curve seems to be attributed to the curve obtained as a function of acidity at constant t because the dissociation constant of water vapor is a function of pressure.[30] $\chi_{am}(X)$ tends to decrease as a function of P until $P = 2.9$ MPa ($T = 508$ K) and the change in $\chi_{am}(X)$ by variation of t is complicated, first decreasing considerably (until $t = 30$ s) and then recovering to a level slightly lower than that of original cellulose, despite the increase in S_a. In contrast, $\chi_{am}(C_3)$ and $\chi_{am}(C_6)$ tend to give very similar curves to S_a as functions of both P and t, particularly for $\chi_{am}(C_3)$. This means that S_a is primarily determined by $\chi_{am}(C_3)$ if the P_v dependence of S_a is neglected. These results also indicate that the crystallinity measured by X-ray diffraction does not directly correspond with the crystallinity estimated by NMR, which has already been pointed out by Kamide et $al.$[28] suggesting that the parameters $\chi_{am}(X)$ and $\chi_{am}(C_k)$ express a different order of structure. Interestingly, $\chi_{am}(C_2)$ and $\chi_{am}(C_6)$ curves show quite a contrasted change as a function of both P and t. This is implausible if the cellulose with cellulose I crystal is allowed to form only $O_3 \cdots O_5'$ and $O_6 \cdots H-O_2'$ types of intramolecular hydrogen bonds, which are particularly characteristic of the cellulose I crystal. As pointed out in our previous work,[1] a molecular packing state must exist where molecules may take the form of $O_2 \cdots H-O_6'$ which is characteristic of the cellulose II crystal.

However, it should be noted that $\chi_{am}(C_6)$ may be also influenced by intermolecular hydrogen bonding at C_6, as pointed out in our previous paper.[12] We conclude, therefore, that $\chi_{am}(X)$ is not a ruling factor of S_a, but rather, $\chi_{am}(C_k)$ can describe the S_a of cellulose, although the P_v dependence of S_a should be considered. Exact dependence of S_a on $X_{am}(C_k)$ should be examined at constant P_v.

In summary, (1) the maximum S_a ($ca.$ 99%) is obtained under conditions of $P = 2.9$ MPa ($T = 508$ K) and $t = 30$ s. (2) t and P contribute equally to the decrease in P_v when $t \leq 180$ s and $P \leq 2.9$ MPa ($T \leq 508$ K) but when P exceeds 2.9 MPa ($T = 508$ K) t is the only dominant factor. (3) Higher water content in the untreated sample gives a lower degree of decrease in P_v and 100% water content may give the best S_a. (4) The amorphous content as determined by X-ray analysis has a tendency to decrease. (5) The structural parameters estimated by the CP/MAS ^{13}C NMR range are sensitively changed by steam explosion, and these may be more closely correlated with S_a than the X-ray analysis parameter.

5.1.3 Effect of crystal forms of original cellulose[31]

All the above experimental results were obtained only with natural cellulose (wood pulp and cotton linter), which only ever has crystal form I. Therefore, in this section, for a better and more thorough understanding of the detailed mechanism of steam explosion, the dependence of the changes in several physical and structural parameters during the steam explosion on the crystal forms of cellulose (I, II, and III) was examined.

Three cellulose samples with having different crystal forms (cellulose I, II, and III crystals) were treated by the steam explosion method. The latter two samples were prepared by applying crystalline transition treatments in the solid state to the cellulose I sample (soft wood pulp cellulose with crystal form I) in order to avoid a possible significant variation in supermolecular structure except the crystal form between the samples.

Cellulose samples

A starting cellulose sample with crystal form I was obtained from Alaska pulp (soft wood pulp manufactured by Alaska Pulp Co., USA: α-cellulose content, 90.1 wt%; $P_v = 1060$; referred to hereafter simply as Cell-I) was used. Cellulose samples with crystal form II (Cell-II) was prepared by one part of Cell-I being immersed in 10 parts of a 17.5 wt% aq. NaOH solution at 298 K for 1 h and the excess aq. alkali squeezed out. To the resultant alkali cellulose mass (alkali cellulose I),[32] 350 parts of water were added to create a slurry, which was neutralized with 5% aq acetic acid, washed with a large amount of water, and dried at room temperature in air. Samples with crystal form III (Cell-III) were prepared by Cell-I samples being dipped in a large amount of liquid ammonia at 239.6 K for 1 h to produce ammonia cellulose I.[33] The ammonia adhering to the cellulose was then vaporized in a hood at 298 K for 3 days.

Preparation procedures are briefly illustrated in Figure 5.1.18.

Steam explosion apparatus and procedure

The steam explosion apparatus as described in Section 5.1.1 was utilized without serious modification. According to the procedure described in the previous papers,[1,27] the original cellulose samples (water content; $100 \times$ water (wt)/dry cellulose (wt) = *ca.* 8%)

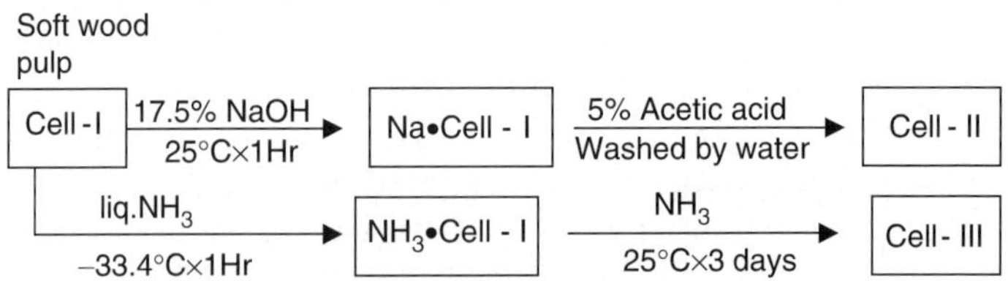

Figure 5.1.18 Preparation of cellulose samples having cellulose II and III crystals from cellulose with cellulose crystal I (soft wood pulp).[31]

were precisely conditioned in advance with great care to have water content of 10% and were subjected to the steam explosion experiments (treatment time $t = 15$–300 s; vapor pressure $P = 2.9$ MPa corresponding to saturated steam temperature $T = 508$ K).

Figure 5.1.19 shows the changes in P_v and S_a of the treated samples of Step 4 (samples obtained by washing and drying the steam-treated samples) as a function of treatment time t at constant water vapor pressure P (2.9 MPa). In the shorter treatment time ($t \leq 15$ s) the depolymerization becomes less remarkable in the following order: Cell-I $\geq$ Cell-III $>$ Cell-II. But the P_v values where the curve levels off for Cell-II and Cell-III (73 and 82, respectively), attained by the longer treatment, are notably lower than that of Cell-I(136). S_a for the treated Cell-I and Cell-III increases, corresponding in part to the lowering of their P_v during the treatment, approaching nearly 100% at $t \geq 30$ s. Note that S_a of the treated Cell-II sample, which is not regenerated cellulose and whose

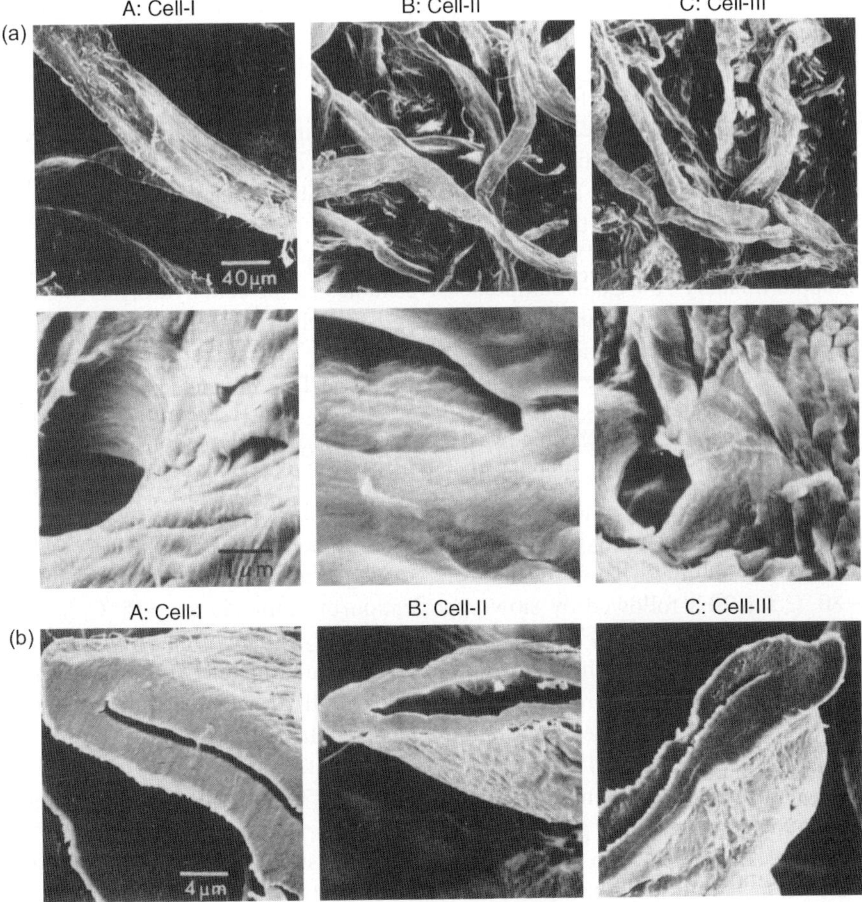

Figure 5.1.19 SEM micrographs of (a) whole feature and (b) cross sectional view of the original cellulose samples subjected to the steam explosion treatment.

morphology and P_v are almost the same as those of the Cell-I sample, remains unexpectedly as low as 45%, even where P_v (73) levels off.

The P_v versus treatment time t curves are similar to the conventional acid hydrolysis curves obtained as a function of hydrolysis time at constant acid strength. Evidently there are three steps of the depolymerization for Cell-I and Cell-III, each corresponding to $t < 15$ s, 15 s $\leq t < 180$ s, and $t > 180$ s, and for Cell-II there are two depolymerization steps at $t < 180$ s and $t \geq 180$ s, respectively. The existence of these depolymerization steps during the steam explosion treatment for each cellulose sample corresponds excellently with the morphological changes during the treatment observed by SEM. The scission of fiber in the direction of the diameter starts from weak points, which allow easy penetration of acidic water vapor (*ca.* pH 3–5) for Cell-I and Cel-III even at $t \leq 5$ s and the scission of fiber seems to be random. In contrast, the depolymerization of Cell-II seems to take place evenly within a fiber making it considerably porous, but without any scission of fiber at $t \leq 180$ s, probably due to a gradual penetration of acidic water vapor, and then the obvious scission of fiber seems to occur. Cell-III shows the same tendency as Cell-II. The rate constant k for each sample at the initial step is highly correlated with the molecular packing density of the samples, which decreases in the order Cell-II > Cell-III > Cell-I. Under the steam explosion conditions used here ($P = 2.9$ MPa), P_v levels off for Cell-I at approximately 140 and for Cell-III and Cell-II, at about 80 and 70, respectively. The experimentally confirmed existence of a 'level-off' P_v means that the proposed first order kinetics theory cannot always explain the whole depolymerization process by a single rate constant k value, especially at the later stage. Nelson and Tripp[34] studied the acid hydrolysis of cotton linter in 0.01 pH (2–2.5 N) aqueous hydrochloric acid at 80–100 °C for a maximum of 600 h, showing the level-off P_v of approximately 200. Later comprehensive studies on the acid hydrolysis in 2.4N aq. HCl at 100 °C for 10 h of natural celluloses (Cell-I) such as ramie, cotton, flax, valonia, and dissolved pulp were carried out by Watanabe *et al.*[35] who also obtained a common level-off P_v of approximately 200 for these celluloses. This is fairly consistent with our present results on Cell-I, although the value estimated here is slightly smaller. Akabori *et al.*[36] prepared cellulose samples with the crystal form of cellulose II by neutralizing the mercerized natural celluloses including ramie, cotton, flax, and valonia in 1% aqueous acetic acid, and hydrolyzing these samples in 2.4 N aq. HCl at 100 °C for 10 h, giving a definite level-off P_v of approximately 80. They also demonstrated that celluloses with the crystal form of cellulose II,[36] which were prepared by treating ramie with liquid ammonia at -80 °C for 20 h followed by slow vaporization of ammonia at -15 °C, yielded a level-off P_v of approximately 80 under the same hydrolysis conditions. The level-off P_v value of approximately 80 which they obtained for cellulose II and III closely coincides with our present results. The reason why Cell-II and Cell-III, which are more resistant to the penetration of acidic water vapor into their molecular space than Cell-I, give lower level-off P_v values than Cell-I is not fully clear at present, but may be due to the ability to retain acidic water once penetrated into the inner space. The retention ability is probably inversely proportional to the penetration ability.

For the samples at Step 4 obtained under the steam explosion conditions of $P = 2.9$ MPa, $t = 30$ s: (1) the peaks in X-ray diffraction curve corresponding to the (002) plane for Cell-I ($2\theta = 22.55°$) and Cell-II ($2\theta = 21.70°$) shift to 22.70° and 22.10°, respectively; (2) the crystal peaks corresponding to the (002) plane for Cell-I and Cell-II are obviously sharpened by the steam explosion treatment; (3) the crystal peak corresponding to the (10$\bar{1}$) plane for Cell-II ($2\theta = 20.00°$) also sharpens; (4) a characteristic peak ($2\theta = 20.70°$) for Cell-III considerably decreases in its intensity and in turn the peak at $2\theta = 22.50°$ responsible for Cell-I becomes very remarkable. These experimental results indicate that the steam explosion treatment causes the distance between crystal lattices of the (002) plane to become shorter and the degree of perfection of the crystal lattice to increase.

$\chi_{am}(X)$ of both Cell-I and Cell-II, treated at $P = 2.9$ MPa, $t = 30$ and 180 s, decreases by nearly 10% with steam explosion ($t = 30$ s) and longer treatment time affords lower $\chi_{am}(X)$ values for these samples. In other words, the crystallinity evaluated by the X-ray method of Cell-I and Cell-II increases by the steam explosion treatment. The extents of conversion of Cell-III to Cell-I, when treated at $P = 2.9$ MPa, $t = 30$ s, and at $P = 2.9$ MPa, $t = 180$ s, are roughly estimated to be *ca.* 70% and *ca.* 100%, respectively. $\chi_{am}(X)$ of the treated Cell-III at Step 4, of which crystalline form is almost cellulose I, obtained under the explosion conditions of $P = 2.9$ MPa, $t = 180$ s, is estimated as 21%, being almost the same as those of the Cell-I and Cell-II samples, treated under the same conditions.

The experimental results obtained for samples steam treated under the conditions of $P = 2.9$ MPa, $t = 30$ s, are summarized as follows. The steam treatment brings about a change in the NMR range of cellulose samples similar to that brought about by simple addition of water to the original samples at room temperature. Especially for Cell-III, the lower magnetic field peak in the C_4 carbon region shifts to a much lower magnetic field by approximately 1 ppm and the C_2, C_3, and C_5 carbon peak region clearly splits into two peaks located at 75.1 and 72.6 ppm. This indicates the creation of cellulose with cellulose I crystal.

The treated samples obtained by washing and drying the steam exploded samples (Step 4) gave the following results. (1) Drying of the steam exploded samples caused insignificant changes in range except for a sharpening of the C_6 carbon peak for Cell-II, compared with the range for the steam exploded samples. (2) The ^{13}C NMR range for the treated Cell-I sample is similar to that for the original Cell-I but the peaks of Cell-II are somewhat broader than those of the original Cell-II.

The values of $\chi_{am}(C_k)$ ($k = 2, 3,$ and 6) for the original samples and the samples treated at $t = 30$ and 180 s under $P = 2.9$ MPa are evaluated: both $\chi_{am}(C_3)$ and $\chi_{am}(C_6)$ of Cell-I increase with only the addition of water at room temperature, and $\chi_{am}(C_3)$ and $\chi_{am}(C_6)$ of the Cell-I samples at Steps 3 and 4 keep the level higher than for the original Cell-I over a wide range of treatment time. $\chi_{am}(C_3)$ of Cell-II increases with simple wet-treatment, but the steam explosion treatment followed by washing and drying considerably lowers $\chi_{am}(C_3)$ of Cell-II. Longer treatment time tends to give the lower $\chi_{am}(C_3)$ value for the treated Cell-II sample at Step 4. This may be closely correlated with the experimental fact that S_a of Cell-II samples is not significantly improved by the steam treatment, even if P_v of the steam-treated sample is as low as 73. The Cell-III samples at Steps 3 and 4 obtained under the conditions of $P = 2.9$ MPa and $t = 30$ s contain cellulose I and cellulose III crystals, as proven by X-ray analysis, and it is difficult to estimate accurate values of $\chi_{am}(C_3)$ due to the lack of a proper method for Cell-III. In contrast to this, the crystal form (cellulose III) of the Cell-III samples at Steps 3 and 4 treated under the steam explosion conditions of $P = 2.9$ MPa and $t = 180$ s is ascertained to be transformed almost completely from cellulose III to cellulose I crystal during the steam explosion treatment. In fact, all $\chi_{am}(C_k)$ values lie in the same range as those of the Cell-I samples at Steps 3 and 4 prepared under the same steam explosion conditions.

All the data points can be represented by two straight lines:

$$S_a = -0.080P_v + 116.6 \qquad \text{for Cell-I and Cell-III} \qquad (5.1.4)$$

$$S_a = -0.031P_v + 45.9 \qquad \text{for Cell-II} \qquad (5.1.5)$$

The original Cell-I and Cell-III, with the same P_v, have almost the same S_a. S_a of Cell-II samples is by far lower than the S_a values for Cell-I and Cell-III samples at the same P_v level. Clearly, the dependence of S_a on P_v should be investigated within the framework of cellulose samples of the same origin with the same crystal forms. In other words, S_a of cellulose is a function of the crystal form as well as of P_v.

The solubility of the cellulose with the crystal form of cellulose II obtained by the alkali cellulose method is much lower than that of celluloses with the crystal forms of cellulose I and III at the same P_v level. The well known and widely accepted experimental fact that cellulose II crystal is energetically more stable than the others[37] is the most probable reason for the results in this study. But the solubility of the so-called regenerated cellulose (Bemberg fiber®), prepared by dissolving cellulose (cotton linter) once into cuprammonium solution and then by regenerating from cellulose cuprammonium complex with acid, was improved from 62 to 100% by the steam explosion treatment. This raises another problem which needs further study.

It is interesting to note that the peaks responsible for the strong intramolecular hydrogen bonds in the ^{13}C NMR range of the original Cell-II appeared more sharply than those for the steam treated sample (after drying). Nevertheless, X-ray diffractometry revealed the higher crystallinity and its higher perfection for the latter sample. This again implies that CP/MAS ^{13}C NMR and X-ray diffraction give different information on the molecular packing of glucopyranose units in cellulose solids and the parameters such as the crystallinity, estimated from the latter method, never govern the solubility of cellulose towards aqueous solution.

In summary, we found that, as judged from changes in morphology, P_v and S_a, the effectiveness of a relatively mild steam explosion treatment on cellulose decreases in the following order: Cell-I > Cell-III > Cell-II. The severe steam explosion treatment gave the lowest level-off P_v to Cell-II. Cellulose III crystal was almost completely converted to cellulose I during the severe steam explosion treatment. The decrease in P_v of all samples by the steam explosion treatment can be reasonably described by two or three steps of the first-order decomposition kinetics with different rate constants, depending on the different morphological changes during the treatment. The solubility, S_a, against aqueous alkali increased remarkably by the steam explosion for Cell-I and Cell-III, but showed only a slight improvement in S_a for Cell-II by the same treatment. The extent of increase in S_a by the treatment is Cell-I $\geq$ Cell-III $\gg$ Cell-II. $\chi_{am}(X)$, as estimated from the X-ray diffraction method, decreased by 10–20% for cellulose having cellulose I or II crystal by the steam explosion. That is, the steam explosion, like other heat-treatment procedures, increases the X-ray crystallinity of cellulose, but concurrently improves the solubility. Although a unique and common structural parameter explaining the solubility of celluloses with different crystal forms has not yet been found, the solubility of Cell-I and Cell-II can be judged from $\chi_{am}(C_3)$, irrespective of their preparative methods. $\chi_{am}(C_3)$ increased significantly by the steam explosion for Cell-I and Cell-III, which showed excellent solubility in aqueous alkali.

REFERENCES

1. T Yamashiki, T Matsui, M Saito, K Okajima, K Kamide and T Sawada, *Br. Polym. J.*, 1990, **22**, 73.
2. C Beadle and HP Stevens, *Eighth Int. Cong. Appl. Chem.*, 1912, **13**, 25.
3. SM Neals, *J. Text. Inst.*, 1920, **20**, T373.
4. H Dillenius, *Kunstseide Zellwolle*, 1940, **22**, 314.
5. E Schwart and W Zimmerman, *Milliands Textilber. Inst.*, 1941, **22**, 525.
6. H Staudinger and RJ Mohr, *Prakt. Chem.*, 1941, **158**, 233.
7. O Eisenluth, *Cell. Chem.*, 1941, **19**, 45.
8. K Kamide, K Okajima, T Matsui and K Kowaska, *Polym. J.*, 1984, **16**, 857.
9. K Tugamura, A Miyazaki, Y Kawashima, T Higuchi and M Tanahashi, *Jpn. J. Zootechn. Sci.*, 1983, Suppl., 45.
10. H Mamers, D Menz and PJ Menz, *Appita*, 1979, **33**, 201.
11. RH Leitheiser, BR Bongner and FC Grant-Acquah, *Proceedings of the Symposium on Wood Adhesives Research, Application and Needs*, USDA Forest Products Laboratory, Madison, 1980, p. 50.
12. K Kowsaka, K Okajima and K Kamide, *International Symposium on New Functionalisation Developments in Cellulosics and Wood, CELLUCON 88*, Kyoto, Japan, abstract, 1988, p. 18.
13. T Sawada, *Proceedings of the International Symposium on Wood and Pulping Chemistry*. Vancouver, Canada, 1985.
14. K Kamide, K Okajima and K Kowsaka, *Polym. J.*, 1992, **24**, 71.
15. L Segal, J Creely, A Martin and C Conra, Jr., *Text. Res. J.*, 1959, **29**, 786.
16. P Scherrer, *Gouinger Nachr.*, 1918, 98.
17. K Kamide, K Okajima and K Kowsaka, *Polym. J.*, 1985, **17**, 701.
18. K Todheid, in *Water—A Comprehensive Treatise*, Vol. 1, (ed. F Franks), Plenum Publishers, New York, 1972, p. 463.
19. RH Marchessault and J St-Pierre, in *Proceedings of the CHEMRAWN Conference*, (eds GR Brown and LE St-Pierre), Toronto, Canada, July 1978, Pergamon Press, New York, 1980, pp. 611–625.
20. M Tanahashi and T Higuchi, *Polym. Appl.*, 1982, **32**(12), 595.
21. F Horii, A Hirai and R Kitamaru, *Macromolecules*, 1987, **20**, 2117.
22. A Hirai, F Horii and R Kitamaru, *Sen-i Gakkai Symposium*, Preprints, 1988, C105.
23. A Sarko and R Muggi, *Macromolecules*, 1974, **7**, 486.
24. K Kamide, K Kowsaka and K Okajima, *Polym. J.*, 1985, **17**, 707.
25. For example, J Hayashi, T Yamada and K Kimura, *Appl. Polym. Symp.*, 1976, **28**, 713.
26. T Yamashiki, K Kamide, K Okajima, K Kowsaka, T Matsui and H Fukase, *Polym. J.*, 1988, **20**, 447.
27. T Yamashiki, T Matsui, M Saito, K Okajima and K Kamide, *Br. Polym. J.*, 1990, **22**, 121.
28. K Kamide, K Okajima, T Matsui and K Kowsaka, *Polym. J.*, 1984, **16**, 857.
29. H Staudinger and M Sorkin, *Ber. Dtsch. Chem. Ges.*, 1937, **70B**, 1565.
30. K Todheid, in *Water—A Comprehensive Treatise,* Vol.1, (ed. F Franks), Plenum Publishers, New York, 1972, p. 463.
31. T Yamashiki, T Matsui, M Saitoh, Y Matsuda, K Okajima and K Kamide, *Br. Polym. J.*, 1990, **22**, 201.
32. H Sobue, H Kiessig and K Hess, *Z. Physik. Chem.*, 1939, **B43**, 309.
33. KT Hess and C Trogus, *Ber. Dtsch. Chem. Ges.*, 1935, **68**, 1986.
34. ML Nelson and VN Tripp, *J. Polym. Sci.*, 1953, **10**, 577.

35. T Watanabe, J Hayashi, A Sueoka and T Akabori, *20th Annual Meeting of Polymer Science, Japan*, preprint, 1971, p. 427.
36. T Akabori, J Hayashi and T Watanabe, *Nihon Kagaku Kaishi*, 1973, **3**, 594.
37. For example, A Sarko, in *Cellulose* (eds RA Young and RM Rowell), Wiley, New York, 1986.

5.2 STRUCTURAL FACTORS GOVERNING DISSOLUTION INTO SOLVENTS

5.2.1 Dissolution of natural cellulose into aq. alkali solution[1]

As is well recognized, cellulose is not soluble either in water or in aqueous aq. alkaline solutions despite the high density of hydroxyl groups in the molecule. However, cellulose chemists only know that cellulose is strongly swollen by aq. alkali solution, resulting in partial dissolution of cellulose.[2] Extensive studies of the swelling phenomena and structural change of cellulose by the action of aq. sodium hydroxide (NaOH) were started after Mercer's discovery (the mercerization process) as early as 1842[3] and have been extensively continued up to now.

Numerous authors[2,4–7] pointed out, without a description of the degree of polymerization (DP), that approximately 8–10 wt% aq. NaOH solutions have the greatest capability to swell towards cellulose and partly dissolve the polymer at temperatures below 10 °C. The solubility, S_a, of various cellulose samples in 9% (w/v) aq. NaOH solution at 'room temperature' was as follows:[4] cellulose fiber obtained by denitration of cellulose nitrate fiber, 90%; viscose rayon, 35–45%; cuprammonium rayon, 30%; wood pulp, 15%; cotton, 6%. Here, great care should be paid to the experimental results on denitrated cellulose fibers in aq. NaOH with a concentration of 8–10 wt% because there is the possibility that very small amounts of residual nitrate groups, which would greatly facilitate solubility towards alkali or water, may remain in the fibers. This effect of residual functional groups has been demonstrated in the case of other cellulose derivatives, such as sodium carboxymethyl cellulose and cellulose acetate, with very low degree of substitution.[8,9] Staudinger *et al.*[4] reported, without a detailed description of the experimental conditions, somewhat different results; that cotton and mercerized cotton could dissolve in a 10% (w/v) aq. NaOH solution when the viscosity-average degree of polymerization P_v was below 400, and that regenerated cellulose fibers with P_v of less than 1200 could also dissolve in 10%(w/v) aq. NaOH. However, those observations were not reproducible, as reported in our previous study.[10] Regarding the solubility of cellulose towards alkali solution, it is worthwhile to note that the soda cellulose Q (Na cell Q, 'Q' is after 'quellung'), which was found by Sobue *et al.*,[11] dissolves in 8%(w/v) aq. NaOH solution at 6 °C when the polymer concentration is less than 0.1 wt%.[12] The observation that a part of cellulose sample dissolves in aq. NaOH solutions at various NaOH concentrations has been used to estimate the 'lateral order distribution', an imaginary distribution function of ordering or hydrogen bond density, of cellulose fibers.[3] In other words, the solubility of cellulose towards aq. NaOH with the specific concentration (*ca.* 8–10 wt%) was believed to be governed by the lateral order distribution, i.e. quantitative distribution of ordering of the hydrogen bonds.

The lateral order distribution concept would be much more useful if the distribution of ordering of hydrogen bonds could be correctly estimated by other methods, and the distribution correlated with the solubility, as compared with the very old and plausible concept of Mark[5] and Hermans,[6] that the solubility or swelling ability of cellulose is controlled by crystallinity as measured by the X-ray diffraction method. The latter concept seems to have lost its basis by recent experimental findings on the solubility of regenerated cellulose by Kamide and his coworkers[10] (Section 4.1). They regenerated cellulose samples, from cuprammonium solution of purified cotton, using three different methods which involved (a) simultaneous coagulation and regeneration in aq. H_2SO_4, (b) partial evaporation of ammonia from the solution followed by regeneration in aq. acid, and (c) coagulation in acetone followed by regeneration in aq. acid. These regenerated cellulose samples had practically the same viscosity-average molecular weight $M_v = ca.$ 8×10^4, but showed different solubilities in the range 52–100% toward 9.1 wt% aq. NaOH solution at 4 °C when the polymer concentration was kept as high as 5 wt%. Kamide *et al.* proved that the solubility of the regenerated cellulose is primarily governed by the degree of break down of O_3–H$\cdots O_5'$ intramolecular hydrogen bond ($\chi_{am}(C_3)$), estimated by solid state cross-polarization magic angle sample spinning (CP/MAS) ^{13}C NMR,[13,14] but not by amorphous content ($\chi_{am}(X)$), estimated from the X-ray crystallinity or by accessibility estimated from deuteration-IR method (Section 4.2). Recently, Yamashiki and co-workers[15] also clarified the specific solvation structure (existence of weakly bonded water layer on strongly solvated water molecules to OH^- anions and Na^+ cations) of aq. NaOH solution with specific concentration (8–10 wt%) as a cellulose solvent at low temperature, by measuring electro-conductivity, adiabatic compressibility, optical rotation, Raman spectroscopy, ^{1}H and ^{23}Na NMR chemical shifts and their spin lattice relaxation times for aq. NaOH and cellobiose/aq. NaOH solutions as a function of alkali concentration (Section 4.3). They suggested that the weakly bonded water layer, which is, of course, surrounded by free water, acts as an energy transfer layer when cellulose dissolves in aq. NaOH solution of a specific concentration.[15] This might be a breakthrough result indicating that complete dissolution of cellulose in solvent can be realized only by an appropriate combination of the specific supermolecular structure of cellulose solid and the specific solvent structure. Discovery of the completely alkali soluble regenerated cellulose motivated their attempt to establish an industrially meaningful method for the preparation of a completely alkali soluble cellulose with the crystal form of cellulose I (hereafter denoted as Cell-I; see, Section 5.1).

Consequently, the research group successfully developed the steam explosion treatment as an appropriate transforming method for the molecular aggregation state or supermolecular structure of natural cellulose (wood pulps) into the alkali soluble form.[15] $\alpha_{s,3}^w$ [15–19] On a soft wood pulp, they disclosed changes in solubility S_a, P_v (by viscosity), morphology (by electron microscope), degree of breakdown in intramolecular hydrogen bonds at C_k positions ($\chi_{am}(C_k)$, $k = 2$, 3, and 6) (by CP/MAS ^{13}C NMR) and amorphous content $\chi_{am}(X)$ (by X-ray diffraction) as function of steam explosion conditions. They found a significant positive correlation between S_a and $\chi_{am}(C_3)$ or $\chi_{am}(C_6)$ although the M_v of the treated soft wood pulps were not the same, but ranged over 8–3.3×10^4 making a correlation between S_a and $\chi_{am}(X)$ impossible. Since steam explosion proved to accelerate acid hydrolysis of cellulose, it is difficult to prepare celluloses with similar P_v values and S_a varies widely in a series of steam explosion experiments.

They established structural parameters which primarily govern the solubility of cellulose with crystal form of cellulose II where the P_v level was the same and confirmed that the solubility of cellulose with the crystal form of cellulose II (including that prepared through Mercerization of Cell-I; hereafter denoted as Cell-II) was mainly governed by $\chi_{am}(C_3)$.[10] It remains necessary to establish supermolecular structural parameters which govern the solubility of a Cell-I solid.

In this section, an attempt has been made to prepare natural celluloses with almost the same P_v value but with a wide range of S_a values using the steam explosion method or conventional acid hydrolysis, and to find the supermolecular structural parameter governing the solubility of Cell-I.[10] For this purpose, IR spectral parameters expressing local molecular vibration corresponding to conformational difference in cellobiose unit were measured, besides $\chi_{am}(C_k)$ and $\chi_{am}(X)$. Structural factors governing solubility of cuprammonium cellulose, cellulose acetate and carboxy-ethyl cellulose will be described in Sections 5.2.2–5.2.4.

Cellulose samples

The starting samples included soft wood (mainly white spruce) pulp Alaska pulp (USA), manufactured by Alaska Pulp Co.: α-cellulose content, 90.1 wt%; viscosity-average molecular weight determined from the limiting viscosity number in cadoxen $M_v = 1.72 \times 10^5$ and a purified cotton linter (*Gossypium hirsutum*; imported from USA and purified by Asahi Chemical Ind. Co., Ltd.: α-cellulose content, 95.7%; $M_v = 2.6 \times 10^5$). Raw cotton linter was purified by 'alkali cooking' (3.5 wt% aq. NaOH; at 172 °C for 90 min), bleaching (at 25 °C for 45 min) with chlorine (270 ppm), and dipping into aq. alkaline solution, followed by washing and drying. The soft wood pulp was cut into small pieces ($10 \times 5 \times 1.5$ mm^3) and subjected to steam explosion, conventional acid hydrolysis, and ball milling (in the dry state).

Steam explosion treatment and acid hydrolysis

The steam explosion treatment of cellulose was carried out using the same apparatus as described in previous studies.[16–19] Steam exploded samples were washed with an excess of water and acetone and finally dried in air as previously described[16–19] (see Figure 5.1.1). The acid hydrolysis of cellulose was carried out in a 5 l round bottom flask with 5N sulfuric acid (cellulose: acid media $= 1{:}20$(w/w)) at various temperatures. To obtain a series of cellulose samples with almost the same M_v ($= ca.$ 7.3×10^4) and different S_a ($= 11$–100%), a considerable number of trial and error steam explosion and acid hydrolysis experiments were carried out. Finally, the procedures, as summarized in the third column of Table 5.2.1, were chosen. The soft wood pulp was ball milled at 25 °C for 4 and 8 h and the samples prepared this way were used as references. Their molecular parameters and S_a are also listed in Table 5.2.1.

Viscosity-average molecular weight M_v of cellulose

M_v was determined from the limiting viscosity nurriber $[\eta]$ ml g^{-1} in cadoxen (cadmium oxide–ethylenediamine–NaOH–H$_2$O $= 5{:}28{:}1.4{:}165.6$, w/w/w/w) at 25 °C

Table 5.2.1

Preparation conditions, M_v, S_a, and structural parameters of cellulose samples[1]

Sample code	Original cellulose	Preparation conditions	M_v (10^4)	S_a (%)	$\chi_{am}(X)$ (%)	$\chi_{am}(C_3 + C_6)$ (%)	A_r (897)	Reference
P-0	–	Soft-wood pulp, as-received	17.2	31	22	44	0.37	
EP-1	P-0	Steam exploded (with 8% NaOH, 2.9 MPa, 30 s)	7.4	99	19	53.5	0.47	
EP-2	P-0	Steam exploded (with water, 2.9 MPa, 15 s)	7.6	84	24	48	0.44	
EP-3	P-0	Steam exploded (with water, 2.4 MPa, 30 s)	7.3	70	24	46.5	0.42	17
EP-4	P-0	Steam exploded (with water, N_2, 5 kg cm^{-2}, 2.9 MPa, 30 s)	5.8	100	–	48	–	
AP-1	P-0	Acid-hydrolyzed ($5N$-H_2SO_4, 60 °C, 120 min)	7.2	64	27	42.5	0.44	
PC-1	P-0	Ball milled (25 °C, 8 h)	1.9	58	92	46.2	–	10
PC-2	P-0	Ball milled (25 °C, 4 h)	9.6	61	73	88.0	0.42	
C-0	–	Purified cotton linter	26.0	25	19	31	0.35	
EC-1	C-0	Steam exploded (with water, 2.9 MPa, 180 s)	9.0	11	17	28	0.34	
EC-2	C-0	Steam exploded (with water, 4.9 MPa, 180 s)	7.1	11	15	33	0.33	16

using the following equation:[20]

$$[\eta] = 3.85 \times 10^{-2} M_w 0.76 \qquad (5.2.1)$$

where M_w is the weight-average molecular weight.

Solubility (S_a) of cellulose

Five grams (g) on a dry basis of cellulose (water content about 8–12 wt%) were dispersed into 95 g of a 9.1 wt% aq. NaOH solution, precooled at 4 °C, and allowed to stand for 8 h with intermittent mixing by a homogenizer. The insoluble part was extracted by centrifugation at $5 \times 10^3 g$ for 60 min. The insoluble alkali-swollen cellulose was neutralized with 1.3 wt% aq. hydrochloric acid in order to precipitate the cellulose completely. The precipitate was washed with pure water and dried in air and *in vacuo* to constant weight (denoted as w in the unit of gram). Solubility S_a was calculated through use of the relationship:[10]

$$S_a = 100 \times (5 - w)/5(\%) \tag{5.2.2}$$

Scanning electron microscopic (SEM) observation

Gold was sputtered onto the samples by a Fine coat ion sputter JFC-1 100, JEOL Ltd, Japan, and the sputtered samples were observed and photographed with a scanning electron microscope (Scanning microscope type JMS-35CF, JEOL Ltd, Japan).

X-ray diffractometry

X-ray diffraction patterns of the samples were recorded on an X-ray diffractometer (Rotor Flex RU-200PL, Rigaku Denki Co., Ltd, Japan) with a position sensitive proportional counter (PSPC) (Type CN5791Pl PSPC-5, Rigaku Denki Co. Ltd, Japan) by a reflection method. The amorphous fraction $\chi_{am}(X)$ of the sample was calculated from the crystallinity $\chi_{am}(X)$, which was estimated by Segal's method,[21] using the relationships:

$$\chi_{am}(X) = 100(I_{002} - I_{am})/I_{002}(\%) \tag{5.2.3}$$

$$\chi_{am}(X) = 100 - \chi_c(X)(\%) \tag{5.2.4}$$

Here, I_{002} and I_{am} are the mean peak intensities, corresponding to (002) plane ($2\theta = 22.6°$) and amorphous region ($2\theta = 19.0°$; see Figure 5.2.3).

^{13}C NMR measurement

Solid state CP/MAS ^{13}C NMR ranges were recorded on an FT NMR spectrometer (FX-200, JEOL Ltd, Japan) under the following operating conditions: data points, 8192 (4096 zero-filling); accumulation, 200–1024; pulse width, 5 μs; contact time, 2 ms; pulse interval, 5 s; spectral width, 20,000 Hz; acquisition time, 102.4 ms. Full peak assignments, established by Kamide *et al.*,[13] of ^{13}C NMR spectra (see Figure 5.2.5) were subsequently employed. Accordingly, seven envelopes numbered from lower magnetic fields were assigned as follows: (1) C_1 carbon; (2) C_4 carbon belonging to cellobiose units which participate in $O_3–H\cdots O_5'$ (denote the neighboring anhydroglucopyranose (AHG) unit) intramolecular hydrogen bond; (3) C_4 carbon which does not participate in intramolecular hydrogen bond; (4) C_5 carbon overlapped with C_3 and C_2 carbons which do not participate in intramolecular hydrogen bonds; (5) C_3 and C_2 carbons belonging to cellobiose units which participate in $O_3–H\cdots O_5'$ and $O_2–H\cdots O_6'$ intramolecular hydrogen bonds, respectively; (6) C_6 carbon which participates in intramolecular hydrogen bond; (7) C_6 carbon belonging to cellobiose units which does not participate in $O_2–H\cdots O_6'$ intramolecular hydrogen bond (see Table 5.2.2).

The degree of breakdown of $O_3–H\cdots O_5'$ intramolecular hydrogen bonds $\chi_{am}(C_3)$, and degree of breakdown of $O_2–H\cdots O_6'$ intramolecular hydrogen bonds $\chi_{am}(C_6)$ were

Table 5.2.2

Chemical shifts, half value of widths and assignments for carbon peaks in
CP/MAS spectrum of cellulose I[1]

No.	δ (ppm)	Δv (Hz)	Assignment
1	105	145	C_1
2	89	130	$C_4(O_3-H\cdots O_5')$
3	83	210	C_4
4	75	150	C_2, C_3, C_5
5	72	160	$C_2(O_2-H\cdots O_6')$, $C_3(O_3-H\cdots O_5')$
6	65	130	$C_6(O_2-H\cdots O_6')$
7	62	180	C_6

independently estimated, analyzing the C_4 carbon region (envelopes No. 2 and 3 in
Figure 5.2.5) and C_6 carbon region (envelopes No. 6 and 7 in Figure 5.2.5) of the
spectrum, respectively, using eqs. (5.2.5) and (5.2.6):

$$\chi_{am}(C_3) = 100\{I_3/(I_2 + I_3)\}(\%) \tag{5.2.5}$$

$$\chi_{am}(C_6) = 100 \times \{I_7/(I_6 + I_7)\}(\%) \tag{5.2.6}$$

Here, I_k ($k = 1-7$) are the intensities of the fractions of envelopes numbered from lower
magnetic field. The parameter $\chi_{am}(C_2)$ represents the degree of breakdown in $O_2-H\cdots O_6'$
intramolecular hydrogen bonds which cannot be estimated directly due to heavy
peak overlapping, but $\chi_{am}(C_2 + C_3)$ [i.e. an average of $\chi_{am}(C_2)$ and $\chi_{am}(C_3)$
($\equiv (\chi_{am}(C_2) + \chi_{am}(C_3))/2$] can be determined from the C_2, C_3, and C_5 carbon regions
(envelopes No. 4 and 5 in Figure 5.2.5) of the spectrum as follows:[16]

$$\chi_{am}(C_2 + C_3) = 100 \times [\{3I_4/(I_4 + I_5)\} - 1]/2(\%) \tag{5.2.7}$$

It is important to note that the estimation of these parameters is only possible for the
Cell-I structure because cellulose I is expected to have a single $O_2-H\cdots O_6'$ type
intramolecular hydrogen bond, while Cell-II is shown to possess two types of $O_6\cdots O_2'$
($O_6-H\cdots O_2'$ and $O_2-H\cdots O_6'$) intramolecular hydrogen bonds, resulting in considerable
overlapping of NMR peaks.[13] Using $\chi_{am}(C_3)$, $\chi_{am}(C_6)$, and $\chi_{am}(C_2 + C_3)$ thus
determined, other structural parameters such as $\chi_{am}(C_2)$, $\chi_{am}(C_2 + C_6)$, $\chi_{am}(C_3 + C_6)$,
and $\chi_{am}(total)$ are defined by the relationships:

$$\chi_{am}(C_2) = \chi_{am}(C_2 + C_3) \times 2 - \chi_{am}(C_3) \tag{5.2.8}$$

$$\chi_{am}(C_2 + C_6)) = \{\chi_{am}(C_6) + \chi_{am}(C_2)\}/2 \tag{5.2.9}$$

$$\chi_{am}(C_3 + C_6) = \{\chi_{am}(C_3) + \chi_{am}(C_6)\}/2 \tag{5.2.10}$$

$$\chi_{am}(total) = \{\chi_{am}(C_2 + C_3) \times 2 + \chi_{am}(C_6)\}/3 \tag{5.2.11}$$

FT-IR measurement

FT-IR spectra of the samples were recorded by a diffusional reflection method on an FT-IR spectrometer (JIR 3505, JEOL, Japan) under operating conditions as follows: resolution, 2 cm^{-1}; accumulation, 100; interpolation, 0; window function, triangle. The relative absorbance for the bands attributable to intra- and intermolecular hydrogen bonds was calculated using the CH stretching band (maximum absorption peak in the wave number 2904–2899 cm^{-1}) as an internal standard.

Figure 5.2.1 shows plots of the solubility S_a in 9 wt% aq. NaOH at 4 °C for all cellulose samples as a function of the viscosity-average molecular weight M_v. The figure is principally constructed on the basis of data in Table I of Ref. 16 and Figure 5 of Ref. 17, to which new data[22] are added. It can be seen from Figure 5.2.1, that a strongly negative correlation between S_a and M_v, as Yamashiki *et al.* first demonstrated, is again confirmed. In the figure, data points of samples having nearly the same M_v ($= 7.1 \times 10^4$–9.0×10^4) are represented as filled circles and such data points scatter over quite a wide range of S_a (from 11 to 99%). This explicitly indicates that M_v is not the sole factor controlling S_a of cellulose. The fourth and fifth columns of Table 5.2.1 show M_v and S_a of the original (untreated) and treated cellulose samples with crystal form of cellulose I. The table also includes literature data for samples EP-3,[17] PC-1,[16] and EC-2.[10] The samples EP-1-3, AP-1, and EC-2 have practically the same M_v values ($7.35 \pm 0.25 \times 10^4$), but M_v, of EC-1 slightly deviates from them. The M_v of PC-1 is lower than those for EP-1-3 and AP-1 by approximately four times. The M_v of PC-2 is in the same range as EC-1. The S_a values for all samples prepared from common original wood pulp in the wet state vary over the range 100–64%, all being far higher than that of the untreated wood pulp (sample P-0; $S_a = 31\%$). In contrast, the S_a ($= 11\%$) values of samples (EC-1 and EC-2) prepared from original cotton linter are less than half of the untreated one (sample code C-0; $S_a = 25\%$) despite considerable depolymerization. Note that S_a of two ball milled

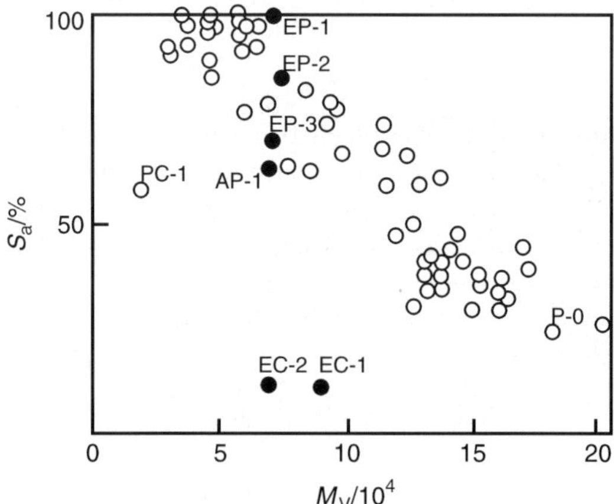

Figure 5.2.1 Plots of solubility S_a against the viscosity-average molecular weight M_v of all cellulose (cellulose I) samples.[1]

celluloses (PC-1 and PC-2) with $M_v = 1.9 \times 10^4$ and 9.6×10^4 are both low (60%) in spite of very low X-ray crystallinity ($\chi_{am}(X) = 92$ and 73%), but their $\chi_{am}(C_3 + C_6)$ values are quite different (46.2 and 88.0%). Ball-milling in a dry state is probably not a reliable method to give an evenly treated sample and therefore these samples should be disregarded for further analysis.

Figure 5.2.2 shows SEM micrographs of samples EP-1-3, AP-1, and EC-1. A comparison of this figure (Figure 5.2.2(a)–(c)) with the morphology of untreated pulp (Figure 3 of Ref. 16) shows that steam explosion treatment on soft wood pulp (P-0) significantly shortens the fiber length of pulp and that fibrillation took place to some extent during treatment. An acid hydrolyzed wood pulp (AP-1) also has morphology similar to the steam exploded pulp. On the other hand, even drastic steam explosion treatment like 2.9 MPa (*ca.* 230 °C) for 180 s on cotton linter (C-0) brings about neither shortening of fiber length nor fibrillation, as previously shown.[16] From this figure we consider that the morphological difference observed among various samples does not seem to be a key factor governing S_a because specific surface areas of samples do not differ greatly.

Figure 5.2.3 shows X-ray diffraction patterns for the original cellulose samples (wood pulp (a) and cotton linter (b)) and treated cellulose samples (c–g). The values of $\chi_{am}(X)$

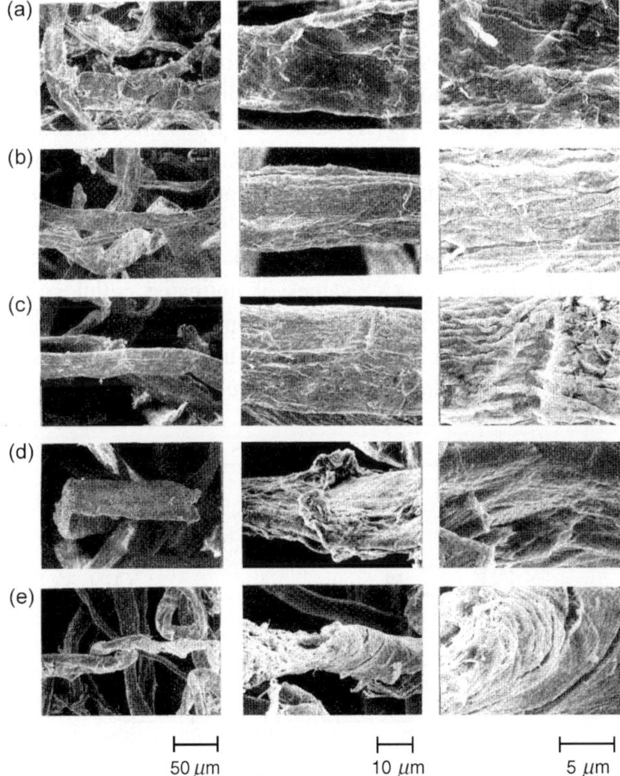

Figure 5.2.2   SEM micrographs of treated cellulose samples:[1] (a) EP-1; (b) EP-2; (c) EP-3; (d) AP-1; (e) EC-1.

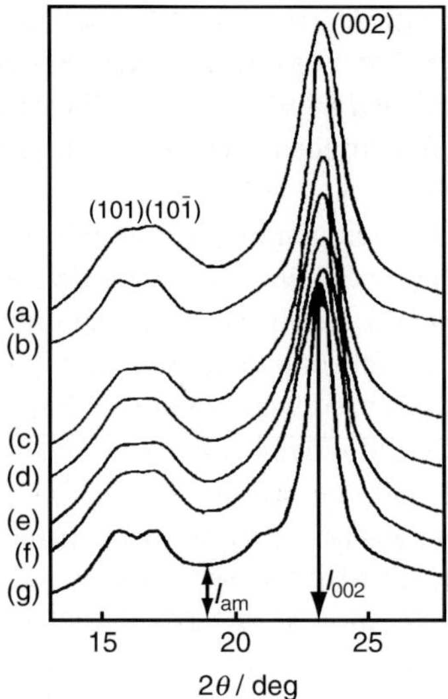

Figure 5.2.3   X-ray diffractograms of cellulose samples:[1] (a) P-0; (b) C-0; (c) EP-1; (d) EP-2; (e) EP-3; (f) AP-1; (g) EC-1.

estimated from the figure are listed in the sixth column of Table 5.2.1, in which the literature data for EP-3, PC-1, and EC-2 are also included. Note that EP-1 and EC-1, with not so different M_v, have similar $\chi_{am}(X)$ values (19 and 11%, respectively). This means that $\chi_{am}(X)$ is not a structural parameter controlling S_a of cellulose, and therefore invalidating the long held hypothesis that S_a is proportional to $\chi_{am}(X)$. This situation is clearly demonstrated in Figure 5.2.4, where the values of S_a of cellulose with the viscosity-average molecular weight ranging from 7.1×10^4 to 9.0×10^4 are plotted against $\chi_{am}(X)$. In the figure, $\chi_{am}(X)$ values of untreated cellulose samples are also plotted as rectangles. Although untreated wood pulp and untreated cotton linter have similar S_a and $\chi_{am}(X)$ values, the steam explosion drastically increases S_a for the former sample, but decreases S_a for the latter. Over an entire $\chi_{am}(X)$ range, there is no significant

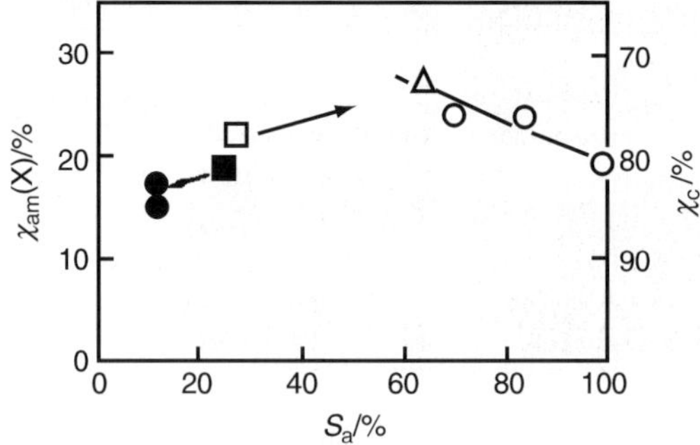

Figure 5.2.4   Plot of $\chi_{am}(X)$ versus S_a of cellulose samples:[1] (□), sample code P-0; (■), sample code C-0; (○), sample codes EP-1-3; (△), sample code AP-1; (●), sample codes EC-1 and 2.

correlation between S_a and $\chi_{am}(X)$ (correlation coefficient $\gamma = 0.616$) and if only the steam exploded wood pulps are considered, there seems to be a significant but negative correlation between S_a and $\chi_{am}(X)$ ($\gamma = 0.931$). That is, the negative correlation suggests that the solubility of cellulose seems proportional to its X-ray crystallinity.

Figure 5.2.5 shows CP/MAS ^{13}C NMR spectra for the original cellulose samples (wood pulp (a) and cotton linter (b)) and treated cellulose samples (c–g). All samples have seven NMR peaks, which are numbered systematically on the spectra from the low magnetic field. Envelopes No. 3 and 7 are very broad. The relative area of envelopes (No. 3, 4, and 7), shadowed in the figure, decreases in the order as follows: EP-1 > EP-2 > EP-3 > AP-1 > EC-1. In this case, the reverse is really true for envelopes No. 2, 5, and 6. Through use of eqs. (5.2.4)–(5.2.11), the structural parameters, such as $\chi_{am}(C_k)$ ($k = 2, 3,$ and 6), $\chi_{am}(C_k + C_l)$ ($k = 2, 3$ and $l = 3, 6$), and $\chi_{am}(\text{total})$ (see eq. (5.2.11)), all related to intramolecular hydrogen bonds, are evaluated and plotted against S_a in Figure 5.2.6, in which the correlation coefficient γ is also shown for each plot. Both $\chi_{am}(C_3)$ and $\chi_{am}(C_6)$ are highly correlated with $S_a(\gamma = 0.989)$. This means that the solubility of Cell-I with almost the same M_v is also governed by the extent of the formation of $O_3–H\cdots O_5'$ intramolecular hydrogen bonds, as previously observed for Cell-II.[10]

Figure 5.2.7(a) illustrates a cellobiose unit constituting a cellulose molecule, in which $O_3–H\cdots O_5'$ and $O_2–H\cdots O_6'$ intramolecular hydrogen bonds, expected to exist for Cell-II,[10] are formed. As is evident from Figure 5.2.7(a), the simultaneous coexistence of the two kinds of intramolecular hydrogen bonds in a naturally occurring cellulose chain makes the chain a kind of very inflexible ladder polymer, building up a rigid planar

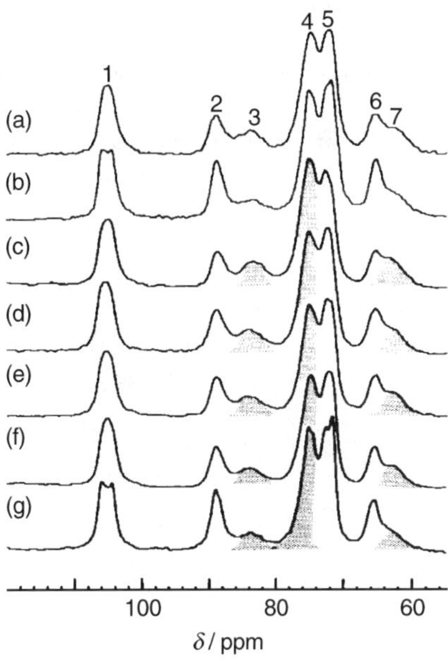

Figure 5.2.5   CP/MAS ^{13}C NMR spectra of cellulose samples:[1] (a) P-0; (b) C-0; (c) EP-1; (d) EP-2; (e) EP-3; (f) AP-1; (g) EC-1.

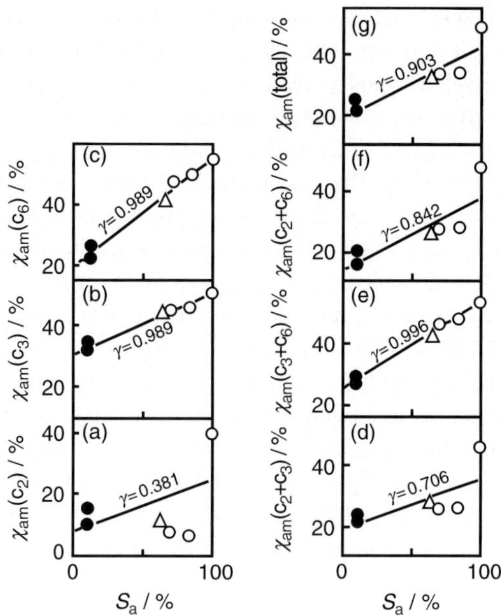

Figure 5.2.6 Plot of $\chi_{am}(C_k)$s versus S_a of cellulose samples:[1] (a) $\chi_{am}(C_2)$; (b) $\chi_{am}(C_3)$; (c) $\chi_{am}(C_6)$; (d) $\chi_{am}(C_{2+}C_3)$; (e) $\chi_{am}(C_{3+}C_6)$; (f) $\chi_{am}(C_{2+}C_6)$; (g) $\chi_{am}(total)$. γ indicates the correlation coefficients.

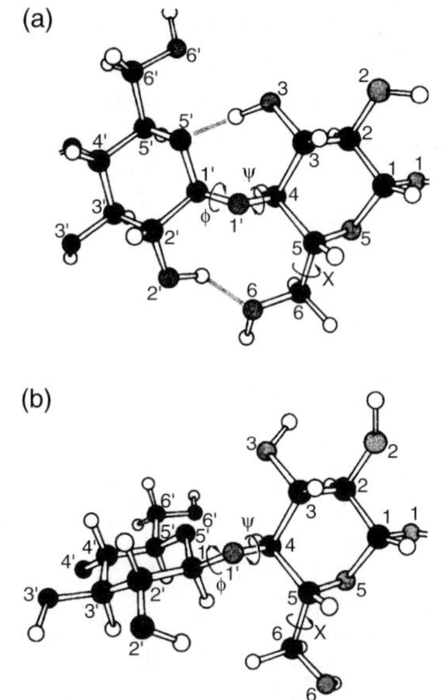

Figure 5.2.7 Schematic representation ('ball and stick' model) of cellobiose units constituting a cellulose molecule in which intramolecular hydrogen bonds in cellobiose units are formed (a) or not formed (b):[1] black balls, carbon atoms; shadowed balls, oxygen atoms; white balls, hydrogen atoms.

structure with fixed values of two torsional angles Ψ and Φ around glucosidic linkage $(C_1'-O-C_4)$ between two AHGS and a torsional angle X around C_5-C_6 linkage in a AHG at room temperature. Manabe $et\ al.$[23] obtained five kinds of glass transition temperatures T_g for cellulose filament, using the method of dynamic mechanical absorption, defining the related absorptions to be: α_{H_2O}, 33–53 °C; α_{sh}, 115–200 °C; $\alpha_{2,2}$, 140–240 °C; $\alpha_{2,1}$, 215–290 °C; α_1 and 285–305 °C. They tentatively assigned these absorptions as follows:[23] the α_{H_2O} absorption corresponds to the cooperative motion of cellulose chains and water molecules in the amorphous region, in which intra- and intermolecular hydrogen bonds are almost completely destroyed. α_{sh}, $\alpha_{2,2}$, and $\alpha_{2,1}$ absorptions are due to micro-Brownian movement of cellulose chains in the amorphous region, in which intra- and intermolecular hydrogen bonds are completely (α_{sh} and $\alpha_{2,2}$) and partially ($\alpha_{2,1}$) destroyed. The α_1 absorption corresponds to the molecular motion of cellulose chains belonging to the amorphous region, in which intra- and intermolecular hydrogen bonds are strongly and densely formed. Since α_{H_2O} can be relatively small in dry wood pulp, T_g is far higher than room temperature. Therefore, at room temperature, the molecular motion in the ladder-like polymer is extremely limited.

Figure 5.2.8 illustrates the top view (a) and side view (b) of three polymer chains packed in Cell-I crystal structure drawn with reference to X-ray diffraction and packing analysis by Stipanovic and Sarko.[24] In the figure, hydrogen atoms are omitted and in the side view, only one AHG unit per single chain is shown. In a Cell-I type solid, an intermolecular hydrogen bond (chain line in the figure) is formed only between the O_3 atom of a chain and O_6 hydroxyl group of the nearest neighboring chain, besides intramolecular hydrogen bonds,[24] which are shown as broken lines in the figure. In Cell-I type solids cellulose chains, located in parallel to each other, form a sheet structure by intermolecular hydrogen bonds. Such numerous sheets are stacked to build Cell-I type crystal structure and even the amorphous phase is not expected to deviate remarkably from the above supermolecular structure. Note that all oxygen atoms lie within a sheet and concern with hydrogen bonds with hydroxyl groups inside the sheet. In this sense,

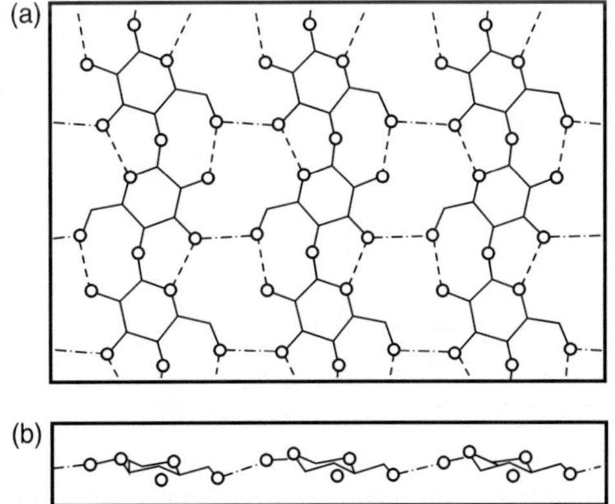

Figure 5.2.8   Schematic representation of the sheet structure of cellulose molecules in Cell-I type crystals:[1] (a) top view; (b) side view; (O), oxygen atoms; broken line, intramolecular hydrogen bonds; chain lines, intermolecular hydrogen bonds.

the AHG unit inside the sheet structure is hydrophobic. When this kind of untreated natural cellulose (Cell-I) solid is dipped into a solvent (in this case, aq. alkali), solvent molecules (in this case, hydrated sodium and hydroxide ions, see Figure 4.3.10[15]) find it difficult to penetrate the gap of the sheets. The solvent molecules which do penetrate, even though few in number, do not dissolve the cellulose strongly enough to destroy the stacking of the sheets. As a result, naturally occurring and untreated cellulose is not completely dissolved by aq. alkali solutions.

Cellulose with neither of the two kinds of intramolecular hydrogen bonds can be prepared by applying high temperature water vapor treatments including steam explosion. This can be interpreted qualitatively as follows: at higher temperatures (steam explosion treatment at the temperatures of $230-260\,°C$), water molecules can penetrate the gap between the hard planes and break the intramolecular hydrogen bonds, which in turn lowers the T_g. As a result, molecular motion is activated, giving rise to great variation of Φ, Ψ, and X values, destruction of the rigid planar structure, and enlargement of intermolecular spacing. When the treated sample is quickly cooled down to room temperature, which is apparently below T_g, the molecular motion is frozen (i.e. the various values of Φ, Ψ, and X, taken at higher temperature in wet state, are fixed) with only partial reformation of intramolecular hydrogen bonds. In the process, recrystallization occurs to some extent. This is demonstrated in Figure 5.2.7(b). Here, the C_4 and C_6 peaks are observed at higher magnetic field positions than those in Figure 5.2.7(a)[13] and at similar positions as those for cellulose/aq. NaOH solution.[10] These peaks (No. 3 and 7, respectively) are broad mainly due to the large scattering of torsional angles Φ, Ψ, and X.

The simultaneous appearance of peaks 2, 3, 6, and 7 in Cell-I solid (see Figure 5.2.5), indicates the existence of two kinds of cellobiose units: one with $O_3-H\cdots O_5'$ and $O_6-H\cdots O_2'$ intramolecular hydrogen bonds and one without intramolecular hydrogen bonds. Following the line of the above discussion, we can consider that $\chi_{am}(C_3)$ estimated by eq. (5.2.5), and $\chi_{am}(C_6)$ by eq. (5.2.6), represent the degree of destruction of $O_3-H\cdots O_5'$ and $O_2-H\cdots O_6'$ intramolecular hydrogen bonds, respectively. The concurrent existence of $O_3-H\cdots O_5'$ and $O_2-H\cdots O_6'$ bonds in a given cellobiose unit is not a prerequisite, but the existence of this kind of bond is expected to accelerate the formation of other kinds of bonds.

When Cell-I solid, whose intramolecular hydrogen bonds are at least partly broken, is contacted with solvent, the solvent molecules penetrate relatively easily into the widen gap between the sheets (i.e. largely deformed sheet) in the amorphous region and the penetration of solvent may result in the dissolving of the chain and may bring about the destruction of the remaining intramolecular hydrogen bonds, which accelerates the microscopic Brownian motion of AHG segment, finally leading to the complete destruction of the crystalline region and total dissolution. Note that it was already pointed out by Kamide $et\ al.$[25] that aq. NaOH solution with specific concentration $(8-9\ wt\%)$ could widen the crystalline lattice of (002) plane of alkali soluble cellulose-I (Section 4.4.1).

Therefore, it is quite understandable that both $\chi_{am}(C_3)$ and $\chi_{am}(C_6)$ are highly correlated with S_a. The previous study[10] demonstrated close correlations between $\chi_{am}(C_3)$ and S_a for Cell-II and if $\chi_{am}(C_6)$ was evaluated for Cell-II it was possible to obtain intimate correlations between $\chi_{am}(C_6)$ and S_a for Cell-II (Section 4.4.1).

As noted previously, there exists a strong correlation between $S_a(C_3)$ and $\chi_{am}(C_6)$ for Cell-I solid making reasonable a correlation of S_a with either $\chi_{am}(C_3)$ or $\chi_{am}(C_6)$ and a correlation between S_a and $\chi_{am}(C_3 + C_6)$, reflecting two kinds of intramolecular hydrogen bond, $O_3-H\cdots O_5'$ and $O_6-H\cdots O_2'$. Figure 5.2.6 seems to support this prediction experimentally. In this figure, the highest correlation coefficient ($\gamma = 0.996$) is obtained between S_a and $\chi_{am}(C_3 + C_6)$. The S_a value of Cell-I can be represented in terms of $\chi_{am}(C_3)$, $\chi_{am}(C_6)$, and $\chi_{am}(C_3 + C_6)$ through the equations:

$$S_a = 4.81\chi_{am}(C_3) - 146.5 \tag{5.2.12}$$

$$S_a = 2.73\chi_{am}(C_6) - 53.7 \tag{5.2.13}$$

$$S_a = 3.54\chi_{am}(C_3 + C_6) - 89.4 \tag{5.2.14}$$

Here, it should be noted that the minimum value of $\chi_{am}(C_3)$, above which Cell-I completely dissolves into 9.1 wt% aq. NaOH at 4 °C, $\chi_{am}^c(C_3)$, is 54%, which is far smaller than that ($= 94\%$) for Cell-II.[10]

Note a slight, but significant difference in the method for determining the solubility S_a between this study for Cell-I and the previous study[10] for Cell-II. In the previous study,[10] dilution by the addition of aq. NaOH to the test solution of Cell-II sample was not made before exclusion of the undissolved part by centrifugation. In this case, it is empirically known that the occurrence of partial gelatinizing of the solution during centrifugation makes the apparent solubility erroneously smaller. In order to prevent such undesirable gelatinization in this study, the solution was diluted with aq. NaOH before centrifugation. Therefore, $\chi_{am}^c(C_3)$ (94%) reported for Cell-II may be considered to be more or less (some few %) overestimated. However, a great difference in $\chi_{am}^c(C_3)$ (94 and 54%) cannot be explained by a slight difference in the method, but may suggest that when intramolecular hydrogen bonds developed to the same degree (i.e. at the same χ_{am} level) the solubility of Cell-I into aq. alkali solution is higher than that of Cell-II. The past study[25] demonstrated that, unlike the sheet structure observed in Cell-I type structure; the intermolecular hydrogen bonds are well formed three dimensionally in the Cell-II solid. For the Cell-II solid, the destruction of larger numbers of intramolecular hydrogen bonds is needed for solubilization.

Since $O_6-H\cdots O_2'$ type intramolecular hydrogen bonds exist, in which C_6 and C_2 hydroxyl groups are concurrently participating (see Figure 5.2.7), $\chi_{am}(C_6)$ should be equal to $\chi_{am}(C_2)$. However, this ($\chi_{am}(C_6) = \chi_{am}(C_2)$) is not the case, as is evident from Figure 5.2.6(a) and (c), in which $\chi_{am}(C_6) > \chi_{am}(C_2)$. Some possible reasons for this were given by Yamashiki et al.[16] and additional reasons are worthwhile. In soda cellulose sodium cations selectively coordinate at the C_2 position. The change in electron density with the formation of $O_2-H\cdots O_6'$, as measured by NMR, may be partly cancelled due to the hydrogen bond ($O_2\cdots H-O-H$) formed by coordination of water molecules, existing unavoidably in the sample dried at room temperature, at the hydroxyl group at the C_2 position.

Figure 5.2.9 shows the FT-IR spectrum of an air-dried EP-1 sample. The absorption bands indicated by arrows in the figure are assigned with reference to literature[26-28] as follows: 3600–3200 cm^{-1}, OH stretching; 2902 cm^{-1} CH stretching; 1429 cm^{-1}, CH$_2$ bending; 1370 cm^{-1}, CH bending; 1318 cm^{-1}, CH$_2$ wagging; 1202 cm^{-1}, OH in-plane

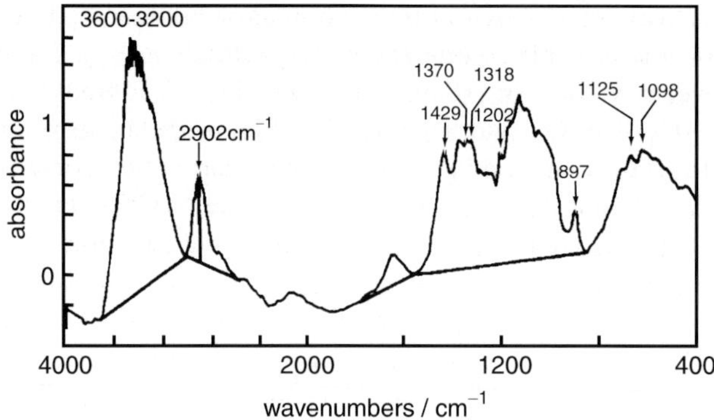

Figure 5.2.9   FT-IR spectrum of sample code EP-1 (air dried).[1]

bending; 1125 cm^{-1} asymmetrical in-phase ring stretching; 1098 cm^{-1}, OH association; 897 cm^{-1}, C_1–H deformation or asymmetrical out of phase ring stretching.

Figure 5.2.10 shows plots of the relative absorbance $A_r(n)$ ($n = 1429$, 1370, 1318, 1202, and 897 cm^{-1}) against S_a. $A_r(897)$ is well correlated to S_a with the correlation coefficient $\gamma = 0.970$, irrespective of sample preparation method and is listed in the eighth column of Table 5.2.1.

Other $A_r(n)$ ($n \neq 897$) always have roughly linear correlations with S_a except for the data point of sample AP-1. Note that only AP-1 is prepared by acid hydrolysis. The S_a values can be empirically expressed as a function of $A_r(897)$ by the following equation:

$$S_a = 699.3A_r(897)29.3 \qquad (5.2.15)$$

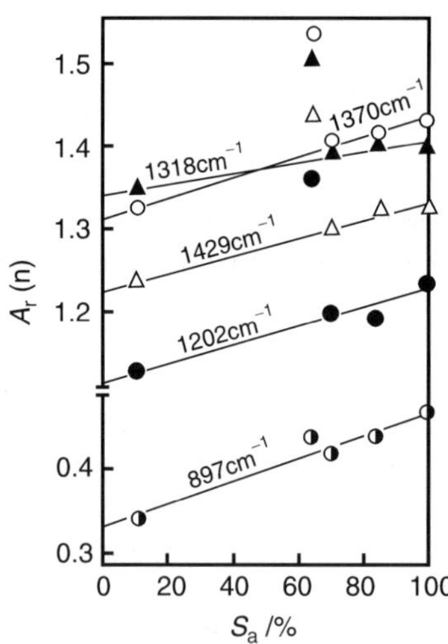

Figure 5.2.10   Plot of the relative absorbance at wave number n, $A_r(n)$s versus S_a of air dried cellulose samples:[1] (●), A_r (1202); (▲), A_r (1318); (○), A_r (1370); (△), A_r (1429); (half black circle), A_r (897).

Nelson and O'Connor[27,28] showed that the intensity of the peak at 897 cm^{-1} was smaller in native cotton (Cell-I) than the corresponding peak in Fortisan (Cell-II; at 893 cm^{-1}) and ball milled cotton (at 897 cm^{-1}; see Figure 2 of Ref. 24). Higgins $et\ al.$[29] demonstrated that the intensity of the band at the wavelength λ of 11.2 μm (i.e. 893 cm^{-1}) reduced by deuteration, although the general absorption in this region was increased, concluding that the intensity of the band at 893 cm^{-1} was sensitive to deuteration and to the nature of the hydrogen bond system. Hayashi $et\ al.$[30] pointed out that the IR absorption band at $ca.$ 897 cm^{-1} is distinctive for celluloses with crystal forms of Cell-III$_I$, Cell-II, Cell-III$_{II}$, and Cell-IV$_{II}$ but not for celluloses with crystal forms of Cell-I and Cell-IV$_I$. From the above literature results, the intensity of the band at 893–897 cm^{-1} can be considered to represent the degree of destruction of hydrogen bonds of cellulose I. A high correlation is experimentally obtained between $A_r(897)$ and $\chi_{am}(C_3 + C_6)$ (see Figure 5.2.11). McKenzie and Higgins[31] showed that the intensities of the bands at 1429, 1111, 990, and 897 (893 in Cell-II[27]) cm^{-1} of various natural celluloses, treated with aq. NaOH, increased with concentration of NaOH and these bands play a role as infrared spectroscopic criteria of transition from cellulose I to cellulose II. The experimental fact that A_r (1429) and A_r (897) are closely correlated with S_a indicates that partial destruction of the supermolecular structure intrinsic to Cell-I is necessary for dissolution of cellulose in solvents. In addition, correlations between the absorbance of the peak at 1318 cm^{-1}, which originates from CH_2 wagging, A_r (1318) and S_a mean that the conformation around C_6 carbon, which can be expressed by torsional angle X, and is related to $O_2-H\cdots O_6'$ intramolecular hydrogen bonds, significantly influences S_a. In short, the results of infrared spectra indirectly confirm the validity of the conclusions, previously reached through NMR analysis, that S_a of cellulose samples is governed by the degree of destruction of $O_3-H\cdots O_5'$ and $O_2-H\cdots O_6'$ intramolecular hydrogen bonds.

Figure 5.2.11 shows correlation coefficients γ between two arbitrarily chosen parameters among S_a, $\chi_{am}(X)$, $\chi_{am}(C_3 + C_6)$, and A_r (897).

S_a has very close correlation with $\chi_{am}(C_3 + C_6)$ and with A_r (897) (and the former has the largest γ). The correlation between $\chi_{am}(C_3 + C_6)$ and A_r (897) is also quite high. In contrast, the correlations of $\chi_{am}(X)$ to any other parameters are very low. Extremely poor correlations between $\chi_{am}(C_3)$ and $\chi_{am}(X)$ confirmed for cellulose with the crystal form of cellulose I and almost the same molecular weight, and for cellulose with crystal

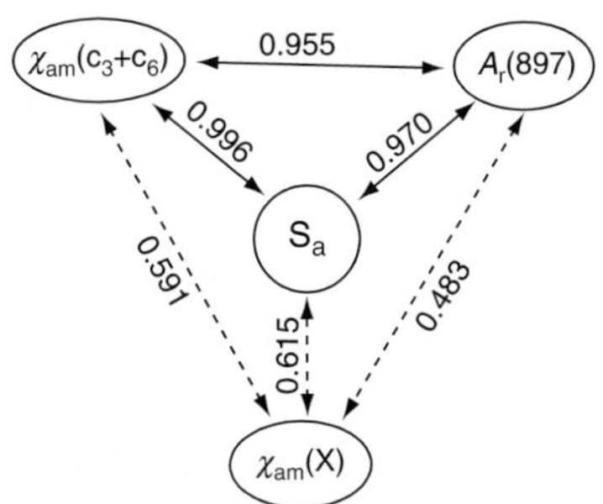

Figure 5.2.11   Correlation coefficients between $\chi_{am}(C_k)$, $\chi_{am}(X)$, $A_r(897)$ and S_a.[1]

form of cellulose II[10] means that an NMR peak component appeared at a lower magnetic field in C_4 carbon, peak region does not originate from the crystalline component as determined by X-ray diffraction. In this regard, our results support the work of Earl and Van der Hart[32] but not that of Horii *et al.*[33]

It can be concluded from Figure 5.2.6 that Cell-I with $\chi_{am}(C_3 + C_6) > 54\%$ is completely soluble, but Cell-I with $\chi_{am}(C_3 + C_6) < 25\%$ is absolutely insoluble in 9.1 wt% aq. NaOH at 4 °C. It is clear from Table 5.2.1 that for all cellulose samples $\chi_{am}(C_3 + C_6)$ is always larger than $\chi_{am}(X)$ by 1.5–2.8 times, strongly suggesting that even in the crystalline region intramolecular hydrogen bonds are not completely organized. This suggestion cannot be so easily understood by considering that cellulose solid consists of a completely crystalline part and a completely amorphous part alone (i.e. the two phase model). Such an oversimplified supermolecular model may not be applicable in a simple way to natural celluloses, in which the ordering of molecular packing depends on the development of intramolecular hydrogen bonds in a very complex manner. $\chi_{am}(C_3)$ and $\chi_{am}(C_3 + C_6)$ should be functions of the way the cellulose is generated in nature. Therefore, it is not too ridiculous to consider the existence of incomplete crystallinity or paracrystallinity, which is regarded as totally crystalline by X-ray diffraction. It is then plausible that intramolecular hydrogen bonds could be broken down to some extent even in the crystalline regions.

Generally, it is well known for CP/MAS ^{13}C NMR measurements that the peak intensity for the nuclei with large spin lattice relaxation time T_1 is smaller than that of the nuclei having relatively small T_1. In this study, the peaks (Nos. 2 and 6) responsible for the carbon nuclei belonging to strong intramolecular hydrogen bonds have relatively larger T_1; hence there is a considerable possibility that $\chi_{am}(C_3 + C_6)$ is overestimated. However, the large disparity between $\chi_{am}(C_3 + C_6)$ and $\chi_{am}(X)$ seems not be explained by this possibility alone.

In summary, cellulose samples with the crystal form of cellulose I and having almost the same degree of polymerization and various solubilities were prepared by the steam explosion treatment and acid hydrolysis on soft wood pulp. Their solubility towards aq. alkali solution proved to be governed by $\chi_{am}(C_3)$ expressing the degree of breakdown in $O_3-H\cdots O_5'$ intramolecular hydrogen bonds, as confirmed for cellulose with the crystal form of cellulose II. $\chi_{am}(C_3 + C_6)$ expressing the degree of breakdown in both $O_3-H\cdots O_5'$ and $O_2-H\cdots O_6'$ intramolecular hydrogen bond proved to be able to explain the solubility of the cellulose more precisely. The degree of destruction of intramolecular hydrogen bonds could be estimated by IR analysis of the samples. Figure 5.2.12 illustrates the underlying principle of the preparation of aq. alkali soluble cellulose and Figure 5.2.13 demonstrates the mechanism of dissolution of cellulose in aq. alkali solution.

5.2.2   Dissolution of cellulose into aq. cuprammonium solution[34]

Even in cases where cellulose or cellulose derivative solid is not easily soluble in solvents, once it is dissolved, the solution is usually stable. Cellulose triacetate (CTA)– acetone,[35] cellulose–aq. NaOH,[10] cellulose–aq. LiOH[36] and cellulose–aq. cuparammonium solution[37] are typical of such cases. Acetone was disclosed to be a thermodynamically

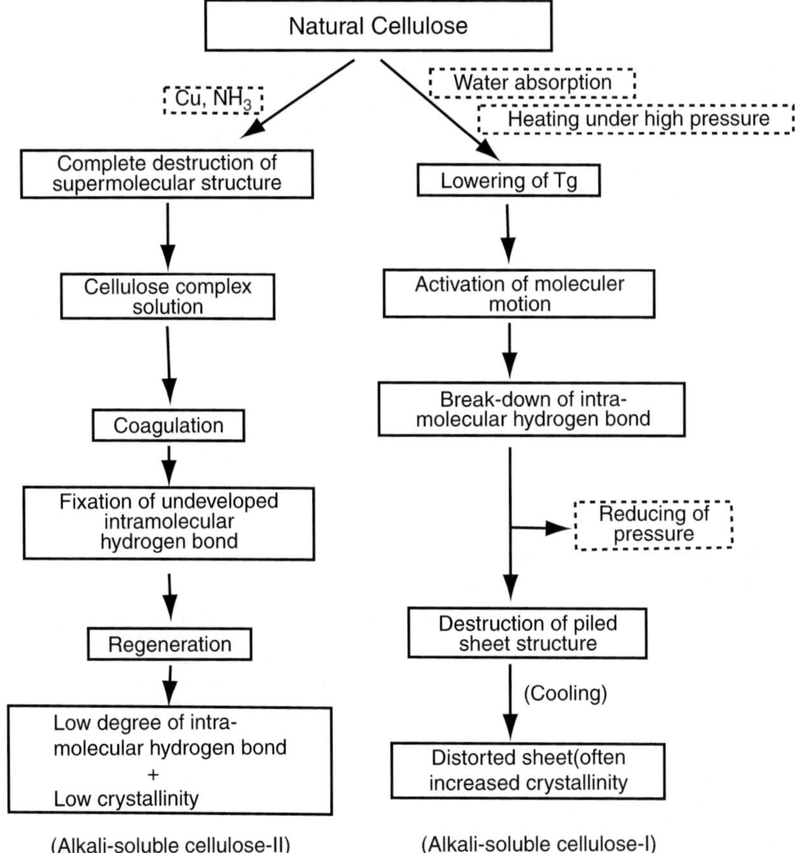

Figure 5.2.12   Underlying principle of the preparation of alkali soluble cellulose.[1]

good solvent for CTA (see, Table 3.3.6).[35] Therefore, the dissolved state as an equilibrium state should be rigorously discriminated from the dissolving process. These two processes had not been previously distinguished in studies of equilibriums of (1) cellulose–copper hydroxide–ammonia–water system[38] and (2) cellulose–copper hydroxide–ammonia–sodium hydroxide–water system[39,40] carried out during 1906–1930. When cellulose solid is mixed with aq. cuprammonium solution, cupric tetramine ion (Cu $(NH_3)_4)^{++}$ reacts with cellulose to form a complex. In this case, first cupric tetramine ion penetrates, in particular, under the coexistence of excess ammonia, into the intervals between two neighboring glucopyranose rings with breakdown of the intermolecular hydrogen bonds and secondly cupric tetramine ions build up a 'cellulose–cuparammonium complex' with breakdown of the intramolecular hydrogen bonds of glucopyranose units. A cellulose molecule, in which some glucopyranose rings are forming complex with cupric tetramine ions, disperses molecularly into the remaining solution. Cuprammonium ions coordinate preferentially with the hydroxyl group at C_2 and C_3, resulting in a chelate form.[41]

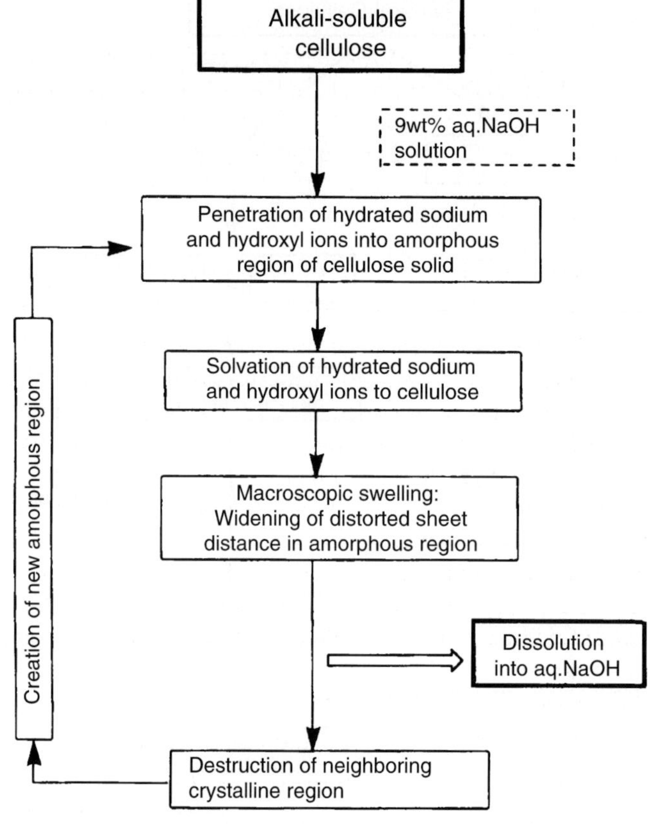

Figure 5.2.13   Dissolution route of cellulose in aq. NaOH solution.[1]

Kamide and his coworkers indicated that cellulose solid, in which the intramolecular hydrogen bonds[14] are broken in advance by some physical methods (for example, see Section 5.1) dissolves into aq. sodium hydroxide or aq. lithium hydroxide completely to yield concentrated cellulose solution.[2] The best composition of dilute alkali as solvent of cellulose is 2.5 mol l^{-1}[42] and cellulose is solvated by alkali molecules.[13] An adequate method for preparing alkali-soluble cellulose is, for example, the steam explosion method.[15-17] The dissolution process of cellulose into aq. NaOH was investigated in detail.[1,25,43] New cellulose fibers were produced from cellulose–aq. NaOH system.[44,45] CTA fibers, with similar properties to commercial products, were spun from CTA–acetone system on a semi-commercial scale.[46] The solubility of sodium salt of carboxyethyl cellulose was explained reasonably in terms of the intramolecular hydrogen bonding concept.[47]

The underlying mechanism for the dissolution of cellulose solid into solvent is very common among cuprammonium solution and dilute alkali solutions. Cupric tetramine ions act as a breaking agent of the intramolecular hydrogen bonds. Miyamoto et al.[41] showed that the apparent degree of breakdown in the intramolecular hydrogen bond at O_3H-O_5' of the undissolved part of the cellulose solid, which was treated with cuprammonium solution and regenerated with dilute acid, increased with the concentration of the hydroxyl group C_{OH}, if $C_{OH} > \sim 0.48$ mol 1^{-1}, where the cuprammonium complex is ionized. Their results are strongly indicative of cuprammonium ion breaking the intramolecular hydrogen bonds and if the treated sample is not separated from the cuprammonium solution, the dissolution may proceed further.

Their treated samples are regenerated for NMR measurements to estimate the degree of breakdown of the intramolecular hydrogen bonds and the possible rebuilding of hydrogen bonding during regeneration may bring about some extent of random crosslinking. From the above discussion it is now clear that the dissolution of cellulose does not prerequisite the formation of 'cellulose-complex with definite composition' (cellulose-complex in the narrow sense of coordinate chemistry). In fact, Miyamoto et al.[48] succeeded to produce the cellulose–cuprammonium solution, which contains 1/2 copper and 1/3 ammonia in their volumes, compared with those used in conventional cellulose–cuprammonium solution. Cuprammonium ion is shown by the experiment to interact strongly with the hydroxyl groups at C_2 and C_3 positions,[8] and 'cellulose complex' in the dissolution process can be similarly considered as a kind of strong solvation, interfering effectively with the association of dispersed cellulose molecules, and strongly solvated solute molecules cannot, of course, be isolated from the solution. Therefore, the key factors governing the dissolution of cellulose is (1) the breakdown of supermolecular (especially, intramolecular hydrogen bonds) of solid, and (2) formation of strong solvation (or interaction).

Figure 5.2.14 shows schema of dissolution of cellulose into aq. cuprammonium solution.

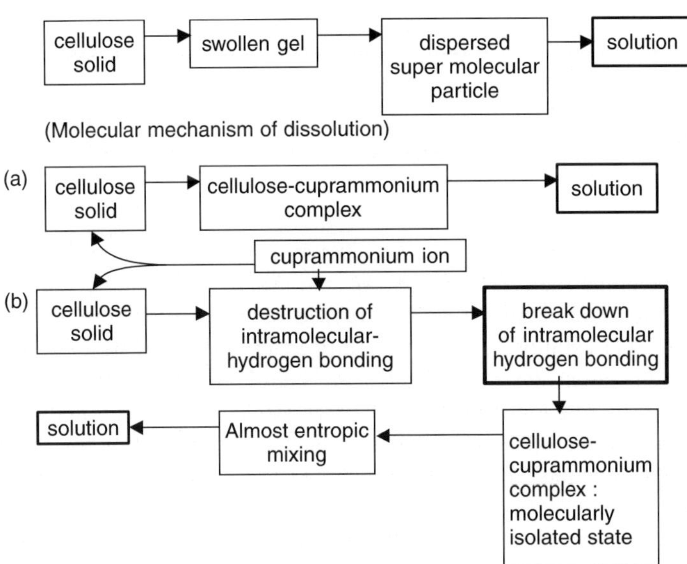

Figure 5.2.14 Concept of dissolution of cellulose into aq. cuprammonium solution:[34] (a) classical concept, (b) new concept.

5.2.3 Cellulose acetate

Average acetyl group distribution on glucopyranose unit[49]

Generally, cellulose acetate with $\langle\!\langle F\rangle\!\rangle = 1.7-2.5$ readily dissolves at room temperature in various organic solvents, including acetone, whereas it never dissolves in water at any temperature. Kamide *et al.*[50] synthesized a series of cellulose acetate with low $\langle\!\langle F\rangle\!\rangle$ by adopting a method of hydrolysis reaction of cellulose acetate with $\langle\!\langle F\rangle\!\rangle = 2.46$ with hydrochloric acid, confirming that cellulose acetate prepared this way is water-soluble at room temperature in the $\langle\!\langle F\rangle\!\rangle$ range of *ca.* 0.4–0.9. They established, for well fractionated cellulose acetate with $\langle\!\langle F\rangle\!\rangle = 0.49$, relationships between the limiting viscosity number $[\eta]$, the radius of gyration $\langle S^2\rangle_z^{1/2}$ and the weight-average molecular weight M_{w}, estimating the Flory's viscosity parameter Φ, the unperturbed chain dimension A and the conformation parameter σ.

Four years later, Miyamoto *et al.*[51] investigated the water solubility of cellulose acetate with low $\langle\!\langle F\rangle\!\rangle$, which are prepared by (1) the hydrolysis of cellulose acetate sample with $\langle\!\langle F\rangle\!\rangle = 2.94$ in aq. acetic acid (D series) or in dimethylsulfoxide (DMSO) (H series) and (2) the acetylation of cellulose dissolved in dimethylacetamide (DMAc)/lithium chloride (LiCl) mixture (A series). They concluded that only cellulose acetate samples with $\langle\!\langle F\rangle\!\rangle$ of 0.5–1.1 prepared by method (1) are completely water soluble at 20 °C. Based on analysis of $\langle\!\langle f_k\rangle\!\rangle$ of the water soluble and insoluble parts, estimated from ^{13}C and ^{1}H NMR spectra, they showed $\langle\!\langle f_6\rangle\!\rangle > \langle\!\langle f_3\rangle\!\rangle$ or $\langle\!\langle f_2\rangle\!\rangle$ for H and A series, but $\langle\!\langle f_2\rangle\!\rangle \approx \langle\!\langle f_3\rangle\!\rangle \approx \langle\!\langle f_6\rangle\!\rangle$ for D series.

In this section we clarify the effects of the preparation conditions on $\langle\!\langle f_k\rangle\!\rangle$ and the relative ratio of various substituted glucopyranose units of cellulose acetate, and obtain the relationships between $\langle\!\langle F\rangle\!\rangle$, $\langle\!\langle f_k\rangle\!\rangle$, ($k = 2$, 3, and 6) and the solubility of cellulose acetate, discussing the dissolution mechanism of cellulose derivatives in solvents. For this purpose, we prepared nine cellulose acetate samples by (1) a direct acetylation method of cellulose solution (one step method) and (2) the acid-hydrolysis of cellulose triacetate with $\langle\!\langle F\rangle\!\rangle = 2.92$ (two step method), as described in the previous paper,[50] and measured ^{13}C NMR spectra of the samples in deuterated dimethyl-sulfoxide (DMSO-d_6).

Preparation of cellulose acetate samples. One step method. Softwood pulp (cellulose I, the viscosity-average molecular weight $M_{\mathrm{v}} = 1.7 \times 10^5$) was immersed in 5N sulfuric acid at 60 °C for 80 min to give a hydrolyzed cellulose (cellulose I, $M_{\mathrm{v}} = 8 \times 10^4$). Five grams of this cellulose sample was dipped in 200 g of DMAc heated at 165 °C for 30 min with reflux and 16 g of LiCl was added at 100 °C, cooled down to 20 °C and maintained at that temperature with vital mechanical agitation to give a visibly clear and colorless transparent solution. Cellulose acetate samples were prepared by adding various amounts of acetic anhydride and pyridine (as a catalyst) to the solution and stirring it at 20 °C for 4–6 h. Table 5.2.3 summarizes the detailed preparation conditions. In the fifth column of the table, M_{v} of the samples, estimated from the hypothetical MHS equation of cellulose acetate/DMAc or cellulose acetate/DMSO system at 25 °C, is compiled. The K_{m} and a values in the MHS equations for estimating M_{v} for cellulose acetate having given $\langle\!\langle F\rangle\!\rangle$ were derived

Table 5.2.3

Acetylation condition, $\langle\langle F \rangle\rangle$, and $\langle\langle f_k \rangle\rangle$ of CAs (one-step)[49]

Sample code	Ac$_2$O[a] (mol)	Pyridine[a] (mol)	Reaction time (h)	$\bar{M}_v$ (10^5)	$\langle\langle F \rangle\rangle$[b]	$\langle\langle f_2 \rangle\rangle + \langle\langle f_3 \rangle\rangle$[c]	$\langle\langle f_6 \rangle\rangle$[c]	$\langle\langle f_6 \rangle\rangle / \langle\langle F \rangle\rangle$[c]	20 °C water soluble part (%)	Solubility[d] 20 °C water	20 °C DMAc	80 °C DMAc	80 °C DMAc
I-1	6.0	2.0	4.0	2.6[e]	0.38	0.13	0.27	0.71	–	×[f]	×	×	O[g]
I-2	8.0	2.0	4.0	2.6[e]	0.57	0.20	0.37	0.65	3.3	×	×	×	O
I-3	10.0	2.0	5.0	2.6[e]	0.68	0.27	0.41	0.60	–	×	Δ	Δ	O
I-4	12.0	3.0	6.0	2.2[h]	0.93	0.36	0.57	0.61	2.9	×	Δ	O	O
I-5	15.0	3.0	6.0	2.8[h]	1.24	0.54	0.70	0.56	2.1	Δ[i]	Δ	O	O

[a]mol mol-1 glucose unit.
[b]From titration.
[c]From NMR.
[d]Judged visually.
[e]From (η) in DMSO.
[f]Insoluble.
[g]Soluble.
[h]From (η) in DMAc.
[i]Highly swelling.

from correlations between $\langle\!\langle F\rangle\!\rangle$ and K_{m}, a values in the MHS equations for various cellulose acetate established by Kamide *et al.*[7,10–12] $\langle\!\langle F\rangle\!\rangle$, determined by the neutralization titration method, of these cellulose acetate samples is also given in the sixth column of the table.

Two step method. An unfractionated cellulose triacetate sample with $\langle\!\langle F\rangle\!\rangle$ of 2.92 was deacetylated in acidic media in the same manner as described in the previous papers[35,50,52,53] to obtain four cellulose acetate samples with different $\langle\!\langle F\rangle\!\rangle$. Table 5.2.3 summarizes the data of M_{w} by the light scattering method and $\langle\!\langle F\rangle\!\rangle$ by the titration method of four cellulose acetate samples prepared by the two step method. In the table, the data on the original CTA sample are also included.

The weight percentage of the water soluble part of cellulose acetate is indicated in the 10th column of Table 5.2.3 and the seventh column of Table 5.2.4. The water solubility was also judged in a qualitative manner visually, as was the solubility of cellulose acetate against DMAc and DMSO. The results are compiled in the last four and three columns of Tables 5.2.3 and 5.2.4, respectively. In Tables 5.2.3 and 5.2.4, $\langle\!\langle f_6\rangle\!\rangle$ and $(\langle\!\langle f_2\rangle\!\rangle + \langle\!\langle f_3\rangle\!\rangle)$ data for all cellulose acetate samples estimated this way are collected.

Figure 5.2.15 shows the plot of $\langle\!\langle f_6\rangle\!\rangle$ against $\langle\!\langle F\rangle\!\rangle$ for two series of CA samples (one step and two step methods). In the figure, the unfilled mark denotes water insoluble, the half filled mark denotes partially soluble and highly swelling in water, and the filled mark indicates completely water soluble. From Tables 5.2.3 and 5.2.4 together with Figure 5.2.15, the following relationships hold: $\langle\!\langle f_6\rangle\!\rangle/\langle\!\langle F\rangle\!\rangle = 0.65 \pm 0.06$ for CA (one step method) and $\langle\!\langle f_6\rangle\!\rangle \approx (\langle\!\langle f_2\rangle\!\rangle + \langle\!\langle f_3\rangle\!\rangle)/2$ for CA (two step method).

In the one step method, the hydroxyl group at C_6 position is 3–4 times reactive against acetic anhydride than that at C_2 or C_3 position. Whereas, in the two step method, the acetyl groups at three different carbon positions are almost equally hydrolyzed when dissolved in acidic media. In other words, the reactivity of acetylation strongly depends

Table 5.2.4

M_{w}, $\langle\!\langle F\rangle\!\rangle$ and $\langle\!\langle f_k\rangle\!\rangle$ of CAs (two-step)[47]

Sample code	$M_{\mathrm{w}}(M_{\mathrm{c}})$ (10^5)	$\langle\!\langle F\rangle\!\rangle^{a}$	$\langle\!\langle f_2\rangle\!\rangle + \langle\!\langle f_3\rangle\!\rangle^{b}$	$\langle\!\langle f_6\rangle\!\rangle^{b}$	$\langle\!\langle f_6\rangle\!\rangle/\langle\!\langle F\rangle\!\rangle^{b}$	20 °C water soluble part (%)	Solubility[c]		
							20 °C water	20 °C DMAc	80 °C DMSO
CTA-0	2.32	2.92	1.89	1.03	0.35	0	×[d]	O[e]	O
II-1	1.20	2.46	1.59	0.86	0.35	0	×	O	O
II-2	0.46[f]	1.75	1.14	0.61	0.35	0	×	O	O
II-3	0.41[f]	0.80	0.53	0.27	0.34	100	O	O	O
II-4	0.80	0.68	0.45	0.23	0.34	100	O	O	O

[a]From titration.
[b]From NMR.
[c]Judged visually.
[d]Insoluble.
[e]Soluble.
[f]M_{v} from $[\eta]$ in DMAc.

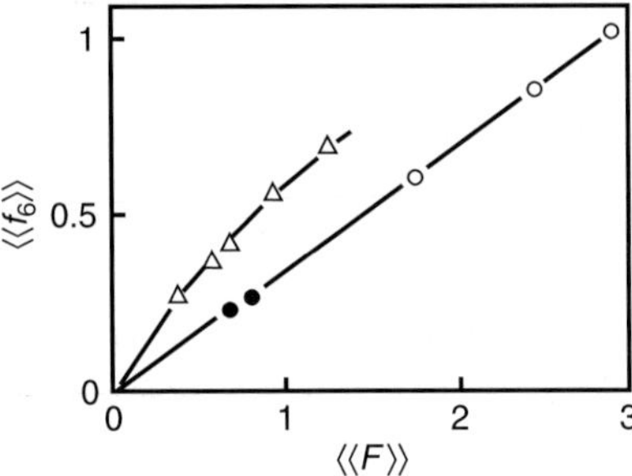

Figure 5.2.15   Plot of $\langle\!\langle f_6\rangle\!\rangle$ of CA against $\langle\!\langle F\rangle\!\rangle$.[49] ($\Delta$), (half filled triangle), one step method;[49] ($\bigcirc$,$\bullet$), two step method; (unfilled marks), water-insoluble; (half filled marks), highly swelling in water; filled marks, water soluble.

on the carbon positions to which the hydroxyl group is located, but that of deacetylation does not significantly depend on whether the acetyl group is derived from the primary hydroxyl or secondary group.

All CA samples used here are soluble in DMSO at 80 °C and cellulose acetate samples prepared in the two step method are also soluble in DMAc at 20 °C. For many years, we have regarded DMAc as a common solvent for cellulose acetate with various $\langle\!\langle F\rangle\!\rangle$, and, in fact, the measurements of the solution properties of cellulose acetate have been carried out in the DMAc system. However, CA samples prepared by the one step method were found not to dissolve in DMAc at 20 °C. This is the first demonstration of the solubility study on CA. It is also evident from the tables that the solubility of cellulose acetate samples against water and DMAc is different from sample to sample. Our previous study[50] revealed that M_w dependence of solubility of CA prepared by the two step method was not found up to $M_w = 1.45 \times 10^5$, showing that M_w is not essential for the solubility of cellulose acetate.

A comparison of sample codes I-1 and II-3, both with the same $\langle\!\langle f_6\rangle\!\rangle$ ($= 0.27$) and different $\langle\!\langle F\rangle\!\rangle$, shows that even if $\langle\!\langle f_6\rangle\!\rangle$ is the same, the solubility against water and DMAc is better for the polymer with larger $(\langle\!\langle f_2\rangle\!\rangle + \langle\!\langle f_3\rangle\!\rangle)$. A comparison of sample codes I-3 and II-4 leads us to the conclusion that CA with larger $(\langle\!\langle f_2\rangle\!\rangle + \langle\!\langle f_3\rangle\!\rangle)$ exhibits a better solubility when compared at the same $\langle\!\langle F\rangle\!\rangle$. Therefore, it seems doubtlessly reasonable to consider that the solubility of CA is predominantly governed by $(\langle\!\langle f_2\rangle\!\rangle + \langle\!\langle f_3\rangle\!\rangle)$. However, CA samples (I-5, II-3) with almost the same $(\langle\!\langle f_2\rangle\!\rangle + \langle\!\langle f_3\rangle\!\rangle)$ (≈ 0.54) have different solubilities against water and DMAc. As mentioned before, CA sample code II-3, which is completely water soluble, has a higher probability of substitution at C_3 or higher fraction of C_3-mono-substituted glucopyranose units than cellulose acetate sample codes I-5. Therefore, the minimum necessary conditions, under which the CA sample dissolves in water and polar solvent such as DMAc, seem to be that the degree of substitution at C_3 should be higher than a given threshold value. The threshold value cannot be exactly estimated. But from $^{13}C\{^1H\}$ NMR spectra in O-acetyl carbonyl carbon region of CTA, CA (one step method) and CA (two step method), the threshold value may be around 0.2 for water. That is, substitution at C_3 is one/five glucopyranose units. The introduction of an hydrophobic group at C_3 may break down the intramolecular hydrogen bond between $O_3 \cdots O_5'$. For two series of the CA sample, the solubility power of DMAc is somewhat larger than water. This may be due to the difference in solvation power to O-acetyl and hydroxyl groups of the solvents. These experimental results agree well with those of Miyamoto $et\ al.$[51] who demonstrated that CA, having higher $\langle\!\langle f_6 \rangle\!\rangle$ than $\langle\!\langle f_2 \rangle\!\rangle$ and $\langle\!\langle f_3 \rangle\!\rangle$, synthesized by the one step method (i.e. A series samples in their paper), does not dissolve in water at room temperature, and they proposed that $\langle\!\langle f_2 \rangle\!\rangle$ and $\langle\!\langle f_3 \rangle\!\rangle$ play key roles in the water solubility of cellulose acetate. However, they did not give any explanation for their results. Note that $(\langle\!\langle f_2 \rangle\!\rangle + \langle\!\langle f_3 \rangle\!\rangle)$ or $\langle\!\langle f_3 \rangle\!\rangle$ or fraction of C_3-mono-substituted glucopyranose units is, of course, not a single parameter controlling water solubility.

Concerning the reactivity of cellulose towards acetic anhydride in DMAc/LiCl medium, Miyamoto $et\ al.$[51] suggested that the strength of hydrogen bonding by C_2 and C_3 hydroxyl groups plays an important role, as compared with that of C_6 hydroxyl group. They based their discussion on Gagnaire $et\ al.$'s claim[54] that the intramolecular hydrogen bonding between the hydroxyl group at the C_3 position and a heterocyclic oxygen atom in neighboring glucopyranose unit of cellulose may exist even in the solution. Note that Gagnaire $et\ al.$'s assertion was formed only indirectly by observing the constancy of ^{1}H NMR spectra peak position of hydroxyl proton at C_3 position (4.6 ppm) of cellobiose (not cellulose!) dissolved in dimethylacetamide (DMAc)/LiCl system. It is well recognized that ^{13}C NMR information on polysaccharides depends significantly on the degree of polymerization (DP) in an extremely low DP region. Kamide and Saito[55] showed experimentally that the unperturbed chain dimension of cellulose derivatives, including cellulose acetate, increases with the polarity of the solvent (dielectric constant ε) and both the substituent groups and remaining hydroxyl group of cellulose derivatives solvate with the solvent and the degree of solvation is closely correlated with ε. The facts explicitly demonstrate the impossibility of the existence of the intramolecular hydrogen bonding of cellulose acetate in the solvents. Accordingly, the selective acetylation at C_6 position for the cellulose in DMAc/LiC1 does not indicate the difference in strength of intra-molecular hydrogen bonds between the hydroxyl groups at C_2, C_3, and C_6 positions and a heterocyclic oxygen atom. The experimental data, summarized in Table 5.2.4, suggest that a primary hydroxyl group at the C_6 position is highly reactive to an acetylation reagent than secondary hydroxyl groups at C_2 and C_3 positions as in the case of low molecular weight organic compounds.[56]

Kamide $et\ al.$[57] demonstrated that, as long as NaCMC does not dissolve in the liquids, the absorbency can be accurately determined by $\langle\!\langle f_6 \rangle\!\rangle$ alone. They explained this fact as follows: the introduction of a bulky substituent (carboxymethyl group) at the C_6 position destroys the intermolecular hydrogen bonds between the hydroxyl group at the C_6 position and a heterocyclic oxygen atom in a neighboring glucopyranose unit and also widens the distance between molecular chains, resulting in the destruction of the intermolecular hydrogen bonds, creating a wide space to readily receive the absorbed liquids (Section 2.8). It should be noted that the results by Kamide $et\ al.$[57] for NaCMC do not mean that the simple introduction of the bulky substituent group into the C_6 position is an important step to make cellulose soluble in water or aqueous solutions. A carboxymethyl group has the same molecular volume as an acetyl group and the introduction of these groups into the C_6 position is then expected to bring about the destruction of intermolecular hydrogen bonding to the same degree. If the destruction of hydrogen bonding associated with the C_6 position is a single universal factor controlling the water solubility, cellulose acetate with high $\langle\!\langle f_6 \rangle\!\rangle$ should be water soluble. Unfortunately, this is not true. It is interesting to note that for NaCMC, the carboxymethyl group is more hydrophilic than the hydroxyl group, and in cellulose acetate the acetyl group is less hydrophilic than the hydroxyl group. Therefore, the affinity of the substituent group against water seems another important factor.

Of course, attempts to directly correlate the chemical structure, as characterized in terms of $\langle\!\langle f_2 \rangle\!\rangle$, $\langle\!\langle f_3 \rangle\!\rangle$, and $\langle\!\langle f_6 \rangle\!\rangle$, of cellulose derivatives with their solubility are too simple and rough. The solubility should be mainly determined by the supermolecular structure of the solid (i.e. the crystalline and amorphous phases, intra- and intermolecular hydrogen bonds) and the latter is vitally controlled by the chemical structure, provided that the heterogeneity of the chemical structure in glucopyranose units can be ignored and the solid is formed by similar procedure. Kamide *et al.*[34] pointed out that the solubility of cellulose in a 10 w% aq. NaOH at 4 °C could generally be correlated to the relative amount of the high magnetic field envelope of C_4 carbon NMR peak (i.e. that of the region where intramolecular hydrogen bonds are at least partly broken).

Supermolecular structure formed by introduction of substituent group
into glucopyranose unit[58]

Recently, Kamide *et al.*[1] found that S_a of the cellulose with crystal form of cellulose I was primarily governed by the degree of breakdown in $O_3-H\cdots O_5'$ and $O_2-H\cdots O_6'$ intramolecular hydrogen bonds $\chi_{am}(C_3 + C_6)$ ([i.e. an average of the degrees of breakdown in $O_3-H\cdots O_5'$ and $O_2-H\cdots O_6'$ intramolecular hydrogen bonds, estimated from C_3 and C_6 carbon regions, respectively, $\chi_{am}(C_3)$ and $\chi_{am}(C_6)$ [$\equiv (\chi_{am}(C_3)$ and $\chi_{am}(C_6))/2$]). In the above studies[10,44] it was affirmed that S_a of the celluloses was not governed by the amorphous content ($\chi_{am}(X)$) calculated from the X-ray crystallinity. These experimental findings definitely explain why solid cellulose does not dissolve in water despite the high density of hydroxyl groups in its structure.

On the other hand, it is well known that almost every cellulose derivative with a low total degree of substitution ($\langle\!\langle F \rangle\!\rangle$) is water soluble irrespective of the hydrophilic or hydrophobic nature of the substituent: The water soluble $\langle\!\langle F \rangle\!\rangle$ ranges of cellulose derivatives with hydrophobic substituent are as follows:[59] cellulose acetate (CA), 0.6–0.8; cellulose nitrate (CN), *ca.* 1; ethyl cellulose (EC), 1.0–1.5; cyanoethyl cellulose (CyEC), 0.68–0.98; methyl cellulose (MC), 1.28–1.95. The comprehensive compilation of literature data[59] also indicates that, except for cellulose acetate, all derivatives described above are alkali soluble in the $\langle\!\langle F \rangle\!\rangle$ range which is lower than the water soluble range. A typical hydrophilic cellulose derivative, the sodium salt of carboxymethyl cellulose (NaCMC), is water soluble in the range $\langle\!\langle F \rangle\!\rangle = 0.3-0.8$, which is a little lower range than those for cellulose derivatives with hydrophobic substituents.[61] NaCMC has been believed to be soluble in 4–8 wt% aq. NaOH solution in the $\langle\!\langle F \rangle\!\rangle$ range 0.05–0.25.[59] Therefore, the solubility behavior of cellulose derivatives with low $\langle\!\langle F \rangle\!\rangle$ cannot simply be explained by the nature of the substituent group introduced into the chain. Consequently, it is reasonable to consider that the introduction of a small number of substituent groups into the glucopyranose rings might give rise to some significant change in the supermolecular structure of the cellulose derivative solids.

Since in most of the above water soluble cellulose derivatives the unsubstituted hydroxyl groups are larger in number than the substituent group, the existence of unsubstituted hydroxyl groups might strongly influence the water solubility. Therefore, the distribution of the substituted hydroxyl groups at the C_k ($k = 2$, 3, and 6) position ($\langle\!\langle f_k \rangle\!\rangle$ ($k = 2$, 3, and 6)) in the glucopyranose ring cannot be ignored when discussing the solubility in water. In fact, $\langle\!\langle f_k \rangle\!\rangle$ has been extensively studied for several cellulose derivatives in connection with their properties.[49,51,60-64] Kamide and his coworkers[49] attempted to explain the solubility behavior of cellulose acetate in water, DMAc and DMSO in terms of $\langle\!\langle f_k \rangle\!\rangle$ and confirmed that cellulose acetate samples with $\langle\!\langle F \rangle\!\rangle =$ 0.38–1.24 and $\langle\!\langle f_6 \rangle\!\rangle > \langle\!\langle f_2 \rangle\!\rangle \approx \langle\!\langle f_3 \rangle\!\rangle$ are not water soluble. Furthermore, they suggested, by analyzing the carbonyl-carbon peak region of ^{13}C NMR spectra of cellulose acetate samples, that the polymer with a large content of C_3-monosubstituted glucopyranose unit, dissolves in water[49] (see 'Average acetyl group distribution on glucopyranose unit').

In this section an attempt is made, as an extension of 'Average acetyl group distribution on glucopyranose unit', to clarify the solubility behavior of a cellulose derivative with hydrophobic substituent (O-acetyl group), cellulose acetate with wide variety of $\langle\!\langle F \rangle\!\rangle$ into several solvents, especially that of cellulose acetate with low $\langle\!\langle F \rangle\!\rangle$ into water, in connection with distribution of substitution and the relating supermolecular structure.

Characterization of cellulose acetate samples and their solubility. Tables 5.2.5 and 5.2.6 summarize M_w, $\langle\!\langle F \rangle\!\rangle$, $\langle\!\langle f_k \rangle\!\rangle$, $\langle\!\langle f_{lmn} \rangle\!\rangle$ of CA-I and CA-II samples and their solubility values against water, water/acetone (1/1, w/w at 20 °C), acetone, and DMSO at 20 °C. For CA-II series an empirical relation $\langle\!\langle f_2 \rangle\!\rangle \approx \langle\!\langle f_3 \rangle\!\rangle \approx \langle\!\langle f_6 \rangle\!\rangle$ roughly holds its validity and for CA-I series $\langle\!\langle f_6 \rangle\!\rangle > \langle\!\langle f_2 \rangle\!\rangle \approx \langle\!\langle f_3 \rangle\!\rangle$ holds. These relations are clearly pictured in Figure 5.2.16 where $\langle\!\langle f_k \rangle\!\rangle$ ($k = 2$, 3, and 6) is plotted against $\langle\!\langle F \rangle\!\rangle$. None of the eight $\langle\!\langle f_{lmn} \rangle\!\rangle$ precisely express the water solubility of CA samples. For example, $\langle\!\langle f_{000} \rangle\!\rangle$ for CA-I-1 and CA-II-9 is almost the same (0.66 and 0.63, respectively) except for their solubility S towards water at 20 °C is 0 and 41%. Likewise, the almost identical $\langle\!\langle f_{100} \rangle\!\rangle$ values (0.13 and 0.12) for CA-II-8 and CA-I-3 give quite different S values (100 and 3%). A similar phenomenon is observed for $\langle\!\langle f_{010} \rangle\!\rangle$–$S$ relationships. From the results in Table 5.2.6, it is not clear whether the substituted and unsubstituted glucopyranose parts are arranged in random or in block for both series of CA samples. However, our recent analysis on a CA-II series sample with $\langle\!\langle F \rangle\!\rangle \approx 0.60$, with the aid of an enzymatic degradation method, revealed that (1) the CA sample contained a small amount of CA with $\langle\!\langle F \rangle\!\rangle \approx 1.76$ in block arrangement; (2) the weight fraction of unsubstituted and substituted glucopyranose units was *ca.* 34:66%; and (3) $\langle\!\langle F \rangle\!\rangle$ of the latter part is generally constant (≈ 1.0)[65,66] (Section 2.6).

Upper and lower limits of the $\langle\!\langle F \rangle\!\rangle$ values for complete dissolution of the CA-II samples differ depending on the solvent employed. Note that all cellulose acetate samples are completely soluble in hot DMSO. From the table, we can conclude that (1) CA-II samples with low $\langle\!\langle F \rangle\!\rangle$ dissolve in hydrophilic solvent, (2) those with high $\langle\!\langle F \rangle\!\rangle$ dissolve in hydrophobic solvent, and (3) those with medium $\langle\!\langle F \rangle\!\rangle$ dissolve in a mixture of these solvents. An amphoteric solvent, such as DMSO, can dissolve cellulose acetate samples with a wide range of $\langle\!\langle F \rangle\!\rangle$. To explain this solubility behavior of cellulose acetate samples,

Table 5.2.5

Molecular weight, degrees of substitution, and solubility towards several solvents of cellulose acetate samples[58]

Sample code	$M_w(M_v)$ (10^4)	$\langle\!\langle F\rangle\!\rangle^a$	$\langle\!\langle f_2\rangle\!\rangle$	$\langle\!\langle f_3\rangle\!\rangle$	$\langle\!\langle f_6\rangle\!\rangle$	Solubility[b] (20°)				Reference
						Water (wt%)	Acetone/water (1/1)	Acetone	DMSO	
CA-0	23.2^a	2.92	0.92^a	1.01^a	0.99^a	0	×	Δ	○	64
CA-II-1	12.0^a	2.46	0.79^a	0.82^a	0.85^a	0	×	○	○	64
CA-II-2	4.6^a	1.75	0.56^a	0.63^a	0.56^a	0	×	○	○	64
CA-II-3	–	1.23	0.42^a	0.40^a	0.40^a	0	○	×	○	64
CA-II-4	–	1.06	0.34^a	0.37^a	0.35^a	3	○	×	○	64
CA-II-5	–	0.95	0.29^a	0.33^a	0.33^a	79	○	×	○	64
CA-II-6	–	0.77	0.26^a	0.24^a	0.27^a	98	Δ	×	○	64
CA-II-7	–	0.69	0.20^a	0.23^a	0.26^a	100	Δ	×	○	64
CA-II-8	–	0.54	0.17^a	0.15^a	0.22^a	100	×	×	○	64
CA-II-9	–	0.43	0.13^a	0.12^a	0.18^a	41	×	×	Δ	64
CA-I-1	26^b	0.38	0.07	0.05	0.27	0	×	×	×	49
CA-I-2	26^b	0.52	0.13	0.05	0.34	3	×	×	Δ	49
CA-I-3	26^b	0.67	0.15	0.12	0.40	3	×	×	Δ	49
CA-I-4	22^b	0.89	0.19	0.16	0.54	3	○	×	○	49
CA-I-5	28^b	1.17	0.31	0.20	0.66	2	○	×	○	49

○, completely soluble; Δ, partially soluble with considerable swelling; ×, insoluble.
[a]Literature data reproduced from our previous papers cited in column 11.
[b]M_v.

Table 5.2.6

Parameters representing distribution of degree of substitution $\langle\langle f_{lmn} \rangle\rangle$, $\langle\langle f_k \rangle\rangle_{calc}$, $\langle\langle F \rangle\rangle_{calc}$ and solubility against water of cellulose acetate samples[58]

Sample code	$\langle\langle f_{111} \rangle\rangle$	$\langle\langle f_{110} \rangle\rangle$	$\langle\langle f_{101} \rangle\rangle$	$\langle\langle f_{011} \rangle\rangle$	$\langle\langle f_{100} \rangle\rangle$	$\langle\langle f_{010} \rangle\rangle$	$\langle\langle f_{001} \rangle\rangle$	$\langle\langle f_{000} \rangle\rangle$	$\langle\langle f_2 \rangle\rangle_{calc}$	$\langle\langle f_3 \rangle\rangle_{calc}$	$\langle\langle f_6 \rangle\rangle_{calc}$	$\langle\langle F \rangle\rangle_{calc}$	S^a (wt%)
CA-II-1	0.48[b]	0.18[b]	0.21[b]	0.16[b]	0[b,c]	0[b,c]	0[b,c]	0[b,c]	0.87[b]	0.82[b]	0.85[b]	2.54[b]	0
CA-II-8	0[b,c]	0.02[b]	0.02[b]	0.07[b]	0.13[b]	0.08[b]	0.13[b]	0.45[b]	0.17[b]	0.17[b]	0.22[b]	0.56[b]	100
CA-II-9	0[b,c]	0.00[b]	0.01[b]	0.05[b]	0.11[b]	0.08[b]	0.12[b]	0.63[b]	0.12[b]	0.13[b]	0.18[b]	0.43[b]	41
CA-I-1	0[c]	0.00	0.01	0.01	0.06	0.05	0.23	0.66	0.07	0.06	0.25	0.38	0
CA-I-2	0[c]	0.00	0.01	0.03	0.09	0.06	0.32	0.49	0.10	0.09	0.36	0.55	3
CA-I-3	0[c]	0.01	0.02	0.03	0.12	0.05	0.38	0.39	0.15	0.09	0.43	0.67	3

[a]Against 20 °C water.
[b]Literature data reproduced from Table V in Ref. 64.
[c]Approximated *a priori*.

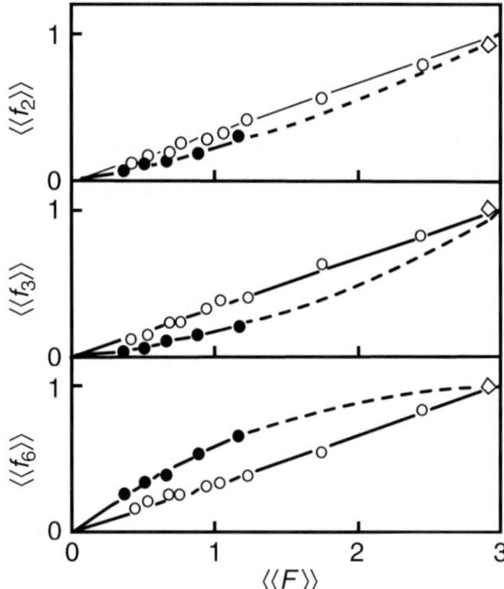

Figure 5.2.16   Plots of $\langle\langle f_k \rangle\rangle$ against $\langle\langle F \rangle\rangle$ of CA samples.[58] (●), CA-I series; (○), CA-II series;[64] (◇), sample code CA-0.

it should be noted that the role of unreacted hydroxyl groups which can strongly solvate with hydrophilic solvents cannot be ignored and must be taken into account as well as the hydrophobic substituent (acetyl) groups in cellulose acetate molecules. In other words, the balance of hydrophilic and hydrophobic interactions between cellulose acetate molecules and solvents is one of the more important factors controlling the solubility of cellulose acetate in solvents.[65,66]

Solubility of cellulose acetate. Water soluble cellulose acetate can be prepared by homogeneous hydrolysis of CA with $\langle\langle F \rangle\rangle = 2.46$ dissolved in acetic acid, and the samples (that is, CA-II series) have $\langle\langle F \rangle\rangle (\approx 3\langle\langle f_2 \rangle\rangle \approx 3\langle\langle f_3 \rangle\rangle \approx 3\langle\langle f_6 \rangle\rangle) = 0.77-0.54$. CA samples (CA-I series) with $\langle\langle F \rangle\rangle = 0.38-1.17$ and $\langle\langle f_6 \rangle\rangle \gg \langle\langle f_2 \rangle\rangle > \langle\langle f_3 \rangle\rangle$, obtained by direct acetylation of wood pulp in DMAc/LiCl are practically insoluble in both water and acetone. DMSO dissolves all the CA samples (CA-I with $\langle\langle F \rangle\rangle \geq 0.38$, CA-II with $\langle\langle F \rangle\rangle \geq$ 0.43) employed here at 80 °C but somewhat different results are observed at 20 °C: CA-II series with $\langle\langle F \rangle\rangle \geq 0.54$ and CA-I series with $\langle\langle F \rangle\rangle \geq 0.89$ are all soluble in DMSO, while CA-I -2 with $\langle\langle F \rangle\rangle = 0.52$ only swells in DMSO at 20 °C. The solubility behavior of CA against water and DMSO clearly indicates that the distribution of substituent is one of the important factors ruling the solubility, as well as solvent nature. The solvent mixture of acetone/water dissolves both the CA-I and CA-I series of samples when their $\langle\langle F \rangle\rangle$ values range from 0.9 to 1.23 (CA-II, 1.23–0.95; CA-I, 1.17–0.89).

Structural parameters governing the solubility of cellulose acetate. Figure 5.2.17 shows the relationships between the solubility (S) of cellulose acetate samples in (A) water, (B) acetone/water, (C) acetone, and (D) DMSO, and their $\langle\langle f_k \rangle\rangle$ values ($k = 3$ and 6) by a thin solid line. In this figure, ($S-\langle\langle f_3 \rangle\rangle$ relation), $\chi_{am}(C_3)-\langle\langle f_3 \rangle\rangle$, and half value width of C_1 carbon peak $\Delta_{1/2}(C_1)-\langle\langle f_3 \rangle\rangle$ relations are plotted as bold solid and broken lines,

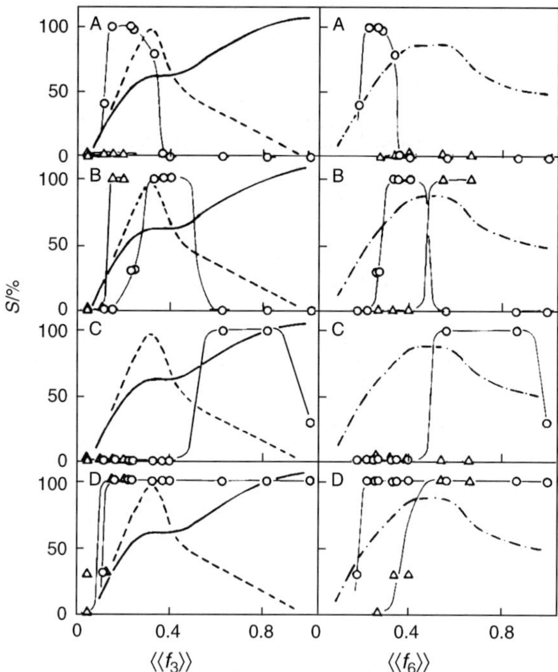

Figure 5.2.17 Change of solubility S against various solvent of CA samples against $\langle\!\langle f_k\rangle\!\rangle$.[58] (A) water; (B) acetone/water (1/1, w/w); (C) acetone; (D) DMSO. (Δ), CA-I series; ($\bigcirc$), CA-II.

respectively. The $\Delta_{1/2}(C_6)-\langle\!\langle f_6\rangle\!\rangle$ relation is also shown as a chain line in the same figure ($S-\langle\!\langle f_6\rangle\!\rangle$ relation).

(a) *Acetone/water mixture.* When $\langle\!\langle f_3\rangle\!\rangle$ is in the range $0.2-0.3 \leq \langle\!\langle f_3\rangle\!\rangle \leq 0.42$, the cellulose acetate sample is soluble in acetone/water (1/1, w/w at 20 °C), and this corresponds to the plateau region of $\chi_{\mathrm{am}}(C_3)$. The absolute value (*ca.* 0.92) of $\chi_{\mathrm{am}}(C_3)$ is an asymptotic value of the first increasing curve in the $\chi_{\mathrm{am}}(C_3)-\langle\!\langle f_k\rangle\!\rangle$ relation. Therefore, in this region, the ladder structure of the cellulose backbone formed by intramolecular hydrogen bonds can be considered to be almost completely broken, making the penetration of the two solvents (acetone and water) into such a solid structure very simple. The value of $\langle\!\langle f_6\rangle\!\rangle$ is not a unique factor controlling the solubility because the value of the $\langle\!\langle f_6\rangle\!\rangle$ range necessary for complete dissolution of CA depends strongly on the method of synthesis of cellulose acetate. Thus, the distribution of the substituent group in CA defined as $\langle\!\langle f_k\rangle\!\rangle$ is not useful for expressing its solubility in acetone/water. Furthermore, the dissolving power of acetone/water depends on the absolute number of hydroxyl groups per glucopyranose unit on the cellulose chain, which can interact with water, and on the number of *O*-acetyl groups, which can interact with acetone. At the lower region of $\langle\!\langle F\rangle\!\rangle$, acetone acts as a non-solvent and at the upper region of $\langle\!\langle F\rangle\!\rangle$, water acts as a non-solvent. The mixture may not behave as an amphoteric solvent, which must have both cationic and anionic sites in a molecule or a molecular associate, as will be

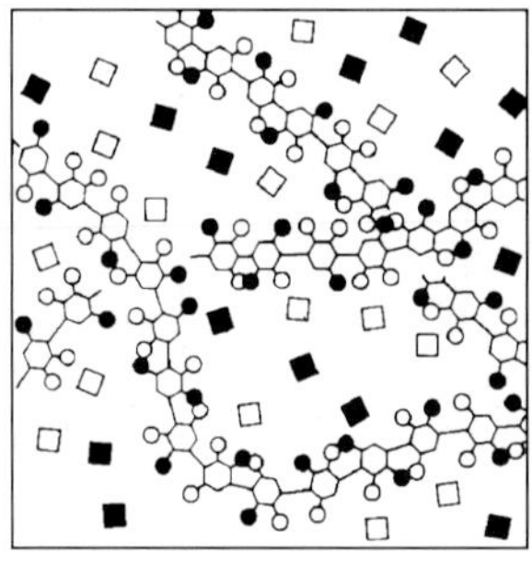

acetone/water-soluble

Figure 5.2.18 Schematic representation of the dissolved state of CA with medium $\langle\!\langle F \rangle\!\rangle$ (≈ 1.2) in acetone/water.[58] (O), Hydroxyl group; (●), acetyl group; (□), water molecule; (■), acetone molecule.

described later for DMSO. A schematic representation of the dissolution mechanism for CA with medium value of $\langle\!\langle F \rangle\!\rangle$ in acetone/water is shown in Figure 5.2.18.

(b) *Acetone*. The acetone soluble $\langle\!\langle f_k \rangle\!\rangle$ ($k = 3$ and 6) range is as follows: $0.63 \leq \langle\!\langle f_3 \rangle\!\rangle \leq 0.82$; $0.56 \leq \langle\!\langle f_6 \rangle\!\rangle \leq 0.85$ for CA-II series. The $\langle\!\langle f_3 \rangle\!\rangle$ region lies in the second increasing stage of $\chi_{am}(C_3)$ but also in the decreasing stage of $\Delta_{1/2}(C_1)$. These observations are contradictory and $\chi_{am}(C_3)$ is no longer meaningful, as described before. The $\langle\!\langle f_6 \rangle\!\rangle$ range also lies in the decreasing stage of half value width of C_6 carbon peak $\Delta_{1/2}(C_6)$. It is not understandable that when the allowed variation of the torsional angles around C_1–O–C_4' (see Figure 5.2.23) is relatively limited and the intermolecular regularity is higher, all the CA samples with $\langle\!\langle f_3 \rangle\!\rangle$ and $\langle\!\langle f_6 \rangle\!\rangle$ in the above ranges are soluble in acetone. It probably means that the solubility of CA is mainly enhanced by the formation of interaction between solvent and side chain (*O*-acetyl group) of CA. A schematic representation of the dissolved state of CA in acetone is illustrated in Figure 5.2.19(b). Acetone-insoluble CA with low $\langle\!\langle F \rangle\!\rangle$ is shown in Figure 5.2.19(a).

However, it should be noted that when both $\Delta_{1/2}(C_1)$ and $\Delta_{1/2}(C_6)$ are at the lowest value, that is, when the conformational regularities around the C_1 and C_6 carbons are highest, the cellulose acetate does not completely dissolve in acetone. This observation suggests the existence of some kind of intramolecular interaction between the *O*-acetyl groups, which act as structural blockage for complete dissolution of the cellulose acetate as demonstrated in Figure 5.2.19(c). The existence of this kind of intramolecular interaction was confirmed by the CP/MAS ^{13}C NMR spectrum for CA-0, where a strong sharp peak for C_1 appeared at 100.0 ppm. This sample was previously found to be soluble in acetone when the dissolution temperature was first lowered to $-40\,^{\circ}$C and then elevated to room temperature.[35] During the dissolution process the structure (regularity of molecular packing) of the cellulose acetate may be destroyed as shown in Figure 5.2.19(c'). This process resembles the steam explosion treatment of cellulose, which is an effective way to obtain the alkali soluble cellulose,[16] and which gives rise to considerable alternations in the supermolecular structure.

(c) *DMSO*. DMSO dissolves CA with $\langle\!\langle f_3 \rangle\!\rangle \geq 0.12$ but the solubility of CA samples with $\langle\!\langle f_6 \rangle\!\rangle = ca.\ 0.2$–$0.4$ towards DMSO differs depending on the method of synthesis, as shown in Table 5.2.5. As is the case for acetone/water, the dissolution of cellulose

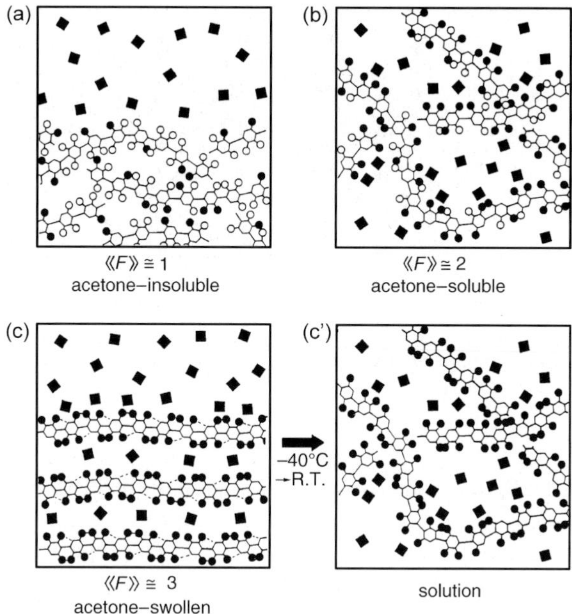

Figure 5.2.19   Schematic representation of the dissolved state of CA with medium $\langle\!\langle F\rangle\!\rangle$ (≈ 2.0) in acetone.[58] (○), Hydroxyl group; (●), acetyl group; (■), acetone molecule.

acetate in DMSO is mainly governed by the absolute number of hydroxyl and O-acetyl groups per anhydroglucose unit which can interact with DMSO as an amphoteric solvent. For the lowest $\langle\!\langle f_3\rangle\!\rangle$ value of CA, which permits dissolution in DMSO, both $\chi_{am}(C_3)$ and $\Delta_{1/2}(C_1)$ are in the first increasing stage. This implies that, at the lower limit of $\langle\!\langle f_3\rangle\!\rangle$, some extent of breakdown in intramolecular hydrogen bond is required. At far higher values of $\langle\!\langle F\rangle\!\rangle$ ($= ca.$ 3), where some kind of interaction exists between the O-acetyl group at the C_3 position in a given glucopyranose unit and ring oxygen in the neighboring unit, DMSO is powerful enough to destroy the interaction, leading to dissolution of the cellulose acetate. A schematic representation of the dissolving mechanism of CA with a wide range of $\langle\!\langle F\rangle\!\rangle$ is given in Figure 5.2.20.

(d) *Water.* Water solubility of the samples seems to depend on both $\langle\!\langle f_3\rangle\!\rangle$ and $\langle\!\langle f_6\rangle\!\rangle$ in their specific ranges: $0.17-0.20 \leq \langle\!\langle f_3\rangle\!\rangle \leq 0.32, \langle\!\langle f_6\rangle\!\rangle \leq 0.27$. The necessary $\langle\!\langle f_3\rangle\!\rangle$ values correspond to the first increasing step of both $\chi_{am}(C_3)$ and $\Delta_{1/2}(C_1)$, suggesting that water solubility of CA is at least governed by the degree of breakdown in $O_3-H\cdots O_5'$ intramolecular hydrogen bonds and by the variation of torsional angles around the C_1-O-C_4' linkage, which is closely correlated with $O_3-H\cdots O_5'$ intramolecular hydrogen bonds. The C_1 carbon chemical shifts of water soluble cellulose acetate are practically the same as that of pure cellulose having cellulose II crystal, and the CP/MAS ^{13}C NMR spectra measured as a function of standing time for CA-II-9 ($\langle\!\langle F\rangle\!\rangle = 0.43$)/water system, in which CA does not completely dissolve, show the characteristic peaks for $O_3-H\cdots O_5'$ hydrogen bonds at 89 ppm. In this connection, $\chi_{am}(C_3)$ for CA-I series is always smaller than that for water-soluble CA-II samples.

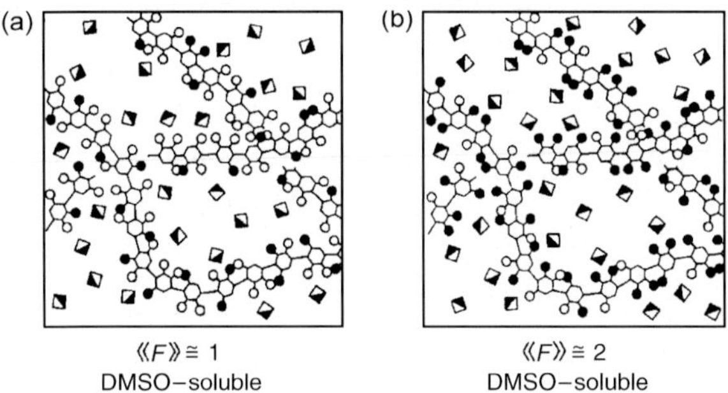

Figure 5.2.20 Schematic representation of the dissolved state of CA.[58] (a) with $\langle\!\langle F \rangle\!\rangle \approx 1$ and (b) with $\langle\!\langle F \rangle\!\rangle \approx 2$ in DMSO. (O), Hydroxyl group; (●), acetyl group; (half-filled rectangle), DMSO molecule.

The $\chi_{am}(C_3)$ values for CA-I series lie in the range 0.80–0.85, which is substantially smaller than the value (0.94) necessary for complete dissolution of cellulose with cellulose crystal II into aqueous alkali. This could be the primary reason why none of CA-II series is water soluble. It is of interest to note that $\chi_{am}(C_3)$ is the main common factor governing the solubility of CA with lower values of $\langle\!\langle f_3 \rangle\!\rangle$ against water and that of pure cellulose against aqueous alkali.

On the basis of the above discussion on water solubility of CA, a schematic representation of dissolution or precipitation behavior of water insoluble CA with relatively low $\langle\!\langle f_3 \rangle\!\rangle$ is demonstrated in Figure 5.2.21(a) and (b). At the lower limit of $\langle\!\langle f_3 \rangle\!\rangle$, a considerable breakdown in intramolecular hydrogen bonds and the conformational variation around C_1-O-C_4' due to this breakdown are required for dissolving in water, but neither $\chi_{am}(C_3)$ nor $\Delta_{1/2}(C_1)$ describe the solubility of CA adequately.

Note that the intermolecular hydrogen bond formed between unsubstituted hydroxyl groups of glucopyranose units in different molecular chains, or the intermolecular interaction between *O*-acetyl groups in different molecular chains, might not strongly influence the solubility of CA samples if the solvent has an ability to swell the samples more or less, because the swelling may occur as a result of rupturing of intermolecular interaction. In other words, intermolecular interaction is more easily broken down by the invasion of a solvent. In Section 2.8 it was shown on NaCMC that the degree of

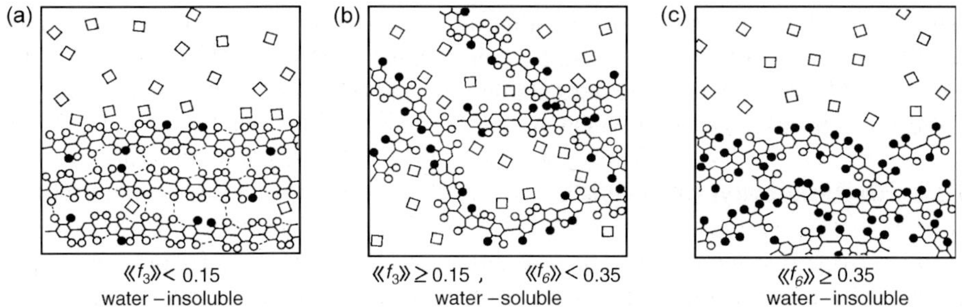

Figure 5.2.21 Schematic representation of CA molecules in water.[58] (O), Hydroxyl group; (●), acetyl group; (□), water molecule.

Table 5.2.7

Solubility of CA in solvents[66]

Type	Sample code	$M_v \times 10^{-4}$	$\langle\!\langle f_2 \rangle\!\rangle$	$\langle\!\langle f_3 \rangle\!\rangle$	$\langle\!\langle f_6 \rangle\!\rangle$	$\langle\!\langle F \rangle\!\rangle$	Solubility at 25 °C		
							DMSO	DMAc	H_2O
C_6-substituted	HCA 05	19.0	0.00	0.00	0.62	0.62	○	×	×
C_2- and C_3-sub.	SCA14-02	5.3	0.39	0.29	0.00	0.68	○	○	○
C_2, C_3, C_6-sub.	RCA02-03	5.4	0.23	0.16	0.25	0.64	○	○	○

introduction of hydrophilic substituent at the C_6 position ($\langle\!\langle f_6 \rangle\!\rangle$) governed its water absorbency.[64] In the present case, the O-acetyl group is hydrophilic and the residual hydroxyl group at the C_6 position ($1 - \langle\!\langle f_6 \rangle\!\rangle$) is expected to be one of the factors controlling the water solubility of cellulose acetate, and conversely, CA with relatively high $\langle\!\langle f_6 \rangle\!\rangle$ is water insoluble (Figure 5.2.21(c)).

In summary,

(1) considerable breakdown in $O_3-H\cdots O_5'$ intramolecular hydrogen bonding and a wide variation of torsional angles around the C_1-O-C_4' glucoside linkage, which closely correlates with breakdown in $O_3-H\cdots O_5'$ intramolecular hydrogen bond, are the least necessary condition for cellulose acetate to dissolve in any kind of solvent,

(2) acetone/water mixture dissolves cellulose acetate depending on the total number of hydroxyl and O-acetyl groups,

(3) acetone solubility of cellulose acetate is mainly governed by solvation of side chains but an intramolecular interaction like $O_3-H\cdots O_5'$ intramolecular hydrogen bond in cellulose tends to lessen its solubility,

(4) DMSO behaves as an amphoteric solvent solvating both hydroxyl and O-acetyl groups in cellulose acetate molecules, and

(5) water solubility of cellulose acetate is mainly governed by the breakdown of the above mentioned hydrogen bond by introducing a substituent at the C_3 position and the number of residual hydroxyl groups at the C_6 position.

Supplement to Section 5.2.2

Table 5.2.7 collects the results of solubility test of C_6-substituted, C_2 and C_3-substituted and C_2, C_3, C_6-substituted CA samples in DMSO, DMAc and water 25 °C.[66] C_6-substituted CA with $\langle\!\langle F \rangle\!\rangle = 0.62$ does not dissolve in DMAc and water, but the other two CA samples dissolve readily into DMAc and water.

5.2.4 Carboxyethyl cellulose[47]

Kowsaka *et al.*[49] disclosed that for cellulose acetate (CA), in which the total degree of substitution ($\langle\!\langle F \rangle\!\rangle$) is less than 0.95, both the degree of breakdown in $O_3-H\cdots O_5'$ intramolecular hydrogen bonds ($\chi_{am}(C_3)$) and the degree of substitution of hydrophobic substituent at the C_6 position heavily influence its solubility in water. It is well known that cellulose derivatives with a relatively low degree of substitution are water soluble irrespective of whether the substituent group is hydrophilic or hydrophobic. Except for cellulose esters, which are generally hydrolyzed in aq. alkali to regenerate cellulose, the many other derivatives are alkali soluble in the $\langle\!\langle F \rangle\!\rangle$ range less than the water soluble $\langle\!\langle F \rangle\!\rangle$ range.[60] Therefore, it is interesting to confirm experimentally whether solubilization of cellulose by introduction of a small amount of the hydrophilic substituent group is explained by the breakdown of the intramolecular hydrogen bond in solid, or not, and, in other words, to ascertain whether the intramolecular hydrogen bonding also governs the solubilities of cellulose derivatives with low $\langle\!\langle F \rangle\!\rangle$, or not. For this purpose, in this section the sodium salt of carboxyethyl cellulose (NaCEC), synthesized by reacting homogeneously acrylonitrile with alkali soluble cellulose[18] dissolved in 9 wt% aq. NaOH, was subjected to the solubility test and NMR measurements.

Samples

Alkali soluble cellulose (10 g) ($M_v = 12.1 \times 10^4$) was regenerated from a cuprammonium solution according to the procedure described in a previous paper,[67] and was dissolved in a mixture of 9 wt% aq. NaOH solution (220 g) and acrylonitrile (AN) (16 g) maintained in advance at 0 °C. The resultant solution was agitated mechanically at 25 °C for various periods of time and the reaction of cellulose with AN was stopped by pouring the solution into excess methanol. The precipitates obtained were dissolved in a 9 wt% aq. NaOH solution and stored at 25 °C for 12 h to convert the cyanoethyl and carbamoylethyl groups substituted with hydroxyl groups in the cellulose chain into carboxyethyl groups completely. The NaCEC sample was recovered from the solution by adding methanol, washed with a mixture of methanol/water (8/2, v/v at 25 °C) and then with methanol, and vacuum-dried at 50 °C for 8 h. In this manner NaCEC samples with various $\langle\!\langle F \rangle\!\rangle$ were synthesized.

^{13}C NMR measurement

The apparent degree of breakdown in intramolecular hydrogen bonding at the C_3 position of NaCEC samples $\chi_{am}(C_3)$ was determined according to the equation proposed by Kamide *et al.*[10] for cellulose solid:

$$\chi_{am}(C_3) = 100 I_h(C_4)/\{I_h(C_4) + I_l(C_4)\}(\%) \tag{5.2.16}$$

Here, $I_h(C_4)$ and $I_l(C_4)$ are the fractions of higher and lower magnetic peaks in the C_4 carbon peak region. Maximum peak positions for C_1, C_4, C_6 carbons, and their half value widths were also analyzed.

   Table 5.2.8 shows values of $\langle\!\langle F \rangle\!\rangle$, $\langle\!\langle f_6 \rangle\!\rangle$, and $\langle\!\langle f_2 \rangle\!\rangle + \langle\!\langle f_3 \rangle\!\rangle$ of NaCEC samples and their solubilities in water and in 2, 5, 9, 13, and 18 wt% aq. NaOH solutions at 20 °C and in 9 wt% aq. NaOH at 4 °C. Solubility of the original alkali soluble cellulose is also compiled in the table. $\langle\!\langle F \rangle\!\rangle$ of the samples employed varies from 0.39 to 0.03. Except for sample code NaCEC-1, $\langle\!\langle f_2 \rangle\!\rangle + \langle\!\langle f_3 \rangle\!\rangle$ values are zero, indicating that all substituent groups were located preferentially at the C_6 position. Only NaCEC-1, whose degrees of substitution are $\langle\!\langle F \rangle\!\rangle = 0.39$ and $\langle\!\langle f_2 \rangle\!\rangle + \langle\!\langle f_3 \rangle\!\rangle = 0.12$, is water soluble. With an increase in alkali concentration, the dissolution power of aq. NaOH solution against NaCEC

Table 5.2.8

Total degree of substitution $\langle\langle F \rangle\rangle$ and degrees of substitution at C_k position $\langle\langle f_k \rangle\rangle$ ($k = 2$, 3, and 6) of sodium carboxyethyl cellulose samples and their solubility against water and aq. NaOH solutions[47]

Sample code	$\langle\langle F \rangle\rangle$	$\langle\langle f_6 \rangle\rangle$	$\langle\langle f_2 \rangle\rangle + \langle\langle f_3 \rangle\rangle$	Solubility against aq. NaOH						at 4 °C
				at 20 °C						
				0 wt%	2 wt%	5 wt%	9 wt%	13 wt%	18 wt%	9 wt%
NaCEC-1	0.39	0.27	0.12	◎	◎	◎	◎	◎	◎	◎
NaCEC-2	0.13	0.13	0.00	×	×	◎	◎	◎	◎	◎
NaCEC-3	0.17	0.07	0.00	×	×	◎	◎	○	Δ	◎
NaCEC-4	0.05	0.05	0.00	×	×	○	◎	–	–	◎
NaCEC-5	0.03	–	–	×	×	×	○	–	–	◎
Cellulose[a]	–	–	–	×	×	×	Δ	Δ	×	◎

[a]Cellulose sample employed as a starting material for NaCEC synthesis in this study.

becomes stronger and then weaker after passing through a maximum. A 9 wt% aq. NaOH solution at 4 °C can dissolve all NaCEC samples, in which the smallest $\langle\!\langle F \rangle\!\rangle$ is 0.03, which was found also most adequate to dissolve alkali soluble cellulose.[10,15] In this sense, NaCEC shows similar solubility behavior to that of cellulose.

Figure 5.2.22 shows CP/MAS ^{13}C NMR range of NaCEC samples. NaCEC with $\langle\!\langle F \rangle\!\rangle \geq 0.13$ (Figure 5.2.22(a) and (b)) gives two broad signals at around 190 and 40 ppm, assigned to the carbonyl carbon and the α-methylene carbon, both in the carboxyethyl group, respectively. For NaCEC samples with $\langle\!\langle F \rangle\!\rangle < 0.13$ (Figure 5.2.22 (c)–(e)) those signals are not detected, probably being hidden in noise, except for the O-acetyl carbonyl carbon peak of sample code NaCEC-3. Two peak envelopes responsible for C_1 and C_6 (especially C_6) carbons gradually become broad with an increase in $\langle\!\langle F \rangle\!\rangle$. On the whole, the NMR range of NaCEC samples with $\langle\!\langle F \rangle\!\rangle \leq 0.05$ are quite similar to those for cellulose having cellulose II crystal (Cell-II). In other words, four sharp peaks at 107, 105, 89, and 88 ppm, all of which are characteristic of Cell-II, were also detected at least partly for sample codes NaCEC-3–5, as marked by arrows in Figure 5.2.22. Kamide et al.[67] demonstrated for Cell-II solid that these sharp peaks at 107, 105, 89, and 88 ppm

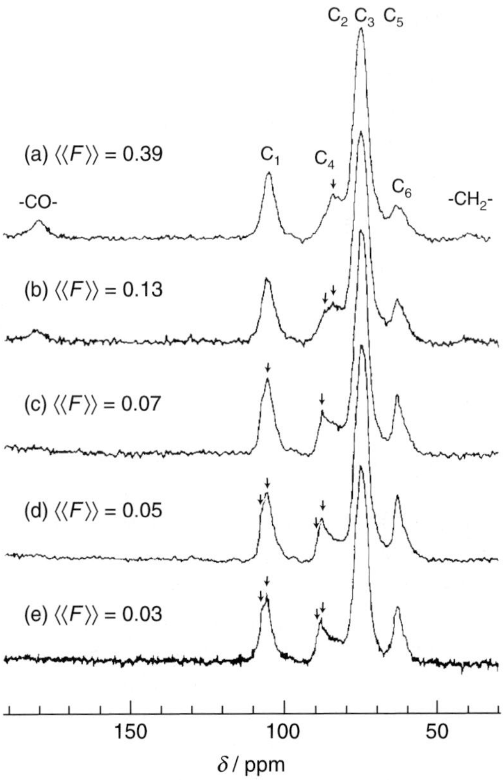

Figure 5.2.22   CP/MAS ^{13}C NMR spectra of NaCEC samples: (a) NaCEC-1; (b) NaCEC-2; (c) NaCEC-3; (d) NaCEC-4; (e) NaCEC-5.

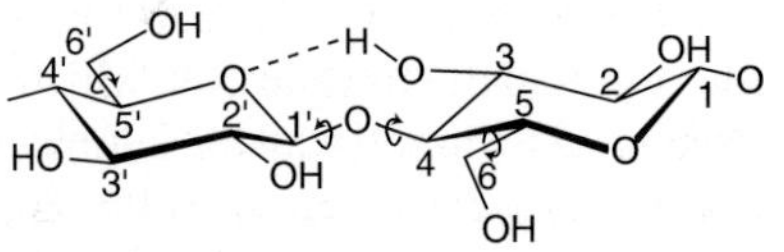

Figure 5.2.23   Molecular structure of a cellobiose unit constituting a cellulose molecule.

are responsible for $O_3-H\cdots O_5'$ intramolecular hydrogen bonds. Therefore, we can conclude that $O_3-H\cdots O_5'$ intramolecular hydrogen bonds exist in cellulose derivative solids when $\langle\!\langle F\rangle\!\rangle$ is far below 0.1. A close inspection of the spectrum for NaCEC with $\langle\!\langle F\rangle\!\rangle = 0.13$ (Figure 5.2.22(b)) also reveals two peaks among those peaks as shoulders. Water soluble NaCEC has the highest values of $\chi_{am}(C_3)$, $\Delta_{1/2}(C_1)$, and $\Delta_{1/2}(C_6)$.

Figure 5.2.23 shows the molecular conformation of a cellobiose unit constituting a cellulose molecule. Chemical shifts of C_1 and C_6 carbons are influenced by the torsional angles around C_1-O-C_4 and C_5-C_6-O linkages, respectively. The wide variation in these torsional angles causes the broadening of C_1 and C_6 carbon peaks in the ^{13}C NMR spectrum. Therefore, it seems reasonable that $\Delta_{1/2}(C_1)$ and $\Delta_{1/2}(C_6)$ represent the wide variation of conformation of molecular chains. Consequently, introduction of substituent groups effectively destroys the ordered structure of cellulose, and, as a result, water soluble NaCEC has a highly disordered structure. Kamide *et al.*[67] suggested that $\chi_{am}(C_3)$ represents the degree of breakdown in $O_3-H\cdots O_5'$ intramolecular hydrogen bonding of cellulose, and they demonstrated[57] that the complete dissolution of cellulose acetate, which is a typical derivative with hydrophobic substituent groups, into water requires $\chi_{am}(C_3) \geq 0.87$. It became clear that NaCEC dissolves completely in water when $\chi_{am}(C_3) \geq 0.76$ and the small difference in the lowest limit of $\chi_{am}(C_3)$ between cellulose acetate and NaCEC probably correlates with the differences in the chemical nature of the substituent group or in the steric hindrance due to spatial volume bulkiness. In other words, there is a critical value of $\chi_{am}(C_3)$, $\chi_{am}^c(C_3)$ above which the polymer dissolves completely in water, and the absolute magnitude of $\chi_{am}^c(C_3)$ differs for different polymers depending on the chemical and physical natures of the substituent group. Except for sample code NaCEC-1, $\chi_{am}(C_3)$ increases with an increase in $\langle\!\langle f_6\rangle\!\rangle$. This is readily understood because an introduction of the bulky substituent at the C_6 position may suppress the formation of the $O_3-H\cdots O_5'$ intramolecular hydrogen bonds to some extent, in spite of the substituent group at the C_6 position supposedly not breaking the intramolecular hydrogen bond at the C_3 position directly (see Figure 5.2.23). $\chi_{am}(C_3)$ ($= 0.46$) obtained for NaCEC-5, which was soluble in 9 wt% aq. NaOH solution at 4 °C and insoluble in water, 2, 5, and 9 wt% aq. NaOH solution at 20 °C, is far smaller than $\chi_{am}^c(C_3)$ ($= 0.94$) found for Cell-II, which was soluble in 9 wt% aq. NaOH at 4 °C.[34] We can conclude that the NaCEC sample, in which intramolecular hydrogen bonds exist at least in part, has higher solubility against aq. NaOH than cellulose. This can be tentatively explained as follows: the carboxyethyl group at the C_6 position, which is highly hydrophilic and very bulky, widens the distance between molecular chains, making it easier for the solvent molecules to penetrate into the space between polymer chains.

As previously reported,[61] the sodium salt of carboxymethyl cellulose (NaCMC) with $\langle\!\langle F\rangle\!\rangle \leq 0.47$, prepared by the heterogeneous reaction of cellulose with sodium monochloroacetate, is water insoluble, and its water absorbency is primarily governed by $\langle\!\langle f_6 \rangle\!\rangle$. Correlation of water absorbency and $\langle\!\langle f_6 \rangle\!\rangle$ of NaCMC can be explained as in the case of NaCEC: the bulky substituent group at the C_6 position widens the distance between molecular chains, creating a wider space to readily receive the absorbed liquids. The IR spectra affords further experimental evidence supporting the existence of intramolecular hydrogen bonds in water insoluble NaCEC.

In summary, at a relatively low degree of substitution the solubility of the derivative in water or aqueous alkali was mainly governed by considerable destruction of the intramolecular hydrogen bonds. These experimental facts support our previous molecular interpretation of the solubility of cellulose acetate (a derivative with hydrophobic substituent) and of cellulose in water or aq. alkali.

REFERENCES

1. K Kamide, K Okajima and K Kowsaka, *Polym. J.*, 1992, **24**, 71.
2. E Parnell, in *Life and Labours of John Mercer*, Green & Co., London, 1886.
3. H Sobue and N Migita (eds), *Cellulose Handbook (in Japanese)*. Asakura-Shoten, Tokyo, 1958.
4. H Staudinger and R Mohr, *J. Pract. Chem.*, 1941, **158**, 233.
5. H Mark, *J. Phys. Chem.*, 1940, **44**, 764.
6. P Hermans, *J. Phys. Chem.*, 1941, **45**, 827.
7. E Ott and M Spurlin (eds), *Cellulose and Cellulose Derivatives*, Vol. II, 2nd Edn., Interscience Publishers, New York, 1954.
8. K Kowsaka, K Okajima and K Kamide, *Polym. Int.*, 1991, **25**, 91.
9. K Kowsaka, K Okajima and K Kamide, *Polym. Int.*, 1992, **29**, 47.
10. K Kamide, K Okajima, T Matsui and K Kowsaka, *Polym. J.*, 1984, **16**, 857.
11. H Sobue, H Kiessig and K Hess, *Z. Phys. Chem.*, 1939, **B34**, 309.
12. H Maeda, H Kawada and T Kawai, *Makromol. Chem.*, 1970, **131**, 139.
13. K Kamide, K Okajima, K Kowsaka and T Matsui, *Polym. J.*, 1985, **17**, 701.
14. K Kamide, K Okajima, K Kowsaka and T Matsui, *Polym. J.*, 1985, **17**, 707.
15. T Yamashiki, K Kamide, K Okajima, K Kowsaka, T Matsui and H Fukase, *Polym. J.*, 1988, **20**, 447.
16. T Yamashiki, T Matsui, M Saitoh, K Okajima, K Kamide and T Sawada, *Br. Polym. J.*, 1990, **22**, 73.
17. T Yamashiki, T Matsui, M Saitoh, K Okajima, K Kamide and T Sawada, *Br. Polym. J.*, 1990, **22**, 121.
18. T Yamashiki, T Matsui, M Saitoh, Y Matsuda, K Okajima, K Kamide and T Sawada, *Br. Polym. J.*, 1990, **22**, 201.
19. K Kamide and K Okajima, US Patent 4,634,470, 1987.
20. W Brown and R Wikström, *Eur. Polym. J.*, 1966, **1**, 1.
21. L Segal, J Creely, A Martin and C Conrad Jr., *Text. Res. J.*, 1959, **29**, 786.
22. K Kamide, K Okajima and T Matsui, unpublished results.
23. S Manabe, M Iwata and K Kamide, *Polym. J.*, 1986, **18**, 1.
24. A Stipanovic and A Sarko, *Macromolecules*, 1976, **9**, 851.
25. K Kamide, K Yasuda, T Matsui, K Okajima and T Yamashiki, *Cell. Chem. Technol.*, 1990, **24**, 23.
26. M Tsuboi, *J. Polym. Sci.*, 1957, **28**, 159.
27. M Nelson and R O'Connor, *J. Appl. Polym. Sci.*, 1964, **8**, 1325.
28. M Nelson, T Robert and R O'Connor, *J. Appl. Polym. Sci.*, 1964, **8**, 1311.
29. H Higgins, C Stewart and K Harrington, *J. Polym. Sci.*, 1961, **51**, 59.
30. J Hayashi, M Sueoka and T Watanabe, *Nippon Kagaku Kaishi*, 1974, **7**, 1320.
31. A McKenzie and H Higgins, *Svensk Paperstidn*, 1958, **61**, 593.

32. W Earl and D Van der Hart, *Macromolecules*, 1981, **14**, 570.
33. F Horii, A Hirai and R Kitamaru, *Polym. Bull.*, 1982, **8**, 163.
34. K Kamide and K Nishiyama, *J. Ind. Econ., Nara Sangyo Univ.*, 2001, **15**(5), 35.
35. K Kamide, Y Miyazaki and T Abe, *Polym. J.*, 1979, **11**, 523.
36. K Kamide and M Saito, *Polym. J.*, 1986, **18**, 569.
37. See, for example, A Munekata, in *Chemical Fibers*, Vol. II, Maruzen, Tokyo, 1956, p. 235.
38. I Sakurada, *Ber. Dtsch. Chem. Ges*, 1930, **63**, 2027.
39. W Norman, *Chem. Ztg.*, 1906, **30**, 584.
40. K Hess and C Trogus, *Ann. Chem.*, 1929, **145**, 401.
41. I Miyamoto, M Imamoto, T Matsui, M Saito and K Okajima, *Polym. J.*, 1995, **27**, 1113.
42. T Yamasiki, K Kamide, K Okajima, K Kowsaka, T Matsui and H Fukase, *Polym. J.*, 1988, **20**, 447.
43. H Yamada, K Kowsaka, T Matsui, K Okajima and K Kamide, *Cell. Chem. Technol.*, 1992, **24**, 141.
44. T Yamashiki, T Matsui, K Kowsaka, M Saitoh, K Okajima and K Kamide, *J. Appl. Polym. Sci.*, 1992, **44**, 691.
45. T Yamashiki, M Saito, K Yasuda, K Okajima and K Kamide, *Cell. Chem. Technol.*, 1990, **24**, 237.
46. K Kamide, T Abe, Y Miyazaki and M Watanabe, *J. Text. Mach. Soc. Japan*, 1891, **34**, T149.
47. K Kowsaka, K Yasuda, K Okajima and K Kamide, *Polym. Intern.*, 1991, **25**, 99.
48. I Miyamoto, T Matsuoka, T Matsui and K Okajima, *Polym. J.*, 1995, **27**, 1123.
49. K Kamide, K Okajima, K Kowsaka and T Matsui, *Polym. J.*, 1987, **19**, 1405.
50. K Kamide, M Saito and T Abe, *Polym. J.*, 1981, **13**, 421.
51. T Miyamoto, Y Sato, T Shibata, M Tanahashi and H Inagaki, *J. Polym. Sci. Polym. Chem. Ed.*, 1985, **23**, 1373.
52. K Kamide, T Terakawa and Y Miyazaki, *Polym. J.*, 1979, **11**, 285.
53. M Saito, *Polym. J.*, 1983, **15**, 249.
54. D Gagnaire, J Saint-Germain and M Vincendon, *J. Appl. Polym. Sci. Polym. Symp.*, 1983, **37**, 261.
55. K Kamide and M Saito, *Eur. Polym. J.*, 1984, **20**, 903.
56. For example, J Roberts, R Stewart and M Caserio, *Organic Chemistry*. W. A. Benjamin, Inc., Menlo Park, CA, 1973.
57. K Kamide, K Okajima, K Kowsaka, T Matsui, S Nomura and K Hikichi, *Polym. J.*, 1985, **17**, 909.
58. K Okajima, K Kowsaka and K Kamide, *Polym. Intern.*, 1992, **29**, 47.
59. J Brundrup and E Immergut (eds), *Polymer Handbook*, 3rd Edn., Wiley, New York, 1989.
60. K Kamide and K Okajima, *Polym. J.*, 1981, **13**, 163.
61. K Kamide, K Okajima, T Matsui, M Ohnishi and H Kobayashi, *Polym. J.*, 1983, **15**, 309.
62. K Kamide, K Okajima and K Kowsaka, *Polym. J.*, 1987, **19**, 231.
63. K Kamide, K Okajima, T Matsui and M Ohnishi, *Polym. J.*, 1987, **19**, 347.
64. K Kowsaka, K Okajima and K Kamide, *Polym. J.*, 1988, **20**, 827.
65. H Iijima, K Kowsaka and K Kamide, *Polym. J.*, 1992, **24**, 1077.
66. K Kamide and M Saito, *Macromol. Symp.*, 1994, **83**, 233.
67. K Kamide, K Yasuda and K Okajima, *Polym. J.*, 1988, **20**, 259.

Subject Index

(number in parenthesis means page)